Springer Series in Computational Mathematics

39

For further volumes:
http://www.springer.com/series/797

Stefan A. Sauter · Christoph Schwab

Boundary Element Methods

Springer

Stefan A. Sauter
Universität Zürich
Institut für Mathematik
Winterthurerstr. 190
8057 Zürich
Switzerland
stas@math.uzh.ch

Christoph Schwab
ETH Zürich
Seminar für Angewandte
Mathematik
Rämistrasse 101
8092 Zürich
Switzerland
christoph.schwab@math.ethz.ch

Translation and expanded edition from the German language edition: *Randelementmethoden*, © Teubner 2004. Vieweg+Teubner is part of Springer Science and Business Media, All Rights Reserved

ISSN 0179-3632
ISBN 978-3-540-68092-5 e-ISBN 978-3-540-68093-2
DOI 10.1007/978-3-540-68093-2
Springer Heidelberg Dordrecht London New York

Mathematics Subject Classification (2010): 45B05, 65F10, 65N38, 65R20, 65D32, 65F50, 78M16

Cover design: deblik, Berlin

Printed on acid-free paper

Springer is part of Springer Science+Business Media (www.springer.com)

We dedicate this book to
Annika, Lina, and Gaby

Preface

The integral equation method is an elegant mathematical way of transforming elliptic partial differential equations (PDEs) into boundary integral equations (BIEs). The focus of this book is the systematic development of efficient numerical methods for the solution of these boundary integral equations and therefore of the underlying differential equations.

The integral equation method has a long history that is closely linked to mathematicians such as I. Fredholm, D. Hilbert, E. Nyström, J. Hadamard, J. Plemelj, J. Radon and many others. Here is a list of some of the original works on the subject: [46, 96, 101, 126, 164, 165, 173, 175–177, 181, 182, 188, 214, 229].

With the introduction of variational methods for partial differential equations at the beginning of the twentieth century, integral equations lost some of their importance for the area of analysis. This was due to the difficulty of formulating precise results on existence and uniqueness by means of classical integral equations.

Since the middle of the twentieth century the need for numerical methods for partial differential methods began to grow. This was reflected also in the rapidly increasing interest in integral equation methods. Some advantages of this approach for certain classes of problems compared to domain methods (difference methods and finite elements) are given in the following:

1. The treatment of equations on spatial domains with a complex geometry is simpler with respect to mesh generation – this is the subdivision of the domain into small geometric elements – for boundary integral equations than for domain methods, since only a surface mesh of the domain has to be generated as opposed to an entire volume mesh.
2. The numerical treatment of problems on unbounded domains is especially simple with integral equation methods, while the treatment by means of domain methods requires the generation of a mesh on an unbounded domain, which is rather problematic.
3. For some parameter dependent problems, for example, from the area of electromagnetism at high frequencies, numerical methods for integral equations remain more stable for extreme parameters than for domain discretizations.

4. The large linear systems of equations that appear in almost every discretization method have a better condition number than the systems of equations for domain discretizations. Basic iterative methods thus converge more rapidly.
5. The drawbacks of the integral equation method, such as the numerical integrations that are necessary to generate and solve the systems of equations, are being resolved with numerical methods that have been under constant development since about 1980.

The Nyström or quadrature formula methods and the collocation method are classical numerical solution methods for integral equations. One of the first textbooks on this topic was written by K.E. Atkinson [7] with an extended new edition [8]. These methods are suited to the solution of boundary integral equations of *the second kind*. These are integral equations with operators of the form $I + K$, where I denotes the identity and K an integral operator. They can be implemented on computers relatively easily, although they do have two significant drawbacks: (a) the Nyström and collocation methods cannot be applied to all boundary integral equations that appear in connection with elliptic boundary value problems and (b) the convergence and stability of the methods can only be shown for very restrictive conditions imposed on the underlying differential equation and the smoothness of the physical domain.

Since about 1980–1990 the Galerkin methods for the discretization of boundary integral equations have been gaining importance for practical problems. From a theoretical point of view this method is superior to the alternatives such as the Nyström and collocation methods: stability, consistency and convergence of the Galerkin method can be shown for a very general class of boundary integral equation. The approach is based on a variational formulation of boundary integral equations as opposed to the pointwise, classical approach. This approach is explained in detail in, for example, [72,74,80,167,171,238] or in the monographs [137,162,170].

The breakthrough for the Galerkin methods for practical, three-dimensional problems was achieved through the development of numerical methods for the approximation of integrals in order to determine the system matrix and through the development of fast algorithms to represent the non-local (boundary integral) operators.

The focus of this book is the systematic development of numerical methods to determine the Galerkin solution of boundary integral equations. All necessary tools from the area of analysis are presented, most of which are proven and derived; some, however, are only cited so that this book does not become too expansive. This book can be used as the basis for a lecture course of four hours a week on the numerics of boundary integral equations, consisting of an intensive short course on functional analysis and with a focus on the numerical methods. Some of the subsections bridge the gap between the textbook and current areas of research or should be seen as complements to the material. They are marked by a star ($\star$). The applications from the area of *electromagnetism* (Maxwell and wave equations, Helmholtz equation for high frequencies), for which integral equation methods are currently being developed intensively, serve as examples. The methods that are dealt with in this book form the basis with which to treat such problems. We will, however, not elaborate on the concrete applications.

So as not to go into too much detail we refrained from representing methods to couple finite elements with boundary elements and domain decomposition methods (see [49, 54, 71, 73, 135, 149]).

First and foremost the aim of this book is to represent and mathematically analyze efficient methods. Its purpose is not the treatment of concrete applications from the area of engineering. For this the books [13, 23, 32] may serve as an introduction.

Other textbooks and monographs from the area of numerical analysis for integral equations include [23, 60, 117, 216].

This book is the translation of the German version [204] and extended by chapters on p-parametric surface approximation and a posteriori error estimates – thanks are due to E. Louw for the translation of the German version. In addition we have corrected some misprints and incorporated additional material at various places.

The authors would like to thank their colleagues Profs. W. Hackbusch, R. Hiptmair and W. Wendland for the numerous discussions concerning the topics of this book, their co-workers L. Banjai, N. Krzebek, M. Rech, N. Stahn, R. Warnke for their support during the reading and correction of the manuscript. We also owe thanks to the Springer-Verlag for their understanding and unproblematic cooperation.

Zürich
July 2010

Stefan Sauter
Christoph Schwab

Contents

Chapter 1
Introduction

Many physical processes can be described by systems of linear and non-linear differential and integral equations. Only in very few special cases can such equations be solved analytically, which is why numerical methods have to be developed for their solution. In light of the complexity of the problems that appear in practice, it is unrealistic to expect to find a numerical method that offers a black-box type of numerical method that is suitable for all these problems. A more reasonable approach is to develop special numerical methods for specific classes of problems in order to take advantage of the characteristic properties of these classes. These numerical methods should then be decomposed into isolated and elementary partial problems. It would then be possible to employ or develop efficient methods for these subproblems.

The fundamental aim of this book is to systematically develop the *boundary element methods* for integral equations. These methods are developed for boundary integral equations that result when elliptic boundary value problems on spatial domains are transformed to integral equations on the boundary of the physical domain. In this introductory chapter we will briefly describe the structure and contents of this book.

1.1 The Concept of the Boundary Element Method

The boundary element method is a method for the solution of integral equations. In this book we will restrict ourselves to integral equations for the solution of elliptic boundary value problems. In order to offer a comprehensive description we will introduce basic terminology and theorems from the theory of partial differential equations and the integral equation method in Chaps. 2 and 3.

1.1.1 Basic Terminology

We consider the following problem. Our aim is to determine a physical quantity u that depends on the spatial variable $\mathbf{x} \in \mathbb{R}^d$. Here d denotes the spatial dimension.

S.A. Sauter and C. Schwab, *Boundary Element Methods*, Springer Series in Computational Mathematics 39, DOI 10.1007/978-3-540-68093-2_1,

We call equations for u that contain partial derivatives

$$\partial_i u = \frac{\partial u}{\partial x_i}$$

$(1 \leq i \leq d)$ or partial derivatives of u of higher order *partial differential equations*. Classical differential operators that contain partial derivatives of a function are the *gradient*, *divergence* and *Laplace* operator. For this let u be a differentiable, scalar function. Then the gradient of u is given by

$$\operatorname{grad} u := \nabla u := (\partial_1 u, \partial_2 u, \ldots, \partial_d u)^\mathsf{T}.$$

The *divergence* of a differentiable vector field $\mathbf{w}$ is defined by

$$\operatorname{div} \mathbf{w} := \nabla \cdot \mathbf{w} = \sum_{i=1}^{d} \partial_i w_i .$$

If u is twice differentiable the Laplace operator can be defined:

$$\Delta u = \sum_{i=1}^{d} \partial_i^2 u.$$

It is easy to verify that

$$\Delta u = \operatorname{div} \operatorname{grad} u.$$

For a differentiable vector field $\mathbf{u} : \mathbb{R}^3 \to \mathbb{R}^3$ the curl operator is given by

$$\operatorname{curl} \mathbf{u} := \nabla \times \mathbf{u} := (\partial_2 u_3 - \partial_3 u_2, \partial_3 u_1 - \partial_1 u_3, \partial_1 u_2 - \partial_2 u_1)^\mathsf{T}.$$

Definition 1.1.1. A subset $\Omega \subset \mathbb{R}^d$ is a domain if it is open and connected.

The domain Ω is called a normal domain if the Gaussian integral theorem holds. Sufficient conditions can be found in, for example, [246], [128, Chap. 4], [142].

Theorem 1.1.2 (Gauss' Integral Theorem). *Let $\Omega \subset \mathbb{R}^d$ be a normal domain with boundary $\partial\Omega$, let $\mathbf{n} : \partial\Omega \to \mathbb{R}^d$ denote the exterior normal field and let $U \supset \Omega$ denote an open subset of $\mathbb{R}^d$. Then for every continuously differentiable vector field $\mathbf{v} : U \to \mathbb{R}^d$:*

$$\int_\Omega \operatorname{div} \mathbf{v}(\mathbf{x})\, d\mathbf{x} = \int_{\partial\Omega} \langle \mathbf{v}(\mathbf{x}), \mathbf{n}(\mathbf{x}) \rangle \, ds_{\mathbf{x}}.$$

1.1.2 A Physical Example

We consider the following physical problem. Find the fields **E**, **D** in the domain $\mathbb{R}^3$ which are characterized by the Maxwell equations:

$$\operatorname{curl} \mathbf{E} = \mathbf{0}, \tag{1.1}$$

$$\operatorname{div} \mathbf{D} = \rho, \tag{1.2}$$

$$\mathbf{D} = \varepsilon \mathbf{E}, \tag{1.3}$$

where ε denotes the electrostatic permeability and ρ the electric charge density. In a vacuum we have, for example:

$$\varepsilon = \varepsilon_0 = \frac{10^{-9}}{36\pi} \text{Farad/meter.}$$

The quantity $\mathbf{E} : \mathbb{R}^3 \to \mathbb{R}^3$ denotes the unknown electrostatic field and $\mathbf{D}$: $\mathbb{R}^3 \to \mathbb{R}^3$ the electrostatic induction. Equation (1.1) implies that there exists a potential $\Phi : \mathbb{R}^3 \to \mathbb{R}$ such that

$$\mathbf{E} = -\operatorname{grad} \Phi. \tag{1.4}$$

If we insert this into (1.3) we obtain

$$\mathbf{D} = -\varepsilon \operatorname{grad} \Phi.$$

Combining this with (1.2) yields a scalar equation for the potential Φ:

$$-\operatorname{div}(\varepsilon \operatorname{grad} \Phi) = \rho \tag{1.5}$$

in $\mathbb{R}^3$.

We now consider a conductor that is described by a bounded domain $\Omega^- \subset \mathbb{R}^3$. The complement or exterior domain is denoted by $\Omega^+ = \mathbb{R}^3 \backslash \overline{\Omega^-}$. We assume that the electrical permeability in Ω^- is given by a positive constant ε^- and, in the exterior Ω^+, is given by a further positive constant ε^+. In the interior Ω^- we obtain the Poisson equation

$$-\Delta \Phi = \frac{\rho}{\varepsilon^-} \qquad \text{in } \Omega^- \tag{1.6}$$

and owing to $\rho \equiv 0$ in the exterior of the conductor we obtain in Ω^+ the *Laplace equation*, i.e., the homogeneous Poisson equation:

$$-\Delta \Phi = 0. \tag{1.7}$$

Remark 1.1.3. *If the electrical permeability ε is only piecewise smooth but discontinuous the assumptions in the definition of the divergence operator [see* (1.5)*] are* *generally not fulfilled. If we only consider the equation in the subdomains* Ω^- *and* Ω^+*, in which* ε *is smooth, applying the* div*-operator is not a problem.*

Through the *potential approach* (1.4) we have reduced the stationary Maxwell equations to a scalar differential equation. Since derivatives of second order do occur but none of higher order, we are dealing with a scalar differential equation of second order.

In order to find unique solutions for differential equations, we still need to prescribe suitable boundary conditions. For example, for $d = 2$ the real and imaginary part of any holomorphic function satisfy (1.7).

Example 1.1.4. *For* $d = 2$ *the functions* $u \equiv 1$, $u = x$, $u = y$, $u = x^2 - y^2$, $u = 2xy, \ldots$ *satisfy the Laplace equation. The functions can also be seen as functions in* $\mathbb{R}^d$ *with* $d > 2$ *(constant in all further variables). They then also satisfy the Laplace equation.*

We assume that the boundary Γ of Ω^- can be oriented and is sufficiently smooth such that a continuous normal field $\mathbf{n} : \Gamma \to \mathbb{R}^3$ can be defined. We assume that $\mathbf{n}(\mathbf{x})$ points in the direction of the exterior Ω^+. Using physical arguments (see [138]) that are formulated in a mathematical manner in Sect. 3.3, we find that the tangential component of the $\mathbf{E}$-field and the normal component of the $\mathbf{D}$-field are continuous across the boundary. We will prove in Chap. 3 that under suitable condition the $\mathbf{E}$ and $\mathbf{D}$-fields can be extended continuously (to one side) to functions $\mathbf{E}^+$, $\mathbf{D}^+ : \overline{\Omega^+} \to \mathbb{R}^3$ and $\mathbf{E}^-$, $\mathbf{D}^- : \overline{\Omega^-} \to \mathbb{R}^3$. Therefore the normal components of $\mathbf{D}$ for $\mathbf{x} \in \Gamma$ can be defined by

$$D_n^+(\mathbf{x}) = \left\langle \mathbf{n}(\mathbf{x}), \mathbf{D}^+(\mathbf{x}) \right\rangle, \qquad D_n^-(\mathbf{x}) = \left\langle \mathbf{n}(\mathbf{x}), \mathbf{D}^-(\mathbf{x}) \right\rangle$$

and the tangential components by

$$\mathbf{E}_t^+(\mathbf{x}) = \mathbf{n}(\mathbf{x}) \times \mathbf{E}^+(\mathbf{x}), \qquad \mathbf{E}_t^-(x) = \mathbf{n}(\mathbf{x}) \times \mathbf{E}^-(\mathbf{x}).$$

The *transmission conditions* are given by

$$\mathbf{E}_t^+(\mathbf{x}) = \mathbf{E}_t^-(\mathbf{x}), \qquad D_n^+(\mathbf{x}) = D_n^-(\mathbf{x}) \qquad \forall \mathbf{x} \in \Gamma.$$

We now insert the potential approach (1.4) into these conditions. Note, however, that the approach (1.4) only determines the potential uniquely up to a constant. It can be chosen such that the transmission conditions for the potential Φ are given by

$$\Phi^+(\mathbf{x}) = \Phi^-(\mathbf{x}) \tag{1.8}$$

$$\varepsilon^+ \frac{\partial \Phi^+}{\partial \mathbf{n}}(\mathbf{x}) = \varepsilon^- \frac{\partial \Phi^-}{\partial \mathbf{n}}(\mathbf{x}) \tag{1.9}$$

for all $\mathbf{x} \in \Gamma$. The quantity $\varepsilon \frac{\partial \Phi}{\partial \mathbf{n}}$ is called the potential flux. Conditions (1.8) and (1.9) imply that the potential and the potential flux are continuous across the boundary Γ.

In summary, we have derived the equations:

$$\begin{aligned} -\Delta\Phi^- &= \tfrac{\rho}{\varepsilon^-} && \text{in } \Omega^- \\ -\Delta\Phi^+ &= 0 && \text{in } \Omega^+ \\ \Phi^+ = \Phi^- \text{ and } \varepsilon^+ \tfrac{\partial\Phi^+}{\partial\mathbf{n}} &= \varepsilon^- \tfrac{\partial\Phi^-}{\partial\mathbf{n}} && \text{on } \Gamma \end{aligned} \tag{1.10}$$

for the electrostatic potential. These equations do not necessarily have a unique solution.

Example 1.1.5. *Let Ω^- be the ball in $\mathbb{R}^3$ with radius 1 around the origin. In (1.10) we choose $\rho = -12\,\|\mathbf{x}\|$ as the right-hand side. By introducing three-dimensional polar coordinates we can show that all functions of the form*

$$\begin{aligned} \Phi^-(\mathbf{x}) &= \frac{1}{\varepsilon^-}\left(\|\mathbf{x}\|^3 + \frac{3+\varepsilon^+ a}{\|\mathbf{x}\|} + a\left(\varepsilon^- - \varepsilon^+\right) + b\varepsilon^- - 4\right) \\ \Phi^+(\mathbf{x}) &= \frac{a}{\|\mathbf{x}\|} + b \end{aligned}$$

satisfy (1.10). If we impose the condition that Φ be regular at the origin we obtain $a = -3/\varepsilon^+$ and we obtain the single-parameter family of solutions:

$$\begin{aligned} \Phi^-(\mathbf{x}) &= \tfrac{\|\mathbf{x}\|^3 - 1}{\varepsilon^-} - \tfrac{3}{\varepsilon^+} + b, \\ \Phi^+(\mathbf{x}) &= -\tfrac{3}{\varepsilon^+\|\mathbf{x}\|} + b. \end{aligned} \tag{1.11}$$

In order to guarantee the unique solvability of partial differential equations on unbounded spatial domains we still need to prescribe suitable *decay conditions* at infinity. For the Laplace equation and spatial dimension $d = 3$ these can be written as

$$\left.\begin{aligned} \left|\Phi^+(\mathbf{x})\right| &\le C\,\|\mathbf{x}\|^{-1} \\ \left\|\operatorname{grad}\Phi^+(\mathbf{x})\right\| &\le C\,\|\mathbf{x}\|^{-2} \end{aligned}\right\} \quad \text{for } \|\mathbf{x}\| \to \infty. \tag{1.12}$$

In the case of the solution of Example 1.1.5 we obtain $b = 0$ as well as the unique solution

$$\Phi(\mathbf{x}) = \begin{cases} \tfrac{\|\mathbf{x}\|^3 - 1}{\varepsilon^-} - \tfrac{3}{\varepsilon^+} & \|\mathbf{x}\| < 1, \\ -\tfrac{3}{\varepsilon^+\|\mathbf{x}\|} & \|\mathbf{x}\| > 1. \end{cases}$$

Differential equations that are posed on all of $\mathbb{R}^3$ are called full space problems. One also often considers differential equations on bounded domains $\Omega \subset \mathbb{R}^3$. For example, if one is only interested in the electrical field inside the conductor Ω^- the differential equation (1.6) is only considered in the domain Ω^-. In place of the transmission conditions (1.8) and (1.9) one now has to deal with boundary conditions that can be obtained through physical measurements. If the potential is measured on Γ we speak of Dirichlet or essential boundary conditions. The associated interior problem reads

$$\begin{aligned} -\Delta\Phi^- &= \rho/\varepsilon^- \text{ in } \Omega^-, \\ \Phi^- &= g_D \quad \text{on } \Gamma. \end{aligned} \tag{1.13}$$

If the fluxes are measured on Γ we speak of Neumann or natural boundary conditions. The interior problem reads

$$\begin{aligned} -\Delta\Phi^- &= \rho/\varepsilon^- \text{ in } \Omega^-, \\ \varepsilon^-\partial\Phi^-/\partial\mathbf{n} &= g_N \quad \text{on } \Gamma. \end{aligned} \tag{1.14}$$

In the case of Neumann boundary conditions the right-hand side ρ/ε has to satisfy suitable compatibility conditions. These can be obtained by integrating the Poisson equation (1.6) over Ω^- and by then applying Gauss' integral theorem (Theorem 1.1.2) to grad Φ

$$\int_{\Omega^-} \frac{\rho(\mathbf{x})}{\varepsilon^-} d\mathbf{x} = -\int_{\Omega^-} \Delta\Phi(\mathbf{x})\, d\mathbf{x} = \int_{\partial\Omega^-} \frac{\partial\Phi}{\partial\mathbf{n}}(\mathbf{x})\, ds_{\mathbf{x}}.$$

It follows that the compatibility condition

$$\int_{\Omega^-} \frac{\rho(\mathbf{x})}{\varepsilon^-} d\mathbf{x} = \int_{\Gamma} \frac{\partial\Phi}{\partial\mathbf{n}}(\mathbf{x})\, ds_{\mathbf{x}},$$

which links the right-hand side in (1.6) with the given Neumann data g_N, is necessary for the solvability of the Neumann boundary value problem.

The exterior problems can be formulated in the same way. The Dirichlet exterior problem consists in solving the problem

$$\begin{aligned} -\Delta\Phi^+ &= \rho/\varepsilon^+ \text{ in } \Omega^+, \\ \Phi^+ &= g_D \quad \text{on } \Gamma \end{aligned} \tag{1.15}$$

and the Neumann exterior problem reads

$$\begin{aligned} -\Delta\Phi^+ &= \rho/\varepsilon^+ \text{ in } \Omega^+, \\ \varepsilon^+\frac{\Phi^+}{\partial\mathbf{n}} &= g_N \quad \text{on } \Gamma. \end{aligned} \tag{1.16}$$

As with the full space problem, suitable decay conditions have to be imposed at infinity to guarantee the unique solvability. For spatial dimensions $d = 3$ these read

$$\begin{aligned} |u(\mathbf{x})| &\leq O(1/\|\mathbf{x}\|), \text{ for } \|\mathbf{x}\| \to \infty, \\ \|\nabla u(\mathbf{x})\| &= O\left(\|\mathbf{x}\|^{-2}\right) \text{ for } \|\mathbf{x}\| \to \infty. \end{aligned} \tag{1.17}$$

1.1.3 Fundamental Solutions

It is our aim to transform the boundary value problems from the previous section into an integral equation on the *boundary* $\Gamma := \partial\Omega^-$ and only then solve it numerically. To transform a partial differential equation into an integral equation one needs the fundamental solution of the underlying differential operator. We again consider the Poisson equation:

$$-\Delta\Phi = \frac{\rho}{\varepsilon} \tag{1.18}$$

in $\mathbb{R}^3$. The function

$$N(\mathbf{x}) = \int_{\mathbb{R}^3} G(\mathbf{x}-\mathbf{y}) \frac{\rho(\mathbf{y})}{\varepsilon} d\mathbf{y} \tag{1.19}$$

with the kernel function

$$G(\mathbf{z}) = \frac{1}{4\pi \|\mathbf{z}\|} \tag{1.20}$$

is called the Newton potential and the function G from (1.20) is called the fundamental solution or the singularity function for the Laplace operator (for $d = 3$). The Newton potential exists for all functions $\rho \in C^0(\mathbb{R}^3)$ with compact support and solves the Poisson equation. For a proof we refer to [115, Chap. 17, Theorem 2]. The fundamental solution satisfies the Laplace equation in $\mathbb{R}^3 \backslash \{0\}$:

$$\Delta G = 0 \qquad \text{in } \mathbb{R}^3 \backslash \{0\}. \tag{1.21}$$

More specifically, we have, in the sense of distributions, the equality $\Delta G = \delta_0$ on $\mathbb{R}^3$ with the delta distribution δ_0 at the point zero. These and further properties of the fundamental solution and of the Newton potential are discussed in Chap. 3.

1.1.4 Potentials and Boundary Integral Operators

The boundary element method can be applied especially efficiently to *homogeneous* boundary value problems. If the equation is inhomogeneous the problem can transformed into a homogeneous problem by using the Newton potential. Since evaluating the Newton potential at one point $\mathbf{x}$ requires an integration over Ω (or $\Omega^{\pm}$), the method becomes very expensive if ρ has a large or, in the most extreme case, unbounded support. For this reason we will generally assume that the inhomogeneous part ρ has compact support: $\operatorname{supp}\rho \subset\subset \mathbb{R}^3$. It can be shown under these conditions that the Newton potential always satisfies the decay conditions (1.17) (see Chap. 3).

The Newton potential solves the Poisson equation. In general this potential will not satisfy the boundary conditions or the jump conditions. It only represents a special solution of the problem with which the Poisson equation can be transformed into the Laplace equation. All solutions of the Poisson equation can be written as the sum of a special solution and a solution of the homogeneous problem

$$\Phi = N + \Phi_0. \tag{1.22}$$

We will first consider the exterior problem with Dirichlet boundary conditions (1.15) and decay conditions (1.17). Evaluating N at one point $\mathbf{x}$ requires an integration over the unbounded exterior domain Ω^+.

The principle of superposition implies that Φ_0 is the solution of the Laplace equation

$$\begin{aligned} \Delta\Phi_0 &= 0, && \text{in } \Omega^+, \\ \Phi_0 &= \tilde{g}_D && \text{on } \Gamma, \\ \left.\begin{aligned} |u(\mathbf{x})| &\to 0 \\ \|\nabla u(\mathbf{x})\| &= O\left(\|\mathbf{x}\|^{-2}\right) \end{aligned}\right\} && && \text{for } \|\mathbf{x}\| \to \infty \end{aligned} \tag{1.23}$$

with the modified boundary conditions $\tilde{g}_D = g_D - N|_\Gamma$. With the help of the fundamental solution we can define an approach for $\mathbf{x} \in \mathbb{R}^3 \backslash \Gamma$ which reads

$$\Phi_0(\mathbf{x}) = \int_\Gamma G(\mathbf{x}-\mathbf{y})\,\sigma(\mathbf{y})\,d\Gamma_\mathbf{y}. \tag{1.24}$$

The function $\sigma : \Gamma \to \mathbb{C}$ has not yet been determined and is called *density*. For continuous densities $\sigma \in C^0(\Gamma)$ the integral in (1.24) exists as a Riemann integral. The right-hand side of (1.24) defines the single layer potential $S(\sigma)$ of the density σ. Since $\mathbf{x} \in \mathbb{R}^3 \backslash \Gamma$ and $\mathbf{y} \in \Gamma$ the differentiation and integration commute and with (1.21) we obtain:

$$\Delta S(\sigma) = 0$$

in $\mathbb{R}^3 \backslash \Gamma$. We will show in Chap. 3 that $S(\sigma)$ satisfies the decay conditions (1.23) for every $\sigma \in C^0(\Gamma)$. The problem (1.23) has thus been solved [and with it so has the initial problem (1.15)] if the density σ can be determined in such a way that the boundary conditions $\Phi_0|_\Gamma = \tilde{g}_D$ are satisfied. It will be shown in Theorem 3.1.16 that the single layer potential $S(\sigma)$ can be continuously extended across the surface Γ by

$$V(\sigma)(\mathbf{x}) = \int_\Gamma G(\mathbf{x}-\mathbf{y})\,\sigma(\mathbf{y})\,d\Gamma_\mathbf{y}, \qquad \text{for } \mathbf{x} \in \Gamma. \tag{1.25}$$

Thereby, the integral for $\sigma \in C^0(\Gamma)$ in (1.25) exists as an improper Riemann integral. The boundary integral equation to determine the density σ then reads:

$$V(\sigma) = \tilde{g}_D, \qquad \text{on } \Gamma, \tag{1.26}$$

or explicitly:

$$\int_\Gamma \frac{\sigma(\mathbf{y})}{4\pi\,\|\mathbf{x}-\mathbf{y}\|}\,d\Gamma_\mathbf{y} = \tilde{g}_D(\mathbf{x}), \qquad \text{for all } \mathbf{x} \in \Gamma.$$

Note that this integral equation represents a *boundary integral equation*, as we have $\mathbf{x}, \mathbf{y} \in \Gamma$ and the functions σ and $\tilde{g}_D$ are mappings from Γ to $\mathbb{C}$. Integral equations where the unknown function only appears under the integral are called integral equations of *the first kind*. The idea is based on the fact that the approach (1.25) satisfies the differential equation in Ω^+ for all densities. This approach for the solution of a differential equation is called the *potential approach method* or the *indirect formulation*.

As a generalization of this approach we note that every derivative of the form

$$k(\mathbf{x}, \mathbf{y}) = \sum_{\nu,\mu} c_{\nu,\mu}(\mathbf{y})\, \partial_{\mathbf{x}}^{\nu} \partial_{\mathbf{y}}^{\mu} G(\mathbf{x} - \mathbf{y})$$

satisfies the Laplace equation $\Delta_{\mathbf{x}} k(\mathbf{x}, \mathbf{y}) = 0$ for $\mathbf{x} \in \Omega^+$ and $\mathbf{y} \in \Gamma$ and therefore also the potential formed with k.

We will introduce the double layer potential for the interior problem with Dirichlet conditions. Once again the Poisson problem can be transformed into the Laplace equation by using the Newton potential:

$$\begin{aligned} \Delta \Phi_0 &= 0 \text{ in } \Omega^-, \\ \Phi_0 &= \tilde{g}_D \text{ on } \Gamma \end{aligned}$$

with $\tilde{g}_D = g_D - N|_{\Gamma}$. We set

$$k(\mathbf{x}, \mathbf{y}) = \langle \mathbf{n}(\mathbf{y}), \nabla_{\mathbf{y}} G(\mathbf{x} - \mathbf{y}) \rangle = -\frac{\langle \mathbf{n}(\mathbf{y}), \mathbf{y} - \mathbf{x} \rangle}{4\pi \|\mathbf{y} - \mathbf{x}\|^3} \tag{1.27}$$

and with this we form the double layer potential

$$D(\theta)(\mathbf{x}) := \int_{\Gamma} k(\mathbf{x}, \mathbf{y})\, \theta(\mathbf{y})\, d\Gamma_{\mathbf{y}} \qquad \text{for all } \mathbf{x} \in \mathbb{R}^3 \backslash \Gamma.$$

Again, for $\mathbf{x} \in \mathbb{R}^3 \backslash \Gamma$ and $\mathbf{y} \in \Gamma$ the differentiation and integration commute and we obtain $\Delta D(\theta) = 0$ in Ω^-. If the unknown density $\theta : \Gamma \to \mathbb{C}$ can be chosen such that

$$\lim_{\mathbf{x} \to \mathbf{x}_0} D(\theta)(\mathbf{x}) = \tilde{g}_D(\mathbf{x}_0), \qquad \text{for all } \mathbf{x}_0 \in \Gamma \tag{1.28}$$

then $\Phi = N + D(\theta)$ solves the interior problem (1.13). We will show in Chap. 3 that $D(\theta)$ can be extended continuously from the interior to the boundary. The extension has the following representation for sufficiently smooth boundaries:

$$-\frac{1}{2}\theta(\mathbf{x}) + K(\theta)(\mathbf{x}) \qquad \mathbf{x} \in \Gamma \tag{1.29}$$

with the boundary integral operator [k as in (1.27)]

$$K(\theta)(\mathbf{x}) := \int_{\Gamma} k(\mathbf{x}, \mathbf{y}) \, \theta(\mathbf{y}) \, d\Gamma_{\mathbf{y}}, \qquad \text{for all } \mathbf{x} \in \Gamma. \tag{1.30}$$

If we insert the representation (1.30) into (1.29) and then into (1.28) we obtain the boundary integral equation

$$-\frac{1}{2}\theta(\mathbf{x}) + K(\theta)(\mathbf{x}) = \tilde{g}_D(\mathbf{x}), \qquad \text{for all } \mathbf{x} \in \Gamma \tag{1.31}$$

in order to determine the unknown density $\theta : \Gamma \to \mathbb{C}$. The integral equation (1.31) is defined on Γ ($\mathbf{x}, \mathbf{y} \in \Gamma$ and $\theta, \tilde{g}_D : \Gamma \to \mathbb{C}$) and, thus, is again a boundary integral equation. Since the unknown function θ appears in the integrand as well as outside of the integrand, (1.31) is called a boundary integral equation of *the second kind*.

If the surface is sufficiently smooth the integral in (1.30) exists as an improper Riemann integral.

In Chap. 3, we will present further possibilities of transforming even more general elliptic differential equations with more general boundary conditions into boundary integral equations.

1.2 Numerical Analysis of Boundary Integral Equations

In Chaps. 4–6, we will deal with the numerics of boundary integral equations. Primarily, we will consider Galerkin boundary element methods for the discretization. Alternatives, such as collocation methods, are considered in examples.

1.2.1 Galerkin Method

In Chap. 4 we will consider the Galerkin boundary element method in its original form.

The basis of the Galerkin method is a finite-dimensional subspace S of the function space H which contains the continuous solution of the boundary integral equation. As an example we consider the boundary integral equation (1.26) for the single layer potential. The construction of the boundary element space S is based on a decomposition of the boundary Γ of Ω into non-overlapping panels which defines the surface mesh $\mathcal{G}$ of Γ. For a panel $\tau \in \mathcal{G}$, $b_\tau : \Gamma \to \{0, 1\}$ denotes the characteristic function on τ. The space S is the span of the basis functions $(b_\tau)_{\tau \in \mathcal{G}}$

$$S := \operatorname{span}\{b_\tau : \tau \in \mathcal{G}\}. \tag{1.32}$$

The dimension of S is denoted by $N := \dim S$. Every function $\sigma \in S$ is uniquely determined by the coefficient vector $(\sigma_\tau)_{\tau \in \mathcal{G}} \in \mathbb{R}^N$ with respect to the basis representation $\sigma = \sum_{\tau \in \mathcal{G}} \sigma_\tau b_\tau$.

Usually one cannot expect the boundary integral equation (1.26) to have a solution in S. Since every function in S is determined by N *degrees of freedom*, in general only N conditions can be imposed to determine the coefficient vector $(\sigma_\tau)_{\tau \in \mathcal{G}}$.

For the Galerkin method (1.26) is multiplied by the basis functions b_τ and then integrated over the boundary Γ. The equations to determine the Galerkin solution then read: Find $\sigma_S \in S$ such that

$$a(\sigma_S, b_\tau) := \int_\Gamma V(\sigma_S) b_\tau ds = \int_\Gamma \tilde{g}_D b_\tau ds =: F(b_\tau) \qquad \forall \tau \in \mathcal{G}. \tag{1.33}$$

It will be shown in Chap. 4 under which conditions the results on existence and uniqueness for the continuous boundary integral equations can be transferred to the Galerkin equations.

Questions concerning convergence and convergence rates are just as important for the evaluation of the method. We will show that under suitable conditions for a sufficiently fine surface mesh $\mathcal{G}$ the Galerkin solution converges quasi-optimally: There exists a constant C which is independent of the right-hand side such that

$$\|\sigma - \sigma_S\|_E \leq C \operatorname{dist}(\sigma, S) \quad \text{with} \quad \operatorname{dist}(\sigma, S) := \inf_{\theta \in S} \|\sigma - \theta\|_E \tag{1.34}$$

holds. The quantity $\operatorname{dist}(\sigma, S)$ depends only on the regularity of the solution σ, the chosen norm $\|\cdot\|_E$ and the boundary element space S.

The quasi-optimality of the Galerkin method, i.e., the error estimate (1.34), is proven under suitable conditions in Chap. 4.

In order to estimate the quantity $\operatorname{dist}(\sigma, S)$, the regularity of the continuous solution σ has to be analyzed. Depending on the smoothness of the boundary Γ and the right-hand side $\tilde{g}_D$ it can shown that the solution $\sigma \in H$ lies in a *smoother* space $W \subset H$.

We use the dimension N of the boundary element space as a parameter to describe the rate of convergence. We would like to estimate the quantity $\operatorname{dist}(\sigma, S)$ depending on N. The regularity of the solution combined with the *approximation property* of the boundary element space leads to the error estimate

$$\operatorname{dist}(\sigma, S) \leq C N^{-\alpha} \|\sigma\|_W,$$

where $\alpha > 0$ denotes the *rate of convergence*, which depends on W and S. In summary, we obtain

$$\|\sigma - \sigma_S\|_E \leq C N^{-\alpha} \|\sigma\|_W. \tag{1.35}$$

1.2.2 *Efficient Methods for the Solution of the Galerkin Equations*

The Galerkin solution is entirely defined by (1.33). However, these equations do not offer a clear idea of how to solve them efficiently. Since on the computer it is not possible to work with continuous (boundary element) functions but only real numbers, we transform (1.33) into a linear system of equations for the coefficient vector $(\sigma_\tau)_{\tau\in\mathcal{G}}$. If we insert the ansatz

$$\sigma_S = \sum_{\tau\in\mathcal{G}} \sigma_\tau b_\tau \tag{1.36}$$

into (1.33) and use the linearity of the operator V we obtain

$$\sum_{t\in\mathcal{G}} \sigma_t \int_\Gamma V(b_t)\, b_\tau ds = \int_\Gamma \tilde{g}_D b_\tau ds \qquad \forall\tau\in\mathcal{G}.$$

We define the system matrix $\mathbf{V} := (\mathbf{V}_{\tau,t})_{\tau,t\in\mathcal{G}}$ by $\mathbf{V}_{\tau,t} := \int_\Gamma V(b_t)\, b_\tau ds$ for τ, $t\in\mathcal{G}$ and the vector $\mathbf{g} := (\mathbf{g}_\tau)_{\tau\in\mathcal{G}}$ by $\mathbf{g}_\tau := \int_\Gamma \tilde{g}_D b_\tau ds$ for all $\tau\in\mathcal{G}$ and obtain a linear system of equations for the coefficient vector $(\sigma_\tau)_{\tau\in\mathcal{G}}$

$$\mathbf{V}\sigma = \mathbf{g}. \tag{1.37}$$

The Galerkin solution σ_S results from the vector σ through (1.36). In order to evaluate the Galerkin solution efficiently, it is thus necessary to develop quadrature methods, specifically designed for each problem, to determine the entries of the system matrix as well as fast methods to solve the system of equations.

1.2.2.1 Quadrature Methods

In the case of the basis functions b_τ from (1.32) the matrix entries for $\mathbf{V}$ are defined by the integrals

$$\mathbf{V}_{\tau,t} := \int_{\tau\times t} \frac{1}{4\pi\,\|\mathbf{x}-\mathbf{y}\|} ds_\mathbf{x} ds_\mathbf{y}. \tag{1.38}$$

For $\tau = t$ these integrals are singular for $\mathbf{x}=\mathbf{y}$ and special quadrature methods have to be developed to approximate them. These consist of a combination of *regularizing* coordinate transformations and Gaussian quadrature formulas. We will illustrate the idea of coordinate transformations by using the simple example of the integration of an integrand with characteristic singular behavior over the triangle with the vertices $(0,0)^\intercal$, $(1,0)^\intercal$, $(1,1)^\intercal$

$$I := \int_0^1 \int_0^{x_1} \frac{f(x_1, x_2)}{\sqrt{x_1^2 + x_2^2}} dx_2 dx_1.$$

The transformation $(\xi_1, \xi_1) = (\eta_1, \eta_1 \eta_2)$ maps the (η_1, η_2)-coordinates of the unit square $(0, 1)^2$ to the triangle $\hat{\tau}$. From this and with the determinant of the Jacobian $\det J(\eta_1, \eta_2) = \eta_1$ we obtain the representation

$$I = \int_0^1 \int_0^1 \eta_1 \frac{f(\eta_1, \eta_1 \eta_2)}{\sqrt{\eta_1^2 + \eta_1^2 \eta_2^2}} d\eta_2 d\eta_1 = \int_0^1 \int_0^1 \frac{f(\eta_1, \eta_1 \eta_2)}{\sqrt{1 + \eta_2^2}} d\eta_2 d\eta_1.$$

The integrand in the last integral is smooth for smooth functions f and the integral can be approximated by using Gaussian quadrature.

In Chap. 5 we will generalize these *Duffy coordinates* (see [83]) to pairs $\tau \times t$ of panels.

The approximation of the entries of the system matrix by means of quadrature methods leads to a perturbed linear system of equations

$$\widetilde{V} \tilde{\sigma} = \mathbf{g} \tag{1.39}$$

as well as a perturbed Galerkin solution $\tilde{\sigma}_S = \sum_{\tau \in \mathcal{G}} \tilde{\sigma}_\tau b_\tau$. The consistency and stability analysis of this perturbation allows us to choose the *order* of the quadrature such that the order of convergence α in (1.35) of the unperturbed Galerkin solution is maintained. In the second part of Chap. 5 this influence will be analyzed.

1.2.2.2 Solving the Linear System of Equations

In Chap. 6 we will study efficient methods for the solution of the linear system of equations (1.39). We will also analyze their convergence.

For a large dimension $N = \dim S$ of the boundary element space, methods such as the LR decomposition cannot be considered as their complexity grows in cubic proportion to the dimension N. Instead, iterative methods are used. The convergence of classical iterative methods is determined by the condition of the matrix $\mathbf{V}$. The integral equations under consideration can be divided into three types:

1. Equations with non-symmetric system matrices and a bounded condition number.
2. Equations with symmetric, positive definite system matrices and a condition number that grows as $N^{1/2}$ in proportion to the dimension N of the boundary element space. The underlying boundary integral operator is smoothing, i.e., the order of differentiability of the image of the function is one order higher than that of the function itself.
3. As in (2) but the boundary integral operator has differentiating properties, i.e., the order of differentiability of the image of a function is one order lower than that of the function itself.

For systems of equations of type 1 minimal residual methods – these are variants of the cg-method for non-symmetric matrices – can be used. The number of iterations necessary to reach a prescribed error tolerance is independent of the dimension N.

For systems of equations of type 2 and 3 the cg-method can be used. We will show in both cases that the number of iterations to reach a prescribed error tolerance grows as $N^{1/4}$ in proportion to the dimension of the system of equations.

Since the equations of type 3 have a differentiating effect they are closely related to Finite Element discretizations of elliptic boundary value problems. The multi-grid methods that are used in connection with these discretizations can be generalized for boundary integral equations of type 3. We will prove in Chap. 6 that the number of multi-grid iterations needed to reach a prescribed tolerance is independent of N.

1.2.2.3 Cluster Method

The complexity for the solution of the linear system of equations (1.37) with an iterative method is the product of the number of iterations and the complexity per iteration. If the iterative methods from Chap. 6 are used we face the following dilemma. The number of iterations is essentially independent of the dimension of the system of equations; however, the complexity per iteration grows quadratically with respect to N. This is due to the fact that the system matrix for integral operators is generally dense [see (1.38)].

In Chap. 7 we study the cluster method, with which a matrix-vector multiplication can be approximated and the complexity of which is proportional to $O\left(N \log^{\kappa} N\right)$ for $\kappa \approx 4$ to 6. Closely related to the cluster method is the fast multipole method (FMM) which was originally developed for N-body particle problems (see [111, 193]). We will briefly discuss this method in Chap. 7 as well.

We will explain the idea of the cluster method by using a simple model problem. For this purpose we assume that the kernel function is degenerate, i.e.,

$$k\left(\mathbf{x}, \mathbf{y}\right) = \sum_{i=1}^{m} \Phi_i\left(\mathbf{x}\right) \Psi_i\left(\mathbf{y}\right) \tag{1.40}$$

for suitable functions $(\Phi_i)_{i=1}^m$ and $(\Psi_i)_{i=1}^m$ with $m \ll N$. Then the coefficients of the system matrix of the associated boundary integral operator are given by

$$\begin{aligned}
\mathbf{V}_{\tau,t} &:= \int_\Gamma \int_\Gamma k\left(\mathbf{x}, \mathbf{y}\right) b_\tau\left(\mathbf{x}\right) b_t\left(\mathbf{y}\right) ds_{\mathbf{x}} ds_{\mathbf{y}} \\
&= \sum_{i=1}^{m} \left(\int_\Gamma \Phi_i\left(\mathbf{x}\right) b_\tau\left(\mathbf{x}\right) ds_{\mathbf{x}} \right) \left(\int_\Gamma \Psi_i\left(\mathbf{y}\right) b_t\left(\mathbf{y}\right) ds_{\mathbf{y}} \right).
\end{aligned}$$

Although this matrix is also dense in general, it can however be stored by using $O(N)$ quantities and it can be multiplied by a vector with a complexity of $O\left(N\right)$

arithmetic operations. For this we define the coefficients

$$\mathbf{L}_{i,\tau} := \int_\Gamma \Phi_i(\mathbf{x})\, b_\tau(\mathbf{x})\, ds_\mathbf{x} \quad \text{and} \quad \mathbf{R}_{i,\tau} := \int_\Gamma \Psi_i(\mathbf{y})\, b_\tau(\mathbf{y})\, ds_\mathbf{y}$$
$$\forall \tau \in \mathcal{G} \quad \forall 1 \le i \le m.$$

Since the support of the basis functions b_τ only consists of the panel τ, the integration over Γ can be reduced to the panel τ. If the functions Φ_i and Ψ_i are sufficiently smooth the assumption that every one of the numbers $\mathbf{L}_{i,\tau}$, $\mathbf{R}_{i,\tau}$ can be evaluated with a complexity of $O(1)$ arithmetic operations, independent of N, is justified. The overall complexity for the computation of all quantities is then given by $O(N)$.

One matrix-vector multiplication $\theta = \mathbf{V}\sigma$ can then be evaluated as follows:

1. Determine the auxiliary quantities $\gamma_i := \sum_{\tau\in\mathcal{G}} \mathbf{R}_{i,\tau}\sigma_\tau$ for $1 \le i \le m$ with an arithmetic complexity of $O(N)$ operations.
2. Determine θ as given by $\theta_\tau := \sum_{i=1}^m \gamma_i \mathbf{L}_{i,\tau}$ for all $\tau \in \mathcal{G}$. Complexity: $O(N)$ arithmetic operations.

In the context of iterative methods to solve linear systems of equations it is often sufficient to have a subroutine at your disposal that evaluates a matrix-vector multiplication. Furthermore, only $O(N)$ matrix entries have to be stored (for example, the diagonal elements for the Jacobi method). We have thus shown that for degenerate kernel functions it is sufficient to compute $O(N)$ `real` numbers in order to evaluate one matrix-vector multiplication with a complexity of $O(N)$.

We would like to emphasize at this point that the kernel functions for integral equations are generally not degenerate, but the approach (1.40) has to be generalized. The matrix-vector multiplication is, in the general case, only approximated and the influence of this additional perturbation on the Galerkin solution will also be analyzed in Chap. 7.

1.2.2.4 Surface Approximation

In practical applications, the description of the "true" physical surface might be very complicated or even not available as an exact analytic function and has to be approximated by using, e.g., pointwise measurements of the surface or some geometric modelling software. In this introduction, we illustrate the concept by the example of the first kind integral equation for the single layer potential on a smooth surface Γ in $\mathbb{R}^3$.

The construction of an approximate surface starts with definition of an interpolating polyhedron Ω^{affine} with surface Γ^{affine}. Let $\mathcal{G}^{\text{affine}} = \{\tau_1^{\text{affine}}, \dots \tau_N^{\text{affine}}\}$ denote a surface mesh of Γ^{affine} consisting of plane triangles with straight edges τ_i^{affine} which interpolate the exact surface Γ in their vertices. The affine pullback of $\tau^{\text{affine}} \in \mathcal{G}^{\text{affine}}$ to the two-dimensional reference triangle $\hat{\tau}$ with vertices $(0,0)^\intercal$, $(1,0)^\intercal$, $(1,1)^\intercal$ is denoted by $\chi_\tau^{\text{affine}} : \hat{\tau} \to \tau^{\text{affine}}$. Let $P : U \to \Gamma$ denote the orthogonal projection of a sufficiently small neighborhood $U \subset \mathbb{R}^3$ of Γ. A surface mesh for Γ is then defined by $\mathcal{G} = \{P(\tau^{\text{affine}}) : \tau^{\text{affine}} \in \mathcal{G}^{\text{affine}}\}$. The pullbacks of $\tau \in \mathcal{G}$ to the reference element is given by $\chi_\tau := P \circ \chi_\tau^{\text{affine}}$.

In principle, the Galerkin boundary element method can be applied directly to the surface mesh by defining the boundary element spaces with respect to $\mathcal{G}$. However, the mapping P is in general complicated and non-linear or even available only by pointwise measurements and hence must be approximated and realized by numerical approximation.

The p-parametric surface approximation is defined by replacing the mapping χ_τ by an (componentwise) interpolating polynomial $\chi_{\tau,p}$ of degree p and approximating the panels $\tau \in \mathcal{G}^{\text{affine}}$ by $\chi_{\tau,p}(\hat{\tau})$. (Note that for the definition of the interpolating polynomial $\chi_{\tau,p}$ only the evaluation of P at the interpolation points are required and not its functional representation.) This leads to the mesh $\mathcal{G}^p := \{\chi_{\tau,p}(\hat{\tau}) : \tau \in \mathcal{G}\}$ and the approximate surface

$$\Gamma^p := \bigcup_{\tau \in \mathcal{G}^p} \overline{\tau}.$$

The corresponding piecewise constant boundary element space is $S_p := \operatorname{span}\{b_\tau : \tau \in \mathcal{G}^p\}$, where $b_\tau : \Gamma^p \to \mathbb{R}$ is the characteristic function for $\tau \in \mathcal{G}^p$. The Galerkin method with piecewise constant boundary elements and p-parametric surface approximation for the single layer equation (1.26) is given by: Find $\sigma_S^p \in S_p$ such that

$$\int_{\Gamma^p \times \Gamma^p} \frac{\sigma_S^p(\mathbf{y})\, b_\tau(\mathbf{x})}{4\pi \|\mathbf{x}-\mathbf{y}\|} ds_{\mathbf{y}} ds_{\mathbf{x}} = \int_{\Gamma^p} g_D^p b_\tau ds \qquad \forall \tau \in \mathcal{G}^p. \tag{1.41}$$

Here, g_D^p is some extension of $\tilde{g}_D$ in (1.26) to Γ^p, e.g., by polynomial interpolation. From the numerical point of view, the problem is substantially simplified because the parametrization of Γ^p is explicitly given by polynomials instead of the complicated projection P.

In order to compare the solution σ_S^p with the continuous solution σ for the error analysis we have to lift σ_S^p to the original surface Γ. For sufficiently small mesh width $h := \max\{\operatorname{diam} \tau : \tau \in \mathcal{G}\}$, we assume that the restriction $P : \Gamma^p \to \Gamma$ is bijective and set $\theta^p = \left(P|_{\Gamma^p}\right)^{-1}$. Let $\check{\sigma}_S^p := \sigma^p \circ \theta^p$ denote the Galerkin solution which is lifted to the surface Γ so that the error $\sigma - \check{\sigma}_S^p : \Gamma \to \mathbb{C}$ is well defined. For the error analysis it is convenient to rewrite (1.41) equivalently as a problem on the true surface Γ. For this, let $\check{S}_p := \{\sigma_S^p \circ \theta^p : \sigma_S^p \in S_p\}$ denote the lifted boundary element space. Then $\check{\sigma}_S^p \in \check{S}_p$ is the solution of

$$\begin{aligned} a_S^p\left(\check{\sigma}_S^p, b_{\check{\tau}}\right) &:= \int_{\Gamma\times\Gamma} \frac{\check{\sigma}_S^p(\mathbf{y})\, b_{\check{\tau}}(\mathbf{x})}{4\pi \|\theta^p(\mathbf{x}) - \theta^p(\mathbf{y})\|} \rho^p(\mathbf{y})\, \rho^p(\mathbf{x})\, ds_{\mathbf{y}} ds_{\mathbf{x}} \\ &= \int_\Gamma \check{g}_D^p b_{\check{\tau}} \rho^p ds = F_S^p(b_{\check{\tau}}) \quad \forall \check{\tau} \in \mathcal{G}, \end{aligned} \tag{1.42}$$

where $\check{g}_D^p := g_D^p \circ \theta^p$ and $\rho^p : \Gamma \to \mathbb{R}$ reflects the change of metric. For any $\tau \in \mathcal{G}$ it is defined by

$$\rho^p|_\tau := \frac{g_{\tau,p} \circ \chi_\tau^{-1}}{g_\tau \circ \chi_\tau^{-1}} \quad \text{with} \quad g_\tau = \sqrt{\det\left((D\chi_\tau)^\intercal D\chi_\tau\right)} \quad \text{and}$$

$$g_{\tau,p} = \sqrt{\det\left(\left(D\chi_{\tau,p}\right)^\intercal D\chi_{\tau,p}\right)}.$$

For the error estimates, one has to compare the bilinear forms and right-hand sides in (1.33) and (1.42) and estimate the quantities δ^{I} and δ^{II} in

$$\begin{aligned} \left|a(u,v) - a_S^p(u,v)\right| &\leq \delta^{\mathrm{I}} \|u\|_H \|v\|_H \quad && \forall u,v \in \check{S}_p, \\ \left|F(u) - F_S^p(u)\right| &\leq \delta^{\mathrm{II}} \|u\|_H && \forall u \in \check{S}_p. \end{aligned}$$

To keep this outline short, we assume that the error related to the right-hand side vanishes, i.e., $\delta^{\mathrm{II}} = 0$. The estimate of δ^{I} can be derived from stability and consistency estimates of the form

$$\left|\frac{1}{\|\mathbf{x}-\mathbf{y}\|} - \frac{1}{\|\theta^p(\mathbf{x}) - \theta^p(\mathbf{y})\|}\right| \leq \frac{C}{\|\mathbf{x}-\mathbf{y}\|} h^{p+1}, \quad \frac{1}{\|\theta^p(\mathbf{x}) - \theta^p(\mathbf{y})\|} \leq \frac{C}{\|\mathbf{x}-\mathbf{y}\|},$$

$$\left|\rho^p(\mathbf{y}) - 1\right| \leq Ch^{p+1}, \qquad \left|\rho^p(\mathbf{y})\right| \leq C. \tag{1.43}$$

These estimates and similar estimates for more general kernel functions and also for only piecewise smooth surfaces will be derived in Chap. 8. It will also been shown that (1.43) implies $\delta^{\mathrm{I}} \leq Ch^{p+1}$ and in the case of full regularity of the solution σ we get

$$\left\|\sigma - \check{\sigma}_S^p\right\|_H \leq C_\sigma \left(h^{3/2} + h^{p+1}\right).$$

This allows us to conclude that, for the boundary integral equation for the single layer potential on a smooth surface discretized by piecewise constant boundary elements, the approximation of Γ by an interpolating polyhedron, i.e., $p = 1$, suffices in order to preserve the convergence rates of the original Galerkin discretization.

1.2.2.5 A Posteriori Error Estimation

The a priori error analysis of Galerkin boundary element methods shows the asymptotic convergence rate of the Galerkin solution by combining (a) the discrete stability of the variational formulation, e.g., in the form of an ellipticity estimate with (b) the regularity analysis of the continuous solution, e.g., by analyzing the smoothness of the solution in dependence of the smoothness of the given data, and (c) approximation properties of the boundary element functions for functions which belong to the regularity class of the exact solution. These estimates can be applied to large (infinite-dimensional) problem classes – however they might be very pessimistic for the concrete problem under consideration. In practical applications, the typical question is to "compute a numerical solution to a prescribed accuracy with minimal cost". However, the exact discretization error is not available in general because it requires the knowledge of the exact solution. Hence the only way to guarantee a prescribed accuracy of the numerical solution is to estimate the error theoretically by quantities which are computable. Since the upper bounds in a priori estimates, in general, are by far too pessimistic and various constants appearing therein cannot be estimated in a sharp, problem-dependent way, the condition: "refine the boundary element space until the upper bound becomes smaller than the given error tolerance" is not practicable because it exceeds the capacities of modern computers as a consequence of these pessimistic estimates.

The a posteriori error analysis allows us to estimate the error by some quantities (denoted as error indicators and error estimators) whose computation uses the computed numerical solution and hence is adapted to the concrete problem under consideration. The advantage is two-fold: (a) the estimates are sharp and can be used to guarantee a prescribed accuracy requirement and (b) the local error indicators provide information concerning the local error distribution and can be used to enrich the boundary element space in an adaptive solution process.

In order to explain the principal idea we consider again the Galerkin discretization (1.33) of the single layer potential. For the error $e_S := \sigma - \sigma_S$ we obtain

$$\|\sigma - \sigma_S\|_H = \left\|V^{-1} V (\sigma - \sigma_S)\right\|_H \leq C_V \left\|V (\sigma - \sigma_S)\right\|_{H'} = C_V \left\|\tilde{g}_D - V\sigma_S\right\|_{H'}. \tag{1.44}$$

Here H is the infinite-dimensional space in which the continuous problem is formulated and $V : H \to H'$ is the boundary integral operator associated to the single layer potential. H' is the dual space of H and

$$C_V := \left\|V^{-1}\right\|_{H \leftarrow H'}$$

is the continuity constant of the inverse operator V^{-1}. Note that (1.44) contains the right-hand side $\tilde{g}_D$ as in (1.26) and the numerically computed solution σ_S but not the exact solution. For the operator V the norm in the space H' is given for $w \in H'$ by

$$\begin{aligned}\|w\|_{H',\Gamma} &:= \sqrt{\|w\|^2_{L^2(\Gamma)} + |w|^2_{H',\Gamma}} : \\ &= \left\{\|w\|^2_{L^2(\Gamma)} + \int_\Gamma \int_\Gamma \frac{|w(\mathbf{x}) - w(\mathbf{y})|^2}{\|\mathbf{x} - \mathbf{y}\|^3} ds_{\mathbf{x}} ds_{\mathbf{y}}\right\}^{1/2}\end{aligned}$$

which is denoted as the Sobolev norm of fractional order $1/2$. The goal is to estimate the $\|\cdot\|_{H',\Gamma}$-norm of the residual $r = \tilde{g}_D - V\sigma_S$ [cf. (1.44)] and use this as a bound for the discretization error. To get local insights on the error distribution, it will also be important to estimate $\|r\|_{H',\Gamma}$ by a sum of local *error indicators*. For this, let $\mathcal{I}$ denote the set of counting indices of the basis functions b_i, $i \in \mathcal{I}$, for the boundary element space S and let

$$\omega_i := \operatorname{supp} b_i$$

denote their support. Then we will show that

$$C_{\text{eff}}\sqrt{\sum_{i\in\mathcal{I}}\eta_i^2} \leq \|\sigma-\sigma_S\|_H \leq C_{\text{rel}}\sqrt{\sum_{i\in\mathcal{I}}\eta_i^2}, \quad \text{where} \quad \eta_i := |r|_{H',\omega_i}. \tag{1.45}$$

A posteriori error estimators which satisfy the upper estimate in (1.45) are called *reliable* and if the lower estimate is satisfied they are called *efficient*. These estimates require the localization of the integral $\int_{\Gamma\times\Gamma}$ in the definition of the fractional order Sobolev norm to a sum of integrals over $\int_{\omega_i\times\omega_i}$. It is relatively simple (but technical) to prove that

$$|v|^2_{H',\Gamma} \leq \sum_{i\in\mathcal{I}}|v|^2_{H',\omega_i} + C\sum_{\tau\in\mathcal{G}}h_\tau^{-1}\|v\|^2_{L^2(\tau)}.$$

Hence it remains to estimate

$$h_\tau^{-1}\|v\|^2_{L^2(\tau)} \leq C\|v\|^2_{H',\omega_\tau}, \quad \text{where} \quad \omega_\tau := \overline{\bigcup_{t\in\mathcal{G}:\bar{t}\cap\bar{\tau}\neq\emptyset} t}. \tag{1.46}$$

Unfortunately, this estimate cannot hold for all functions $v \in H'$ as can be easily seen by considering the function $v = 1$. However, for functions which satisfy a certain orthogonality relation with respect to the boundary element space, estimate (1.46) can be proved. Furthermore, we will prove that the residual r satisfies this orthogonality relation and hence the reliability of the error estimator can be concluded.

Notation 1.2.1. *Throughout the book C and c denote generic positive constants which may vary from inequality to inequality.*

Chapter 2
Elliptic Differential Equations

Integral equations occur in many physical applications. We encounter some of the most important ones when we try to solve elliptic differential equations. These can be transformed into integral equations and can then be solved numerically by means of the boundary element method. The subject of this chapter is the formulation and analysis of scalar, elliptic boundary value problems.

2.1 Elementary Functional Analysis

In this chapter we will present a few fundamental results from functional analysis that we will need at a later stage. It is not intended as an introduction to functional analysis; instead we will refer to other textbooks or we will give schematic proofs if we think this might help the reader's understanding of the subject. The presentation is based on the book [115, Chap. 6]. A detailed introduction to linear functional analysis can be found in, e.g., [3, 62, 98, 195, 243].

2.1.1 Banach and Hilbert Spaces

2.1.1.1 Normed Spaces

We denote by X a normed, linear space over the coefficient field $\mathbb{K} \in \{\mathbb{R}, \mathbb{C}\}$. A norm $\|\cdot\| : X \to [0, \infty)$ is a mapping with the properties

$$\forall x \in X : \|x\| = 0 \Longrightarrow x = 0, \tag{2.1a}$$

$$\forall \lambda \in \mathbb{K} : \|\lambda x\| = |\lambda| \, \|x\|, \tag{2.1b}$$

$$\forall x, y \in X : \|x + y\| \leq \|x\| + \|y\|. \tag{2.1c}$$

We will use the notation $\|\cdot\|_X$ if the space X is not clear from the context. We call the pair $(X, \|\cdot\|)$ a normed space.

S.A. Sauter and C. Schwab, *Boundary Element Methods*, Springer Series in Computational Mathematics 39, DOI 10.1007/978-3-540-68093-2_2,

With $\|\cdot\|_X$ we have defined a topology on X: a subset $A \subset X$ is open if there exists a constant $\varepsilon > 0$ for all $x \in A$ such that the ball $\{y \in X : \|x-y\|_X < \varepsilon\} \subset A$. For a sequence $(x_n)_n \subset X$ we write $x_n \to x$ if

$$x = \lim_{n\to\infty} x_n \Longleftrightarrow \lim_{n\to\infty} \|x - x_n\|_X = 0.$$

Remark 2.1.1. *Every norm $\|\cdot\| : X \to [0,\infty)$ is continuous since we have from (2.1c) the reverse triangle inequality*

$$\forall x, y \in X : \ |\|x\| - \|y\|| \leq \|x - y\|. \tag{2.2}$$

We can define several different norms on X. Two norms $\|\cdot\|_1$, $\|\cdot\|_2$ on X are *equivalent* if and only if

$$\exists C > 0 : \quad C^{-1} \|x\|_1 \leq \|x\|_2 \leq C \|x\|_1 \qquad \forall x \in X. \tag{2.3}$$

Equivalent norms induce the same topology on X.

2.1.1.2 Linear Operators

Let X and Y be normed spaces with the respective norms $\|\cdot\|_X$ and $\|\cdot\|_Y$. A linear mapping $T : X \to Y$ is called an operator. An operator $T : X \to Y$ is called *bounded* if

$$\|T\|_{Y\leftarrow X} := \sup\left\{\|Tx\|_Y / \|x\|_X : 0 \neq x \in X\right\} < \infty. \tag{2.4}$$

Here $\|T\|_{Y\leftarrow X}$ is the *operator norm*. The set of all bounded linear operators $T : X \to Y$ is denoted by $L(X, Y)$ and together with

$$(T_1 + T_2)x := T_1 x + T_2 x, \ (\lambda T_1)x = T_1(\lambda x), \ \lambda \in \mathbb{K}, \tag{2.5}$$

constitutes a normed, linear space $(L(X, Y), \|\cdot\|_{Y\leftarrow X})$. If $X = Y$ we write $L(X)$ instead of $L(X, X)$. $L(X)$ is an algebra if we set

$$\forall T_1, T_2 \in L(X, X) : \ (T_1 T_2)x := T_1(T_2 x).$$

For a normed space X, $I_X \in L(X)$ denotes the identity on X. A mapping $T^{-1} \in L(Y, X)$ is the *inverse* of the mapping $T \in L(X, Y)$ if we have $TT^{-1} = I_Y$ and $T^{-1}T = I_X$.

Exercise 2.1.2. *(a) Show that for all $x \in X$ and $T \in L(X, Y)$ we have*

$$\|Tx\|_Y \leq \|T\|_{Y\leftarrow X} \|x\|_X. \tag{2.6}$$

(b) Show that for $T_1 \in L(Y, Z)$, $T_2 \in L(X, Y)$ we have $T_1T_2 \in L(X, Z)$ and

$$\|T_1T_2\|_{Z\leftarrow X} \le \|T_1\|_{Z\leftarrow Y}\, \|T_2\|_{Y\leftarrow X}. \tag{2.7}$$

Definition 2.1.3. The sequence $(T_n)_n \subset L(X, Y)$ converges to T if

$$T_n \to T \iff \|T - T_n\|_{Y\leftarrow X} \to 0 \text{ for } n \to \infty.$$

It converges pointwise to T if

$$\forall x \in X : \|T_n x - Tx\|_Y \to 0 \text{ for } n \to \infty.$$

2.1.1.3 Banach Spaces

The sequence $\{x_n\} \subset X$ is called *Cauchy sequence* if $\sup\{\|x_n - x_m\|_X : n, m \ge k\} \to 0$ for $k \to \infty$. X is called *complete* if all Cauchy sequences converge to an $x \in X$. A complete, normed, linear space is called a *Banach space*.

Proposition 2.1.4. *Let X be a normed space and Y a Banach space. Then $L(X, Y)$ is a Banach space.*

Proposition 2.1.5. *Let X be a Banach space and $Z \subset X$ a closed subspace. The* quotient space *X/Z consists of the classes $\tilde{x} := \{x + z : z \in Z\}$ for all $x \in X$. The quotient space X/Z with the norm $\|\tilde{x}\| := \inf\{\|x + z\|_X : z \in Z\}$ is a Banach space.*

We call the set $A \subset X$ dense in X if we have for the closure $\overline{A} = X$. More specifically, this means that for all $x \in X$ there exists a sequence $(x_n)_n \subset A$ with $x_n \to x$. If $(X, \|\cdot\|_X)$ is normed but not complete, then the Banach space $(\widetilde{X}, \|\cdot\|_{\widetilde{X}})$ is the *completion* of X if X is dense in $\widetilde{X}$, $\widetilde{X}$ is complete and we have $\|x\|_{\widetilde{X}} = \|x\|_X$ for all $x \in X$.

The Banach space X is called *separable* if there exists a countable, dense subset $A = \{a_n : n \in \mathbb{N}\} \subset X$.

The completion $\widetilde{X}$ is unique up to isomorphism. The continuous extension of a linear operator $T \in L(X, Y)$ from a dense subset $X_0 \subset X$ to X is also uniquely determined. The following proposition explains this in more detail.

Proposition 2.1.6. *Let X_0 be a dense subset of $(X, \|\cdot\|_X)$. An operator $T_0 \in L(X_0, Y)$ with*

$$\|T_0\|_{Y\leftarrow X_0} = \sup\{\|T_0 x\|_Y / \|x\|_X : 0 \ne x \in X_0\} < \infty$$

has a unique extension $T \in L(X, Y)$ that satisfies the following conditions:

1. *For all $x \in X_0$ we have $Tx = T_0 x$.*
2. *For all sequences $(x_n)_n \subset X_0$ with $x_n \to x \in X$ we have $Tx = \lim_{n\to\infty} T_0 x_n$.*
3. *$\|T\|_{Y\leftarrow X} = \|T_0\|_{Y\leftarrow X_0}$.*

The following theorem and corollary are both a result of the open mapping theorem (see [243, Theorem 6.6]).

Theorem 2.1.7. *Let X, Y be Banach spaces, let $T \in L(X, Y)$ be injective ($Tx = Ty \Longrightarrow x = y$) and surjective (for all $y \in Y$ there exists an $x \in X$ with $Tx = y$). Then the mapping $T^{-1} \in L(Y, X)$ exists.*

Corollary 2.1.8. *Let X, Y be Banach spaces and let $T \in L(X, Y)$ be injective. Then the following conditions are equivalent:*

(a) $Y_0 := \{Tx : x \in X\}$ with $\|\cdot\|_Y$ is a closed subspace of Y.
(b) T^{-1} exists on Y_0 and $T^{-1} \in L(Y_0, X)$.

2.1.1.4 Embeddings

Let X, Y be Banach spaces with $X \subset Y$. The injection (or embedding) $I : X \to Y$ is defined by $Ix = x$ for all $x \in X$ and clearly is linear. If I is bounded, that is,

$$\forall x \in X : \; \|x\|_Y \leq C \|x\|_X, \tag{2.8}$$

we have $I \in L(X, Y)$. If X is also dense in Y, we call X densely and continuously embedded in Y.

2.1.1.5 Hilbert Spaces

Let X be a vector space. A mapping $(\cdot, \cdot) : X \times X \to \mathbb{K}$ is called an *inner product* on X if

$$(x, x) > 0 \;\; \forall x \in X \setminus \{0\}, \tag{2.9a}$$

$$(\lambda x + y, z) = \lambda(x, z) + (y, z) \quad \forall \lambda \in \mathbb{K},\ x, y, z \in X, \tag{2.9b}$$

$$(x, y) = \overline{(y, x)} \quad \forall x, y \in X. \tag{2.9c}$$

A Banach space $(X, \|\cdot\|_X)$ is called a *Hilbert space* if there exists an inner product on X such that $\|x\|_X = (x, x)^{1/2}$ for all $x \in X$.

Furthermore, from (2.9) we have the Cauchy–Schwarz inequality

$$|(x, y)| \leq \|x\| \; \|y\| \quad \forall x, y \in X. \tag{2.10}$$

Two vectors $x, y \in X$ are *orthogonal* if $(x, y) = 0$. We denote this by $x \perp y$. For $A \subset X$, $A^{\perp} := \{x \in X \mid \forall a \in A : (x, a) = 0\}$ is a closed subspace of X.

Proposition 2.1.9. *Let X be a Hilbert space and $U \subset X$ a closed subspace. Then we have $X = U \oplus U^{\perp}$, i.e.,*

$$\forall x \in X : x = u + v,\ u \in U,\ v \in U^{\perp},\ \|x\|^2 = \|u\|^2 + \|v\|^2.$$

A system of orthonormal vectors $(v_j)_{j\in\mathcal{I}}$ in a Hilbert space X is an *orthonormal basis* of X if, for every $x \in X$, the Fourier expansion

$$x = \sum_{j\in\mathcal{I}} (x, v_j)\, v_j$$

converges.

Theorem 2.1.10. *For every Hilbert space, there exists an orthonormal basis.*

A proof can be found, e.g., in [131, Theorem 65.1], [141].

2.1.2 *Dual Spaces*

2.1.2.1 Dual Space of a Normed, Linear Space

Let X be a normed, linear space over $\mathbb{K} \in \{\mathbb{R}, \mathbb{C}\}$. The dual space X' of X is the space of all bounded, linear mappings (functionals)

$$X' = L(X, \mathbb{K}).$$

X' is a Banach space with norm

$$\|x'\|_{X'} := \|x'\|_{\mathbb{K}\leftarrow X} = \sup\left\{\left|x'(x)\right| / \|x\|_X : x \in X\backslash\{0\}\right\}. \tag{2.11}$$

For $x'(x)$ one can also write

$$\langle x, x'\rangle_{X\times X'} = \langle x', x\rangle_{X'\times X} = x'(x), \tag{2.12}$$

where $\langle\cdot,\cdot\rangle_{X\times X'}$, $\langle\cdot,\cdot\rangle_{X'\times X}$ are called *dual forms* or *duality pairings*.

Lemma 2.1.11. *Let $X \subset Y$ be continuously embedded. Then $Y' \subset X'$ is continuously embedded.*

Proof. Since $X \subset Y$, any $y' \in Y'$ is defined on X. We therefore have $Y' \subset X'$. Since $X \subset Y$, we have, due to (2.8),

$$\|y'\|_{Y'} = \sup_{x\in Y\backslash\{0\}} \left\{|y'(x)|/\|x\|_Y\right\} \geq C^{-1} \sup_{x\in X\backslash\{0\}} \left\{|y'(x)|/\|x\|_X\right\} = C^{-1}\,\|y'\|_{X'}$$

and therefore $\|y'\|_{X'} \leq C\|y'\|_{Y'}$. This proves that the embedding $Y' \subset X'$ is continuous. □

The *bidual space* X'' of X is defined as

$$X'' = L(X', \mathbb{K}).$$

In general we have the strict inclusion $X \subset X''$. However, in many cases X is isomorphic to X'', i.e., every $x'' \in X''$ can be identified with an $x \in X$. We write $X \cong X''$. In this case we call X *reflexive*. In particular, all Hilbert spaces are reflexive.

2.1.2.2 Dual Operator

One of the most general principles in functional analysis is the extension of continuous linear operators which are defined on some subspace of a Banach space to the whole Banach space. We will need here the version of the Hahn–Banach extension theorem in Banach spaces.

Theorem 2.1.12. *Let X be a Banach space, M a subspace of X and f_0 a continuous linear functional defined on M. Then there exists a continuous linear functional f defined on X such that (i) f is an extension of f_0 and (ii) $\|f_0\|_{\mathbb{C}\leftarrow M} = \|f\|_{\mathbb{C}\leftarrow X}$.*

The proof can be found, e.g., in [243, Chap. IV, Sect. 5].

Corollary 2.1.13. *Let X be a Banach space and $x_0 \in X\backslash\{0\}$. Then there exists a continuous linear functional f_0 on X such that*

$$f_0(x_0) = \|x_0\|_X \quad \textit{and} \quad \|f_0\|_{X'} = 1.$$

Proof. Let $M := \operatorname{span}\{x_0\}$ and define $f_0 : M \to \mathbb{R}$ by

$$f_0(\alpha x_0) := \alpha \|x_0\|_X \quad \forall \alpha \in \mathbb{C}.$$

Then f is linear on M and $|f_0(\alpha x_0)| = |\alpha \|x_0\|_X| = \|\alpha x_0\|_X$, i.e., $\|f_0\|_{\mathbb{C}\leftarrow M} = 1$. Theorem 2.1.12 implies that there is a continuous linear functional f defined on X such that

$$f(x) = f_0(x) \quad \forall x \in M \quad \text{and} \quad \|f\|_{\mathbb{C}\leftarrow X} = \|f_0\|_{\mathbb{C}\leftarrow M} = 1.$$

□

Proposition 2.1.14. *Let X, Y be Banach spaces and let $T \in L(X, Y)$. For $y' \in Y'$,*

$$\langle Tx, y' \rangle_{Y\times Y'} = \langle x, x' \rangle_{X\times X'} \quad \forall x \in X \tag{2.13}$$

defines a unique $x' \in X'$. The mapping $y' \to x'$ is linear and defines the dual operator *$T' : Y' \to X'$ as given by $T'y' = x'$. Furthermore, we have $T' \in L(Y', X')$*

and

$$\|T'\|_{X' \leftarrow Y'} = \|T\|_{Y \leftarrow X}. \tag{2.14}$$

Proof. The relation given in (2.13) can be written as $y'(Tx) = x'(x)$ or $x' = y' \circ T$. It follows from $y' \in L(Y, \mathbb{K})$ and $T \in L(X, Y)$ that $x' \in L(X, \mathbb{K})$ (Exercise 2.1.2).

From the defining relation $\langle Tx, y' \rangle_{Y \times Y'} = \langle x, T'y' \rangle_{X \times X'}$ we obtain

$$\begin{aligned}
\left\|T'y'\right\|_{X'} &= \sup_{x \in X \setminus \{0\}} \frac{|\langle T'y', x \rangle_{X' \times X}|}{\|x\|_X} = \sup_{x \in X \setminus \{0\}} \frac{|\langle Tx, y' \rangle_{Y \times Y'}|}{\|x\|_X} \\
&\leq \left\|y'\right\|_{Y'} \sup_{x \in X \setminus \{0\}} \frac{\|Tx\|_Y}{\|x\|_X} = \left\|y'\right\|_{Y'} \|T\|_{Y \leftarrow X}.
\end{aligned}$$

Hence $\|T'\|_{X' \leftarrow Y'} \leq \|T\|_{Y \leftarrow X}$. The reverse inequality is proved next. Corollary 2.1.13 implies that for any $x_0 \in X \setminus \{0\}$ there exists a functional $f_0 \in Y'$ such that $\|f_0\|_{Y'} = 1$ and $f_0(Tx_0) = \langle Tx_0, f_0 \rangle_{Y \times Y'} = \|Tx_0\|_Y$. Thus $f_0' := T' f_0 \in X'$ satisfies

$$\langle x_0, f_0' \rangle_{X \times X'} = \|Tx_0\|_Y$$

and so

$$\|Tx_0\|_Y = \langle x_0, T' f_0 \rangle_{X \times X'} \leq \left\|T'\right\|_{X' \leftarrow Y'} \|x_0\|_X \|f_0\|_{Y'} = \left\|T'\right\|_{X' \leftarrow Y'} \|x_0\|_X.$$

Note that this estimate trivially holds for $x_0 = 0$. We conclude that $\|T\|_{Y \leftarrow X} \leq \|T'\|_{X' \leftarrow Y'}$ and (2.14) follows.

□

Conclusion 2.1.15. *For two operators $S \in L(X, Y)$, $T \in L(Y, Z)$ we have:*

(i) $(TS)' = S'T'$.
(ii) *S is surjective $\Longrightarrow$ $S' \in L(Y', X')$ is injective.*

Definition 2.1.16. Let X be a Banach space over $\mathbb{K} \in \{\mathbb{R}, \mathbb{C}\}$. A function $f : X \to Y$ is *conjugate linear* if

$$f(\alpha u + \beta v) = \bar{\alpha} f(u) + \bar{\beta} f(v) \qquad \forall u, v \in X \text{ and } \alpha, \beta \in \mathbb{K}.$$

2.1.2.3 Adjoint Operator

Let X be a Hilbert space over $\mathbb{K} \in \{\mathbb{R}, \mathbb{C}\}$. For all $y \in X$,

$$f_y(\cdot) := (\cdot, y)_X : X \to \mathbb{K}$$

is continuous and linear. Thus we have $f_y(\cdot) \in X'$ and $\|f_y\|_{X'} = \|y\|_X$. The converse is a result of Riesz' theorem.

Theorem 2.1.17 (Riesz Representation Theorem). *Let X be a Hilbert space. For all $f \in X'$ there exists a unique $y_f \in X$ such that*

$$\|f\|_{X'} = \|y_f\|_X \quad \text{and} \quad f(x) = (x, y_f)_X \qquad \forall x \in X.$$

Conclusion 2.1.18. *Let X be a Hilbert space. We use the same notation as in Theorem 2.1.17:*

(a) There exists a bounded, invertible conjugate linear mapping $J_X : X \to X'$ with $J_X y = f_y$, $J_X^{-1} f = y_f$. The mapping J_X is an isometry: $\|J_X\|_{X' \leftarrow X} = \|J_X^{-1}\|_{X \leftarrow X'} = 1$.
(b) X' is a Hilbert space with inner product $(x', y')_{X'} := \overline{(J_X^{-1} x', J_X^{-1} y')_X}$.
(c) $\|x'\|_{X'}$ in (2.11) is equal to $(x', x')_{X'}^{1/2}$.
(d) $X \cong X''$ with $x(x') := x'(x)$ and we identify X with X''. In particular, we have $J_{X'} = J_X^{-1}$, $J_X = (J_X)'$, $T'' = T$ for $T \in L(X, Y)$ if $Y = Y''$ and if both are Hilbert spaces.
(e) If $\mathbb{K} = \mathbb{R}$, the spaces X and X' can be identified with each other by means of the isomorphism J_X. Then we have $X := X' \Longrightarrow J_X = I$.
(f) Let $\mathbb{K} = \mathbb{C}$. According to Theorem 2.1.10 we may choose a basis $(v_i)_{i \in \mathcal{I}}$ in X and define the complex conjugation by

$$Cx := \overline{x} := \sum_{j \in \mathcal{I}} \overline{(x, v_j)_X} v_j \tag{2.15}$$

which satisfies $C^{-1} = C$ and C, C^{-1} are conjugate linear isometries. Hence $\bar{J}_X := J_X C$ is an isometric isomorphism and we may identify any Hilbert space with its dual by means of $\bar{J}_X$.

Definition 2.1.19. Let X, Y be Hilbert spaces and $T \in L(X, Y)$. The adjoint operator of T is given by $T^* := J_X^{-1} T' J_Y \in L(Y, X)$.

We have

$$\|T\|_{Y \leftarrow X} = \|T^*\|_{X \leftarrow Y} \quad \text{and} \quad (Tx, y)_Y = (x, T^* y)_X \quad \forall x \in X, y \in Y. \tag{2.16}$$

Definition 2.1.20.

(a) $T \in L(X)$ is self adjoint if $T = T^*$.
(b) $T \in L(X)$ is a projection if $T^2 = T$.

Proposition 2.1.21. *Let $X_0 \subset X$ be a closed subspace of the Hilbert space X. For $x \in X$ there exists a unique $x_0(x) \in X_0$ with*

$$\|x - x_0\|_X = \min\{\|x - y\|_X : y \in X_0\}. \tag{2.17}$$

The mapping $x \to x_0 =: Px$ is an orthogonal projection.

Proof. Existence and uniqueness: The decomposition $x = x_0 + x_\perp$, $x_0 \in X_0$, $x_\perp \in X_0^\perp$ is unique. We will show that $y = x_0$ minimizes the right-hand side in (2.17). If we take $x_\perp \perp (x_0 - z)$ into consideration, we have for every $z \in X_0$

$$\begin{aligned}\|x-z\|_X^2 &= \|x-x_0+x_0-z\|_X^2 = \|x-x_0\|_X^2 - 2\,\mathrm{Re}\,(x-x_0, x_0-z)_X + \|x_0-z\|_X^2 \\ &= \|x-x_0\|_X^2 + \|x_0-z\|_X^2 \ge \|x-x_0\|_X^2 . \end{aligned} \tag{2.18}$$

This means that x_0 minimizes as required. The inequality in (2.18) only becomes an equality for $x_0 = z$, which gives us the uniqueness.

Projection property: For $x \in X_0$ the first part of the proof implies $Px = x$ and therefore $P^2 = P$.

Orthogonality: Let $P^* \in L(X)$ be the adjoint operator of P. For $x, y \in X$ with $x_0 := Px$ and $y_0 = Py$ we have $P = P^*$, since

$$\begin{aligned}\left(x, P^*y\right)_X &= (Px, y)_X = (x_0, y)_X = (x_0, y_0 + y_\perp)_X \\ &= (x_0, y_0)_X = (x_0 + x_\perp, y_0)_X = (x, Py)_X .\end{aligned}$$

The assertion follows from the fact that, for all $y \in X_0$, we have

$$\begin{aligned}(x - Px, y)_X &= (x - Px, Py)_X = \left(P^*x - P^*Px, y\right)_X \\ &= \left(Px - P^2x, y\right)_X = (Px - Px, y)_X = 0.\end{aligned}$$

□

2.1.2.4 Gelfand Triple

In this section V and U will always denote Hilbert spaces with a continuous and dense embedding $V \subset U$.

Proposition 2.1.22. *We have*

$$U' \subset V' \text{ is continuously and densely embedded.} \tag{2.20}$$

Proof. The continuity of the embedding $U' \subset V'$ follows from Lemma 2.1.11. The fact that the embedding is dense follows from the auxiliary result: $(U')^\perp = \{0\}$ in V'. In order to prove this we choose a $v' \in V' \backslash \{0\}$ and set $u := J_V^{-1} v' \in V \subset U$. The function $u' := J_U u \in U' \subset V'$ satisfies $u'(x) = (x, u)_U$ for all $x \in U$. By choosing $x := u = J_V^{-1} v'$ we obtain

$$(v', u')_{V'} = \overline{(J_V^{-1} v', J_V^{-1} u')_V} = \overline{(u, J_V^{-1} u')_V} = \overline{u'(u)} = \overline{(u, u)_U} > 0.$$

Therefore for all $0 \neq v' \in V'$ there exists a $u' \in U'$ with $(u', v')_{V'} \neq 0$. From this we have $(U')^\perp = \{0\} \subset V'$ and therefore U' is dense in V'. □

We identify U and U' (cf. Conclusion 2.1.18e,f) and obtain the Gelfand triple

$$V \subset U \subset V' \quad (V \subset U \text{ continuous and dense}). \tag{2.21}$$

Proposition 2.1.23. *In the Gelfand triple (2.21), V and U are also continuously and densely embedded in V'.*

Remark 2.1.24. *(a) In (2.21) one could also choose $V = V'$, which would result in $U' \subset V' = V \subset U$. For $U \neq V$ it is not possible to set $U = U'$ and $V = V'$ simultaneously, as, for $x, y \in U$, we would have $x(y) = \langle y, x\rangle_{U\times U'} = (y, x)_U = (y, x)_V$, which is a contradiction for $U \neq V$.*

(b) Since $U = U'$, $(x, y)_U$ can also be interpreted as $\langle x, y\rangle_{U\times U'}$. For $x \in V \subset U$, we have $y(x) = \langle x, y\rangle_{V\times V'} = (x, y)_U$ for all $y \in U \subset V'$. Since $U \subset V'$ is dense and continuous, $(\cdot,\cdot)_U$ can be extended continuously to $V \times V'$ as the dual form $\langle\cdot,\cdot\rangle_{V\times V'}$.

2.1.2.5 Weak Convergence

The Bolzano–Weierstrass theorem states that in $\mathbb{K} \in \{\mathbb{R}, \mathbb{C}\}$ every bounded sequence has at least one accumulation point. This statement only holds in a weaker form when considering infinite-dimensional function spaces. First we will need to define the concept of weak convergence.

Definition 2.1.25. Let B be a Banach space and let B' be its dual space. A sequence $(u_\ell)_{\ell\in\mathbb{N}}$ in B converges *weakly* to an element $u \in B$ if

$$\lim_{\ell\to\infty} \|f(u) - f(u_\ell)\|_{B'} = 0 \qquad \forall f \in B'.$$

Theorem 2.1.26. *Let the Banach space B be reflexive and let $(u_\ell)_{\ell\in\mathbb{N}}$ be a bounded sequence in B:*

$$\sup_{\ell\in\mathbb{N}_0} \|u_\ell\|_B \le C < \infty.$$

Then there exists a subsequence $\left(u_{\ell_j}\right)_{j\in\mathbb{N}}$ that converges weakly to a $u \in B$.

The proof can be found in, e.g., [141, V, Sect. 7, Theorem 7], [131, Theorem 60.6]. In order to distinguish the weak convergence of a sequence $(u_\ell)_{\ell\in\mathbb{N}}$ to an element u from the usual (strong) convergence, we use the notation

$$u_\ell \rightharpoonup u.$$

2.1.3 Compact Operators

Definition 2.1.27. The subset $U \subset X$ of the Banach space X is called precompact if every sequence $(x_n)_{n\in\mathbb{N}} \subset U$ has a convergent subsequence $\left(x_{n_i}\right)_{i\in\mathbb{N}}$. It is compact if, furthermore, $x = \lim_{i\to\infty} x_{n_i} \in U$.

Definition 2.1.28. Let X, Y be Banach spaces. $T \in L(X,Y)$ is called compact if $\{Tx : x \in X,\ \|x\|_X \leq 1\}$ is precompact in Y.

We will often consider operators that consist of several other operators.

Lemma 2.1.29. *Let X, Y, Z be Banach spaces, let $T_1 \in L(X,Y)$, $T_2 \in L(Y,Z)$ and let at least one of the operators T_i be compact. Then $T = T_2T_1 \in L(X,Z)$ is also compact.*

Lemma 2.1.30. *$T \in L(X,Y)$ compact $\Longrightarrow T' \in L(Y', X')$ compact.*

Definition 2.1.31. Let Y be a Banach space and $X \subset Y$ a subspace that is continuously embedded. The embedding is compact if the injection $I \in L(X,Y)$ is compact. We denote this by $X \subset\subset Y$.

Conclusion 2.1.32. *$X \subset\subset Y$ if every sequence $(x_i)_{i\in\mathbb{N}} \subset X$ with $\|x_i\|_X \leq 1$ has a subsequence that converges in Y.*

Lemma 2.1.33. *Let $V \subset U \subset V'$ be a Gelfand triple and let the embedding $V \subset\subset U$ be compact. For $T \in L(V', V)$ the restrictions $T \in L(V', V')$, $T \in L(U, U)$, $T \in L(V, V)$, $T \in L(V', U)$ and $T \in L(U, V)$ are all compact.*

Proof. According to the assumptions the embedding $I \in L(V, U)$ is compact, and therefore so is $I \in L(U, V')$ (see Lemma 2.1.30). $T \in L(U, V)$ is the composition of the (compact) embedding $I \in L(U, V')$ with $T \in L(V', V)$ and thus it is also compact (see Lemma 2.1.29). □

Remark 2.1.34. *For dim$(X) < \infty$ or dim$(Y) < \infty$, $T \in L(X,Y)$ is compact.*

The following lemma will be needed later for existence theorems when dealing with variational problems.

Lemma 2.1.35. *Let $X \subset Y \subset Z$ be Banach spaces with continuous embeddings and let $X \subset\subset Y$. Then for all $\varepsilon > 0$ there exists a constant $C_\varepsilon > 0$ with*

$$\forall x \in X:\ \|x\|_Y \leq \varepsilon \|x\|_X + C_\varepsilon \|x\|_Z.$$

2.1.4 Fredholm–Riesz–Schauder Theory

Let X be a Banach space and let $T \in L(X)$ be a compact operator. In the following theorem we will establish the connection between the spectrum

$$\sigma(T) := \{\lambda \in \mathbb{C} : (T - \lambda I)^{-1} \notin L(X)\} \tag{2.22}$$

and the eigenvalues of T.

Theorem 2.1.36. *(i) For all* $\lambda \in \mathbb{C}\backslash\{0\}$ *we have one of the alternatives* *(a)* $(T - \lambda I)^{-1} \in L(X)$ *or (b)* λ *is an eigenvalue of* T.
The following alternatives are equivalent: (a′) The equation

$$Tx - \lambda x = y$$

has a unique solution $x \in X$ *for all* $y \in X$. *(b′) There exists a finite-dimensional eigenspace*

$$E(\lambda, T) = \{v \in X \mid Tv = \lambda v\}$$
$$\forall 0 \neq x \in E(\lambda, T):\ Tx = \lambda x.$$

(ii) $\sigma(T)$ *consists of all eigenvalues of* T *and it includes* $\lambda = 0$ *if* $T^{-1} \notin L(X, X)$. *There are at most countably many eigenvalues* $\{\lambda_j\}$ *and the only possible accumulation point is zero.*

(iii)
$$\lambda \in \sigma(T) \Longleftrightarrow \overline{\lambda} \in \sigma(T'). \tag{2.23}$$

(iv) We have
$$\dim(E(\lambda, T)) = \dim(E(\overline{\lambda}, T')) < \infty. \tag{2.24}$$

(v) For $\lambda \in \sigma(T)\backslash\{0\}$ *the equation*

$$(T - \lambda I)x = y$$

has at least one solution if and only if the compatibility condition

$$\langle y, x' \rangle_{X \times X'} = 0 \quad \forall x' \in E(\overline{\lambda}, T') \tag{2.25}$$

is satisfied.

The following corollary is a result of Theorem 2.1.36 and will play a significant role in later applications.

Corollary 2.1.37. *Let* X *be a Banach space and let* $T \in L(X)$ *be a compact operator. Then we have the following equivalence:*

$$I + T \text{ is injective} \Longleftrightarrow I + T \text{ is an isomorphism.}$$

2.1.5 Bilinear and Sesquilinear Forms

Let H_1, H_2 be Hilbert spaces with norms $\|\cdot\|_{H_1}, \|\cdot\|_{H_2}$ over $\mathbb{K}$. A mapping $a(\cdot,\cdot):$ $H_1 \times H_2 \to \mathbb{K}$ is called a sesquilinear form if

$$\forall u_1, u_2 \in H_1, v_1, v_2 \in H_2, \lambda \in \mathbb{K} : a(u_1 + \lambda u_2, v_1) = a(u_1, v_1) + \lambda a(u_2, v_1),$$
$$a(u_1, v_1 + \lambda v_2) = a(u_1, v_1) + \overline{\lambda} a(u_1, v_2). \tag{2.26}$$

For a sesquilinear form $a : H \times H \to \mathbb{C}$ the adjoint sesquilinear form $a^* : H \times H \to \mathbb{C}$ is defined by

$$a^*(u, v) = \overline{a(v, u)} \qquad \forall u, v \in H. \tag{2.27}$$

It is called *Hermitian* if $a = a^*$.

If $\mathbb{K} = \mathbb{R}$ we speak of a bilinear form. The bilinear form $a : H \times H \to \mathbb{R}$ is called symmetric if

$$a(u, v) = a(v, u) \qquad \forall u, v \in H.$$

A sesquilinear form $a(\cdot, \cdot) : H_1 \times H_2 \to \mathbb{K}$ is *continuous* (or *bounded*) if there exists a constant $C < \infty$ with

$$|a(u, v)| \le C \|u\|_{H_1} \|v\|_{H_2}, \tag{2.28}$$

for all $u \in H_1$, $v \in H_2$. The smallest C in (2.28) is the norm of $a(\cdot, \cdot)$ and we write

$$\|a\| := \sup_{u \in H_1 \setminus \{0\}} \sup_{v \in H_2 \setminus \{0\}} \frac{|a(u, v)|}{\|u\|_{H_1} \|v\|_{H_2}}. \tag{2.29}$$

We can identify sesquilinear forms with linear operators.

Lemma 2.1.38. *Let H_1, H_2 be Hilbert spaces over $\mathbb{K}$:*

(a) *For every sesquilinear form $a(\cdot, \cdot)$: $H_1 \times H_2 \to \mathbb{C}$ there exists a unique $A \in L(H_1, H_2')$ such*[1] *that*

$$a(u, v) = \langle Au, v \rangle_{H_2' \times H_2} \qquad \forall u \in H_1, v \in H_2. \tag{2.30}$$

It satisfies

$$\|A\|_{H_2' \leftarrow H_1} \le \|a\|. \tag{2.31}$$

(b) *Let S_1, S_2 be dense in H_1, H_2 and let the sesquilinear form $a(\cdot, \cdot)$ be defined on $S_1 \times S_2$. We assume that (2.28) holds for all $u_1 \in S_1$, $v_1 \in S_2$. Then $a(\cdot, \cdot)$ can be uniquely and continuously extended on $H_1 \times H_2$ and (2.28) holds on $H_1 \times H_2$ with the same constant $C = \|a\|$.*

---

[1] More precisely, $A \in L(H_1, H_2^*)$, where the *anti-dual space* H_2^* contains all bounded conjugate linear forms on H_2. A linear operator is defined by $\langle \widetilde{A}u, \overline{v} \rangle_{H_2' \times H_2} = a(u, v)$ for all $u \in H_1$, $v \in H_2$. [Recall that complex conjugation in Hilbert spaces is well defined; see (2.15).] Note that $A = C'\widetilde{A}$, where C' is the dual operator for the complex conjugation operator C as in (2.15). If no confusion is possible, we do not distinguish in the notation the dual space from the anti-dual space and always write H_2'.

Proof. (a) For $u \in H_1$, $\varphi_u(v) := a(u,v)$ defines a linear functional $\varphi_u(\cdot) \in H_2'$ with $\|\varphi_u\|_{H_2'} \le C\|u\|_{H_1}$. We set $Au := \varphi_u$ for all $u \in H_1$. We then have $\|Au\|_{H_2'} \le C\|u\|_{H_1}$ and as a consequence (2.31). It also follows that

$$\langle Au, v\rangle_{H_2'\times H_2} = \langle \varphi_u, v\rangle_{H_2'\times H_2} = \varphi_u(v) = a(u,v).$$

Conversely, let $A \in L(H_1, H_2')$. Then $a(u,v) := \langle Au, v\rangle_{H_2'\times H_2}$ is a sesquilinear form with

$$|\langle Au, v\rangle_{H_2'\times H_2}| \le \|Au\|_{H_2'}\|v\|_{H_2} \le \|A\|_{H_2'\leftarrow H_1}\|u\|_{H_1}\|v\|_{H_2}.$$

(b) According to Proposition 2.1.6, for the above argument we only need to consider the definition of A on dense subspaces $S_1 \subset H_1$, $S_2 \subset H_2$ to extend the form $a(u,v)$ to $H_1 \times H_2$. $\langle Au, v\rangle_{H_2'\times H_2}$ then denotes the extension. □

The operator A from Lemma 2.1.38 is called the associated operator of $a(\cdot,\cdot)$.

Remark 2.1.39. *The results from Lemma 2.1.38 can be analogously transferred to bilinear forms* $a : H_1 \times H_2 \to \mathbb{R}$.

Let H be a Hilbert space and let $A \in L(H, H')$. Then $\overline{A} \in L(H, H')$ is defined by

$$\langle \overline{A}u, v\rangle_{H'\times H} := \overline{\langle A\overline{u}, \overline{v}\rangle_{H'\times H}},$$

where complex conjugation in Hilbert spaces is introduced in (2.15). An operator $A \in L(H, H')$ is said to be Hermitian if $A = \overline{A'}$. In the case that $\mathbb{K} = \mathbb{R}$ we use the term "symmetric".

Remark 2.1.40. *Let H be a Hilbert space and $a : H \times H \to \mathbb{C}$ a sesquilinear form with associated operator A. The statements (i) and (ii) are equivalent:*

(i) $a(\cdot,\cdot)$ *is Hermitian.*
(ii) $\widetilde{A}$ *is Hermitian, where*

$$\langle \widetilde{A}u, v\rangle_{H'\times H} := \langle Au, \overline{v}\rangle_{H'\times H} = a(u, \overline{v}) \qquad \forall u, v \in H. \tag{2.32}$$

Proof. By using the definition of $\widetilde{A}$, A', and the complex conjugation in Hilbert spaces we obtain

$$a^*(u,v) = \overline{a(v,u)} = \overline{\langle \widetilde{A}v, \overline{u}\rangle_{H'\times H}} = \overline{\langle \widetilde{A}'\overline{u}, v\rangle_{H'\times H}} = \left\langle \overline{\widetilde{A}'}u, \overline{v}\right\rangle_{H'\times H}.$$

On the other hand, we have

$$a(u,v) = \langle \widetilde{A}u, \overline{v}\rangle_{H'\times H}.$$

Hence the equivalence is proved:

$$a = a^* \iff \widetilde{A} = \overline{\widetilde{A}'}.$$

□

Exercise 2.1.41. *If A is associated with the form $a(\cdot,\cdot)$ then $C'\overline{A'}C$ is associated with the form a^*, where C' is the dual operator for the complex conjugation in Hilbert spaces. Similarly, if $\widetilde{A}$ is associated with $a(\cdot,\cdot)$ as in (2.32), then $\widetilde{A}'$ corresponds to a^*.*

A sesquilinear or bilinear form $b(\cdot,\cdot) : H_1 \times H_2 \to \mathbb{C}$ is compact if the associated operator $T \in L(H_1, H_2')$ with $\langle Tu, v\rangle_{H_2' \times H_2} := b(u,v)$ is compact.

Example 2.1.42. *Let $H_1 = H_2 = \mathbb{R}^n$ with inner product $(\mathbf{x},\mathbf{y}) = \sum_{i=1}^n \mathbf{x}_i \mathbf{y}_i$. Then every matrix $\mathbf{A} \in \mathbb{R}^{n\times n}$ induces a bilinear form according to the relation $a(\mathbf{x},\mathbf{y}) = (\mathbf{A}\mathbf{x},\mathbf{y}) : \mathbb{R}^n \times \mathbb{R}^n \to \mathbb{R}$. The form is symmetric if and only if $\mathbf{A}$ is symmetric, i.e., if $\mathbf{A}_{ij} = \mathbf{A}_{ji}$, $1 \le i,j \le n$.*

Example 2.1.43. *We call a matrix $\mathbf{A} \in \mathbb{R}^{n\times n}$ positive definite if it is symmetric and if we have*

$$(\mathbf{A}\mathbf{x},\mathbf{x}) > 0 \qquad \forall \mathbf{x} \in \mathbb{R}^n \backslash \{0\}.$$

For positive definite matrices, $a(\mathbf{x},\mathbf{y}) := (\mathbf{A}\mathbf{x},\mathbf{y})$ defines an inner product on $\mathbb{R}^n$.

2.1.6 Existence Theorems

Differential and integral equations can often be formulated as *variational problems*. In this section we will define abstract variational problems and prove the existence and uniqueness of solutions under suitable conditions. As standard references and additional material we refer, e.g., to [9, Chap. 5], [151, 166, 174].

For this let H_1, H_2 be Hilbert spaces, let $a(\cdot,\cdot)\colon H_1 \times H_2 \to \mathbb{K}$ be a continuous sesquilinear form and $\ell : H_2 \to \mathbb{K}$ a continuous, linear functional. We consider the abstract problem: Find $u \in H_1$ with

$$a(u,v) = \ell(v) \quad \forall v \in H_2. \tag{2.33}$$

The form $a(\cdot,\cdot)$ satisfies the *inf–sup condition* if

$$\inf_{u\in H_1\backslash\{0\}} \sup_{v\in H_2\backslash\{0\}} \frac{|a(u,v)|}{\|u\|_{H_1}\|v\|_{H_2}} \ge \gamma > 0, \tag{2.34a}$$

$$\forall v \in H_2\backslash\{0\} : \quad \sup_{u\in H_1\backslash\{0\}} |a(u,v)| > 0. \tag{2.34b}$$

Theorem 2.1.44. *The following statements are equivalent:*

(a) For every $\ell \in (H_2)'$ the abstract problem (2.33) has a unique solution $u \in H_1$ and we have

$$\|u\|_{H_1} \leq \frac{1}{\gamma} \|\ell\|_{H_2'} . \tag{2.35}$$

(b) The sesquilinear form $a(\cdot,\cdot)$ satisfies the inf-sup *condition (2.34).*

Proof. (b) $\Longrightarrow$ (a):
For the proof we will proceed in several steps:

(i) We choose an arbitrary $u \in H_1$. Then the functional $\phi_u \in H_2'$, which is given by $\phi_u := a(u,\cdot)$, is continuous and linear on H_2. This follows from the continuity of $a(\cdot,\cdot) : H_1 \times H_2 \to \mathbb{K}$ since

$$|\phi_u(v)| = |a(u,v)| \leq \|a\| \, \|u\|_{H_1} \, \|v\|_{H_2} = C(a,u) \, \|v\|_{H_2} .$$

Let A be the associated operator of $a(\cdot,\cdot)$. The mapping $u \to \phi_u$ can therefore be written as $\phi_u = Au$. The operator $A : H_1 \to H_2'$ is continuous and linear, since

$$\|Au\|_{H_2'} = \sup_{v \in H_2 \setminus \{0\}} \frac{|\phi_u(v)|}{\|v\|_{H_2}} \leq \|a\| \, \|u\|_{H_1} < \infty .$$

(ii) We will show that the image of H_1 under A is closed in H_2'. We have for all $u \in H_1$

$$\|Au\|_{H_2'} = \sup_{v \in H_2 \setminus \{0\}} \frac{|\phi_u(v)|}{\|v\|_{H_2}} = \sup_{v \in H_2 \setminus \{0\}} \frac{|a(u,v)|}{\|v\|_{H_2}} \overset{(2.34a)}{\geq} \gamma \, \|u\|_{H_1} .$$

Now, let $(u_n)_n \subset H_1$ be such that $(Au_n)_n$ is a Cauchy sequence in H_2. Then $(u_n)_n$ is a Cauchy sequence in H_1, since

$$\|Au_m - Au_n\|_{H_2'} = \|A(u_n - u_m)\|_{H_2'} \geq \gamma \, \|u_n - u_m\|_{H_1} .$$

Therefore the image of H_1 under A is closed in H_2'.

(iii) We claim that $A(H_1) = H_2'$. If this were not true, we would have

$$A(H_1) = \overline{A(H_1)}^{\|\cdot\|_{H_2'}} \neq H_2',$$

and the image of H_1 under A would be a closed proper subset of H_2'. Then, according to the Hahn–Banach theorem (see, e.g., [141], [3, Theorem 4.1]), there exists a $v_0 \in H_2'' \setminus \{0\}$ with $v_0(r) = 0$ for all $r \in A(H_1)$.

As H_2 is reflexive, it follows that $v_0 \in H_2 \cong H_2''$ and therefore with $r = Au$, we have the equation

$$0 = v_0(r) = r(v_0) = (Au)(v_0) = a(u, v_0) \quad \forall u \in H_1.$$

This is a contradiction to (2.34b). Therefore we have $A(H_1) = H_2'$ and then $A : H_1 \to H_2'$ is surjective. This means that (2.33) has a unique solution for all $\ell \in H_2'$.

(iv) We will show the a priori estimate. For every $\ell \in H_2'$ the equation $Au = \ell$ has a unique solution $u_0 = A^{-1}\ell$ and we have

$$|a(u_0, v)| = |\ell(v)| \le \|\ell\|_{H_2'} \|v\|_{H_2} \quad \forall v \in H_2.$$

It follows that

$$\begin{aligned}
\|\ell\|_{H_2'} &= \sup_{v\in H_2\setminus\{0\}} \frac{|a(u_0, v)|}{\|v\|_{H_2}} = \|u_0\|_{H_1} \sup_{v\in H_2\setminus\{0\}} \frac{|a(u_0, v)|}{\|u_0\|_{H_1} \|v\|_{H_2}} \\
&\ge \|u_0\|_{H_1} \inf_{u\in H_1\setminus\{0\}} \sup_{v\in H_2\setminus\{0\}} \frac{|a(u, v)|}{\|u\|_{H_1} \|v\|_{H_2}} \\
&\ge \gamma \|u_0\|_{H_1}.
\end{aligned}$$

(a) $\Longrightarrow$ *(b):*

If $A^{-1} \in L(H_2', H_1)$ exists, $A \in L(H_1, H_2')$ is bijective and the associated form $a(u, v) = \langle Au, v\rangle_{H_2'\times H_2}$ satisfies (2.34b). Now, we will show (2.34a): Since $A \in L(H_1, H_2')$ is bijective, with (2.35) we have

$$\begin{aligned}
\inf_{u\in H_1\setminus\{0\}} \sup_{v\in H_2\setminus\{0\}} \frac{\langle Au, v\rangle_{H_2'\times H_2}}{\|u\|_{H_1} \|v\|_{H_2}} &= \inf_{w\in H_2'\setminus\{0\}} \sup_{v\in H_2\setminus\{0\}} \frac{\langle w, v\rangle_{H_2'\times H_2}}{\|A^{-1}w\|_{H_1} \|v\|_{H_2}} \\
&\ge \inf_{w\in H_2'\setminus\{0\}} \sup_{v\in H_2\setminus\{0\}} \frac{\langle w, v\rangle_{H_2'\times H_2}}{\gamma^{-1}\|w\|_{H_2'} \|v\|_{H_2}} =: *.
\end{aligned}$$

According to Conclusion 2.1.18 (a) there exists an isometry $J_{H_2} : H_2 \to H_2'$. We therefore have

$$\begin{aligned}
* &= \gamma \inf_{\breve{w}\in H_2\setminus\{0\}} \sup_{v\in H_2\setminus\{0\}} \frac{\langle J_{H_2}\breve{w}, v\rangle_{H_2'\times H_2}}{\|J_{H_2}\breve{w}\|_{H_2'} \|v\|_{H_2}} \\
&= \gamma \inf_{\breve{w}\in H_2\setminus\{0\}} \frac{1}{\|\breve{w}\|_{H_2}} \sup_{v\in H_2\setminus\{0\}} \frac{\langle J_{H_2}\breve{w}, v\rangle_{H_2'\times H_2}}{\|v\|_{H_2}} \\
&= \gamma \inf_{\breve{w}\in H_2\setminus\{0\}} \frac{1}{\|\breve{w}\|_{H_2}} \|J_{H_2}\breve{w}\|_{H_2'} = \gamma.
\end{aligned}$$

□

Remark 2.1.45. *The adjoint conditions*

$$\inf_{v\in H_2\backslash\{0\}} \sup_{u\in H_1\backslash\{0\}} \frac{|a(u,v)|}{\|u\|_{H_1}\|v\|_{H_2}} \geq \gamma' > 0 \tag{2.36a}$$

$$\forall u \in H_1\backslash\{0\}: \quad \sup_{v\in H_2\backslash\{0\}} |a(u,v)| > 0 \tag{2.36b}$$

are equivalent to (2.34).

Proof. (see [115, Lemma 6.5.3]).

1. (2.34) $\Longrightarrow$ (2.36).
   From Condition (2.34a) we clearly have (2.36b). In the following we will show (2.36a). Let $A : H_1 \to H_2'$ be the associated operator of $a(\cdot,\cdot)$. From Theorem 2.1.44 we have
   $$\left\|A^{-1}\right\|_{H_1\leftarrow H_2'} \leq 1/\gamma. \tag{2.37}$$
   Proposition 2.1.14 gives us $\left\|(A')^{-1}\right\|_{H_2\leftarrow H_1'} \leq 1/\gamma$. We denote the left-hand side of (2.36a) by I. We then have
   $$\begin{aligned} I &= \inf_{v\in H_2\backslash\{0\}} \sup_{u\in H_1\backslash\{0\}} \frac{\left|\langle Au, v\rangle_{H_2'\times H_2}\right|}{\|u\|_{H_1}\|v\|_{H_2}} = \inf_{v\in H_2\backslash\{0\}} \sup_{u\in H_1\backslash\{0\}} \frac{\left|\langle u, A'v\rangle_{H_1\times H_1'}\right|}{\|u\|_{H_1}\|v\|_{H_2}} \\ &= \inf_{v'\in H_1'\backslash\{0\}} \|(A')^{-1}v'\|_{H_2}^{-1} \sup_{u\in H_1\backslash\{0\}} \frac{\left|\langle u, v'\rangle_{H_1\times H_1'}\right|}{\|u\|_{H_1}} \\ &= \inf_{v'\in H_1'\backslash\{0\}} \|(A')^{-1}v'\|_{H_2}^{-1} \left\|v'\right\|_{H_1'} \\ &= \frac{1}{\sup_{v'\in H_1'\backslash\{0\}} \left\|(A')^{-1}v'\right\|_{H_2} / \|v'\|_{H_1'}} = \left\|(A')^{-1}\right\|_{H_2\leftarrow H_1'}^{-1} \geq \gamma. \end{aligned}$$
   This is the same as (2.36a) with $\gamma' = \gamma > 0$.
2. (2.36) $\Longrightarrow$ (2.34).
   The proof of the converse is analogous to the first part. □

Remark 2.1.46. *(i) Let $A \in L(H_1, H_2')$ be the operator that is associated with the form $a(\cdot,\cdot)$ (see Lemma 2.1.38). Let (2.34) hold. Then $A^{-1} \in L(H_2', H_1)$ exists and*

$$\|A^{-1}\|_{H_1\leftarrow H_2'} \leq \gamma^{-1}. \tag{2.38}$$

(ii) Conversely, if $A^{-1} \in L(H_2', H_1)$ exists and (2.38) holds, then we have (2.34).

Proof. Part (i) follows from (2.37) and Part (ii) from Theorem 2.1.44. □

Remark 2.1.47. *The following statement is equivalent to the* inf-sup *condition (2.34a): There exists a constant $\gamma > 0$ with*

$$\forall u \in H_1 \backslash \{0\} : \sup_{v \in H_2 \backslash \{0\}} \frac{|a(u,v)|}{\|v\|_{H_2}} \geq \gamma \|u\|_{H_1}. \tag{2.39}$$

Since $a(u,v) = \langle Au, v \rangle_{H_2' \times H_1}$ (see Lemma 2.1.38), (2.39) is also equivalent to

$$\forall u \in H_1 \backslash \{0\} : \|Au\|_{H_2'} \geq \gamma \|u\|_{H_1}. \tag{2.40}$$

Remark 2.1.48. *In order to prove the* inf-sup *condition (2.34a) we will use the following method: Let $u \in H_1$ be arbitrary and given. If we can find a $v_u \in H_2$ with the following properties:*

$$\|v_u\|_{H_2} \leq C_1 \|u\|_{H_1}, \quad |a(u, v_u)| \geq C_2 \|u\|_{H_1}^2, \tag{2.41}$$

where C_1, C_2 are independent of u and v_u, then we have (2.34a) with $\gamma = C_2/C_1$.

Proof. Let $u \in H_1$ and $v_u \in H_2$ such that (2.41) holds. Then

$$\begin{aligned} \inf_{u \in H_1 \backslash \{0\}} \sup_{v \in H_2 \backslash \{0\}} \frac{|a(u,v)|}{\|u\|_{H_1} \|v\|_{H_2}} &\geq \inf_{u \in H_1 \backslash \{0\}} \frac{|a(u,v_u)|}{\|u\|_{H_1} \|v_u\|_{H_2}} \\ &\geq \inf_{u \in H_1 \backslash \{0\}} \frac{C_2 \|u\|_{H_1}^2}{\|u\|_{H_1} C_1 \|u\|_{H_1}} = \frac{C_2}{C_1} > 0. \end{aligned}$$

Remark 2.1.49. *Theorem 2.1.44 also holds for reflexive Banach spaces H_1, H_2.*

Now let

$$H_1 = H_2 = H$$

and let $a : H \times H \to \mathbb{C}$ be a sesquilinear form. In this case, the associated variational problem reads: For a given $\ell \in H'$ find $u \in H$ with

$$a(u,v) = \ell(v) \qquad \forall v \in H. \tag{2.42}$$

The sesquilinear form $a(\cdot,\cdot)$ is called H-elliptic if there exists a constant $\gamma > 0$ and a $\sigma \in \mathbb{C}$ with $|\sigma| = 1$ such that

$$\forall u \in H : \operatorname{Re}(\sigma a(u,u)) \geq \gamma \|u\|_H^2. \tag{2.43}$$

Remark 2.1.50. *1. Let $a : H \times H \to \mathbb{K}$ be a continuous and H-elliptic sesquilinear form. Then σ in (2.43) can be chosen in such a way that we have* $\operatorname{Re} \sigma \neq 0$.

2. Let H be a real Hilbert space and $a(\cdot,\cdot)$ a (real) bilinear form. Then, in (2.43), we can choose $\sigma \in \{-1, 1\}$.

3. *The H-ellipticity implies*

$$\forall u \in H: \quad |a(u,u)| \geq \gamma \|u\|_H^2. \tag{2.44}$$

Proof. of 1: If, for σ, we have in (2.43) $\operatorname{Re}\sigma \neq 0$, nothing needs to be shown. Therefore in the following we assume that $\operatorname{Re}\sigma = 0$ and choose $\tilde{\sigma} \in \mathbb{C}\backslash\{\sigma\}$ with $|\tilde{\sigma}| = 1$, so that we have

$$C_c |\sigma - \tilde{\sigma}| \leq \gamma/2 \quad \text{and} \quad \sigma \neq -\tilde{\sigma}$$

with the continuity constant C_c of $a(\cdot,\cdot)$. It then follows that

$$\operatorname{Re}(\tilde{\sigma} a(u,u)) = \operatorname{Re}(\sigma a(u,u)) + \operatorname{Re}((\sigma - \tilde{\sigma}) a(u,u)).$$

The continuity of $a(\cdot,\cdot)$ gives us

$$\operatorname{Re}((\sigma - \tilde{\sigma}) a(u,u)) \leq C_c |\sigma - \tilde{\sigma}| \|u\|_H^2 \leq \gamma/2 \|u\|_H^2,$$

from which we have the assertion with $\gamma \leftarrow \gamma/2$:

$$\operatorname{Re}(\tilde{\sigma} a(u,u)) \geq \gamma \|u\|_H^2 - C_c |\sigma - \tilde{\sigma}| \|u\|_H^2 \geq \gamma/2 \|u\|_H^2.$$

of 2: Let σ be as in (2.43). The assumptions give us that

$$\forall u \in H: \gamma \|u\|_H^2 \leq \operatorname{Re}(\sigma a(u,u)) = (\operatorname{Re}\sigma) a(u,u). \tag{2.45}$$

From this we have $\operatorname{Re}\sigma \neq 0$. For $\operatorname{Re}\sigma > 0$, (2.45) gives us the estimate $a(u,u) \geq 0$ for all $u \in H$. It follows that $(\operatorname{Re}\sigma) a(u,u) \leq a(u,u)$ and $\sigma = 1$ satisfies (2.43). The case $\operatorname{Re}\sigma < 0$ can be proven analogously with $\sigma = -1$.

of 3: We have $\gamma \|u\|_H^2 \leq \operatorname{Re}(\sigma a(u,u)) \leq |\sigma a(u,u)| = |a(u,u)|$. □

Lemma 2.1.51 (Lax–Milgram). *Let H be a Hilbert space. Let the sesquilinear form $a: H \times H \to \mathbb{C}$ be H-elliptic. Then (2.34) holds and the variational problem (2.42) has a unique solution $u \in H$ for all $\ell \in H'$ with*

$$\|u\|_H \leq \frac{1}{\gamma} \|\ell\|_{H'}. \tag{2.46}$$

Proof. We will show (2.34a) as in Remark 2.1.48: for $u \in H$ we choose $v_u = u \in H$. Then, due to $\|v_u\|_H = \|u\|_H$ and (2.44), we have the inequality

$$|a(u, v_u)| = |a(u,u)| \geq \gamma \|u\|_H^2.$$

From this we have (2.34a). We can prove the inequality (2.34b) in a similar way. Thus, let $0 \neq v \in H$. Then

$$\sup_{u \in H} |a(u,v)| \geq |a(v,v)| \geq \gamma \|v\|_H^2 > 0.$$

The statement follows from Theorem 2.1.44. □

Remark 2.1.52. *The Lax–Milgram lemma still holds if we replace Condition (2.43) by Condition (2.44) (see [137, Theorem 5.2.3]).*

Note that in (2.43) we do not impose any conditions on the symmetry of $a(\cdot,\cdot)$. If $a(\cdot,\cdot)$ is symmetric the solution of (2.42) can be characterized as a minimum.

Proposition 2.1.53. *Let the form $a : H \times H \to \mathbb{K}$ be symmetric and H-elliptic with $\sigma = 1$ in (2.43) [see Remark 2.1.50(2)]. Then for all $\ell \in H'$ the unique solution of problem (2.42) is also a solution of the problem of finding a minimizer of the quadratic functional*

$$\Pi(v) = \tfrac{1}{2} a(v,v) - \ell(v). \tag{2.47}$$

If, on the other hand, $u \in V$ minimizes $\Pi(\cdot)$ then u solves (2.42).

Proof. Let u be the solution of (2.42) and $v \in H \setminus \{0\}$. Then

$$\begin{aligned} 2\Pi(u+v) &= a(u+v,u+v) - 2\ell(u+v) \\ &= a(u,u) + 2a(u,v) + a(v,v) - 2\ell(u) - 2\ell(v) \\ &= 2\Pi(u) + a(v,v) + 2(a(u,v) - \ell(v)) \\ &= 2\Pi(u) + a(v,v) \geq 2\Pi(u) + \gamma \|v\|_V^2 \geq 2\Pi(u), \end{aligned}$$

and therefore u solves Problem (2.47).

Now, let u be the solution of (2.47). Then

$$\begin{aligned} \forall v \in V: \quad 0 &= \tfrac{d}{d\varepsilon}(\Pi(u+\varepsilon v))|_{\varepsilon=0} \\ &= \tfrac{d}{d\varepsilon}\left(\tfrac{1}{2}a(u+\varepsilon v, u+\epsilon v) - \ell(u+\varepsilon v)\right)\Big|_{\varepsilon=0} \\ &= \tfrac{d}{d\varepsilon}\left(\tfrac{1}{2}a(u,u) + \varepsilon a(u,v) + \tfrac{1}{2}\varepsilon^2 a(v,v) - \ell(u) - \varepsilon\ell(v)\right)\Big|_{\varepsilon=0} \\ &= a(u,v) - \ell(v) \end{aligned}$$

and thus u solves (2.42). □

In some of the applications that we are going to study later on, we will often encounter sesquilinear forms $a(\cdot,\cdot)$ that do not satisfy (2.43) but only a weaker condition, the *H-coercivity*.

Definition 2.1.54. Let the Hilbert spaces U, H constitute a Gelfand triple $H \subset U \subset H'$ with the continuous and dense embedding $H \subset U$. The sesquilinear form $a(\cdot,\cdot)\colon H \times H \to \mathbb{C}$ is said to be H-coercive if there exist constants $\gamma > 0$, $C_U \in \mathbb{R}$ and $\sigma \in \mathbb{C}$ with $|\sigma| = 1$ such that

$$\forall u \in H : \operatorname{Re}(\sigma a(u,u)) \geq \gamma \|u\|_H^2 - C_U \|u\|_U^2 . \tag{2.48}$$

Remark 2.1.55. *The elliptic and coercive forms that we are going to study in this book will always satisfy the inequalities (2.43) or (2.48) with $\sigma = 1$. However, in other applications, for example from the area of electromagnetism, there are forms that have an imaginary principal part and therefore do not allow setting $\sigma = 1$ (see [40, 44]).*

Remark 2.1.50(1), (3) can be applied to H-coercive sesquilinear forms appropriately.

Remark 2.1.56. *1. Let $a : H \times H \to \mathbb{K}$ be a continuous and H-coercive sesquilinear form. Then σ in (2.48) can be chosen so that* $\operatorname{Re}\sigma > 0$.
2. The H-coercivity implies that

$$\forall u \in H : \quad |a(u,u)| \geq \gamma \|u\|_H^2 - C_U \|u\|_U^2 . \tag{2.49}$$

H-coercive forms $a(\cdot,\cdot)$ remain H-coercive when perturbed by the addition of suitable forms $b(\cdot,\cdot)$ which are either "small"with respect to the form $a(\cdot,\cdot)$ or compact.

Lemma 2.1.57. *Let the Hilbert spaces U, H constitute a Gelfand triple $H \subset U \subset H'$ with a dense and continuous embedding $H \subset U$. Let the sesquilinear form $a(\cdot,\cdot) : H \times H \to \mathbb{C}$ be H-coercive and let $b : H \times H \to \mathbb{C}$ be continuous. Then the form $a(\cdot,\cdot) + b(\cdot,\cdot)$ is again H-coercive if one of the following conditions is satisfied:*

(i) For all $\varepsilon > 0$ there exists a constant $C(\varepsilon) > 0$ with

$$\forall u \in H : \quad |b(u,u)| \leq \varepsilon \|u\|_H^2 + C(\varepsilon) \|u\|_U^2 . \tag{2.50}$$

(ii) X, Y are Hilbert spaces with continuous embeddings $H \subset X \subset U$, $H \subset Y \subset U$. One of the embeddings $H \subset X$, $H \subset Y$ is compact. Furthermore,

$$\forall u, v \in H : \quad |b(u,v)| \leq C_b \|u\|_X \|v\|_Y . \tag{2.51}$$

(iii) The embeddings $H \subset X \subset U$, $H \subset Y \subset U$ are continuous and (2.51) holds, as well as:

For all $\varepsilon > 0$ there exists a constant $C(\varepsilon) > 0$ such that for all $u \in H$

$$\|u\|_Y \leq \varepsilon \|u\|_H + C(\varepsilon) \|u\|_U \quad \text{or} \quad \|u\|_X \leq \varepsilon \|u\|_H + C(\varepsilon) \|u\|_U . \tag{2.52}$$

Proof. We will show (a): (i) implies the coercivity of $a(\cdot,\cdot) + b(\cdot,\cdot)$, (b): (ii) $\Rightarrow$ (iii) and (c): (iii) $\Rightarrow$(i):

(a) In (2.50) we set $\varepsilon = \gamma/2$ with γ from (2.48). Then for all $u \in H$

$$\begin{aligned}\operatorname{Re}\left(\sigma\{a(u,u)+b(u,u)\}\right) &= \operatorname{Re}\left(\sigma a(u,u)\right)+\operatorname{Re}\left(\sigma b(u,u)\right)\\ &\geq \gamma\,\|u\|_H^2 - C_U\,\|u\|_U^2 - \tfrac{\gamma}{2}\,\|u\|_H^2 - C(\varepsilon)\|u\|_U^2\\ &= \tfrac{\gamma}{2}\,\|u\|_H^2 - \left(C_U + C(\varepsilon)\right)\|u\|_U^2,\end{aligned}$$

which is (2.48) for $a(\cdot,\cdot)+b(\cdot,\cdot)$.

(b) As an example, we consider the case $H \subset\subset X \subset U$. Lemma 2.1.35 implies

$$\forall\varepsilon > 0\ \exists C(\varepsilon) > 0: \quad \forall u \in H: \|u\|_X \leq \varepsilon\|u\|_H + C(\varepsilon)\,\|u\|_U, \tag{2.53}$$

from which we have (2.52).

(c) The embedding $H \subset Y$ is continuous and therefore there exists a constant $C_Y < \infty$ with

$$\forall u \in H: \ \|u\|_Y \leq C_Y\,\|u\|_H.$$

As an example, we assume that the right-hand inequality in (2.52) is satisfied. Then, by (2.51), we have for all $\varepsilon > 0$ the inequality

$$\begin{aligned}\forall u \in H: \ |b(u,u)| &\leq C_b\,\|u\|_X\,\|u\|_Y\\ &\leq C_b C_Y \left(\varepsilon\,\|u\|_H^2 + C(\varepsilon)\,\|u\|_U\,\|u\|_H\right)\\ &\leq C_b C_Y \left(2\varepsilon\|u\|_H^2 + \tfrac{(C(\varepsilon))^2}{4\varepsilon}\,\|u\|_U^2\right)\\ &= \varepsilon'\,\|u\|_H^2 + C'(\varepsilon)\,\|u\|_U^2\end{aligned}$$

and thus we have (2.50). □

Remark 2.1.58. *Lemma 2.1.57 still holds if Condition (2.48) is replaced by the Gårding inequality: There exists a compact operator $T: H \to H'$ such that*

$$\forall u \in H: |a(u,u) + (Tu,u)_{H'\times H}| \geq \gamma\,\|u\|_H^2.$$

For a proof of this remark, we refer to [137, Remark 5.3.2]. The following special case of Lemma 2.1.57 is particularly important when dealing with boundary integral equations.

Corollary 2.1.59. *We assume that U, H, H' form a Gelfand triple $H \subset U \subset H'$ with a compact embedding $H \subset\subset U$. Let $a(\cdot,\cdot)$ be H-coercive and let either $b(\cdot,\cdot)$: $H \times U \to \mathbb{C}$ or $b(\cdot,\cdot)$: $U \times H \to \mathbb{C}$ be continuous. Then $a(\cdot,\cdot) + b(\cdot,\cdot)$ is H-coercive.*

The following theorem is an application of the Fredholm–Riesz–Schauder theory to H-coercive sesquilinear forms $a: H \times H \to \mathbb{C}$.

Theorem 2.1.60. *Let $H \subset U \subset H'$ be a Gelfand triple with a compact and dense embedding $H \subset\subset U$. Let the sesquilinear form $a(\cdot,\cdot) : H \times H \to \mathbb{C}$ that is associated with the operator $A \in L(H, H')$ be H-coercive.*

Let I denote the embedding: $I : H \to H'$. Then we have for all $\lambda \in \mathbb{C}$ either

$$(A - \lambda I)^{-1} \in L(H', H) \quad \textit{and} \quad (A' - \overline{\lambda} I)^{-1} \in L(H', H) \tag{2.54}$$

or

$$\lambda \textit{ is an eigenvalue of } A. \tag{2.55}$$

Should (2.54) hold, then the variational problems: Find $x, x^* \in H$ such that

$$\begin{aligned} a(x,y) - \lambda(x,y)_U &= \langle f, y\rangle_{H'\times H} \quad \textit{and} \quad a(y,x^*) - \overline{\lambda}(x^*,y)_U \\ &= \langle f, y\rangle_{H'\times H} \qquad \forall y \in H \end{aligned} \tag{2.56}$$

have a unique solution for all $f \in H'$. Should, however, (2.55) hold, then the eigenspaces

$$\{0\} \neq E(\lambda) = \ker(A - \lambda I), \ \{0\} \neq E'(\lambda) = \ker(A' - \overline{\lambda} I)$$

are finite-dimensional and we have for all $y \in H$

$$x \in E(\lambda): \quad a(x,y) = \lambda(x,y)_U, \tag{2.57}$$

$$x^* \in E'(\lambda): \quad a(y,x^*) = \overline{\lambda}(x^*,y)_U. \tag{2.58}$$

The spectrum $\sigma(A)$ of A consists of at most countably many eigenvalues $\{\lambda_i\}$ and the only possible accumulation point is at infinity. Furthermore, we have

$$\lambda \in \sigma(A) \Longleftrightarrow \overline{\lambda} \in \sigma(A').$$

For $\lambda \in \sigma(A)$ the variational problem

$$x \in H: \ a(x,y) - \lambda(x,y)_U = \langle f, y\rangle_{H'\times H} \quad \forall y \in H \tag{2.59}$$

has at least one solution if and only if $f \perp E'(\lambda)$, i.e., if and only if $f \in H'$ satisfies the compatibility condition

$$\forall x^* \in E'(\lambda): \quad \langle f, x^*\rangle_{H'\times H} = 0. \tag{2.60}$$

Proof. The statements follow from Theorem 2.1.36. Here we check the assumptions.

If $H \subset\subset U$ then we also have $H \subset\subset H'$ and the embedding $I : H \to H'$ is compact. Due to Remark 2.1.56(1) we can assume that $\operatorname{Re}\sigma \neq 0$ and set $\widetilde{C} := C_U / \operatorname{Re}\sigma$.

The sesquilinear form $a(\cdot,\cdot) + \widetilde{C}\,\|\cdot\|_U^2$ is H-elliptic since, by (2.48), we have for all $u \in H$

$$\operatorname{Re}\left(\sigma\{a(u,u) + \widetilde{C}\,\|u\|_H^2\}\right) = \operatorname{Re}(\sigma a(u,u)) + C_U\,\|u\|_U^2 \geq \gamma\,\|u\|_H^2$$

with $\gamma > 0$. According to Lemma 2.1.51, $(A + \widetilde{C}I)^{-1} \in L(H', H)$ exists. Lemma 2.1.29 states that $K := (A + \widetilde{C}I)^{-1}I : H \to H$ is compact, and therefore Theorem 2.1.36 can be applied to the operator $K - \mu I$. With

$$\begin{aligned} K - \mu I &= -\mu(I - \mu^{-1}K) = -\mu(A + \widetilde{C}I)^{-1}(A + \widetilde{C}I - \frac{1}{\mu}I) \\ &= -\mu(A + \widetilde{C}I)^{-1}(A - \lambda I) \end{aligned}$$

and $\lambda = \mu^{-1} - \widetilde{C}$ we obtain Theorem 2.1.60 for

$$A - \lambda I = -\frac{1}{\mu}(A + \widetilde{C}I)(K - \mu I)$$

as a consequence of the statements of Theorem 2.1.36 for $K - \mu I$.

□

The combination of Corollary 2.1.59 and Theorem 2.1.60 gives us the following existence theorem, which is often used in the variational formulation of integral equations.

Corollary 2.1.61. *We assume that U, H, H' form a Gelfand triple $H \subset U \subset H'$ with compact embedding $H \subset\subset U$. Let $a(\cdot,\cdot)$ and $b(\cdot,\cdot)$: $H \times H \to \mathbb{C}$ be continuous sesquilinear forms, let $a(\cdot,\cdot)$ be H-coercive and assume that $b(\cdot,\cdot)$ satisfies*

$$\forall u, v \in H : \ |b(u,v)| \leq C_b\,\|u\|_U\,\|v\|_H \ \ \text{or} \ \ |b(u,v)| \leq C_b\,\|u\|_H\,\|v\|_U.$$

Furthermore, let the form $c(\cdot,\cdot) := a(\cdot,\cdot) + b(\cdot,\cdot)$ be injective:

$$\forall v \in H : \ c(u,v) = 0 \Longrightarrow u = 0. \tag{2.61}$$

Then for every $f \in H'$ the variational problem

$$u \in H : \quad a(u,v) + b(u,v) = \langle f, v\rangle_{H' \times H} \quad \forall v \in H \tag{2.62}$$

has a unique solution u.

Proof. According to Corollary 2.1.59, $c(\cdot,\cdot)$ is H-coercive and satisfies (2.48). According to (2.61), $\lambda = 0$ is not an eigenvalue of $a(\cdot,\cdot) + b(\cdot,\cdot)$ and so Problem (2.62) has a unique solution, as stated by Theorem 2.1.60. □

2.1.7 Interpolation Spaces*

When dealing with the variational formulation of boundary integral equations as well as the error analysis of boundary element methods, it is very useful to study function spaces that describe the differentiability of a function. In classical analysis differentiation is only defined for integer orders. However, by using "interpolation spaces" it is possible to formulate properties concerning the smoothness of functions for non-integer orders of differentiability. There are different, not necessarily equivalent interpolation methods. Here we will only introduce the "real interpolation method". For a detailed discussion on interpolation spaces as well as proofs we refer to [22] and [155].

Let X_0, X_1 be two Banach spaces with continuous embedding $X_1 \hookrightarrow X_0$ (this property is not strictly necessary; however, in the cases that are of interest to us it is always given). For $u \in X_0$ and all $t > 0$ we define the "K-functional" as

$$K(t,u) := \inf_{v \in X_1} (\|u - v\|_{X_0} + t\,\|v\|_{X_1}). \tag{2.63}$$

Clearly, we have for $u \in X_1$

$$K(t,u) \le t\,\|u\|_{X_1}, \quad K(t,u) \le \|u\|_{X_0}.$$

For $0 \le \theta \le 1$ and $1 \le p < \infty$ we define the norm

$$\|u\|_{[X_0,X_1]_{\theta,p}} := \left(\int_0^\infty t^{-\theta p}\, K(t,u)^p \,\frac{dt}{t} \right)^{1/p}. \tag{2.64a}$$

For $p = \infty$ we define

$$\|u\|_{[X_0,X_1]_{\theta,\infty}} := \sup_{0<t<\infty} t^{-\theta}\, K(t,u). \tag{2.64b}$$

Then the set

$$[X_0, X_1]_{\theta,p} = X_{\theta,p} := \left\{ u \in X_0 : \|u\|_{[X_0,X_1]_{\theta,p}} < \infty \right\}$$

is a Banach space with norm (2.64).

Let $X_i, Y_i, i = 0, 1$, be two pairs of Banach spaces as given above, with $X_i \subset Y_i$. We then have

$$X_{\theta,p} \subset Y_{\theta,p}, \; X_1 \subset X_{\theta,p} \subset X_0, \; X_{\theta,p} \subset X_{\theta,\infty}$$

* This section should be read as a complement to the core material of this book.

and $X_{\theta,1} \subset X_{\theta,p}$ for all $1 \le p \le \infty$. The spaces $X_{\theta,p}$ form a scale:

$$X_{\theta_2,p} \subset X_{\theta_1,p} \quad \text{for } 1 \le p \le \infty,\ \theta_1 \le \theta_2.$$

Proposition 2.1.62. *Let X_i, Y_i be two pairs of Banach spaces and let $T \in L(Y_i, X_i)$, $i = 0, 1$. Then we have*

$$T \in L(Y_{\theta,p}, X_{\theta,p}) \quad \text{for } 0 \le \theta \le 1,\ 1 \le p \le \infty \tag{2.65}$$

and

$$\|T\|_{X_{\theta,p} \leftarrow Y_{\theta,p}} \le \|T\|_{X_0 \leftarrow Y_0}^{1-\theta} \, \|T\|_{X_1 \leftarrow Y_1}^{\theta}. \tag{2.66}$$

Another important result is the "re-iteration theorem". It states that one cannot obtain "new" interpolation spaces by repeated interpolation.

Proposition 2.1.63. *We have for all $0 \le \theta_0 < \theta_1 \le 1$, $1 \le p_0, p_1, q \le \infty$ and $0 < \lambda < 1$:*

$$\left[[X_0, X_1]_{\theta_0,p_0}, [X_0, X_1]_{\theta_1,p_1} \right]_{\lambda,q} = [X_0, X_1]_{(1-\lambda)\theta_0 + \lambda\theta_1, q}.$$

The dual spaces of interpolation spaces are isomorphic to the interpolation spaces of the respective dual spaces. The following proposition will clarify the details.

Proposition 2.1.64. *Let X_1 be dense in X_0. Then we have for all $0 < \theta < 1$, $1 \le p < \infty$, $\frac{1}{p} + \frac{1}{p'} = 1$,*

$$[X_0, X_1]'_{\theta,p} = [X'_1, X'_0]_{1-\theta,p'} = [X'_0, X'_1]_{\theta,p'}.$$

For functions from X_1, the square of the norm of the interpolation space $[X_0, X_1]_{\theta,p}$ can be estimated by the product of the norms in X_0 and X_1. We only need this result in the case $p = 2$.

Proposition 2.1.65. *There exists a constant $c > 0$ such that for all $u \in X_1$ the inequality*

$$\|u\|_{[X_0,X_1]_{\theta,2}} \le c\, \|u\|_{X_0}^{1-\theta} \, \|u\|_{X_1}^{\theta}$$

is satisfied.

We refer to [230], [22] and [155] for proofs of these statements as well as further reading.

2.2 Geometric Tools

2.2.1 Function Spaces

Boundary integral equations are formulated on the surfaces of domains in $\mathbb{R}^d$. In order to define the relevant function spaces on the boundaries one has to characterize the smoothness of the boundaries. For this one needs Hölder continuous parametrizations that have to be introduced first. Let $k \in \mathbb{N}_0$ and let $\Omega \subset \mathbb{R}^d$ be a domain. The space of all k times continuously differentiable functions on Ω is denoted by

$$C^k\left(\overline{\Omega}\right) := \{f : \Omega \to \mathbb{C} : f \text{ is } k \text{ times continuously differentiable and } \partial^\alpha f \text{ can be extended as a continuous function on } \overline{\Omega} \text{ for all } 0 \leq |\alpha| \leq k\}.$$

Here $\alpha \in \mathbb{N}_0^d$ is a multi-index and we use the following conventions. For $\mu \in \mathbb{N}_0^d$ we set

$$\begin{aligned} &\mu! := \prod_{i=1}^d \mu_i!, \quad |\mu|_1 := |\mu| := \sum_{i=1}^d \mu_i, |\mu|_\infty := \max_{1\leq i\leq d} |\mu_i|, \\ &\forall \mathbf{v} = (v_i)_{i=1}^d \in \mathbb{C}^d : \mathbf{v}^\mu := \prod_{i=1}^d v_i^{\mu_i}, \quad \partial^\mu f(\mathbf{x}) := \partial_{\mathbf{x}}^\mu f(\mathbf{x}) := \frac{\partial^{|\mu|} f(\mathbf{x})}{\partial_{x_1}^{\mu_1} \partial_{x_2}^{\mu_2} \dots \partial_{x_d}^{\mu_d}}. \end{aligned} \tag{2.67}$$

On the vector space $C^k\left(\overline{\Omega}\right)$ we can define the following norms

$$\|\varphi\|_{C^0(\overline{\Omega})} := \sup_{\mathbf{x}\in\Omega} \{|\varphi(\mathbf{x})|\}, \quad \|\varphi\|_{C^k(\overline{\Omega})} := \max_{0\leq|\alpha|\leq k} \left\{\|\partial^\alpha \varphi\|_{C^0(\overline{\Omega})}\right\}.$$

A function $\varphi \in C^0\left(\overline{\Omega}\right)$ is Hölder continuous of order $\lambda \in]0,1]$ in Ω, if

$$|\varphi|_{C^{0,\lambda}(\overline{\Omega})} := \sup_{\mathbf{x},\mathbf{y}\in\Omega} \frac{|\varphi(\mathbf{x}) - \varphi(\mathbf{y})|}{\|\mathbf{x}-\mathbf{y}\|^\lambda} < \infty.$$

The set of all Hölder continuous functions is given by $C^{0,\lambda}\left(\overline{\Omega}\right)$. The space $C^{k,\lambda}\left(\overline{\Omega}\right)$ contains all functions on Ω with $\partial^\alpha \varphi \in C^{0,\lambda}\left(\overline{\Omega}\right)$ for all $|\alpha| \leq k$. On $C^{k,\lambda}\left(\overline{\Omega}\right)$ a norm is given by

$$\|\varphi\|_{C^{k,\lambda}(\overline{\Omega})} := \|\varphi\|_{C^k(\overline{\Omega})} + \max_{|\alpha|=k} |\partial^\alpha \varphi|_{C^{0,\lambda}(\overline{\Omega})}.$$

Remark 2.2.1. *For all $k \in \mathbb{N}_0$ and $0 < \lambda \leq 1$, $C^{k,\lambda}\left(\overline{\Omega}\right)$ is a Banach space.*

Exercise 2.2.2. *Let $\mu \in [0,1]$. Determine the maximum $\lambda \in]0,1]$ so that the function $f : (-1,1) \to \mathbb{R}$, $f(x) = |x|^\mu$ lies in the space $C^{0,\lambda}([-1,1])$.*

The space of all infinitely differentiable functions is given by

$$C^\infty\left(\overline{\Omega}\right) := \bigcap_{k\in\mathbb{N}_0} C^k\left(\overline{\Omega}\right).$$

All the functions we have considered so far are scalar, i.e., they map points from a domain Ω to $\mathbb{C}$. The definitions can, however, be generalized for vector-valued functions $\Phi = (\Phi_i)_{i=1}^d : \Omega_1 \to \Omega_2$ on domains $\Omega_1, \Omega_2 \subset \mathbb{R}^d$. We set

$$\mathbf{C}^{k,\lambda}\left(\overline{\Omega_1}, \overline{\Omega_2}\right) := \left\{\Phi : \Omega_1 \to \Omega_2 \mid \forall 1 \leq i \leq d : \Phi_i \in C^{k,\lambda}\left(\overline{\Omega_1}\right)\right\}. \quad (2.68)$$

If the condition $\Phi_i \in C^{k,\lambda}\left(\overline{\Omega_1}\right)$ in (2.68) is replaced by $\Phi_i \in C^k\left(\overline{\Omega_1}\right)$ we obtain the space $\mathbf{C}^k\left(\overline{\Omega_1}, \overline{\Omega_2}\right)$. For $\Omega_1 = \Omega_2$ we use the notation $\mathbf{C}^{k,\lambda}\left(\overline{\Omega_1}\right) := \mathbf{C}^{k,\lambda}\left(\overline{\Omega_1}, \overline{\Omega_2}\right)$ and similarly $\mathbf{C}^k\left(\overline{\Omega_1}\right) := \mathbf{C}^k\left(\overline{\Omega_1}, \overline{\Omega_2}\right)$.

Definition 2.2.3. Let Ω_1, $\Omega_2 \subset \mathbb{R}^d$ be two domains and let $k \in \mathbb{N}_0 \cup \{\infty\}$. A mapping $\Phi : \Omega_1 \to \Omega_2$ is a C^k-diffeomorphism if it satisfies the conditions (a)–(c):

(a) $\Phi \in \mathbf{C}^k\left(\overline{\Omega_1}, \overline{\Omega_2}\right)$.
(b) The inverse mapping $\Phi^{-1} : \Omega_2 \to \Omega_1$ exists and satisfies

$$\Phi^{-1} \in \mathbf{C}^k\left(\overline{\Omega_2}, \overline{\Omega_1}\right).$$

(c) There exists a constant $0 < c < \infty$ such that the Jacobian $D\Phi = \left(\frac{\partial \Phi_i}{\partial \mathbf{x}_j}\right)_{1 \leq i,j \leq d}$ satisfies the inequality

$$\forall \mathbf{x} \in \Omega_1 : 0 < c \leq |\det (D\Phi(\mathbf{x}))| \leq 1/c. \quad (2.69)$$

Remark 2.2.4 , follows from the inverse mapping theorem (see, e.g., [245, Sect. 8.6], [95, Chap. 8]).

Remark 2.2.4. *If $\Omega \subset \mathbb{R}^d$ is bounded, (a) and (b) imply (c).*
If $k \geq 1$ and Φ is surjective, (a) and (c) imply (b).

Definition 2.2.5. A function $\Phi : \Omega_1 \to \Omega_2$ is *bi-Lipschitz continuous* if in Definition 2.2.3 we have $\mathbf{C}^{0,1}\left(\overline{\Omega_i}, \overline{\Omega_j}\right)$ instead of $\mathbf{C}^k\left(\overline{\Omega_i}, \overline{\Omega_j}\right)$ and

$$0 < c \leq \sup_{\substack{\mathbf{x},\mathbf{y} \in \Omega_1 \\ \mathbf{x} \neq \mathbf{y}}} \frac{|\Phi(\mathbf{x}) - \Phi(\mathbf{y})|}{\|\mathbf{x} - \mathbf{y}\|} \leq 1/c \quad (2.70)$$

instead of (2.69).

The space of all Lebesgue measurable functions that are bounded almost everywhere on Ω is denoted by $L^\infty(\Omega)$. The term "almost everywhere" always refers to everywhere except on sets with Lebesgue measure 0.

Proposition 2.2.6. *Let $\Omega \subset \mathbb{R}^d$ be bounded and let $\varphi \in C^{0,1}\left(\overline{\Omega}\right)$, $d \geq 2$. Then we have:*

(a) For all $\varphi \in C^{0,1}\left(\overline{\Omega}\right)$, the partial derivatives $(\partial\varphi/\partial x_i)_{i=1}^d$ exist almost everywhere in Ω, they are measurable and bounded almost everywhere, i.e., $\partial^\alpha \varphi \in L^\infty(\Omega)$ for all $|\alpha| = 1$.
(b) We have the more general property for $k \in \mathbb{N}_0$

$$\varphi \in C^{k,1}\left(\overline{\Omega}\right) \Rightarrow \partial^\alpha \varphi \in L^\infty(\Omega) \qquad \forall\, |\alpha| \leq k+1.$$

2.2.2 Smoothness of Domains

In order to describe the smoothness of domains one uses local as well as global criteria. Lipschitz domains represent a reasonably general class of domains for whose boundaries integral equations can be defined. Lipschitz domains are given by the existence of an atlas which consists of bi-Lipschitz continuous charts. In Chap. 4 we examine Galerkin boundary element methods in order to numerically solve integral equations, for which it is necessary to decompose the surface into curved triangles and rectangles. To be able to do this the surface needs to locally satisfy a greater degree of smoothness.

In general, we assume that $\Omega \subset \mathbb{R}^d$ is a domain with compact boundary $\Gamma = \partial\Omega$. For $r > 0$, B_r denotes the open ball in $\mathbb{R}^d$ with radius r around the origin. We set

$$\begin{aligned} B_r^+ &:= \{\xi \in B_r : \xi_d > 0\}, \qquad B_r^- := \{\xi \in B_r : \xi_d < 0\}, \\ B_r^0 &:= \{\xi \in B_r : \xi_d = 0\}. \end{aligned} \tag{2.71}$$

Definition 2.2.7. A domain $\Omega \subset \mathbb{R}^d$ is a Lipschitz domain $\left(\Omega \in C^{0,1}\right)$ if there exists a finite cover $\mathcal{U}$ of open subsets in $\mathbb{R}^d$ such that the associated bijective mappings[2] $\left\{\chi_U : \overline{B_2} \to \overline{U}\right\}_{U \in \mathcal{U}}$ have the following properties:

1. $\chi_U \in C^{0,1}\left(\overline{B_2}, \overline{U}\right), \quad \chi_U^{-1} \in C^{0,1}\left(\overline{U}, \overline{B_2}\right)$.
2. $\chi_U\left(B_2^0\right) = U \cap \Gamma$.
3. $\chi_U\left(B_2^+\right) = U \cap \Omega$.
4. $\chi_U\left(B_2^-\right) = U \cap \mathbb{R}^d \backslash \overline{\Omega}$.

Let $k \in \mathbb{N} \cup \{\infty\}$. A domain Ω is a C^k-domain if Property 1 can be replaced by

$$\chi_U \in C^k\left(\overline{B_2}, \overline{U}\right), \quad \chi_U^{-1} \in C^k\left(\overline{U}, \overline{B_2}\right).$$

Remark 2.2.8. *Properties 2–4 in Definition 2.2.7 express the fact that Ω is locally situated on one side of the boundary $\partial\Omega$.*

---

[2] The choice of the radius $r = 2$ for the ball K_2 is arbitrary but will slightly reduce the technicalities in the definition of boundary element meshes because the *master element* then is contained in K_2.

In order to describe the local smoothness of the surface we use surface meshes. For this, let $q \in \mathbb{N}$,

$$\widehat{S}_q := \left\{\xi \in \mathbb{R}^q : 0 < \xi_1 < \xi_2 < \ldots \xi_{q-1} < \xi_q < 1\right\}$$

be the *unit simplex* and

$$\widehat{Q}_q := (0,1)^q$$

be the *unit cube*. In the following these domains will be called *reference elements* and will be abbreviated by $\hat{\tau}_q$. If there is no confusion with respect to the dimension q we will simply write $\hat{\tau}$.

Definition 2.2.9. Let $\Omega \subset \mathbb{R}^d$ (for $d = 2, 3$) be a bounded domain with boundary Γ:

1. A subset $\tau \subset \Gamma$ is called a boundary element or panel of smoothness $k \in \mathbb{N}_0 \cup \{\infty\}$ – in short a C^k-element – if there exists a C^k-diffeomorphism $\chi_\tau : \hat{\tau} \to \tau$ which can be extended to a C^k-diffeomorphism $\chi_\tau^\star : \hat{\tau}^\star \to \tau^\star$. Here $\hat{\tau}^\star \subset \mathbb{R}^{d-1}$ signifies a neighborhood of $\overline{\hat{\tau}}$ and $\overline{\tau} \subset \tau^\star$.
2. A set $\mathcal{G}$ is called a surface mesh (of smoothness $k \in \mathbb{N}_0$) if:
   (a) All $\tau \in \mathcal{G}$ panels are of smoothness k.
   (b) The elements of $\mathcal{G}$ are open and disjoint.
   (c) $\Gamma = \bigcup_{\tau \in \mathcal{G}} \overline{\tau}$.
3. A surface mesh $\mathcal{G}$ does not have any hanging nodes if the intersection $\overline{\tau} \cap \overline{t}$ of all non-identical elements $\tau, t \in \mathcal{G}$ is either the empty set, a common point or – if $d = 3$ – a common edge.

Definition 2.2.10. A bounded domain $\Omega \subset \mathbb{R}^d$, $d = 2, 3$, is piecewise smooth with the index $k \in \mathbb{N} \cup \{\infty\}$, in short $\Omega \in C^k_{\mathrm{pw}}$, if:

1. There exists a surface mesh $\mathcal{G}$ of smoothness k.
2. Ω is a Lipschitz domain, where the mapping χ_U from Definition 2.2.7 can be chosen in such a way that $\chi_U|_\tau = \chi_\tau$.

Similarly, the boundary $\Gamma = \partial\Omega$ of a bounded C^k_{pw}-domain $\Omega \subset \mathbb{R}^d$, $d = 2, 3$, is also called piecewise smooth with the index $k \in \mathbb{N} \cup \{\infty\}$ and is denoted by $\Gamma \in C^k_{\mathrm{pw}}$.

The definition of C^k_{pw}-domains that we have presented here has been chosen in such a way that we will not need to introduce a new notation for the discretization.

Exercise 2.2.11. *Show that polygonal domains (domains whose boundaries are described by polygonal curves) are Lipschitz domains.*

Show that the bounded (cusp-) domain $\Omega \subset \mathbb{R}^2$ which is bounded by the boundary segments $\{0\} \times [0,1]$, $[0, 1/2] \times \{1\}$, $\{(t, t^s) : 0 \leq t \leq 1/2\}$, $\{1/2\} \times [2^{-s}, 1]$ *is not a Lipschitz domain for all* $s \in (0,1)$ *(see Fig. 2.1).*

A surface mesh allows us to define *piecewise* smooth functions on surfaces.

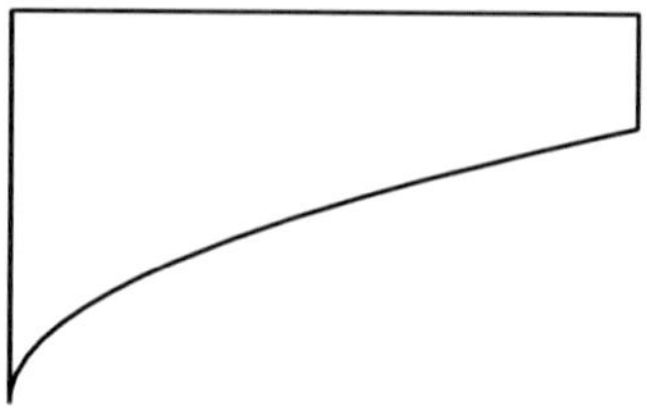

Fig. 2.1 Cusp domain as in Exercise 2.2.11

Definition 2.2.12. Let $k \in \mathbb{N}_0 \cup \{\infty\}$ and $\Gamma \in C^k_{\mathrm{pw}}$. A function $f : \Gamma \to \mathbb{C}$ is called k times piecewise differentiable if there exists a surface mesh $\mathcal{G}$ of smoothness k with

$$f \circ \chi_\tau \in C^k\left(\overline{\hat{\tau}}\right) \qquad \forall \tau \in \mathcal{G}.$$

The set of all k times piecewise differentiable mappings on Γ is denoted by $C^k_{\mathrm{pw}}(\Gamma)$.

2.2.3 *Normal Vector*

Let $\Omega \subset \mathbb{R}^d$ with $d = 2, 3$ be a bounded domain of the type C^1_{pw} and let $\mathcal{G}$ be the surface mesh from Definition 2.2.9 of smoothness $k \geq 1$. The sphere in $\mathbb{R}^d$ is denoted by $\mathbb{S}_{d-1}$. For $\mathbf{x} \in \tau \in \mathcal{G}$ we define a normal vector $\mathbf{n}(\mathbf{x}) \in \mathbb{S}_{d-1}$ by

$$\tilde{\mathbf{n}}(\mathbf{x}) := \left\{ \begin{array}{ll} \left(\chi'_\tau(\xi)\right)^\perp & d = 2 \\ \\ \left(\partial \chi_\tau(\xi) / \partial \xi_1\right) \times \left(\partial \chi_\tau(\xi) / \partial \xi_2\right) & d = 3 \end{array} \right\}$$

$$\text{with } \xi = \chi_\tau^{-1}(\mathbf{x}) \quad \text{and } \mathbf{v}^\perp = \begin{pmatrix} v_2 \\ -v_1 \end{pmatrix},$$

$$\mathbf{n}(\mathbf{x}) := \tilde{\mathbf{n}}(\mathbf{x}) / \|\tilde{\mathbf{n}}(\mathbf{x})\|. \tag{2.72}$$

In general, we assume that the orientation of the charts χ_τ is chosen in such a way that the normal vector points towards the unbounded space outside of Ω:

Remark 2.2.13. *1. For domains of type C^1_{pw} the set*

$$\left\{ \mathbf{x} \in \Gamma : \mathbf{x} \notin \bigcup_{\tau \in \mathcal{G}} \tau \right\}$$

has zero surface measure. Therefore (2.72) defines an outer normal field on Γ almost everywhere.

2. For domains of type C^1 there exists a normal vector for all $\mathbf{x} \in \Gamma$.

Lemma 2.2.14. *Let τ be a C^2-element. Then there exists a constant $0 < C_{\mathbf{n}} < \infty$ such that*

$$|\langle \mathbf{n}(\mathbf{y}), \mathbf{y}-\mathbf{x}\rangle| \le C_{\mathbf{n}} \|\mathbf{y}-\mathbf{x}\|^2 .$$

Proof. Let $\chi_\tau : \hat{\tau} \to \tau$ be the C^2-diffeomorphism given in Definition 2.2.9. For two points $\mathbf{x}, \mathbf{y} \in \tau$ we set

$$\hat{\mathbf{x}} := \chi_\tau^{-1}(\mathbf{x}) \quad \text{and} \quad \hat{\mathbf{y}} := \chi_\tau^{-1}(\mathbf{y}).$$

The mean value theorem guarantees the existence of a point ξ on the line $[\hat{\mathbf{x}}, \hat{\mathbf{y}}]$ such that

$$\langle \mathbf{n}(\mathbf{y}), \mathbf{y}-\mathbf{x}\rangle = \langle \mathbf{n}(\mathbf{y}), \chi_\tau(\hat{\mathbf{y}}) - \chi_\tau(\hat{\mathbf{x}})\rangle = \langle \mathbf{n}(\mathbf{y}), (\mathbf{J}_\tau(\xi))(\hat{\mathbf{y}}-\hat{\mathbf{x}})\rangle ,$$

where $\mathbf{J}_\tau := \mathbf{D}\chi_\tau \in \mathbb{R}^{d\times(d-1)}$ is the Jacobian. Since $\mathbf{n}(\mathbf{y})$ is perpendicular to the column vectors of $\mathbf{J}_\tau(\hat{\mathbf{y}})$, we have

$$\langle \mathbf{n}(\mathbf{y}), \mathbf{y}-\mathbf{x}\rangle = \langle \mathbf{n}(\mathbf{y}), (\mathbf{J}_\tau(\xi) - \mathbf{J}_\tau(\hat{\mathbf{y}}))(\hat{\mathbf{y}}-\hat{\mathbf{x}})\rangle .$$

The assumptions we made concerning the smoothness of τ imply that the matrix $\mathbf{J}_\tau$ is continuously differentiable (componentwise). This, however, proves the assertion since

$$\begin{aligned}|\langle \mathbf{n}(\mathbf{y}), (\mathbf{J}_\tau(\xi) - \mathbf{J}_\tau(\hat{\mathbf{y}}))(\hat{\mathbf{y}}-\hat{\mathbf{x}})\rangle| &\le C_1 \|\xi - \hat{\mathbf{y}}\| \sum_{i=1}^{3}\sum_{j=1}^{2} |\mathbf{n}_i(\mathbf{y})| \, |\hat{\mathbf{y}}-\hat{\mathbf{x}}|_j \\ &\le C_2 \|\mathbf{n}(\mathbf{y})\| \left\|\chi_\tau^{-1}(\mathbf{y}) - \chi_\tau^{-1}(\mathbf{x})\right\|^2 \le C_3 \|\mathbf{y}-\mathbf{x}\|^2 .\end{aligned}$$

□

2.2.4 Boundary Integrals

Let τ be a C^1-panel with the parametrization $\chi_\tau : \hat{\tau} \to \tau$ and let $f : \tau \to \mathbb{C}$ be a measurable function. Then the surface integral of f over τ can be written as

$$\int_\tau f(\mathbf{x})\, ds_{\mathbf{x}} = \int_{\hat{\tau}} \hat{f}(\hat{\mathbf{x}}) \sqrt{g(\hat{\mathbf{x}})} d\hat{\mathbf{x}} \qquad \text{with} \quad \hat{f} := f \circ \chi_\tau . \tag{2.73}$$

Here g signifies the Gram determinant, which is defined as follows. The Jacobian of the parametrization χ_τ is denoted by $\mathbf{J}_\tau := D\chi_\tau = \left(\frac{\partial \chi_i}{\partial \hat{x}_j}\right)_{\substack{1\le i\le d\\ 1\le j\le d-1}}$. The Gram matrix is given by

$$G(\hat{\mathbf{x}}) := \mathbf{J}_\tau^{\intercal}(\hat{\mathbf{x}})\, \mathbf{J}_\tau(\hat{\mathbf{x}}) \in \mathbb{R}^{(d-1)\times(d-1)}.$$

The surface element $\sqrt{g(\hat{\mathbf{x}})}$ in (2.73) is the square root of the determinant of the Gram matrix

$$g(\hat{\mathbf{x}}) := \det G(\hat{\mathbf{x}}).$$

More generally, for piecewise smooth boundaries $\Gamma \in C^1_{\mathrm{pw}}$ and measurable functions $f : \Gamma \to \mathbb{C}$ we have

$$\int_\Gamma f(\mathbf{x})\, ds_\mathbf{x} := \sum_{\tau \in \mathcal{G}} \int_{\hat{\tau}} \hat{f}_\tau(\hat{\mathbf{x}}) \sqrt{g_\tau(\hat{\mathbf{x}})} d\hat{\mathbf{x}}$$

where $\hat{f}_\tau := f \circ \chi_\tau$ and $\sqrt{g_\tau}$ is the surface element of the parametrization χ_τ.

For a measurable subset γ of a surface Γ we denote the surface measure by $|\gamma| := \int_\gamma 1 ds_\mathbf{x}$. For measurable subsets $\omega \subset \mathbb{R}^d$ we use the same notation and set $|\omega| := \int_\omega 1 d\mathbf{x}$.

2.3 Sobolev Spaces on Domains Ω

Results concerning existence and uniqueness can be formulated for elliptic boundary value problems by using Sobolev spaces on domains. We will briefly review some of the properties of the function space $L^2(\Omega)$, after which we will introduce Sobolev spaces. The relevant proofs can be found in [242], for example.

We consider an open subset $\Omega \subset \mathbb{R}^d$. $L^2(\Omega)$ denotes all Lebesgue measurable functions $f : \Omega \to \mathbb{C}$ that satisfy $\int_\Omega |f|^2 d\mathbf{x} < \infty$. We do not distinguish between two functions u, v if they differ on a set of zero measure.

Theorem 2.3.1. *$L^2(\Omega)$ is a Hilbert space with inner product*

$$(u, v)_{0,\Omega} := (u, v)_{L^2(\Omega)} := \int_\Omega u(\mathbf{x}) \overline{v(\mathbf{x})} d\mathbf{x}$$

and norm $\|u\|_{0,\Omega} := \|u\|_{L^2(\Omega)} := (u, u)^{1/2}_{0,\Omega}$.

If there is no cause for confusion we will simply write $(u, v)_0$ and $\|u\|_0$ instead of $(u, v)_{0,\Omega}$ and $\|u\|_{0,\Omega}$.

It is not possible to define classical derivatives (e.g., pointwise as the limit of difference quotients) for functions from $L^2(\Omega)$. In order to define a generalized derivative we use the fact that every function from $L^2(\Omega)$ can be approximated by smooth functions. For a continuous function $u \in C^0(\Omega)$,

$$\operatorname{supp}(u) := \overline{\{x \in \Omega : u(x) \neq 0\}} \tag{2.74}$$

denotes the support of the function u. The space of all infinitely differentiable functions on Ω is denoted by $C^\infty(\Omega)$ and we set

$$C_0^\infty(\Omega) := \{u \in C^\infty(\Omega) : \operatorname{supp}(u) \subset\subset \Omega\}.$$

The space of all functions from $C^\infty(\Omega)$ with compact support is defined as

$$C^\infty_{\text{comp}}(\Omega) := C_0^\infty\left(\mathbb{R}^d\right)\Big|_\Omega := \left\{u|_\Omega : u \in C_0^\infty\left(\mathbb{R}^d\right)\right\}. \tag{2.75}$$

Remark 2.3.2. *It should be noted that the support of a function $u \in C^\infty_{\text{comp}}(\Omega)$ is compact in $\mathbb{R}^d$ but in general is not compact in Ω. Therefore we have $C^\infty_{\text{comp}}(\Omega) \neq C_0^\infty(\Omega)$ for domains $\emptyset \neq \Omega \neq \mathbb{R}^d$.*

Lemma 2.3.3. *The spaces $C^\infty(\Omega) \cap L^2(\Omega)$ and $C_0^\infty(\Omega)$ are dense in $L^2(\Omega)$.*

Definition 2.3.4. A function $u \in L^2(\Omega)$ has a weak derivative $g := \partial_w^\alpha u \in L^2(\Omega)$ if the property

$$(v, g)_{0,\Omega} = (-1)^{|\alpha|}(\partial^\alpha v, u)_{0,\Omega}, \qquad \forall v \in C_0^\infty(\Omega)$$

is satisfied.

We denote the weak derivative by ∂_w.

Remark 2.3.5. *If u has a weak derivative $\partial_w^\alpha u \in L^2(\Omega)$ and the classical derivative $\partial^\alpha u$ exists on $\omega \subset \Omega$ then these two derivatives coincide on ω (almost everywhere). For this reason, in the following we will omit the index w in ∂_w^α.*

Definition 2.3.6. Let $\Omega \subset \mathbb{R}^d$ be a bounded domain. For $\ell = 0, 1, 2, \ldots$ the Sobolev space $H^\ell(\Omega)$ is given by

$$H^\ell(\Omega) := \left\{\varphi \in L^2(\Omega) : \partial^\alpha \varphi \in L^2(\Omega) \text{ for all } |\alpha| \le \ell\right\}. \tag{2.76}$$

On the space $H^\ell(\Omega)$ we define the inner product

$$(\varphi, \psi)_\ell := \sum_{|\alpha| \le \ell} (\partial^\alpha \varphi, \partial^\alpha \psi)_0 = \sum_{|\alpha| \le \ell} \int_\Omega \partial^\alpha \varphi \overline{\partial^\alpha \psi}\, d\mathbf{x} \tag{2.77}$$

and the norm

$$\|\varphi\|_\ell := (\varphi, \varphi)_\ell^{1/2}. \tag{2.78}$$

The space $H^\ell(\Omega)$ is sometimes denoted by $W^{\ell,2}(\Omega)$. We will also need the space

$$W^{\ell,\infty}(\Omega) := \{\varphi \in L^\infty(\Omega) : \partial^\alpha \varphi \in L^\infty(\Omega) \text{ for all } |\alpha| \le \ell\}$$

which is equipped with the norm

$$\forall \varphi \in W^{\ell,\infty}(\Omega) \qquad \|\varphi\|_{W^{\ell,\infty}(\Omega)} := \max_{|\alpha| \le \ell} \|\partial^\alpha \varphi\|_{L^\infty(\Omega)}.$$

If in (2.77) we only sum over those multi-indices with $|\alpha| = \ell$ we can define a seminorm on $H^\ell(\Omega)$ by

$$|\varphi|_\ell^2 := \sum_{|\alpha|=\ell} \int_\Omega |\partial^\alpha \varphi|^2 \, d\mathbf{x}. \tag{2.79}$$

Sobolev spaces can also be defined for non-integer exponents. For $\ell \in \mathbb{R}$, $\lfloor \ell \rfloor$ denotes the greatest integer for which $\lfloor \ell \rfloor \le \ell$. For a non-integer $\ell \ge 0$, i.e., $\ell = \lfloor \ell \rfloor + \lambda$ with $\lambda \in (0,1)$, we define

$$\begin{aligned}(\varphi, \psi)_\ell := &\sum_{|\alpha| \le \lfloor \ell \rfloor} (\partial^\alpha \varphi, \partial^\alpha \psi)_0 \\ &+ \sum_{|\alpha| \le \lfloor \ell \rfloor} \int_{\Omega \times \Omega} \frac{(\partial^\alpha \varphi(\mathbf{x}) - \partial^\alpha \varphi(\mathbf{y})) \overline{(\partial^\alpha \psi(\mathbf{x}) - \partial^\alpha \psi(\mathbf{y}))}}{\|\mathbf{x} - \mathbf{y}\|^{d+2\lambda}} d\mathbf{x} d\mathbf{y}\end{aligned} \tag{2.80}$$

and

$$\|\varphi\|_\ell := (\varphi, \varphi)_\ell^{1/2}. \tag{2.81}$$

For a non-integer ℓ the Sobolev space $H^\ell(\Omega)$ is defined as the closure of

$$\{u \in C^\infty(\Omega) : \|u\|_\ell < \infty\} \tag{2.82}$$

with respect to the norm $\|\cdot\|_\ell$ from (2.81).

Proposition 2.3.7. *The space $H^\ell(\Omega)$ is a separable Hilbert space, i.e., $H^\ell(\Omega)$ has a countable basis (see Sect. 2.1.1.3). We can define an inner product by (2.77), (2.80) and a norm by (2.78), (2.81).*

The fact that certain smooth function spaces are dense in $H^\ell(\Omega)$ becomes very helpful with respect to techniques used in proofs concerning Sobolev spaces.

Definition 2.3.8. $H_0^\ell(\Omega)$ is the closure of the space $C_0^\infty(\Omega)$ with respect to the $\|\cdot\|_\ell$ norm.

Proposition 2.3.9. *We have*

$$H^0(\Omega) = H_0^0(\Omega) = L^2(\Omega), \quad H^\ell\left(\mathbb{R}^d\right) = H_0^\ell\left(\mathbb{R}^d\right).$$

Proposition 2.3.10. *Let $\Omega \subset \mathbb{R}^d$ be open and let $\ell \ge 0$. Then the space $H^\ell(\Omega) \cap C^\infty(\Omega)$ is dense in $H^\ell(\Omega)$.*

The proofs of Proposition 2.3.9 and 2.3.10 can be found in [242, Theorems 3.3–3.6, Conclusion 3.1], for example.

The Sobolev spaces $H^\ell(\Omega)$ of non-integer order $\ell = \lfloor \ell \rfloor + \lambda$ can also be characterized by interpolation. We have

Proposition 2.3.11. *Let $k \in \mathbb{N}_0$ and $0 < \lambda < 1$. For a bounded domain Ω with a Lipschitz boundary we have*

$$H^{k+\lambda}(\Omega) = [H^k(\Omega), H^{k+1}(\Omega)]_{\lambda,2}. \tag{2.83}$$

A proof can be found in, e.g., [230] or [155].

2.4 Sobolev Spaces on Surfaces Γ

In order to define boundary integral equations one needs Sobolev spaces on boundaries $\Gamma := \partial\Omega$ of domains. These are defined with the help of Sobolev spaces on Euclidean (parameter-) domains by means of "lifting".

2.4.1 *Definition of Sobolev Spaces on* Γ

Let $\Omega \subset \mathbb{R}^d$ be a bounded Lipschitz domain. We can define coordinates on the boundary $\Gamma := \partial\Omega$ with the help of a surface mesh $\mathcal{G}$. For this we use the notation from Definition 2.2.7 and introduce the following restrictions

$$\chi_{U,0} : B_2^0 \to U_0 := U \cap \Gamma, \qquad \chi_{U,0} := \chi_U|_{B_2^0}.$$

Now we can define a coordinate system $a = \left(U_0, \chi_{U,0}^{-1}\right)_{U \in \mathcal{U}}$ as well as a subordinate partition of unity $\{\beta_U : \Gamma \to \mathbb{R}\}_{U \in \mathcal{U}}$ by

$$1 = \sum_{U \in \mathcal{U}} \beta_U \text{ on } \Gamma, \qquad \operatorname{supp}(\beta_U) \subset U_0, \qquad \beta_U \circ \chi_{U,0} \in C_0^{0,1}\left(\overline{B_2^0}\right).$$

Functions $\varphi : \Gamma \to \mathbb{C}$ can be localized with the help of this partition. The function

$$\varphi_U = \varphi\beta_U : \Gamma \to \mathbb{C} \quad \text{satisfies} \quad \operatorname{supp}(\varphi_U) \subset U_0.$$

If Ω is a C^k-domain with $k \geq 1$ we can carry out the localization in an analogous way, in which case the functions $\chi_{U,0}$ are C^k-diffeomorphisms. The smoothness of a function on the surface Γ is characterized by the smoothness of the pullback of the localized function to the parameter domain. In this light, we define

$$\widehat{\varphi_U} := \varphi_U \circ \chi_{U,0} : B_2^0 \to \mathbb{C}, \qquad U \in \mathcal{U}.$$

Therefore it is obvious that the maximal smoothness of the domain Ω is an upper bound for the order of differentiability of the Sobolev spaces on Γ. More specifically, for $C^{0,1}$ or C^k-domains, only Sobolev spaces $H^\ell(\Gamma)$ with a maximal order of differentiability ℓ, invariant under the choice of the coordinate system, that satisfy

$$\begin{aligned} &\ell \leq 1 \text{ for Lipschitz domains } \Omega, \\ &\ell \leq k \text{ for } C^k\text{-domains } \Omega \end{aligned} \tag{2.84}$$

can be defined. We use the previously introduced notation for the following definition.

Definition 2.4.1. Let $\Omega \subset \mathbb{R}^d$ be a bounded $C^{0,1}$ or C^k-domain with $k \geq 1$. We assume that (2.84) is satisfied for $\ell \in \mathbb{R}_{\geq 0}$. The space $H^\ell(\Gamma)$ contains all functions $\varphi : \Gamma \to \mathbb{C}$ that satisfy $\widehat{\varphi_U} \in H_0^\ell(B_2^0)$ for all $U \in \mathcal{U}$.

In the same way as in (2.80), we can define a norm on $H^\ell(\Gamma)$ for $\varphi \in H^\ell(\Gamma)$ by setting $\lambda := \ell - \lfloor \ell \rfloor$ and

$$\|\varphi\|_{\ell,\Gamma}^2 := \begin{cases} \sum_{|\alpha|\leq \ell} \|\varphi_\alpha\|_{L^2(\Gamma)}^2 & \text{if } \ell \in \mathbb{N}_0, \\ \sum_{|\alpha|\leq \lfloor \ell \rfloor} \|\varphi_\alpha\|_{L^2(\Gamma)}^2 + \sum_{|\alpha|\leq \lfloor \ell \rfloor} \int_{\Gamma\times\Gamma} \frac{|\varphi_\alpha(\mathbf{x}) - \varphi_\alpha(\mathbf{y})|^2}{\|\mathbf{x}-\mathbf{y}\|^{d-1+2\lambda}} ds_{\mathbf{x}} ds_{\mathbf{y}} & \text{if } \ell \in \mathbb{R}\backslash\mathbb{N}_0, \end{cases} \tag{2.85}$$

where the functions $\varphi_\alpha : \Gamma \to \mathbb{C}$ are given by

$$\varphi_\alpha(\mathbf{x}) := \sum_{U\in\mathcal{U}} \partial_\xi^\alpha (\widehat{\varphi_U})(\xi) \quad \text{with } \mathbf{x} = \chi_{U,0}(\xi) \tag{2.86}$$

and ∂_ξ^α denotes the differentiation with respect to the variable ξ.

Formally, the Sobolev space $H^\ell(\Gamma)$ depends on the chosen coordinates. Should it be necessary, we will write $H_a^\ell(\Gamma)$ instead of $H^\ell(\Gamma)$. It can, however, be shown that $H^\ell(\Gamma)$ is defined invariantly on Γ under the condition that there is a suitable relation between the order of differentiation ℓ and the smoothness of the boundary.

Proposition 2.4.2. *Let Ω be a bounded Lipschitz domain or a C^k-domain with $k \geq 1$. We assume that the index of differentiation ℓ satisfies (2.84). Let a_1, a_2 be two coordinate systems on Γ. Then the spaces $H_{a_1}^\ell(\Gamma)$ and $H_{a_2}^\ell(\Gamma)$ are equivalent, i.e., they contain the same set of functions, and the norms are equivalent.*

The proof of this proposition can be found in [242, Theorem 4.2].

Sobolev spaces $H^\ell(\Gamma)$ of non-integer order can also be characterized by interpolation. We have the following theorem, which is analogous to Proposition 2.3.11:

Proposition 2.4.3. *Let $\Omega \subset \mathbb{R}^d$ be a bounded Lipschitz or C^k-domain with $k \geq 1$ and $\Gamma := \partial\Omega$: Furthermore, let $\ell \in \mathbb{N}_0$ be such that $\ell + 1$ satisfies Condition (2.84). Then for $0 < \lambda < 1$*

$$H^{\ell+\lambda}(\Gamma) = \left(H^\ell(\Gamma), H^{\ell+1}(\Gamma)\right)_{\lambda,2}. \tag{2.87}$$

More generally, if ℓ_1, ℓ_2 satisfy (2.84) then

$$H^\ell(\Gamma) = \left(H^{\ell_1}(\Gamma),\ H^{\ell_2}(\Gamma)\right)_{\lambda,2} \tag{2.88}$$

for $\ell = \lambda\ell_1 + (1-\lambda)\ell_2$ with $0 \leq \lambda \leq 1$.

We have introduced Sobolev spaces with non-negative differentiation indices for domains Ω and their boundaries Γ. The dual spaces of these Sobolev spaces contain all the continuous linear functionals defined thereon. Let X be either a domain Ω or a surface Γ. Then the following notation is used for the dual space:

$$H^{-\ell}(X) := \left(H_0^{\ell}(X)\right)', \quad \ell \geq 0. \tag{2.89}$$

Note that in the case of closed surfaces ($X = \Gamma$) the boundary of X is the empty set and therefore $H_0^{\ell}(X) = H^{\ell}(X)$.

All the results for $H_0^{\ell}(X)$ concerning density can be directly transferred to the dual spaces. Due to the Riesz representation theorem (Theorem 2.1.17) for every $F \in H^{-\ell}(X)$ there exists an element $f \in H_0^{\ell}(X)$ with

$$F(v) = (v, f)_{H^{\ell}(\Omega)} \qquad \forall v \in H_0^{\ell}(X).$$

If a space U is dense in $H_0^{\ell}(X)$, then for every functional $F \in H^{-\ell}(X)$ there exists a sequence of elements $(f_i)_{i \in \mathbb{N}_0}$ in U such that

$$\lim_{i \to \infty} (\cdot, f_i)_{H^{\ell}(\Omega)} = F.$$

2.4.2 Sobolev Spaces on $\Gamma_0 \subset \Gamma$

In order to formulate integral equations on domains with "cracks" we need Sobolev spaces on open manifolds with boundary conditions. In the following we will briefly discuss the most important definitions and properties and refer to [162, Sect. 3], for example, for a more detailed discussion.

Let $\Gamma_0 \subset \Gamma$ be a measurable subset of the boundary with $|\Gamma_0| > 0$. The Sobolev space $\widetilde{H}^s(\Gamma_0)$, $s \in [0, 1]$, is defined by

$$\widetilde{H}^s(\Gamma_0) := \left\{u \in H^s(\Gamma) : \operatorname{supp}(u) \subset \overline{\Gamma_0}\right\}. \tag{2.90}$$

The norm on $\widetilde{H}^s(\Gamma_0)$ is given by

$$\|u\|_{\widetilde{H}^s(\Gamma_0)} := \left\|u^\star\right\|_{H^s(\Gamma)}, \tag{2.91}$$

where $u^\star$ denotes the extension of u on Γ by zero.

Exercise 2.4.4. *Let $\Gamma = (-1, 2)$ and $\Gamma_0 = (0, 1)$. Show that the characteristic function*

$$u(x) := \begin{cases} 1 & x \in \Gamma_0, \\ 0 & \text{otherwise} \end{cases}$$

is in $\widetilde{H}^s(\Gamma_0)$ for $s < 1/2$ but not for $s \geq 1/2$.

The spaces with a negative index are again defined as dual spaces: $\widetilde{H}^{-s}(\Gamma_0) := (H^s(\Gamma_0))'$ for $s \in [0,1]$. Conversely, we have: $H^{-s}(\Gamma_0) = \left(\widetilde{H}^s(\Gamma_0)\right)'$ for $s \in [0,1]$. Note that for closed surfaces Γ the spaces $H^s(\Gamma)$ and $\widetilde{H}^s(\Gamma)$ are isomorphic.

2.5 Embedding Theorems

The spaces $H^\ell(\Omega)$, $H^\ell(\Gamma)$ are nested for a continuous *scale* of indices ℓ.

Theorem 2.5.1. *There holds*

$$\left.\begin{array}{l} H^{\ell_1}(\Omega) \subset H^{\ell_2}(\Omega) \\ H^{\ell_1}(\Gamma) \subset H^{\ell_2}(\Gamma) \end{array}\right\} \qquad \ell_1 \geq \ell_2 \geq 0. \tag{2.92}$$

In the case of a surface we require that Condition (2.84) is satisfied with $\ell = \ell_1$.

The Sobolev spaces with a positive differentiation index ℓ together with $L^2(\Omega)$ and their dual spaces form a Gelfand triple.

Proposition 2.5.2. *For $\ell > 0$ the triples*

$$H^\ell(\Omega) \subset L^2(\Omega) \subset \left(H^\ell(\Omega)\right)'$$
$$H_0^\ell(\Omega) \subset L^2(\Omega) \subset (H_0^\ell(\Omega))'$$
$$H^\ell(\Gamma) \subset L^2(\Gamma) \subset \left(H^\ell(\Gamma)\right)'$$

are Gelfand triples, whereby we again require that Condition (2.84) be satisfied in the case of a surface. The inner product $(\cdot,\cdot)_{L^2(\Omega)}$ can therefore be continuously extended to dual pairings on $H^\ell(\Omega) \times \left(H^\ell(\Omega)\right)'$, $\left(H^\ell(\Omega)\right)' \times H^\ell(\Omega)$, $H_0^\ell(\Omega) \times \left(H_0^\ell(\Omega)\right)'$ and $\left(H_0^\ell(\Omega)\right)' \times H_0^\ell(\Omega)$. Analogously, the inner product $(\cdot,\cdot)_{L^2(\Gamma)}$ can be continuously extended to dual pairings on $H^\ell(\Gamma) \times \left(H^\ell(\Gamma)\right)'$ and $\left(H^\ell(\Gamma)\right)' \times H^\ell(\Gamma)$.

Notation 2.5.3. *Assuming the same conditions as in Proposition 2.5.2 we again denote the extensions by $(\cdot,\cdot)_{L^2(\Omega)}$ and $(\cdot,\cdot)_{L^2(\Gamma)}$, in case the relevant function spaces can be determined from the arguments. If the domain Ω is clear from the context, we simply write $(\cdot,\cdot)_0$.*

It is interesting to know under which conditions every function (equivalence class) $\varphi \in H^\ell(\Omega)$ has a continuous representative. This question is answered by the Sobolev embedding theorem.

Theorem 2.5.4 (Sobolev Embedding Theorem). *Let $\Omega \subset \mathbb{R}^d$ be a bounded Lipschitz domain. Then for $\ell > d/2$:*

$$H^{\ell}(\Omega) \subset C^{0}\left(\overline{\Omega}\right).$$

For bounded C^k-domains Ω, $k > \ell$, functions (equivalence classes) $\varphi \in H^{\ell}(\Omega)$ have an m times continuously differentiable representative for integer $m < \ell - d/2$:

$$H^{\ell}(\Omega) \subset C^{m}\left(\overline{\Omega}\right)$$

with a continuous embedding:

$$\|\varphi\|_{C^m\left(\overline{\Omega}\right)} \leq C\, \|\varphi\|_{H^{\ell}(\Omega)}, \qquad \forall \varphi \in H^{\ell}(\Omega).$$

In order to determine whether a function from $H^{\ell_2}(\Omega)$ can be approximated with respect to a (weaker) norm $\|\cdot\|_{\ell_1,\Omega}$, i.e., $\ell_1 < \ell_2$, the compactness (see Definition 2.1.31) of the embedding $I : H^{\ell_1}(\Omega) \to H^{\ell_2}(\Omega)$ will be crucial.

Theorem 2.5.5 (Rellich). *Let $\Omega \subset \mathbb{R}^d$ be a bounded Lipschitz domain. Then the first of the embeddings in (2.92) for $\ell_1 > \ell_2$ is compact. To guarantee the compactness of the second embedding, Condition (2.84) also has to be satisfied.*

The proof can be found in, e.g., [3, 242].

In some proofs, results are first proven for dense subspaces of Sobolev spaces. They are then applied to the Sobolev spaces by considering Cauchy sequences and their limits. Keeping this in mind we will use Rellich's embedding theorem (see, e.g., [3, Sect. 5.9(4), A 5.4], [242]).

Theorem 2.5.6. *Let $\Omega \subset \mathbb{R}^d$ be a bounded domain with a Lipschitz boundary. Then the embedding $H^1(\Omega) \hookrightarrow L^2(\Omega)$ is compact, i.e., there exists for every bounded sequence in $H^1(\Omega)$ a subsequence that converges with respect to the norm in $L^2(\Omega)$.*

The Poincaré inequalities are useful consequences of these compact embeddings.

Theorem 2.5.7. *Let $\Omega \subset \mathbb{R}^d$ be bounded and $\ell = 1, 2, \ldots$. Then for all $\varphi \in H_0^{\ell}(\Omega)$*

$$\|\varphi\|_{\ell,\Omega}^2 \leq C\,(1 + \operatorname{diam}\Omega)^{2\ell} \sum_{|\alpha|=\ell} \int_{\Omega} |\partial^{\alpha}\varphi|^2\, d\mathbf{x}. \tag{2.93}$$

The inequality in (2.93) is also referred to as the Friedrichs inequality.

Corollary 2.5.8. *Let Γ_D be a subset of the boundary Γ with a positive $(d-1)$-dimensional surface measure. Theorem 2.5.7 remains valid if the space $H_0^{\ell}(\Omega)$ is replaced by the space $\overline{\left\{\varphi \in C^{\infty}\left(\overline{\Omega}\right) : \varphi = 0 \text{ on } \Gamma_D\right\}}^{\|\cdot\|_{\ell,\Omega}}$.*

For $\varphi \in H^{\ell}(\Omega)$, this assertion is only true in the modified form (2.94).

Theorem 2.5.9. *Let $\Omega \subset \mathbb{R}^d$ be a Lipschitz domain. Then for all $\varphi \in H^\ell(\Omega)$*

$$\|\varphi\|_{\ell,\Omega}^2 \leq C \left\{ \sum_{|\alpha|=\ell} \int_\Omega |\partial^\alpha \varphi|^2 \, d\mathbf{x} + \sum_{|\alpha|<\ell} \left| \int_\Omega \partial^\alpha \varphi \, d\mathbf{x} \right|^2 \right\}. \tag{2.94}$$

The inequalities (2.93) and (2.94) are called the first and second Poincaré inequalities and are proven in [166, Theorems 1.1 and 1.5].

Corollary 2.5.10. *Let $\Omega \subset \mathbb{R}^d$ be a Lipschitz domain. Then there exists a constant $c_\Omega > 0$ such that for all $\varphi \in H^1(\Omega)$*

$$\inf_{z \in \mathbb{R}} \|\varphi - z\|_{1,\Omega} \leq c_\Omega |\varphi|_{1,\Omega}.$$

Proof. Choose $\alpha := \int_\Omega \varphi d\mathbf{x} / |\Omega|$ and define $\varphi_\alpha := \varphi - \alpha$. It then follows from (2.94) that

$$\inf_{z \in \mathbb{R}} \|\varphi - z\|_{1,\Omega} \leq \|\varphi_\alpha\|_{1,\Omega} \leq C \left(|\varphi_\alpha|_{1,\Omega} + \left| \int_\Omega \varphi_\alpha d\mathbf{x} \right| \right) = C |\varphi|_{1,\Omega}.$$

□

For convex, polygonal domains $\Omega \subset \mathbb{R}^2$, the constant C is known explicitly.

Theorem 2.5.11. *Let $\Omega \subset \mathbb{R}^2$ be a convex and polygonal domain. For any function $w \in H^1(\Omega)$ with $\int_\Omega w \, d\mathbf{x} = 0$, there holds the estimate*

$$\|w\|_{L^2(\Omega)} \leq \frac{1}{\pi} (\operatorname{diam} \Omega) |w|_{H^1(\Omega)}.$$

For the proof we refer to [180].

Theorem 2.5.11 can be generalized to neighborhoods of convex sets.

Corollary 2.5.12. *Let $\Omega \subset \Omega^\star \subset \mathbb{R}^2$ denote two convex and polygonal domains. For $u \in H^1(\Omega^\star)$, let $\Pi_\Omega(u) := \int_\Omega u d\mathbf{x} / |\Omega|$. Then*

$$\|u - \Pi_\Omega(u)\|_{L^2(\Omega^\star)} \leq \left(1 + \sqrt{\frac{|\Omega^\star|}{|\Omega|}} \right) \frac{\operatorname{diam} \Omega^\star}{\pi} |u|_{H^1(\Omega^\star)}.$$

Proof. For $u \in H^1(\Omega^\star)$, let $\Pi_{\Omega^\star}(u) := \int_{\Omega^\star} u d\mathbf{x} / |\Omega_{\Omega^\star}|$. The projection property of Π_Ω leads to

$$u - \Pi_\Omega u = (I - \Pi_\Omega)(u - \Pi_{\Omega^\star} u),$$

where I is the identity. Hence

$$|u - \Pi_\Omega u|_{L^2(\Omega^\star)} \leq \left(1 + \sup_{v \in L^2(\Omega^\star) \setminus \{0\}} \frac{\|\Pi_\Omega v\|_{L^2(\Omega^\star)}}{\|v\|_{L^2(\Omega^\star)}} \right) \|u - \Pi_{\Omega^\star} u\|_{L^2(\Omega^\star)}. \tag{2.95}$$

The supremum can be estimated by using the Cauchy–Schwarz inequality for the L^2-scalar product

$$\|\Pi_\Omega v\|_{L^2(\Omega^\star)}^2 = \int_{\Omega^\star} \left(\frac{1}{|\Omega|} \int_\Omega v d\mathbf{x} \right)^2 d\mathbf{y} \leq \int_{\Omega^\star} \frac{1}{|\Omega|} \|v\|_{L^2(\Omega)}^2 d\mathbf{y}$$
$$= \frac{|\Omega^\star|}{|\Omega|} \|v\|_{L^2(\Omega)}^2 \leq \frac{|\Omega^\star|}{|\Omega|} \|v\|_{L^2(\Omega^\star)}^2 .$$

Applying Theorem 2.5.11 to the right-hand side in (2.95) results in

$$|u - \Pi_\Omega u|_{L^2(\Omega^\star)} \leq \left(1 + \sqrt{\frac{|\Omega^\star|}{|\Omega|}} \right) \frac{\operatorname{diam} \Omega^\star}{\pi} |u|_{H^1(\Omega^\star)} .$$

□

2.6 Trace Operators

The trace (restriction) of a function $u \in H^\ell(\Omega)$ on the boundary $\partial\Omega$ can be given a reasonable definition if the differentiation index of the Sobolev space and the regularity of the surfaces are both sufficiently large. The main result is summarized in Theorems 2.6.8 and 2.6.9. The Sobolev norm of the trace of a sufficiently smooth function $u : \Omega \to \mathbb{C}$ can be estimated by the Sobolev norm of u in a local neighborhood of $\partial\Omega$. Therefore traces can also be defined for functions that are only locally in $H^\ell(\Omega)$. The relevant space is called $H^\ell_{\mathrm{loc}}(\Omega)$. For later applications we will also introduce the associated dual space $H^s_{\mathrm{comp}}(\Omega)$. Although norms cannot be defined on these spaces, a metric and therefore a topology can be defined. We refer to [243] and [85, pp. 48, 114 ff] for details. However, for our applications we only need the results given in Theorem 2.6.7, which provides us with criteria with which we can prove the continuity of mappings from and to these spaces.

Definition 2.6.1. Let Ω be a (possibly unbounded) domain. The space $H^\ell_{\mathrm{loc}}(\Omega)$ contains all continuous, linear functionals (distributions) on $C^\infty_{\mathrm{comp}}(\Omega)$, in short $u \in C^\infty_{\mathrm{comp}}(\Omega)'$, with the property that $\varphi u \in H^\ell(\Omega)$ for all $\varphi \in C^\infty_{\mathrm{comp}}(\Omega)$.

Remark 2.6.2. *(a) The definition of the space $H^\ell_{\mathrm{loc}}(\Omega)$ does not contain any restrictions with respect to the growth of the function towards infinity. For example, every polynomial as well as the exponential function are in $H^\ell_{\mathrm{loc}}(\mathbb{R})$ for an arbitrary $\ell \geq 0$.*

(b) By choosing $\varphi \equiv 1$ we see that for bounded domains Ω, $H^\ell_{\mathrm{loc}}(\Omega)$ and $H^\ell(\Omega)$ coincide (see Remark 2.3.2).

(c) Let $\Omega^- \subset \mathbb{R}^d$ be a bounded domain with boundary $\Gamma := \partial\Omega^-$ and $\Omega^+ := \mathbb{R}^d \backslash \overline{\Omega^-}$. Then the growth of functions from $H^\ell_{\mathrm{loc}}(\Omega^+)$ is not restricted towards

infinity. However, it is restricted in every bounded neighborhood of Γ*: For* $u \in H^{\ell}_{\text{loc}}(\Omega^{+})$ *we have* $u|_U \in H^{\ell}(U)$ *for every bounded subdomain* $U \subset \Omega^{+}$.

(d) *In some literature, in the definition of* $H^{\ell}_{\text{loc}}(\Omega)$ *the condition* $\varphi \in C^{\infty}_{\text{comp}}(\Omega)$ *is sometimes replaced by* $\varphi \in C^{\infty}_{0}(\Omega)$*. In this case, using the conditions in (c), the growth of the functions in a local neighborhood of* Γ *is also not restricted.*

In order to define the dual space of $H^{\ell}_{\text{loc}}(\Omega)$, we need to extend the definition of the support of a function [see (2.74)] to Sobolev functions.

Definition 2.6.3. Let $\ell \geq 0$. The restriction of a function $u \in H^{\ell}(\Omega)$ to an open subset $U \subset \Omega$ is the zero function $u|_U = 0$ if

$$(u, w)_{H^{\ell}(\Omega)} = 0$$

for all $w \in C^{\infty}(\Omega)$ with $\operatorname{supp} w \subset U$.

For $\ell < 0$ the condition $(u, w)_{H^{\ell}(\Omega)} = 0$ has to be replaced by $u(w) = 0$.

Definition 2.6.4. Let $u \in H^{\ell}(\Omega)$. The support $\operatorname{supp}(u)$ is the largest, relatively closed set $V \subset \Omega$ for which u is the zero function on $\Omega \backslash V$.

Definition 2.6.5. Let $\ell \in \mathbb{R}$ and $\Omega \subset \mathbb{R}^d$ be open. The space $H^{\ell}_{\text{comp}}(\Omega)$ is given by

$$H^{\ell}_{\text{comp}}(\Omega) := \bigcup_{K} \left\{ u \in H^{\ell}_{\text{loc}}(\Omega) : \operatorname{supp}(u) \subset K \right\},$$

where the union is taken over all relatively compact subsets $K \subset \Omega$.

Remark 2.6.6. *Note that for bounded domains,* $H^{\ell}_{\text{comp}}(\Omega)$ *coincides with* $H^{\ell}(\Omega)$ *and* $H^{\ell}_{\text{loc}}(\Omega)$.

Theorem 2.6.7. (a) *For every* $s \in \mathbb{R}$ *the bilinear form* $\langle \cdot, \cdot \rangle : C^{\infty}(\Omega) \times C^{\infty}_{\text{comp}}(\Omega) \to \mathbb{K}$*:*

$$\langle u, v \rangle = \int_{\Omega} uv dx \tag{2.96}$$

can be extended to a dual pairing $\langle \cdot, \cdot \rangle : H^{s}_{\text{loc}}(\Omega) \times H^{-s}_{\text{comp}}(\Omega) \to \mathbb{K}$.

(b) *Let E be a normed space. A linear mapping* $A : H^{s}_{\text{comp}}(\Omega) \to E$ *is continuous if and only if the restriction* $A : \left\{ u \in H^{s}_{\text{loc}}(\Omega) : \operatorname{supp}(u) \subset K \right\} \to E$ *is continuous for all compact sets* $K \subset \Omega$*. A linear mapping* $A : H^{s}_{\text{loc}}(\Omega) \to E$ *is continuous if and only if there exist* $\varphi \in C^{\infty}_{\text{comp}}(\Omega)$ *and a constant* $C < \infty$ *such that*

$$\|Au\|_E \leq C \|\varphi u\|_{H^s(\Omega)} \qquad \forall u \in H^{s}_{\text{loc}}(\Omega).$$

(c) *A linear mapping* $A : E \to H^{s}_{\text{loc}}(\Omega)$ *is continuous if and only if for all* $\varphi \in C^{\infty}_{\text{comp}}(\Omega)$ *there exists a* $C < \infty$ *such that*

$$\|\varphi (Au)\|_{H^s(\Omega)} \leq C \|u\|_E \qquad \forall u \in E.$$

(d) *A linear mapping* $A : H^s_{\text{comp}}(\Omega) \to H^t_{\text{loc}}(\Omega)$ *is continuous if and only if for all compact sets* $K \subset \Omega$ *and all* $\varphi \in C^\infty_{\text{comp}}(\Omega)$ *there exists a* $C < \infty$ *such that*

$$\|\varphi(Au)\|_{H^t(\Omega)} \le C \|u\|_{H^s(\Omega)} \qquad \forall u \in H^s_{\text{comp}}(\Omega) : \operatorname{supp}(u) \subset K.$$

(e) *A linear mapping* $A : H^s_{\text{loc}}(\Omega) \to H^t_{\text{loc}}(\Omega)$ *is compact if and only if for all cut-off functions* $\varphi, \psi \in C^\infty_{\text{comp}}(\Omega)$ *the restriction* $u \to \varphi A(\psi u) : H^s(\Omega) \to H^t(\Omega)$ *is compact.*

Theorem 2.6.8. *Let* Ω^- *be a bounded Lipschitz domain with boundary* Γ *and* $\Omega^+ := \mathbb{R}^d \backslash \overline{\Omega^-}$.

(a) *For* $1/2 < \ell < 3/2$ *there exists a continuous, linear trace operator* $\gamma_0 : H^\ell_{\text{loc}}(\mathbb{R}^d) \to H^{\ell-1/2}(\Gamma)$ *with*

$$\gamma_0\varphi = \varphi\,|_\Gamma \qquad \textit{for all } \varphi \in C^0\left(\mathbb{R}^d\right).$$

(b) *For* $s \in \{+,-\}$ *there exist one-sided, continuous, linear trace operators* $\gamma_0^s : H^\ell_{\text{loc}}(\Omega^s) \to H^{\ell-1/2}(\Gamma)$ *with*

$$\gamma_0^s\varphi = \varphi\,|_\Gamma \qquad \textit{for all } \varphi \in C^0\left(\overline{\Omega^s}\right)$$

and

$$\gamma_0^+ u = \gamma_0^- u = \gamma_0 u \qquad \textit{almost everywhere}$$

for all $u \in H^\ell_{\text{loc}}(\mathbb{R}^d)$.

This result can be generalized for smoother domains.

Theorem 2.6.9. *Let* $\Omega^- \subset \mathbb{R}^d$ *be a bounded* C^k*-domain,* $k \in \mathbb{N} \cup \{\infty\}$ *and* $\Omega^+ := \mathbb{R}^d \backslash \overline{\Omega^-}$. *Let the differentiation index* ℓ *satisfy the condition* $1/2 < \ell \le k$. *Then the trace operator from Theorem 2.6.8 is a continuous operator* $\gamma_0 : H^\ell_{\text{loc}}(\mathbb{R}^d) \to H^{\ell-1/2}(\Gamma)$ *which satisfies the property*

$$\gamma_0\varphi = \varphi\,|_\Gamma \qquad \forall \varphi \in C_0^\infty\left(\mathbb{R}^d\right).$$

The proof is based on a localization of the statement with the help of a C^k or $C^{0,1}$-atlas of Γ and a subordinate partition of unity. In doing so, the trace theorem can be reduced to a trace theorem in the half-space and can then be proven by characterizing Sobolev norms in terms of Fourier transforms (see [162] and [72]). A direct result is the fact that the trace of a function is solely determined by its local behavior in a neighborhood of Γ.

Remark 2.6.10. *With the same conditions as in Theorem 2.6.8 we have for all* $v \in H^\ell_{\text{loc}}(\mathbb{R}^d)$, $1/2 < \ell < 3/2$, *and all cut-off functions* $\chi \in C_0^\infty(\mathbb{R}^d)$ *that satisfy* $\chi \equiv 1$ *in a neighborhood of* Γ

$$\gamma_0 (\chi v) = \gamma_0 (v) .$$

For $v^+ \in H^\ell_{\text{loc}} (\Omega^+)$ *and* $v^- \in H^\ell_{\text{loc}} (\Omega^-)$ *we have*

$$\gamma_0^+ (\chi v^+) = \gamma_0^+ (v^+) \quad \textit{and} \quad \gamma_0^- (\chi v^-) = \gamma_0^- (v^-) .$$

The trace theorem answers the question under which conditions functions from a Sobolev space $H^\ell (\Omega)$ can be restricted to surfaces Γ. It turns out that this is possible for sufficiently smooth surfaces. Furthermore, the differentiability is reduced by half an order.

There are a number of applications in which the inverse of this question plays a significant role. Can functions from $H^{\ell-1/2} (\Gamma)$, that are given on surfaces, be extended to $H^\ell (\Omega)$?

Theorem 2.6.11. *Let* Ω^- *be a bounded Lipschitz domain with surface* Γ *and* $\Omega^+ := \mathbb{R}^d \backslash \overline{\Omega^-}$. *Then for* $1/2 < \ell < 3/2$ *there exists a linear, continuous extension operator* $Z : H^{\ell-1/2} (\Gamma) \to H^\ell_{\text{comp}} (\mathbb{R}^d)$ *with* $(\gamma_0 \circ Z) (\varphi) = \varphi$ *on* $H^{\ell-1/2} (\Gamma)$. *For* $\Omega \in \{\Omega^-, \Omega^+\}$ *the composition*

$$Z_\Omega := R_\Omega Z : H^{\ell-1/2} (\Gamma) \to H^\ell_{\text{comp}} (\Omega)$$

is continuous. Here, R_Ω *denotes the restriction of a function in* $H^\ell_{\text{comp}} (\mathbb{R}^d)$ *to* Ω.

The proof is given in, e.g., [242, Theorem 8.8].

Notation 2.6.12. *Alternatively,* Z_{Ω^+} *is denoted by* Z_+ *and* Z_{Ω^-} *by* Z_-.

2.7 Green's Formulas and Normal Derivatives

Classically, elliptic boundary value problems consist of a differential equation for the unknown function on the domain Ω and associated boundary conditions. We formulate the Laplace problem with Dirichlet boundary conditions as a prototype for a linear elliptic differential equation: Let $f \in C^0 (\overline{\Omega})$ and $g_D \in C^0 (\Gamma)$ be given. Find $u \in C^2 (\Omega) \cap C^0 (\overline{\Omega})$ such that

$$-\Delta u = f \quad \text{in } \Omega, \qquad u = g_D \quad \text{on } \Gamma. \tag{2.97}$$

In general, for equations of second order, either the trace or the normal derivative of the unknown function is given on the boundary.

In this section we will define the conormal derivative to the general linear, elliptic differential operator with constant coefficients. This operator has the form

$$Lu := -\operatorname{div} (\mathbf{A} \operatorname{grad} u) + 2 \langle \mathbf{b}, \operatorname{grad} u \rangle + cu, \tag{2.98}$$

where we generally suppose that $\mathbf{A} \in \mathbb{R}^{d \times d}$ is positive definite, $\mathbf{b} \in \mathbb{R}^d$ and $c \in \mathbb{R}$. The smallest eigenvalue of $\mathbf{A}$ is denoted by $a_{\min}$ and the largest by $a_{\max}$. We always assume that we have

$$0 < a_{\min} \leq a_{\max} < \infty. \tag{2.99}$$

The Laplace problem (2.97) results if we choose $\mathbf{A} = \mathbf{I}$, $\mathbf{b} = \mathbf{0}$ and $c = 0$.

In order to define the conormal derivative we will multiply the operator L in (2.98) by suitable functions and integrate by parts over Ω. The equations that result from this are called Green's formulas and also form part of this chapter. Let $\mathbf{L}^\infty(\Gamma) := (L^\infty(\Gamma))^d$.

Theorem 2.7.1 (Rademacher). *Let Ω be a bounded Lipschitz domain with boundary Γ. Then there exists an outer normal vector almost everywhere on Γ which satisfies $\mathbf{n} \in \mathbf{L}^\infty(\Gamma)$.*

A proof of this theorem can be found in, e.g., [241, 11A, p. 272]. Next we will introduce some conventions which will be used frequently.

Convention 2.7.2. *Let Ω^- be a bounded Lipschitz domain with boundary Γ and let $\Omega^+ := \mathbb{R}^d \backslash \overline{\Omega^-}$. We assume that each of these domains is connected and, furthermore, that the orientation of the normal field $\mathbf{n} : \Gamma \to \mathbb{S}_{d-1}$ is chosen in the direction of Ω^+. In the following Ω denotes one of the domains Ω^-, Ω^+, and the algebraic sign function σ_Ω is given by*

$$\sigma_\Omega := \begin{cases} 1 & \text{for } \Omega = \Omega^-, \\ -1 & \text{for } \Omega = \Omega^+. \end{cases}$$

Therefore $\sigma_\Omega \mathbf{n}$ is the outer normal relative to Ω.

The principal part of the operator L in (2.98) is given by $\operatorname{div}(\mathbf{A} \operatorname{grad} \cdot)$. Gauss' theorem deals with integration by parts of integrands in "divergence form".

Theorem 2.7.3 (Gauss' Theorem). *Let $\Omega \in \{\Omega^-, \Omega^+\}$. For all $\mathbf{F} \in H^1(\Omega, \mathbb{R}^d)$ we have*

$$\int_\Omega (\operatorname{div} \mathbf{F})\, d\mathbf{x} = \int_\Gamma \langle \sigma_\Omega \mathbf{n}, \mathbf{F} \rangle \, ds_{\mathbf{x}}.$$

The proof can be found in, e.g., [162, Theorem 3.34, Lemma 4.1]. A direct result of Gauss' theorem is the first of Green's formulas.

Theorem 2.7.4. *Let $\mathbf{A} \in \mathbb{R}^{d \times d}$ be symmetric and positive definite. Then we have for all $u \in H^2(\Omega)$ and $v \in H^1(\Omega)$ Green's first formula*

$$\int_\Omega \operatorname{div}(\mathbf{A} \operatorname{grad} u)\, v d\mathbf{x} = -\int_\Omega \langle \mathbf{A} \operatorname{grad} u, \operatorname{grad} v \rangle \, d\mathbf{x} + \sigma_\Omega \int_\Gamma \langle \mathbf{A}\mathbf{n}, \operatorname{grad} u \rangle \, v ds_{\mathbf{x}}. \tag{2.100}$$

For $v \in H^2(\Omega)$ one obtains Green's second formula

$$\int_\Omega \operatorname{div}(\mathbf{A}\operatorname{grad} u)\, v d\mathbf{x} - \int_\Omega u \operatorname{div}(\mathbf{A}\operatorname{grad} v)\, d\mathbf{x}$$
$$= \sigma_\Omega \left(\int_\Gamma \langle \mathbf{A}\mathbf{n}, \operatorname{grad} u\rangle\, v ds_\mathbf{x} - \int_\Gamma u \langle \mathbf{A}\mathbf{n}, \operatorname{grad} v\rangle\, ds_\mathbf{x} \right). \tag{2.101}$$

A proof can be found in, e.g., [162, Chap. 4].

For $u, v \in H^1(\Omega)$ we can define the sesquilinear form

$$B(u,v) := \int_\Omega \left(\langle \mathbf{A}\operatorname{grad} u, \overline{\operatorname{grad} v}\rangle + 2\langle \mathbf{b}, \operatorname{grad} u\rangle \overline{v} + cu\overline{v} \right) d\mathbf{x} \tag{2.102}$$

and for $u \in H^2(\Omega)$ we can define the conormal derivative

$$\gamma_1 u := \langle \mathbf{A}\mathbf{n}, \gamma_0 \operatorname{grad} u\rangle . \tag{2.103}$$

A direct consequence of (2.100) is the representation

$$\int_\Omega (Lu)\overline{v} d\mathbf{x} = B(u,v) - \sigma_\Omega \int_\Gamma (\gamma_1 u)\overline{(\gamma_0 v)} ds_\mathbf{x} \tag{2.104}$$

for all $u \in H^2(\Omega)$ and $v \in H^1(\Omega)$.

The *formal adjoint* operator of L is given by

$$L^* v := -\operatorname{div}(\mathbf{A}\operatorname{grad} v) - 2\langle \mathbf{b}, \operatorname{grad} v\rangle + cv. \tag{2.105}$$

The term "*formal adjoint*" refers to the property (whose proof is achieved by means of integration by parts)

$$(Lu, v)_{L^2(\mathbb{R}^d)} = \left(u, L^* v\right)_{L^2(\mathbb{R}^d)}$$

for all $u, v \in C^\infty\left(\mathbb{R}^d\right)$ which have the property that one of the two functions u, v has compact support. In general, this relation does not hold for bounded domains. Through integration by parts we have the following representation

$$\int_\Omega u\overline{(L^* v)} d\mathbf{x} = B(u,v) - \sigma_\Omega \int_\Gamma (\gamma_0 u)\overline{(\widetilde{\gamma_1} v)} ds_\mathbf{x} \tag{2.106}$$

for all $u \in H^1(\Omega)$ and $v \in H^2(\Omega)$ with the modified conormal derivative

$$\widetilde{\gamma_1} v := \langle \mathbf{n}\gamma_0, \mathbf{A}\operatorname{grad} v + 2\mathbf{b}v\rangle = \gamma_1 v + 2\langle \mathbf{b}, \mathbf{n}\rangle \gamma_0 v. \tag{2.107}$$

Remark 2.7.5. *The boundary differential operators γ_1 and $\widetilde{\gamma_1}$ are continuous mappings from $H^2(\Omega)$ to $H^{1/2}(\Gamma)$.*

The formulas (2.104) and (2.106) are called Green's first formulas for the operator L. The domain of the conormal derivatives γ_1 and $\widetilde{\gamma_1}$ can, for $s \geq 1$, be extended to the space

$$H_L^s(\Omega) := \left\{u \in H_{\mathrm{loc}}^s(\Omega) : Lu \in L_{\mathrm{comp}}^2(\Omega)\right\} \tag{2.108}$$

by using the relations (2.104) and (2.106).

Definition 2.7.6. Let $\Omega \in \{\Omega^-, \Omega^+\}$ be as in Convention 2.7.2 and let Z_Ω be the extension operator from Theorem 2.6.11. Then the (weak) conormal derivative $\gamma_1 : H_L^1(\Omega) \to H^{-1/2}(\Gamma)$ is characterized by

$$(\gamma_1 u, \psi)_{L^2(\Gamma)} = \sigma_\Omega \left(B(u, Z_\Omega \psi) - (Lu, Z_\Omega \psi)_{L^2(\Omega)}\right) \qquad \forall \psi \in H^{1/2}(\Gamma). \tag{2.109}$$

The modified (weak) conormal derivative $\widetilde{\gamma_1} : H_L^1(\Omega) \to H^{-1/2}(\Gamma)$ is given by

$$(\psi, \widetilde{\gamma_1} v)_{L^2(\Gamma)} = \sigma_\Omega \left(B(Z_\Omega \psi, v) - \left(Z_\Omega \psi, L^* v\right)_{L^2(\Omega)}\right) \qquad \forall \psi \in H^{1/2}(\Gamma).$$

Theorem 2.7.7. *Let $\Omega \in \{\Omega^-, \Omega^+\}$ be as in Convention 2.7.2. The conormal derivative $\gamma_1 : H_L^1(\Omega) \to H^{-1/2}(\Gamma)$ is continuous. Let $u \in H_L^1(\Omega)$, $v \in H_{\mathrm{loc}}^1(\Omega)$ and let one of the two functions u, v have compact support, then*

$$(\gamma_1 u, \gamma_0 v)_{L^2(\Gamma)} = \sigma_\Omega \left\{B(u, v) - (Lu, v)_{L^2(\Omega)}\right\}. \tag{2.110}$$

For $u \in H^2(\Omega)$, $\gamma_1 u$ coincides with the conormal derivative from (2.103) almost everywhere on Γ.

The modified conormal operator $\widetilde{\gamma_1} : H_L^1(\Omega) \to H^{-1/2}(\Gamma)$ is continuous. Let $v \in H_L^1(\Omega)$, $u \in H_{\mathrm{loc}}^1(\Gamma)$ and let one of the two functions u, v have compact support. Then

$$(\gamma_0 u, \widetilde{\gamma_1} v)_{L^2(\Gamma)} = \sigma_\Omega \left\{B(u, v) - \left(u, L^* v\right)_{L^2(\Omega)}\right\}. \tag{2.111}$$

For $v \in H^2(\Omega)$, $\widetilde{\gamma_1}$ coincides with the modified conormal operator defined in (2.107) almost everywhere on Γ.

A proof of this theorem can be found in [162, Lemma 4.3].

Theorem 2.7.7 implies that the definitions of γ_1 and $\widetilde{\gamma_1}$ are independent of the choice of the trace extension Z_Ω. In order to see this we consider another continuous trace extension $\widetilde{Z_\Omega} : H^{1/2}(\Gamma) \to H_{\mathrm{comp}}^1(\Omega)$. For $u \in H_L^1(\Omega)$ we define $g := \gamma_1 u$ as in (2.109) and we define $\tilde{g}$ by replacing Z_Ω in (2.109) by $\widetilde{Z_\Omega}$. Then

$$(\tilde{g}, \psi)_{L^2(\Gamma)} = \sigma_\Omega \left(B\left(u, \widetilde{Z_\Omega} \psi\right) - \left(Lu, \widetilde{Z_\Omega} \psi\right)_{L^2(\Omega)}\right) \qquad \forall \psi \in H^{1/2}(\Gamma).$$

For $\psi \in H^{1/2}(\Gamma)$ we define $v := Z_\Omega \psi$ and $\tilde{v} := \widetilde{Z_\Omega} \psi$ and note that $\gamma_0 v = \gamma_0 \tilde{v} = \psi$ (see Theorem 2.6.11). In (2.110) we set $\gamma_1 u = g$, $\gamma_0 v = \psi$ and $v = Z_\Omega \psi$. From this we subtract (2.110) while setting $\gamma_1 u = \widetilde{G}$, $\gamma_0 \widetilde{V} = \psi$ and $\widetilde{V} = \widetilde{Z_\Omega} \psi$ and thus we obtain

$$\sigma_\Omega (g - \tilde{g}, \psi)_{L^2(\Gamma)} = B\left(u, \left(Z_\Omega - \widetilde{Z_\Omega}\right)\psi\right) - \left(Lu, \left(Z_\Omega - \widetilde{Z_\Omega}\right)\psi\right)$$
$$\overset{(2.110)}{=} \sigma_\Omega \left(g, \gamma_0 \left(Z_\Omega - \widetilde{Z_\Omega}\right)\psi\right)_{L^2(\Gamma)} = 0.$$

Since $\psi \in H^{1/2}(\Gamma)$ was arbitrary, we have as a consequence $g = \tilde{g}$. The corresponding result for $\widetilde{\gamma_1}$ can be proven analogously.

Corollary 2.7.8. *Let $u \in H^1_L(\Omega)$. Then, for every cut-off function $\chi \in C_0^\infty(\mathbb{R}^d)$ that satisfies $\chi \equiv 1$ in a neighborhood of Γ, the equation $\gamma_1(\chi u) = \gamma_1 u$ holds.*

Proof. Let U be a neighborhood of Γ with $\chi \equiv 1$ on U. We choose a second cut-off function χ_2 that satisfies $\operatorname{supp} \chi_2 \subset U$ and $\chi_2 \equiv 1$ in another neighborhood U_2 of Γ. We define $\widetilde{Z_\Omega} := \chi_2 Z_\Omega$ and $\gamma_1 u$ in (2.109) by using $\widetilde{Z_\Omega}$ instead of Z_Ω. (The definition of $\gamma_1 u$ is independent of the choice of the trace extension, as was shown above.) By using $\chi \equiv 1$ on $\operatorname{supp}\left(\widetilde{Z_\Omega}\varphi\right)$ we obtain for all $\varphi \in H^{1/2}(\Gamma)$

$$(\gamma_1 u, \varphi)_{L^2(\Gamma)} = \sigma_\Omega \left\{ B\left(u, \widetilde{Z_\Omega}\varphi\right) - \left(Lu, \widetilde{Z_\Omega}\varphi\right)_{L^2(\Omega)} \right\}$$
$$= \sigma_\Omega \left\{ B\left(\chi u, \widetilde{Z_\Omega}\varphi\right) - \left(L\chi u, \widetilde{Z_\Omega}\varphi\right)_{L^2(\Omega)} \right\} = (\gamma_1(\chi u), \varphi)_{L^2(\Gamma)}.$$

□

Remark 2.7.9. *The definition of the conormal derivative depends only on the principal part of the operator L from (2.98), since the sum of the terms of lower order on the right-hand side in (2.110) equals zero [see (2.102)].*

Remark 2.7.10. *In order to distinguish whether γ_1 is applied to functions in Ω^+ or Ω^-, we write γ_1^+, γ_1^- or $\widetilde{\gamma_1^+}$, $\widetilde{\gamma_1^-}$. Analogously, the notations $B_+(\cdot,\cdot)$ and $B_-(\cdot,\cdot)$ indicate whether the sesquilinear form B is defined with respect to Ω^+ or Ω^-.*

Remark 2.7.11. *By combining the formulas (2.110), (2.111) for the two functions $u, v \in H^1_L(\Omega)$, one of which has compact support, we obtain Green's second formula*

$$(Lu, v)_{L^2(\Omega)} - (u, L^* v)_{L^2(\Omega)} = \sigma_\Omega \left\{ (\gamma_0 u, \widetilde{\gamma_1} v)_{L^2(\Gamma)} - (\gamma_1 u, \gamma_0 v)_{L^2(\Gamma)} \right\}. \quad (2.112)$$

Green's third formula appears in connection with transmission problems. Here the goal is to find a function u whose restrictions $u^+ := u|_{\Omega^+}$ and $u^- := u|_{\Omega^-}$ satisfy the equations

$$\begin{aligned} Lu^+ &= f^+ && \text{in } \Omega^+ \\ Lu^- &= f^- && \text{in } \Omega^-. \end{aligned}$$

Apart from this, the behavior of the traces of u^+ and u^- has to be prescribed on the boundary Γ. In the remainder of this section we will derive a Green's representation formula which is relevant to the study of this question.

For $u \in L^2(\mathbb{R}^d)$ we first introduce the abbreviations $u^+ := u|_{\Omega^+}$ and $u^- := u|_{\Omega^-}$ in order to define the space

$$H_L^1\left(\mathbb{R}^d \backslash \Gamma\right) := \left\{u \in L^2\left(\mathbb{R}^d\right) \mid u^+ \in H_L^1\left(\Omega^+\right) \wedge u^- \in H_L^1\left(\Omega^-\right)\right\}, \quad (2.113)$$

where the spaces $H_L^1(\Omega^-)$, $H_L^1(\Omega^+)$ are given as in (2.108).

Remark 2.7.12. *In general, for a function $u \in H_L^1(\mathbb{R}^d \backslash \Gamma)$, we do not have $Lu \in L^2(\mathbb{R}^d)$. An easy counter-example is obtained by setting $u^- \equiv 1$ and $u^+ \equiv 0$.*

For $u \in H_L^1(\mathbb{R}^d \backslash \Gamma)$ the function $L_\pm u \in L^2(\mathbb{R}^d)$ is defined as

$$L_\pm u := \begin{cases} Lu^- \text{ in } \Omega^- \\ Lu^+ \text{ in } \Omega^+ \end{cases}. \quad (2.114)$$

For a given $f \in L^2_{\text{comp}}(\mathbb{R}^d)$ we now consider functions $u \in H_L^1(\mathbb{R}^d \backslash \Gamma)$ that satisfy

$$L_\pm u = f \qquad \text{in } \mathbb{R}^d \backslash \Gamma. \quad (2.115)$$

For $u \in H_L^1(\mathbb{R}^d \backslash \Gamma)$ the application Lu can be defined as a functional (distribution) on $C_0^\infty(\mathbb{R}^d)$

$$(Lu, v)_{L^2(\mathbb{R}^d)} := \left(u, L^* v\right)_{L^2(\mathbb{R}^d)} \qquad \forall v \in C_0^\infty\left(\mathbb{R}^d\right). \quad (2.116)$$

We use $u \in L^2(\mathbb{R}^d)$, $L^* v \in C_0^\infty(\mathbb{R}^d)$, (2.111) and (2.110) to maintain the separation

$$(Lu, v)_{L^2(\mathbb{R}^d)} = \sum_{s \in \{+,-\}} \left(u, L^* v\right)_{L^2(\Omega^s)} = \sum_{s \in \{+,-\}} \left(B_s(u, v) - \sigma_{\Omega^s}\left(\gamma_0^s u, \widetilde{\gamma_1^s} v\right)_{L^2(\Gamma)}\right) \quad (2.117)$$

$$= \left(L_\pm u, v\right)_{L^2(\mathbb{R}^d)} + \sum_{s \in \{+,-\}} \sigma_{\Omega^s}\left(\left(\gamma_1^s u, \gamma_0^s v\right)_{L^2(\Gamma)} - \left(\gamma_0^s u, \widetilde{\gamma_1^s} v\right)_{L^2(\Gamma)}\right) \quad (2.118)$$

on the right-hand side in (2.116). The traces and conormal derivatives of u are, in general, discontinuous across the boundary Γ. As an abbreviation we use the following notation

$$[u] := \gamma_0^+ u - \gamma_0^- u \quad \text{and} \quad [\gamma_1 u] := \gamma_1^+ u - \gamma_1^- u. \tag{2.119}$$

The smoothness of $v \in C_0^\infty(\mathbb{R}^d)$ implies $[\widetilde{\gamma_1} v] = [v] = 0$ and therefore (2.118) implies the equation

$$(Lu, v)_{L^2(\mathbb{R}^d)} = (f, v)_{L^2(\mathbb{R}^d)} + ([u], \widetilde{\gamma_1} v)_{L^2(\Gamma)} - ([\gamma_1 u], \gamma_0 v)_{L^2(\Gamma)} \tag{2.120}$$

for all $u \in H_L^1(\mathbb{R}^d \backslash \Gamma)$ and $v \in C_0^\infty(\mathbb{R}^d)$.

By using the dual mappings of $\widetilde{\gamma_1}$ and γ_0, (2.120) can be expressed without the use of test functions. Theorem 2.6.8 implies that the trace mapping $\gamma_0 : H^1_{\text{loc}}(\mathbb{R}^d) \to H^{1/2}(\Gamma)$ is continuous and that the dual mapping $\gamma_0' : H^{-1/2}(\Gamma) \to H^{-1}_{\text{comp}}(\mathbb{R}^d)$ is characterized by

$$(w, \gamma_0 v)_{L^2(\Gamma)} = (\gamma_0' w, v)_{L^2(\mathbb{R}^d)} \qquad \forall v \in H^1_{\text{loc}}\left(\mathbb{R}^d\right), w \in H^{-1/2}(\Gamma). \tag{2.121}$$

For functions $v \in C^\infty(\mathbb{R}^d)$ the modified conormal derivative can be written as

$$\widetilde{\gamma_1} v = \langle \mathbf{n}\gamma_0, \mathbf{A} \operatorname{grad} v + 2\mathbf{b}v \rangle$$

and we have $\widetilde{\gamma_1} v \in L^\infty(\Gamma)$. As a consequence, $\widetilde{\gamma_1}'$ can be defined on $(L^\infty(\Gamma))' = L^1(\Gamma)$ by

$$(w, \widetilde{\gamma_1} v)_{L^2(\Gamma)} = (\widetilde{\gamma_1}' w, v)_{L^2(\mathbb{R}^d)} \qquad \forall w \in L^1(\Gamma), \qquad v \in C^\infty\left(\mathbb{R}^d\right)$$

and therefore $\widetilde{\gamma_1}' w$ describes a functional on $C^\infty(\mathbb{R}^d)$. With this result and (2.120) we obtain the third of Green's formulas

$$Lu = f + \widetilde{\gamma_1}'([u]) - \gamma_0'([\gamma_1 u]) \tag{2.122}$$

for $u \in H_L^1(\mathbb{R}^d \backslash \Gamma)$ as a functional on $C_0^\infty(\mathbb{R}^d)$.

The derivation of (2.122) can be done in complete analogy for $u \in H_L^1(\mathbb{R}^d)$ with compact support and $v \in C^\infty(\mathbb{R}^d)$.

Proposition 2.7.13. *If $u \in H_L^1(\mathbb{R}^d \backslash \Gamma)$ has compact support, (2.122) still holds when considered as a functional on $C^\infty(\mathbb{R}^d)$.*

2.8 Solution Operator

Let Ω be a *bounded* Lipschitz domain with L as in (2.98). Then the modified differential operator $\widetilde{L}$ can be defined by

$$\widetilde{L} := L + \lambda, \tag{2.123}$$

where $\lambda \geq 0$ will be determined at a later stage. $B(\cdot,\cdot)$ denotes the sesquilinear form from (2.102) and $\widetilde{B} = B + \lambda (\cdot,\cdot)_{L^2(\Omega)}$. For given boundary values $\varphi \in H^{1/2}(\Gamma)$ we consider the homogeneous Dirichlet problem: Find $u \in H^1(\Omega)$ with $\gamma_0 u = \varphi$ and

$$\widetilde{B}(u,v) = 0 \qquad \forall v \in H_0^1(\Omega). \tag{2.124}$$

Proposition 2.8.1. *Every solution of (2.124) that satisfies* $u \in H_L^1(\Omega)$ *solves the homogeneous equation*

$$\widetilde{L}u = 0 \qquad \text{in } \Omega. \tag{2.125}$$

Solutions of (2.125) are called $\widetilde{L}$-harmonic.

Proof. From $Lu \in L^2_{\text{comp}}(\Omega)$ and $u \in L^2_{\text{loc}}(\Omega) = L^2(\Omega)$ (see Remark 2.6.2.b) we have $\widetilde{L}u = Lu + \lambda u \in L^2_{\text{comp}}(\Omega)$ and therefore $u \in H^1_{\widetilde{L}}(\Omega)$. By using (2.110) with $L \leftarrow \widetilde{L}$ and $\gamma_0 v \equiv 0$ for all $v \in H_0^1(\Omega)$ we obtain the assertion. □

By using the trace extension Z_Ω from Definition 2.6.11 the problem (2.124) can be transformed into an inhomogeneous Dirichlet problem with homogeneous boundary conditions. We set $u_1 := Z_\Omega \varphi$ and $u_0 = u - u_1$. We apply this approach in (2.124) and thus obtain the equation that determines u_0: Find $u_0 \in H_0^1(\Omega)$ with

$$\widetilde{B}(u_0,v) = -\widetilde{B}(u_1,v) \qquad \forall v \in H_0^1(\Omega). \tag{2.126}$$

We will show in Lemma 2.10.1 that

$$\operatorname{Re} B(u,u) \geq c \|u\|^2_{H^1(\Omega)} - C \|u\|^2_{L^2(\Omega)}$$

holds with constants $c > 0$, $C \in \mathbb{R}$ that do not depend on $u \in H_0^1(\Omega)$. The choice of $\lambda > C$ therefore implies that the modified sesquilinear form $\widetilde{B}$ is elliptic in $H_0^1(\Omega)$. We will also prove the continuity in Lemma 2.10.1. According to the Lax–Milgram lemma (Lemma 2.1.51) problem (2.126) has a unique solution u_0, which satisfies the inequality

$$\|u_0\|_{H^1(\Omega)} \leq C \|u_1\|_{H^1(\Omega)}.$$

By using the continuity of the extension operator (see Theorem 2.6.11) one can deduce that

$$\|u\|_{H^1(\Omega)} \leq \|u_0\|_{H^1(\Omega)} + \|u_1\|_{H^1(\Omega)} \leq (1+C)\|Z_\Omega\varphi\|_{H^1(\Omega)} \leq \widetilde{C}\|\varphi\|_{H^{1/2}(\Gamma)}.$$

With these results, the existence of a continuous solution operator $T : H^{1/2}(\Gamma) \to H^1(\Omega)$ that maps the Dirichlet data $\varphi \in H^{1/2}(\Gamma)$ to the solution u of problem (2.124) has been shown. It satisfies

$$\|T\|_{H^1(\Omega) \leftarrow H^{1/2}(\Gamma)} \leq \widetilde{C}.$$

Since $\widetilde{L}Tu \equiv 0$ the mapping $T : H^{1/2}(\Gamma) \to H_L^1(\Omega)$ is also continuous. It follows from the mapping properties of T that the operator $\gamma_1 T$ is well defined.

This operator maps the trace of an $\widetilde{L}$-harmonic function to its conormal derivative and is called Steklov–Poincaré operator. The mapping properties of γ_1 and T imply that

$$\gamma_1 T : H^{1/2}(\Gamma) \to H^{-1/2}(\Gamma).$$

This result can be generalized to a larger scale of orders of differentiation.

Theorem 2.8.2. *Let Ω be a bounded Lipschitz domain. Then for $-1/2 \leq s \leq 1/2$, both the Steklov–Poincaré operator*

$$\gamma_1 T : H^{1/2+s}(\Gamma) \to H^{-1/2+s}(\Gamma)$$

and, for $-1/2 < s < 1/2$, the solution operator T

$$T : H^{1/2+s}(\Gamma) \to H^{1+s}(\Omega)$$

are continuous.

A proof of the first assertion can be found in [166, Chap. 5, Theorem 1.3, Lemma 1.4] (see also [72, Lemma 3.7] or [162, Theorem 4.25]). The second assertion is proven in [72, Lemma 4.2].

The solution operator allows us to show that the conormal operator is also continuous for a scale of Sobolev spaces.

Theorem 2.8.3. *We use the notations from Convention 2.7.2. The conormal trace operators*

$$\gamma_1^- : H_L^{s+1}(\Omega^-) \to H^{s-1/2}(\Gamma),$$
$$\gamma_1^+ : H_L^{s+1}(\Omega^+) \to H^{s-1/2}(\Gamma)$$

are continuous for $-1/2 < s < 1/2$.

Proof. Due to Remark 2.7.9 we only need to consider the case $Lu = -\operatorname{div}(\mathbf{A} \operatorname{grad} u)$.

We begin by considering the interior problem. For $u \in H_L^1(\Omega^-)$ and $\varphi \in H^{1/2}(\Gamma)$ we set $v := T\varphi$. We then use Remark 2.7.11 in which we replace L by $\widetilde{L}$. With $\widetilde{L}^* = \widetilde{L}$, $\widetilde{L}T\varphi = 0$ and $\widetilde{\gamma_1^-} = \gamma_1^-$ we get

$$\left(\gamma_1^- u, \varphi\right)_{L^2(\Gamma)} = \left(\gamma_0 u, \gamma_1^- T\varphi\right)_{L^2(\Gamma)} - \left(\widetilde{L}u, T\varphi\right)_{L^2(\Omega^-)}. \tag{2.127}$$

By using the dual operators T' and $(\gamma_1 T)'$ we obtain the representation

$$\gamma_1^- = \left(\gamma_1^- T\right)' \gamma_0 - T'\widetilde{L}.$$

To obtain the first assertion we combine $\gamma_0 : H_{\mathrm{loc}}^{1+s}(\Omega^-) \to H^{1/2+s}(\Gamma)$ with $(\gamma_1 T)' : H^{1/2+s}(\Gamma) \to H^{s-1/2}(\Gamma)$. For the second term we use $\widetilde{L} : H_L^{1+s}(\Omega^-)$

$\to L^2(\Omega^-)$, $L^2(\Omega^-) \subset (H^{1-s}(\Omega^-))'$ for all $-1/2 < s < 1/2$ with a continuous and dense embedding and finally $T' : (H^{1-s}(\Omega^-))' \to H^{-1/2+s}(\Gamma)$.

The continuity of γ_1^+ for the exterior problem is proven by localization (see Theorem 2.6.7b). Let $\chi_1 \in C^\infty_{\text{comp}}(\Omega^+)$ be an arbitrary cut-off function that is equal to 1 in a neighborhood of Γ. We need to show the following:

$$\left\| \gamma_1^+(\chi_1 w) \right\|_{H^{s-1/2}(\Gamma)} \le C \left\| \chi_1 w \right\|_{H_L^{s+1}(\Omega^+)} \qquad \forall w \in H_L^{s+1}(\Omega^+). \tag{2.128}$$

We choose a sufficiently large, open ball B_R with

$$\operatorname{supp} \chi_1 \cup \overline{\Omega^-} \subset B_R.$$

Let $\chi_2 \in C_0^\infty(\mathbb{R}^d)$ be another cut-off function with $\operatorname{supp} \chi_2 \subset B_R$ and $\chi_2 \equiv 1$ on $\operatorname{supp} \chi_1$.

We set $\Omega_R^+ := B_R \cap \Omega^+$ and define the solution operator T with respect to the Lipschitz domain Ω_R^+ with prescribed boundary conditions $\varphi \in H^{1/2}(\Gamma)$ and zero boundary conditions on ∂B_R.

As before, we choose $\varphi \in H^{1/2}(\Gamma)$ and in this case set $v := \chi_2 T\varphi$. For an arbitrary $w \in H_L^1(\Omega^+)$ we define $u := \chi_1 w$ and observe that $u, v \in H_L^1(\Omega^+)$.

It should be noted that, taking Remark 2.7.11 into consideration, the traces of u and v are zero on the outer boundary ∂B_R. Furthermore, we have

$$\begin{aligned}(u, \widetilde{L}^* v)_{L^2(\Omega^+)} &= (\chi_1 w, \widetilde{L}(\chi_2 T\varphi))_{L^2(\Omega^+)} = (\chi_1 w, \widetilde{L}(\chi_2 T\varphi))_{L^2(\Omega^+ \cap \operatorname{supp} \chi_1)} \\ &= (\chi_1 w, \widetilde{L} T\varphi)_{L^2(\Omega^+ \cap \operatorname{supp} \chi_1)} = 0.\end{aligned}$$

and

$$(\widetilde{L}u, v)_{L^2(\Omega^+)} = (\widetilde{L}u, T\varphi)_{L^2(\Omega^+ \cap \operatorname{supp} \chi_1)} = (\widetilde{L}u, T\varphi)_{L^2(\Omega_R^+)} = (T'\widetilde{L}u, \varphi)_{L^2(\Gamma)}.$$

By Remark 2.7.11 we then obtain

$$(\gamma_1^+ u, \gamma_0 v)_{L^2(\Gamma)} = (\gamma_0 u, \gamma_1^+(\chi_2 T\varphi))_{L^2(\Gamma)} + (T'\widetilde{L}u, \varphi)_{L^2(\Gamma)}.$$

By using $\gamma_1^+ \chi_2 T\varphi = \gamma_1^+ T\varphi$ (see Corollary 2.7.8) we obtain the representation

$$\gamma_1^+ u = T'\widetilde{L}u + (\gamma_1^+ T)' \gamma_0 u$$

as a functional on $H^{1/2}(\Gamma)$. In the same way as was done for the interior problem we can now deduce the estimate (2.128) for $u = \chi_1 w$. □

In order to prove the mapping properties of boundary integral operators we will need dense subspaces of the Sobolev spaces $H^s(\Gamma)$. The relevant results are summarized in the following lemma.

Lemma 2.8.4. *We assume the same conditions as in Theorem 2.6.8:*

(a) The trace mapping γ_0 maps $C_0^\infty\left(\mathbb{R}^d\right)$ to a dense subspace of $H^{1/2}(\Gamma)$.
(b) The trace mapping (γ_0, γ_1) maps $C_0^\infty\left(\mathbb{R}^d\right)$ to a dense subspace of $H^{1/2}(\Gamma) \times H^{-1/2}(\Gamma)$.

The proof can be found in, e.g., [72, Lemma 3.5].

2.9 Elliptic Boundary Value Problems

We introduced boundary value problems for the Laplace operator in Chap. 1. In this section we will treat the boundary value problems related to the operator

$$Lu = -\operatorname{div}(\mathbf{A} \operatorname{grad} u) + 2\langle \mathbf{b}, \operatorname{grad} u\rangle + cu$$

[see (2.98)]. In general, we will assume that the space is of dimension $d = 3$. The trace and conormal operators γ_0, γ_1 on the boundary Γ of a bounded Lipschitz domain $\Omega^- \subset \mathbb{R}^3$ were introduced in Theorem 2.6.8 and (2.103) as well as Definition 2.7.6.

For sufficiently smooth functions $u \in C^0\left(\overline{\Omega}\right)$ and $v \in C^1\left(\overline{\Omega}\right)$ these can be written as

$$\gamma_0 u = u|_\Gamma \quad \text{and} \quad \gamma_1 v = \langle \mathbf{A}\mathbf{n}, (\operatorname{grad} v)|_\Gamma\rangle.$$

2.9.1 *Classical Formulation of Elliptic Boundary Value Problems*

First we will present the classical (or strong) formulation of elliptic boundary value problems, after which we will introduce the relevant variational formulations. We will use the notation from Convention 2.7.2.

2.9.1.1 Interior Dirichlet Problem (IDP)

For a given $f \in C^0(\Omega^-)$ and $g_D \in C^0(\Gamma)$ find $u \in C^2(\Omega^-) \cap C^0\left(\overline{\Omega^-}\right)$ such that

$$\begin{aligned} Lu &= f \quad \text{in } \Omega^-, \\ u &= g_D \quad \text{on } \Gamma. \end{aligned} \tag{2.129}$$

2.9.1.2 Interior Neumann Problem (INP)

For a given $f \in C^0(\Omega^-)$ and $g_N \in C^0(\Gamma)$ find $u \in C^2(\Omega^-) \cap C^1\left(\overline{\Omega^-}\right)$ such that

$$\begin{aligned} Lu &= f \quad \text{in } \Omega^-, \\ \gamma_1 u &= g_N \quad \text{on } \Gamma. \end{aligned} \tag{2.130}$$

2.9.1.3 Interior Mixed Boundary Value Problem (IMP)

Let Γ be partitioned into relatively open, non-empty subsets Γ_D and Γ_N, i.e.,

$$\Gamma = \overline{\Gamma_D} \cup \overline{\Gamma_N}, \quad \Gamma_D \cap \Gamma_N = \emptyset,$$

where we assume that the surface measure $|\Gamma_D| > 0$. For a given right-hand side $f \in C^0(\Omega^-)$ and boundary data $g_D \in C^0(\Gamma_D)$ and $g_N \in C^0(\Gamma_N)$, find $u \in C^2(\Omega^-) \cap C^0\left(\overline{\Omega^-}\right) \cap C^1(\Omega^- \cup \Gamma_N)$ such that

$$\begin{aligned} Lu &= f \quad && \text{in } \Omega^-, \\ u &= g_D && \text{on } \Gamma_D, \\ \gamma_1 u &= g_N && \text{on } \Gamma_N. \end{aligned} \tag{2.131}$$

2.9.1.4 Exterior Dirichlet Problem (EDP)

In order to formulate the exterior boundary value problem one needs to prescribe the boundary conditions on Γ as well as the behavior of the solution at infinity. For the strong formulation we will only present these decay or radiation conditions, which depend on the coefficients of the differential operator L, for the case $c \geq 0$ [see (2.98)] and for the Helmholtz equation.

Let $c \geq 0$ for the coefficient c in L. Then the decay conditions are given by

$$|u(\mathbf{x})| \leq C \|\mathbf{x}\|^{-1} \qquad \text{for } \|\mathbf{x}\| \to \infty. \tag{2.132}$$

We consider the Helmholtz operator $Lu = -\Delta u - k^2 u$ and a positive wave number $k > 0$. For these we impose *Sommerfeld's radiation conditions*

$$\left.\begin{aligned} |u(\mathbf{x})| &\leq C \|\mathbf{x}\|^{-1} \\ \left|\frac{\partial u}{\partial r} - iku\right| &\leq C \|\mathbf{x}\|^{-2} \end{aligned}\right\} \quad \text{for } \|\mathbf{x}\| \to \infty. \tag{2.133}$$

Here $\partial u / \partial r = \langle \mathbf{x}/\|\mathbf{x}\|, \nabla u \rangle$ denotes the radial derivative.

The radiation condition (2.133) describes outgoing waves. Time-harmonic incoming waves can be described analogously. In this case in (2.133) k is simply replaced by $-k$.

For a given $f \in C^0\left(\Omega^+\right)$ and $g_D \in C^0(\Gamma)$ find $u \in C^2\left(\Omega^+\right) \cap C^0\left(\Omega^+ \cup \Gamma\right)$ such that

$$\begin{aligned} Lu &= f && \text{in } \Omega^+, \\ u &= g_D && \text{on } \Gamma, \\ u \text{ satisfies } &\begin{cases} (2.132) & \text{if } c \geq 0, \\ (2.133) & \text{for the Helmholtz problem.} \end{cases} \end{aligned} \tag{2.134}$$

2.9.1.5 Exterior Neumann Problem (ENP)

For a given $f \in C^0\left(\Omega^+\right)$ and $g_N \in C^0(\Gamma)$ find $u \in C^2\left(\Omega^+\right) \cap C^1\left(\Omega^+ \cup \Gamma\right)$ such that

$$\begin{aligned} Lu &= f && \text{in } \Omega^+, \\ \gamma_1 u &= g_N && \text{on } \Gamma, \\ u \text{ satisfies } &\begin{cases} (2.132) & \text{if } c \geq 0, \\ (2.133) & \text{for the Helmholtz problem.} \end{cases} \end{aligned} \tag{2.135}$$

2.9.1.6 Exterior Mixed Boundary Value Problem (EMP)

Let Γ be partitioned into relatively open, disjoint, non-empty subsets Γ_D and Γ_N, i.e.,

$$\Gamma = \overline{\Gamma_D} \cup \overline{\Gamma_N} \quad \text{and} \quad |\Gamma_N| > 0,\ |\Gamma_D| > 0.$$

For a given right-hand side $f \in C^0\left(\Omega^+\right)$ and boundary data $g_D \in C^0(\Gamma_D)$ and $g_N \in C^0(\Gamma_N)$ find $u \in C^2\left(\Omega^+\right) \cap C^0\left(\Omega^+ \cup \Gamma_D\right) \cap C^1\left(\Omega^+ \cup \Gamma_N\right)$ such that

$$\begin{aligned} Lu &= f && \text{in } \Omega^+, \\ u &= g_D && \text{on } \Gamma_D, \\ \gamma_1 u &= g_N && \text{on } \Gamma_N, \\ u \text{ satisfies } &\begin{cases} (2.132) & \text{if } c \geq 0, \\ (2.133) & \text{for the Helmholtz problem.} \end{cases} \end{aligned} \tag{2.136}$$

2.9.1.7 Transmission Problem (TP)

Finally, we want to formulate the transmission problem. The differential equation is considered in both the interior and the exterior domain and appropriate transmission conditions are imposed on the common boundary $\Gamma = \partial\Omega^+ = \partial\Omega^-$. The differential operators in the interior and exterior domain need not be the same, which is why we denote the differential operator in the interior domain by L^- and for the exterior domain Ω^+ by L^+. The relevant coefficients for $s \in \{-,+\}$ are denoted by $\mathbf{A}^s, \mathbf{b}^s, c^s$.

For a given right-hand side $f = \left(f^-, f^+\right)$ with $f^s \in C^0(\Omega^s)$ for $s \in \{-,+\}$ and transmission data $g_D \in C^0(\Gamma_D)$, $g_N \in C^0(\Gamma_N)$, the aim is to find $u = \left(u^-, u^+\right)$ with $u^s \in C^2(\Omega^s) \cap C^1(\Omega^s \cup \Gamma)$ for $s \in \{-,+\}$ such that

$$\begin{aligned} L^s u^s &= f^s && \text{in } \Omega^s \text{ for } s \in \{-,+\} \\ [u] &= g_D && \text{on } \Gamma, \\ [\gamma_1 u] &= g_N && \text{on } \Gamma, \\ u^+ \text{ satisfies} & \begin{cases} (2.132) \\ (2.133) \end{cases} && \begin{array}{l} \text{if } c \geq 0, \\ \text{for the Helmholtz problem.} \end{array} \end{aligned} \tag{2.137}$$

2.9.2 Variational Formulation of Elliptic Boundary Value Problems

One of the decisive disadvantages of the strong formulation of boundary value problems is that questions concerning existence and uniqueness cannot be answered in a satisfactory way. We can overcome these difficulties by choosing a *variational formulation* in appropriate function spaces instead. In order to do this we multiply the differential equation by a test function and then integrate over Ω. If we then use integration by parts, the boundary conditions can directly be incorporated in the variational formulation. The solution of the variational problem is called a *weak* solution. In contrast to the strong formulation, the solutions are either sought or given in Sobolev spaces. We will show in Sect. 2.9.3 that if the solution to the variational problem is sufficiently smooth, it coincides with the classical solution. We briefly review the formal definition of the sesquilinear form B from (2.102), which will appear in the variational formulation

$$B(u, v) = \int_\Omega \left(\left\langle \mathbf{A} \operatorname{grad} u, \overline{\operatorname{grad} v}\right\rangle + 2 \left\langle \mathbf{b}, \operatorname{grad} u\right\rangle \overline{v} + cu\overline{v}\right) d\mathbf{x}.$$

For some of the boundary value problems that are to follow, the coefficients $\mathbf{A}$, $\mathbf{b}$ and c will have to satisfy additional conditions.

2.9.2.1 Interior Dirichlet Problem (IDP)

First we suppose that the function u in (2.129) is sufficiently smooth. More specifically, this means that $u \in H_L^1(\Omega^-)$ and therefore $f \in L^2(\Omega^-)$. We multiply (2.129) by the functions $v \in C_0^\infty(\Omega^-)$ and integrate over Ω^-. The conditions we imposed on u and v allow us to apply Green's formula (2.110)

$$B(u, v) - (\gamma_1 u, \gamma_0 v)_{L^2(\Gamma)} = (f, v)_{L^2(\Omega^-)} \,.$$

If we set $\gamma_0 v = 0$ we obtain

$$B(u, v) = (f, v)_{L^2(\Omega)} \qquad \forall v \in H_0^1(\Omega^-)\,. \tag{2.138}$$

Since

$$F(v) = (f, v)_{L^2(\Omega^-)} \tag{2.139}$$

the right-hand side in (2.138) defines a functional on $H_0^1(\Omega^-)$. Equation (2.138) is also valid for functions $u, v \in H^1(\Omega^-)$. This leads us to the

Variational formulation of the interior Dirichlet problem (2.129): For a given $F \in H^{-1}(\Omega^-)$ and $g_D \in H^{1/2}(\Gamma)$ find $u \in H^1(\Omega^-)$ with $\gamma_0^- u = g_D$ on Γ such that

$$B(u, v) = F(v) \qquad \forall v \in H_0^1(\Omega^-). \tag{2.140}$$

Solutions of (2.140) that are not in $C^2(\Omega^-)$ are called weak solutions. Conversely, solutions that are in $u \in C^2(\Omega^-) \cap H_0^1(\Omega^-)$ are called strong solutions. By using the trace extension Z_- (see Remark 2.6.12) the problem can be formulated as a homogeneous Dirichlet problem. We set $u_1 := Z_- g_D$ and suppose that $u = u_0 + u_1$. Then the unknown function u_0 is the solution to the problem: Find $u_0 \in H_0^1(\Omega^-)$ such that

$$B(u_0, v) = F(v) - B(u_1, v) \qquad \forall v \in H_0^1(\Omega^-). \tag{2.141}$$

2.9.2.2 Interior Neumann Problem (INP)

In the case of Neumann boundary conditions it is not the trace but the conormal derivative of the solution that is given. Therefore we use the function space $H^1(\Omega^-)$. We then multiply by a test function and apply Green's formulas which gives us

$$B(u, v) = F(v) \tag{2.142}$$

for all $v \in H^1(\Omega^-)$. In general, F is a given functional from $\left(H^1(\Omega^-)\right)'$. If the boundary data in (2.130) is given, with $f \in \left(H^1(\Omega^-)\right)'$ and $g_N \in H^{-1/2}(\Gamma)$, the associated functional is given by

$$F(v) := (f, v)_{L^2(\Omega^-)} + (g_N, \gamma_0 v)_{L^2(\Gamma)}. \tag{2.143}$$

2.9.2.3 Interior Mixed Boundary Value Problem (IMP)

In the case of the mixed boundary value problem, the trace of the solution is given on the Dirichlet boundary $\Gamma_D \subset \Gamma$ with $|\Gamma_D| > 0$. This fact is the motivation behind the definition of the Sobolev space

$$H_D^1(\Omega^-) := \left\{v \in H^1(\Omega^-) : v = 0 \text{ on } \Gamma_D \text{ in the sense of traces}\right\}.$$

Again we multiply the differential equation by test functions v from $H_D^1(\Omega^-)$ and integrate over Ω^-. By applying Green's formula (2.110) we obtain

$$B(u,v) = (f,v)_{L^2(\Omega^-)} + (\gamma_1 u, \gamma_0 v)_{L^2(\Gamma)}\,.$$

Since v vanishes on Γ_D and $\gamma_1 u = g_N$ on Γ_N, we obtain

$$(\gamma_1 u, \gamma_0 v)_{L^2(\Gamma)} = (g_N, \gamma_0 v)_{L^2(\Gamma_N)}\,.$$

The variational formulation for the interior mixed boundary value problem reads: Find $u \in H^1(\Omega)$ with $\gamma_0 u = g_D$ on Γ_D such that

$$B(u,v) = F(v) \qquad \forall v \in H^1_D(\Omega^-)\,. \tag{2.144}$$

For given data $f \in \left(H^1_D(\Omega)\right)'$ and $g_N \in H^{-1/2}(\Gamma_N)$, F is defined by

$$F(v) := (f,v)_{L^2(\Omega^-)} + (g_N, \gamma_0 v)_{L^2(\Gamma_N)}\,. \tag{2.145}$$

By using an arbitrary trace extension $u_1 \in H^1(\Omega^-)$, i.e., $\gamma_0 u_1 = g_D$ on Γ_D, we can apply the approach $u = u_0 + u_1$. The function u_0 is the solution of the equation with homogeneous Dirichlet boundary conditions: Find $u_0 \in H^1_D(\Omega^-)$ with

$$B(u_0,v) = F(v) - B(u_1,v) \qquad \forall v \in H^1_D(\Omega^-)\,. \tag{2.146}$$

2.9.2.4 Function Spaces for Exterior Problems

We now move on to exterior problems. In principle we approach the problem in the same way as for interior problems. However, now we have to consider the decay conditions, which have to be formulated within the definition of the function spaces in a suitable way. We achieve this by introducing suitable weight functions in the definition of the Sobolev spaces. These characterize the behavior of the functions at infinity and depend on the differential operator L under consideration. The notation $H^1\left(L,\Omega^+\right)$ makes this dependency evident. The applied weight function should, on the one hand, guarantee the existence and uniqueness of the solution of the variational problem. On the other hand it should also imply that the weak solution also solves the strong formulation of the boundary value problem, possibly after imposing certain conditions on the smoothness. In the following we will specify the relevant function spaces for the general differential operator with a positive reaction component ($c > 0$) and for the Laplace and Helmholtz operators. The variational formulation requires different trial and test spaces for the Helmholtz problem. The trial space will always be denoted by $H^1(L,\Omega)$ and the test space by $H^1_T(L,\Omega)$.

General Differential Operator with $c > 0$

If, in the general differential operator L from (2.98) the component $c > 0$, then the weighted and non-weighted Sobolev spaces coincide:

$$H^1(L,\Omega^+) := H^1(\Omega^+) \quad \text{and} \quad H_0^1(L,\Omega^+) := H_0^1(\Omega^+).$$

We define the norm appropriately $\|\cdot\|_{H^1(L,\Omega^+)} := (\cdot,\cdot)^{1/2}_{H^1(L,\Omega^+)}$ with the usual inner product

$$(u,v)_{H^1(L,\Omega^+)} := \int_{\Omega^+} (\langle \nabla u, \nabla \overline{v}\rangle + u\overline{v})\, d\mathbf{x}. \tag{2.147}$$

The trial and test spaces coincide: $\|\cdot\|_{H_T^1(L,\Omega^+)} := \|\cdot\|_{H^1(L,\Omega^+)}$, $H_T^1(L,\Omega^+) := H^1(L,\Omega^+)$ and $H_{T,0}^1(L,\Omega^+) := H_0^1(L,\Omega^+)$.

Laplace Operator

The differential equation for the Laplace operator $L = -\Delta$ leads to the Poisson equation

$$-\Delta u = f \qquad \text{in } \Omega^+.$$

For sufficiently smooth functions $u, v \in C^\infty_{\text{comp}}(\Omega^+)$ [see (2.75)] we can define the inner product

$$(u,v)_{H^1(L,\Omega^+)} := \int_{\Omega^+} \left(\langle \nabla u, \nabla \overline{v}\rangle + \frac{u\overline{v}}{1+\|\mathbf{x}\|^2}\right) d\mathbf{x} \tag{2.148}$$

as well as the norm $\|u\|_{H^1(L,\Omega^+)} := (u,u)^{1/2}_{H^1(L,\Omega^+)}$. For $L = -\Delta$ the weighted Sobolev spaces $H^1(L,\Omega^+)$ and $H_0^1(L,\Omega^+)$ are given by the closures of the spaces $C^\infty_{\text{comp}}(\Omega^+)$ and $C_0^\infty(\Omega^+)$ respectively, with respect to the norm $\|\cdot\|_{H^1(L,\Omega^+)}$ in (2.148).

The trial and test spaces coincide for the Laplace problem: $\|\cdot\|_{H_T^1(L,\Omega^+)} := \|\cdot\|_{H^1(L,\Omega^+)}$, $H_T^1(L,\Omega^+) := H^1(L,\Omega^+)$ and $H_{T,0}^1(L,\Omega^+) := H_0^1(L,\Omega^+)$.

Remark 2.9.1. *Since $H^1(\Omega^+) \subset H^1(L,\Omega^+)$ in the case of the Laplace operator, one could also formulate the exterior boundary value problem in $H^1(\Omega^+)$. However, as functions that satisfy the classical decay conditions (2.132) in general are not in $H^1(\Omega^+)$, the solutions of the variational problem would be "unphysical". In contrast, the space $H^1(L,\Omega^+)$ allows for solutions with a physically correct behavior $O\left(\|\mathbf{x}\|^{-1}\right)$ for $\|\mathbf{x}\| \to \infty$.*

For $a_{\min} c > \|\mathbf{b}\|^2$ [see (2.99)] the physically relevant solutions show exponential decay for $\|\mathbf{x}\| \to \infty$ (see Lemma 3.1.9) and so the Sobolev space $H^1(\Omega^+)$ can be used for the formulation of the relevant variational problem.

Remark 2.9.2. *For every bounded, open domain $\omega \subset \Omega^+$ the spaces $H^1(L,\omega)$ and $H^1(\omega)$ coincide as sets and the norms are equivalent. We have the same assertion for the spaces $H_0^1(L,\omega)$ and $H_0^1(\omega)$.*

Helmholtz Equation

The Helmholtz equation is given by

$$L_k u := -\Delta u - k^2 u = f \qquad \text{in } \Omega^+.$$

Let $\rho(r) := 1 + r^2$ and $\tilde{\rho} := \rho^{-1}$. For sufficiently smooth functions $u, v \in C^\infty_{\text{comp}}(\Omega^+)$ the inner product can be defined as

$$(u,v)_{\rho,H^1(L,\Omega+)} := \int_{\Omega+} \left(\frac{\langle \nabla u, \nabla \overline{v} \rangle + u\overline{v}}{\rho(\|\mathbf{x}\|)} + \left(\frac{\partial u}{\partial r} - iku \right) \overline{\left(\frac{\partial v}{\partial r} - ikv \right)} \right) d\mathbf{x} \tag{2.149}$$

(see [154, 170]). The norm is given by $\|u\|_{H^1(L,\Omega+)} := (u,u)^{1/2}_{\rho,H^1(L,\Omega+)}$. The weighted Sobolev spaces $H^1(L,\Omega^+)$ and $H^1_0(L,\Omega^+)$ are the closures of the spaces $C^\infty_{\text{comp}}(\Omega^+)$ and $C^\infty_0(\Omega^+)$ respectively, with respect to the norm $\|\cdot\|_{H^1(L,\Omega+)}$.

The associated test spaces $H^1_T(L,\Omega^+)$ and $H^1_{T,0}(L,\Omega^+)$ are the closures of the spaces $C^\infty_{\text{comp}}(\Omega^+)$ and $C^\infty_0(\Omega^+)$ respectively, with respect to the norm $\|u\|_{H^1_T(L,\Omega+)} := (u,u)^{1/2}_{H^1_T(L,\Omega+)} := (u,u)^{1/2}_{\tilde{\rho},H^1(L,\Omega+)}$.

Remark 2.9.3. *We have introduced the weighted Sobolev spaces*

$$H^1\left(L,\Omega^+\right) := \overline{C^\infty_{\text{comp}}(\Omega+)}^{\|\cdot\|_{H^1(L,\Omega+)}} \quad \text{and} \quad H^1_0\left(L,\Omega^+\right) := \overline{C^\infty_0(\Omega+)}^{\|\cdot\|_{H^1(L,\Omega+)}}$$

$$H^1_T\left(L,\Omega^+\right) := \overline{C^\infty_{\text{comp}}(\Omega+)}^{\|\cdot\|_{H^1_T(L,\Omega+)}} \quad \text{and} \quad H^1_{T,0}\left(L,\Omega^+\right) := \overline{C^\infty_0(\Omega+)}^{\|\cdot\|_{H^1_T(L,\Omega+)}}$$

for the general differential operator L from (2.98) with $c > 0$, as well as for the Laplace and Helmholtz operators. The norms are the square roots of the inner products from:

- *Equation (2.147) for the general differential operator L from (2.98) with $c > 0$*
- *Equation (2.148) for the Laplace operator*
- *Equation (2.149) for the Helmholtz operator*

2.9.2.5 Exterior Dirichlet Problem (EDP)

Let L again be the general differential operator from (2.98) with $c > 0$ or the Laplace or Helmholtz operator and let $H^1(L,\Omega^+)$, $H^1_T(L,\Omega^+)$ and $H^1_0(L,\Omega^+)$, $H^1_{T,0}(L,\Omega^+)$ be as in Remark 2.9.3.

We obtain the variational formulation of the exterior Dirichlet problem by multiplying the differential equation in the strong formulation (2.134) by test functions $v \in C^\infty_0(\Omega^+)$. We can apply Green's formula (Theorem 2.7.7) under the condition that $u \in H^1_L(\Omega^+)$ and we then have

$$B(u,v) = \int_{\Omega^+} f\bar{v}d\mathbf{x}.$$

The sesquilinear form B can be extended to $H^1(L,\Omega^+) \times H^1_{T,0}(L,\Omega^+)$. The variational formulation of the exterior Dirichlet problem then reads: Let $F \in \left(H^1_{T,0}(L,\Omega^+)\right)'$ and $g_D \in H^{1/2}(\Gamma)$ be given. Find $u \in H^1(L,\Omega^+)$ with $u = g_D$ on Γ such that

$$B(u,v) = F(v) \qquad \forall v \in H^1_{T,0}(L,\Omega). \tag{2.150}$$

If we consider the strong formulation (2.134) the functional F is defined as

$$F(v) := (f,v)_{L^2(\Omega^+)}. \tag{2.151}$$

Here we assume that f is sufficiently smooth, so that the right-hand side in (2.151) exists for all $v \in H^1_0(L,\Omega^+)$.

We can also transform this problem into one with homogeneous boundary conditions by applying a trace extension. We do this by choosing a function $u_1 \in H^1(L,\Omega)$ with $u_1 = g_D$ and by then applying the approach $u = u_0 + u_1$. The unknown function u_0 is then the solution of the homogeneous Dirichlet problem: Find $u_0 \in H^1_0(L,\Omega^+)$ such that

$$B(u_0,v) = F(v) - B(u_1,v) \qquad \forall v \in H^1_{T,0}(L,\Omega^+). \tag{2.152}$$

2.9.2.6 Exterior Neumann Problem (ENP)

Let L again be the general differential operator from (2.98) with $c > 0$ or the Laplace or Helmholtz operator and let $H^1(L,\Omega^+)$, $H^1_T(L,\Omega^+)$ and $H^1_0(L,\Omega^+)$, $H^1_{T,0}(L,\Omega^+)$ be as in Remark 2.9.3.

In this case we multiply the differential equation in the strong formulation (2.135) by test functions from $C^\infty_{\text{comp}}(\Omega^+)$ and then apply Green's second formula from Theorem 2.7.7. We then have

$$B(u,v) = \int_{\Omega^+} f\bar{v}d\mathbf{x} + \int_\Gamma g_N\bar{v}ds_{\mathbf{x}},$$

where $\gamma_1 u$ has already been replaced by the Neumann data g_N. The sesquilinear form B can be extended to $H^1(L,\Omega^+) \times H^1_T(L,\Omega^+)$. The variational problem then becomes: For a given $F \in (H^1_T(L,\Omega^+))'$ find $u \in H^1(L,\Omega^+)$ such that

$$B(u,v) = F(v) \qquad \forall v \in H^1_T(L,\Omega^+). \tag{2.153}$$

Combined with sufficiently smooth data g_N and an f [see (2.135)] which is known, the functional F is given by

$$F(v) := \int_{\Omega^+} f\bar{v}d\mathbf{x} + \int_{\Gamma} g_N \bar{v} ds_{\mathbf{x}}. \qquad (2.154)$$

2.9.2.7 Exterior Mixed Boundary Value Problem (EMP)

Let L again be the general differential operator from (2.98) with $c > 0$ or the Laplace or Helmholtz operator and let $H^1(L, \Omega^+)$, $H_T^1(L, \Omega^+)$ and $H_0^1(L, \Omega^+)$, $H_{T,0}^1(L, \Omega^+)$ be as in Remark 2.9.3.

Let Γ be decomposed into Γ_D and Γ_N as in (2.136). The relevant function space is given by

$$H_D^1(L, \Omega^+) := \left\{ u \in H^1(L, \Omega^+) : u = 0 \text{ on } \Gamma_D \text{ in the sense of traces} \right\}.$$

The variational formulation of the exterior mixed boundary value problem then reads: For $F \in \left(H_{T,D}^1(L, \Omega^+)\right)'$ find $u \in H^1(L, \Omega^+)$ with $u = g_D$ on Γ_D such that

$$B(u, v) = F(v) \qquad \forall v \in H_{T,D}^1(L, \Omega^+). \qquad (2.155)$$

For the strong formulation the functional F is given by

$$F(v) := \int_{\Omega^+} f\bar{v}d\mathbf{x} + \int_{\Gamma_N} g_N \bar{v} ds_{\mathbf{x}} \qquad \forall v \in H_{T,D}^1(L, \Omega^+). \qquad (2.156)$$

By means of a trace extension $u_0 \in H^1(L, \Omega^+)$ that satisfies $u_0 = g_D$ on Γ_D, this problem can again be transformed into a homogeneous boundary value problem, although we will not go into detail here.

2.9.2.8 Transmission Problem (TP)

For $s \in \{-, +\}$ the differential operator L^s refers to the domain Ω^s (see Sect. 2.9.1.7). Let L^+ be the general differential operator from (2.98) with $c^+ > 0$ or the Laplace or Helmholtz operator and let $H^1(L^+, \Omega^+)$, $H_T^1(L^+, \Omega^+)$ and $H_0^1(L^+, \Omega^+)$, $H_{T,0}^1(L^+, \Omega^+)$ be as in Remark 2.9.3.

In order to derive the variational formulation we multiply (2.137) by functions $v \in C_0^\infty(\mathbb{R}^d)$ and integrate over $\Omega^- \cup \Omega^+$. The conditions for u and v allow us to apply Green's formula (2.110)

$$B_-(u, v) + B_+(u, v) = (f, v)_{L^2(\mathbb{R}^d)} - (g_N, \gamma_0 v)_{L^2(\Gamma)}. \qquad (2.157)$$

The sesquilinear form is well defined for functions $u = (u^-, u^+)$, $v = (v^-, v^+) \in H^1(\Omega^-) \times H^1(L, \Omega^+)$ that satisfy $[v] = 0$. We define the closed subspace $W \subset H^1(\Omega^-) \times H^1_T(L, \Omega^+)$ and the associated test space $\widetilde{W}$ as

$$\begin{aligned} W &:= \left\{u = (u^-, u^+) \in H^1(\Omega^-) \times H^1(L, \Omega^+) : [u] = 0\right\}, \\ \widetilde{W} &:= \left\{v = (v^-, v^+) \in H^1_T(\Omega^-) \times H^1_T(L, \Omega^+) : [v] = 0\right\}. \end{aligned} \tag{2.158}$$

The norm on W is given by $\|u\|_W := \left\{\|u^-\|^2_{H^1(\Omega^-)} + \|u^+\|^2_{H^1(L,\Omega^+)}\right\}^{1/2}$ and the norm $\|u\|_{\widetilde{W}}$ is defined analogously.

Then the variational formulation of the transmission problem (2.137) reads: Let $F \in \widetilde{W}'$ and $g_D \in H^{1/2}(\Gamma)$ be given. Find $u \in H^1(\Omega^-) \times H^1(L, \Omega^+)$ with $[u] = g_D$ such that

$$B_-(u, v) + B_+(u, v) = F(v) \qquad \forall v \in \widetilde{W}. \tag{2.159}$$

For the strong formulation (2.137) the functional F is defined by the right-hand side of (2.157) for sufficiently smooth f and g.

2.9.3 Equivalence of Strong and Weak Formulation

The variational formulation was derived from the strong formulation by multiplying it by test functions, integrating and then integrating by parts.

In this section we will discuss whether the solution of the variational problem is also a solution of the strong formulation of the boundary value problem. To do this we have to integrate back by parts in the variational formulation. However, the conditions in Theorem 2.7.7 require that $u \in H^1_L(\Omega)$. In general, a weak solution does not satisfy these conditions. If the weak solution is to solve the strong formulation an additional regularity condition $u \in H^1_L(\Omega)$ has to be fulfilled.

2.9.3.1 Interior Problems

Let u be the solution of one of the interior problems: IDP (2.140) with right-hand side (2.139), INP (2.142) with right-hand side (2.143) or IMP (2.144) with right-hand side (2.145). In the derivation we make the following assumption.

Assumption 2.9.4. *For the variational problem we have:*

(a) *Ω^- is a Lipschitz domain.*
(b) *The weak solution satisfies $u \in H^1_L(\Omega^-)$ and in the case of a Neumann problem $\gamma_1 u \in L^2(\Gamma)$.*
(c) *The functional on the right-hand side is defined by (2.139), (2.143) or (2.145) with $f \in L^2(\Omega^-)$ and $g_N \in L^2(\Gamma)$.*

The condition $u \in H^1_L(\Omega^-)$ allows us to apply Theorem 2.7.7, which in turn allows us to undo the integration by parts. Therefore the weak solution u satisfies

$$(Lu - f, v)_{L^2(\Omega^-)} = 0 \qquad \forall v \in V,$$

with

$$V := \begin{cases} H^1_0(\Omega^-) & \text{for (2.140)}, \\ H^1(\Omega^-) & \text{for (2.142)}, \\ H^1_D(\Omega^-) & \text{for (2.144)}. \end{cases}$$

Assumptions 2.9.4.a and b imply that $Lu - f \in L^2(\Omega^-)$. Since $H^1_0(\Omega^-)$ is densely embedded into $L^2(\Omega^-)$ (see Proposition 2.5.2) the embedding $V \subset L^2(\Omega^-)$ is also dense. Therefore there exists a sequence $(v_n)_n \subset V \subset L^2(\Omega^-)$ with $v_n \to Lu - f$ in $L^2(\Omega^-)$ that satisfies

$$0 = \lim_{n\to\infty} (Lu - f, v_n)_{L^2(\Omega^-)} = \left(Lu - f, \lim_{n\to\infty} v_n\right)_{L^2(\Omega^-)} = \|Lu - f\|^2_{L^2(\Omega^-)}.$$

With this we have shown that if the weak solution u satisfies the additional conditions $u \in H^1_L(\Omega^-)$ and $f \in L^2(\Omega^-)$ it also satisfies the differential equation $Lu = f$ almost everywhere.

For the IDP and the IMP the Dirichlet boundary conditions $\gamma_0 u = g_D$ on Γ_D are required explicitly. Here we will only consider the INP, the proof for the IMP can be done in the same way.

From $Lu = f$ in $L^2(\Omega^-)$, (2.142) and Theorem 2.7.7 we have

$$\begin{aligned} 0 &= B(u, v) - (f, v)_{L^2(\Omega^-)} - (g_N, \gamma_0 v)_{L^2(\Gamma)} \\ &= (Lu - f, v)_{L^2(\Omega^-)} + (\gamma_1 u - g_N, \gamma_0 v)_{L^2(\Gamma)} \\ &= (\gamma_1 u - g_N, \gamma_0 v)_{L^2(\Gamma)} \end{aligned}$$

for all $v \in H^1(\Omega^-)$. Lemma 2.8.4 implies that the image of $H^1(\Omega^-)$ under γ_0 is dense in $H^{1/2}(\Gamma)$ and therefore is also dense in $L^2(\Gamma)$. It follows that

$$(\gamma_1 u - g_N, w)_{L^2(\Gamma)} = 0 \qquad \forall w \in L^2(\Gamma).$$

With Assumptions 2.9.4(b) and 2.9.4.c we have $\gamma_1 u = g_N$ in $L^2(\Gamma)$.

2.9.3.2 Exterior Problems

In principle, the argument for exterior problems is the same as for interior problems. However, Assumption 2.9.4 does not guarantee that the conditions of Theorem 2.7.7 are satisfied. In general, neither the weak solution nor the test function $v \in H^1_T(L, \Omega^+)$ has compact support. We will however show that under suitable conditions Green's second formula remains valid for functions that do not have compact support. For this we consider the following abstract situation, which has been chosen to fit Remark 2.9.3.

Assumption 2.9.5. *$V := \overline{C^{\infty}_{\text{comp}}(\Omega+)}^{\|\cdot\|_V}$ and $V_T := \overline{C^{\infty}_{\text{comp}}(\Omega+)}^{\|\cdot\|_{V_T}}$ are the closures of the sets of smooth functions with compact support with respect to the norms $\|\cdot\|_V$ and $\|\cdot\|_{V_T}$ respectively, so that:*

(a) V, V_T are complete and satisfy $V, V_T \subset H^1_{\text{loc}}(\Omega^+)$.
(b) $B : V \times V_T \to \mathbb{C}$ is continuous.

Theorem 2.9.6. *Let Assumption 2.9.5 hold for the spaces V, V_T. Then Green's second formula is applicable for all $u \in V \cap H^1_L(\Omega^+)$ and $v \in V_T$.*

Proof. (a) Since $V_T \subset H^1_{\text{loc}}(\Omega^+)$ there exists a continuous trace operator $\gamma_0 : V_T : H^{1/2}(\Gamma)$ (see Theorem 2.6.8).

(b) For all $u \in V \cap H^1_L(\Omega^+)$ and $v \in C^{\infty}_{\text{comp}}(\Omega^+)$ we have, with Theorem 2.7.7, Green's second formula

$$(Lu, v)_{L^2(\Omega+)} - B(u, v) = (\gamma_1 u, \gamma_0 v)_{L^2(\Gamma)}\,.$$

The mapping $(Lu, \cdot)_{L^2(\Omega)} : V_T \to \mathbb{C}$ is continuous, since we assume that Lu has compact support and that $V_T \subset H^1_{\text{loc}}(\Omega^+)$. Due to Assumption 2.9.5.b, the sesquilinear form $B : V \times V_T \to \mathbb{C}$ is continuous. If, for an arbitrary $v \in V_T$, we choose a Cauchy sequence $(v_n)_n \subset C^{\infty}_{\text{comp}}$ that converges towards v with respect to the V_T-norm we have

$$\lim_{n\to\infty} \left\{(Lu, v_n)_{L^2(\Omega+)} - B(u, v_n)\right\} = (Lu, v)_{L^2(\Omega+)} - B(u, v)\,. \tag{2.160}$$

On the other hand, according to Theorem 2.7.7, $\gamma_1 u$ defines a continuous functional on $H^{1/2}(\Gamma)$. Therefore because of (a) we have

$$\lim_{n\to\infty} (\gamma_1 u, \gamma_0 v_n)_{L^2(\Gamma)} = (\gamma_1 u, \gamma_0 v)_{L^2(\Gamma)}\,,$$

which with (2.160) proves the statement. □

In order to apply this result to the exterior problem we use the function spaces that were defined in Remark 2.9.3. We set $V := H^1(L, \Omega^+)$, $V_T := H^1_T(L, \Omega^+)$ and check the conditions from Assumption 2.9.5.

Continuity:

In Lemma 2.10.1 and Theorem 2.10.10 we will show the continuity of the sesquilinear form B on $H^1(\Omega^+)$ and on $H^1(-\Delta, \Omega^+)$. The continuity for the Helmholtz problem is investigated in Corollary 2.10.3.

Embedding:

The embedding from Assumption 2.9.5.a follows for all considered differential operators, as the weight functions only influence the behavior of the functions $f \in W$ at infinity.

Therefore Assumption 2.9.5 is also satisfied and Green's second formula is applicable for $u \in H_L^1(\Omega) \cap H^1(L,\Omega)$ and $v \in H_T^1(L,\Omega)$. The way in which we treated the interior problems, in this case deriving the strong formulation from the weak formulation, can be repeated identically for the exterior problems.

Decay Condition:

In the following chapters, the solutions of the boundary value problems that have been discussed above will be represented by means of a surface integral, from which the decay conditions for the Laplace operator, Helmholtz operator and the general elliptic operator with $c > 0$ can be deduced immediately [see (3.22) and (3.23)]. The Sommerfeld radiation conditions for the solution of the Helmholtz problem are discussed with the help of the integral representation in Exercise 3.1.15.

2.10 Existence and Uniqueness

In the previous section we formulated interior and exterior elliptic boundary value problems as variational problems. We will now give the most important results on existence and uniqueness. Since the focus of this book is on *integral equations* for elliptic boundary value problems, we will not elaborate the analysis of elliptic differential equations. Instead we will refer to the appropriate textbooks. First we will prove the continuity and coercivity of the sesquilinear form B for the interior problem.

Lemma 2.10.1. *Let $\Omega \in \{\Omega^-, \Omega^+\}$. The sesquilinear form $B(\cdot,\cdot)$ as in (2.102) is continuous and there exist positive constants C_1, C_2 such that*

$$\operatorname{Re} B(u,u) \geq C_1 \|u\|_{H^1(\Omega)}^2 - C_2 \|u\|_{L^2(\Omega)}^2 \qquad \forall u \in H^1(\Omega).$$

Therefore B is coercive on $H^1(\Omega)$ for $\Omega = \Omega^-$.

Proof. Let $a_{\max}$ ($a_{\min}$) be the largest (smallest) eigenvalue of the matrix $\mathbf{A}$. Then, with the notation (2.79), we have for all $u, v \in H^1(\Omega)$

$$\begin{aligned}|B(u,v)| &\leq \int_\Omega \left(a_{\max} \|\nabla u\| \|\nabla v\| + 2\|\mathbf{b}\| \|\nabla u\| |v| + |c| |u| |v|\right) dx \\ &\leq a_{\max} |u|_{H^1(\Omega)} |v|_{H^1(\Omega)} + 2\|\mathbf{b}\| |u|_{H^1(\Omega)} \|v\|_{L^2(\Omega)} \\ &\quad + |c| \|u\|_{L^2(\Omega)} \|v\|_{L^2(\Omega)} \\ &\leq 3\max\{a_{\max}, 2\|\mathbf{b}\|, |c|\} \|u\|_{H^1(\Omega)} \|v\|_{H^1(\Omega)}.\end{aligned}$$

For the proof of the coercivity we use for arbitrary $\delta, \varepsilon > 0$

$$\begin{aligned}\operatorname{Re} B(u,u) &\geq a_{\min} |u|^2_{H^1(\Omega)} - \|\mathbf{b}\| \left(\delta |u|^2_{H^1(\Omega)} + \delta^{-1} \|u\|^2_{L^2(\Omega)} \right) + c \|u\|^2_{L^2(\Omega)} \\ &\geq (a_{\min} - \delta \|\mathbf{b}\|) |u|^2_{H^1(\Omega)} + \varepsilon \|u\|^2_{L^2(\Omega)} + \left(c - \|\mathbf{b}\| \delta^{-1} - \varepsilon\right) \|u\|^2_{L^2(\Omega)} .\end{aligned}$$

We now choose $0 < \varepsilon < a_{\min}$ and set $\delta := (a_{\min} - \varepsilon) / \|\mathbf{b}\|$ if $\|\mathbf{b}\| \neq 0$ and $\delta := +\infty$ otherwise. We then have

$$\operatorname{Re} B(u,u) \geq \varepsilon \|u\|^2_{H^1(\Omega)} + \left(c - \|\mathbf{b}\| \delta^{-1} - \varepsilon\right) \|u\|^2_{L^2(\Omega)} . \tag{2.161}$$

If $\Omega = \Omega^-$ the compact embedding $L^2(\Omega) \hookrightarrow H^1(\Omega)$ (see Theorem 2.5.5) implies the coercivity. □

In Corollary 2.10.2 the quotient space $H^1(\Omega^-)/\mathbb{K}$ appears. Its equivalence classes consist of functions in $H^1(\Omega^-)$ that only differ by a constant. A norm on this space is given by

$$\|u\|_{H^1(\Omega^-)/\mathbb{K}} := \inf_{c \in \mathbb{R}} \|u - c\|_{H^1(\Omega^-)} . \tag{2.162}$$

Corollary 2.10.2. *(a) The result from Lemma 2.10.1 also holds for every subspace of $H^1(\Omega^-)$.*
(b) The sesquilinear form is elliptic on $H^1(\Omega^-)$ if we have $a_{\min} c > \|\mathbf{b}\|^2$.
(c) For $\mathbf{b} = \mathbf{0}$ and $c = 0$, the sesquilinear form B_- is elliptic on $H^1_0(\Omega^-)$.
(d) For $\mathbf{b} = \mathbf{0}$ and $c = 0$, the sesquilinear form B_- is elliptic on $H^1(\Omega^-)/\mathbb{K}$.
(e) The inequality (2.161) also holds for the exterior problem, i.e., $\Omega = \Omega^+$ and $B = B_+$. Therefore, for $a_{\min} c > \|\mathbf{b}\|^2$, the sesquilinear form B_+ is elliptic on $H^1(\Omega^+)$ and, consequently, on every subspace.

Proof. Statements (a), (b) and (e) follow directly from the proof of the previous lemma.

(c): Here we use Theorem 2.5.7 and obtain

$$\operatorname{Re} B(u,u) \geq a_{\min} |u|^2_{H^1(\Omega^-)} \geq C \|u\|^2_{H^1(\Omega^-)} \qquad \forall u \in H^1_0(\Omega^-) .$$

Statement (d) follows from the second Poincaré inequality (Corollary 2.5.10)

$$\|u\|^2_{H^1(\Omega^-)/\mathbb{K}} \leq C |u|^2_{H^1(\Omega^-)} \leq C/a_{\min} \operatorname{Re} B(u,u) \qquad \forall u \in H^1(\Omega^-)/\mathbb{K}.$$

□

Corollary 2.10.3. *The sesquilinear form of the exterior Helmholtz problem is continuous.*

Proof. If we use the explicit representation of the sesquilinear form of the Helmholtz equation and apply the Cauchy–Schwarz inequality, we obtain for all $u \in H^1$

(L, Ω^+) and $v \in H_T^1(L, \Omega^+)$ the continuity

$$B(u, v) = \int_{\Omega^+} \left(\langle \nabla u, \nabla \bar{v} \rangle - k^2 u \bar{v} \right) d\mathbf{x} \leq |u|_{H^1(\Omega^+)} |v|_{H^1(\Omega^+)}$$
$$+ k^2 \left\| u \rho^{-1/2} \right\|_{L^2(\Omega^+)} \left\| v \rho^{1/2} \right\|_{L^2(\Omega^+)}$$
$$\leq \left(1 + k^2\right) \|u\|_{H^1(L,\Omega^+)} \|v\|_{H_T^1(L,\Omega^+)} .$$

□

By combining Lemma 2.10.1 and Corollary 2.10.2 with the results from Sect. 2.1.6 one obtains the existence and uniqueness of the solution for the boundary value problems from Sect. 2.9.2.

2.10.1 Interior Problems

We first prove the results on existence and uniqueness for interior problems.

2.10.1.1 Interior Dirichlet Problem

The following theorem will demonstrate that the Fredholm alternative always holds for the Dirichlet interior problem. Furthermore, if the coefficients of the differential operator satisfy suitable conditions, the Lax–Milgram lemma guarantees the existence and uniqueness of the solution.

Theorem 2.10.4. *We consider the IDP [see (2.140)] and assume that the functional F is defined as in (2.139) and that we have $g_D \in H^{1/2}(\Gamma)$ for the boundary data:*

1. *The Fredholm alternative is applicable: Either, for every right-hand side $F \in H_0^1(\Omega^-)'$ and boundary data $g_D \in H^{1/2}(\Gamma)$, the problem (2.140) has a unique solution $u \in H_0^1(\Omega^-)$ that depends continuously on the right-hand side, i.e.,*

$$\|u\|_{H^1(\Omega^-)} \leq C \left(\|F\|_{H_0^1(\Omega^-)'} + \|g_D\|_{H^{1/2}(\Gamma)} \right)$$

   or zero is an eigenvalue of the operator associated with B that corresponds to a finite-dimensional eigenspace.
2. *The condition $a_{\min} c > \|\mathbf{b}\|^2$ implies that the first case will always apply in the above-mentioned alternative.*
3. *Statement (2) remains true if the condition $a_{\min} c > \|b\|^2$ is replaced by $c = \|\mathbf{b}\| = 0$.*

Proof. For 1: We use the approach $u = u_0 + u_1$ with the trace extension $u_1 := Z_{-} g_D$. The right-hand side in (2.141) defines a continuous functional on $H_0^1(\Omega)$:

$$\begin{aligned}|F(v) - B(u_1, v)| &\leq \left(\|F\|_{H^{-1}(\Omega^-)} + C_1 \|u_1\|_{H^1(\Omega^-)}\right) \|v\|_{H^1(\Omega^-)} \\ &\leq \left(\|F\|_{H^{-1}(\Omega^-)} + C_2 \|g_D\|_{H^{1/2}(\Gamma)}\right) \|v\|_{H^1(\Omega^-)},\end{aligned}$$

so that the Fredholm alternative (see Theorem 2.1.60) is applicable for (2.141). The assertion for the problem (2.140) then follows from $\|u\|_{H^1(\Omega^-)} \leq \|u_0\|_{H^1(\Omega^-)} + C \|g_D\|_{H^{1/2}(\Gamma)}$.

For 2,3: The proof is done in the same way as in the first part. Here we combine Corollary 2.10.2.(b),(c) with Lemma 2.1.51. □

2.10.1.2 Interior Neumann Problem

In the following theorem we will formulate results on existence and uniqueness for the interior Neumann problem.

Theorem 2.10.5. *We consider the INP (2.142), (2.143) and assume that* $f \in \left(H^1(\Omega^-)\right)'$, $g_N \in H^{-1/2}(\Gamma)$:

1. *The Fredholm alternative is applicable: Either, for every right-hand side* $F \in \left(H^1(\Omega^-)\right)'$ *the problem (2.142) has a unique solution* $u \in H^1(\Omega^-)$ *that depends continuously on the right-hand side [see* (2.89)*], i.e.,*

$$\|u\|_{H^1(\Omega^-)} \leq C \|F\|_{H^1(\Omega^-)'} \leq C \left(\|f\|_{(H^1(\Omega^-))'} + \|g_N\|_{H^{-1/2}(\Gamma)}\right)$$

   or zero is an eigenvalue of the operator associated with B *that corresponds to a finite-dimensional eigenspace.*
2. *The condition* $a_{\min} c > \|\mathbf{b}\|^2$ *implies that the first case will always apply in the above-mentioned alternative.*
3. *Let* $c = \|\mathbf{b}\| = 0$*. Then there exists a solution* $u \in H^1(\Omega^-)$ *if and only if* f *and* g *satisfy the relation*

$$\langle f, 1\rangle_{L^2(\Omega^-)} + \langle g_N, 1\rangle_{L^2(\Gamma)} = 0. \tag{2.163}$$

   The solution is unique up to a constant function. Therefore, if we restrict the solution space to $H^1(\Omega^-)/\mathbb{K}$*, there exists, for all* $f \in \left(H^1(\Omega^-)\right)'$ *and* $g_N \in H^{-1/2}(\Gamma)$ *that satisfy (2.163), a unique solution in* $H^1(\Omega^-)/\mathbb{K}$ *that depends continuously on the data.*

Proof. The proofs of (1) and (2) are similar to the proof of Theorem 2.10.4. The proof of (3) can be found in, e.g., [162, Theorem 8.19]. □

2.10.1.3 Interior Mixed Boundary Value Problem

We now move on to existence and uniqueness results for the interior mixed boundary value problem [see (2.9.2.3)].

Theorem 2.10.6. *The results from Theorem 2.10.4 also hold for the interior mixed boundary value problem. Here Γ has to be replaced by Γ_D and $H_0^1(\Omega^-)$ by $H_D^1(\Omega^-)$.*

Proof. The proof of this statement is analogous to the proof of Theorem 2.10.4. To prove the third statement we use Corollary 2.5.8. □

2.10.2 Exterior Problems

Results on existence and uniqueness for exterior problems require the use of the weighted Sobolev spaces $H^1(L, \Omega^+)$ that were introduced in the previous section (see Remark 2.9.3).

2.10.2.1 General Elliptic Operator with $a_{\min} c > \|\mathbf{b}\|^2$

If we combine Corollary 2.10.2.e with the Lax–Milgram lemma we get existence and uniqueness for the exterior boundary value problems under the condition that $a_{\min} c > \|\mathbf{b}\|^2$. Note that in this case we have $H^1(L, \Omega^+) = H_T^1(L, \Omega^+)$ which is why we can omit the "$\sim$" notation.

Theorem 2.10.7. *Let $a_{\min} c > \|\mathbf{b}\|^2$.*

(a) EDP (2.150). For all $F \in H^{-1}(\Omega^+)$ and $g_D \in H^{1/2}(\Gamma)$ there exists a unique solution $u \in H^1(\Omega^+)$ that depends continuously on the data, i.e.,

$$\|u\|_{H^1(\Omega^+)} \leq C \left(\|F\|_{H^{-1}(\Omega^+)} + \|g_D\|_{H^{1/2}(\Gamma)} \right).$$

(b) ENP (2.153). For all $F \in (H^1(\Omega^+))'$ there exists a unique solution $u \in H^1(\Omega^+)$ that depends continuously on the data, i.e.,

$$\|u\|_{H^1(\Omega^+)} \leq C \, \|F\|_{(H^1(\Omega^+))'}.$$

(c) EMP (2.155). For all $F \in (H_D^1(\Omega^+))'$ and $g_D \in H^{1/2}(\Gamma_D)$ there exists a unique solution $u \in H_D^1(\Omega^+)$ that depends continuously on the data

$$\|u\|_{H^1(\Omega^+)} \leq C \left(\|F\|_{(H_D^1(\Omega^+))'} + \|g_D\|_{H^{1/2}(\Gamma_D)} \right).$$

(d) TP (2.159). Let W be as in (2.158). For all $F \in W'$ and $g_D \in H^{1/2}(\Gamma)$ there exists a unique solution $u \in W$ that depends continuously on the data

$$\|u\|_W \le C\left(\|F\|_{W'} + \|g_D\|_{H^{1/2}(\Gamma)}\right).$$

Proof. The proofs for the interior problem can be directly transferred to this particular case. □

2.10.2.2 Laplace Operator

The proof of existence and uniqueness results for the Laplace operator is more elaborate than for the differential operator from Sect. 2.10.2.1 because of the use of the weighted Sobolev spaces.

We begin with a few auxiliary results on the function spaces $H^1\left(-\Delta, \Omega^+\right)$ and $H_0^1\left(-\Delta, \Omega^+\right)$ (see [80, Chap. XI, Part B]). For the Laplace operator the inner product and the norm on $H^1\left(-\Delta, \Omega^+\right)$ and $H_0^1\left(-\Delta, \Omega^+\right)$ are defined by (2.148). We will first show that the norms [see (2.148), (2.79)] $\|\cdot\|_{H^1(-\Delta,\Omega^+)}$ and $|\cdot|_{H^1(\Omega^+)}$ are equivalent on $H_0^1\left(-\Delta, \Omega^+\right)$ for exterior problems. For the proof we again denote the ball with radius $a > 0$ around the origin by $B_a := \left\{\mathbf{x} \in \mathbb{R}^3 : \|\mathbf{x}\| < a\right\}$ and the exterior complement of B_a by $B_a^+ := \mathbb{R}^d \backslash \overline{B_a}$.

Proposition 2.10.8. *For $a > 0$, the norms $|\cdot|_{H^1\left(B_a^+\right)}$ and $\|\cdot\|_{H^1\left(-\Delta, B_a^+\right)}$ are equivalent on $H_0^1\left(-\Delta, B_a^+\right)$.*

Proof. As $C_0^\infty\left(B_a^+\right)$ is dense in $H_0^1\left(-\Delta, B_a^+\right)$ it suffices to show the equivalence for smooth functions.

(i) The inequality $|u|_{H^1\left(B_a^+\right)} \le \|u\|_{H^1\left(-\Delta, B_a^+\right)}$ is obvious.

(ii) We will now show that $\|u\|_{H^1\left(-\Delta, B_a^+\right)} \le C\,|u|_{H^1\left(B_a^+\right)}$. Obviously, we only need to show that

$$\int_{B_a^+} \frac{|u(\mathbf{x})|^2}{1+\|\mathbf{x}\|^2} d\mathbf{x} \le C\,|u|^2_{H^1\left(B_a^+\right)}.$$

We introduce spherical coordinates $\mathbf{x} = r\zeta$ with $\zeta = \mathbf{x}/\|\mathbf{x}\| \in \mathbb{S}_2$, where $\mathbb{S}_2$ denotes the unit sphere in $\mathbb{R}^3$. We then have

$$\int_{B_a^+} \frac{|u(\mathbf{x})|^2}{1+\|\mathbf{x}\|^2} d\mathbf{x} \le \int_a^\infty \int_{\mathbb{S}_2} \frac{|u(r\zeta)|^2}{1+r^2} r^2 d\zeta dr.$$

For functions $f(r)$ that vanish for sufficiently large r and satisfy $f(a) = 0$, we have, with the help of integration by parts:

$$\int_a^\infty |f(r)|^2 \, dr = -\int_a^\infty 2r \operatorname{Re}\left(f(r)\, \partial_r \overline{f}(r)\right) dr \qquad (2.164)$$
$$\leq 2 \left(\int_a^\infty |f(r)|^2 \, dr \right)^{1/2} \left(\int_a^\infty |\partial_r f(r)|^2 \, r^2 dr \right)^{1/2},$$

i.e.,

$$\int_a^\infty |f(r)|^2 \, dr \leq 4 \int_a^\infty |\partial_r f(r)|^2 \, r^2 dr.$$

Since u has compact support, we can choose $f(r) = u(r\zeta)$ in (2.164). We obtain the required inequality by integrating over $\mathbb{S}_2$:

$$\int_{\mathbb{S}_2} \int_a^\infty \frac{|u(r\zeta)|^2}{1+r^2} r^2 dr d\zeta \leq \int_{\mathbb{S}_2} \int_a^\infty |u(r\zeta)|^2 \, dr d\zeta$$
$$\leq 4 \int_{\mathbb{S}_2} \int_a^\infty |\partial_r u(r\zeta)|^2 \, r^2 dr d\zeta$$
$$= 4 \int_{B_a^+} \left| \left\langle \frac{\mathbf{x}}{\|\mathbf{x}\|}, \nabla u(\mathbf{x}) \right\rangle \right|^2 d\mathbf{x} \leq 4 \, |u|^2_{H^1\left(B_a^+\right)}.$$

□

Note that the equivalence constant is independent of a.

Remark 2.10.9. *The spaces $H_0^1\left(-\Delta, \Omega^+\right)$ and $H^1\left(-\Delta, \Omega^+\right)$ are complete by definition. They become Hilbert spaces once we define the inner product from (2.148) on them.*

From Proposition 2.10.8 we directly have an inequality of Poincaré type for $H^1\left(-\Delta, \Omega^+\right)$.

Theorem 2.10.10. *$|\cdot|_{H^1(\Omega^+)}$ defines a norm on $H^1\left(-\Delta, \Omega^+\right)$ and $H_0^1\left(-\Delta, \Omega^+\right)$ that is equivalent to the $\|\cdot\|_{H^1(-\Delta,\Omega^+)}$ norm.*

Proof. Since $H_0^1\left(-\Delta, \Omega^+\right) \subset H^1\left(-\Delta, \Omega^+\right)$ we only need to prove the statement for $H^1\left(-\Delta, \Omega^+\right)$.

(i) We obviously have $|u|_{H^1(\Omega^+)} \leq \|u\|_{H^1(-\Delta,\Omega^+)}$.
(ii) We prove $\|u\|_{H^1(-\Delta,\Omega^+)} \leq C\,|u|_{H^1(\Omega^+)}$ indirectly. We assume that there exists a sequence $(u_n)_{n\in\mathbb{N}}$ in $H^1\left(-\Delta, \Omega^+\right)$ such that

$$|u_n|_{H^1(\Omega^+)} \leq n^{-1}, \qquad \|u_n\|_{H^1(-\Delta,\Omega^+)} = 1. \qquad (2.165)$$

For sufficiently large $a > 0$ we have $\Omega \subset B_a$. We choose a cut-off function $\phi, \psi \in C^\infty\left(\Omega^+\right)$ (see Fig. 2.2) with

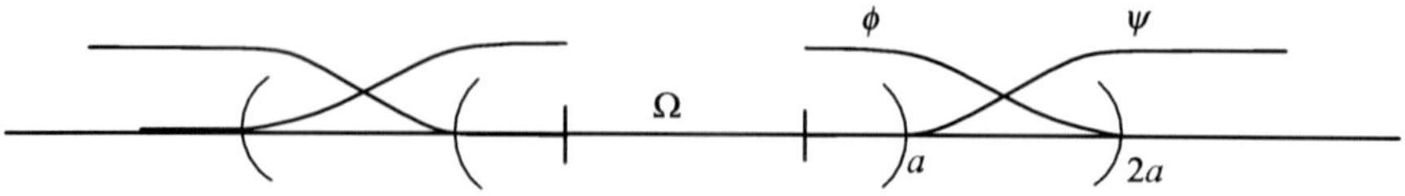

Fig. 2.2 Cutoff functions ϕ and ψ for the proof of Theorem 2.10.10

$$\phi, \psi \geq 0, \quad \phi + \psi = 1 \text{ in } \Omega^+, \quad \phi(\mathbf{x}) = 0 \quad \text{for all } \|\mathbf{x}\| \geq 2a \quad \text{and}$$
$$\psi(\mathbf{x}) = 0 \quad \text{for all } \|\mathbf{x}\| < a.$$

We obviously have $u_n = \phi u_n + \psi u_n$. By differentiating the products ψv and ϕv for $v \in H^1(-\Delta, \Omega^+)$ we obtain

$$\begin{aligned} |\psi v|_{H^1(B_a^+)} &\leq |v|_{H^1(B_a^+)} + c\,\|v\|_{L^2(B_{2a}\cap\Omega^+)}, \\ |\phi v|_{H^1(\Omega^+)} &\leq |v|_{H^1(B_{2a}\cap\Omega^+)} + C\,\|v\|_{L^2(B_{2a}\cap\Omega^+)}. \end{aligned} \tag{2.166}$$

For every $a > 0$, $H^1(B_{2a} \cap \Omega^+)$ is compactly embedded in $L^2(B_{2a} \cap \Omega^+)$ (see Theorem 2.5.5). Therefore there exists a subsequence $(u_{n_j})_{j\in\mathbb{N}}$ (see Theorem 2.5.6) that satisfies

$$u_{n_j} \to u \qquad \text{in } L^2(B_{2a} \cap \Omega^+).$$

From (2.165) and (2.166), (ψu_{n_j}) and (ϕu_{n_j}) are Cauchy sequences in $H^1(B_a^+)$ and $H^1(\Omega^+)$. Since $\psi u_{n_j} \in H_0^1(B_a^+)$, Proposition 2.10.8 is applicable and implies convergence in the $H^1(-\Delta, B_a^+)$-norm as well. Since ϕu_{n_j} has compact support the norms in $H^1(B_a^+)$ and $H^1(-\Delta, B_a^+)$ are equivalent and hence

$$\begin{aligned} \phi u_{n_j} &\to w_1 \qquad \text{with respect to } \|\cdot\|_{H^1(-\Delta,\Omega^+)}, \\ \psi u_{n_j} &\to w_2 \qquad \text{with respect to } |\cdot|_{H^1(B_a^+)}. \end{aligned}$$

Therefore $u_{n_j} = (\phi + \psi)\, u_{n_j}$ converges to some $w \in H^1(-\Delta, \Omega^+)$. Assumption (2.165) implies that $\nabla w = 0$ and therefore that w is constant. Finally, from $w \in H^1(-\Delta, \Omega^+)$ we have $w \equiv 0$, which is a contradiction to (2.165). □

Exterior Dirichlet Problem

We now come to the theorem on existence and uniqueness for the exterior Dirichlet problem.

Theorem 2.10.11. *We consider the (EDP) [see (2.150)]. Let F be defined as in (2.151). Then, for every $f \in (H_0^1(-\Delta, \Omega^+))'$ and $g_D \in H^{1/2}(\Gamma)$, the exterior*

Dirichlet problem has a unique solution $u \in H_0^1\left(-\Delta,\Omega^+\right)$ *that satisfies the inequality*

$$\|u\|_{H^1(-\Delta,\Omega+)} \leq \|F\|_{(H_0^1(-\Delta,\Omega+))'} + C\,\|g_D\|_{H^{1/2}(\Gamma)}\,.$$

Proof. We choose $a > 0$ with $\Omega \subset B_a$. By using the trace extension operator (see Theorem 2.6.11) $Z_a := Z_{B_a} : H^{1/2}(\Gamma) \rightarrow H_0^1(B_a)$ the Dirichlet data g_D can be extended:

$$\|Z_a g_D\|_{H^1(-\Delta,\Omega+)} \leq C_1\,\|Z_a g_D\|_{H^1(B_a)} \leq C_2\,\|g_D\|_{H^{1/2}(\Gamma)}. \tag{2.167}$$

We use the approach $u = u_0 + Z_a g_D$ in (2.150) and thus obtain an equation of the form (2.152) with $u_1 = Z_a g_D$ for $u_0 \in H_0^1\left(-\Delta,\Omega^+\right)$.

As in the proof of Theorem 2.10.4, part 1, it can be shown that the right-hand side in (2.152) defines a continuous linear functional on $H_0^1\left(-\Delta,\Omega^+\right)$. Theorem 2.10.10 implies that the sesquilinear form B is elliptic on $H^1\left(-\Delta,\Omega^+\right)$ (and therefore also on $H_0^1\left(-\Delta,\Omega^+\right)$). Thus

$$B(v,v) = \int_{\Omega+} \|\nabla v\|^2\,d\mathbf{x} \geq C\,\|v\|^2_{H^1(-\Delta,\Omega+)} \qquad \forall v \in H^1\left(-\Delta,\Omega^+\right). \tag{2.168}$$

The continuity of B follows from Theorem 2.10.10. The Lax–Milgram Lemma 2.1.51 becomes applicable and as a consequence the problem (2.152) has a unique solution $u_0 \in H_0^1\left(-\Delta,\Omega^+\right)$ with

$$\begin{aligned}\|u_0\|_{H^1(-\Delta,\Omega+)} &\leq \|F\|_{(H_0^1(-\Delta,\Omega+))'} + C\,\|Z_a g_D\|_{H^1(B_a)}\\ &\leq \|F\|_{(H_0^1(-\Delta,\Omega+))'} + C(\Omega)\,\|g_D\|_{H^{1/2}(\Gamma)}\,.\end{aligned}$$

□

Exterior Neumann Problem

We now consider the exterior Neumann problem for the Laplace operator.

Theorem 2.10.12. *For every* $F \in \left(H^1\left(-\Delta,\Omega^+\right)\right)'$ *there exists a unique solution* $u \in H^1\left(-\Delta,\Omega^+\right)$ *of the exterior Neumann problem (2.153) that depends continuously on* F*:*

$$\|u\|_{H^1(-\Delta,\Omega+)} \leq C\,\|F\|_{(H^1(-\Delta,\Omega+))'}\,.$$

Proof. We combine Corollary 2.10.2.(e) and (2.168) with the Lax–Milgram lemma 2.1.51 and thus obtain the statement. □

Exterior Mixed Boundary Value Problem

Results concerning existence and uniqueness can be found for the EMP analogously [see (2.155)].

Theorem 2.10.13. *The results from Theorem 2.10.11 hold for the exterior mixed boundary value problem. Here* Γ *has to be replaced by* Γ_D *and* $H_0^1(L,\Omega^+)$ *by* $H_D^1(L,\Omega^+)$.

Transmission Problem

We finally turn our attention to the transmission problem (2.157).

Theorem 2.10.14. *We consider the TP (2.159) with* W *as in (2.158). For every* $F \in W'$ *and* $g_D \in H^{1/2}(\Gamma)$ *there exists a unique solution* $u \in H^1(\Omega^-) \times H^1(-\Delta,\Omega^+)$ *of the transmission problem (2.159). It depends continuously on* F *and* g_D*:*

$$\|u\|_W \le C \|F\|_{W'} + \|g_D\|_{H^{1/2}(\Gamma)}.$$

Proof. We use the approach $u = u_0 + u_1$ with $u_1|_{\Omega^+} = 0$ and $u_1|_{\Omega^-} := -Z_- g_D$. Then we have $u_1 \in H^1(\Omega^-) \times H^1(-\Delta,\Omega^+)$ and $[u_1] = g_D$. Then u_0 is the solution of the problem: Find $u_0 \in W$ such that

$$B_{\Omega^-\cup\Omega^+}(u_0,v) := B_-(u_0,v) + B_+(u_0,v) = F(v) - B_-(u_1,v) \qquad \forall v \in W. \tag{2.169}$$

The continuity of $B_{\Omega^-\cup\Omega^+}$ and the right-hand side in (2.169) follow from Lemma 2.10.1.

For the Lax–Milgram lemma we still need to show the ellipticity of $B_{\Omega^-\cup\Omega^+}$.

For an arbitrary function $v \in W$ we set $v^- := v|_{\Omega^-}$ and $v^+ := v|_{\Omega^+}$. Since $u_0 \in W \subset H^1(\Omega^-) \times H^1(-\Delta,\Omega^+)$, the ellipticity of B_+ on $H^1(-\Delta,\Omega^+)$ gives us [see (2.168)]

$$B_+(v^+,v^+) \ge C_0 \|v^+\|^2_{H^1(-\Delta,\Omega^+)}. \tag{2.170}$$

For v^- we set $g := \gamma_0^- v^-$ and $w^- := Z_- g$ with the trace extension $Z_- : H^{1/2}(\Gamma) \to H^1(\Omega^-)$ from Theorem 2.6.11. Then we have $v^- - w^- \in H_0^1(\Omega^-)$ and due to the Friedrichs inequality (see Theorem 2.5.7)

$$\begin{aligned}
\|v^-\|^2_{H^1(\Omega^-)} &\le 2\|v^- - w^-\|^2_{H^1(\Omega^-)} + 2\|w^-\|^2_{H^1(\Omega^-)} \\
&\le 2C\,|v^- - w^-|^2_{H^1(\Omega^-)} + 2\|w^-\|^2_{H^1(\Omega^-)} \\
&\le 4C\,|v^-|^2_{H^1(\Omega^-)} + (2+4C)\|w^-\|^2_{H^1(\Omega^-)} \\
&= C_1 B_-(v^-,v^-) + C_2\|w^-\|^2_{H^1(\Omega^-)}.
\end{aligned}$$

The continuity of the trace extension and the trace operator as well as the condition $[v] = 0$ give us

$$\begin{aligned}\|w^-\|_{H^1(\Omega^-)} &= \|Z_- g\|_{H^1(\Omega^-)} \le C_3 \|g\|_{H^{1/2}(\Gamma)} = C_3 \|\gamma_0^- v^-\|_{H^{1/2}(\Gamma)} \\ &= C_3 \left\|\gamma_0^+ v^+\right\|_{H^{1/2}(\Gamma)} \le C_4 \left\|v^+\right\|_{H^1(-\Delta,\Omega^+)} .\end{aligned}$$

From this we obtain

$$\|v^-\|^2_{H^1(\Omega^-)} \le C_1 B_-(v^-, v^-) + C_2 C_4^2 \left\|v^+\right\|^2_{H^1(-\Delta,\Omega^+)} .$$

If we combine this result with (2.170) we obtain the assertion

$$\begin{aligned}\|v\|_W^2 = \|v^-\|^2_{H^1(\Omega^-)} + \left\|v^+\right\|^2_{H^1(-\Delta,\Omega^+)} &\le C_1 B_-(v^-, v^-) \\ &+ \left(1 + C_2 C_4^2\right) C_0^{-1} B_+\left(v^+, v^+\right) .\end{aligned} \tag{2.171}$$

□

In the case of the general elliptic operator L from (2.98), especially when $L \neq -\Delta$ and $a_{\min} c < \|\mathbf{b}\|^2$, proving the results on existence and uniqueness becomes far more complicated.

2.10.2.3 Helmholtz Equation

The Helmholtz equation in the interior space

$$L_k u = -\Delta u - k^2 u = f \text{ in } \Omega^- \tag{2.172}$$

with Dirichlet or Neumann boundary conditions $\gamma_0^- u = g_D \in H^{1/2}(\Gamma)$ or $\gamma_1^- u = g_N \in H^{-1/2}(\Gamma)$ and with the sesquilinear form $B_-(u,v) = \int_{\Omega^-} (\langle \nabla u, \nabla v\rangle - k^2 uv)\, d\mathbf{x}$ has, according to Theorem 2.10.4.1 and Theorem 2.10.5.1, a unique solution if and only if k^2 is not an eigenvalue of the IDP or INP.

For the Helmholtz equation in the exterior space

$$L_k u := -\Delta u - k^2 u = 0 \quad \text{in } \Omega^+ \tag{2.173}$$

we are looking for solutions that satisfy the Sommerfeld radiation conditions (2.133).

Theorem 2.10.15. *The variational problem (2.150) that is associated with the EDP from (2.173), has a unique solution* $u \in H^1(L_k, \Omega^+)$ *for every* $g_D \in H^{1/2}(\Gamma)$.

For a proof we refer to, e.g., [154, 170]. Note that the additional term $(\partial_r u - iku, \partial_r v - ikv)_{L^2}$ in the sesquilinear form (2.149) represents the analogy to the Sommerfeld radiation condition. We have a corresponding result for the Neumann problem:

Theorem 2.10.16. *The variational problem (2.153) that is associated with the ENP from (2.172) has a unique solution in $H^1(L_k, \Omega^+)$ for every $g_N \in H^{-1/2}(\Gamma)$.*

Chapter 3
Elliptic Boundary Integral Equations

Homogeneous, linear elliptic boundary value problems with constant coefficients can be transformed into boundary integral equations by using the *integral equation method*. In this chapter we will introduce the relevant boundary integral operators and we will derive the most important mapping properties and representations. We will also present the boundary integral equations for the boundary value problems from the previous chapter. Finally, we will prove the appropriate results on existence and uniqueness for these boundary integral equations.

3.1 Boundary Integral Operators

We consider the differential operator L from (2.98)

$$Lu = -\operatorname{div}(\mathbf{A} \operatorname{grad} u) + 2 \langle \mathbf{b}, \operatorname{grad} u \rangle + cu. \tag{3.1}$$

Our goal is to solve the homogeneous differential equation

$$Lu = 0 \qquad \text{in } \Omega \tag{3.2}$$

for this operator with appropriate boundary conditions. Solutions of these differential equations can be formulated with the help of potentials that are closely linked to the fundamental solution of the operator L, which in turn can be formulated explicitly. In general, we assume that the coefficients of L satisfy $\mathbf{A} \in \mathbb{R}^{d\times d}$ positive definite, $\mathbf{b} \in \mathbb{R}^d$ and $c \in \mathbb{R}$. With the help of the matrix $\mathbf{A}$ we can define an inner product and a norm on $\mathbb{R}^d$

$$\langle \mathbf{x}, \mathbf{y} \rangle_{\mathbf{A}} = \mathbf{x}^{\mathsf{T}} \mathbf{A}^{-1} \mathbf{y} \qquad \text{and } \|\mathbf{x}\|_{\mathbf{A}} = \langle \mathbf{x}, \mathbf{x} \rangle_{\mathbf{A}}^{1/2}.$$

We set $\vartheta := c + \|\mathbf{b}\|_{\mathbf{A}}^2$ and $\lambda = \sqrt{\vartheta}$ for $\vartheta \geq 0$ and $\lambda = -i\sqrt{|\vartheta|}$ otherwise. The fundamental solution $G(\mathbf{x} - \mathbf{y})$ then has the following form

S.A. Sauter and C. Schwab, *Boundary Element Methods*, Springer Series in Computational Mathematics 39, DOI 10.1007/978-3-540-68093-2_3,

$$G(\mathbf{z}) = \begin{cases} \dfrac{e^{\langle \mathbf{b},\mathbf{z}\rangle_{\mathbf{A}}}}{2\pi\sqrt{\det \mathbf{A}}} \log \dfrac{1}{\|\mathbf{z}\|_{\mathbf{A}}} & \text{for } d = 2 \text{ and } \lambda = 0, \\[2ex] \dfrac{e^{\langle \mathbf{b},\mathbf{z}\rangle_{\mathbf{A}}}}{4\sqrt{\det \mathbf{A}}} i H_0^{(1)}\left(i\lambda \|\mathbf{z}\|_{\mathbf{A}}\right) & \text{for } d = 2 \text{ and } \lambda \neq 0, \\[2ex] \dfrac{1}{4\pi\sqrt{\det \mathbf{A}}} \dfrac{e^{\langle \mathbf{b},\mathbf{z}\rangle_{\mathbf{A}} - \lambda\|\mathbf{z}\|_{\mathbf{A}}}}{\|\mathbf{z}\|_{\mathbf{A}}} & \text{for } d = 3. \end{cases} \tag{3.3}$$

The function G is singular for $\mathbf{z} = 0$ and analytic for $\mathbf{z} \neq 0$. By choosing $\mathbf{A} = \mathbf{I}$, $\mathbf{b} = \mathbf{0}$ and $c = 0$ we obtain the Laplace operator $L = -\Delta$ and the fundamental solution to the Laplace operator.

With the help of the fundamental solution we can introduce the single layer and double layer potentials for $v \in L^1(\Gamma)$. We recall the notation γ_0 for the trace operator (see Theorem 2.6.8), γ_1 for the conormal derivative (see (2.103) or Definition 2.7.6) and $\widetilde{\gamma_1}$ for the modified conormal derivative (see (2.107) or Definition 2.7.6). In order to explicitly state whether the trace is applied from the interior or exterior domain, we use the indices "−" for the interior and "+" for the exterior domain.

Single Layer Potential:

$$(Sv)(\mathbf{x}) := \int_\Gamma G(\mathbf{x}-\mathbf{y})\, v(\mathbf{y})\, ds_{\mathbf{y}} \qquad \mathbf{x} \in \mathbb{R}^d \backslash \Gamma. \tag{3.4}$$

Double Layer Potential:

$$(Dv)(\mathbf{x}) := \int_\Gamma \widetilde{\gamma_{1,\mathbf{y}}} G(\mathbf{x}-\mathbf{y})\, v(\mathbf{y})\, ds_{\mathbf{y}} \qquad \mathbf{x} \in \mathbb{R}^d \backslash \Gamma, \tag{3.5}$$

where the subscript $\mathbf{y}$ in $\widetilde{\gamma_{1,\mathbf{y}}}$ indicates that the modified conormal derivative $\widetilde{\gamma_1}$ is applied with respect to the $\mathbf{y}$-variable. Since the fundamental solution $G(\mathbf{x}-\mathbf{y})$ is regular for $\mathbf{x} \neq \mathbf{y}$, the single and double layer potentials are both well defined.

Theorem 3.1.1. *Let* $v \in L^1(\Gamma)$.

(a) We have

$$(LSv)(\mathbf{x}) = (LDv)(\mathbf{x}) = 0 \qquad \textit{for all } \mathbf{x} \in \mathbb{R}^d \backslash \Gamma.$$

(b) The functions Sv *and* Dv *are infinitely differentiable in* $\mathbb{R}^d \backslash \Gamma$.

Proof. For (a): We set $k_S(\mathbf{x},\mathbf{y}) := G(\mathbf{x}-\mathbf{y})$ and $k_D(\mathbf{x},\mathbf{y}) := \widetilde{\gamma_{1,\mathbf{y}}} G(\mathbf{x}-\mathbf{y})$. Let $\mathbf{x}_0 \in \mathbb{R}^3 \backslash \Gamma$. Then there exists a compact neighborhood U_0 of $\mathbf{x}$ that is entirely contained in $\Omega \in \{\Omega^-, \Omega^+\}$ and therefore has a positive distance to Γ. The restrictions $k_S, k_D : U_0 \times \Gamma \to \mathbb{C}$ are then bounded and are differentiable for almost every $\mathbf{y} \in \Gamma$ on U_0. For all $\mathbf{x} \in U_0$, k_S and k_D are integrable over Γ. The theorem on dominated convergence then implies that differentiation and integration may be interchanged. $L_{\mathbf{x}}$ denotes the application of L with respect to the $\mathbf{x}$ variable. This gives us the assertion from $L_{\mathbf{x}}G(\mathbf{x}-\mathbf{y}) = L_{\mathbf{x}}\left(\widetilde{\gamma_{1,\mathbf{y}}}G(\mathbf{x}-\mathbf{y})\right) = \widetilde{\gamma_{1,\mathbf{y}}}L_{\mathbf{x}}G(\mathbf{x}-\mathbf{y}) = 0$.

For (b): The statement then follows from repeated application of the arguments given in (a) by means of induction. □

In order to solve problem (3.2) we can therefore consider the *ansatz* Sv or Dv. As a result of Theorem 3.1.1, for every *boundary density* v, this ansatz satisfies the homogeneous differential equation (3.2). Therefore the problem is reduced to the question of whether the *boundary density* v can be determined in such a way that the boundary traces of these potentials satisfy the boundary conditions.

Formally, the boundary integral operators V, K, K', W can be defined by means of the introduced conormal operators γ_0, γ_1^+, γ_1^-. For $\sigma \in \{-,+\}$ we set

$$\begin{aligned} Vv &:= \gamma_0 Sv, \quad K_\sigma\psi := \gamma_0^\sigma D\psi, \\ K'_\sigma\psi &:= \gamma_1^\sigma S\psi, \quad Wu := -\gamma_1(Du). \end{aligned} \tag{3.6}$$

The index + or − indicates that the trace operators $\gamma_0(w)$ and $\gamma_1(w)$ are applied to the restrictions $w|_{\Omega^+}$ and $w|_{\Omega^-}$ respectively. We will show (see Theorem 3.3.1) that we have $\gamma_0^+ Sv = \gamma_0^- Sv$ and $\gamma_1^+ Du = \gamma_1^- Du$, which is why we have already omitted the indices ± in the definition of V and W.

3.1.1 Newton Potential

Before we turn our attention to the mapping properties of the above-mentioned potentials and boundary integral operators, we will consider the converse problem. If the Dirichlet and Neumann data of a function u that satisfies $Lu = f$ are known, it can be formulated in terms of the boundary data and f *explicitly*. The associated formula is called *Green's representation formula*. The derivation of this formula uses the mapping properties of the Newton potential, which in turn are proven by means of a Fourier analysis. Here we will restrict ourselves to an overview of the required properties, and for proofs we refer to [133, 184] and [162, Theorem 6.1].

For a given $f \in L^2_{\text{comp}}\left(\mathbb{R}^d\right)$ we consider the functions $u \in H^1_L\left(\mathbb{R}^d\backslash\Gamma\right)$ [see (2.113)] with

$$L_\pm u = f \qquad \text{in } \mathbb{R}^d\backslash\Gamma \tag{3.7}$$

[see (2.114)]. The Newton potential

$$\mathcal{N}f(\mathbf{x}) := \int_{\mathbb{R}^d} G(\mathbf{x}-\mathbf{y}) f(\mathbf{y})\, d\mathbf{y} \qquad \forall \mathbf{x} \in \mathbb{R}^d \tag{3.8}$$

will play a significant role in the representation of the solutions of this equation. We will first need a few mapping properties of the Newton potential. For functions $f \in C_0^\infty(\mathbb{R}^d)$, $\mathcal{N} f$ is defined as an improper integral: $\mathcal{N} : C_0^\infty(\mathbb{R}^d) \to C^\infty(\mathbb{R}^d)$. We use Fubini's theorem to obtain a representation of the dual operator. The extension of the inner product $(\cdot,\cdot)_{L^2(\mathbb{R}^d)}$ to $L^2_{\text{loc}}(\mathbb{R}^d) \times L^2_{\text{comp}}(\mathbb{R}^d)$ is again denoted by $(\cdot,\cdot)_{L^2(\mathbb{R}^d)}$. For $f, g \in C_0^\infty(\mathbb{R}^d)$ we then have

$$
\begin{aligned}
(\mathcal{N} f, g)_{L^2(\mathbb{R}^d)} &= \int_{\mathbb{R}^d} \left(\int_{\mathbb{R}^d} G(\mathbf{x}-\mathbf{y}) f(\mathbf{y})\, d\mathbf{y} \right) \overline{g(\mathbf{x})}\, d\mathbf{x} \\
&= \int_{\mathbb{R}^d} f(\mathbf{y}) \overline{\left(\int_{\mathbb{R}^d} \overline{G}(\mathbf{x}-\mathbf{y})\, g(\mathbf{y})\, d\mathbf{y} \right)} d\mathbf{x},
\end{aligned}
$$

and thus we obtain the representation

$$
(\mathcal{N}' g)(\mathbf{y}) = \int_{\mathbb{R}^d} \overline{G}(\mathbf{x}-\mathbf{y})\, g(\mathbf{x})\, d\mathbf{x} \qquad \forall \mathbf{y} \in \mathbb{R}^d. \tag{3.9}
$$

The mapping property $\mathcal{N}' : C_0^\infty(\mathbb{R}^d) \to C^\infty(\mathbb{R}^d)$ of the dual Newton potential can be shown in the same way. The domain of the Newton potential can be extended to functionals $f \in (C^\infty(\mathbb{R}^d))'$ by means of the dual mapping. The functional $\mathcal{N} f \in (C_0^\infty(\mathbb{R}^d))'$ is characterized by

$$
(\mathcal{N} f, v)_{L^2(\mathbb{R}^d)} = (f, \mathcal{N}' v)_{L^2(\mathbb{R}^d)} \qquad \forall v \in C_0^\infty\left(\mathbb{R}^d\right).
$$

The Newton potential can also be defined for functions in Sobolev spaces (see [216, Sect. 6.1]).

Theorem 3.1.2. *For the Newton potential, the mapping*

$$
\mathcal{N} : H^s_{\text{comp}}\left(\mathbb{R}^d\right) \to H^{s+2}_{\text{loc}}\left(\mathbb{R}^d\right)
$$

is continuous for all $s \in \mathbb{R}$.

Remark 3.1.3. *Problems in acoustics and electromagnetism can often be described by the Helmholtz equation with the operator*

$$
L_k u = -\Delta u - k^2 u, \qquad k \in \mathbb{R}.
$$

The associated fundamental solution [see (3.3)] for $d = 3$ *is given by*

$$
G_k(\mathbf{z}) = \frac{e^{ik\|\mathbf{z}\|}}{4\pi \|\mathbf{z}\|},
$$

and the associated Newton potential is denoted by $\mathcal{N}_k$. *Then* $\mathcal{N}_0$ *is the Newton potential (Coulomb potential) for the Laplace operator. The expansion*

$$G_k(\mathbf{z}) = G_k^{\mathrm{I}}(\mathbf{z}) + G_k^{\mathrm{II}}(\mathbf{z}) + G_k^{\mathrm{III}}(\mathbf{z}) := \frac{1}{4\pi\|\mathbf{z}\|} + \frac{ik}{4\pi} + O(k^2\|\mathbf{z}\|)$$

shows that the kernel $G_k - G_0$ is continuous at $\|\mathbf{z}\| = 0$ and has bounded derivatives that are discontinuous in $\mathbf{z} = \mathbf{0}$. With the help of the calculus of pseudo-differential operators we obtain the mapping property

$$\mathcal{N}_k - \mathcal{N}_0 : H^s_{\mathrm{comp}}(\mathbb{R}^3) \to H^{s+4}_{\mathrm{loc}}(\mathbb{R}^3) \qquad \forall s \in \mathbb{R}$$

by reasoning as follows. From [137, Definition 7.1.1 and (7.1.2)], we conclude that G_k^{III} has a pseudohomogeneous expansion of degree 1 so that, according to [137, Theorem 7.1.1], the associated volume potential $\mathcal{N}_k^{\mathrm{III}} u = \int_{\mathbb{R}^3} G_k^{\mathrm{III}}(\cdot - y)\, u(y)\, dy$ belongs to $\mathcal{L}^{-4}_{\mathrm{cl}}(\mathbb{R}^3)$, i.e., to the classical symbol class as defined, e.g., in [137, Definition 6.1.6]. From [137, Definition 6.1.12], we conclude that $\mathcal{N}_k^{\mathrm{III}} : H^s_{\mathrm{comp}}(\mathbb{R}^3) \to H^{s+4}_{\mathrm{loc}}(\mathbb{R}^3)$ is continuous for all $s \in \mathbb{R}$. For the volume potential associated to $G_k^{\mathrm{II}}(\mathbf{z})$ this mapping property follows trivially because the kernel function is constant.

The formal adjoint operator $L^* : C^\infty(\mathbb{R}^d) \to C^\infty(\mathbb{R}^d)$ from (2.105) satisfies

$$(Lu, v)_{L^2(\mathbb{R}^d)} = (u, L^* v)_{L^2(\mathbb{R}^d)} \qquad \forall u \in C_0^\infty\left(\mathbb{R}^d\right) \quad \forall v \in C^\infty\left(\mathbb{R}^d\right).$$

Thus the domain of L can be extended to $(C^\infty(\mathbb{R}^d))'$ as well as $(C_0^\infty(\mathbb{R}^d))'$:

$$(Lf, g)_{L^2(\mathbb{R}^d)} := (f, L^* g)_{L^2(\mathbb{R}^d)} \qquad \forall f \in \left(C^\infty\left(\mathbb{R}^d\right)\right)' \quad \forall g \in C^\infty\left(\mathbb{R}^d\right),$$
$$(Lf, g)_{L^2(\mathbb{R}^d)} := (f, L^* g)_{L^2(\mathbb{R}^d)} \qquad \forall f \in \left(C_0^\infty\left(\mathbb{R}^d\right)\right)' \quad \forall g \in C_0^\infty\left(\mathbb{R}^d\right).$$

The following theorem shows that the Newton potential constitutes a right and left inverse of the operator L.

Theorem 3.1.4. *For all functionals $u \in (C^\infty(\mathbb{R}^d))'$ we have*

$$L\mathcal{N}u = u = \mathcal{N}Lu \qquad \text{in } \left(C_0^\infty\left(\mathbb{R}^d\right)\right)'.$$

The explicit representations (3.4) and (3.5) of the operators S and D are only suited to locally integrable functions $v \in L^1(\Gamma)$. The domain of the single and double layer potential can be substantially enlarged.

Definition 3.1.5. The single layer potential S and the double layer potential D are given by

$$S := \mathcal{N}\gamma_0', \qquad D := \mathcal{N}\widetilde{\gamma_1}'.$$

Theorem 3.1.6 deals with the connection between the abstract Definition 3.1.5 and the explicit representations (3.4) and (3.5). The jumps $[u]$, $[\gamma_1 u]$ of a function $u \in H^1_L(\mathbb{R}^d \backslash \Gamma)$ across Γ were introduced in (2.119).

Theorem 3.1.6. *(a) For functions $u \in H_L^1\left(\mathbb{R}^d\backslash\Gamma\right)$ with compact support and $f = L_\pm u$ [see (3.7)] we have Green's representation formula*

$$u = \mathcal{N}f - S\left([\gamma_1 u]\right) + D\left([u]\right) \tag{3.10}$$

as a functional on $C_0^\infty\left(\mathbb{R}^d\right)$.

(b) The operators $S = \mathcal{N}\gamma_0'$ and $D = \mathcal{N}\widetilde{\gamma_1}'$ have the representations (3.4) and (3.5) for $u \in L^1(\Gamma)$ on $\mathbb{R}^d\backslash\Gamma$.

Proof. Functions $u \in H_L^1\left(\mathbb{R}^d\backslash\Gamma\right)$ with compact support can be interpreted as functionals on $C^\infty\left(\mathbb{R}^d\right)$ according to

$$U := (u, \cdot)_{L^2(\mathbb{R}^d)}\,.$$

Applying Theorem 3.1.4, we then have

$$\begin{aligned}(L\mathcal{N}u, v)_{L^2(\mathbb{R}^d)} &:= \left(u, \mathcal{N}'L'v\right)_{L^2(\mathbb{R}^d)} = U\left(\mathcal{N}'L'v\right) = (L\mathcal{N}U)(v)\\ &= U(v) = (u, v)_{L^2(\mathbb{R}^d)}\end{aligned}$$

for all $v \in C^\infty\left(\mathbb{R}^d\right)$ and subsequently the equality $L\mathcal{N}u = u$ in the sense of a functional on $C^\infty\left(\mathbb{R}^d\right)$. $\mathcal{N}Lu = u$ can be shown analogously.

The operator $\mathcal{N}$ can then be applied to Green's third formula (2.122). With Theorem 3.1.4 we then obtain the representation

$$u = \mathcal{N}f - \mathcal{N}\gamma_0'\left([\gamma_1 u]\right) + \mathcal{N}\widetilde{\gamma_1}'\left([u]\right).$$

For $u \in C^\infty\left(\mathbb{R}^d\right)$ we have $[\gamma_1 u] \in L^1(\Gamma)$. Under these conditions we will show that for $\mathbf{x} \in \mathbb{R}^d\backslash\Gamma$ we have the representation (3.4) for $\mathcal{N}\gamma_0'$. In the following, let S and D be defined by the right-hand sides in (3.4) and (3.5). The representation (3.9) gives us, along with Fubini's theorem, for all $v \in L^1(\Gamma)$ and $w \in C_0^\infty\left(\mathbb{R}^d\right)$

$$\begin{aligned}\left(\mathcal{N}\gamma_0' v, w\right)_{L^2(\mathbb{R}^d)} &= \left(v, \gamma_0\mathcal{N}'w\right)_{L^2(\Gamma)} = \int_\Gamma v(\mathbf{y})\overline{\left(\int_{\mathbb{R}^d}\overline{G}(\mathbf{x}-\mathbf{y})\,w(\mathbf{x})\,d\mathbf{x}\right)}ds_{\mathbf{y}}\\ &= \int_{\mathbb{R}^d}\overline{w}(\mathbf{x})\left(\int_\Gamma v(\mathbf{y})\,G(\mathbf{x}-\mathbf{y})\,ds_{\mathbf{y}}\right)d\mathbf{x} = (Sv, w)_{L^2(\mathbb{R}^d)}\,.\end{aligned}$$

Let $\mathbf{x} \in \Omega^+$ and $\mathcal{U} \subset \Omega^+$ be an arbitrary, compact neighborhood of $\mathbf{x}$. Then $Sv|_{\mathcal{U}} \in C^\infty(\mathcal{U})$ (see Theorem 3.1.1). Since the restriction of $C_0^\infty\left(\mathbb{R}^d\right)$ on $\mathcal{U}$ is dense in $L^2(\mathcal{U})$, we have the equality $\mathcal{N}\gamma_0' = S$ in $\mathbf{x}$. The assertion for $\mathbf{x} \in \Omega^-$ can be shown analogously.

In order to prove the representation (3.5) for $\mathcal{N}\widetilde{\gamma_1}'$ for $\mathbf{x} \in \mathbb{R}^d\backslash\Gamma$, we again first consider the case $\mathbf{x} \in \Omega^+$ and a compact neighborhood $\mathcal{U} \subset \Omega^+$ of $\mathbf{x}$. Let $\chi \in C_0^\infty\left(\mathbb{R}^d\right)$ be a test function with supp $\chi \subset \Omega^+$ and $\chi \equiv 1$ on $\mathcal{U}$. For $v \in L^1(\Gamma)$ and $w \in C_0^\infty\left(\mathbb{R}^d\right)$ we obtain

$$\left(\mathcal{N}\widetilde{\gamma_1}'v, \chi w\right)_{L^2(\mathbb{R}^d)} = \left(v, \widetilde{\gamma_1}\mathcal{N}'\chi w\right)_{L^2(\Gamma)}$$
$$= \int_\Gamma v(\mathbf{y}) \overline{\left(\widetilde{\gamma_{1,\mathbf{y}}}\left(\int_{\operatorname{supp}\chi} \overline{G}(\mathbf{x}-\mathbf{y})\,\chi w(\mathbf{x})\,d\mathbf{x}\right)\right)}\,ds_{\mathbf{y}}.$$

Since the distance between (supp χ) and Γ is positive, the kernel function $\overline{G}(\mathbf{x}-\mathbf{y})$ is smooth and the differentiation and integration can be interchanged. Fubini's theorem then gives us

$$\left(\mathcal{N}\widetilde{\gamma_1}'v, \chi w\right)_{L^2(\mathbb{R}^d)} = \int_{\operatorname{supp}\chi} \overline{\chi w}(\mathbf{x}) \left(\int_\Gamma v(\mathbf{y}) \left(\widetilde{\gamma_{1,\mathbf{y}}} G(\mathbf{x}-\mathbf{y})\right) ds_{\mathbf{y}}\right) d\mathbf{x}$$
$$= (Dv, \chi w)_{L^2(\mathbb{R}^d)}.$$

We obtain the equality $\mathcal{N}\widetilde{\gamma_1}'v = Dv$ on Ω^+ because the restriction $\chi C_0^\infty(\mathbb{R}^d)$ on $\mathcal{U}$ coincides with the restriction $C_0^\infty(\mathbb{R}^d)\big|_{\mathcal{U}}$, which is itself dense in $L^2(\mathcal{U})$. □

We will generalize Theorem 3.1.1 by proving that $L_\pm Sv \equiv 0$ for all $v \in H^{-1/2}(\Gamma)$.

Proposition 3.1.7. *Let* $-1/2 < s < 1/2$ *and* $v \in H^{-1/2+s}(\Gamma)$. *Then* $LSv \equiv 0$ *on* $\mathbb{R}^d \backslash \Gamma$.

Proof. We use Definition 3.1.5 and obtain

$$LSv = L\mathcal{N}\gamma_0'v.$$

The mapping properties of the trace operator $\gamma_0 : H_{\mathrm{loc}}^{1+s}(\mathbb{R}^d) \to H^{1/2+s}(\Gamma)$ imply the continuity of the dual operator

$$\gamma_0' : H^{-1/2-s}(\Gamma) \to H_{\mathrm{comp}}^{-1-s}(\mathbb{R}^d) = \left(H_{\mathrm{loc}}^{1+s}(\mathbb{R}^d)\right)' \subset \left(C^\infty(\mathbb{R}^d)\right)'.$$

Therefore Theorem 3.1.4 becomes applicable and we obtain

$$LSv = \gamma_0'v \tag{3.11}$$

in the sense of a functional on $C^\infty(\mathbb{R}^d)$. Let $\psi \in C^\infty(\mathbb{R}^d\backslash\Gamma)$ with $\operatorname{supp}\psi \subset \mathbb{R}^d\backslash\Gamma$. Without loss of generality we assume that $\operatorname{supp}\psi \subset \Omega^+$. From this we have $\gamma_0\psi \equiv 0$ and

$$\left(\gamma_0'v, \psi\right)_{L^2(\mathbb{R}^d\backslash\Gamma)} = (v, \gamma_0\psi)_{L^2(\Gamma)} = 0$$

and as a consequence $LS \equiv 0$ on $\mathbb{R}^d\backslash\Gamma$. □

Green's representation formula (3.10) was shown for functions u with compact support. We will prove a modified form of Green's representation formula for functions that satisfy the characteristic physical decay condition but do not have compact support. We will restrict ourselves to functions that satisfy $Lu = 0$ in $\mathbb{R}^3 \backslash \Gamma$. In the following section we will anticipate a result and use the mapping properties of the potentials S and D from Theorem 3.1.16:

$$S : H^{-1/2}(\Gamma) \to H^1_{\text{loc}}(\Gamma) \quad \text{and} \quad D : H^{1/2}(\Gamma) \to H^1_L\left(\mathbb{R}^d \backslash \Gamma\right) \text{ are continuous.} \tag{3.12}$$

First, we will choose a sufficiently large $a > 0$ with $\overline{\Omega^-} \subset B_a$. Let $u \in H^1_L\left(\mathbb{R}^d \backslash \Gamma\right)$ with $Lu \equiv 0$ in $\mathbb{R}^d \backslash \{\Gamma\}$. For the boundary of the intersection of the domains $\Omega_a := \Omega^+ \cap B_a$ we have $\partial\Omega_a = \Gamma \cup \Gamma_a$ with $\Gamma_a := \partial B_a$. Let the normals on Γ again point in the direction of Ω^+ and those on Γ_a in the direction of $B_a^+ := \mathbb{R}^d \backslash \overline{B_a}$. The function $u_a := u$ in B_a and $u_a \equiv 0$ in B_a^+ satisfies $u_a \in H^1_L\left(\mathbb{R}^d \backslash \partial\Omega_a\right)$ and has compact support. Therefore we can apply Green's representation formula (3.10), which gives us

$$\begin{aligned} u &= -S\left[\gamma_1 u\right] + D\left[u\right] + v \quad \text{in } B_a \backslash \Gamma, \\ 0 &= -S\left[\gamma_1 u\right] + D\left[u\right] + v \quad \text{in } B_a^+ \end{aligned} \tag{3.13}$$

with

$$v := S_a\left(\left(\gamma_1 u\right)|_{\Gamma_a}\right) - D_a\left(\left(\gamma_0 u\right)|_{\Gamma_a}\right) \qquad \text{in } B_a \cup B_a^+. \tag{3.14}$$

Here, in (3.14), S_a and D_a denote the single and the double layer potentials for Γ_a, while S and D in (3.13) denote those for Γ. We define

$$w(x) := \begin{cases} v(x) & \text{in } B_a, \\ v(x) + u(x) & \text{in } B_a^+. \end{cases} \tag{3.15}$$

Combining the first equation in (3.13) with the first equation in (3.15) gives us

$$w = u + S\left[\gamma_1 u\right] - D\left[u\right] \qquad \text{in } B_a \backslash \Gamma. \tag{3.16}$$

The mapping properties of S and D [see (3.12)], the boundedness of B_a and Proposition 3.1.7 imply that

$$w|_{B_a} = v|_{B_a} \in H^1\left(B_a\right) \quad \text{and} \quad Lw \equiv 0 \quad \text{in } B_a. \tag{3.17}$$

Combining the second equation in (3.13) and (3.15) gives us, together with (3.16),

$$w = u + S\left[\gamma_1 u\right] - D\left[u\right] \qquad \text{in } \mathbb{R}^d \backslash \partial\Omega_a. \tag{3.18}$$

It follows from (3.18) together with the mapping properties of S and D that $w|_{\Omega^+} \in H^1_{\text{loc}}\left(\Omega^+\right)$ and $Lw = 0$ in Ω^+. With (3.17) this gives us $w \in H^1_{\text{loc}}\left(\mathbb{R}^d\right)$ and $Lw = 0$ in $\mathbb{R}^d$. These ideas are summarized in the following theorem.

Theorem 3.1.8. *Let $u \in H_L^1\left(\mathbb{R}^d \backslash \Gamma\right)$ with $Lu \equiv 0$ in $\Omega^- \cup \Omega^+$. Then*

$$u = -S\left[\gamma_1 u\right] + D\left[\gamma_0 u\right] + w \qquad \text{in } \Omega^- \cup \Omega^+ \tag{3.19}$$

with an L-harmonic function $w \in H_{\mathrm{loc}}^1\left(\mathbb{R}^d\right)$.

Theorem 3.1.8 generalizes Green's representation formula so that it applies to functions that have unbounded support. However, the space $H_{\mathrm{loc}}^1\left(\Omega^+\right)$ also contains functions with an unphysical behavior for $\|\mathbf{x}\| \to \infty$. Ideally, for $\|\mathbf{x}\| \to \infty$ the required behavior of u at infinity should imply $w \equiv 0$. For such functions u Green's representation formula remains valid unchanged. We will carry out these ideas and use the Laplace and Helmholtz operators as examples. We will also consider the operator L under the condition that the coefficients satisfy $a_{\min} c > \|\mathbf{b}\|^2$. Here $a_{\min}$ again denotes the smallest eigenvalue of the matrix $\mathbf{A}$ [see (2.99)].

Lemma 3.1.9. *Let $d = 3$ and $a_{\min} c > \|\mathbf{b}\|^2$. Then, for all $\varphi \in H^{-1/2}(\Gamma)$ and $\psi \in H^{1/2}(\Gamma)$, there exist positive constants C_1, C_2 such that*

$$|S\varphi(\mathbf{x})| + |D\psi(\mathbf{x})| + \|\nabla(S\varphi)(\mathbf{x})\| + \|\nabla(D\varphi)(\mathbf{x})\| \leq C_1 e^{-C_2\|\mathbf{x}\|}$$

for all $\mathbf{x} \in \mathbb{R}^3$ with $\|\mathbf{x}\| \geq a$. Here $a > 0$ is chosen so that $\Gamma \subset\subset B_a$ and

$$\inf_{(\mathbf{x},\mathbf{y}) \in \Gamma \times \partial B_a} \|\mathbf{x} - \mathbf{y}\| \geq 1. \tag{3.20}$$

Proof. It follows from $a_{\min} c > \|\mathbf{b}\|^2$ that we have for the exponent of the fundamental solution in (3.3)

$$\langle \mathbf{b}, \mathbf{z}\rangle_{\mathbf{A}} - \lambda\|\mathbf{z}\|_{\mathbf{A}} \leq \|\mathbf{b}\|_{\mathbf{A}}\|\mathbf{z}\|_{\mathbf{A}} - \|\mathbf{z}\|_{\mathbf{A}}\sqrt{c + \|\mathbf{b}\|_{\mathbf{A}}^2} \leq -\gamma\|\mathbf{z}\|$$

with

$$\gamma := \left(\sqrt{c + \|\mathbf{b}\|_{\mathbf{A}}^2} - \|\mathbf{b}\|_{\mathbf{A}}\right) / \sqrt{a_{\max}} > 0.$$

From this we have for the fundamental solution under consideration

$$|G(\mathbf{z})| \leq C_2 e^{-\gamma\|\mathbf{z}\|} / \|\mathbf{z}\|.$$

This gives us the estimate

$$\begin{aligned} |G(\mathbf{x}-\mathbf{y})| &\leq C_2 e^{-\gamma\|\mathbf{x}-\mathbf{y}\|} = C_2 e^{\gamma(\|\mathbf{x}\| - \|\mathbf{x}-\mathbf{y}\|)} e^{-\gamma\|\mathbf{x}\|} \leq C_2\left(\max_{\mathbf{y}\in\Gamma} e^{\gamma\|\mathbf{y}\|}\right) e^{-\gamma\|\mathbf{x}\|} \\ &=: C_3 e^{-\gamma\|\mathbf{x}\|} \end{aligned}$$

for all $\mathbf{x} \in \mathbb{R}^3 \backslash B_a$ and $\mathbf{y} \in \Gamma$. Now let $\varphi \in H^{-1/2}(\Gamma)$ and $\|\mathbf{x}\| \geq a$. Then

$$
\begin{aligned}
|S\varphi(\mathbf{x})| &\le \int_\Gamma |G(\mathbf{x}-\mathbf{y})|\,|\varphi(\mathbf{y})|\,ds_\mathbf{y} \le C_3 e^{-\gamma\|\mathbf{x}\|}\int_\Gamma |\varphi(\mathbf{y})|\,ds_\mathbf{y} \\
&\le C_3\|1\|_{H^{1/2}(\Gamma)}e^{-\gamma\|\mathbf{x}\|}\|\varphi\|_{H^{-1/2}(\Gamma)} = C_4 e^{-\gamma\|\mathbf{x}\|}\|\varphi\|_{H^{-1/2}(\Gamma)}.
\end{aligned}
$$

The result concerning the double layer potential and the gradients of the potentials can be proven analogously. □

The following lemma shows that L-harmonic functions on $\mathbb{R}^3$ are always polynomials for the coefficients under consideration.

Lemma 3.1.10. *Let $d = 3$ and $a_{\min}c > \|\mathbf{b}\|^2$ or $c = \|\mathbf{b}\| = 0$. Then every $w \in H^1_{\mathrm{loc}}(\mathbb{R}^3)$ with $Lw \equiv 0$ on $\mathbb{R}^3$ is a polynomial. If $c = \|\mathbf{b}\| = 0$ the statement remains valid for the space $H^1(L,\mathbb{R}^3)$.*

The assertion of this lemma follows from [80, Chap. XI, Part B, Sect. 2, Theorem 1].

Theorem 3.1.11. *Let $d = 3$ and $a_{\min}c > \|\mathbf{b}\|^2$. Let the function $u \in H^1(\mathbb{R}^3\backslash\Gamma)$ satisfy $Lu = 0$ in $\Omega^- \cup \Omega^+$. Then the representation formula (3.19) holds with $w \equiv 0$.*

Proof. The mapping properties of S, D imply $S[\gamma_1 u], D[u] \in H^1_{\mathrm{loc}}(\mathbb{R}^3\backslash\Gamma)$. Lemma 3.1.9 gives us the stronger statement $S[\gamma_1 u]$, $D[u] \in H^1(\mathbb{R}^3\backslash\Gamma)$. According to the conditions, the left-hand side in (3.19) is in $H^1(\mathbb{R}^3\backslash\Gamma)$ and therefore the right-hand side is also in $H^1(\mathbb{R}^3\backslash\Gamma)$. Since the only polynomial for which $w \in H^1(\mathbb{R}^3)$ is the zero polynomial, we have $w \equiv 0$. □

Theorem 3.1.12. *Let $d = 3$ and $L = -\Delta$. Let the function $u \in H^1(\Omega^-) \times H^1(-\Delta,\Omega^+)$ satisfy $\Delta u = 0$ in $\Omega^+ \cup \Omega^-$. Then the representation formula holds with $w = 0$.*

Proof. Choose a as in (3.20). For $\|\mathbf{b}\| = c = 0$, the fundamental solution and its derivatives satisfy the inequalities

$$
\left.\begin{aligned}
|G(\mathbf{z})| &\le C_1\|\mathbf{z}\|^{-1} \\
\|\nabla G(\mathbf{z})\| &\le C_1\|\mathbf{z}\|^{-2}
\end{aligned}\right\} \qquad \forall\mathbf{z} \in \mathbb{R}^3\backslash\{0\}.
$$

It follows for all $\mathbf{y} \in \Gamma$ and $\|\mathbf{x}\| \ge a$ with $C_\Gamma := \max_{\mathbf{y}\in\Gamma}(1+\|\mathbf{y}\|)$ that the following inequalities hold

$$
\begin{aligned}
|G(\mathbf{x}-\mathbf{y})| &\le C_1\|\mathbf{x}-\mathbf{y}\|^{-1} \le C_1 C_\Gamma\|\mathbf{x}\|^{-1} \\
\|\nabla_\mathbf{x} G(\mathbf{x}-\mathbf{y})\| &\le C_1 C_\Gamma^2\|\mathbf{x}\|^{-2}, \\
\left|\partial G(\mathbf{x}-\mathbf{y})/\partial\mathbf{n}_\mathbf{y}\right| &\le C_1\|\mathbf{x}-\mathbf{y}\|^{-2} \le C_1 C_\Gamma^2\|\mathbf{x}\|^{-2}, \\
\left\|\nabla_\mathbf{x}\partial G(\mathbf{x}-\mathbf{y})/\partial\mathbf{n}_\mathbf{y}\right\| &\le C_2\|\mathbf{x}-\mathbf{y}\|^{-3} \le C_2 C_\Gamma^3\|\mathbf{x}\|^{-3}.
\end{aligned} \tag{3.21}
$$

From this one deduces, as in the proof of Lemma 3.1.9, that for all $\varphi \in H^{-1/2}(\Gamma)$ and $\|\mathbf{x}\| \ge a$

$$\begin{aligned}|S\varphi(\mathbf{x})| &\le \int_\Gamma |G(\mathbf{x}-\mathbf{y})|\,|\varphi(\mathbf{y})|\,ds_\mathbf{y} \le C_1 C_\Gamma \frac{1}{\|\mathbf{x}\|}\int_\Gamma |\varphi(\mathbf{y})|\,ds_\mathbf{y} \\ &\le C_3\,\|\mathbf{x}\|^{-1}\,\|\varphi\|_{H^{-1/2}(\Gamma)}\,. \end{aligned} \tag{3.22}$$

In a similar way one can prove for all $\varphi \in H^{-1/2}(\Gamma)$, $\psi \in H^{1/2}(\Gamma)$ and $\|\mathbf{x}\| \ge a$ the inequalities

$$\begin{aligned}\|\nabla(S\varphi)(\mathbf{x})\| &\le C_4\,\|\mathbf{x}\|^{-2}\,\|\varphi\|_{H^{-1/2}(\Gamma)}\,,\\ |D\psi(\mathbf{x})| &\le C_5\,\|\mathbf{x}\|^{-2}\,\|\psi\|_{H^{1/2}(\Gamma)}\,,\\ \|\nabla(D\psi)(\mathbf{x})\| &\le C_6\,\|\mathbf{x}\|^{-3}\,\|\psi\|_{H^{1/2}(\Gamma)}\,.\end{aligned} \tag{3.23}$$

Note that the constants $C_1,\dots C_6$ are independent of a. Theorem 3.1.16 implies [see (3.12)] that $S\varphi, D\psi \in H^1_{\mathrm{loc}}(\mathbb{R}^3\backslash\Gamma)$. From the boundedness of Ω^- and $\Omega_a := \Omega^+ \cap B_a$ one deduces

$$S\varphi|_{B_a\backslash\Gamma}\,,\ D\psi|_{B_a\backslash\Gamma} \in H^1(B_a\backslash\Gamma)\,.$$

Let $B_a^+ := \mathbb{R}^3\backslash\overline{B_a}$. If one combines the inequalities (3.22), (3.23) with the definition of the $H^1(L, B_a^+)$-norm, it follows that (see Exercise 3.1.14)

$$S\varphi|_{B_a^+} \in H^1(L, B_a^+) \quad \text{and} \quad D\psi|_{B_a^+} \in H^1(L, B_a^+)\,.$$

Finally, by the equivalence of the norms in $H^1(\Omega_a)$ and $H^1(L,\Omega_a)$ on the bounded domain Ω_a one obtains the property

$$S\varphi, D\psi \in H^1(\Omega^-) \times H^1(L,\Omega^+)\,. \tag{3.24}$$

The condition $u \in H^1(\Omega^-)\times H^1(L,\Omega^+)$ combined with (3.24), Theorem 3.1.8 and Lemma 3.1.10 gives us that the L-harmonic function w is a polynomial with $w \in H^1(\Omega^-) \times H^1(L,\Omega^+)$. Hence $w \equiv 0$. □

Theorem 3.1.13. *Let $d = 3$ and $Lu := -\Delta u - k^2 u$ with a positive wave number*[1] *$k > 0$. Let the space $H^1(L,\Omega)$ be defined as in Remark 2.9.3. Let the function $u \in H^1(\Omega^-) \times H^1(L,\Omega^+)$ satisfy $Lu = 0$ in $\Omega^+ \cup \Omega^-$. Then the representation formula holds with $w = 0$.*

Proof. The statement follows from [80, Chap. XI, Part B, Sect. 3], as no plain wave $e^{i\langle\mathbf{k},\mathbf{x}\rangle}$ with $\|\mathbf{k}\| = k$ is contained in $H^1(L,\Omega^+)$. □

Exercise 3.1.14. *Let $\varphi \in H^{-1/2}(\Gamma)$ and $\psi \in H^{1/2}(\Gamma)$. Show that the single and double layer potentials for the Laplace problem satisfy*

$$S\varphi \in H^1(L, B_a^+) \quad \textit{and} \quad D\psi \in H^1(L, B_a^+)\,.$$

Exercise 3.1.15. *Let $\varphi \in H^{-1/2}(\Gamma)$ and $\psi \in H^{1/2}(\Gamma)$, and let S be the single and D the double layer potential for the Helmholtz problem. Show that $S\varphi$ and $D\psi$ then satisfy the Sommerfeld radiation conditions (2.133).*

[1] The *wave number* is a scalar quantity which characterizes the oscillatory behavior of time periodic waves. It is proportional to the reciprocal of the wave length.

3.1.2 Mapping Properties of the Boundary Integral Operators

In this section we will derive the mapping properties of the potentials and the boundary integral operators. The definitions of S, D, V, $K_{\pm}$, $K'_{\pm}$, W can be found in Definition 3.1.5 and (3.6).

Theorem 3.1.16. *Let $\Omega \subset \mathbb{R}^3$ be a bounded Lipschitz domain with boundary $\Gamma := \partial\Omega$. The operators S, D, V, K_+, K_-, K'_+, K'_- and W are continuous for $|s| < 1/2$:*

(i) $S : H^{-1/2+s}(\Gamma) \to H^{1+s}_{\mathrm{loc}}(\mathbb{R}^3)$.
(ii) $D : H^{1/2+s}(\Gamma) \to H^{1+s}_L(\mathbb{R}^3 \backslash \Gamma)$.
(iii) $V : H^{-1/2+s}(\Gamma) \to H^{1/2+s}(\Gamma)$.
(iv) $\sigma \in \{-,+\}$: $\quad K_\sigma : H^{1/2+s}(\Gamma) \to H^{1/2+s}(\Gamma)$.
(v) $\sigma \in \{-,+\}$: $\quad K'_\sigma : H^{-1/2+s}(\Gamma) \to H^{-1/2+s}(\Gamma)$.
(vi) $W : H^{1/2+s}(\Gamma) \to H^{-1/2+s}(\Gamma)$.

Proof. By Definition 3.1.5, finding the mapping properties of S reduces to finding the mapping properties of $\mathcal{N}$ and γ'_0. The trace theorem implies the continuity of $\gamma_0 : H^{1-s}_{\mathrm{loc}}(\mathbb{R}^3) \to H^{1/2-s}(\Gamma)$ for $|s| < 1/2$. This in turn implies the continuity of the dual operator $\gamma'_0 : H^{s-1/2}(\Gamma) \to H^{-1+s}_{\mathrm{comp}}(\mathbb{R}^3)$. Combined with Theorem 3.1.2 this gives us

$$S : H^{s-1/2}(\Gamma) \to H^{1+s}_{\mathrm{loc}}(\mathbb{R}^3).$$

The mapping properties of the operator V follow directly from the mapping properties of the trace operator $\gamma_0 : H^{1+s}_{\mathrm{loc}}(\mathbb{R}^3) \to H^{s+1/2}(\Gamma)$.

We now consider the double layer potential. This case can be reduced to the previous case by representing the double layer potential in terms of the single layer potential. We use the solution operator T from Sect. 2.8 for the interior problem and define, for given boundary data $v \in H^{1/2+s}(\Gamma)$, the function $u \in H^1_L(\mathbb{R}^3 \backslash \Gamma)$ by

$$u := \begin{cases} Tv \text{ in } \Omega^-, \\ 0 \ \text{ in } \Omega^+. \end{cases}$$

Note that we have for the jumps of u and $\gamma_1 u$ across Γ

$$[u] = u^+ - u^- = -v \quad \text{and} \quad [\gamma_1 u] = \gamma_1^+ u - \gamma_1^- u = -\gamma_1^- u.$$

We define $f \in L^2_{\mathrm{comp}}(\mathbb{R}^3)$ by

$$f := L_{\pm} u = \begin{cases} -\lambda T v & \text{in } \Omega^-, \\ 0 & \text{in } \Omega^+. \end{cases}$$

Green's formula (3.10) can be applied, as u has compact support, which gives us the relation

$$u = \mathcal{N} f + S \gamma_1^- u - D v.$$

If we solve for Dv we obtain

$$Dv = \mathcal{N}\begin{pmatrix} -\lambda T v \\ 0 \end{pmatrix} + S\left(\gamma_1^- T\right) v - \begin{pmatrix} T v \\ 0 \end{pmatrix}, \tag{3.25}$$

where $(v_-, v_+)^\intercal$ is an abbreviation for $v_-\chi_- + v_+\chi_+$ with the characteristic functions χ_-, χ_+ for the domains Ω^-, Ω^+. The mapping properties of $S, \mathcal{N}$ and $\gamma_1^- T$ (see Theorem 2.8.2) imply that

$$\begin{aligned} &H^{1/2+s}(\Gamma) \overset{\binom{-\lambda T}{0}}{\to} L^2_{\mathrm{comp}}\left(\mathbb{R}^3\right) \overset{\mathcal{N}}{\to} H^2_{\mathrm{loc}}\left(\mathbb{R}^3\right), \\ &H^{1/2+s}(\Gamma) \overset{\gamma_1^- T}{\longrightarrow} H^{-1/2+s}(\Gamma) \overset{S}{\to} H^{1+s}_{\mathrm{loc}}\left(\mathbb{R}^3\right), \\ &H^{1/2+s}(\Gamma) \overset{\binom{T}{0}}{\to} H^{1+s}\left(\mathbb{R}^3 \backslash \Gamma\right). \end{aligned}$$

Combined this gives us $D : H^{1/2+s}(\Gamma) \to H^{1+s}_{\mathrm{loc}}\left(\mathbb{R}^3\backslash\Gamma\right)$. By Proposition 3.1.7 one obtains $S : H^{-1/2+s}(\Gamma) \to H^{1+s}_L\left(\mathbb{R}^3\backslash\Gamma\right)$. Let $\widetilde{L}$ be as in (2.123). From $\widetilde{L}T \equiv 0$ in Ω^- it follows that $(T, 0)^\intercal : H^{1/2+s}(\Gamma) \to H^{1+s}_L\left(\mathbb{R}^3\backslash\Gamma\right)$, which proves that $D : H^{1/2+s}(\Gamma) \to H^{1+s}_L\left(\mathbb{R}^3\backslash\Gamma\right)$.

The continuity of the operators $K_\pm$, $K'_\pm$, W follows from the continuity of the trace operators $\gamma_0^\pm : H^{1+s}_{\mathrm{loc}}\left(\Omega^\pm\right) \to H^{s+1/2}(\Gamma)$ and $\gamma_1^\pm : H^{1+s}_L\left(\Omega^\pm\right) \to H^{s-1/2}(\Gamma)$ (see Theorem 2.6.8 and Theorem 2.8.3) combined with the mapping properties of S and D. □

Corollary 3.1.17. *From the proof of the mapping properties for the double layer potential D and Proposition 3.1.7 we have $LDv = 0$ in $\mathbb{R}^3\backslash\Gamma$ for all $v \in H^{1/2+s}(\Gamma)$ and $-1/2 < s < 1/2$.*

We close this section with a remark on the optimality of the interval $|s| < 1/2$ in Theorem 3.1.16.

Remark 3.1.18. *(a) The restriction $|s| < 1/2$ in Theorem 3.1.16 is a result of the representation $S = \mathcal{N}\gamma_0'$ of the single layer operator, used in the previous proof, as well as the range of the trace operator γ_0 for Lipschitz domains (see Theorem 2.6.8). In general, this interval cannot be enlarged for Lipschitz domains, which means that the interval $|s| < 1/2$ cannot be determined more accurately with the method chosen for the proof (see [72, 162]).*

(b) The continuity of the operators in Theorem 3.1.16 can also be shown for $s = \pm 1/2$ for the Laplace operator. The proof requires methods from the area of harmonic analysis and goes beyond the scope of this book. It can, however, be found in, e.g., [143, 231].

(c) If the Lipschitz boundary Γ is globally smooth, $\Gamma \in C^\infty$, the trace operator $\gamma_0 : H^\ell_{\mathrm{loc}}\left(\mathbb{R}^d\right) \to H^{\ell-1/2}(\Gamma)$ is continuous on the entire range $\ell > 1/2$ and Theorem 3.1.16 is then valid for all $s > -1/2$ (see [170, Chap. 4]).

(d) *If the Lipschitz boundary* $\Gamma = \partial\Omega$ *is piecewise smooth, more specifically, if there exists a finite number of disjoint, relatively open surface patches* $\Gamma_j \subset \Gamma$ *with* $1 \le j \le q$ *that are smooth,* $\Gamma_j \subset C^\infty$*, and that satisfy* $\Gamma = \bigcup_{j=1}^{q} \overline{\Gamma_j}$*, then the upper bound of the interval on which the trace operator in Theorem 2.6.8 is continuous can be extended beyond the interval of indices given (see [39, 79]). From this we have the mapping properties in Theorem 3.1.16 for all* $-1/2 < s \le s_0$ *with an* $s_0 > 1/2$.

3.2 Regularity of the Solutions of the Boundary Integral Equations

The derivation of a priori convergence rates for discretizations of the boundary integral equations is based on regularity properties of the continuous solutions. More specifically, this means that the solution of the associated integral equation should not only exist in the energy space ($s = 0$ in Theorem 3.1.16) but should also be sufficiently smooth. The regularity theory for boundary integral equations follows from the regularity of the solutions of the associated partial differential equations. Both of these questions go beyond the scope of this book and we will only present the relevant results. Appropriate proofs can be found in, e.g., [72, Theorem 3], [162, Theorems 7.16, 7.17], [79, 113, 145].

For the formulation of regularity results we distinguish between the following cases: a globally smooth surface, a piecewise smooth Lipschitz polyhedron and a general Lipschitz surface.

Definition 3.2.1. A domain $\Omega \subset \mathbb{R}^3$ is a Lipschitz polyhedron if $\Omega \in C^{0,1}$ and there exist finitely many disjoint, relatively open surface patches $\Gamma_j \subset \Gamma$, $1 \le j \le q$ that are smooth, $\Gamma_j \in C^\infty$, and satisfy $\Gamma = \bigcup_{j=1}^{q} \overline{\Gamma_j}$.

Theorem 3.2.2. *Let* $\Omega \subset \mathbb{R}^3$ *be a bounded domain with a globally smooth boundary* $\Gamma = \partial\Omega \in C^\infty$*:*

(a) *Let* $\varphi \in H^{-1/2}(\Gamma)$ *and* $V\varphi = f \in H^{1/2+s}(\Gamma)$ *for an arbitrary* $s \ge 0$*. Then* $\varphi \in H^{-1/2+s}(\Gamma)$ *and*

$$\|\varphi\|_{H^{-1/2+s}(\Gamma)} \le C\left(\|f\|_{H^{1/2+s}(\Gamma)} + \|\varphi\|_{H^{-1/2}(\Gamma)}\right).$$

(b) *Let* $\varphi \in H^{-1/2}(\Gamma)$ *and* $K'_+\varphi = f \in H^{-1/2+s}(\Gamma)$ *for an arbitrary* $s \ge 0$*. Then* $\varphi \in H^{-1/2+s}(\Gamma)$ *and*

$$\|\varphi\|_{H^{-1/2+s}(\Gamma)} \le C\left(\|f\|_{H^{-1/2+s}(\Gamma)} + \|\varphi\|_{H^{-1/2}(\Gamma)}\right).$$

The analogous result holds for the operator K'_-.

(c) *Let* $\psi \in H^{1/2}(\Gamma)$ *and* $K_+\psi = f \in H^{1/2+s}(\Gamma)$ *for an arbitrary* $s \geq 0$. *Then* $\psi \in H^{1/2+s}(\Gamma)$ *and*

$$\|\psi\|_{H^{1/2+s}(\Gamma)} \leq C\left(\|f\|_{H^{1/2+s}(\Gamma)} + \|\psi\|_{H^{1/2}(\Gamma)}\right).$$

The analogous result holds for the operator K_-.

(d) *Let* $\psi \in H^{1/2}(\Gamma)$ *and* $W\psi = f \in H^{-1/2+s}(\Gamma)$ *for an arbitrary* $s \geq 0$. *Then* $\psi \in H^{1/2+s}(\Gamma)$ *and*

$$\|\psi\|_{H^{1/2+s}(\Gamma)} \leq C\left(\|f\|_{H^{-1/2+s}(\Gamma)} + \|\psi\|_{H^{1/2}(\Gamma)}\right).$$

The constants C *in the above-mentioned inequalities depend on* s.

In the following theorem we consider surfaces of Lipschitz domains and bounded Lipschitz polyhedra.

Theorem 3.2.3. (a) *Let* Γ *be the surface of a bounded Lipschitz polyhedron* $\Omega \subset \mathbb{R}^3$. *Then there exists some* $s_0 = s_0(\Gamma) > 1/2$ *such that the regularity and the a priori estimates from Theorem 3.2.2 hold for all* $0 \leq s < s_0$.

(b) *For general Lipschitz domains this statement only holds for* $s_0 = 1/2$.

3.3 Jump Relations of the Potentials and Explicit Representation Formulas

In this section we will first derive the jump properties of the potentials and their conormal derivatives on an abstract level, after which we will give an explicit representation. The approach we have chosen here avoids the calculus of pseudo-differential operators, which is used in [144], for example, for the derivation of one-sided jump relations.

3.3.1 Jump Properties of the Potentials

The single and double layer potentials have characteristic jump properties on the surface Γ.

Theorem 3.3.1. *Let* Ω *be a bounded Lipschitz domain with boundary* $\Gamma = \partial\Omega$. *The single and double layer potentials satisfy for all* $\varphi \in H^{-1/2}(\Gamma)$ *and* $\psi \in H^{1/2}(\Gamma)$ *the jump relations*

$$\begin{aligned}
[S\varphi] &= 0, \qquad [D\psi] = \psi \quad \text{in } H^{1/2}(\Gamma),\\
[\gamma_1 S\varphi] &= -\varphi, \quad [\gamma_1 D\psi] = 0 \text{ in } H^{-1/2}(\Gamma).
\end{aligned}$$

Proof. The first jump relation $[S\varphi] = 0$ is a direct result of Theorem 2.6.8 and the mapping properties of S (see Theorem 3.1.16).

To deal with the jump of the normal of S, we let $\varphi \in H^{-1/2}(\Gamma)$ and set $u = S\varphi$. Theorem 3.1.16 combined with Proposition 3.1.7 imply that $u \in H^1_L(\mathbb{R}^d \backslash \Gamma)$ and $Lu = 0$ in $\mathbb{R}^d \backslash \Gamma$. By using Green's second formula (2.112) with $v \in C_0^\infty(\mathbb{R}^d)$ we obtain, for $\Omega \in \{\Omega^-, \Omega^+\}$,

$$-\left(u, L'v\right)_{L^2(\Omega)} = \sigma_\Omega \left\{(\gamma_0 u, \widetilde{\gamma_1} v)_{L^2(\Gamma)} - (\gamma_1 u, \gamma_0 v)_{L^2(\Gamma)}\right\}.$$

If we add both equations (for $\Omega = \Omega^-$ and $\Omega = \Omega^+$) while using $[v] = [\widetilde{\gamma_1} v] = 0$ and $u \in L^2_{\text{loc}}(\mathbb{R}^d)$ we obtain

$$-\left(u, L'v\right)_{L^2(\mathbb{R}^d)} = -\left([u], \widetilde{\gamma_1} v\right)_{L^2(\Gamma)} + \left([\gamma_1 u], \gamma_0 v\right)_{L^2(\Gamma)}.$$

We have already shown that $[u] = [S\varphi] = 0$ and thus obtain

$$\left([\gamma_1 u], \gamma_0 v\right)_{L^2(\Gamma)} = -\left(u, L'v\right)_{L^2(\mathbb{R}^d)}. \tag{3.26}$$

Combining (2.116) with the definition of S (Definition 3.1.5) and Theorem 3.1.4 we obtain

$$\begin{aligned}\left(u, L'v\right)_{L^2(\mathbb{R}^d)} &= (Lu, v)_{L^2(\mathbb{R}^d)} = (LS\varphi, v)_{L^2(\mathbb{R}^d)} = \left(L\mathcal{N}\gamma_0'\varphi, v\right)_{L^2(\mathbb{R}^d)} \\ &= \left(\gamma_0'\varphi, v\right)_{L^2(\mathbb{R}^d)} = (\varphi, \gamma_0 v)_{L^2(\Gamma)}\end{aligned} \tag{3.27}$$

and combined with (3.26)

$$\left([\gamma_1 S\varphi], \gamma_0 v\right)_{L^2(\Gamma)} = -(\varphi, \gamma_0 v)_{L^2(\Gamma)}.$$

The assertion follows, as $\gamma_0 C_0^\infty(\mathbb{R}^d)$ is dense in $H^{1/2}(\Gamma)$ (see Lemma 2.8.4).

To deal with the jump properties of D, we start in the same way as for the single layer potential with an arbitrary function $\varphi \in H^{1/2}(\Gamma)$ and set $u = D\varphi$. As before, this time by using Corollary 3.1.17, Green's second formula gives

$$-\left(u, L'v\right)_{L^2(\mathbb{R}^d)} = -\left([u], \widetilde{\gamma_1} v\right)_{L^2(\Gamma)} + \left([\gamma_1 u], \gamma_0 v\right)_{L^2(\Gamma)}$$

for all $v \in C_0^\infty(\mathbb{R}^d)$. With the definition of D (Definition 3.1.5) we obtain

$$\begin{aligned}\left(u, L'v\right)_{L^2(\mathbb{R}^d)} &= (Lu, v)_{L^2(\mathbb{R}^d)} = (LD\varphi, v)_{L^2(\mathbb{R}^d)} = \left(L\mathcal{N}\widetilde{\gamma_1}'\varphi, v\right)_{L^2(\mathbb{R}^d)} \\ &= \left(\widetilde{\gamma_1}'\varphi, v\right)_{L^2(\mathbb{R}^d)} = (\varphi, \widetilde{\gamma_1} v)_{L^2(\Gamma)}\end{aligned}$$

and therefore

$$\left([\gamma_1 D\varphi], \gamma_0 v\right)_{L^2(\Gamma)} = \left([D\varphi] - \varphi, \widetilde{\gamma_1} v\right)_{L^2(\Gamma)}. \tag{3.28}$$

Since $\gamma_0 C_0^\infty(\mathbb{R}^d) \times \widetilde{\gamma_1} C_0^\infty(\mathbb{R}^d)$ is dense in $H^{1/2}(\Gamma) \times H^{-1/2}(\Gamma)$ (see Lemma 2.8.4) each of the two sides in (3.28) has to be equal to zero. This, however, gives us exactly the two stated jump relations for the double layer potentials.

□

3.3.2 Explicit Representation of the Boundary Integral Operator V

Theorem 3.3.1 demonstrates that the one-sided limits of the potentials and their conormal derivatives in general define different functions on Γ. For the numerical solution of boundary integral equations it is essential that the integral operators can be evaluated on the surface Γ. For this, the *representations* of the integral operators as one-sided traces of potentials, i.e., as one-sided limits, prove to be unsuitable. We have seen for the single and double layer potentials that for sufficiently smooth data v the explicit representation (3.4) and the abstract definition 3.1.5 coincide.

The functions that we use for the discretization are always bounded, i.e., in $L^\infty(\Gamma)$. Under this condition the integral operators have an explicit representation on piecewise smooth surfaces. In order to determine the limits of the potentials we need estimates of the fundamental solution G [see (3.3)] and its derivatives. These depend on the coefficients $\mathbf{A}$, $\mathbf{b}$ and c in the definition of the differential operator L [see (3.1)] as given in (3.3). With regard to (3.3) we introduce the function

$$g : \mathbb{R}^d \to \mathbb{C}, \qquad g(\mathbf{z}) := \exp\left(\langle \mathbf{b}, \mathbf{z}\rangle_{\mathbf{A}} - \lambda \|\mathbf{z}\|_{\mathbf{A}}\right)$$

with λ as in (3.3). The behavior of this function depends on the coefficients $\mathbf{A}$, $\mathbf{b}$ and c.

Lemma 3.3.2. *Let $a_{\min}$ again be the smallest and $a_{\max}$ the largest eigenvalue of* $\mathbf{A}$*:*

1. $c > 0$. *Then for all* $\mathbf{z} \in \mathbb{R}^d$

$$|g(\mathbf{z})| \leq e^{-\rho\|\mathbf{z}\|_{\mathbf{A}}} \qquad \textit{with } \rho := \frac{c}{\sqrt{c} + 2\|\mathbf{b}\|_{\mathbf{A}}}.$$

2. $c = 0$. *For all* $\mathbf{z} \in \mathbb{R}^d$ *we have*

$$|g(\mathbf{z})| \leq 1.$$

3. $c < 0$ *and* $\mathbf{b} = \mathbf{0}$*: Then we have*

$$|g(\mathbf{z})| = 1 \qquad \forall \mathbf{z} \in \mathbb{R}^d.$$

4. $c < 0$ *and* $\mathbf{b} \neq \mathbf{0}$. *Then the function g diverges exponentially in the direction of* $\mathbf{b}$*; more precisely, for all* $\alpha > 0$ *we have*

$$|g(\alpha\mathbf{b})| \geq e^{\alpha \min\left\{|c|/2, \|\mathbf{b}\|_{\mathbf{A}}^2\right\}}.$$

Proof. For case 1. we have

$$\lambda = \sqrt{\|\mathbf{b}\|_{\mathbf{A}}^2 + c} = \|\mathbf{b}\|_{\mathbf{A}} + \frac{c}{\|\mathbf{b}\|_{\mathbf{A}} + \sqrt{\|\mathbf{b}\|_{\mathbf{A}}^2 + c}} \geq \|\mathbf{b}\|_{\mathbf{A}} + \frac{c}{\sqrt{c} + 2\|\mathbf{b}\|_{\mathbf{A}}}$$

and thus

$$\langle \mathbf{b}, \mathbf{z} \rangle_{\mathbf{A}} - \lambda \|\mathbf{z}\|_{\mathbf{A}} \leq \|\mathbf{b}\|_{\mathbf{A}} \|\mathbf{z}\|_{\mathbf{A}} - (\|\mathbf{b}\|_{\mathbf{A}} + \rho) \|\mathbf{z}\|_{\mathbf{A}} = -\rho \|\mathbf{z}\|_{\mathbf{A}} < 0.$$

Case 2: For $c = 0$ it follows from the Cauchy–Schwarz inequality that

$$g(\mathbf{z}) = e^{\langle \mathbf{b}, \mathbf{z} \rangle_{\mathbf{A}} - \|\mathbf{b}\|_{\mathbf{A}} \|\mathbf{z}\|_{\mathbf{A}}} \leq e^0 = 1.$$

Case 3: For $c < 0$ and $\mathbf{b} = 0$ we have

$$|g(\mathbf{z})| = \left| e^{-i\sqrt{|c|}\|\mathbf{z}\|_{\mathbf{A}}} \right| = 1.$$

Case 4: First let $-\|\mathbf{b}\|_{\mathbf{A}}^2 < c < 0$. We choose $\mathbf{z} = \alpha \mathbf{b}$ with $\alpha \in \mathbb{R}_{>0}$. In the same way as before one can show that

$$\lambda = \sqrt{\|\mathbf{b}\|_{\mathbf{A}}^2 + c} = \|\mathbf{b}\|_{\mathbf{A}} + \frac{c}{\|\mathbf{b}\|_{\mathbf{A}} + \sqrt{\|\mathbf{b}\|_{\mathbf{A}}^2 + c}} \leq \|\mathbf{b}\|_{\mathbf{A}} + \frac{c}{2\|\mathbf{b}\|_{\mathbf{A}}}.$$

From this, with $\mathbf{z} = \alpha \mathbf{b}$, it follows that

$$g(\mathbf{z}) = g(\alpha \mathbf{b}) = e^{\alpha \|\mathbf{b}\|_{\mathbf{A}} (\|\mathbf{b}\|_{\mathbf{A}} - \lambda)} \geq e^{\alpha |c|/2},$$

and the function diverges exponentially for $\alpha \to \infty$.

Now let $c \leq -\|\mathbf{b}\|_{\mathbf{A}}^2 < 0$. In this case we have $\lambda = -i\sqrt{|c| - \|\mathbf{b}\|_{\mathbf{A}}^2}$ and therefore for all $\mathbf{z} \in \mathbb{R}^d$

$$|g(\mathbf{z})| = \left| e^{\langle \mathbf{b}, \mathbf{z} \rangle} \right| = \left| e^{\alpha \|\mathbf{b}\|_{\mathbf{A}}^2} \right|.$$

□

Fundamental solutions, the coefficients of which correspond to case 4 in Corollary 3.3.2, induce potentials with exponential growth in certain directions for $\|\mathbf{x}\| \to \infty$. From a physical point of view, these potentials are not very important and will no longer be examined.

Lemma 3.3.3. *Let* $\mathbf{b} = \mathbf{0}$ *or* $c \geq 0$.

(a) For all $0 < \varepsilon < 1$ *and* $R > 0$ *there exists a constant* $C < \infty$ *such that for all* $\mathbf{y} \in \Gamma$ *and all* $\mathbf{x} \in B_R(\mathbf{y})$ *we have*

$$|G(\mathbf{x}-\mathbf{y})| \leq \frac{C}{\|\mathbf{x}-\mathbf{y}\|^{d-2+\varepsilon}}. \tag{3.29}$$

(b) *Under the additional assumption "$\Gamma \in C^2$ in a local neighborhood of* $\mathbf{y}$" *we have*

$$\left|\widetilde{\gamma_{1,\mathbf{y}}} G(\mathbf{x}-\mathbf{y})\right| + \left|\gamma_{1,\mathbf{y}} G(\mathbf{y}-\mathbf{x})\right| \leq C\left(\frac{|\langle \mathbf{n}_\mathbf{y}, \mathbf{x}-\mathbf{y}\rangle|}{\|\mathbf{x}-\mathbf{y}\|^d} + \frac{1}{\|\mathbf{x}-\mathbf{y}\|^{d-2+\varepsilon}}\right)$$

for all $\mathbf{y} \in \Gamma$ *and* $\mathbf{x} \in U_\mathbf{y}$ *where* $U_\mathbf{y}$ *denotes an arbitrary, bounded neighborhood of* $\mathbf{y}$.

Remark 3.3.4. *For $d \geq 3$ we can choose $\varepsilon = 0$ in Lemma 3.3.3. In two spatial dimensions the kernel function has a logarithmic singularity and $\varepsilon = 0$ is not admissible.*

Proof of Lemma 3.3.3: We will only prove the assertion for $d = 3$ and refer to [102] for the general case.
Corollary 3.3.2 implies the uniform boundedness

$$\frac{\left|e^{\langle \mathbf{b}, \mathbf{z}\rangle_\mathbf{A} - \lambda \|\mathbf{z}\|_\mathbf{A}}\right|}{\sqrt{\det \mathbf{A}}} \leq C \qquad \forall \mathbf{z} \in \mathbb{R}^d$$

for the considered values of the coefficients. From this follows (3.29).

We will now study the derivatives of the fundamental solution. We have

$$\nabla_\mathbf{z} G(\mathbf{z}) = -\frac{1}{4\pi\sqrt{\det \mathbf{A}}} \frac{\mathbf{A}^{-1}\mathbf{z}}{\|\mathbf{z}\|_\mathbf{A}^d} + \mathbf{R}_1(\mathbf{z}). \tag{3.30}$$

The remainder can be estimated by

$$\|\mathbf{R}_1(\mathbf{z})\| \leq \frac{C}{\|\mathbf{z}\|^{d-2+\varepsilon}}$$

with a suitable $\varepsilon \in]0,1[$. The definition of the modified conormal derivative (2.107) leads us to the decomposition

$$\widetilde{\gamma_{1,\mathbf{y}}} G(\mathbf{x}-\mathbf{y}) = \gamma_{1,\mathbf{y}} G(\mathbf{x}-\mathbf{y}) + 2\langle \mathbf{n}, \mathbf{b}\rangle G(\mathbf{x}-\mathbf{y}), \tag{3.31}$$

and because of (3.29) it suffices to consider the first summand. Using (3.30) this gives us

$$\left|\gamma_{1,\mathbf{y}} G(\mathbf{x}-\mathbf{y})\right| \leq C\left(\frac{|\langle \mathbf{n}_\mathbf{y}, \mathbf{x}-\mathbf{y}\rangle|}{\|\mathbf{x}-\mathbf{y}\|^d} + \frac{1}{\|\mathbf{x}-\mathbf{y}\|^{d-2+\varepsilon}}\right).$$

□

We use the representation of the single layer potential from (3.4), and for $\varphi \in L^{\infty}(\Gamma)$ we define the extension of S on $\mathbb{R}^d$ by

$$(S\varphi)(\mathbf{x}) := \int_{\Gamma} G(\mathbf{x}-\mathbf{y})\,\varphi(\mathbf{y})\,ds_{\mathbf{y}} \qquad \forall \mathbf{x} \in \mathbb{R}^d. \tag{3.32}$$

Theorem 3.3.5. *Let Γ be the surface of a bounded Lipschitz domain $\Omega \subset \mathbb{R}^d$. For any $\varphi \in L^{\infty}(\Gamma)$, the integral in (3.32) exists as an improper integral and defines a continuous function $S\varphi$ on $\mathbb{R}^d$.*

Proof. The continuity of $S\varphi$ in $\mathbb{R}^d \backslash \Gamma$ has already been determined in Theorem 3.1.1.

We use the notation from Definition 2.2.7. Let $\mathbf{x} \in \Gamma$ and $U_{\mathbf{x}}$ be a d-dimensional neighborhood of $\mathbf{x}$ for which there exists a bi-Lipschitz continuous mapping $\chi : B_2 \to U_{\mathbf{x}}$. Here B_2 again denotes the d-dimensional ball with radius 2 around the origin, and B_2^0 is defined as in (2.71). Without loss of generality we assume that $\chi(0) = \mathbf{x}$. The integral in (3.32) is split into $\int_{\Gamma \backslash U_{\mathbf{x}}} + \int_{\Gamma \cap U_{\mathbf{x}}}$ and, for the first part of the integral, the continuity in $\mathbf{x}$ follows as in the proof of Theorem 3.1.1.

In local coordinates the second part of the integral gives us

$$S_2(\mathbf{x}) := \int_{\Gamma \cap U_{\mathbf{x}}} G(\mathbf{x}-\mathbf{y})\,\varphi(\mathbf{y})\,ds_{\mathbf{y}} = \int_{B_2^0} G(\chi(0) - \chi(\hat{\mathbf{y}}))\,\hat{\varphi}(\hat{\mathbf{y}})\,d\hat{\mathbf{y}}$$

with $\hat{\varphi}(\hat{\mathbf{y}}) := g(\hat{\mathbf{y}})(\varphi \circ \chi)(\hat{\mathbf{y}}) \in L^{\infty}(B_2^0)$ and the surface element $g(\hat{\mathbf{y}})$ (see 2.2.4). The Lipschitz continuity of the surface implies the existence of a constant $C_1 > 1$ that depends only on Γ, with

$$C_1^{-1}\|\hat{\mathbf{x}} - \hat{\mathbf{y}}\| \le \|\chi(\hat{\mathbf{x}}) - \chi(\hat{\mathbf{y}})\| \le C_1\|\hat{\mathbf{x}} - \hat{\mathbf{y}}\| \qquad \forall \hat{\mathbf{x}}, \hat{\mathbf{y}} \in B_2.$$

Combined with (3.29) we obtain the following estimate for an arbitrary $0 < \varepsilon < 1$

$$|G(\chi(0) - \chi(\hat{\mathbf{y}}))| \le \frac{C_2}{\|\hat{\mathbf{y}}\|^{d-2+\varepsilon}} \tag{3.33}$$

for all $\hat{\mathbf{y}} \in B_2^0 \backslash \{0\}$, where C_2 depends only on C_1, d, ε and $\mathbf{A}$. This proves that

$$S_2(\mathbf{x}) \le C_2 \int_{B_2^0} \frac{|\hat{\varphi}(\hat{\mathbf{y}})|}{\|\hat{\mathbf{y}}\|^{d-2+\varepsilon}} d\hat{\mathbf{y}}.$$

The regularity of the parameterization combined with $\varphi \in L^{\infty}(\Gamma)$ result in the existence of a constant $M < \infty$ such that

$$\sup_{\hat{\mathbf{y}} \in B_2^0} |\hat{\varphi}(\hat{\mathbf{y}})| \le M.$$

With this we have

$$S_2(\mathbf{x}) \le C_2 M \int_{B_2^0} \frac{1}{\|\hat{\mathbf{y}}\|^{d-2+\varepsilon}} d\hat{\mathbf{y}}.$$

The integrand on the right-hand side defines an integrable upper bound (see Exercise 3.3.6) so that the right-hand side in (3.32) exists as an improper integral.

In order to prove the continuity of S_2 in $\mathbf{x}$ we consider a sequence of points $(\mathbf{x}_n)_{n\in\mathbb{N}}$ in $U_{\mathbf{x}}$ that converges to $\mathbf{x}$. The associated sequence $(\hat{\mathbf{x}}_n)_{n\in\mathbb{N}}$ in B_2 converges to zero. Without loss of generality we assume that $(\hat{\mathbf{x}}_n)_{n\in\mathbb{N}} \subset B_1$. Let $\lambda \in C_0^\infty(\mathbb{R}^d)$ be a cut-off function with $0 \le \lambda \le 1$ and

$$\lambda \equiv 1 \text{ on } B_2^0 \quad \text{and} \quad \lambda \equiv 0 \text{ on } \mathbb{R}^d \backslash B_3.$$

We have to show that

$$\lim_{n\to\infty} \int_{\mathbb{R}^{d-1}} \frac{\lambda(\hat{\mathbf{y}})}{\|\hat{\mathbf{x}}_n - \hat{\mathbf{y}}\|^{d-2+\varepsilon}} d\hat{\mathbf{y}} = \int_{\mathbb{R}^{d-1}} \frac{\lambda(\hat{\mathbf{y}})}{\|\hat{\mathbf{y}}\|^{d-2+\varepsilon}} d\hat{\mathbf{y}}.$$

The integrand on the right-hand side defines the function f. The integral on the left-hand side can be written as

$$\int_{\mathbb{R}^{d-1}} \frac{\lambda(\hat{\mathbf{y}} + \hat{\mathbf{x}}_n)}{\|\hat{\mathbf{y}}\|^{d-2+\varepsilon}} d\hat{\mathbf{y}} =: \int_{\mathbb{R}^{d-1}} f_n(\hat{\mathbf{y}})\, d\hat{\mathbf{y}}.$$

Note the inclusion

$$\bigcup_{n\in\mathbb{N}} \operatorname{supp} \lambda(\cdot + \hat{\mathbf{x}}_n) \subset \overline{B_4}.$$

If we use the fact that f_n is integrable, converges to f almost everywhere and is bounded above by the integrable function

$$g(\hat{\mathbf{y}}) := \begin{cases} \|\hat{\mathbf{y}}\|^{-(d-2+\varepsilon)} & \text{if } \hat{\mathbf{y}} \in \overline{B_4} \\ 0 & \text{otherwise} \end{cases}$$

we can apply Lebesgue's theorem of dominated convergence and thus prove the statement. □

Exercise 3.3.6. *Let B_1 be the $(d-1)$-dimensional unit ball and $0 < \varepsilon < 1$. Show that*

$$\int_{B_1} \frac{1}{\|\hat{\mathbf{y}}\|^{d-2+\varepsilon}} d\hat{\mathbf{y}}$$

exists as an improper integral. Hint: Use polar coordinates.

3.3.3 Explicit Representation of the Boundary Integral Operators K and K′

We will now turn our attention to the one-sided traces of the double layer potential. The derivation of explicit representations of the traces $\gamma_0^+ D$, $\gamma_0^- D$, $\gamma_1^+ S$, $\gamma_1^- S$ is, technically speaking, more complex than for the boundary integral operator V. We will limit this discussion to the essential arguments and refer to [102] for details.

Explicit representations have a great significance, especially for the numerical solution of the integral equations. For the numerical treatment, the boundary Γ is always assumed to be piecewise smooth and usually one uses piecewise smooth functions for the discretization. Keeping this in mind, it is therefore only necessary to derive the explicit representation formulas under these conditions. These assumptions will greatly reduce the technical difficulties in the following section. For a more general discussion we refer to [162, Chap. 7].

Assumption 3.3.7. *The surface Γ belongs to the class C^2_{pw} (see Definition 2.2.10).*

This assumption on the smoothness implies that the conormal derivative of the fundamental solution G is improperly integrable.

Lemma 3.3.8. *Let $\Gamma \in C^2_{\mathrm{pw}}$ and $\varphi \in L^\infty(\Gamma)$:*

(a) The function $\varphi(\mathbf{y})\,\widetilde{\gamma_{1,\mathbf{y}}}G(\mathbf{x}-\mathbf{y})$ is improperly integrable on Γ with respect to $\mathbf{y}$.
(b) If Γ is smooth in $\mathbf{x} \in \Gamma$ then the function $\varphi(\mathbf{y})\,\gamma_{1,\mathbf{x}}G(\mathbf{x}-\mathbf{y})$ is improperly integrable with respect to $\mathbf{y}$.

Proof. Lemma 3.3.3 implies the existence of an $\varepsilon \in\,]0,1[$ with

$$\left|\widetilde{\gamma_{1,\mathbf{y}}}G(\mathbf{x}-\mathbf{y})\right| \le C\left(\frac{|\langle \mathbf{n}_\mathbf{y}, \mathbf{x}-\mathbf{y}\rangle|}{\|\mathbf{x}-\mathbf{y}\|^d} + \frac{1}{\|\mathbf{x}-\mathbf{y}\|^{d-2+\varepsilon}}\right).$$

It suffices to consider a local neighborhood $U_\mathbf{x} \subset \mathbb{R}^d$ of $\mathbf{x}$ to determine whether the function is integrable or not. If we choose a sufficiently small $U_\mathbf{x}$ there exist parts $\Gamma_i \subset \Gamma$, $1 \le i \le q$, with $\Gamma_i \in C^2$ and

$$U_\mathbf{x} \cap \Gamma = \bigcup_{i=1}^{q} (U_\mathbf{x} \cap \Gamma_i).$$

On each of these parts Γ_i we have (see Lemma 2.2.14)

$$|\langle \mathbf{n}(\mathbf{y}), \mathbf{x}-\mathbf{y}\rangle| \le C\,\|\mathbf{x}-\mathbf{y}\|^2,$$

and therefore

$$\left|\varphi(\mathbf{y})\,\widetilde{\gamma_{1,\mathbf{y}}}G(\mathbf{x}-\mathbf{y})\right| \le CM\,\|\mathbf{x}-\mathbf{y}\|^{-(d-2+\varepsilon)} \qquad \text{with} \quad M := \sup_{\mathbf{z}\in\Gamma} |\varphi(\mathbf{z})|.$$

Hence, as was shown in the proof of Theorem 3.3.5, we have found an improperly integrable upper bound.

In case (b) we need the smoothness of Γ in $\mathbf{x}$ to be able to define the conormal derivative $\gamma_{1,\mathbf{x}}$ pointwise. In this case we can use Lemma 3.3.3 in the same way as above to find an improperly integrable upper bound. □

This lemma shows that

$$\begin{aligned}(K\varphi)(\mathbf{x}) &:= \int_\Gamma \widetilde{\gamma_{1,\mathbf{y}}} G(\mathbf{x}-\mathbf{y})\,\varphi(\mathbf{y})\,ds_\mathbf{y} \quad \mathbf{x}\in\Gamma,\\ (K'\varphi)(\mathbf{x}) &:= \int_\Gamma \gamma_{1,\mathbf{x}} G(\mathbf{x}-\mathbf{y})\,\varphi(\mathbf{y})\,ds_\mathbf{y} \quad \Gamma \text{ smooth in } \mathbf{x}\in\Gamma.\end{aligned} \tag{3.34}$$

are well defined under the condition that $\varphi \in L^\infty(\Gamma)$.

Corollary 3.3.9. *Let $\Gamma \in C^2_{\mathrm{pw}}$. The mappings K and K' from (3.34) map $L^\infty(\Gamma)$ to $L^\infty(\Gamma)$ continuously. The continuous embedding $L^\infty(\Gamma) \subset L^2(\Gamma)$ results in the continuity of the operators*

$$\begin{aligned}&K : L^\infty(\Gamma) \to L^\infty(\Gamma), \quad K : L^\infty(\Gamma) \to L^2(\Gamma),\\ &K' : L^\infty(\Gamma) \to L^\infty(\Gamma), \quad K' : L^\infty(\Gamma) \to L^2(\Gamma).\end{aligned}$$

Proof. Let $\varphi \in L^\infty(\Gamma)$. As in the proof of Lemma 3.3.8 one can deduce that

$$|K\varphi(\mathbf{x})| \le \|\varphi\|_{L^\infty(\Gamma)} \int_\Gamma \frac{1}{\|\mathbf{x}-\mathbf{y}\|^{d-2+\varepsilon}} ds_\mathbf{y} \tag{3.35}$$

for all $\mathbf{x} \in \Gamma$. From the proof of Theorem 3.3.5 we have the boundedness of the integral

$$\int_\Gamma \frac{1}{\|\mathbf{x}-\mathbf{y}\|^{d-2+\varepsilon}} ds_\mathbf{y} \le C$$

with a constant that does not depend on $\mathbf{x}$.

The operator K' can be estimated in the same way in all smooth points $\mathbf{x} \in \Gamma$ by the right-hand side in (3.35). Since the set of all non-smooth points has zero measure, the mapping properties from the assertion result in this case as well. □

The existence of $K\varphi$ and $K'\varphi$ on the surface of Γ does not in any way mean that these functions are limits of the potential $D\varphi$ as a transition from $\Omega^\pm$ on Γ. In the following these one-sided limits will be put into relation with one another.

The utilized geometric construction is illustrated in Fig. 3.1. Let $S(\mathbf{x}, r)$ be the surface of the d-dimensional sphere around $\mathbf{x} \in \mathbb{R}^d$ with radius $r > 0$. We set $H(\mathbf{x}, r) := S(\mathbf{x}, r) \cap \Omega^-$. The functional $J : \mathbb{R}^d \to [0, 1]$ is related to the principal part of the differential operator L [see (2.98)] and is defined by

$$J(\mathbf{x}) := \lim_{r\to 0} \frac{r}{\omega_d \sqrt{\det \mathbf{A}}} \int_{H(\mathbf{x},r)} \frac{1}{\|\mathbf{x}-\mathbf{y}\|^d_\mathbf{A}} ds_\mathbf{y}. \tag{3.36}$$

Fig. 3.1 Intersection $H(x,r) = S(x,r) \cap \Omega^-$ and corresponding choice of the orientation of the normal vector with respect to the local coordinate system

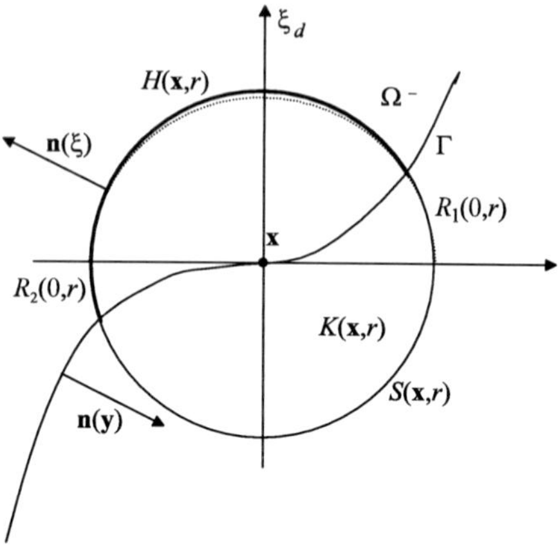

Here ω_d denotes the surface measure of the unit sphere in $\mathbb{R}^d$, i.e., $\omega_2 = 2\pi$, $\omega_3 = 4\pi$, etc. Before we present the properties of J we will need a preparatory lemma.

Lemma 3.3.10. *For all nonsingular matrices* $\mathbf{B} \in \mathbb{R}^{d\times d}$ *we have*

$$\int_{\mathbb{S}_{d-1}} \frac{1}{\|\mathbf{Bx}\|^d} ds_{\mathbf{x}} = \frac{\omega_d}{|\det \mathbf{B}|}. \tag{3.37}$$

Proof. We define the function $\varphi : \mathbb{R} \to \mathbb{R}$ by $r \to r^2 \exp\left(-r^2\right)$. On the one hand we have

$$|\det \mathbf{B}| \int_{\mathbb{R}^d} \frac{\varphi(\|\mathbf{x}\|)}{\|\mathbf{Bx}\|^d} d\mathbf{x} = |\det \mathbf{B}| \int_0^\infty \int_{\mathbb{S}_{d-1}} \frac{\varphi(r)\, r^{d-1}}{r^d \|\mathbf{By}\|^d} ds_{\mathbf{y}} dr$$
$$= |\det \mathbf{B}| \int_{\mathbb{S}_{d-1}} \frac{1}{\|\mathbf{By}\|^d} ds_{\mathbf{y}} \int_0^\infty \frac{\varphi(r)}{r} dr = \frac{1}{2} |\det \mathbf{B}| \int_{\mathbb{S}_{d-1}} \frac{1}{\|\mathbf{By}\|^d} ds_{\mathbf{y}}$$

and on the other

$$|\det \mathbf{B}| \int_{\mathbb{R}^d} \frac{\varphi(\|\mathbf{x}\|)}{\|\mathbf{Bx}\|^d} d\mathbf{x} = \int_{\mathbb{R}^d} \frac{\varphi\left(\|\mathbf{B}^{-1}\mathbf{x}\|\right)}{\|\mathbf{x}\|^d} d\mathbf{x} = \int_{\mathbb{S}_{d-1}} \int_0^\infty \frac{\varphi\left(r\,\|\mathbf{B}^{-1}\mathbf{y}\|\right)}{\|r\mathbf{y}\|^d\, r^{1-d}} dr ds_{\mathbf{y}}$$
$$= \int_{\mathbb{S}_{d-1}} \int_0^\infty \frac{\varphi\left(r\,\|\mathbf{B}^{-1}\mathbf{y}\|\right)}{r} dr ds_{\mathbf{y}} = \frac{\omega_d}{2},$$

which proves the assertion. □

From this lemma we have the estimate

$$0 \leq \frac{r}{\omega_d \sqrt{\det \mathbf{A}}} \int_{H(\mathbf{x},r)} \frac{1}{\|\mathbf{x}-\mathbf{y}\|_{\mathbf{A}}^d} ds_{\mathbf{y}}$$
$$\leq \frac{r}{\omega_d \sqrt{\det \mathbf{A}}} \int_{S(\mathbf{x},r)} \frac{1}{\|\mathbf{x}-\mathbf{y}\|_{\mathbf{A}}^d} ds_{\mathbf{y}} = \frac{1}{\omega_d \sqrt{\det \mathbf{A}}} \int_{S(\mathbf{0},1)} \frac{1}{\|\mathbf{y}\|_{\mathbf{A}}^d} ds_{\mathbf{y}} = 1. \quad (3.38)$$

We will show in Lemma 3.3.11 that the limit in the definition of $J(\mathbf{x})$ exists and that we therefore have $J(\mathbf{x}) \in [0,1]$.

The conormal derivative of the principal part of the double layer potential, applied to the unit function, coincides with the functional J up to a sign. The principal part of the fundamental solution is defined by

$$G_0(\mathbf{z}) := \begin{cases} \dfrac{1}{2\pi\sqrt{\det \mathbf{A}}} \log \dfrac{1}{\|\mathbf{z}\|_{\mathbf{A}}} & \text{for } d = 2, \\ \dfrac{1}{(d-2)\,\omega_d \sqrt{\det \mathbf{A}}} \dfrac{1}{\|\mathbf{z}\|_{\mathbf{A}}^{d-2}} & \text{for } d \geq 3 \end{cases} \quad (3.39)$$

and the principal part of the double layer potential is defined by

$$(D_0 v)(\mathbf{x}) := \int_\Gamma \gamma_{1,\mathbf{y}} G_0(\mathbf{x}-\mathbf{y})\, v(\mathbf{y})\, ds_{\mathbf{y}} \qquad \mathbf{x} \in \mathbb{R}^d \backslash \Gamma. \quad (3.40)$$

Lemma 3.3.11. *The functional J has the representation*

$$J(\mathbf{x}) = -\int_\Gamma \gamma_{1,\mathbf{y}} G_0(\mathbf{x}-\mathbf{y})\, ds_{\mathbf{y}}, \quad (3.41)$$

while for γ_1 in (3.41) one can choose γ_1^+ as well as γ_1^-.

Proof. (Schematic proof) The proof for $\mathbf{x} \in \mathbb{R}^d \backslash \Gamma$ follows from Green's formulas by a suitable choice of the functions u and v. We refer to [102] for the details and restrict ourselves to the more interesting case $\mathbf{x} \in \Gamma$.

The normal field to $H(\mathbf{x}, r)$ is chosen according to

$$\mathbf{n}(\xi) := \frac{1}{r}(\xi - \mathbf{x}) \qquad \forall \xi \in H(\mathbf{x}, r)$$

and thus γ_1 is defined on $H(\mathbf{x}, r)$ by $\gamma_1(\cdot) := \langle \mathbf{A}\mathbf{n}, \gamma_0 \operatorname{grad} \cdot \rangle$. Let $\widetilde{\Omega^-} := \Omega^- \backslash \overline{B_r(\mathbf{x})}$. By applying Gauss' theorem (for sufficiently small $r > 0$) to $\widetilde{\Omega^-}$ and by using the equality $L_0 G_0(\mathbf{x} - \cdot) = 0$ in $\widetilde{\Omega^-}$, we obtain

$$\int_{\Gamma \backslash B_r(\mathbf{x})} \gamma_{1,\mathbf{y}} G_0(\mathbf{x}-\mathbf{y})\, ds_{\mathbf{y}} = \int_{H(\mathbf{x},r)} \gamma_{1,\mathbf{y}} G_0(\mathbf{x}-\mathbf{y})\, ds_{\mathbf{y}}.$$

The theorem of dominated convergence implies that we can let $r \to 0$, i.e.,

$$\int_\Gamma \gamma_{1,\mathbf{y}} G_0(\mathbf{x}-\mathbf{y})\, ds_\mathbf{y} = \lim_{r\to 0}\int_{H(\mathbf{x},r)} \gamma_{1,\mathbf{y}} G_0(\mathbf{x}-\mathbf{y})\, ds_\mathbf{y}$$
$$= -\lim_{r\to 0}\frac{r}{\omega_d\sqrt{\det \mathbf{A}}}\int_{H(\mathbf{x},r)}\frac{1}{\|\mathbf{y}-\mathbf{x}\|_\mathbf{A}^d}ds_\mathbf{y} = -J(\mathbf{x}).$$

□

Corollary 3.3.12. *We have*

$$J(\mathbf{x}) = \begin{cases} 0 & \mathbf{x}\in\Omega^+, \\ 1 & \mathbf{x}\in\Omega^-, \\ \frac{1}{2} & \mathbf{x}\in\Gamma \text{ and } \Gamma \text{ is smooth in } \mathbf{x}. \end{cases}$$

Let $(\mathbf{x}_n^+)_{n\in\mathbb{N}}$ *and* $(\mathbf{x}_n^-)_{n\in\mathbb{N}}$ *be sequences of points in* Ω^+ *and* Ω^- *respectively that converge to a point* $\mathbf{x}\in\Gamma$. *Then we have for* $\sigma\in\{-,+\}$ *the relation*

$$\lim_{n\to\infty} J(\mathbf{x}_n^\sigma) = -\int_\Gamma \gamma_{1,\mathbf{y}} G_0(\mathbf{x}-\mathbf{y})\, ds_\mathbf{y} - \left(\frac{\sigma 1 - 1}{2} + J(\mathbf{x})\right). \tag{3.42}$$

Proof. For $\mathbf{x}\in\Omega^+$ and a sufficiently small $r>0$ we have $H(\mathbf{x},r)=\emptyset$ and the integral in (3.36) equals zero. For $\mathbf{x}\in\Omega^-$ and a sufficiently small $r>0$ we have $H(\mathbf{x},r)=S(\mathbf{x},r)$ and the result follows from (3.38). We now need to consider the remaining case $\mathbf{x}\in\Gamma$ and Γ smooth at $\mathbf{x}$. Since the integral in (3.36) is invariant under rotation and translation of the coordinate system, we can choose a Cartesian coordinate system $(\xi_i)_{i=1}^d$ with origin $\mathbf{x}$ and the first $d-1$ coordinates in the tangent plane at $\mathbf{x}\in\Gamma$. The component ξ_d points towards Ω^-. Let $T(\mathbf{0},r) := \{\xi\in S(\mathbf{0},r) : \xi_d > 0\}$ be the upper half sphere and

$$R_1(\mathbf{0},r) := T(\mathbf{0},r)\backslash H(\mathbf{0},r), \qquad R_2(\mathbf{0},r) := H(\mathbf{0},r)\backslash T(\mathbf{0},r).$$

The smoothness of the surface at $\mathbf{x}$ implies that

$$|R_1(\mathbf{0},r)| + |R_2(\mathbf{0},r)| \le Cr^d.$$

From this we have

$$\frac{r}{\omega_d\sqrt{\det\mathbf{A}}}\int_{H(\mathbf{x},r)}\frac{1}{\|\mathbf{x}-\mathbf{y}\|_\mathbf{A}^d}ds_\mathbf{y} = \frac{r}{\omega_d\sqrt{\det\mathbf{A}}}\int_{H(\mathbf{0},r)}\frac{1}{\|\xi\|_\mathbf{A}^d}ds_\xi$$
$$= \frac{r}{\omega_d\sqrt{\det\mathbf{A}}}\int_{T(\mathbf{0},r)}\frac{1}{\|\xi\|_\mathbf{A}^d}ds_\xi \tag{3.43}$$
$$+\frac{r}{\omega_d\sqrt{\det\mathbf{A}}}\int_{R_2(\mathbf{0},r)}\frac{1}{\|\xi\|_\mathbf{A}^d}ds_\xi - \frac{r}{\omega_d\sqrt{\det\mathbf{A}}}\int_{R_1(\mathbf{0},r)}\frac{1}{\|\xi\|_\mathbf{A}^d}ds_\xi.$$

It follows from the symmetry of $\mathbb{S}_{d-1}$ that the integral (3.37) over the half sphere $T(\mathbf{0},1)$ is equal to $\omega_d/(2\det\mathbf{B})$. As in the proof of (3.38) it follows that the first term in the right-hand side of (3.43) is equal to $1/2$. The other terms tend towards zero for $r\to 0$ since

$$\frac{r}{\omega_d\sqrt{\det\mathbf{A}}}\left\{\left|\int_{R_2(\mathbf{0},r)}\frac{1}{\|\xi\|_{\mathbf{A}}^d}ds_\xi\right|+\left|\int_{R_1(\mathbf{0},r)}\frac{1}{\|\xi\|_{\mathbf{A}}^d}ds_\xi\right|\right\}\leq Cr.$$

One obtains the second equality, i.e., (3.42), by using Lemma 3.3.11. □

Lemma 3.3.8 implies that the double layer potential is defined as an improper integral on the surface of Γ. The one-sided limits $K_\pm$ and $K'_\pm$ also exist, as given in (3.6). The relation between these three functions on Γ is given in the following result.

Theorem 3.3.13. *Let $\Gamma\in C^2_{\mathrm{pw}}$ and let $\left(\mathbf{x}_n^+\right)_{n\in\mathbb{N}}$ and $\left(\mathbf{x}_n^-\right)_{n\in\mathbb{N}}$ be sequences of points in Ω^+ and Ω^- respectively that converge to $\mathbf{x}\in\Gamma$. Let the density $\varphi\in L^\infty(\Gamma)$ be continuous at $\mathbf{x}$. Then we have for the double layer potential and $\sigma\in\{-,+\}$ the jump relations*

$$\lim_{n\to\infty}(D\varphi)\left(\mathbf{x}_n^\sigma\right)=\int_\Gamma\varphi(\mathbf{y})\,\widetilde{\gamma_{1,\mathbf{y}}}G(\mathbf{x}-\mathbf{y})\,ds_\mathbf{y}+\left(\frac{\sigma 1-1}{2}+J(\mathbf{x})\right)\varphi(\mathbf{x}). \quad (3.44)$$

Proof. Let $\sigma\in\{-,+\}$. We first consider the statement for the principal part G_0 of the fundamental solution G [see (3.39)] and the associated double layer potential D_0 [see (3.40)].

For $\mathbf{x}\in\Gamma$ and $\xi\in\mathbb{R}^d$ we define

$$\psi_0(\xi):=\int_\Gamma(\varphi(\mathbf{y})-\varphi(\mathbf{x}))\,\gamma_{1,\mathbf{y}}G_0(\xi-\mathbf{y})\,ds_\mathbf{y}$$

and for $\xi\notin\Gamma$ obtain the representation

$$(D_0\varphi)(\xi)=-\varphi(\mathbf{x})\,J(\xi)+\psi_0(\xi). \quad (3.45)$$

Next, we replace D and G in (3.44) by D_0 and G_0 with, consequently, $\widetilde{\gamma_1}=\gamma_1$. By Corollary 3.3.12 and (3.45) the resulting right and left-hand sides of (3.44) have the representations

$$\psi_0(\mathbf{x})+\varphi(\mathbf{x})\left(\int_\Gamma\gamma_{1,\mathbf{y}}G_0(\mathbf{x}-\mathbf{y})\,ds_\mathbf{y}+\frac{\sigma 1-1}{2}+J(\mathbf{x})\right)=\psi_0(\mathbf{x})-\varphi(\mathbf{x})\lim_{n\to\infty}J(\mathbf{x}_n^\sigma)$$

and

$$-\varphi(\mathbf{x})\lim_{n\to\infty}J\left(\mathbf{x}_n^\sigma\right)+\lim_{n\to\infty}\psi_0\left(\mathbf{x}_n^\sigma\right)$$

respectively. Therefore it suffices to show that $\lim_{n\to\infty}\psi_0\left(\mathbf{x}_n^\sigma\right)\to\psi_0(\mathbf{x})$.

For the difference function we obtain the estimate

$$\left|\psi_0(\mathbf{x}) - \psi_0\left(\mathbf{x}_n^\sigma\right)\right| \le \frac{1}{\omega_d \sqrt{\det \mathbf{A}}} \int_\Gamma k\left(\mathbf{x}_n^\sigma, \mathbf{x}, \mathbf{y}\right) ds_{\mathbf{y}}$$

with

$$k(\xi, \mathbf{x}, \mathbf{y}) := \left| \frac{\langle \mathbf{n}(\mathbf{y}), \mathbf{x} - \mathbf{y} \rangle}{\|\mathbf{x} - \mathbf{y}\|_{\mathbf{A}}^d} - \frac{\langle \mathbf{n}(\mathbf{y}), \xi - \mathbf{y} \rangle}{\|\xi - \mathbf{y}\|_{\mathbf{A}}^d} \right| |\varphi(\mathbf{y}) - \varphi(\mathbf{x})|.$$

The limit $\mathbf{x}_n^\sigma \to \mathbf{x}_n$ is determined by the behavior of the kernel function k for $\xi \to \mathbf{x}$. For this, the integral over Γ is split into an integral over $\Gamma \backslash B_\delta(\mathbf{x})$ and $\Gamma \cap B_\delta(\mathbf{x})$ with a sufficiently small $\delta > 0$. For $\xi \to \mathbf{x}$ and $\mathbf{y} \in \Gamma \backslash B_\delta(\mathbf{x})$ we clearly have $\lim_{\xi \to \mathbf{x}} k(\xi, \mathbf{x}, \mathbf{y}) = 0$ and thus

$$\lim_{\xi \to \mathbf{x}} \int_{\Gamma \backslash B_\delta(\mathbf{x})} k(\xi, \mathbf{x}, \mathbf{y}) \, ds_{\mathbf{y}} = 0.$$

The domain of integration $\Gamma \cap B_\delta(\mathbf{x})$ is further split into smooth parts. The following construction is illustrated in Fig. 3.2. For this let τ be a panel with $\mathbf{x} \in \overline{\tau}$ and a smooth extension $\tau^\star$ (see Definition 2.2.9). For the discussion on convergence we may assume without loss of generality that all $\mathbf{x}_n^\sigma$ are contained in $B_\delta(\mathbf{x})$ and that for every $\xi \in \{\mathbf{x}_n^\sigma : n \in \mathbb{N}\}$ there exists an orthogonal element $\xi_\perp \in \tau^\star$ such that

$$\xi - \xi_\perp = \sigma \|\xi - \xi_\perp\| \, \mathbf{n}(\xi_\perp). \tag{3.46}$$

The proof is given once the

Auxiliary Assumption:

$$I(\xi) := \int_{\tau \cap B_\delta(\mathbf{x})} k(\xi, \mathbf{x}, \mathbf{y}) \, ds_{\mathbf{y}} \to 0 \qquad \text{for } \xi \to \mathbf{x}$$

has been proven.

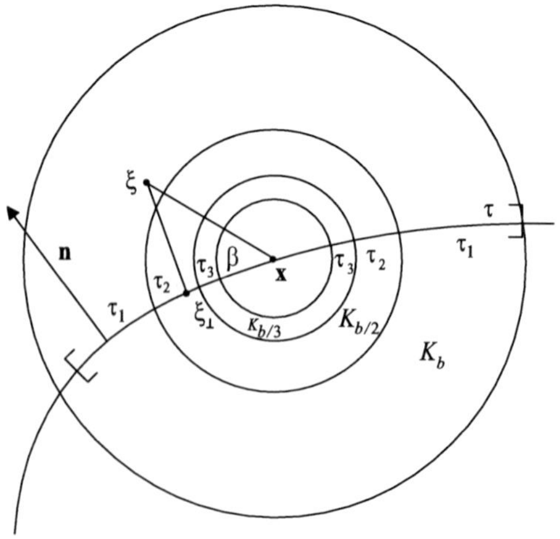

Fig. 3.2 Point ξ which converges towards $\mathbf{x} \in \Gamma$. The angle condition implies that the angle β is bounded from below away from 0. The intersection of the concentric circles $K_{b/i}$ with τ defines the subsets τ_i

Case a: We will first consider sequences that satisfy an angle condition, i.e., there exists some $\alpha \in]0,1[$ with

$$\|\xi - \xi_\perp\| \geq \alpha \|\xi - \mathbf{x}\| \qquad \forall \xi \in \{\mathbf{x}_n^\sigma : n \in \mathbb{N}\}. \tag{3.47}$$

For sufficiently large $n \in \mathbb{N}$ there exists some $m = m(n) \in \mathbb{N}$ with

$$\frac{b}{4m} \leq \|\mathbf{x}_n^\sigma - \mathbf{x}\| \leq \frac{b}{2m}, \tag{3.48}$$

where $b > 0$ is the smallest number such that $\tau \subset B_b(\mathbf{x})$. In the next step $\tau \cap B_\delta(\mathbf{x})$ is decomposed into m parts

$$\begin{aligned} \tau_m &:= \tau \cap B_{\frac{b}{m}}(\mathbf{x}), \\ \tau_i &:= \tau \cap \left(B_{\frac{b}{i}}(\mathbf{x}) \backslash B_{\frac{b}{i+1}}(\mathbf{x})\right) \qquad 1 \leq i \leq m-1. \end{aligned} \tag{3.49}$$

We then have

$$\tau \cap B_\delta(\mathbf{x}) = \bigcup_{i=1}^m \tau_i, \qquad |\tau_m| \leq C m^{1-d}, \qquad |\tau_i| \leq C i^{-d} \quad \text{for } 1 \leq i < m-1. \tag{3.50}$$

The integrals over these parts are denoted by

$$T_i := \int_{\tau_i} k\left(\mathbf{x}_n^\sigma, \mathbf{x}, \mathbf{y}\right) ds_{\mathbf{y}},$$

while we use the convention that integrals over sets of zero measure and over empty sets are equal to zero. Finally, we require the parameter

$$\rho(\beta) := \sup\{|\varphi(\eta) - \varphi(\mathbf{x})| : \|\eta - \mathbf{x}\| \leq \beta\}, \tag{3.51}$$

which converges to zero for $\beta \to 0$ because of the continuity of φ in $\mathbf{x}$. It follows that

$$I(\xi) \leq \sum_{i=1}^m T_i.$$

We begin by estimating the kernel function on τ_m and use Lemma 2.2.14, (3.51) and the equivalence of the norms $\|\cdot\|$, $\|\cdot\|_{\mathbf{A}}$ to obtain

$$k(\xi, \mathbf{x}, \mathbf{y}) \leq C\rho\left(\frac{b}{m}\right)\left(\left|\frac{1}{\|\mathbf{x}-\mathbf{y}\|_{\mathbf{A}}^{d-2}}\right| + \left|\frac{1}{\|\xi-\mathbf{y}\|_{\mathbf{A}}^{d-1}}\right|\right). \tag{3.52}$$

We estimate the denominator of the second summand

$$\|\xi - \mathbf{y}\|_{\mathbf{A}}^2 \geq c\,\|\xi - \mathbf{y}\|^2 \geq c\left(\|\xi - \xi_\perp\|^2 - 2\,|\langle \xi - \xi_\perp, \xi_\perp - \mathbf{y}\rangle| + \|\xi_\perp - \mathbf{y}\|^2\right). \tag{3.53}$$

The middle summand in (3.53) can be estimated by using Lemma 2.2.14 and (3.46), so that

$$|\langle \xi-\xi_\perp, \xi_\perp-\mathbf{y}\rangle| = \|\xi-\xi_\perp\|\,|\langle \mathbf{n}(\xi_\perp), \xi_\perp-\mathbf{y}\rangle| \leq C\,\|\xi-\mathbf{x}\|\,\|\xi_\perp-\mathbf{y}\|^2 \leq C\delta\,\|\xi_\perp-\mathbf{y}\|^2.$$

Therefore, for $\delta \leq (4C)^{-1}$, we have shown that

$$\|\xi-\mathbf{y}\|_{\mathbf{A}}^2 \geq c\,\|\xi-\mathbf{y}\|^2 \geq \frac{c}{2}\left(\|\xi-\xi_\perp\|^2 + \|\xi_\perp-\mathbf{y}\|^2\right) \overset{(3.47)}{\geq} \frac{c}{2}\alpha^2\,\|\xi - \mathbf{x}\|^2 \geq \frac{c}{2}\left(\frac{b\alpha}{m}\right)^2.$$

Finally, we obtain the inequality

$$k(\xi, \mathbf{x}, \mathbf{y}) \leq \widetilde{C}\rho\left(\frac{b}{m}\right)\left(\left|\frac{1}{\|\mathbf{x}-\mathbf{y}\|_{\mathbf{A}}^{d-2}}\right| + \left(\frac{m}{\alpha}\right)^{d-1}\right).$$

This means that

$$T_m = \int_{\tau_m} k\left(\mathbf{x}_n^\sigma, \mathbf{x}, \mathbf{y}\right) ds_{\mathbf{y}} \leq \widetilde{C}\rho\left(\frac{b}{m}\right)\left\{\int_\Gamma \frac{1}{\|\mathbf{x}-\mathbf{y}\|_{\mathbf{A}}^{d-2}} ds_{\mathbf{y}} + |\tau_m|\left(\frac{m}{\alpha}\right)^{d-1}\right\}.$$

As has already been shown in Theorem 3.3.5, the integral in the above inequality is bounded. Since Cm^{1-d} forms an upper bound for the surface $|\tau_m|$, from $\rho(b/m) \to 0$ it follows that $T_m \to 0$ for $m \to \infty$.

For the remaining terms T_i, $1 \leq i \leq m-1$, we use the decomposition

$$\begin{aligned}
&\frac{\langle \mathbf{n}(\mathbf{y}), \mathbf{x}-\mathbf{y}\rangle}{\|\mathbf{x}-\mathbf{y}\|_{\mathbf{A}}^d} - \frac{\langle \mathbf{n}(\mathbf{y}), \xi-\mathbf{y}\rangle}{\|\xi-\mathbf{y}\|_{\mathbf{A}}^d} \\
&= \frac{\langle \mathbf{n}(\mathbf{y}), \mathbf{x}-\xi\rangle}{\|\xi-\mathbf{y}\|_{\mathbf{A}}^d} + \left(\frac{1}{\|\mathbf{x}-\mathbf{y}\|_{\mathbf{A}}} - \frac{1}{\|\xi-\mathbf{y}\|_{\mathbf{A}}}\right)\sum_{k=1}^{d}\frac{\langle \mathbf{n}(\mathbf{y}), \mathbf{x}-\mathbf{y}\rangle}{\|\mathbf{x}-\mathbf{y}\|_{\mathbf{A}}^{d-k}\,\|\xi-\mathbf{y}\|_{\mathbf{A}}^{k-1}} \\
&=: S_1 + S_2.
\end{aligned}$$

The second summand can be estimated by the reverse triangle inequality

$$\begin{aligned}
|S_2| &\leq \frac{|\|\xi-\mathbf{y}\|_{\mathbf{A}} - \|\mathbf{x}-\mathbf{y}\|_{\mathbf{A}}|}{\|\mathbf{x}-\mathbf{y}\|_{\mathbf{A}}\,\|\xi-\mathbf{y}\|_{\mathbf{A}}}\sum_{k=1}^{d}\frac{\|\mathbf{x}-\mathbf{y}\|}{\|\mathbf{x}-\mathbf{y}\|_{\mathbf{A}}^{d-k}\,\|\xi-\mathbf{y}\|_{\mathbf{A}}^{k-1}} \\
&\leq C\sum_{k=1}^{d}\frac{\|\mathbf{x}-\xi\|_{\mathbf{A}}}{\|\mathbf{x}-\mathbf{y}\|_{\mathbf{A}}^{d-k}\,\|\xi-\mathbf{y}\|_{\mathbf{A}}^{k}},
\end{aligned}$$

so that on τ_i

$$|S_1 + S_2| \le \frac{\|\mathbf{x} - \xi\|}{\|\xi - \mathbf{y}\|_{\mathbf{A}}^d} + \widetilde{C} \sum_{k=1}^{d} \frac{\|\mathbf{x} - \xi\|_{\mathbf{A}}}{\|\mathbf{x} - \mathbf{y}\|_{\mathbf{A}}^{d-k} \|\xi - \mathbf{y}\|_{\mathbf{A}}^k} =: S^{(i)}$$

is proven. If we apply the inequalities (3.48)–(3.50) and

$$\frac{1}{\|\xi - \mathbf{y}\|^k} \le \frac{1}{|\|\mathbf{x} - \mathbf{y}\| - \|\xi - \mathbf{x}\||^k} \le \frac{1}{\left(\frac{b}{i+1} - \frac{b}{2m}\right)^k} \le \left(\frac{i+1}{b}\right)^k \frac{1}{\left(1 - \frac{i+1}{2m}\right)^k}$$
$$\le \left(2\frac{i+1}{b}\right)^k$$

to τ_i we obtain

$$S^{(i)} \le C \frac{(i+1)^d}{m}$$

with a constant C that depends only on $\mathbf{A}$, b and d. From this we have, with (3.50),

$$\sum_{i=1}^{m} T_i \le \frac{C}{m} \sum_{i=1}^{m} \rho\left(\frac{b}{i}\right) \frac{(i+1)^d}{i^d} \le \frac{\widehat{C}}{m} \sum_{i=1}^{m} \rho\left(\frac{b}{i}\right). \tag{3.54}$$

Since $\rho(b/i)$ is a null sequence the right-hand side in (3.54) converges to zero for $m \to \infty$. Because of (3.48) we also have $m \to \infty$ from $\mathbf{x}_n^\sigma \to \mathbf{x}$.

Case b: The proof of the auxiliary assumption for sequences that do not satisfy an angle condition of the form (3.47) requires a more complicated decomposition of the surface element τ and will not be carried out here. Instead we refer to [102, Theorem 18] for details.

The proof for the general double layer potential is based on the fact that the singularity of the difference function

$$\gamma_{1,\mathbf{y}} \left(G_0 (\mathbf{x} - \mathbf{y}) - G (\mathbf{x} - \mathbf{y})\right) \tag{3.55}$$

is reduced and therefore that the operator which is associated with the difference kernel (3.55) can be continuously extended to $\mathbb{R}^d$. Again we will not elaborate and refer to [117, Lemma 8.1.5] and [102, Theorem 41]. □

We have already mentioned that the explicit representation of the boundary integral operators is essential for their numerical solution. In Chap. 4 we will focus on discretization methods, for which the boundary integral operators have to be applied to bounded, piecewise smooth functions. The resulting functions will always be interpreted as L^2-functions, the values of which are always determined up to a set of zero measure. Under these conditions Theorem 3.3.13 can be simplified. We will use the notation from Definitions 2.2.9 and 2.2.12.

Corollary 3.3.14. *Let* $\Gamma \in C^2_{\mathrm{pw}}$ *and* $\varphi \in C^1_{\mathrm{pw}}(\Gamma)$. *Then (3.44) can be simplified for* $\sigma \in \{-, +\}$ *as an equality*

$$\gamma_0^\sigma D\varphi = \sigma \frac{1}{2}\varphi + K\varphi \tag{3.56}$$

in $L^2(\Gamma)$ *with* K *as given in (3.34).*

Proof. Let $\varphi \in C^1_{\text{pw}}(\Gamma)$ be arbitrary. Corollary 3.3.9 implies the continuity of the operators $K : C^1_{\text{pw}}(\Gamma) \to L^2(\Gamma)$ and $K' : C^1_{\text{pw}}(\Gamma) \to L^2(\Gamma)$. From this we have $K\varphi \in L^2(\Gamma)$. As the set of non-smooth points $\mathbf{x} \in \Gamma$ has zero measure, these have no effect on the equality in $L^2(\Gamma)$ and the assertion follows from Theorem 3.3.13. □

In the next theorem we will present the conormal trace of the single layer potential.

Theorem 3.3.15. *Let* $\Gamma \in C^2_{\text{pw}}$ *and* $\varphi \in C^1_{\text{pw}}(\Gamma)$. *Then we have for* $\sigma \in \{-,+\}$

$$\gamma_1^\sigma S\varphi = -\left(\sigma \frac{1}{2}\varphi - K'\varphi\right) \qquad \text{a.e. on } \Gamma \tag{3.57}$$

with K' *as given in (3.34).*

The jump relation (3.57) is first proven at smooth surface points $\mathbf{x}$, similarly to Theorem 3.3.13. Furthermore, the difference $\mathbf{n}(\mathbf{y}) - \mathbf{n}(\mathbf{x})$ has to be estimated in a neighborhood of $\mathbf{x}$. Details can be found in [102, Theorem 21] and will not be presented here.

3.3.4 Explicit Representation of the Boundary Integral Operator W

We will now turn our attention to the operator W. We have already shown in Theorem 3.3.1 that $[\gamma_1 D] = 0$ holds. This statement, however, does not contain any explicit representation of the boundary integral operator W. We will note in advance that the integral over the function $\gamma_{1,\mathbf{x}}\widetilde{\gamma_{1,\mathbf{y}}}G(\mathbf{x}-\mathbf{y})\varphi(\mathbf{y})$ in general does not exist as an improper integral over $\Gamma \times \Gamma$ and therefore the differentiation cannot be interchanged with the integration (see Remark 4.1.35). There do, however, exist different representations of the trace $\gamma_1 D$ as an improper integral. Here we choose the representation by means of *integration by parts*, which possesses favorable stability properties with respect to numerical discretization and is due to [159], [171] and [136]. Alternatively, one can also define the integral for the function $\gamma_{1,\mathbf{x}}\widetilde{\gamma_{1,\mathbf{y}}}G(\mathbf{x}-\mathbf{y})\varphi(\mathbf{y})$ by means of a generalized form of integration (Cauchy principal value, part-fini integral). We will deal with the Cauchy principal value in Sect. 5.1.2. For an approach via hypersingular or part-fini integrals we refer to [201] and [211].

In this section we will use elementary properties of Fourier analysis that are proven in, e.g., [243].

For $\varphi \in H^{1/2}(\Gamma)$ the double layer potential $D\varphi$ is defined by (3.5) or Definition 3.1.5. We set

$$X := \left\{ D\varphi : \varphi \in H^{1/2}(\Gamma) \right\}$$

and recall the relations

$$[u] = \varphi, \quad [\gamma_1 u] = 0 \quad \text{and} \quad L\left(u|_{\Omega^- \cup \Omega^+}\right) = 0 \quad \text{on } \Omega^- \cup \Omega^+ \tag{3.58}$$

for all $u = D\varphi \in X$ and L as in (3.1). We set $\mathbf{L}^2\left(\mathbb{R}^3\right) := \left(L^2\left(\mathbb{R}^3\right)\right)^3$. As the matrix $\mathbf{A}$ is positive definite, another positive definite matrix $\mathbf{A}^{1/2}$ can be uniquely defined by $\mathbf{A}^{1/2}\mathbf{A}^{1/2} = \mathbf{A}$. The function u is, in general, discontinuous across Γ so that the gradient ∇u cannot be defined in the classical sense, but instead has to be interpreted as a distribution. The "function part" $\mathbf{g} \in \mathbf{L}^2\left(\mathbb{R}^3\right)$ of $\mathbf{A}^{1/2}\nabla u$ is defined by

$$\mathbf{g}|_{\Omega^\sigma} := \mathbf{A}^{1/2}\nabla\, u|_{\Omega^\sigma} \qquad \sigma \in \{-,+\} \tag{3.59}$$

on $\Omega^- \cup \Omega^+$.

Lemma 3.3.16. *Let* $\mathbf{g}$ *be as in (3.59) and let the coefficients in* L *be denoted, as in (3.1), by* $\mathbf{A}$, $\mathbf{b}$ *and* c. *Then*

$$\operatorname{div}\left(\mathbf{A}^{1/2}\mathbf{g}\right) - 2\left\langle \mathbf{A}^{-1/2}\mathbf{b}, \mathbf{g}\right\rangle - cu = 0$$

in the sense of distributions on $C_0^\infty\left(\mathbb{R}^3\right)$.

Proof. The jump relations (3.58), the definition of the weak derivative and integration by parts on the subdomains Ω^-, Ω^+ together give us for all $w \in C_0^\infty\left(\mathbb{R}^3\right)$

$$\begin{aligned}
\operatorname{div}\left(\mathbf{A}^{1/2}\mathbf{g}\right)(w) &= -\int_{\mathbb{R}^3} \left\langle \mathbf{A}^{1/2}\mathbf{g}, \nabla w\right\rangle d\mathbf{x} = -\int_{\Omega^- \cup \Omega^+} \left\langle \mathbf{A}^{1/2}\mathbf{g}, \nabla w\right\rangle d\mathbf{x} \\
&= \int_{\Omega^- \cup \Omega^+} \operatorname{div}\left(\mathbf{A}^{1/2}\mathbf{g}\right) w\, d\mathbf{x} + \int_\Gamma [\gamma_1 u]\, w\, ds \\
&= \int_{\Omega^- \cup \Omega^+} \left((-Lu) + 2\left\langle \mathbf{A}^{-1/2}\mathbf{b}, \mathbf{g}\right\rangle + cu\right) w\, d\mathbf{x}.
\end{aligned}$$

□

In the following lemma we will be using the surface distribution δ_Γ, which is defined by

$$\mathbf{v}\delta_\Gamma(\mathbf{w}) := \int_\Gamma \langle \mathbf{v}, \mathbf{w}\rangle(\mathbf{x})\, ds_{\mathbf{x}} \tag{3.60}$$

for sufficiently smooth test functions $\mathbf{v}$, $\mathbf{w}$.

Lemma 3.3.17. *Let* $u = D\varphi \in X$ *and* $\mathbf{g}$ *be as in (3.59). Then*

$$\mathbf{A}^{1/2}\nabla u - \mathbf{g} = \varphi\mathbf{A}^{1/2}\mathbf{n}\delta_\Gamma$$

in the sense of distributions on $\mathbf{C}_0^\infty\left(\mathbb{R}^3\right)$.

Proof. For $\mathbf{w} \in \mathbf{C}_0^\infty(\mathbb{R}^3)$ we have

$$\left(\mathbf{A}^{1/2}\nabla u\right)(\mathbf{w}) = -\int_{\mathbb{R}^3} u \operatorname{div}\left(\mathbf{A}^{1/2}\mathbf{w}\right) d\mathbf{x} = \int_{\Omega^- \cup \Omega^+} \left\langle \mathbf{A}^{1/2}\nabla u, \mathbf{w}\right\rangle d\mathbf{x}$$
$$+ \int_\Gamma \varphi \left\langle \mathbf{A}^{1/2}\mathbf{n}, \mathbf{w}\right\rangle ds = \int_{\Omega^- \cup \Omega^+} \langle \mathbf{g}, \mathbf{w}\rangle \, d\mathbf{x} + \varphi \mathbf{A}^{1/2}\mathbf{n}\delta_\Gamma(\mathbf{w}).$$

□

Therefore we have derived the system of equations

$$\begin{aligned} \operatorname{div}\left(\mathbf{A}^{1/2}\mathbf{g}\right) - 2\left\langle \mathbf{A}^{-1/2}\mathbf{b}, \mathbf{g}\right\rangle - cu &= 0 \\ \mathbf{A}^{1/2}\nabla u - \mathbf{g} &= \varphi \mathbf{A}^{1/2}\mathbf{n}\delta_\Gamma \end{aligned} \tag{3.61}$$

for the functions $u, \mathbf{g}$.

Case 1: $c \neq 0$.

Elimination of u in the first equation gives us, with the abbreviation $\nabla^\intercal := \operatorname{div}$,

$$\mathbf{L}\mathbf{g} := \frac{1}{c}\left(\mathbf{A}^{1/2}\nabla\nabla^\intercal \mathbf{A}^{1/2}\mathbf{g} - 2\mathbf{A}^{1/2}\nabla \mathbf{b}^\intercal \mathbf{A}^{-1/2}\mathbf{g}\right) - \mathbf{g} = \varphi \mathbf{A}^{1/2}\mathbf{n}\delta_\Gamma. \tag{3.62}$$

We set

$$\Sigma := -cG\mathbf{I} - \operatorname{curl}_{\mathbf{A},-2\mathbf{b}} \operatorname{curl}_{\mathbf{A},\mathbf{0}}(G\mathbf{I}) \tag{3.63}$$

with

$$\operatorname{curl}_{\mathbf{A},\mathbf{v}} \mathbf{w} := \left(\mathbf{A}^{1/2}\nabla + \mathbf{A}^{-1/2}\mathbf{v}\right) \times \mathbf{w}$$

and the fundamental solution G from (3.3). The application of the differential operator $\operatorname{curl}_{\mathbf{A},\mathbf{v}}$ to a matrix is defined columnwise:

$$\Sigma w = -cG\mathbf{w} - \operatorname{curl}_{\mathbf{A},-2\mathbf{b}} \operatorname{curl}_{\mathbf{A},\mathbf{0}}(G\mathbf{w}).$$

Lemma 3.3.18. *Let $c \neq 0$. The function Σ in (3.63) is a fundamental solution of the operator $\mathbf{L}$ in (3.62), i.e.,*

$$\mathbf{L}\Sigma = \delta \mathbf{I}. \tag{3.64}$$

Proof. The Fourier transform of equation (3.64) combined with the substitutions $\zeta := \mathbf{A}^{1/2}\xi$ and $\tilde{\mathbf{b}} := \mathbf{A}^{-1/2}\mathbf{b}$ gives us

$$\left(-\frac{1}{c}\zeta\zeta^\intercal - \frac{2i}{c}\zeta\tilde{\mathbf{b}}^\intercal - \mathbf{I}\right)\hat{\Sigma} = \mathbf{I}. \tag{3.65}$$

If we insert the Fourier transform of Σ

$$\hat{\Sigma} = \frac{-c\mathbf{I}}{\|\zeta\|^2 + 2i\tilde{\mathbf{b}}^\intercal\zeta + c} - \left(i\zeta - 2\tilde{\mathbf{b}}\right) \times \left(i\zeta \times \frac{\mathbf{I}}{\|\zeta\|^2 + 2i\tilde{\mathbf{b}}^\intercal\zeta + c}\right) \tag{3.66}$$

into (3.65) and use the statement of Exercise 3.3.19 we obtain the assertion. □

Exercise 3.3.19. *Show that the function $\hat{\Sigma}$ in (3.66) satisfies (3.65).*

By means of a fundamental solution Σ, the solution of (3.62) can be written as a convolution

$$\mathbf{g} = \Sigma \star \left(\varphi \mathbf{A}^{1/2} \mathbf{n} \delta_\Gamma\right). \tag{3.67}$$

Case 2: $c = 0$.

We eliminate the function u from the second equation in (3.61) by applying the operator $\operatorname{curl}_{\mathbf{A},\mathbf{0}}$ and by then using $\operatorname{curl}_{\mathbf{A},\mathbf{0}} \left(\mathbf{A}^{1/2} \operatorname{grad} u\right) = 0$. This gives us

$$-\operatorname{curl}_{\mathbf{A},\mathbf{0}} \mathbf{g} = \operatorname{curl}_{\mathbf{A},\mathbf{0}} \left(\varphi \mathbf{A}^{1/2} \mathbf{n} \delta_\Gamma\right). \tag{3.68}$$

Applying the operator $\mathbf{A}^{1/2} \nabla$ to the first equation (3.61) (with $c = 0$) and using the relation

$$\mathbf{A}^{1/2} \nabla \left(\operatorname{div} \mathbf{A}^{1/2} \mathbf{g}\right) = \operatorname{div}(\mathbf{A}\nabla)\, \mathbf{g} + \operatorname{curl}_{\mathbf{A},\mathbf{0}} \operatorname{curl}_{\mathbf{A},\mathbf{0}} \mathbf{g}$$

gives us the equation

$$\operatorname{div}(\mathbf{A}\nabla)\, \mathbf{g} + \operatorname{curl}_{\mathbf{A},\mathbf{0}} \operatorname{curl}_{\mathbf{A},\mathbf{0}} \mathbf{g} - 2\mathbf{A}^{1/2} \nabla \left\langle \mathbf{A}^{-1/2} \mathbf{b}, \mathbf{g} \right\rangle = 0. \tag{3.69}$$

Elementary tensor algebra then results in the two relations:

$$\begin{aligned} \operatorname{curl}_{\mathbf{A},\mathbf{0}} \operatorname{curl}_{\mathbf{A},\mathbf{0}} \mathbf{g} &= \operatorname{curl}_{\mathbf{A},-2\mathbf{b}} \operatorname{curl}_{\mathbf{A},\mathbf{0}} \mathbf{g} + 2\left(\mathbf{A}^{-1/2} \mathbf{b}\right) \times \operatorname{curl}_{\mathbf{A},\mathbf{0}} \mathbf{g} \\ 2\left(\mathbf{A}^{-1/2} \mathbf{b}\right) \times \operatorname{curl}_{\mathbf{A},\mathbf{0}} \mathbf{g} - 2\mathbf{A}^{1/2} \nabla \left\langle \mathbf{A}^{-1/2} \mathbf{b}, \mathbf{g} \right\rangle &= -2 \left\langle \mathbf{b}, \nabla \right\rangle \mathbf{g}. \end{aligned} \tag{3.70}$$

By combining (3.68)–(3.70) we obtain

$$\operatorname{div}(\mathbf{A}\nabla)\, \mathbf{g} - 2 \left\langle \mathbf{b}, \nabla \right\rangle \mathbf{g} = \operatorname{curl}_{\mathbf{A},-2\mathbf{b}} \operatorname{curl}_{\mathbf{A},\mathbf{0}} \left(\varphi \mathbf{n} \delta_\Gamma\right).$$

The solution of this equation is

$$\mathbf{g} = -\left(G\mathbf{I}\right) \star \operatorname{curl}_{\mathbf{A},-2\mathbf{b}} \operatorname{curl}_{\mathbf{A},\mathbf{0}} \left(\varphi \mathbf{n} \delta_\Gamma\right).$$

As differential operators commute with convolutions, we have

$$\mathbf{g} = -\operatorname{curl}_{\mathbf{A},-2\mathbf{b}} \operatorname{curl}_{\mathbf{A},\mathbf{0}} \left(G\mathbf{I}\right) \star \left(\varphi \mathbf{n} \delta_\Gamma\right).$$

This means that the representations (3.63) and (3.67) remain valid for $c = 0$ as well.

Elementary properties of convolutions (see, e.g., [243, Chap. VI.3]) combined with (3.60) yield

$$\mathbf{g}(\mathbf{x}) = \left(\varphi \mathbf{A}^{1/2}\mathbf{n}\delta_\Gamma \star \Sigma\right)(\mathbf{x}) = \int_\Gamma \varphi(\mathbf{y}) \left\langle \mathbf{A}^{1/2}\mathbf{n}_\mathbf{y}, \Sigma(\mathbf{x}-\mathbf{y})\cdot\right\rangle ds_\mathbf{y}. \tag{3.71}$$

Since in (3.58) we set $u = D\varphi \in X$ for an arbitrary $\varphi \in H^{1/2}(\Gamma)$. We use the relation (3.59) between $\mathbf{g}$ and $\mathbf{A}^{1/2}\nabla u$ and note the continuity of $\left\langle \mathbf{A}^{1/2}\mathbf{n}_\mathbf{x}, \mathbf{g}(\mathbf{x})\right\rangle$ over Γ [see (3.58)]. Then we have for all $\psi \in H^{1/2}(\Gamma)$

$$\int_\Gamma (\gamma_1 u)\,\psi ds = \int_\Gamma \langle \mathbf{A}\mathbf{n}, \nabla u\rangle\,\psi ds = \int_\Gamma \left\langle \mathbf{A}^{1/2}\mathbf{n}_\mathbf{x}, \mathbf{g}(\mathbf{x})\right\rangle \psi(\mathbf{x})\,ds_\mathbf{x} \tag{3.72}$$

$$= \int_\Gamma\int_\Gamma \varphi(\mathbf{y})\,\psi(\mathbf{x}) \left\langle \mathbf{A}^{1/2}\mathbf{n}_\mathbf{y}, \Sigma(\mathbf{x}-\mathbf{y})\,\mathbf{A}^{1/2}\mathbf{n}_\mathbf{x}\right\rangle ds_\mathbf{y}ds_\mathbf{x}. \tag{3.73}$$

The definition of the function Σ in (3.63) is somewhat unwieldy. We will therefore simplify the integrand in (3.73) somewhat in the next step. The first summand of Σ in (3.63) is the motivation behind the definition of the integral

$$I_1 := -c\int_\Gamma\int_\Gamma G(\mathbf{x}-\mathbf{y})\,\varphi(\mathbf{y})\,\psi(\mathbf{x}) \left\langle \mathbf{A}^{1/2}\mathbf{n}_\mathbf{y}, \mathbf{A}^{1/2}\mathbf{n}_\mathbf{x}\right\rangle ds_\mathbf{y}ds_\mathbf{x}. \tag{3.74}$$

Hence the right-hand side in (3.73) is equal to $I_1 + I_2$ with

$$I_2 := \int_\Gamma \left\langle -\operatorname{curl}_{\mathbf{A},-2\mathbf{b}} \operatorname{curl}_{\mathbf{A},0} G\mathbf{I} \star \left(\varphi \mathbf{A}^{1/2}\mathbf{n}\delta_\Gamma\right), \psi \mathbf{A}^{1/2}\mathbf{n}\right\rangle ds.$$

Since differentiation and convolutions commute, we have

$$\begin{aligned} I_2 &= -\int_\Gamma \left\langle \operatorname{curl}_{\mathbf{A},-2\mathbf{b}} G\mathbf{I} \star \left(\operatorname{curl}_{\mathbf{A},0}\left(\varphi \mathbf{A}^{1/2}\mathbf{n}\delta_\Gamma\right)\right), \psi \mathbf{A}^{1/2}\mathbf{n}\right\rangle ds \\ &= -\left(\operatorname{curl}_{\mathbf{A},-2\mathbf{b}} G\mathbf{I} \star \left(\operatorname{curl}_{\mathbf{A},0}\left(\varphi \mathbf{A}^{1/2}\mathbf{n}\delta_\Gamma\right)\right)\right)\left(\psi \mathbf{A}^{1/2}\mathbf{n}\delta_\Gamma\right). \end{aligned}$$

Lemma 3.3.20. *The integral I_2 has the representation*

$$I_2 = \left\langle G\mathbf{I} \star \left(\operatorname{curl}_{\mathbf{A},0}\left(\varphi \mathbf{A}^{1/2}\mathbf{n}\delta_\Gamma\right)\right), \operatorname{curl}_{\mathbf{A},2\mathbf{b}}\left(\psi \mathbf{A}^{1/2}\mathbf{n}\delta_\Gamma\right)\right\rangle.$$

Proof. Let $\mathbf{v} \in \mathbb{R}^3$, $\mathbf{q} := G\mathbf{I}\star\left(\operatorname{curl}_{\mathbf{A},0}\left(\varphi \mathbf{A}^{1/2}\mathbf{n}\delta_\Gamma\right)\right)$ and $\mathbf{w} := \psi \mathbf{A}^{1/2}\mathbf{n}\delta_\Gamma$. It follows from Parseval's equation that

$$\begin{aligned} \left(\operatorname{curl}_{\mathbf{A},\mathbf{v}}(\mathbf{q}), \mathbf{w}\right)_{L^2(\mathbb{R}^3)} &= \frac{1}{(2\pi)^3}\left(\widehat{\operatorname{curl}_{\mathbf{A},\mathbf{v}}(\mathbf{q})}, \widehat{\mathbf{w}}\right)_{L^2(\mathbb{R}^3)} \\ &= \frac{1}{(2\pi)^3}\int_{\mathbb{R}^3} \left\langle \left(\mathbf{A}^{1/2} i\xi + \mathbf{A}^{-1/2}\mathbf{v}\right) \times \mathbf{q}(\xi), \overline{\widehat{\mathbf{w}}}(\xi)\right\rangle d\xi, \end{aligned}$$

where $\langle\cdot,\cdot\rangle$ is defined without complex conjugation. Elementary tensor algebra gives us

$$\begin{aligned}(\operatorname{curl}_{\mathbf{A},\mathbf{v}}(\mathbf{q}),\mathbf{w})_{L^2(\mathbb{R}^3)} &= \frac{1}{(2\pi)^3}\int_{\mathbb{R}^3}\left\langle \mathbf{q}(\xi),\overline{\left(\mathbf{A}^{1/2}i\xi-\mathbf{A}^{-1/2}\mathbf{v}\right)\times\hat{\mathbf{w}}(\xi)}\right\rangle d\xi \\ &= (\mathbf{q},\operatorname{curl}_{\mathbf{A},-\mathbf{v}}(\mathbf{w}))_{L^2(\mathbb{R}^3)},\end{aligned}$$

from which we have the assertion. □

Finally, we will apply $\operatorname{curl}_{\mathbf{A},\mathbf{v}}$ to the distribution $\psi\mathbf{A}^{1/2}\mathbf{n}\delta_\Gamma$. For this we define, for $\lambda \in H^{1/2}(\Gamma)$, the boundary differential operator

$$\operatorname{curl}_{\Gamma,\mathbf{A},\mathbf{v}}\lambda := \left(\mathbf{A}^{1/2}\operatorname{grad}\lambda^\star + \lambda\mathbf{A}^{-1/2}\mathbf{v}\right)\times\mathbf{A}^{1/2}\mathbf{n}, \tag{3.75}$$

where $\lambda^\star := Z_-\lambda \in H^1(\Omega^-)$ denotes the trace extension of λ in Ω^- (see Remark 2.6.12).

Lemma 3.3.21. *Let $\lambda \in H^{1/2}(\Gamma)$. Then*

$$\operatorname{curl}_{\mathbf{A},\mathbf{v}}\left(\lambda\mathbf{A}^{1/2}\mathbf{n}\delta_\Gamma\right) = (\operatorname{curl}_{\Gamma,\mathbf{A},\mathbf{v}}\lambda)\,\delta_\Gamma$$

in the sense of distributions on $\mathbf{C}_0^\infty(\mathbb{R}^3)$.

Proof. Let $\mathbf{w} \in \mathbf{C}_0^\infty(\mathbb{R}^3)$. Then

$$\begin{aligned}\left(\operatorname{curl}_{\mathbf{A},\mathbf{v}}\left(\lambda\mathbf{A}^{1/2}\mathbf{n}\delta_\Gamma\right)\right)(\mathbf{w}) &= \int_\Gamma\left\langle\lambda\mathbf{A}^{1/2}\mathbf{n},\operatorname{curl}_{\mathbf{A},-\mathbf{v}}\mathbf{w}\right\rangle ds \\ &= \int_\Gamma\left\langle\mathbf{n},\lambda\mathbf{A}^{1/2}\operatorname{curl}_{\mathbf{A},-\mathbf{v}}\mathbf{w}\right\rangle ds.\end{aligned} \tag{3.76}$$

Let $\lambda^\star := Z_-\lambda$. One can easily verify that $\operatorname{div}\left(\mathbf{A}^{1/2}\operatorname{curl}_{\mathbf{A},\mathbf{0}}(\cdot)\right) = 0$. Thus with Gauss' theorem we obtain

$$\int_\Gamma\left\langle\mathbf{n},\mathbf{A}^{1/2}\operatorname{curl}_{\mathbf{A},\mathbf{0}}(\lambda\mathbf{w})\right\rangle ds = \int_{\Omega^-}\operatorname{div}\left(\mathbf{A}^{1/2}\operatorname{curl}_{\mathbf{A},\mathbf{0}}(\lambda^\star\mathbf{w})\right)d\mathbf{x} = 0. \tag{3.77}$$

On the other hand, elementary tensor algebra gives us

$$\mathbf{A}^{1/2}\operatorname{curl}_{\mathbf{A},\mathbf{0}}(\lambda^\star\mathbf{w}) = \mathbf{A}^{1/2}\left(\left(\mathbf{A}^{1/2}\operatorname{grad}\lambda^\star\right)\times\mathbf{w} + \lambda^\star\operatorname{curl}_{\mathbf{A},\mathbf{0}}\mathbf{w}\right). \tag{3.78}$$

By combining (3.76)–(3.78) we obtain

$$\begin{aligned}
\operatorname{curl}_{\mathbf{A},\mathbf{v}}\left(\lambda\mathbf{A}^{1/2}\mathbf{n}\delta_\Gamma\right)(\mathbf{w}) &= \int_\Gamma \left\langle \mathbf{n}, \lambda\mathbf{A}^{1/2}\operatorname{curl}_{\mathbf{A},\mathbf{0}}\mathbf{w}\right\rangle - \left\langle \mathbf{A}^{1/2}\mathbf{n}, \lambda\left(\mathbf{A}^{-1/2}\mathbf{v}\right)\times\mathbf{w}\right\rangle ds \\
&= -\int_\Gamma \left\langle \mathbf{A}^{1/2}\mathbf{n}, \left(\mathbf{A}^{1/2}\operatorname{grad}\lambda^\star\right)\times\mathbf{w}\right\rangle \\
&\quad + \left\langle \mathbf{A}^{1/2}\mathbf{n}, \lambda\left(\mathbf{A}^{-1/2}\mathbf{v}\right)\times\mathbf{w}\right\rangle ds \\
&= \int_\Gamma \left\langle\left(\mathbf{A}^{1/2}\operatorname{grad}\lambda^\star + \lambda\mathbf{A}^{-1/2}\mathbf{v}\right)\times\mathbf{A}^{1/2}\mathbf{n}, \mathbf{w}\right\rangle ds.
\end{aligned}$$

□

We have thus derived the following representation of I_2

$$I_2 = -\int_{\Gamma\times\Gamma} G(\mathbf{x}-\mathbf{y})\left\langle \operatorname{curl}_{\Gamma,\mathbf{A},\mathbf{0}}\varphi(\mathbf{x}), \operatorname{curl}_{\Gamma,\mathbf{A},2\mathbf{b}}\psi(\mathbf{y})\right\rangle ds_{\mathbf{x}}ds_{\mathbf{y}}.$$

If we combine the previous representations we obtain the proof of the following theorem.

Theorem 3.3.22. *Let $\varphi,\psi\in H^{1/2}(\Gamma)$ and $u = D\varphi$. Then*

$$\begin{aligned}
b_W(\varphi,\psi) &:= \int_{\Gamma\times\Gamma} G(\mathbf{x}-\mathbf{y})\left\langle \operatorname{curl}_{\Gamma,\mathbf{A},\mathbf{0}}\overline{\psi}(\mathbf{x}), \operatorname{curl}_{\Gamma,\mathbf{A},2\mathbf{b}}\varphi(\mathbf{y})\right\rangle ds_{\mathbf{x}}ds_{\mathbf{y}} \\
&\quad + c\int_{\Gamma\times\Gamma} G(\mathbf{x}-\mathbf{y})\overline{\psi}(\mathbf{x})\varphi(\mathbf{y})\left\langle \mathbf{A}^{1/2}\mathbf{n}(\mathbf{x}), \mathbf{A}^{1/2}\mathbf{n}(\mathbf{y})\right\rangle ds_{\mathbf{x}}ds_{\mathbf{y}} \\
&= -\int_\Gamma (\gamma_1 u)\overline{\psi}\,ds.
\end{aligned}$$

Remark 3.3.23. *As $\gamma_1 u = \gamma_1 D\varphi = -W\varphi$, we have the representation*

$$b_W(\varphi,\psi) = (W\varphi,\psi)_{L^2(\Gamma)} \qquad \forall\varphi,\psi\in H^{1/2}(\Gamma).$$

Corollary 3.3.24. *For the Laplace operator "$-\Delta$", the bilinear form b_W is explicitly given by*

$$b_W(\varphi,\psi) := \int_{\Gamma\times\Gamma} \frac{\left\langle \operatorname{curl}_\Gamma\varphi(\mathbf{y}), \operatorname{curl}_\Gamma\psi(\mathbf{x})\right\rangle}{4\pi\|\mathbf{x}-\mathbf{y}\|} ds_{\mathbf{x}}ds_{\mathbf{y}},$$

where

$$\operatorname{curl}_\Gamma\lambda := \operatorname{grad}\lambda^\star\times\mathbf{n} \quad \textit{and} \quad \lambda^\star := Z_-\lambda.$$

For the Helmholtz operator "$-\Delta - k^2$", we obtain

$$\begin{aligned}
b_W(\varphi,\psi) := \int_{\Gamma\times\Gamma} \frac{\mathrm{e}^{\mathrm{i}k\|\mathbf{x}-\mathbf{y}\|}}{4\pi\|\mathbf{x}-\mathbf{y}\|} &\left\{\left\langle \operatorname{curl}_\Gamma\varphi(\mathbf{y}), \operatorname{curl}_\Gamma\overline{\psi}(\mathbf{x})\right\rangle\right. \\
&\left. - k^2\left\langle \mathbf{n}(\mathbf{x}), \mathbf{n}(\mathbf{y})\right\rangle\varphi(\mathbf{y})\overline{\psi}(\mathbf{x})\right\} ds_{\mathbf{x}}ds_{\mathbf{y}}.
\end{aligned}$$

The implementation of the Galerkin discretization of boundary integral equations requires the transformation of the operator ($\gamma_0 \operatorname{grad} Z_{-}u$) in (3.75) to two-dimensional parameter domains. The following remark provides us with the relevant transformation formula (see [170, Chap. 2]).

Exercise 3.3.25. *Let $\tau \subset \Gamma$ be a panel which is transformed back to a two-dimensional parameter domain $\hat{\tau} \subset \mathbb{R}^2$ by means of a $\mathbf{C}^1$-diffeomorphism $\chi_\tau : \hat{\tau} \to \tau$. Let $\hat{u} : \hat{\tau} \to \mathbb{K}$ be a sufficiently smooth function and let $u : \tau \to \mathbb{K}$ be defined by lifting, i.e., $u := \hat{u} \circ \chi_\tau^{-1}$. Then we have*

$$(\gamma_0 \operatorname{grad} Z_{-}u) \circ \chi_\tau = \mathbf{J}_\tau \mathbf{G}_\tau^{-1} \widehat{\nabla} \hat{u}, \tag{3.79}$$

where $\widehat{\nabla}$ denotes the gradient with respect to the parameter variables, $J_\tau : \hat{\tau} \to \mathbb{R}^{3\times 2}$ denotes the Jacobian of the transformation χ_τ and $\mathbf{G}_\tau := \mathbf{J}_\tau^\intercal \mathbf{J}_\tau \in \mathbb{R}^{2\times 2}$ is the Gram matrix.

3.4 Integral Equations for Elliptic Boundary Value Problems

In Sect. 2.9 we formulated elliptic boundary value problems as variational problems. These problems can be transformed into integral equations, which are derived in this chapter.

We will present two methods with which one can formulate elliptic boundary value problems as boundary integral equations. The *indirect method* uses an ansatz consisting of potentials. The unknown density functions are then determined by the given boundary data. The *direct method* uses an ansatz where the given boundary data is inserted into Green's representation formula, which in turn is solved for the unknown boundary data. Formulating elliptic boundary value problems as integral equations is very advantageous from a numerical point of view if the right-hand side in the differential equation equals zero. Therefore we will always assume, unless explicitly stated otherwise, that the right-hand sides f in (2.129)–(2.137) are all equal to zero. All formulations can be modified by adding Newton potentials $\mathcal{N} f$ should the source term f not be equal to zero.

The sesquilinear forms associated with the operators V, K_+, K_-, K'_+, K'_-, W [see (3.6)] are, for $\sigma \in \{-, +\}$, given by

$$\begin{array}{ll}
b_V : H^{-1/2}(\Gamma) \times H^{-1/2}(\Gamma) \to \mathbb{C} & b_V(\varphi, \psi) := (V\varphi, \psi)_{L^2(\Gamma)} \\
b_K^\sigma : H^{1/2}(\Gamma) \times H^{-1/2}(\Gamma) \to \mathbb{C} & b_K^\sigma(\varphi, \psi) := \sigma \frac{1}{2}(\varphi, \psi)_{L^2(\Gamma)} + b_K(\varphi, \psi) \\
b_{K'}^\sigma : H^{-1/2}(\Gamma) \times H^{1/2}(\Gamma) \to \mathbb{C} & b_{K'}^\sigma(\varphi, w) := -\sigma \frac{1}{2}(\varphi, \psi)_{L^2(\Gamma)} + b_{K'}(\varphi, \psi) \\
b_W : H^{1/2}(\Gamma) \times H^{1/2}(\Gamma) \to \mathbb{C} & b_W(\varphi, \psi) := (W\varphi, \psi)_{L^2(\Gamma)}
\end{array}$$

with

$$b_K(\varphi, \psi) := (K\varphi, \psi)_{L^2(\Gamma)} \text{ and } b_{K'}(\varphi, \psi) := (K'\varphi, \psi)_{L^2(\Gamma)}$$

[see (3.34)], where $(\cdot,\cdot)_{L^2(\Gamma)}$ again denotes the extension of the $L^2(\Gamma)$ inner product to $H^{1/2}(\Gamma) \times H^{-1/2}(\Gamma)$ or $H^{-1/2}(\Gamma) \times H^{1/2}(\Gamma)$. If the sesquilinear form – as in the case of b_K^σ, $b_{K'}^\sigma$ – contains one summand which is of the form $\pm 1/2\,(u, \varphi)_{L^2(\Gamma)}$ it is called an integral operator of the *second kind*. Otherwise, it is called – in the case b_V and b_W – an operator of the *first kind*. The associated integral equations are appropriately called equations of the first and the second kind.

3.4.1 The Indirect Method

For functions $\varphi \in H^{-1/2}(\Gamma)$ and $\psi \in H^{1/2}(\Gamma)$ we can define the potentials

$$\begin{aligned} u_- &:= S\varphi \quad \text{on } \Omega^-, & w_- &:= D\psi \quad \text{on } \Omega^-, \\ u_+ &:= S\varphi \quad \text{on } \Omega^+, & w_+ &:= D\psi \quad \text{on } \Omega^+. \end{aligned} \tag{3.80}$$

The principle of the indirect method consists in first determining the unknown *density function* φ as a solution of a boundary integral equation by means of the given boundary conditions. Then we can insert it into the associated potentials, which gives us the solution of the boundary value problem. The following proposition recalls the properties of the potentials $u_\pm$ and $w_\pm$.

Proposition 3.4.1. *The functions $u_\pm$, $w_\pm$ from (3.80) satisfy*

1. $u_- \in H^1(\Omega^-)$ *and* $Lu_- = 0$ *in* Ω^-,
2. $u_+ \in H^1_{\text{loc}}(\Omega^+)$ *and* $Lu_+ = 0$ *in* Ω^+,
3. $w_- \in H^1(\Omega^-)$ *and* $Lw_- = 0$ *in* Ω^-,
4. $w_+ \in H^1_{\text{loc}}(\Omega^+)$ *and* $Lw_+ = 0$ *in* Ω^+.

We will formulate the integral equations for the boundary value problems given in Sect. 2.9. The formulation as an integral equation is by no means unique. The single layer potential can just as well be used as the double layer potential.

3.4.1.1 Interior Problems

IDP:

Single Layer Potential: Let $g_D \in H^{1/2}(\Gamma)$ be given. Find $\varphi \in H^{-1/2}(\Gamma)$ such that

$$b_V(\varphi, \eta) = (g_D, \eta)_{L^2(\Gamma)} \qquad \forall \eta \in H^{-1/2}(\Gamma).$$

Double Layer Potential: Let $g_D \in H^{1/2}(\Gamma)$ be given. Find $\psi \in H^{1/2}(\Gamma)$ such that

$$-\frac{1}{2}(\psi,\eta)_{L^2(\Gamma)} + b_K(\psi,\eta) = (g_D,\eta)_{L^2(\Gamma)} \qquad \forall \eta \in H^{-1/2}(\Gamma). \tag{3.81}$$

INP:

Single Layer Potential: Let $g_N \in H^{-1/2}(\Gamma)$ be given. Find $\varphi \in H^{-1/2}(\Gamma)$ such that

$$\frac{1}{2}(\varphi,\eta)_{L^2(\Gamma)} + b_{K'}(\varphi,\eta) = (g_N,\eta)_{L^2(\Gamma)} \quad \forall \eta \in H^{1/2}(\Gamma).$$

Double Layer Potential: Let $g_N \in H^{-1/2}(\Gamma)$ be given. Find $\psi \in H^{1/2}(\Gamma)$ such that

$$b_W(\psi,\eta) = -(g_N,\eta)_{L^2(\Gamma)} \qquad \forall \eta \in H^{1/2}(\Gamma).$$

IDNP:

In order to formulate integral equations for mixed boundary value problems, we need to use Sobolev spaces on the Dirichlet and Neumann parts of the boundary Γ. Here we will only introduce the relevant function spaces and summarize the required theorems. For a detailed analysis we refer to [162, p. 231 ff].

Let $\Gamma_0 \subset \Gamma$ be a measurable subset of the boundary with $|\Gamma_0| > 0$. The Sobolev space $\widetilde{H}^s(\Gamma_0)$, $s \in [0,1]$ was defined in (2.90) as

$$\widetilde{H}^s(\Gamma_0) = \left\{u \in H^s(\Gamma) : \operatorname{supp}(u) \subset \overline{\Gamma_0}\right\}. \tag{3.82}$$

The norm on $\widetilde{H}^s(\Gamma_0)$ is defined by

$$\|u\|_{H^s(\Gamma_0)} := \left\|u^\star\right\|_{H^s(\Gamma)}, \tag{3.83}$$

where $u^\star$ denotes the extension of u on Γ by zero.

The spaces with negative indices are again defined as dual spaces: $\widetilde{H}^{-s}(\Gamma_0) := (H^s(\Gamma_0))'$ for $s \in [0,1]$. Conversely, we have: $H^{-s}(\Gamma_0) = \left(\widetilde{H}^s(\Gamma_0)\right)'$ for $s \in [0,1]$.

For the mixed boundary value problem our aim is to find the Dirichlet data on Γ_N where the Neumann data is given and vice-versa on Γ_D. This requires the localization of the boundary integral operators on Γ_N and Γ_D in the range as well as the domain. For functions φ, ψ on Γ with $\operatorname{supp}(\varphi) \subset \overline{\Gamma_D}$ and $\operatorname{supp}(\psi) \subset \overline{\Gamma_N}$ we set

$$V_{DD}\varphi := (V\varphi)|_{\Gamma_D}, \quad K'_{ND}\varphi := (K'\varphi)|_{\Gamma_N},$$

$$K_{DN}\psi := (K\psi)|_{\Gamma_D}, \; W_{NN}\psi := (W\psi)|_{\Gamma_N},$$

where the operators K, K' are given by

$$K = -\frac{1}{2}I + \gamma_0^+ D = \frac{1}{2}I + \gamma_0^- D$$
$$K' = \frac{1}{2}I + \gamma_1^+ S = -\frac{1}{2}I + \gamma_1^- S$$

and the two equalities on the right-hand side follow from (3.56) and (3.57). For sufficiently smooth φ, ψ they have the representation (3.34). The mapping properties of these operators are given in the following theorem.

Theorem 3.4.2. *We have*

$$V_{DD} : \widetilde{H}^{-1/2}(\Gamma_D) \to H^{1/2}(\Gamma_D),\ K'_{ND} : \widetilde{H}^{-1/2}(\Gamma_D) \to H^{-1/2}(\Gamma_N),$$

$$K_{DN} : \widetilde{H}^{1/2}(\Gamma_N) \to H^{1/2}(\Gamma_D),\ \ W_{NN} : \widetilde{H}^{1/2}(\Gamma_N) \to H^{-1/2}(\Gamma_N).$$

For $\varphi \in \widetilde{H}^{-1/2}(\Gamma_D)$ and $\psi \in \widetilde{H}^{1/2}(\Gamma_N)$ we use, in Ω^σ, $\sigma \in \{-,+\}$, the ansatz

$$u^\sigma = S\varphi - D\psi \qquad \text{in } \Omega^\sigma.$$

If we then form the traces we obtain

$$\gamma_0^\sigma u^\sigma = V\varphi - \left(\frac{\sigma 1}{2}I + K\right)\psi,$$
$$\gamma_1^\sigma u^\sigma = \left(-\frac{\sigma 1}{2}I + K'\right)\varphi + W\psi.$$

If we consider the first integral equation on Γ_D and the second on Γ_N and if we apply the given data $\left(\gamma_0^\sigma u^\sigma\right)\big|_{\Gamma_D} = g_D$ and $\left(\gamma_1^\sigma u^\sigma\right)\big|_{\Gamma_N} = g_N$, we obtain a system of integral equations for $\varphi \in \widetilde{H}^{-1/2}(\Gamma_D)$ and $\psi \in \widetilde{H}^{1/2}(\Gamma_N)$

$$g_D = V_{DD}\varphi - \left(\frac{\sigma 1}{2}I_{DN} + K_{DN}\right)\psi \qquad \text{on } \Gamma_D, \tag{3.84}$$

$$g_N = \left(-\frac{\sigma 1}{2}I_{ND} + K'_{ND}\right)\varphi + W_{NN}\psi \qquad \text{on } \Gamma_N. \tag{3.85}$$

If we then combine the operators in (3.84) and (3.85) to form a 2×2-system of operators we formally obtain for $(\varphi, \psi) \in \widetilde{H}^{-1/2}(\Gamma_D) \times \widetilde{H}^{1/2}(\Gamma_N)$ the equation

$$\begin{bmatrix} V_{DD} & -\left(\sigma\frac{1}{2}I_{DN} + K_{DN}\right) \\ \left(-\frac{\sigma 1}{2}I_{ND} + K'_{ND}\right) & W_{NN} \end{bmatrix}\begin{pmatrix}\varphi\\ \psi\end{pmatrix} = \begin{pmatrix} g_D \\ g_N\end{pmatrix} \tag{3.86}$$

in $H^{1/2}(\Gamma_D) \times H^{-1/2}(\Gamma_N)$. If we multiply (from the right-hand side) by $(\eta, \kappa) \in \widetilde{H}^{-1/2}(\Gamma_D) \times \widetilde{H}^{1/2}(\Gamma_N)$ and integrate over Γ_D and Γ_N we obtain

$$\begin{aligned} b_{V_{DD}}(\varphi,\eta) &- b_{K_{DN}}(\psi,\eta) + b_{K'_{ND}}(\varphi,\kappa) + b_{W_{NN}}(\psi,\kappa) \\ &= (g_D,\eta)_{L^2(\Gamma_D)} + (g_N,\kappa)_{L^2(\Gamma_N)}, \end{aligned} \tag{3.87}$$

where each sesquilinear form is defined by localization analogously

$$b_{V_{DD}} : \widetilde{H}^{-1/2}(\Gamma_D) \times \widetilde{H}^{-1/2}(\Gamma_D) \to \mathbb{C} \quad b_{V_{DD}}(\varphi,\eta) := (V_{DD}\varphi,\eta)_{L^2(\Gamma_D)}$$

$$b_{K_{DN}} : \widetilde{H}^{-1/2}(\Gamma_D) \times \widetilde{H}^{1/2}(\Gamma_N) \to \mathbb{C} \quad b_{K_{DN}}(\psi,\eta) := (K_{DN}\psi,\eta)_{L^2(\Gamma_D)}$$

$$b_{K'_{ND}} : \widetilde{H}^{1/2}(\Gamma_N) \times \widetilde{H}^{-1/2}(\Gamma_D) \to \mathbb{C} \quad b_{K'_{ND}}(\varphi,\kappa) := \left(K'_{ND}\psi,\kappa\right)_{L^2(\Gamma_N)}$$

$$b_{W_{NN}} : \widetilde{H}^{1/2}(\Gamma_N) \times \widetilde{H}^{1/2}(\Gamma_N) \to \mathbb{C} \quad b_{W_{NN}}(\psi,\kappa) := (W_{NN}\psi,\kappa)_{L^2(\Gamma_N)}.$$

Note that the contribution of the identity operators I_{DN} and I_{ND} in (3.86) vanishes, as Γ_N and Γ_D have a disjoint interior. We obtain a more compact representation if we use the left-hand side of (3.87) to define the sesquilinear form b_{mixed} on $\mathbf{H} \times \mathbf{H}$, with

$$\mathbf{H} := \widetilde{H}^{-1/2}(\Gamma_D) \times \widetilde{H}^{1/2}(\Gamma_N). \tag{3.88}$$

The direct method for the interior mixed problem as an integral equation then reads:

In (2.144) and (2.145) let $f \equiv 0$ and $(g_D, g_N) \in H^{1/2}(\Gamma_D) \times H^{-1/2}(\Gamma_N)$ be given. Find $(\varphi,\psi) \in \mathbf{H}$ such that

$$b_{mixed}\left(\begin{pmatrix}\varphi\\ \psi\end{pmatrix}, \begin{pmatrix}\eta\\ \kappa\end{pmatrix}\right) = (g_D,\eta)_{L^2(\Gamma_D)} + (g_N,\kappa)_{L^2(\Gamma_N)} \qquad \forall (\eta,\kappa) \in \mathbf{H}. \tag{3.89}$$

Remark 3.4.3. *Mixed boundary value problems are usually only continuous and regular for a small range of Sobolev indices. The operator on the left-hand side of (3.86) maps* $\widetilde{H}^{-1/2+s}(\Gamma_D) \times \widetilde{H}^{1/2+s}(\Gamma_N)$ *to* $H^{1/2+s}(\Gamma_D) \times H^{-1/2+s}(\Gamma_N)$ *continuously for all* $|s| < s_0(\Gamma) < 1/2$*. We have:*

(a) $s_0 = 1/4$ *for general Lipschitz domains.*
(b) $1/4 < s_0 \leq 1/2$ *for Lipschitz polyhedra (see Definition 3.2.1) and also for globally smooth domains.*

The range of regularity for the mixed boundary value problem is smaller compared to the range of regularity for the pure Dirichlet and Neumann problems (see Theorem 3.2.2 and 3.2.3). Essentially, these theorems can be applied to the present case; however, the range of regularity is given by $|s| < s_0$ *where* s_0 *is defined as in (a) and (b).*

3.4.1.2 Exterior Problems

EDP:

Single Layer Potential: Let $g_D \in H^{1/2}(\Gamma)$ be given. Find $\varphi \in H^{-1/2}(\Gamma)$ such that

$$b_V(\varphi, \eta) = (g_D, \eta)_{L^2(\Gamma)} \qquad \forall \eta \in H^{-1/2}(\Gamma).$$

Double Layer Potential: Let $g_D \in H^{1/2}(\Gamma)$ be given. Find $\psi \in H^{1/2}(\Gamma)$ such that

$$\frac{1}{2}(\psi, \eta)_{L^2(\Gamma)} + b_K(\psi, \eta) = (g_D, \eta)_{L^2(\Gamma)} \qquad \forall \eta \in H^{-1/2}(\Gamma).$$

ENP:

Single Layer Potential: Let $g_N \in H^{-1/2}(\Gamma)$ be given. Find $\varphi \in H^{-1/2}(\Gamma)$ such that

$$-\frac{1}{2}(\varphi, \eta)_{L^2(\Gamma)} + b_{K'}(\varphi, \eta) = (g_N, \eta)_{L^2(\Gamma)} \qquad \forall \eta \in H^{1/2}(\Gamma).$$

Double Layer Potential: Let $g_N \in H^{-1/2}(\Gamma)$ be given. Find $\psi \in H^{1/2}(\Gamma)$ such that

$$b_W(\psi, \eta) = -(g_N, \eta)_{L^2(\Gamma)} \qquad \forall \eta \in H^{1/2}(\Gamma).$$

EDNP:

In (2.155) and (2.156) let $f \equiv 0$ and $(g_D, g_N) \in H^{1/2}(\Gamma_D) \times H^{-1/2}(\Gamma_N)$ be given. Then the associated formulation by integral equations for the indirect method reads: Find $(\varphi, \psi) \in \mathbf{H}$ such that

$$b_{mixed}\left(\begin{pmatrix}\varphi\\ \psi\end{pmatrix}, \begin{pmatrix}\eta\\ \kappa\end{pmatrix}\right) = (g_D, \eta)_{L^2(\Gamma_D)} + (g_N, \kappa)_{L^2(\Gamma_N)} \qquad \forall (\eta, \kappa) \in \mathbf{H}. \tag{3.90}$$

3.4.1.3 Transmission Problem

For $\varphi \in H^{-1/2}(\Gamma)$ and $\psi \in H^{1/2}(\Gamma)$ we use the ansatz

$$u = S\varphi + D\psi \qquad \text{in } \Omega^- \cup \Omega^+$$

and note that $Lu = 0$ in $\Omega^- \cup \Omega^+$. The jump relations from Theorem 3.3.1 give us the two relations

$$g_D = [D\psi] = \psi, \qquad g_N = [\gamma_1 S\varphi] = -\varphi.$$

This means that we can formulate the solution to the TP explicitly:

$$u = -Sg_N + Dg_D \qquad \text{in } \Omega^- \cup \Omega^+.$$

With this we have given formulations by integral equations for all boundary value problems for the indirect method. Once the unknown density functions have been determined they have to be inserted into the associated single or double layer potentials, which results in a function u that satisfies the boundary conditions and the equation $Lu = 0$ in the appropriate domain. We will discuss existence and uniqueness theorems for these integral equations in Sect. 3.5.

3.4.2 *The Direct Method*

The direct method is based on Green's formulas (Theorem 3.1.6). In general, we will again assume that the differential equation is homogeneous, i.e., that we always have $f \equiv 0$ in Ω. In the event that $f \neq 0$ we can use the Newton potential to revert to the case $f = 0$ [see (1.22) and Theorem 3.1.6].

3.4.2.1 Interior Problems

We shall again begin with interior problems. The extension of a function $u \in H_L^1(\Omega^-)$ to Ω^+ by zero will again be denoted by u and satisfies $u \in H_L^1(\mathbb{R}^d)$. Thereby, Green's representation formula (3.10) becomes applicable and gives us

$$u = S(\gamma_1^- u) - D(\gamma_0^- u) \qquad \text{in } \Omega^-. \tag{3.91}$$

This means that the function u in Ω^- is determined as soon as the boundary values $\gamma_0^- u$ or, as is necessary, the values of the conormal derivative $\gamma_1^- u$ are known. By applying γ_0^- or γ_1^- to (3.91) we obtain two boundary integral equations, i.e., a relation between the Dirichlet and Neumann data. We set $u_D := \gamma_0^- u$ and $u_N := \gamma_1^- u$ and obtain

$$\begin{aligned} u_D &= V u_N - \left(K - \frac{1}{2} I\right) u_D \\ u_N &= \left(K' + \frac{1}{2} I\right) u_N + W u_D. \end{aligned} \tag{3.92}$$

By means of these two equations, the interior problem can be transformed into one boundary integral equation of the first kind and another of the second kind.

IDP:

Equation of the first kind: Let $g_D \in H^{1/2}(\Gamma)$ be given. Find $u_N \in H^{-1/2}(\Gamma)$ such that

$$b_V(u_N, \varphi) = \frac{1}{2}(g_D, \varphi)_{L^2(\Gamma)} + b_K(g_D, \varphi) \qquad \forall \varphi \in H^{-1/2}(\Gamma).$$

Equation of the second kind: Let $g_D \in H^{1/2}(\Gamma)$ be given. Find $u_N \in H^{-1/2}(\Gamma)$ such that

$$\frac{1}{2}(u_N,\varphi)_{L^2(\Gamma)} - b_{K'}(u_N,\varphi) = b_W(g_D,\varphi) \qquad \forall\varphi \in H^{1/2}(\Gamma).$$

INP:

Equation of the first kind: Let $g_N \in H^{-1/2}(\Gamma)$ be given. Find $u_D \in H^{1/2}(\Gamma)$ such that

$$b_W(u_D,\varphi) = \frac{1}{2}(g_N,\varphi)_{L^2(\Gamma)} - b_{K'}(g_N,\varphi) \qquad \forall\varphi \in H^{1/2}(\Gamma).$$

Equation of the second kind: Let $g_N \in H^{-1/2}(\Gamma)$ be given. Find $u_D \in H^{1/2}(\Gamma)$ such that

$$\frac{1}{2}(u_D,\varphi)_{L^2(\Gamma)} + b_K(u_D,\varphi) = b_V(g_N,\varphi) \qquad \forall\varphi \in H^{-1/2}(\Gamma).$$

IDNP:

For the mixed boundary value problem we use the first equation in (3.92) on Γ_D and the second equation on Γ_N. In so doing, we obtain the 2×2 system of boundary integral equations

$$\begin{bmatrix} V_{DD} & -K_{DN} \\ K'_{ND} & W_{NN} \end{bmatrix}\begin{pmatrix} u_N \\ u_D \end{pmatrix} = \begin{bmatrix} -V_{DN} & \frac{1}{2}I + K_{DD} \\ \frac{1}{2}I - K'_{NN} & -W_{ND} \end{bmatrix}\begin{pmatrix} g_N \\ g_D \end{pmatrix}.$$

If we then multiply by $(\varphi,\psi) \in \mathbf{H}$ and integrate over the respective surface parts Γ_D and Γ_N we obtain the variational formulation: Find $(u_N,u_D) \in \mathbf{H}$ [cf. (3.88)] such that

$$b_{mixed}\left(\begin{pmatrix} u_N \\ u_D \end{pmatrix},\begin{pmatrix} \varphi \\ \psi \end{pmatrix}\right) = \frac{1}{2}\left\{(g_D,\varphi)_{L^2(\Gamma_D)} + (g_N,\psi)_{L^2(\Gamma_N)}\right\} + b_{mixed}^{rhs}\left(\begin{pmatrix} g_N \\ g_D \end{pmatrix},\begin{pmatrix} \varphi \\ \psi \end{pmatrix}\right)$$

for all $(\varphi,\psi) \in \mathbf{H}$ with

$$b_{mixed}^{rhs}\left(\begin{pmatrix} g_N \\ g_D \end{pmatrix},\begin{pmatrix} \varphi \\ \psi \end{pmatrix}\right) := -b_{V_{DN}}(g_N,\varphi) + b_{K_{DD}}(g_D,\varphi) - b_{K'_{NN}}(g_N,\psi) - b_{W_{ND}}(g_D,\psi). \tag{3.93}$$

Remark 3.4.3, which concerns the mapping and regularity properties of the integral operators of the indirect method as applied to the mixed boundary value problem, also holds for the direct method.

3.4.2.2 Exterior Problems

Green's representation formula as given in (3.19) was proven for exterior problems with an additive, L-harmonic extra term. However, with Theorems 3.1.11, 3.1.12 and 3.1.13, we know that, for the Laplace and Helmholtz problems as well as for the positive definite case $a_{\min} c > \|\mathbf{b}\|^2$, this extra term vanishes and that Green's representation formula holds unchanged. In this section we will assume that there exists a subspace $V \subset H^1_{\text{loc}}\left(\mathbb{R}^d \backslash \Gamma\right)$ such that, for all $u \in V$ with $Lu = 0$ in $\Omega^+ \cup \Omega^-$, the representation

$$u = -S\left[\gamma_1 u\right] + D\left[\gamma_0 u\right] \qquad \text{in } \Omega^- \cup \Omega^+ \tag{3.94}$$

and the trace theorems for γ_0 and γ_1 in V all remain valid unchanged. We set $u^- \equiv 0$ on Ω^- and only consider (3.94) in the exterior space

$$u^+ = -S\gamma_1^+ u^+ + D\gamma_0^+ u^+ \qquad \text{in } \Omega^+.$$

If we form the traces we obtain the equations

$$\begin{aligned} u_D &= -V u_N + \left(K u_D + \tfrac{1}{2} u_D\right), \quad \text{in } H^{1/2}(\Gamma), \\ u_N &= -\left(K' u_N - \tfrac{1}{2} u_N\right) - W u_D, \quad \text{in } H^{-1/2}(\Gamma). \end{aligned} \tag{3.95}$$

In the following we will use these results to derive the boundary integral equations, achieving this by using the known, and solving for the unknown, boundary data.

EDP:

Equation of the first kind: Let $g_D \in H^{1/2}(\Gamma)$ be given. Find $u_N \in H^{-1/2}(\Gamma)$ such that

$$b_V(u_N, \varphi) = -\frac{1}{2}(g_D, \varphi)_{L^2(\Gamma)} + b_K(g_D, \varphi) \qquad \forall \varphi \in H^{-1/2}(\Gamma).$$

Equation of the second kind: Let $g_D \in H^{1/2}(\Gamma)$ be given. Find $u_N \in H^{-1/2}(\Gamma)$ such that

$$\frac{1}{2}(u_N, \psi)_{L^2(\Gamma)} + b_{K'}(u_N, \psi) = -b_W(g_D, \psi) \qquad \forall \psi \in H^{1/2}(\Gamma).$$

ENP:

Equation of the second kind: Let $g_N \in H^{-1/2}(\Gamma)$ be given. Find $u_D \in H^{1/2}(\Gamma)$ such that

$$b_W(u_D, \psi) = -\frac{1}{2}(g_N, \psi)_{L^2(\Gamma)} - b_{K'}(g_N, \psi) \qquad \forall \psi \in H^{1/2}(\Gamma).$$

Equation of the second kind: Let $g_N \in H^{-1/2}(\Gamma)$ be given. Find $u_D \in H^{1/2}(\Gamma)$ such that

$$\frac{1}{2}(u_D, \varphi)_{L^2(\Gamma)} - b_K(u_D, \varphi) = -b_V(g_N, \varphi) \qquad \forall \varphi \in H^{-1/2}(\Gamma). \tag{3.96}$$

EDNP:

Let $g_N \in H^{-1/2}(\Gamma_N)$ and $g_D \in H^{1/2}(\Gamma_N)$ be given. Find $u_D \in \widetilde{H}^{1/2}(\Gamma_N)$ and $u_N \in \widetilde{H}^{-1/2}(\Gamma_D)$ such that

$$b_{mixed}\left(\begin{pmatrix} u_N \\ u_D \end{pmatrix}, \begin{pmatrix} \varphi \\ \psi \end{pmatrix}\right) = -\frac{1}{2}\left\{(g_D, \varphi)_{L^2(\Gamma_D)} + (g_N, \psi)_{L^2(\Gamma_N)}\right\}$$
$$+ b_{mixed}^{rhs}\left(\begin{pmatrix} g_N \\ g_D \end{pmatrix}, \begin{pmatrix} \varphi \\ \psi \end{pmatrix}\right)$$

for all $(\varphi, \psi) \in \mathbf{H}$, where b_{mixed} and b_{mixed}^{rhs} are defined as in (3.90) and (3.93).

3.4.3 Comparison Between Direct and Indirect Method

It is either the integral operators V, W, $\pm I + K$ and $\pm I + K'$ or their localizations on the boundary parts Γ_D and Γ_N that appear in the direct and indirect formulations. This raises the question which of the two formulations is more suited to concrete applications. In the following we will discuss some of the merits and drawbacks of the direct and indirect formulations:

1. The right-hand side of the integral equation for the direct formulation is defined by an integral operator [see, e.g., (3.96)]. In contrast, in order to generate the right-hand side of the indirect formulation one simply needs to evaluate the $L^2(\Gamma)$ inner product of the boundary data and the test functions.
2. When one solves the direct formulation one obtains the Dirichlet and Neumann boundary data explicitly. However, solving the indirect formulation only produces an abstract auxiliary function that subsequently has to be evaluated by means of potentials.
3. The solution of the underlying differential equation in the interior is defined as a representation formula in both cases. For the indirect formulation an integral over the boundary Γ has to be evaluated for every point of the domain [see, e.g.,

(3.4)], while the direct formulation requires the evaluation of two integrals over Γ [see, e.g., Theorem 3.1.6 (with $f = 0$)].

4. For non-smooth surfaces the solutions of the integral equations contain characteristic singularities at the vertices and edges of the surfaces, [237]. With the direct method the solutions are exactly the Cauchy data of the underlying boundary value problem. Therefore the singularities of the solutions of the boundary integral equations *with the direct method* are exactly the Cauchy data of the singularities of the solutions of the boundary value problem. With the indirect method, however, the solutions of the boundary integral equations are jumps of the Cauchy data of the interior as well as the exterior problem. On non-smooth surfaces the solutions of the boundary integral equations, obtained by the indirect method, contain traces of the singularities of solutions of the interior *and* exterior problem, which may in turn reduce the regularity considerably: If Ω^- is a convex polyhedron the solutions of the Dirichlet problem for the Laplace equation are in $H^2(\Omega^-)$ within Ω^-. However, the solutions of the associated exterior problem are, in general, only in $H^s_{\text{loc}}(\Omega^+)$ with $s < 2$ because of the reentering vertices and edges within Ω^+.

These comparisons only give us a rough indication as to which of the two formulations is more suited to practical applications. Note that, in principle, solving the integral equations has the same complexity for both formulations, since in both cases the same operators appear. We will consider two typical applications:

1. If the purpose is simply to determine the unknown boundary data then the direct formulation is more suitable than the indirect formulation.
2. For applications in which the solution of the underlying differential equation has to be evaluated in many points of the domain, the indirect formulation is more suitable.

3.5 Unique Solvability of the Boundary Integral Equations

In this section we will show the coercivity of the integral operators V and W under suitable conditions. Combined with the injectivity of the boundary integral operator we can then deduce the unique solvability of the variational boundary integral equation of the first kind.

3.5.1 Existence and Uniqueness for Closed Surfaces and Dirichlet or Neumann Boundary Conditions

First we will consider closed surfaces $\Gamma = \partial\Omega$ and the case that either Dirichlet or Neumann boundary conditions are prescribed on all of Γ.

We begin with the ellipticity of V and W in the case of the Laplace operator. For this, we will need to generalize Green's formula (2.110) for functions with unbounded support.

Lemma 3.5.1. *Let $a_{\min} c > \|\mathbf{b}\|^2$ or $d = 3$ and $L = -\Delta$. For all $\varphi, \psi \in H^{-1/2}(\Gamma)$ with $u := (S\varphi)|_{\Omega^+}$, $v := (S\psi)|_{\Omega^+}$ we have*

$$\left(\gamma_1^+ u, \gamma_0 v\right)_{L^2(\Gamma)} = -B_{\Omega^+}(u, v). \tag{3.97}$$

Proof. We will first consider the case $L = -\Delta$ and $d = 3$. Let $a > 0$ with $\overline{\Omega^-} \subset K_a$ and

$$\inf_{(\mathbf{x},\mathbf{y}) \in \Gamma \times \partial B_a} \|\mathbf{x} - \mathbf{y}\| \geq 1.$$

We apply Green's formula (2.110) for the bounded domain $\Omega_a := \Omega^+ \cap B_a$ and obtain

$$(\gamma_1 u, \gamma_0 v)_{L^2(\Gamma)} + (\gamma_1 u, \gamma_0 v)_{L^2(\Gamma_a)} = -B_{\Omega_a}(u, v) \tag{3.98}$$

with $\Gamma_a = \partial B_a$. The normal vectors on Γ point, as usual, in the direction of Ω^+ and those on Γ_a in the direction of Ω_a. With this we have

$$\left| \int_{\Gamma_a} \frac{\partial u}{\partial \mathbf{n}} v ds_{\mathbf{y}} \right| \leq \|\varphi\|_{H^{-1/2}(\Gamma)} \|\psi\|_{H^{-1/2}(\Gamma)}$$
$$\int_{\Gamma_a} \|G(\mathbf{x} - \cdot)\|_{H^{1/2}(\Gamma)} \left\| \frac{\partial}{\partial \mathbf{n}} G(\mathbf{x} - \cdot) \right\|_{H^{1/2}(\Gamma)} ds_{\mathbf{x}}.$$

It follows from (3.21) for $\mathbf{x} \in \Gamma_a$ that

$$\|G(\mathbf{x} - \cdot)\|_{H^{1/2}(\Gamma)} \leq C \|G(\mathbf{x} - \cdot)\|_{H^1(\Gamma)} \leq C a^{-1}$$

and

$$\left\| \frac{\partial}{\partial \mathbf{n}} G(\mathbf{x} - \cdot) \right\|_{H^{1/2}(\Gamma)} \leq C \left\| \frac{\partial}{\partial \mathbf{n}} G(\mathbf{x} - \cdot) \right\|_{H^1(\Gamma)} \leq C a^{-2}.$$

With $|\Gamma_a| \leq C a^2$ we obtain

$$\left| \int_{\Gamma_a} \frac{\partial u}{\partial \mathbf{n}} v ds_{\mathbf{y}} \right| \leq C a^{-1} \|\varphi\|_{H^{-1/2}(\Gamma)} \|\psi\|_{H^{-1/2}(\Gamma)}.$$

Therefore, in (3.98), we can let $a \to \infty$ where the second term on the left-hand side goes to zero. This proves the assertion for $L = -\Delta$.

Now let $a_{\min} c > \|\mathbf{b}\|^2$. Lemma 3.1.9 shows that, in this case, the potentials exhibit an exponential decay so that the same arguments as in the previous case can be used to prove the statement. □

Exercise 3.5.2. *Let $a_{\min} c > \|\mathbf{b}\|^2$ or $d = 3$ and $L = -\Delta$. For all $\varphi, \psi \in H^{1/2}(\Gamma)$ with $u := (D\varphi)|_{\Omega^+}$, $v := (D\psi)|_{\Omega^+}$ we have*

$$\left(\gamma_1^+ u, \gamma_0 v\right)_{L^2(\Gamma)} = -B_{\Omega^+}(u, v).$$

Theorem 3.5.3. *Let $d = 3$ and $L = -\Delta$. Then the associated sesquilinear forms $b_V : H^{-1/2}(\Gamma) \times H^{-1/2}(\Gamma) \to \mathbb{C}$ and $b_W : H^{1/2}(\Gamma)/\mathbb{K} \times H^{1/2}(\Gamma)/\mathbb{K}$ are elliptic:*

$$\begin{aligned} b_V(\varphi,\varphi) &\geq c_V \|\varphi\|^2_{H^{-1/2}(\Gamma)} & \forall \varphi \in H^{-1/2}(\Gamma), \\ b_W(\psi,\psi) &\geq c_W \|\psi\|^2_{H^{1/2}(\Gamma)/\mathbb{K}} & \forall \varphi \in H^{1/2}(\Gamma)/\mathbb{K}. \end{aligned}$$

Proof. We begin with the sesquilinear form b_V. For $\varphi \in H^{-1/2}(\Gamma)$ we set $u = S\varphi$. It follows from (3.24) that $u \in H^1(L, \mathbb{R}^d)$. The jump relations give us

$$[\gamma_1 u] = -\varphi \text{ in } H^{-1/2}(\Gamma) \quad \text{and} \quad [u] = 0 \text{ in } H^{1/2}(\Gamma). \tag{3.99}$$

Green's representation formulas (2.110) and (3.97) give us, because of $Lu \equiv 0$ in $\Omega^- \cup \Omega^+$,

$$\begin{aligned} \left(\gamma_1^- u, \gamma_0^- u\right)_{L^2(\Gamma)} &= B_-(u,u), \\ \left(\gamma_1^+ u, \gamma_0^+ u\right)_{L^2(\Gamma)} &= -B_+(u,u). \end{aligned}$$

By subtracting the first equation from the second and by using $[u] = 0$ we obtain, with $B_{\Omega^-\cup\Omega^+} := B_- + B_+$,

$$([\gamma_1 u], \gamma_0 u)_{L^2(\Gamma)} = -B_{\Omega^-\cup\Omega^+}(u,u).$$

Combining (3.99) with the $H^1(L, \mathbb{R}^d)$-ellipticity of $B_{\Omega^-\cup\Omega^+}$ [see (2.171)] gives us

$$b_V(\varphi,\varphi) = (\varphi, V\varphi)_{L^2(\Gamma)} = (\varphi, \gamma_0 u)_{L^2(\Gamma)} = B_{\Omega^-\cup\Omega^+}(u,u) \geq c\,\|u\|^2_{H^1(L,\mathbb{R}^d)}. \tag{3.100}$$

By Definition 2.7.6 and the fact that $Lu \equiv 0$ on $\Omega^- \cup \Omega^+$, we have for the one-sided conormal derivatives (where $\sigma \in \{-,+\}$)

$$\begin{aligned} \|\gamma_1^\sigma u\|_{H^{-1/2}(\Gamma)} &= \sup_{\psi \in H^{1/2}(\Gamma)\backslash\{0\}} \frac{\left|\left(\gamma_1^\sigma u, \psi\right)_{L^2(\Gamma)}\right|}{\|\psi\|_{H^{1/2}(\Gamma)}} = \sup_{\psi \in H^{1/2}(\Gamma)\backslash\{0\}} \frac{|B_\sigma(u, Z_{\Omega^\sigma}\psi)|}{\|\psi\|_{H^{1/2}(\Gamma)}} \\ &\leq C \sup_{\psi \in H^{1/2}(\Gamma)\backslash\{0\}} \frac{\|u\|_{H^1(L,\Omega^\sigma)} \|Z_{\Omega^\sigma}\psi\|_{H^1(L,\Omega^\sigma)}}{\|\psi\|_{H^{1/2}(\Gamma)}}. \end{aligned}$$

Since $Z_{\Omega^\sigma}\psi \in H^1_{\text{comp}}(\Omega^\sigma)$ there exists a ball B_a with $\operatorname{supp}(Z_{\Omega^\sigma}\psi) \subset B_a$. With this and by the equivalence of the norms in $H^1(B_a)$ and $H^1(L, B_a)$ we obtain

$$\|Z_{\Omega^\sigma}\psi\|_{H^1(L,\Omega^\sigma)} \leq \|Z_{\Omega^\sigma}\psi\|_{H^1(B_a\cap\Omega^\sigma)} \leq C\,\|\psi\|_{H^{1/2}(\Gamma)}.$$

From this and with (3.100) we have the ellipticity

$$\begin{aligned} \|\varphi\|^2_{H^{-1/2}(\Gamma)} &= \|[\gamma_1 u]\|^2_{H^{-1/2}(\Gamma)} \leq \|\gamma_1^- u\|^2_{H^{-1/2}(\Gamma)} + \|\gamma_1^+ u\|^2_{H^{-1/2}(\Gamma)} \\ &\leq C\,\|u\|^2_{H^1(L,\mathbb{R}^d)} \leq \frac{C}{c} b_V(\varphi,\varphi). \end{aligned}$$

Next we will prove the ellipticity of the sesquilinear form b_W on $H^{1/2}(\Gamma)/\mathbb{K}$. For $\psi \in H^{1/2}(\Gamma)/\mathbb{K}$ we define $u := D\psi$. From (3.24) we have $u \in H^1(\Omega^-) \times H^1(L, \Omega^+)$. The jump relations give us

$$[u] = \psi, \qquad [\gamma_1 u] = 0. \tag{3.101}$$

Green's representation formulas (2.110) and (3.97) give us, by taking into consideration that $Lu \equiv 0$ on $\Omega^- \cup \Omega^+$,

$$\begin{aligned}(\gamma_1^- u, \gamma_0^- u)_{L^2(\Gamma)} &= B_-(u,u)\\ (\gamma_1^+ u, \gamma_0^+ u)_{L^2(\Gamma)} &= -B_+(u,u).\end{aligned}$$

If we subtract the first equation from the second and use $[\gamma_1 u] = 0$ we obtain

$$(\gamma_1 u, [u])_{L^2(\Gamma)} = -(B_-(u,u) + B_+(u,u)). \tag{3.102}$$

The right-hand side in (3.102) defines, as in (2.169), the sesquilinear form $B_{\Omega^- \cup \Omega^+}$. By combining $\psi = [u]$ with (3.6) we obtain with (3.102)

$$b_W(\psi,\psi) = (\psi, W\psi)_{L^2(\Gamma)} = ([u], -\gamma_1 u)_{L^2(\Gamma)} = B_{\Omega^- \cup \Omega^+}(u,u). \tag{3.103}$$

The continuity and the second Poincaré inequality (Corollary 2.5.10 and Theorem 2.10.10) give us

$$\begin{aligned}\|\psi\|^2_{H^{1/2}(\Gamma)/\mathbb{K}} &= \inf_{c\in\mathbb{R}} \|[u] - c\|^2_{H^{1/2}(\Gamma)}\\ &\le \left(\left\|\gamma_0^+ u\right\|_{H^{1/2}(\Gamma)} + \inf_{c\in\mathbb{R}} \|\gamma_0^-(u-c)\|_{H^{1/2}(\Gamma)}\right)^2\\ &\le 2C\left(\|u\|^2_{H^1(L,\Omega^+)} + \inf_{c\in\mathbb{R}} \|u-c\|^2_{H^1(\Omega^-)}\right)\\ &\le 2C\left(|u|^2_{H^1(L,\Omega^+)} + |u|^2_{H^1(\Omega^-)}\right) \le \widetilde{C} B_{\Omega^- \cup \Omega^+}(u,u) = \widetilde{C} b_W(\psi,\psi).\end{aligned}$$

□

We now turn our attention to the ellipticity of the boundary integral operators V and W for the general elliptic differential equation.

Theorem 3.5.4. *Let $a_{\min} c > \|\mathbf{b}\|^2$. Then the associated sesquilinear forms b_V: $H^{-1/2}(\Gamma) \times H^{-1/2}(\Gamma) \to \mathbb{C}$ and $b_W : H^{1/2}(\Gamma) \times H^{1/2}(\Gamma)$ are elliptic.*

Proof. The proof is analogous to the proof of the previous theorem. We begin with the sesquilinear form b_V. For $\varphi \in H^{-1/2}(\Gamma)$ we set $u = S\varphi$. It follows from Theorem 3.1.16 and Lemma 3.1.9 that $u \in H^1(\mathbb{R}^d)$. As before, the jump relations and Green's representation formulas (2.110) and (3.97) give us

$$([\gamma_1 u], \gamma_0 u)_{L^2(\Gamma)} = -B_{\Omega^- \cup \Omega^+}(u,u).$$

It follows with the $H^1(\mathbb{R}^d)$-ellipticity of $B_{\Omega^- \cup \Omega^+}$ and the continuity of the trace operator that

$$b_V(\varphi, \varphi) = (\varphi, V\varphi)_{L^2(\Gamma)} = (\varphi, \gamma_0 u)_{L^2(\Gamma)} = B_{\Omega^- \cup \Omega^+}(u, u)$$
$$\geq c \|u\|^2_{H^1(L, \mathbb{R}^d)} \geq c \|\varphi\|^2_{H^{-1/2}(\Gamma)}.$$

The proof of the ellipticity of b_W in $H^{1/2}(\Gamma)$ is similar to, yet simpler than, the proof of Theorem 3.5.3. This is due to the fact that, since the interior Neumann problem is uniquely solvable in the case $a_{\min} c > \|\mathbf{b}\|^2$, we need not consider quotient spaces. □

We have now shown that the integral operators V and W are elliptic in suitable Sobolev spaces for elliptic boundary value problems with $L = -\Delta$ or $a_{\min} c > \|\mathbf{b}\|^2$. Thus the unique solvability is a direct consequence of the Lax–Milgram lemma.

Considering the general elliptic operator L in (2.98), we can prove a Gårding inequality for the integral operators V and W and therefore the Riesz–Schauder theory from Sect. 2.1.4 becomes applicable. The details can be found in the following Proposition.

Proposition 3.5.5. *Let G be the fundamental solution [see (3.3)] defined in combination with the operator L from (2.98) and let V and W be the boundary integral operators defined thereby. These satisfy a Gårding inequality in $H^{-1/2}(\Gamma)$ and $H^{1/2}(\Gamma)$. More specifically, there exist compact operators $T_V : H^{-1/2}(\Gamma) \to H^{1/2}(\Gamma)$ and $T_W : H^{1/2}(\Gamma) \to H^{-1/2}(\Gamma)$ such that*

$$((V + T_V) u, u)_{L^2(\Omega)} \geq c_V \|u\|^2_{H^{-1/2}(\Gamma)} \qquad \text{for all } u \in H^{-1/2}(\Gamma), \tag{3.104}$$
$$((W + T_W) v, v)_{L^2(\Omega)} \geq c_W \|v\|^2_{H^{1/2}(\Gamma)} \qquad \text{for all } v \in H^{1/2}(\Gamma).$$

The proof can be found in [72]. The Gårding inequality does not yet provide us with the existence of solutions; we still need to determine the injectivity of V and W.

3.5.2 Existence and Uniqueness for the Mixed Boundary Value Problem*

Let $\Omega^- \subset \mathbb{R}^3$ again be a bounded Lipschitz domain with boundary Γ and let $\Gamma_D, \Gamma_N \subset \Gamma$ be relatively open boundary pieces with $\Gamma_D \cap \Gamma_N = \emptyset$ and

$$\Gamma = \overline{\Gamma_D} \cup \overline{\Gamma_N}. \tag{3.105}$$

---

* This section should be read as a complement to the core material of this book.

For simplicity's sake, we assume that Γ_D and Γ_N are simply connected and we recall the definition (3.82) of the relevant function spaces $H^s(\Gamma_0)$, $\widetilde{H}^s(\Gamma_0)$ on $\Gamma_0 \subset \Gamma$.

We consider the mixed boundary value problem for the Laplace operator:

$$\Delta u = 0 \quad \text{in } \Omega, \qquad u = g_D \quad \text{on } \Gamma_D, \qquad \partial u/\partial \mathbf{n} = g_N \quad \text{on } \Gamma_N \tag{3.106}$$

for given boundary data $g_D \in H^{1/2}(\Gamma_D)$, $g_N \in H^{-1/2}(\Gamma_N)$ and we refer to Sect. 2.9.2.3 for the associated variational formulation.

A (weak) solution $u \in H^1(-\Delta, \Omega^-)$ can be written in terms of its Cauchy data $\varphi = u|_\Gamma$ and $\sigma = \partial u/\partial \mathbf{n}|_\Gamma$ according to

$$u(\mathbf{x}) = (S\sigma)(\mathbf{x}) - (D\varphi)(\mathbf{x}), \qquad \mathbf{x} \in \Omega^- \tag{3.107}$$

(see Theorem 3.1.12).

In order to determine u, we need to determine the missing Cauchy data $(u|_{\Gamma_N}, (\partial u/\partial \mathbf{n})|_{\Gamma_D})$, which can be obtained as a solution of the integral equation (3.89).

In this section we study the ellipticity of the sesquilinear form $b_{mixed} : \mathbf{H} \times \mathbf{H} \to \mathbb{K}$ from (3.89) with $\mathbf{H} = \widetilde{H}^{-1/2}(\Gamma_D) \times \widetilde{H}^{1/2}(\Gamma_N)$. We restrict ourselves to the Laplace problem.

For $(\varphi, \sigma) = (\psi, \eta) \in \mathbf{H}$ we have

$$b_{mixed}\left(\begin{pmatrix}\varphi \\ \sigma\end{pmatrix}, \begin{pmatrix}\varphi \\ \sigma\end{pmatrix}\right) = (V_{DD}\varphi, \varphi)_{L^2(\Gamma_D)} + (W_{NN}\sigma, \sigma)_{L^2(\Gamma_N)}.$$

The following lemma deals with the ellipticity of the single layer operator as well as the normal derivative of the double layer operator.

Lemma 3.5.6. *Let $\Omega \subset \mathbb{R}^3$ be a bounded Lipschitz domain and let Γ_D, Γ_N be a partition of the boundary Γ into simply connected pieces with positive surface measure that satisfy (3.105). Then there exists a constant $\gamma(\Gamma_D, \Gamma_N) > 0$ such that*

$$\forall \varphi \in \widetilde{H}^{-1/2}(\Gamma_D) : \ (V_{DD}\varphi, \varphi)_{L^2(\Gamma_D)} \geq c_V \|\varphi\|^2_{\widetilde{H}^{-1/2}(\Gamma_D)}$$

$$\forall \sigma \in \widetilde{H}^{1/2}(\Gamma_N) : \ (W_{NN}\sigma, \sigma)_{L^2(\Gamma_N)} \geq \gamma \|\sigma\|^2_{\widetilde{H}^{1/2}(\Gamma_N)}.$$

Proof. For the first estimate we use $\varphi \in \widetilde{H}^{-1/2}(\Gamma_D) \Longrightarrow \varphi^\star \in H^{-1/2}(\Gamma)$ with the zero extension $\varphi^\star$ of φ on Γ. It follows from (3.83) that $\|\varphi\|_{\tilde{H}^{-1/2}(\Gamma_D)} = \|\varphi^\star\|_{H^{-1/2}(\Gamma)}$ and the ellipticity of V on Γ (see Theorem 3.5.3) gives us

$$(V_{DD}\varphi, \varphi)_{L^2(\Gamma_D)} = \left(V\varphi^\star, \varphi^\star\right)_{L^2(\Gamma)} \geq c_V \|\varphi^\star\|^2_{H^{-1/2}(\Gamma)} = c_V \|\varphi\|^2_{\tilde{H}^{-1/2}(\Gamma_D)}. \tag{3.108}$$

The estimate for the operator W follows from

$$\sigma \in \widetilde{H}^{1/2}(\Gamma_N) \Longrightarrow \sigma^\star \in H^{1/2}(\Gamma), \qquad \|\sigma\|_{\tilde{H}^{1/2}(\Gamma_N)} = \|\sigma^\star\|_{H^{1/2}(\Gamma)}.$$

From this and from the ellipticity of the hypersingular operator W on Γ (see Theorem 3.5.3) it follows that

$$(W_{NN}\sigma,\sigma)_{L^2(\Gamma_N)} = \left(W\sigma^\star,\sigma^\star\right)_{L^2(\Gamma)} \geq c_W \|\sigma^\star\|^2_{H^{1/2}(\Gamma)/\mathbb{K}}$$
$$= c_W \min_{c\in\mathbb{R}} \|\sigma^\star - c\|^2_{H^{1/2}(\Gamma)}. \tag{3.109}$$

We recall the definition of the $H^{1/2}(\Gamma)$-norm

$$\|\varphi\|^2_{H^{1/2}(\Gamma)} = \|\varphi\|^2_{L^2(\Gamma)} + |\varphi|^2_{H^{1/2}(\Gamma)},$$
$$|\varphi|^2_{H^{1/2}(\Gamma)} = \int_\Gamma\int_\Gamma \frac{|\varphi(\mathbf{x})-\varphi(\mathbf{y})|^2}{\|\mathbf{x}-\mathbf{y}\|^3} ds_{\mathbf{y}} ds_{\mathbf{x}}. \tag{3.110}$$

Therefore we have $|c|_{H^{1/2}(\Gamma)} = 0$ and thus

$$\|\sigma^\star\|^2_{H^{1/2}(\Gamma)/\mathbb{K}} = |\sigma^\star|^2_{H^{1/2}(\Gamma)} + \min_{c\in\mathbb{R}} \|\sigma^\star - c\|^2_{L^2(\Gamma)}.$$

The minimum is attained for the mean value $c := |\Gamma|^{-1}\int_\Gamma \sigma^\star ds = |\Gamma|^{-1}\int_{\Gamma_N}\sigma ds$. This leads to $|c|^2 \leq |\Gamma|^{-2}|\Gamma_N|\|\sigma\|^2_{L^2(\Gamma_N)}$ and

$$\begin{aligned}
\min_{\alpha\in\mathbb{R}} \|\sigma^\star - \alpha\|^2_{L^2(\Gamma)} &= \|\sigma\|^2_{L^2(\Gamma_N)} + \|c\|^2_{L^2(\Gamma)} - 2\int_{\Gamma_N} \sigma c\, ds \\
&= \|\sigma\|^2_{L^2(\Gamma_N)} + |\Gamma||c|^2 - 2c\int_{\Gamma_N} \sigma\, ds \\
&= \|\sigma\|^2_{L^2(\Gamma_N)} + \frac{1}{|\Gamma|}\left(\int_{\Gamma_N} \sigma ds\right)^2 - \frac{2}{|\Gamma|}\left(\int_{\Gamma_N} \sigma ds\right)^2 \\
&= \|\sigma\|^2_{L^2(\Gamma_N)} - \frac{1}{|\Gamma|}\left(\int_{\Gamma_N} \sigma ds\right)^2 \geq \|\sigma\|^2_{L^2(\Gamma_N)}\left(1 - \frac{|\Gamma_N|}{|\Gamma|}\right).
\end{aligned}$$

Therefore we have the following estimate for all $\sigma \in \widetilde{H}^{1/2}(\Gamma_N)$:

$$\begin{aligned}
\|\sigma^\star\|^2_{H^{1/2}(\Gamma)/\mathbb{K}} &\geq \left(1 - \tfrac{|\Gamma_N|}{|\Gamma|}\right)\left(\|\sigma^\star\|^2_{L^2(\Gamma)} + |\sigma^\star|^2_{H^{1/2}(\Gamma)}\right) \\
&= \left(1 - \tfrac{|\Gamma_N|}{|\Gamma|}\right)\|\sigma^\star\|^2_{H^{1/2}(\Gamma)} = \left(1 - \tfrac{|\Gamma_N|}{|\Gamma|}\right)\|\sigma\|^2_{\widetilde{H}^{1/2}(\Gamma_N)}.
\end{aligned} \tag{3.111}$$

This proves part 2 of the assertion. □

From (3.108) and (3.109) we have the ellipticity of b_{mixed}:

Corollary 3.5.7. *For* $|\Gamma_N| < |\Gamma|$ *there exists a constant* $c > 0$ *such that for all* $(\varphi,\sigma) \in \widetilde{H}^{-1/2}(\Gamma_D)\times\widetilde{H}^{1/2}(\Gamma_N)$ *we have*

$$b_{mixed}\left(\begin{pmatrix}\varphi\\ \sigma\end{pmatrix},\begin{pmatrix}\varphi\\ \sigma\end{pmatrix}\right) \geq c\left(\|\varphi\|^2_{\widetilde{H}^{-1/2}(\Gamma_D)} + \|\sigma\|^2_{\widetilde{H}^{1/2}(\Gamma_N)}\right). \tag{3.112}$$

Theorem 3.5.8. *The systems of boundary integral equations (3.89) and (3.86) associated with the mixed boundary value problem have a unique solution* $(\varphi, \sigma) \in \mathbf{H}$ *for all* $g_D \in H^{1/2}(\Gamma_D)$ *and* $g_N \in H^{-1/2}(\Gamma_N)$.

The variational problem (2.144), (2.145) associated with the mixed boundary value problem (3.106) also has a unique solution $u \in H^1_D(\Omega^-)$. *This solution can be written as in (3.107).*

Proof. The unique solvability of the boundary integral equations follows from the $\mathbf{H}$-ellipticity of the bilinear form b_{mixed} and the Lax–Milgram lemma.

The unique solvability of the variational problem (2.144), (2.145) associated with the mixed boundary value problem (3.106) follows from Theorem 2.10.6.

The representation formula was proven in Theorem 3.1.12. □

3.5.3 Screen Problems*

The boundary value problems that we have considered so far always lead to integral equations on closed surfaces. The computation of electric fields and potentials or of stress fields near cracks often requires the solution of screen problems. Here fields are induced by thin charged plates or screens. Screens are modelled as hypersurfaces $\Gamma_0 \subset \mathbb{R}^3$ which, in general, are not closed surfaces, i.e., $\partial\Gamma_0 \neq \emptyset$. The potential equation is formulated on the exterior $\mathbb{R}^3\backslash\Gamma_0$ and the boundary integral equations reduce the problem to the screen Γ_0.

The energy spaces for integral equations on boundary pieces were given by $\widetilde{H}^{-1/2}(\Gamma_D)$ and $\widetilde{H}^{1/2}(\Gamma_N)$. These spaces allow us to consider boundary integral equations on open surface pieces Γ as well.

Therefore, in this subsection we assume that an open surface piece Γ_0 is given, which can be extended to a closed Lipschitz surface Γ in $\mathbb{R}^3$ so that

$$\Gamma = \Gamma_0 \cup \Gamma_0^c, \tag{3.113}$$

where $\Gamma_0^c = \Gamma\backslash\overline{\Gamma}_0$. In order to avoid technical difficulties we assume, as in the previous subsection, that both Γ_0 and Γ_0^c be simply connected.

Dirichlet Screen Problem: Find, for a given $g_D \in H^{1/2}(\Gamma_0)$, the function $u \in H^1_{\mathrm{loc}}(\mathbb{R}^3\backslash\overline{\Gamma}_0)$ such that

$$\Delta u = 0 \quad \text{in } \mathbb{R}^3\backslash\overline{\Gamma}_0, \quad u = g_D \text{ on } \Gamma_0, \qquad |u(\mathbf{x})| = O\left(\|\mathbf{x}\|^{-1}\right) \text{ for } \|\mathbf{x}\| \to \infty. \tag{3.114}$$

Neumann Screen Problem: Find, for a given $g_N \in H^{-1/2}(\Gamma_0)$, the function $u \in H^1_{\mathrm{loc}}(\mathbb{R}^3\backslash\overline{\Gamma}_0)$ such that

* This section should be read as a complement to the core material of this book.

$$\Delta u = 0 \quad \text{in } \mathbb{R}^3 \backslash \overline{\Gamma}_0, \quad \partial u / \partial \mathbf{n} = g_N \text{ on } \Gamma_0 \, |u(\mathbf{x})| = O\left(\|\mathbf{x}\|^{-1}\right) \text{ for } \|\mathbf{x}\| \to \infty. \tag{3.115}$$

By using the ansatz via potentials, these screen problems can be transformed into boundary integral equations and we again formulate these as variational problems.

Dirichlet Screen Problem: Find, for a given $g_D \in H^{1/2}(\Gamma_0)$, the function $\varphi \in \widetilde{H}^{-1/2}(\Gamma_0)$ such that

$$(V\varphi, \eta)_{L^2(\Gamma_0)} = (g_D, \eta)_{L^2(\Gamma_0)} \qquad \forall \eta \in \widetilde{H}^{-1/2}(\Gamma_0). \tag{3.116}$$

Neumann Screen Problem: Find, for a given $g_N \in H^{-1/2}(\Gamma_0)$, the function $\sigma \in \widetilde{H}^{1/2}(\Gamma_0)$ such that

$$(W\sigma, \kappa)_{L^2(\Gamma_0)} = (g_N, \kappa) \qquad \forall \kappa \in \widetilde{H}^{1/2}(\Gamma_0). \tag{3.117}$$

Here we identify, as usual, the inner product $(\cdot,\cdot)_{L^2(\Gamma_0)}$ with its extension on $H^{1/2}(\Gamma_0) \times \widetilde{H}^{-1/2}(\Gamma_0)$ or on $H^{-1/2}(\Gamma_0) \times \widetilde{H}^{1/2}(\Gamma_0)$.

Theorem 3.5.9. *The operators $V : \widetilde{H}^{-1/2}(\Gamma_0) \to H^{1/2}(\Gamma_0)$ and $W : \widetilde{H}^{1/2}(\Gamma_0) \to H^{-1/2}(\Gamma_0)$ are continuous and positive, that is, there exist c_V and $c_W > 0$ such that*

$$\forall \varphi \in \widetilde{H}^{-1/2}(\Gamma_0) : (V\varphi, \varphi)_{L^2(\Gamma_0)} \geq c_V \|\varphi\|^2_{\widetilde{H}^{-1/2}(\Gamma_0)}, \tag{3.118}$$

$$\forall \sigma \in \widetilde{H}^{\frac{1}{2}}(\Gamma_0) : \ (W\sigma, \sigma)_{L^2(\Gamma_0)} \geq c_W \|\sigma\|^2_{\widetilde{H}^{1/2}(\Gamma_0)}. \tag{3.119}$$

Theorem 3.5.9 is a direct consequence of Lemma 3.5.6 with $\Gamma_0 = \Gamma_D$ or $\Gamma_0 = \Gamma_N$, while c_V, c_W depend on Γ_0.

3.6 Calderón Projector*

The direct method results in boundary integral equations for the unknown boundary data of the boundary value problem. In this section we will again turn our attention to the identities (3.92) and (3.95) and derive some useful conclusions.

For $\sigma \in \{-,+\}$ we define the space Y_σ by

$$Y_\sigma := \left\{u \in H^1_L(\Omega^\sigma) : Lu = 0 \text{ in } \Omega^\sigma\right\}.$$

On $Y_- \times Y_+$ Green's representation formula is, in general, only valid in the form (3.19) with an extra additive term.

* This section should be read as a complement to the core material of this book.

For the operator L in (3.1) with $a_{\min}c > \|\mathbf{b}\|^2$ we have shown in Theorem 3.1.11 that the extra term in Green's representation formula disappears for functions from $Y_- \times Y_+$.

This also holds for the Laplace operator $-\Delta$ (see Theorem 3.1.12) on the subspace

$$\left\{u \in H^1\left(\Omega^-\right) : \Delta u = 0\right\} \times \left\{u \in H^1\left(-\Delta, \Omega^+\right) : \Delta u = 0\right\}$$

and for the Helmholtz operator $L_k := -\Delta - k^2$ (see Theorem 3.1.13) on the subspace

$$\left\{u \in H^1\left(\Omega^-\right) : L_k u = 0\right\} \times \left\{u \in H^1\left(L_k, \Omega^+\right) : L_k u = 0\right\}$$

[see (2.149)].

Assumption 3.6.1. *For the given differential operator L we can choose a subspace $X_- \times X_+ \subset Y_- \times Y_+$ such that the extra term in (3.19) vanishes and Green's representation formula holds in the form*

$$u = -S\left[\gamma_1 u\right] + D\left[\gamma_0 u\right] \qquad \text{in } \Omega^- \cup \Omega^+$$

for all $u \in X_- \times X_+$, where $u \in X_- \times X_+$ is an abbreviation for $(u_-, u_+) \in X_- \times X_+$ with the convention $u|_{\Omega^\sigma} = u_\sigma$, $\sigma \in \{-, +\}$.

The direct boundary integral equations are based on the identities (3.92) and (3.95) between the Dirichlet data $u_D \in H^{\frac{1}{2}}(\Gamma)$ and the Neumann data $u_N \in H^{-\frac{1}{2}}(\Gamma)$. For all $u \in X_- \times X_+$, we have with $(u_D, u_N) = \left(\gamma_0^- u, \gamma_1^- u\right)$ in the *inner domain* Ω^-:

$$\begin{pmatrix} \gamma_0^- u \\ \gamma_1^- u \end{pmatrix} = \begin{pmatrix} u_D \\ u_N \end{pmatrix} = \begin{pmatrix} \frac{1}{2}I - K & V \\ W & \frac{1}{2}I + K' \end{pmatrix} \begin{pmatrix} u_D \\ u_N \end{pmatrix} =: P_- \begin{pmatrix} u_D \\ u_N \end{pmatrix}, \tag{3.120}$$

and with $(u_D, u_N) = \left(\gamma_0^+ u, \gamma_1^+ u\right)$ in the *outer domain* Ω^+:

$$\begin{pmatrix} \gamma_0^+ u \\ \gamma_1^+ u \end{pmatrix} = \begin{pmatrix} u_D \\ u_N \end{pmatrix} = \begin{pmatrix} \frac{1}{2}I + K & -V \\ -W & \frac{1}{2}I - K' \end{pmatrix} \begin{pmatrix} u_D \\ u_N \end{pmatrix} =: P_+ \begin{pmatrix} u_D \\ u_N \end{pmatrix}. \tag{3.121}$$

This is the motivation behind the definition of the *Calderón operator* A on the boundary Γ by

$$A = \tfrac{1}{2}(P_- - P_+) = \begin{pmatrix} -K & V \\ W & K' \end{pmatrix}. \tag{3.122}$$

Therefore the compact form of (3.120) and (3.121) reads

$$\begin{pmatrix} \gamma_0^\sigma u \\ \gamma_1^\sigma u \end{pmatrix} = -\left(\sigma A - \tfrac{1}{2} I\right) \begin{pmatrix} \gamma_0^\sigma u \\ \gamma_1^\sigma u \end{pmatrix} \qquad \sigma \in \{-,+\}. \tag{3.123}$$

For $\sigma \in \{-,+\}$ we then have

$$P_\sigma = -\left(\sigma A - \frac{1}{2} I\right).$$

Proposition 3.6.2. *The operators P_σ have the following properties:*

(i) P_σ can be continuously extended to $H^{1/2}(\Gamma) \times H^{-1/2}(\Gamma) \to H^{1/2}(\Gamma) \times H^{-1/2}(\Gamma)$.
(ii) P_σ is a projection of $H^{1/2}(\Gamma) \times H^{-1/2}(\Gamma) \to H^\sigma_{\text{Cauchy}}(\Gamma)$ where the space $H^\sigma_{\text{Cauchy}}(\Gamma)$ is defined as

$$H^\sigma_{\text{Cauchy}}(\Gamma) := \{(\gamma_0^\sigma u, \gamma_1^\sigma u) : u \in X_\sigma\}. \tag{3.124}$$

(iii) We have

$$P_+ + P_- = I \quad \textit{and} \quad P_+ P_- = P_- P_+ \quad \textit{on} \quad H^{1/2}(\Gamma) \times H^{-1/2}(\Gamma).$$

Proof. Statement (i) is a direct result of the mapping properties of the boundary integral operator in the Calderón operator A.

We prove (ii): Let $\varphi \in H^{-1/2}(\Gamma)$, $\psi \in H^{1/2}(\Gamma)$ be arbitrary and given. Then the potential $u = S\varphi - D\psi \in Y_- \times Y_+$ and therefore satisfies the homogeneous differential equation.

We use the notation $u^\sigma = u|_{\Omega^\sigma}$. If we apply the trace operators $\gamma_0^\sigma, \gamma_1^\sigma$ to u^σ we obtain

$$\gamma_0^\sigma u^\sigma = \gamma_0^\sigma S\varphi - \gamma_0^\sigma D\psi = V\varphi - \left(\frac{\sigma 1}{2} + K\right)\psi$$

and

$$\gamma_1^\sigma u^\sigma = \gamma_1^\sigma S\varphi - \gamma_1^\sigma D\psi = \left(-\frac{\sigma 1}{2} I + K'\right)\varphi + W\psi,$$

so that we have

$$P_\sigma \begin{pmatrix} \varphi \\ \psi \end{pmatrix} = \begin{pmatrix} \gamma_0^\sigma u \\ \gamma_1^\sigma u \end{pmatrix} \in H^\sigma_{\text{Cauchy}}(\Gamma).$$

We obtain statement (iii) via the relations

$$P_+ P_- = (A - \frac{1}{2} I)(-A - \frac{1}{2} I) = (-A - \frac{1}{2} I)(A - \frac{1}{2} I) = P_- P_+.$$

□

Definition 3.6.3. For $\sigma \in \{-,+\}$, P_σ are the *Calderón projectors* for the elliptic system (L, γ_0, γ_1) in Ω^σ.

Let $\gamma_0 X := \gamma_0^- X_- + \gamma_0^+ X_+$ and $\gamma_1 X := \gamma_1^- X_- + \gamma_1^+ X_+$. (Note that for all standard cases we have $\gamma_0 X = H^{1/2}(\Gamma)$ and $\gamma_1 X = H^{-1/2}(\Gamma)$.)

Proposition 3.6.4. *We have the* Calderón identities

$$KV = VK' \text{ on } \gamma_1 X, \qquad WK = K'W \text{ on } \gamma_0 X \tag{3.125a}$$

and

$$VW = \tfrac{1}{4}I - K^2 \text{ on } \gamma_0 X, \quad WV = \tfrac{1}{4}I - K'^2 \text{ on } \gamma_1 X. \tag{3.125b}$$

Proof. The identity $(P_\sigma)^2 = P_\sigma$ on $H^\sigma_{\text{Cauchy}}(\Gamma)$ can be written componentwise

$$\begin{pmatrix} \sigma K + K^2 + VW + \frac{1}{4}I & -(\sigma V + KV - VK') \\ -(\sigma W + WK - K'W) & -(\sigma K' - K'^2 - WV - \frac{1}{4}I) \end{pmatrix} = \begin{pmatrix} \sigma K + \frac{1}{2}I & -(\sigma V) \\ -(\sigma W) & -(\sigma K' - \frac{1}{2}I) \end{pmatrix}.$$

By comparing the coefficients we obtain (3.125). □

Remark 3.6.5. *The Calderón identities (3.125) have different implications.*

(a) Relation (3.125a) implies that V and W symmetrize the operators K and K' respectively, more specifically, we have $(KV)' = KV$ and $(WK)' = WK$.
(b) The operators V and W have the order 1 *and* -1 *respectively. We have from (3.125b) that applying the operators successively, i.e., VW and WV, defines operators of order zero. This property can be used advantageously to precondition the linear system of equations that results from the Galerkin discretization of the boundary integral operators V and W (see Chap. 6, [217], [218] and [67]).*

3.7 Poincaré–Steklov Operator*

We consider the Dirichlet interior problem [see (2.140)]: For a given $g_D \in H^{1/2}(\Gamma)$, find $u \in H^1(\Omega^-)$ with $\gamma_0^- u = g_D$ on Γ such that

$$B(u, v) = 0 \qquad \forall v \in H_0^1(\Omega^-). \tag{3.126}$$

In this section we assume that the Dirichlet problem has a unique solution.

Assumption 3.7.1. *Problem (3.126) has, for all $g_D \in H^{1/2}(\Gamma_D)$, a unique solution that depends continuously on g_D:*

$$\|u\|_{H^1(\Omega^-)} \le C \|g_D\|_{H^{1/2}(\Gamma)}.$$

* This section should be read as a complement to the core material of this book.

We will study necessary and sufficient conditions for the coefficients of the differential operator L to satisfy Assumption 3.7.1 in Sect. 2.10.

The mapping $g_D \rightarrow \gamma_1^- u$, which assigns Neumann data to a solution of the Dirichlet problem in Ω^-, defines the Dirichlet-to-Neumann mapping. The associated *Poincaré–Steklov operator* is given by

$$P_S g_D := \gamma_1^- u. \tag{3.127}$$

For $u \in X_-$ (see Assumption 3.6.1) this mapping can also be written as

$$P_S \left(\gamma_0^- u\right) = \gamma_1^- u.$$

Clearly, $P_S : H^{\frac{1}{2}}(\Gamma) \rightarrow H^{-\frac{1}{2}}(\Gamma)$ is continuous. With (3.120) and (3.125) we can write P_S explicitly as

$$P_S = V^{-1}\left(\tfrac{1}{2}I + K\right) = W + \left(\tfrac{1}{2}I + K'\right) V^{-1}\left(\tfrac{1}{2}I + K\right). \tag{3.128}$$

The operator P_S coincides with the composition $\gamma_1 \circ T$, where T denotes the solution operator from Sect. 2.8 (for the choice $\lambda = 0$) and whose existence is guaranteed by Assumption 3.7.1.

We will now consider the Neumann interior problem [see (2.142), (2.143)]. For a given $g_N \in H^{-1/2}(\Gamma)$ find $u \in H^1(\Omega^-)$ such that

$$B(u, v) = (g_N, \gamma_0 v)_{L^2(\Gamma)} \qquad \forall v \in H^1(\Omega^-). \tag{3.129}$$

In this case we also assume the existence of a unique solution.

Assumption 3.7.2. *Problem (3.129) has, for all $g_N \in H^{-1/2}(\Gamma)$, a unique solution that depends continuously on g_N:*

$$\|u\|_{H^1(\Omega^-)} \leq C\, \|g_N\|_{H^{-1/2}(\Gamma)}.$$

We will discuss necessary and sufficient conditions for the coefficients of the differential operator L to satisfy Assumption 3.7.2 in Sect. 2.10. In case that the existence of a unique solution on $H^1(\Omega^-)$ is not ensured (Example: $L = -\Delta$), we may consider suitable subspaces of $H^1(\Omega^-)$ and $H^{-1/2}(\Gamma)$ in order to satisfy the assumption.

The Neumann-to-Dirichlet mapping $S_P : H^{-1/2}(\Gamma) \rightarrow H^{1/2}(\Gamma)$ for $u \in X_-$ is defined by

$$S_P(\gamma_1^- u) = \gamma_0^- u \tag{3.130}$$

and is called *Steklov–Poincaré operator*. Assuming that $W : H^{1/2}(\Gamma) \rightarrow H^{-1/2}(\Gamma)$ is bijective, we obtain the explicit representation

$$S_P = W^{-1}\left(\tfrac{1}{2}I - K'\right) = V + \left(\tfrac{1}{2}I - K\right) W^{-1}\left(\tfrac{1}{2}I - K'\right). \tag{3.131}$$

Remark 3.7.3. *The conditions placed on W can be reduced. Let*

$$R := \left\{ \frac{1}{2}\varphi - K'\varphi : \varphi \in H^{-1/2}(\Gamma) \right\} \quad \text{and} \quad U := \left\{ \varphi \in H^{1/2}(\Gamma) : W\varphi \in R \right\}.$$

Then, under the assumption that $W : U \to R$ is bijective, (3.131) still holds.

Proposition 3.7.4. *The operators P_S, S_P in (3.127), (3.130) are continuous. If the underlying differential operator is self-adjoint, then P_S and S_P are Hermitian: $P_S = P_S'$ and $S_P = S_P'$.*

Proof. Follows from the representations (3.128) and (3.131). □

Remark 3.7.5. *Let Assumptions 3.7.1 and 3.7.2 be satisfied.*

The operators $P_S : H^{1/2}(\Gamma) \to H^{-1/2}(\Gamma)$ and $S_P : H^{-1/2}(\Gamma) \to H^{1/2}(\Gamma)$ are invertible and satisfy

$$P_S S_P = I \quad \text{on } H^{-1/2}(\Gamma), \qquad S_P P_S = I \quad \text{on } H^{1/2}(\Gamma).$$

3.8 Invertibility of Boundary Integral Operators of the Second Kind*

In Sect. 3.4 we have derived integral equations of the second kind in order to solve boundary value problems. The relevant boundary integral operators $\frac{1}{2} I \pm K$, $\frac{1}{2} I \pm K'$ have order 0 and, in general, are not self-adjoint. Therefore it seems logical to choose the function space $L^2(\Gamma)$ for a variational formulation. However, the Calderón identities (3.125) and the mapping properties of K, K' (see Theorem 3.1.16) demonstrate that the function spaces $H^{\pm 1/2}(\Gamma)$ provide a more natural choice for such a formulation. For the existence of solutions on non-smooth boundaries Γ this choice becomes essential.

Assumption 3.8.1. *The single layer operator $V : H^{-1/2}(\Gamma) \to H^{1/2}(\Gamma)$ is Hermitian, continuous and positive: there exists some $c_V > 0$ such that*

$$(\sigma, V\sigma)_{L^2(\Gamma)} \geq c_V \, \|\sigma\|^2_{H^{-1/2}(\Gamma)} \qquad \forall \sigma \in H^{-1/2}(\Gamma). \tag{3.132}$$

Combined with the boundedness of V the expression

$$\|\sigma\|_V := (\sigma, V\sigma)^{1/2}_{L^2(\Gamma)} \tag{3.133}$$

* This section should be read as a complement to the core material of this book.

therefore constitutes a norm on $H^{-1/2}(\Gamma)$ and is equivalent to the $H^{-1/2}(\Gamma)$-norm. In the same way V^{-1}: $H^{1/2}(\Gamma) \to H^{-1/2}(\Gamma)$ is continuous, Hermitian and we have

$$\|\varphi\|_{V^{-1}}^2 := \left(\varphi, V^{-1}\varphi\right)_{L^2(\Gamma)} \leq \frac{1}{c_V} \|\varphi\|_{H^{1/2}(\Gamma)}^2 \qquad \forall \varphi \in H^{1/2}(\Gamma). \tag{3.134}$$

Therefore $\|\varphi\|_{V^{-1}}$ defines a norm that is equivalent to $\|\cdot\|_{H^{1/2}(\Gamma)}$. We can formulate analogous definitions for the hypersingular operator W, which, for example for the Laplace problem, is only positive on quotient spaces.

Definition 3.8.2. For the homogeneous Neumann problem

$$Lu = 0 \quad \text{in } \Omega^-, \qquad \gamma_1 u = 0 \quad \text{on } \partial\Omega^- \tag{3.135}$$

the space of the traces of the solutions is given by

$$\mathcal{N} = \{\gamma_0 u : u \in H^1(\Omega^-) \text{ solves (3.135)}\}.$$

Remark 3.8.3. *(a) The Riesz–Schauder theory (see Sect. 2.1.4) implies that $\mathcal{N} \subset H^{1/2}(\Gamma)$ is finite-dimensional.*
(b) If the operator associated with the boundary value problem (3.135) is injective we have $\mathcal{N} = \{0\}$.
(c) For $L = -\Delta$, $\gamma_1 = \frac{\partial}{\partial n}$ we have $\mathcal{N} = \operatorname{span}\{1\}$.

Remark 3.8.4. *For $\sigma \in \{-, +\}$ the quotient spaces $H^{\sigma 1/2}(\Gamma)/\mathcal{N}$ are given by the classes*

$$\{u\} := \{u + v : v \in \operatorname{span} \mathcal{N}\}, \qquad u \in H^{\sigma 1/2}(\Gamma).$$

These can be identified with the representatives $u_0 = u_0(u) := u + v$, where $v = v(u)$ is chosen so that

$$\forall v \in \mathcal{N} : (u_0, v)_{L^2(\Gamma)} = 0.$$

Thus $H^{\sigma 1/2}(\Gamma)/\mathcal{N}$ is isomorphic to

$$H_{\mathcal{N}}^{\sigma 1/2}(\Gamma) := \{u_0(u) : u \in H^{\sigma 1/2}(\Gamma)\}, \tag{3.136}$$

and the quotient norm $H^{\sigma 1/2}(\Gamma)/\mathcal{N}$ is equivalent to the $H^{\sigma 1/2}(\Gamma)$-norm on $H_{\mathcal{N}}^{\sigma 1/2}(\Gamma)$.

Assumption 3.8.5. *There exists a constant $c_W > 0$ such that*

$$(\varphi, W\varphi)_{L^2(\Gamma)} \geq c_W \|\varphi\|_{H^{1/2}(\Gamma)}^2 \quad \forall \varphi \in H_{\mathcal{N}}^{1/2}(\Gamma). \tag{3.137}$$

In Theorem 3.5.3 we have shown that Assumptions 3.8.1 and 3.8.5 are satisfied for the integral operators V and W associated with $L = -\Delta$.

Exercise 3.8.6. *Prove that*

$$\forall \varphi \in \mathcal{N}: \; W\varphi = \left(\tfrac{1}{2}I + K\right)\varphi = 0 .$$

Hint: In Sect. 3.9.2 the same statement is proven for the Helmholtz operator.

Theorem 3.8.7. *Assuming that Assumptions 3.8.1 and 3.8.5 hold, we have for the product of the constants in (3.132) and (3.137) the estimate $c_V c_W \leq 1/4$ and for $\sigma \in \{-,+\}$*

$$(1 - c_K)\|u\|_{V^{-1}} \leq \left\|\left(\sigma K + \tfrac{1}{2}I\right)u\right\|_{V^{-1}} \leq c_K\|u\|_{V^{-1}} \qquad \forall u \in H^{1/2}_{\mathcal{N}}(\Gamma) . \tag{3.138}$$

with

$$0 < c_K = \tfrac{1}{2} + \sqrt{\tfrac{1}{4} - c_V c_W} < 1 .$$

Proof. The Calderón identity (3.123) combined with (3.128) gives us

$$\begin{aligned}
\left\|\left(\tfrac{1}{2}I + K\right)u\right\|^2_{V^{-1}} &= \left(V^{-1}\left(\tfrac{1}{2}I + K\right)u,\; \left(\tfrac{1}{2}I + K\right)u\right)_{L^2(\Gamma)} \\
&= \left(\left(\tfrac{1}{2}I + K'\right)V^{-1}\left(\tfrac{1}{2}I + K\right)u, u\right)_{L^2(\Gamma)} \\
&= (P_S u, u)_{L^2(\Gamma)} - (Wu, u)_{L^2(\Gamma)} .
\end{aligned}$$

By Assumption 3.8.1, $V^{-1} : H^{1/2}(\Gamma) \to H^{-1/2}(\Gamma)$ has a square root: More precisely, there exist a complete orthonormal system $(e_i)_{i\in\mathbb{N}}$ in $L^2(\Gamma)$ and positive numbers $(\lambda_i)_{i\in\mathbb{N}}$ with $V^{-1}e_i = \lambda_i e_i$ for all $i \in \mathbb{N}$. Thus the *square root* $V^{-1/2}$ of V^{-1} can be defined for all $u \in H^{1/2}(\Gamma)$ as

$$V^{-1/2}u = \sum_{i\in\mathbb{N}} \lambda_i^{1/2}(u, e_i)_{L^2(\Gamma)}\, e_i$$

and satisfies $\left(V^{-1}u, u\right)_{L^2(\Gamma)} = \left(V^{-1/2}u, V^{-1/2}u\right)_{L^2(\Gamma)}$ (see, e.g., [162, Theorem 2.37, Corollary 2.38]).

Hence we have

$$\begin{aligned}
(P_S u, u)_{L^2(\Gamma)} &= \left(V^{-1}VP_S u, u\right)_{L^2(\Gamma)} = \left(V^{-1/2}VP_S u, V^{-1/2}u\right)_{L^2(\Gamma)} \\
&\leq \|V^{-\frac{1}{2}}VP_S u\|_{L^2(\Gamma)}\|V^{-\frac{1}{2}}u\|_{L^2(\Gamma)} \\
&= \|VP_S u\|_{V^{-1}}\,\|u\|_{V^{-1}} \\
&= \left\|\left(\tfrac{1}{2}I + K\right)u\right\|_{V^{-1}}\|u\|_{V^{-1}} .
\end{aligned}$$

For $u \in H^{1/2}_{\mathcal{N}}(\Gamma)$ we have by Assumption 3.8.5 with (3.134)

$$(Wu, u)_{L^2(\Gamma)} \geq c_W\,\|u\|^2_{H^{1/2}(\Gamma)} \geq c_W c_V\left(V^{-1}u, u\right)_{L^2(\Gamma)} = c_V c_W\|u\|^2_{V^{-1}} .$$

It follows for all $u \in H^{1/2}_{\mathcal{N}}(\Gamma)$ that

$$\left\|\left(\tfrac{1}{2}I + K\right)u\right\|^2_{V^{-1}} \leq \left\|\left(\tfrac{1}{2}I + K\right)u\right\|_{V^{-1}} \|u\|_{V^{-1}} - c_V c_W \|u\|^2_{V^{-1}} .$$

This inequality has the form

$$a^2 \leq ab - c_V c_W\, b^2.$$

The case $0 = b = \|u\|_{V^{-1}}$ implies that $u = 0$ and $0 = \left\|\left(\tfrac{1}{2}I + K\right)u\right\|_{V^{-1}} = a$ and therefore is trivial. For $b \neq 0$ the inequality is equivalent to

$$a^2 b^{-2} - ab^{-1} + c_V c_W \leq 0 \Longleftrightarrow 1 - c_K \leq \frac{a}{b} \leq c_K \wedge c_V c_W \leq 1/4,$$

from which we have (3.138) with "+".

We prove the "−" estimate. With the inequality for "+" it follows for $u \in H^{1/2}_{\mathcal{N}}(\Gamma)$ that

$$\begin{aligned}
\|u\|_{V^{-1}} &= \left\|\left(\tfrac{1}{2}I + K + \tfrac{1}{2}I - K\right)u\right\|_{V^{-1}} \\
&\leq \left\|\left(\tfrac{1}{2}I - K\right)u\right\|_{V^{-1}} + \left\|\left(\tfrac{1}{2}I + K\right)u\right\|_{V^{-1}} \\
(1 - c_K)\|u\|_{V^{-1}} &\leq \left\|\left(\tfrac{1}{2}I - K\right)u\right\|_{V^{-1}} .
\end{aligned}$$

The proof for the upper bound uses (3.128):

$$\begin{aligned}
\left\|\left(\tfrac{1}{2}I - K\right)u\right\|^2_{V^{-1}} &= \|u\|^2_{V^{-1}} + \left\|\left(\tfrac{1}{2}I + K\right)u\right\|^2_{V^{-1}} - 2\left(V^{-1}\left(\tfrac{1}{2}I + K\right)u, u\right)_{L^2(\Gamma)} \\
&= \|u\|^2_{V^{-1}} + \left\|\left(\tfrac{1}{2}I + K\right)u\right\|^2_{V^{-1}} - 2\left(P_S u, u\right)_{L^2(\Gamma)} \\
&= \|u\|^2_{V^{-1}} - \left\|\left(\tfrac{1}{2}I + K\right)u\right\|^2_{V^{-1}} - 2\left(W u, u\right)_{L^2(\Gamma)} \\
&\leq \left(1 - (1 - c_K)^2 - 2c_V c_W\right)\|u\|^2_{V^{-1}} \\
&= c_K^2 \|u\|^2_{V^{-1}} .
\end{aligned}$$

□

If the operator that is associated with the boundary value problem (3.135) is injective Theorem 3.8.7 implies that the equation of the second kind

$$\left(\tfrac{1}{2}I - K\right)\varphi = -g_D \ \text{ in } H^{1/2}(\Gamma) \tag{3.139}$$

has a unique solution for the interior Dirichlet problem

$$Lu = 0 \quad \text{in } \Omega \ , \qquad \gamma_0 u = g_D \text{ in } H^{1/2}(\Gamma) \tag{3.140}$$

by the double layer ansatz with the indirect method.

Exercise 3.8.8. *Show that for the solution φ of the equation of the second kind (3.139) we have the representation*

$$\varphi = -\left(\tfrac{1}{2}I - K\right)^{-1} g_D = -\sum_{\nu=0}^{\infty}\left(\tfrac{1}{2}I + K\right)^{\nu} g_D, \tag{3.141}$$

and that the Neumann series converges in $H^{1/2}(\Gamma)$.

Analogously, we have for the Neumann problem

$$Lu = 0 \quad \text{in } \Omega^-, \qquad \gamma_1^- u = g_N \qquad \text{in } H_{\mathcal{N}}^{-1/2}(\Gamma), \tag{3.142}$$

the integral equation of the second kind

$$\left(\tfrac{1}{2}I + K'\right)\psi = g_N \qquad \text{in } H^{-1/2}(\Gamma) \tag{3.143}$$

by means of the single layer ansatz of the indirect method. Formally, the solution is given by the Neumann series

$$\psi = \left(\tfrac{1}{2}I + K'\right)^{-1} g_N = \sum_{\nu=0}^{\infty}\left(\tfrac{1}{2}I - K'\right)^{\nu} g_N. \tag{3.144}$$

The Neumann series (3.144) converges in the $\|\cdot\|_V$-norm. It also converges in $H^{-1/2}(\Gamma)$ because the $\|\cdot\|_V$ and the $\|\cdot\|_{H^{-1/2}(\Gamma)}$ norms are equivalent, as the following result, which is analogous to Theorem 3.8.7, shows.

Theorem 3.8.9. *With $c_K \in (0,1)$ in (3.138) we have for $\sigma \in \{-,+\}$*

$$(1-c_K)\|u\|_V \le \left\|\left(\sigma K' - \tfrac{1}{2}I\right)u\right\|_V \le c_K\|u\|_V \qquad \forall u \in H_{\mathcal{N}}^{-1/2}(\Gamma). \tag{3.145}$$

Proof. It follows from the boundedness of $\frac{1}{2}I - K' : H_{\mathcal{N}}^{-1/2}(\Gamma) \to H_{\mathcal{N}}^{-1/2}(\Gamma)$ and (3.138) that for $\sigma \in \{-,+\}$ we have

$$\begin{aligned}
\left\|\left(\sigma K' - \tfrac{1}{2}I\right)u\right\|_V^2 &= \left(\left(\sigma K' - \tfrac{1}{2}I\right)u,\ V\left(\sigma K' - \tfrac{1}{2}I\right)u\right)_{L^2(\Gamma)} \\
&= \left(V\left(\sigma K' - \tfrac{1}{2}I\right)u,\ V^{-1}\left(\sigma K - \tfrac{1}{2}I\right)Vu\right)_{L^2(\Gamma)} \\
&= \left(\left(\sigma K - \tfrac{1}{2}I\right)Vu,\ V^{-1}\left(\sigma K - \tfrac{1}{2}I\right)Vu\right)_{L^2(\Gamma)} \\
&= \left\|\left(\sigma K - \tfrac{1}{2}I\right)Vu\right\|_{V^{-1}}^2 \\
&\le c_K^2\,\|Vu\|_{V^{-1}}^2 = c_K^2\|u\|_V^2.
\end{aligned}$$

The left-hand inequality is proven in the same way. □

The Neumann series (3.141) and (3.144) for the representation of solutions of the integral equations (3.139) and (3.143) motivate the representation by series for discretized integral equations as well. (Note that in order to evaluate (3.141) and (3.144) one only has to apply the matrices of the discretized boundary integral operators.) However, since there already exist efficient, iterative methods for the solving of (discretized) equations of the second kind (see Chap. 6), we do not advise the use of Neumann series for the numerical solution of integral equations.

The algorithmic realization of equations of the second kind in $H^{\pm 1/2}(\Gamma)$ is technically involved, as the discretization is based on the $H^{\pm 1/2}$ inner-product. We will therefore provide criteria in the remainder of this section, which permit the formulation of equations of the second kind in $L^2(\Gamma)$. As an example we consider the abstract equation of the second kind:

Let $g \in H^{1/2}(\Gamma)$ be given. Find $\varphi \in H^{1/2}(\Gamma)$ such that

$$-\frac{1}{2}(\varphi,\eta)_{L^2(\Gamma)} + b_K(\varphi,\eta) = (g,\eta)_{L^2(\Gamma)} \quad \forall \eta \in H^{-1/2}(\Gamma). \tag{3.146}$$

Here $(\cdot,\cdot)_{L^2(\Gamma)}$ again denotes the continuous extension of the $L^2(\Gamma)$ inner-product to the dual pairing $\langle\cdot,\cdot\rangle_{H^{1/2}(\Gamma)\times H^{-1/2}(\Gamma)}$. The equation

$$\gamma_0^- D\varphi = g \tag{3.147}$$

in $H^{1/2}(\Gamma)$ is equivalent to (3.146), with double layer potential D, where we have used the relation $\gamma_0^- D = -\frac{1}{2}I + K$ [see (3.56)]. According to the Riesz representation theorem, every functional $\eta \in H^{-1/2}(\Gamma)$ has a unique representative $\psi \in H^{1/2}(\Gamma)$ such that

$$\langle v,\eta\rangle_{H^{1/2}(\Gamma)\times H^{-1/2}(\Gamma)} = (v,\psi)_{H^{1/2}(\Gamma)} \qquad \forall v \in H^{1/2}(\Gamma).$$

With this, (3.146) can be equivalently formulated as follows:

Find $\varphi \in H^{1/2}(\Gamma)$ such that

$$\left(\gamma_0^- D\varphi,\psi\right)_{H^{1/2}(\Gamma)} = (g,\psi)_{H^{1/2}(\Gamma)} \qquad \forall \psi \in H^{1/2}(\Gamma). \tag{3.148}$$

The existence and uniqueness of the solution φ and the continuous dependency on the data g is guaranteed by the assumption: $\gamma_0^- D : H^{1/2}(\Gamma) \to H^{1/2}(\Gamma)$ is an isomorphism.

The numerical implementation of the $H^{1/2}(\Gamma)$ inner-product is technically involved. Thus we will discuss in the following under which additional conditions the $H^{1/2}(\Gamma)$ inner-product in (3.148) can be replaced by the $L^2(\Gamma)$ inner-product (see, for example, [86, Corollaries A.2 and A.5] and [160]).

Assumption 3.8.10. *The operator $\gamma_0^- D : H^s(\Gamma) \to H^s(\Gamma)$ is an isomorphism for $s \in \{0, 1/2\}$.*

The following remark shows that Assumption 3.8.10 is satisfied for the Laplace operator.

Remark 3.8.11. *In the case of the Laplace operator, in [86, Corollaries A.2 and A.5] a Gårding inequality is shown for the operator $\gamma_0^- D$ on $H^{1/2}(\Gamma)$. Furthermore, $\gamma_0^- D$ satisfies Assumption 3.8.10.*

The properties that have been described in the previous remark could be directly transferred to the numerical discretization if the operator $\gamma_0^- D$ were seen as an operator in $H^{1/2}(\Gamma)$. However, this is usually avoided because of the increase in complexity for the discretization of the non-local $H^{1/2}(\Gamma)$ inner-product. Instead, $\gamma_0^- D$ is interpreted as an operator in $L^2(\Gamma)$. Assumption 3.8.10 allows us to formulate the integral equation in $L^2(\Gamma)$ as well: Let $g \in L^2(\Gamma)$ be given. Find $\tilde{\varphi} \in L^2(\Gamma)$ such that

$$\left(\gamma_0^- D\tilde{\varphi}, \psi\right)_{L^2(\Gamma)} = (g, \psi)_{L^2(\Gamma)}, \qquad \forall \psi \in L^2(\Gamma). \tag{3.149}$$

Here $(\cdot,\cdot)_{L^2(\Gamma)}$ denotes the usual inner product in $L^2(\Gamma)$ (and not the extension to dual pairings). Assumption 3.8.10 guarantees the existence of a solution in $L^2(\Gamma)$. Under the additional condition that $g \in H^{1/2}(\Gamma)$, the solution satisfies $\tilde{\varphi} \in H^{1/2}(\Gamma)$. Since $\gamma_0^- D : H^{1/2}(\Gamma) \to H^{1/2}(\Gamma)$ is, according to the our assumptions, an isomorphism, it follows that $\tilde{\varphi} = \varphi$.

Remark 3.8.12. *(a) Let Assumption 3.8.10 be satisfied and let $g \in H^{1/2}(\Gamma)$. Then the solutions from (3.148) and (3.149) coincide.*
(b) The statement "$\gamma_0^- D : L^2(\Gamma) \to L^2(\Gamma)$ is an isomorphism" cannot, in general, be transferred to the numerical discretization of (3.149). In general, the operator $\gamma_0^- D$ does not satisfy any Gårding inequality in $L^2(\Gamma)$ and the stability of the discretization has to be analyzed with special methods for concrete situations.

3.9 Boundary Integral Equations for the Helmholtz Equation

3.9.1 Helmholtz Equation

Thus far, we have always assumed $a_{\min} c > \|\mathbf{b}\|^2$ for the solvability of the boundary integral equations or we have considered the Laplace problem. In this section we will discuss physical applications from the areas of time-harmonic acoustics and electromagnetism that are given by the Helmholtz equation with positive wave number $k > 0$

$$L_k u := -\Delta u - k^2 u = f. \tag{3.150}$$

In terms of the spatial dimension we will assume $d = 3$. As usual, these equations require suitable boundary conditions. For the exterior problem we impose the *Sommerfeld radiation conditions* [see (2.133)]

$$\left.\begin{aligned} |u(\mathbf{x})| &\le C\,\|\mathbf{x}\|^{-1} \\ \left|\frac{\partial u}{\partial r} - iku\right| &\le C\,\|\mathbf{x}\|^{-2} \end{aligned}\right\} \quad \text{for } \|\mathbf{x}\| \to \infty. \tag{3.151}$$

Here $\partial u/\partial r = \langle \mathbf{x}/\|\mathbf{x}\|, \nabla u\rangle$ denotes the radial derivative.

As the coefficients of the Helmholtz operator L_k do not satisfy with the above-mentioned conditions, special methods have to be developed for the analysis.

The fundamental solution for the operator L_k is given by [see (3.3)]

$$G_k(\mathbf{z}) = \frac{e^{ik\|\mathbf{z}\|}}{4\pi\,\|\mathbf{z}\|}. \tag{3.152}$$

In Exercise 3.1.15 it had to be shown that the single and double layer potentials for the Helmholtz equation

$$(S_k\varphi)(\mathbf{x}) = \int_{\mathbf{y}\in\Gamma} G_k(\mathbf{x}-\mathbf{y})\,\varphi(\mathbf{y})ds_{\mathbf{y}}, \qquad (D_k\psi)(\mathbf{x}) = \int_{\mathbf{y}\in\Gamma} \gamma_{1,\mathbf{y}}\, G_k(\mathbf{x}-\mathbf{y})\,\psi(\mathbf{y})ds_{\mathbf{y}} \tag{3.153}$$

satisfy the Sommerfeld radiation conditions (3.151).

For a given $g_D \in H^{1/2}(\Gamma)$ the *exterior Dirichlet problem* (EDP) for the Helmholtz equation reads [see (2.134)]:

$$\begin{aligned} &L_k u = 0 \text{ in } \Omega^+, \qquad \gamma_0^+ u = g_D \text{ on } \Gamma, \\ &u \text{ satisfies the Sommerfeld radiation conditions.} \end{aligned} \tag{3.154}$$

For given data $g_N \in H^{-1/2}(\Gamma)$ the *exterior Neumann problem* (ENP) for the Helmholtz equation [see (2.135)] is given by

$$\begin{aligned} &L_k u = 0 \text{ in } \Omega^+, \qquad \gamma_1^+ u = g_N \text{ on } \Gamma, \\ &u \text{ satisfies the Sommerfeld radiation conditions.} \end{aligned} \tag{3.155}$$

3.9.2 Integral Equations and Resonances

In this section we will give necessary and sufficient conditions for the existence of a unique solution of the integral equations for the interior problems of the Helmholtz equation. For the exterior problem the radiation conditions guarantee that the EDP and ENP for the Helmholtz equation has a unique solution for every k. Some of the integral equations that appear during the boundary reduction of the interior and exterior problems are identical. It follows that, although the boundary value problem has a unique solution, the integral operators of the exterior problem are *not* invertible in the natural Sobolev spaces on Γ for every wave number.

It follows from Theorem 3.1.1, Proposition 3.1.7 and Exercise 3.1.15 that for $\varphi \in H^{-1/2}(\Gamma)$ the single layer potential $u = S_k\varphi$ satisfies the homogeneous differential equation $L_k\, S_k\varphi = 0$ in $\Omega^+ \cup \Omega^-$ as well as the Sommerfeld radiation condition (3.151). According to Theorem 3.3.1 we have $[S_k\varphi] = 0$ and the single layer operator

$$V_k\varphi = \gamma_0^+\, S_k\varphi = \gamma_0^-\, S_k\varphi : H^{-1/2}(\Gamma) \to H^{1/2}(\Gamma)$$

is well defined. For the IDP

$$L_k\varphi = 0 \text{ in } \Omega^-, \quad \gamma_0^- u = g_D \in H^{1/2}(\Gamma)$$

we obtain the integral equation of the first kind: Find $\varphi \in H^{-1/2}(\Gamma)$ such that

$$(V_k\varphi, \psi)_{L^2(\Gamma)} = (g_D, \psi)_{L^2(\Gamma)} \quad \forall\psi \in H^{-1/2}(\Gamma). \tag{3.156}$$

In the following theorem we will discuss the invertibility of the operator V_k.

Theorem 3.9.1. *The single layer operator V_k for the Helmholtz problem is invertible on $H^{-1/2}(\Gamma)$ if and only if k^2 is not an eigenvalue of the IDP for the operator $-\Delta$:*

$$-\Delta u = k^2 u \text{ in } \Omega^-, \quad \gamma_0^- u = 0 \Longrightarrow u = 0 \text{ in } \Omega^-.$$

The null space of V_k is given by

$$\operatorname{span}\left\{\gamma_1^- v : -\Delta v = k^2 v \text{ in } \Omega^- \wedge \gamma_0^- v = 0 \text{ on } \Gamma\right\}.$$

Proof. Let v be an eigenfunction of the IDP for the Laplace operator with eigenvalue k^2, i.e., $-\Delta v = k^2 v$ in Ω^-, $\gamma_0^- v = 0$. The single layer potential $S_k\gamma_0^- v$ is identical to zero on $\mathbb{R}^3$. We denote the zero extension of v to all of $\mathbb{R}^3$ by w. Then we have $[\gamma_0 w] = 0$ and $[\gamma_1 w] = -\gamma_1^- v$. Therefore the representation formula (3.10) for the Helmholtz operator is applicable and gives us $v = -S_k\,[\gamma_1 w] = S_k\gamma_1^- v$ in Ω^-. As v is an eigenfunction of the IDP for the Laplace operator with eigenvalue k^2, we have $L_k v = 0$ in Ω^- and $\gamma_0^- v = 0$ and so we have

$$0 = \gamma_0^- v = \gamma_0^-\left(S_k\gamma_1^- v\right) = V_k\gamma_1^- v.$$

Hence $\gamma_1^- v$ is in the zero space of V_k.

Assume k^2 is not an eigenvalue of the IDP for the Laplace operator in Ω^- and let $w \neq 0$ be in the kernel of V_k. Then the single layer potential $v = S_k w$ satisfies the equation $L_k v = 0$ in Ω^- and from $w \in \operatorname{Kern}(V_k)$ we have $\gamma_0^- v = 0$. From this we obtain $L_k v = -\Delta v - k^2 v = 0$ in Ω^-, $\gamma_0^- v = 0$. Since k^2 is not an eigenvalue of the IDP for the Laplace operator in Ω^-, it follows that $v = 0$ in Ω^- and with $w = -\gamma_1^- v = 0$ we arrive at a contradiction.

Corollary 3.9.2. *Although the EDP (3.154) has a unique solution for all $k \geq 0$, the first kind integral equation (3.156) for the EDP that results from the direct method does not have a solution for all $g_D \in H^{1/2}(\Gamma)$ if k^2 is an eigenvalue of the IDP for the operator $-\Delta$.*

We have a similar result for equations of the second kind. If we use the single layer potential ansatz $u = S_k\varphi$ for the ENP we obtain the problem: For given data $g_N \in H^{-1/2}(\Gamma)$ find $\varphi \in H^{-1/2}(\Gamma)$ such that

$$-\frac{1}{2}(\varphi, \eta)_{L^2(\Gamma)} + (K_k'\varphi, \eta)_{L^2(\Gamma)} = (g_N, \eta)_{L^2(\Gamma)} \quad \forall \eta \in H^{1/2}(\Gamma). \tag{3.157}$$

Theorem 3.9.3. *For every eigenvalue k^2 of the IDP for the operator $-\Delta$ in Ω^-, $-\frac{1}{2}I + K_k'$ is not injective.*

Proof. The proof is similar to the proof of Theorem 3.9.1. Let $0 \neq w \in H_0^1(\Omega^-)$ be an eigenfunction of the IDP for $-\Delta$ in Ω^- and let w^* be the zero extension of this eigenfunction to Ω^+. Then w^* solves the homogeneous equation $L_k w^* = 0$ in $\Omega^+ \cup \Omega^-$ and the radiation condition (3.151) holds. Therefore $0 = S_k(\gamma_1^- w)$ is in Ω^+ [see (3.1.13), (3.19)]. It follows that

$$\left(-\frac{1}{2}I + K_k'\right)(\gamma_1^- w) = 0. \tag{3.158}$$

□

Corollary 3.9.4. *The ENP (3.155) cannot be solved for all $g_N \in H^{-1/2}(\Gamma)$ by the integral equation (3.157) if k^2 is an eigenvalue of the IDP.*

We have an analogous result for the integral operators $\frac{1}{2}I + K_k$ and W_k.

Exercise 3.9.5. *Let k^2 be an eigenvalue of the INP for the Laplace equation and let $0 \neq w \in H^1(\Omega^-)$ be an associated eigenfunction. Then we have*

$$\left(-\frac{1}{2}I + K_k\right)(\gamma_1^- w) = 0, \quad W_k(\gamma_0^- w) = 0, \tag{3.159}$$

and the integral operators in the boundary integral equations

$$\left(\frac{1}{2}I + K_k\right)\varphi = g_D, \quad W_k\varphi = g_N, \tag{3.160}$$

for the EDP and ENP of the Helmholtz equation (both of which have a unique solution) are not invertible in this case.

Remark 3.9.6. *Note that the statements from Theorems 3.9.1 and 3.9.3, Corollaries 3.9.2 and 3.9.4 and of Exercise 3.9.5 remain valid unchanged for the operators V_{-k}, K_{-k}, K_{-k}' and W_{-k}, as the associated eigenvalue equation $-\Delta u = k^2 u$ in Ω^- does not depend on the sign of k.*

These observations raise the following dilemma: The solutions of the exterior problems (3.154) and (3.155) are uniquely determined for all k; however, the (standard) boundary integral equations (3.156) and (3.157) *cannot* be solved for the resonant frequencies of the interior problems for arbitrary boundary data $g_D \in H^{1/2}(\Gamma)$, $g_N \in H^{-1/2}(\Gamma)$. In Sect. 3.9.4 we will introduce modified boundary integral equations which avoid this difficulty.

3.9.3 Existence of Solutions of the Exterior Problem

In this section we will provide a proof of existence for the solutions of the Helmholtz exterior problem. In the classical point of view the boundary integral operators for the Helmholtz operator are seen as a compact perturbation of the operators for the Laplace operator. The disadvantage of this approach is the fact that the boundary integral equations do not have a solution for certain critical wave numbers, even though the associated boundary value problem has a unique solution.

One stabilized formulation without critical frequencies is due to Panich and will be introduced in Sect. 3.9.4.

The easiest situation occurs when the boundary Γ is smooth. In this case $K_k : H^{1/2}(\Gamma) \to H^{1/2}(\Gamma)$ is compact, since the associated kernel function is then weakly singular, as the following exercise demonstrates.

Exercise 3.9.7. *Let $\Gamma \in C^2$. Then there exists some $C(\Gamma) > 0$ such that*

$$\left|G_k(\mathbf{x}-\mathbf{y})\right| + \left|\gamma_{1,\mathbf{y}} G_k(\mathbf{x}-\mathbf{y})\right| \leq \frac{C(\Gamma)}{\|\mathbf{x}-\mathbf{y}\|} \qquad \forall \mathbf{x}, \mathbf{y} \in \Gamma,\ \mathbf{x} \neq \mathbf{y}. \tag{3.161}$$

(Hint: Use Lemma 2.2.14.)

If K is compact the integral equation (3.139) becomes a Fredholm equation of the second kind in $H^{1/2}(\Gamma)$. Furthermore, the injectivity of the integral operator in (3.139) implies the existence of a unique solution $\varphi \in H^{1/2}(\Gamma)$, according to Theorem 2.1.36, provided that k^2 does not lie in the spectrum of the INP for the Laplace equation.

If we decompose W_k into a definite operator and a compact perturbation we may apply Theorem 2.1.36 to the integral equation of the first kind, $W_k \varphi = g_N$, which arises during the indirect boundary reduction of the ENP. The following lemma provides the details.

Lemma 3.9.8. *Let Γ be a Lipschitz boundary in $\mathbb{R}^3$ and let $k \in \mathbb{R}$. Then the following operators are compact:*

$$\begin{aligned} V_k - V_0 : &\quad H^{-1/2}(\Gamma) \to H^{1/2}(\Gamma) \\ K_{\sigma,k} - K_{\sigma,0} : &\quad H^{1/2}(\Gamma) \to H^{1/2}(\Gamma), \quad \sigma \in \{+,-\} \end{aligned}$$

$$K'_{\sigma,k} - K'_{\sigma,0} : H^{-1/2}(\Gamma) \to H^{-1/2}(\Gamma), \quad \sigma \in \{+,-\}$$
$$W_k - W_0 : \quad H^{1/2}(\Gamma) \to H^{-1/2}(\Gamma).$$

Proof. We consider the Newton potential for the Helmholtz operator $(\mathcal{N}_k\varphi)(\mathbf{x}) = \int_{\mathbb{R}^3} G_k(\mathbf{x}-\mathbf{y})\,\varphi(\mathbf{y})ds_{\mathbf{y}}$. Then $\mathcal{N}_k - \mathcal{N}_0$ is the potential for the kernel function $G_k(\mathbf{z}) - G_0(\mathbf{z}) = \frac{e^{ik\|\mathbf{z}\|}-1}{4\pi\|\mathbf{z}\|}$ and

$$\mathcal{N}_k - \mathcal{N}_0 : H^{\ell}_{\text{comp}}(\mathbb{R}^3) \to H^{\ell+4}_{\text{loc}}(\mathbb{R}^3) \qquad \forall \ell \in \mathbb{R}$$

is continuous (see Remark 3.1.3). We use the representation $V_k - V_0 = \gamma_0(\mathcal{N}_k - \mathcal{N}_0)\gamma'_0$ [see (3.1.6)]. The continuity of $\gamma_0 : H^1_{\text{loc}}(\mathbb{R}^3) \to H^{1/2}(\Gamma)$ implies the continuity of $\gamma'_0 : H^{-1/2}(\Gamma) \to H^{-1}_{\text{comp}}(\mathbb{R}^3)$. From this we have the compactness of $V_k - V_0 : H^{-1/2}(\Gamma) \to H^{1/2}(\Gamma)$ by considering the composition of the following mappings

$$H^{-1/2}(\Gamma) \underset{\gamma'_0}{\longrightarrow} H^{-1}_{\text{comp}}(\mathbb{R}^3) \underset{\mathcal{N}_k-\mathcal{N}_0}{\longrightarrow} H^3_{\text{loc}}(\mathbb{R}^3) \underset{c}{\hookrightarrow} H^1_{\text{loc}}(\mathbb{R}^3) \underset{\gamma_0}{\longrightarrow} H^{1/2}(\Gamma).$$

Here the compactness of the embedding $H^3_{\text{loc}}(\mathbb{R}^3) \underset{c}{\hookrightarrow} H^1_{\text{loc}}(\mathbb{R}^3)$ follows directly from the compact embedding $H^3(\Omega) \underset{c}{\hookrightarrow} H^1(\Omega)$ for every compact domain Ω (see Theorem 2.6.7). We have $K_{\sigma,k} - K_{\sigma,0} = \gamma_0^\sigma(\mathcal{N}_k - \mathcal{N}_0)(\gamma_1^\sigma)'$ for $\sigma \in \{-,+\}$ (see Definition 3.1.5).

In order to analyze the mapping properties of $K_{\sigma,k} - K_{\sigma,0}$, we use the same approach as in the proof of Theorem 3.1.16. We apply the solution operator T from Sect. 2.8 for the interior problem (with $L \leftarrow L_k$ and $\lambda \leftarrow k^2$) and define the function $u \in H^1_L(\mathbb{R}^3)$ for given boundary data $v \in H^{1/2}(\Gamma)$ by

$$u := \begin{cases} Tv & \text{in } \Omega^-, \\ 0 & \text{in } \Omega^+. \end{cases}$$

We define $f_k \in L^2_{\text{comp}}(\mathbb{R}^3)$ by

$$f_k := (L_k)_{\pm}\, u = \begin{cases} -k^2 Tv & \text{in } \Omega^-, \\ 0 & \text{in } \Omega^+. \end{cases}$$

Green's formula (3.10) may be applied thanks to the compact support of u and gives us the relation

$$u = \mathcal{N}_k f_k + S_k \gamma_1^- u - D_k v,$$
$$u = \mathcal{N}_0 f_0 + S_0 \gamma_1^- u - D_0 v.$$

We clearly have $f_0 = 0$ and by subtracting one equation from the other we obtain

$$(D_k - D_0)\,v = \mathcal{N}_k f_k + (S_k - S_0)\,\gamma_1^- T v. \tag{3.162}$$

In the following we will use the same notation as in the proof of Theorem 3.1.16. The mapping properties of $S_k - S_0$, $\mathcal{N}_k$ and $\gamma_1^- T$ (see Theorem 2.8.2) imply

$$H^{1/2}(\Gamma) \overset{\binom{-k^2T}{0}}{\rightarrow} L^2_{\text{comp}}\left(\mathbb{R}^3\right) \overset{\mathcal{N}_k}{\rightarrow} H^2_{\text{loc}}\left(\mathbb{R}^3\right) \underset{c}{\hookrightarrow} H^1_{\text{loc}}(\mathbb{R}^3),$$

$$H^{1/2}(\Gamma) \overset{\gamma_1^- T}{\longrightarrow} H^{-1/2}(\Gamma) \overset{S_k - S_0}{\rightarrow} H^3_{\text{loc}}\left(\mathbb{R}^3\right) \underset{c}{\hookrightarrow} H^1_{\text{loc}}(\mathbb{R}^3).$$

By combining these results we obtain the compactness of the mapping $D_k - D_0 : H^{1/2}(\Gamma) \to H^1_{\text{loc}}\left(\mathbb{R}^3\right)$. The continuity of the trace operators $\gamma_0^\pm : H^1_{\text{loc}}\left(\Omega^\pm\right) \to H^{1/2}(\Gamma)$ therefore gives us the compactness of the difference mapping $K_{\sigma,k} - K_{\sigma,0} : H^{1/2}(\Gamma) \to H^{1/2}(\Gamma)$.

The right-hand side in (3.162) can be decomposed into $\mathcal{N}_k f_k + S_k \gamma_1^- T v \in H^1_{L_k}\left(\mathbb{R}^3 \backslash \Gamma\right)$ and $S_0 \gamma_1^- T v \in H^1_{L_0}\left(\mathbb{R}^3 \backslash \Gamma\right)$. Hence $\gamma_1^\sigma : H^1_L(\Omega^\sigma) \to H^{-1/2}(\Gamma)$ can be applied to every one of these summands, which yields the compactness of $W_k - W_0 = -\gamma_1 D : H^{1/2}(\Gamma) \to H^{-1/2}(\Gamma)$. □

We now use Lemma 3.9.8 to show existence for the EDP for the Helmholtz equation.

Theorem 3.9.9. *For every $g_D \in H^{1/2}(\Gamma)$ the EDP (3.154) has a unique solution.*

Proof. We transform the EDP with the representation formula $u(x) = S_k u_N - D_k g_D$ and the direct method to the equivalent boundary integral equation: Find $u_N \in H^{-1/2}(\Gamma)$ such that

$$(V_k u_N, \eta)_{L^2(\Gamma)} = \left((-\frac{1}{2} I + K_k) g_D, \eta \right)_{L^2(\Gamma)} \qquad \forall \eta \in H^{-1/2}(\Gamma). \tag{3.163}$$

According to Lemma 3.9.8 there exists a constant $C > 0$ with

$$(V_k \varphi, \varphi)_{L^2(\Gamma)} \geq C \|\varphi\|^2_{H^{-1/2}(\Gamma)} - c(\varphi, \varphi) \qquad \forall \varphi \in H^{-1/2}(\Gamma)$$

and a compact form $c(\cdot,\cdot)$ on $H^{-1/2}(\Gamma)$ (given by the sesquilinear form associated with $V_k - V_0$). The Fredholm alternative is applicable to (3.163) and the injectivity of V_k implies that (3.163) has a unique solution. According to Theorem 3.9.1, V_k is injective on $H^{-1/2}(\Gamma)$ if and only if k^2 is not an eigenvalue of the IDP for the operator $-\Delta$. Then (3.163) has a unique solution $u_N \in H^{-1/2}(\Gamma)$ for all $g_D \in H^{1/2}(\Gamma_D)$.

If k^2 is an eigenvalue of the IDP for the operator $-\Delta$ then, according to the Fredholm alternative, the integral equation (3.163) can be solved if and only if the right-hand side vanishes on the kernel of the adjoint operator of V_k.

The adjoint operator of V_k is given by $V_k^*(x) = \int_\Gamma \frac{e^{-ik\|x-y\|}}{4\pi\|x-y\|} u(y)\, ds_y$. By virtue of Remark 3.9.6, the kernel of V_k^* is spanned by $\gamma_1^- v$ with

$$-\Delta v - k^2 v = 0 \text{ in } \Omega^-, \qquad \gamma_0^- v = 0 \text{ on } \Gamma. \tag{3.164}$$

Let v be a solution of (3.164). Then, with Green's second formula [see (2.112)], we have for this v in Ω^-:

$$\begin{aligned}\left(\gamma_1^- v, (-\frac{1}{2}I + K_k) g_D\right)_{L^2(\Gamma)} &= \left(\gamma_1^- v, \gamma_0^-(D_k g_D)\right)_{L^2(\Gamma)} \\ &= \left(\gamma_0^- v, \gamma_1^-(D_k g_D)\right)_{L^2(\Gamma)} - (L_k v, D_k g_D)_{L^2(\Omega^-)} \\ &\quad + (v, L_k D_k g_D)_{L^2(\Omega^-)} .\end{aligned}$$

All terms on the right-hand side vanish, which is why, according to the Fredholm alternative, the integral equation (3.163) has a solution. The solutions are unique up to elements from the kernel of V_k, i.e., up to $\gamma_1^- v$ for eigenfunctions v of the IDP for the operator $-\Delta$ with eigenvalue k^2.

We have shown that the integral equation (3.163) has a solution u_N for every k and every $g_D \in H^{1/2}(\Gamma)$. By means of $S_k u_N - D_k g_D$ we have therefore shown the existence of a solution of the EDP. □

We can show the existence of solutions for the ENP for all wave numbers with similar methods.

3.9.4 *Modified Integral Equations*

Finding a stable numerical solution of the boundary integral equations for the Helmholtz equation is substantially complicated by the problem of resonant frequencies. Therefore we are interested in modified integral equations that have unique solutions for all wave numbers.

There are several approaches to transform the exterior problems (3.154) and (3.155) into modified integral equations that have unique solutions for all wave numbers. We will present two.

The classical approach is due to Brakhage and Werner [28] and it consists of using a combined single and double layer ansatz. For globally smooth surfaces it can be shown that the resulting boundary integral equations have unique solutions for all wave numbers. We consider the EDP (3.154) and use the indirect method. Let $\eta \in \mathbb{R}$ with

$$\eta \operatorname{Re} k > 0. \tag{3.165}$$

In Ω^+ we set

$$u = D_k \varphi - i\eta S_k \varphi. \tag{3.166}$$

This ansatz satisfies $L_k u = 0$ in Ω^+ as well as the Sommerfeld radiation conditions (3.151). The jump relations give us the boundary integral equation

$$g_D = \gamma_0^+ u = \left(\frac{1}{2}I + K_k\right)\varphi - i\eta V_k \varphi, \tag{3.167}$$

an integral equation of the second kind for the unknown φ. Assuming that Γ is globally smooth it can be shown, as, for example, in [147, Theorems 3.33, 3.34], that the integral operator $\frac{1}{2}I + K_k - i\eta V_k$ is bijective for all wave numbers k. The proof is similar to the proof of Theorem 3.9.1, where, instead of the null space of the IDP, we have the null space of the interior problem

$$-\Delta u - k^2 u = 0 \quad \text{in } \Omega^-, \qquad \gamma_1^- u + i\eta\gamma_0^- u = 0,$$

in which for all $\eta \neq 0$ the term k^2 is not an eigenvalue.

The proof of the bijectivity uses the global smoothness of the boundary, which, according to Exercise 3.9.7, implies the weak singularity of the kernels of K_k and K_k'. Therefore the integral operator in (3.167) is a Fredholm integral operator of the second kind and is thus boundedly invertible for all wave numbers k.

With the Brakhage-Werner regularization (3.167), the question whether $\frac{1}{2}I + K_k - i\eta\, V_k$ is also bijective for piecewise smooth or general Lipschitz boundaries Γ, remains unanswered. In this case the ansatz (3.166) becomes problematic, since the domains of D_k and S_k do not coincide on non-smooth boundaries.

This problem is solved by an approach due to Panich [179], which guarantees the existence of a unique solution for all wave numbers for Lipschitz boundaries as well. We assume that there exists an isomorphism

$$R : H^{-1/2+s}(\Gamma) \to H^{1/2+s}(\Gamma) \qquad \forall |s| \leq 1/2 \tag{3.168}$$

on general Lipschitz boundaries Γ that is Hermitian for $s = 0$.

For the solution of the EDP with Lipschitz boundary we use the ansatz by means of potentials

$$u(x) = D_k\varphi + i\eta\, S_k\, R^{-1}\varphi \qquad \forall \varphi \in H^{1/2}(\Gamma). \tag{3.169}$$

Then we have $L_k u = 0$ in Ω^+ and (3.151) for all $\varphi \in H^{1/2}(\Gamma)$. The unknown data is the solution of the boundary integral equation

$$g_D = \gamma_0^+ u = B_k\varphi := \left(\frac{1}{2}\, I + K_k\right)\varphi + i\eta\, V_k\, R^{-1}\varphi \text{ in } H^{1/2}(\Gamma). \tag{3.170}$$

Furthermore, we have for the potential u in (3.169)

$$[\gamma_0 u] = \varphi \quad \text{and} \quad [\gamma_1 u] = -i\eta\, R^{-1}\varphi. \tag{3.171}$$

If we eliminate the density φ in (3.171) we obtain that u in (3.169) is a solution of the interior problem

$$L_k u = 0 \text{ in } \Omega^-, \quad i\eta\gamma_0^- u + R(\gamma_1^- u) = i\eta g_D + R\gamma_1^+ u. \tag{3.172}$$

Proposition 3.9.10. *For $\eta \neq 0$ the integral operator B_k in (3.170) is injective for all k.*

Proof. Let $0 \neq \varphi \in H^{1/2}(\Gamma)$ be a solution of $B_k\varphi = 0$. Then $u := D_k\varphi + i\eta\, S_k\, R^{-1}\varphi$ is a solution of the EDP (3.154) in Ω^+ with $g_D = 0$. The fact that the EDP has a unique solution implies that $u = 0$ in Ω^+. From this we have that $-\gamma_0^- u = [\gamma_0 u] = \varphi$ and $-\gamma_1^- u = [\gamma_1 u] = -i\eta\, R^{-1}\varphi$. Green's formula in Ω^- yields

$$\|\nabla u\|^2_{L^2(\Omega^-)} - k^2\, \|u\|^2_{L^2(\Omega^-)} = (\gamma_1^- u, \gamma_0^- u)_{L^2(\Gamma)} = -(i\eta R^{-1}\varphi, \varphi)_{L^2(\Gamma)}.$$

Since R is Hermitian, it follows for $0 \neq \eta \in \mathbb{R}$ that the right-hand side of this identity is purely imaginary, from which we have $(R^{-1}\varphi, \varphi)_{L^2(\Gamma)} = 0$. By virtue of (3.168) we then have $\varphi = 0$, i.e., we have the injectivity of B_k. □

The existence of a unique solution of the integral equation (3.170) for all wave numbers follows from the injectivity and a Gårding inequality for B_k in $H^{1/2}(\Gamma)$ that is uniform in k.

Remark 3.9.11. *The choice of R is not unique. In [45], the definition of R is based on the strongly smoothing integral operator $\int_\Gamma e^{-\|\mathbf{x}-\mathbf{y}\|}\varphi(\mathbf{y})\, ds_{\mathbf{y}}$. It is also explicitly analyzed in how far the Galerkin discretization depends on the wave number k. In [132] the inverse of the Laplace–Beltrami operator for the stabilization of the integral equation is proposed.*

For further readings we refer, e.g., to [55, 56, 70].

3.10 Bibliographical Remarks on Variational BIEs

In this and the preceding chapter, we presented elements of variational formulations of boundary integral equations on Lipschitz domains $\Omega \subset \mathbb{R}^3$. We also established the well-posedness of these variational boundary integral equations by proving coercivity of the boundary integral operators in scales of Sobolev spaces on the boundary $\Gamma = \partial\Omega$.

The use of integral equation methods to analyze the existence and unique solvability of elliptic boundary value problems is not recent: it dates back to the nineteenth century in the work of Fredholm, Radon, Neumann. Ideas from the analysis of integral equations entered also in an essential fashion into the development of functional analysis at the beginning of the 20th century. However, in these works the boundary integral operators were analyzed as mappings between Hölder spaces.

In our presentation, we paid particular attention to the proof of coercivity in suitable function spaces of Sobolev type on the surface Γ. A key step in the approach presented here consists in establishing Gårding inequalities in function spaces on Γ by "transfer of coercivity" from the elliptic problem in Ω to spaces of traces and normal derivatives on Γ. This avenue to formulating and to studying well-posedness of BIEs is relatively recent: it originates in work of J.C. Nédélec and Planchard [171] and, independently, of G.C. Hsiao and W.L. Wendland in [136]. A particularly lucid exposition of the approach for boundary value problems of linear, scalar, second order strongly elliptic differential operators in or exterior to bounded Lipschitz domains Ω is [72] due to M. Costabel.

The formal approach has, however, a much wider scope than scalar, second order elliptic problems: it has been extended to quite general classes of boundary value problems for systems of partial differential equations which are elliptic in the sense of Agmon, Douglis and Nirenberg in smooth domains as was shown by M. Costabel and W. Wendland in [74]. This result covers in particular the Lamé–Navier system of linearized elasticity in Lipschitz domains Ω where the analogs of the screen problems of Sect. 3.5.3 are the first kind boundary integral equation reformulations of elastic fracture mechanics. Their well-posedness was first established by E.P. Stephan in [219]. These formulations have the appeal that they require only discretizations of the fracture, and not of the ambient, possibly infinitely large, linearly elastic medium. We note in passing that the formulation and the analysis of problems from both elastostatics as well as elastodynamic problems interior or exterior to bounded domains $\Omega \subset \mathbb{R}^3$ in function spaces of Hölder type is quite mature and classical by now, see, e.g. [23, 150] and the references there. For specific variational formulations of these boundary value problems for vector-valued functions we refer to Chap. 2 of [137]. Importantly, *all concepts presented in Chap. 3 of the present monograph for scalar, second order elliptic problems carry over to problems of elastostatics* verbatim.

Also due to M. Costabel is the application of the variational formalism to the derivation of *coercive coupled variational boundary integral formulations*. There, a variational formulation of a (possibly nonlinear) elliptic boundary value problem in a bounded domain is coupled to a linear elliptic exterior problem by means of a one-sided boundary reduction in the exterior domain. This results, in effect, in variational formulations of elliptic PDEs with nonlocal and, possibly, nonlinear boundary integral operators in their variational form. These coupled formulations constitute a nonlocal *exact artificial boundary condition* for the artificial reduction of boundary value problems on unbounded domains to a bounded, truncated domain.

Once again, strong ellipticity of coupled variational formulations on the boundary can be established, as was first explained by M. Costabel in [73]. The variational approach for deriving well-posed, i.e. strongly elliptic, coupled formulations of boundary value and transmission problems has subsequently found many applications, in particular in *contact problems in elasticity* (see, e.g. [61, 158] for a formulation, and for asymptotic convergence estimates for some Galerkin discretizations of such coupled formulations).

In problems of linearized elasticity, this leads once again to *boundary integral variational inequalities* on the (a priori unknown) contact surface. Again, the boundary integral operators appearing in these inequalities are shown to be strongly elliptic by the variational method; also here, the function spaces $\widetilde{H}^{\pm 1/2}(\Gamma_0)$ enter in an essential fashion.

A particularly interesting feature of boundary integral reformulations of boundary value problems of homogeneous, isotropic three-dimensional, linearized elasticity is the fact that the fundamental solution and the *mapping properties of all boundary integral operators remain unchanged in the incompressible limit*, i.e. in the passage from the Lamé–Navier equations to the Stokes equations. This is in stark contrast to the variational formulations of the domain problems, where in the incompressible limit the function spaces must change, due to the appearance of the "incompressibility constraint" $\operatorname{div} u = 0$ in Ω. This constraint is accounted for exactly by the fundamental solution (which, in this case, is a matrix-valued integral operator with kernels derived from the so-called *Kelvin fundamental tensor*. As a consequence, *variational boundary integral formulations for linearized, three-dimensional elastostatics* do not suffer from the so-called *locking effect* which is a well-known problem for domain based Finite Element discretizations of the PDE.

A further important class of elliptic problems for which the use of boundary integral equations is fertile for efficient numerical treatment is *computational electromagnetism*. This class of problems has not been discussed in the present monograph since, unlike the above-mentioned problems, on nonsmooth domains (such as Lipschitz domains Ω), the Maxwell system does *not* fit straightforwardly into the variational framework of Chaps. 2 and 3. One approach to overcoming this consists in reformulation of the governing equations in terms of vector potentials; indeed, in this setting, the governing Maxwell equations can be, at least on smooth domains Ω, recast into elliptic systems which allow for a strongly elliptic bilinear form (see, e.g. [170] and the references there).

For time-harmonic electromagnetic wave propagation problems, a basic problem in computational electromagnetism is the numerical solution of the time-harmonic Maxwell governing equations subject either to "electric" or to "magnetic" boundary conditions (taking formally the place of Dirichlet and Neumann boundary conditions, but being different from these) in the exterior of a bounded Lipschitz polyhedron $\Omega \subset \mathbb{R}^3$. Degenerate domains like screens, or wires, are again of particular interest in connection with electromagnetic fields in antenna design. Here, a *direct* approach towards variational boundary integral equations on nonsmooth domains has been developed in recent years. It was initiated by electrical engineers, starting from the so-called Stratton–Chu representation formula for electromagnetic fields as potentials of so-called surface currents. The principal issues, definitions and mathematical results in well-posed variational formulations of BIEs on Lipschitz polyhedra is recapitulated in the survey [41] and the references there. In particular, the variational functional framework of the associated BIEs (such as, e.g., the so-called "Electric Field Integral Equation (EFIE)"), is quite distinct from that presented in Chaps. 2 and 3 of the present volume. The structure of the trace spaces on Lipschitz surfaces were only recently characterized in [39]. Likewise, coercive variational formulations of the variational BIEs have been obtained by boundary reduction from corresponding statements of the domain problems in [40]. Many features of the BIEs of electromagnetics are reminiscent of the (simpler) Helmholtz equation which was described in Sect. 3.9 of the present volume: in particular the occurrence of resonance frequencies at which the homogeneous, interior boundary value problems admit nontrivial solutions causes instabilities of the boundary integral equations; these can be overcome by the use of the so-called "combined field integral equation (CFIE)", see for example [42, 43] for details. The numerical analysis of the BIEs obtained by the direct boundary reduction is complicated in that the function spaces which are natural for the BIE have an infinite dimensional null space; accordingly, the standard convergence framework of Galerkin BEM which we presented in the present chapter does not apply anymore. This problem was overcome first by S. Christiansen in [64] and later extended to the case of screen like conductors in [37].

The above references underline the wide applicability of the variational approach to the systematic derivation of stable boundary integral equation formulations, in particular on nonsmooth domains Ω. Still, the variational approach is historically rather recent, and relies on transfer of strong ellipticity of a related partial differential equation together with the existence of a fundamental solution of the differential operator.

We mention several other approaches for proving stability of boundary integral operators which do not draw upon the variational approach. Most if not all of these approaches require, however, surfaces Γ which are considerably more regular than Lipschitz which was sufficient for the variational approach.

One of the earliest approaches is the proof of bounded invertibility of the double layer potential for the Laplacian by means of showing convergence of the so-called Neumann series representation of the inverse in classes of Hölder continuous functions on Γ. We refer to [160] and the references there for more on this topic. In particular, the bounded invertibility of the boundary integral operators for elasticity in classes of Hölder continuous functions has been investigated in [150].

A second, general approach to the analysis of boundary integral operators is by interpreting them as particular instances of *pseudodifferential operators*. In this way, powerful tools from the theory of these operators can be brought to bear. This requires, however, boundaries Γ which are smooth, closed manifolds in $\mathbb{R}^3$. Still, in this case strong ellipticity in the form of coercivity of the boundary integral operators can be established *directly*, i.e. without resorting to ellipticity of a partial differential operator in the domain Ω bounded by Γ. This is done by proving a *Gårding inequality* for the principal part of the boundary integral operator A. The principal part of the operator A at a point $\mathbf{x} \in \Gamma$ coincides with the restriction of this operator on the tangent bundle to Γ at $\mathbf{x}$. A key result from the theory of pseudodifferential operators on manifolds states that the *Gårding inequality* for A follows from the *positivity of the real part of the principal symbol of the boundary integral operator* A. For many boundary integral operators, the principal symbol is easily calculated. The verification of its positivity is then elementary. The mathematical details of this direct approach to strong ellipticity of boundary integral operators are nicely laid out in the recent monograph of G.C. Hsiao and W.L. Wendland [137].

So far, all boundary integral equations considered were posed on *bounded surfaces* $\Gamma \subset \mathbb{R}^3$. There are several cases of practical interest where the surfaces Γ are unbounded. We mention only acoustic or electromagnetic scattering on a halfspace or the integral equations which arise in the modelling of water waves. We refer to the recent papers [58, 59, 186] for more on the formulation and the bounded invertibility of integral operators on noncompact boundaries Γ.

Another area of active current research is the *uniformly bounded invertibility of parametric boundary integral operators*. This pertains in particular to acoustic and electromagnetic scattering problems at high frequencies where the parameter is the (nondimensional) wave number. In our considerations, the boundary integral equations for Helmholtz problems were always considered at fixed wave number κ; however, all constants in the stability estimates for the boundary integral operators which are obtained by the abstract error analysis depend on the wave number κ in an unspecific way. In recent years, considerable progress was made in establishing stability bounds which are explicit in the wave number κ for boundary integral operators for acoustics and electromagnetics (i.e. for the Helmholtz and Maxwell equations) (see, e.g., [15, 45, 55, 56, 146, 156]).

Chapter 4
Boundary Element Methods

In Chap. 3 we transformed strongly elliptic boundary value problems of second order in domains $\Omega \subset \mathbb{R}^3$ into boundary integral equations. These integral equations were formulated as variational problems on a Hilbert space H:

$$\text{Find } u \in H: \qquad b(u, v) = F(v) \qquad \forall v \in H, \tag{4.1}$$

which, in the simplest cases, was chosen as one of the Sobolev spaces $H^s(\Gamma)$, $s = -1/2, 0, 1/2$. The functional $F \in H'$ denotes the given right-hand side, which, in the case of the direct method (see Sect. 3.4.2), may again contain integral operators. The sesquilinear form $b(\cdot, \cdot)$ has the abstract form

$$b(u, v) = (Bu, v)_{L^2(\Gamma)}$$

with the integral operator

$$(Bu)(\mathbf{x}) = \lambda_1(\mathbf{x})\, u(\mathbf{x}) + \lambda_2(\mathbf{x}) \int_\Gamma k(\mathbf{x}, \mathbf{y}, \mathbf{y} - \mathbf{x})\, u(\mathbf{y})\, ds_{\mathbf{y}} \qquad \mathbf{x} \in \Gamma \text{ a.e.} \tag{4.2}$$

Convention 4.0.1. *The inner product $(\cdot, \cdot)_{L^2(\Gamma)}$ is again identified with the continuous extension on $H^{-s}(\Gamma) \times H^s(\Gamma)$.*

The coefficients λ_1, λ_2 are bounded. For $\lambda_1 = 0$, a.e., one speaks of an integral operator of the first kind, otherwise of the second kind. In some applications the kernel function is not improperly integrable, and the integral is defined by means of a suitable regularization (see Theorem 3.3.22).

The sesquilinear form in (4.1) associated with the boundary integral operator in (4.2) satisfies a Gårding inequality: There exist a $\gamma > 0$ and a compact operator $T : H \to H'$ such that

$$\forall u \subset H : |b(u, u) + \langle Tu, u \rangle_{H' \times H}| \geq \gamma \|u\|_H^2 . \tag{4.3}$$

S.A. Sauter and C. Schwab, *Boundary Element Methods*, Springer Series in Computational Mathematics 39, DOI 10.1007/978-3-540-68093-2_4,

The variational formulation (4.1) of the integral equations forms the basis of the numerical solution thereof, by means of finite element methods on the boundary $\Gamma = \partial\Omega$, the so-called boundary element methods. They are abbreviated by "BEM".

Note: Readers who are familiar with the concept of finite element methods will recognize it here. One essential conceptual difference between the BEM and the finite element method is the fact that, in the BEM, the resulting finite element meshes *usually* consist of curved elements and therefore, in general, *no affine* parametrization over a reference element can be found.

Primarily, we consider the Galerkin BEM, which is the most natural method for the variational formulation (4.1) of the boundary integral equation. In Sect. 4.1 we will describe the Galerkin BEM for the boundary value problems of the Laplace equation with Dirichlet, Neumann and mixed boundary conditions, all of which lead to boundary integral equations of the first kind with positive definite bilinear forms. We obtain quasi-optimal approximations and prove asymptotic convergence rates for the Galerkin BEM. In Sect. 4.2 we will then study Galerkin methods in an abstract form for operators that are only positive with a compact perturbation. We will also present a general framework for the convergence analysis of Galerkin methods. In Sect. 4.3 we will finally prove the approximation properties of the boundary element spaces.

4.1 Boundary Elements for the Potential Equation in $\mathbb{R}^3$

We will first introduce the Galerkin BEM for integral equations of the classical potential problem in $\mathbb{R}^3$ and derive relevant error estimates for the simplest boundary elements.

4.1.1 Model Problem 1: Dirichlet Problem

Let $\Omega^- \subset \mathbb{R}^3$ be a bounded polyhedral domain, the boundary $\Gamma = \partial\Omega^-$ of which consists of finitely many, disjoint, plane faces Γ^j, $j = 1, \dots, J$: $\Gamma = \bigcup_{j=1}^J \overline{\Gamma^j}$.

In the exterior $\Omega^+ = \mathbb{R}^3 \backslash \overline{\Omega^-}$ we consider the Dirichlet problem

$$\Delta u = 0 \text{ in } \Omega^+, \tag{4.4a}$$

$$u = g_D \text{ on } \Gamma, \tag{4.4b}$$

$$|u(\mathbf{x})| = O(\|\mathbf{x}\|^{-1}) \text{ for } \|\mathbf{x}\| \to \infty. \tag{4.4c}$$

In Chap. 2 (Theorem 3.5.3) we have shown the unique solvability of Problem (4.4).

Proposition 4.1.1. *For all* $g_D \in H^{1/2}(\Gamma)$ *Problem (4.4) has a unique solution* $u \in H^1(L, \Omega^+)$ *with* $L = -\Delta$.

Proof. Theorem 2.10.11 implies the unique solvability of the variational formulation associated with (4.4) in $H^1\left(L, \Omega^+\right)$ with $L = -\Delta$. In Sect. 2.9.3 we have shown that the solution also solves (4.4a) and (4.4b) almost everywhere.

Decay Condition: Theorem 3.5.3 provides us with the unique solvability of the boundary integral equation that results from (4.4) (with the single layer ansatz) in $H^{-1/2}(\Gamma)$. The associated single layer potential is in $H^1\left(L, \Omega^+\right)$ (see Exercise 3.1.14) and, thus, is the unique solution.

Finally, in (3.22) we have shown that the single layer potential satisfies the decay condition (4.4c). □

We will now reduce (4.4) to a boundary integral equation of the first kind. We ensure that (4.4a), (4.4c) are satisfied by means of the single layer ansatz (see Chap. 3)

$$u(\mathbf{x}) = (S\varphi)(\mathbf{x}) = \int_\Gamma \frac{\varphi(\mathbf{y})}{4\pi\,\|\mathbf{x}-\mathbf{y}\|}\,ds_\mathbf{y}, \qquad \mathbf{x} \in \Omega^+. \tag{4.5}$$

The unknown density φ from (4.5) is the solution of the boundary integral equation

$$V\varphi = g_D \qquad \text{on } \Gamma \tag{4.6}$$

with the single layer operator

$$(V\varphi)(\mathbf{x}) := \int_\Gamma \frac{\varphi(\mathbf{y})}{4\pi\,\|\mathbf{x}-\mathbf{y}\|}\,ds_\mathbf{y} \qquad \mathbf{x} \in \Gamma. \tag{4.7}$$

(4.6) defines a boundary integral equation of the first kind. The Galerkin boundary element method is based on the variational formulation of the integral equation. Instead of imposing (4.6) for all $\mathbf{x} \in \Gamma$, we multiply (4.6) by a "test function" and integrate over Γ. This gives us: Find $\varphi \in H^{-1/2}(\Gamma)$ such that

$$\begin{aligned}\int_\Gamma (V\varphi)\eta\,ds_\mathbf{x} &= \int_\Gamma \left(\int_\Gamma \frac{\varphi(\mathbf{y})}{4\pi\,\|\mathbf{x}-\mathbf{y}\|}\,ds_\mathbf{y}\right)\eta(\mathbf{x})ds_\mathbf{x} \\ &= \int_\Gamma g_D(\mathbf{x})\,\eta(\mathbf{x})\,ds_\mathbf{x} \qquad \forall\eta \in H^{-1/2}(\Gamma).\end{aligned} \tag{4.8}$$

For the Laplace operator we only consider vector spaces over the field $\mathbb{R}$ and not over $\mathbb{C}$, so that in (4.8) there is no complex conjugation.

The "integrals" in (4.8) should be interpreted as duality pairings in $H^{\frac{1}{2}}(\Gamma) \times H^{-\frac{1}{2}}(\Gamma)$ in the following way. For $\varphi \in H^{-1/2}(\Gamma)$ we have $V\varphi \in H^{1/2}(\Gamma)$ and, by Convention 4.0.1, we can write (4.8) as

$$\text{Find } \varphi \in H^{-1/2}(\Gamma):\ (V\varphi, \eta)_{L^2(\Gamma)} = (g_D, \eta)_{L^2(\Gamma)} \qquad \forall\eta \in H^{-1/2}(\Gamma). \tag{4.9}$$

The left-hand side in (4.9) defines a bilinear form $b(\cdot,\cdot)$ on the Hilbert space $H = H^{-1/2}(\Gamma)$ with

$$b(\varphi, \eta) := (V\varphi, \eta)_{L^2(\Gamma)}, \tag{4.10}$$

and the right-hand side defines a linear functional on $H^{-1/2}(\Gamma)$:

$$F(\eta) := (g_D, \eta)_{L^2(\Gamma)}. \tag{4.11}$$

Keeping the duality of $H^{-1/2}(\Gamma)$ and $H^{1/2}(\Gamma)$ in mind, it follows from

$$|F(\eta)| \leq \left(\sup_{\mu \in H^{-1/2}(\Gamma)\setminus\{0\}} \frac{|(g_D, \mu)_{L^2(\Gamma)}|}{\|\mu\|_{H^{-1/2}(\Gamma)}} \right) \|\eta\|_{H^{-1/2}(\Gamma)} = \|g_D\|_{H^{1/2}(\Gamma)} \|\eta\|_{H^{-1/2}(\Gamma)}$$

that F is continuous on $H^{-1/2}(\Gamma)$.

For sufficiently smooth functions φ, η in (4.10) we have, by virtue of Fubini's theorem,

$$b(\varphi, \eta) = \int_\Gamma \int_\Gamma \frac{\eta(\mathbf{x})\varphi(\mathbf{y})}{4\pi \|\mathbf{x}-\mathbf{y}\|} \, ds_{\mathbf{y}} \, ds_{\mathbf{x}} = b(\eta, \varphi) \tag{4.12}$$

and therefore the form $b(\cdot,\cdot)$ is symmetric. Furthermore, it is also $H^{-1/2}$-elliptic (see Theorem 3.5.3). According to the Lax–Milgram lemma (see Sect. 2.1.6), Problem (4.9) has a unique solution $\varphi \in H^{-1/2}(\Gamma)$ for all $g_D \in H^{1/2}(\Gamma)$. In the representational formula (4.5) this φ gives us the unique solution u of the exterior problem (4.4).

The *discretization* of the boundary integral equation consists in the approximation of the unknown density function φ in (4.6) by means of a function $\tilde{\varphi}$ which is defined by finitely many coefficients $(\alpha_i)_{i=1}^N$ in the basis representation. In the Galerkin boundary element method, this is achieved by restricting φ, η in the variational form (4.9) to finite-dimensional subspaces, the boundary element spaces, which we will now construct.

4.1.2 Surface Meshes

Almost all boundary elements are based on a surface mesh $\mathcal{G}$ of the boundary Γ. A surface mesh is the finite union of curved triangles and quadrilaterals on the boundary Γ, which satisfy suitable compatibility conditions. A general element of $\mathcal{G}$ is called a "panel".

For the definition we introduce the reference elements

$$\begin{aligned} &\textit{Unit triangle:} \quad \widehat{S}_2 := \left\{ (\xi_1, \xi_2) \in \mathbb{R}^2 : 0 < \xi_2 < \xi_1 < 1 \right\} \\ &\textit{Unit square:} \quad \widehat{Q}_2 := (0,1)^2. \end{aligned} \tag{4.13}$$

Our generic notation for the reference element is $\hat{\tau}$.

Definition 4.1.2. A surface mesh $\mathcal{G}$ of the boundary Γ is a decomposition of Γ into finitely relatively open, disjoint elements $\tau \subset \Gamma$ that satisfy the following conditions:

(a) $\mathcal{G}$ is a covering of Γ:

$$\Gamma = \overline{\bigcup_{\tau\in\mathcal{G}} \tau}.$$

(b) Every element $\tau \in \mathcal{G}$ is the image of a reference element $\hat{\tau}$ under a regular reference mapping χ_τ. Then χ_τ is called regular if the Jacobian $\mathbf{J}_\tau = D\chi_\tau$ satisfies the condition

$$0 < \lambda_{\min} \leq \inf_{\hat{\xi}\in\hat{\tau}} \inf_{\substack{\mathbf{v}\in\mathbb{R}^2 \\ \|\mathbf{v}\|=1}} \left\langle \mathbf{J}_\tau\left(\hat{\xi}\right)\mathbf{v}, \mathbf{J}_\tau\left(\hat{\xi}\right)\mathbf{v}\right\rangle \leq \sup_{\hat{\xi}\in\hat{\tau}} \sup_{\substack{\mathbf{v}\in\mathbb{R}^2 \\ \|\mathbf{v}\|=1}} \left\langle \mathbf{J}_\tau\left(\hat{\xi}\right)\mathbf{v}, \mathbf{J}_\tau\left(\hat{\xi}\right)\mathbf{v}\right\rangle$$
$$\leq \lambda_{\max} < \infty.$$

(c) For a plane triangle $\tau \in \mathcal{G}$ with straight edges and vertices $\mathbf{P}_0$, $\mathbf{P}_1$ and $\mathbf{P}_2$, the regular mapping χ_τ is affine:

$$\chi_\tau\left(\hat{\xi}\right) = \mathbf{P}_0 + \hat{\xi}_1\left(\mathbf{P}_1 - \mathbf{P}_0\right) + \hat{\xi}_2\left(\mathbf{P}_2 - \mathbf{P}_1\right). \tag{4.14}$$

For a plane quadrilateral $\tau \in \mathcal{G}$ with straight edges and vertices $\mathbf{P}_0$, $\mathbf{P}_1$, $\mathbf{P}_2$ and $\mathbf{P}_3$ (the numbering is counterclockwise) the mapping is bilinear:

$$\chi_\tau\left(\hat{\xi}\right) = \mathbf{P}_0 + \hat{\xi}_1\left(\mathbf{P}_1 - \mathbf{P}_0\right) + \hat{\xi}_2\left(\mathbf{P}_3 - \mathbf{P}_0\right) + \hat{\xi}_1\hat{\xi}_2\left(\mathbf{P}_2 - \mathbf{P}_3 + \mathbf{P}_0 - \mathbf{P}_1\right). \tag{4.15}$$

Figure 4.1 illustrates Definition 4.1.2 for a triangular and a quadrilateral element.

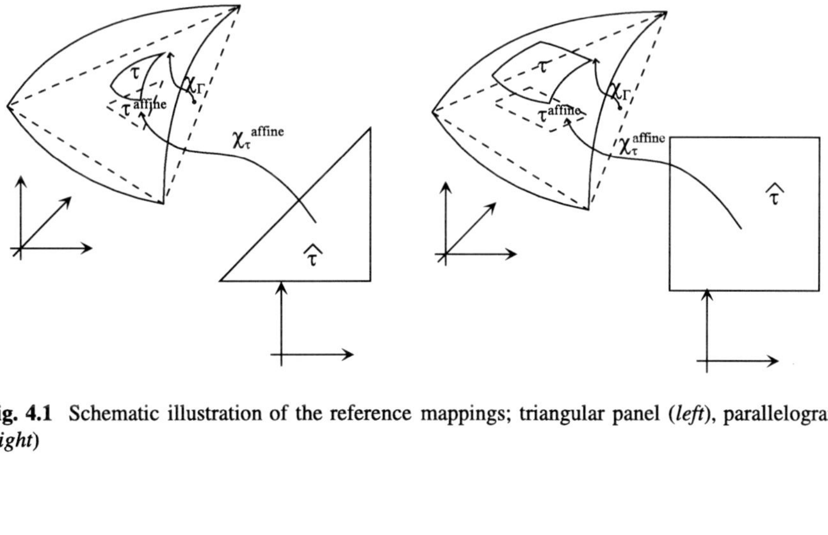

Fig. 4.1 Schematic illustration of the reference mappings; triangular panel (*left*), parallelogram (*right*)

Exercise 4.1.3. *Show the following:*

(a) The affine mapping χ_τ in (4.14) is regular if and only if $\mathbf{P}_0, \mathbf{P}_1, \mathbf{P}_2$ are vertices of a non-degenerate (plane) triangle τ, i.e., they are not colinear. Find an estimate for the constants $\lambda_{\min}, \lambda_{\max}$ from Definition 4.1.2(b) in terms of the interior angles of τ.

(c) Let $\mathbf{P}_0, \mathbf{P}_1, \mathbf{P}_2, \mathbf{P}_3$ be the vertices of a plane quadrilateral τ with straight edges. The mapping χ_τ from (4.15) is regular if all interior angles are smaller than π and larger than 0.

In some cases we will impose a compatibility condition for the intersection of two panels.

Definition 4.1.4. A surface mesh $\mathcal{G}$ of Γ is called *regular* if:

(a) The intersection of two different elements $\tau, \tau' \in \mathcal{G}$ is either empty, a common vertex or a common side.

(b) The parametrizations of the panel edges of neighboring panels coincide: For every pair of different elements $\tau, \tau' \in \mathcal{G}$ with common edge $e = \overline{\tau} \cap \overline{\tau'}$ we have

$$\chi_\tau|_{\hat{e}} = \chi_{\tau'} \circ \gamma_{\tau,\tau'}|_{\hat{e}},$$

where $\hat{e} := \chi_\tau^{-1}(e)$ and $\gamma_{\tau,\tau'} : \hat{\tau} \to \hat{\tau}$ is a suitable affine bijection.

Remark 4.1.5. *Throughout this section we assume that the boundary Γ is Lipschitz and admits a regular surface mesh in the sense of Definitions 4.1.2 and 4.1.4. This is a true restriction since not every Lipschitz surface admits a regular surface mesh.*

For later error estimates we will introduce a few geometric parameters, which represent a measure for the distortion of the panels as well as bounds for their diameters.

Assumption 4.1.6. *There exist open subsets $U, V \subset \mathbb{R}^3$ and a diffeomorphism $\chi_\Gamma : U \to V$ with the following properties:*

(a) $\Gamma \subset U$.

(b) For every $\tau \in \mathcal{G}$, there exists a regular reference mapping $\chi_\tau : \hat{\tau} \to \tau$ of the form

$$\chi_\tau = \chi_\Gamma \circ \chi_\tau^{\text{affine}} : \hat{\tau} \to \tau,$$

where $\chi_\tau^{\text{affine}} : \mathbb{R}^2 \to \mathbb{R}^3$ is a regular, affine mapping.

Example 4.1.7.

1. Let Γ be a piecewise smooth surface that has a bi-Lipschitz continuous parametrization over the polyhedral surface $\hat{\Gamma}$: $\chi_\Gamma : \hat{\Gamma} \to \Gamma$. Let $\mathcal{G}^{\text{affine}} := \{\tau_i^{\text{affine}} : 1 \leq i \leq N\}$ be a regular surface mesh of $\hat{\Gamma}$ with the associated reference mappings $\chi_{\tau^{\text{affine}}}^{\text{affine}} : \hat{\tau} \to \tau^{\text{affine}}$. Then $\mathcal{G} := \{\chi_\Gamma(\tau^{\text{affine}}) : \tau^{\text{affine}} \in \mathcal{G}^{\text{affine}}\}$ defines a regular surface mesh of Γ which satisfies Assumption 4.1.6.

2. *For the unit sphere* $\Gamma := \left\{\mathbf{x} \in \mathbb{R}^3 : \|\mathbf{x}\| = 1\right\}$ *one can choose the inscribed double pyramid with vertices* $(\pm 1, 0, 0)^\intercal$, $(0, \pm 1, 0)^\intercal$, $(0, 0, \pm 1)^\intercal$ *as a polyhedral surface* $\hat{\Gamma}$, *while* $\chi_\Gamma : \hat{\Gamma} \to \Gamma$ *is defined by* $\chi_\Gamma(\mathbf{x}) := \mathbf{x}/\|\mathbf{x}\|$. *By means of* χ_Γ, *regular surface meshes on* Γ *can then be generated through lifting of regular surface meshes of the polyhedral surface* $\hat{\Gamma}$.

In order to construct a sequence of refined surface meshes for Γ, in many cases the procedure is as follows.

Remark 4.1.8. *Let* Γ *be the surface of a bounded Lipschitz domain* $\Omega \subset \mathbb{R}^3$. *In the first step we construct a polyhedron* $\hat{\Gamma}$ *along a bi-Lipschitz continuous mapping* $\chi_\Gamma : \hat{\Gamma} \to \Gamma$ *(see Example 4.1.7). Let* $\mathcal{G}_0^{\text{affine}}$ *be a (very coarse) surface mesh of* $\hat{\Gamma}$. *Then* $\mathcal{G}_0 := \left\{\tau = \chi_\Gamma\left(\tau^{\text{affine}}\right) : \tau^{\text{affine}} \in \mathcal{G}_0^{\text{affine}}\right\}$ *defines a coarse surface mesh of* Γ. *We can obtain a sequence* $\left(\mathcal{G}_\ell^{\text{affine}}\right)_\ell$ *of finer surface meshes if, during each refinement, we decompose every panel in* $\mathcal{G}_0^{\text{affine}}$ *into new panels by means of a fixed refinement method. For triangular elements, for example, we interconnect the midpoints of the sides and for quadrilateral elements we connect both pairs of opposite midpoints. This gives us a sequence of surface meshes by* $\mathcal{G}_\ell := \left\{\tau = \chi_\Gamma\left(\tau^{\text{affine}}\right) : \tau^{\text{affine}} \in \mathcal{G}_\ell^{\text{affine}}\right\}$.

Convention 4.1.9. *If* τ *and* τ^{affine} *appear in the same context the relation between the two is given by* $\tau = \chi_\Gamma\left(\tau^{\text{affine}}\right)$.

The following definition is illustrated in Fig. 4.2.

Definition 4.1.10. Let Assumption 4.1.6 be satisfied. The constants $c_{\text{affine}} > 0$ ($C_{\text{affine}} > 0$) are the maximal (minimal) constants in

$$c_{\text{affine}} \|\mathbf{x} - \mathbf{y}\| \leq \|\chi_\Gamma(\mathbf{x}) - \chi_\Gamma(\mathbf{y})\| \leq C_{\text{affine}} \|\mathbf{x} - \mathbf{y}\| \quad \forall \mathbf{x}, \mathbf{y} \in \tau^{\text{affine}}, \forall \tau^{\text{affine}} \in \mathcal{G}^{\text{affine}}$$

and describe the distortion of curved panels τ compared to their affine pullbacks τ^{affine}.

The diameter of a panel $\tau \in \mathcal{G}$ is given by

$$h_\tau := \sup_{\mathbf{x}, \mathbf{y} \in \tau} \|\mathbf{x} - \mathbf{y}\|$$

and the inner width ρ_τ by the incircle diameter of τ^{affine}.

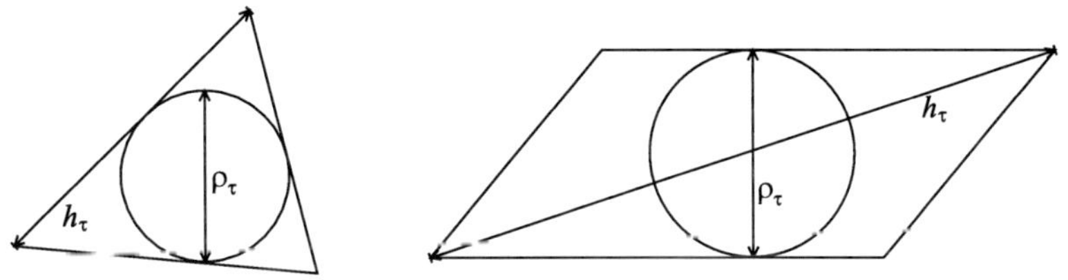

Fig. 4.2 Diameter of a panel and incircle diameter; triangular panel (*left*), parallelogram (*right*)

The *mesh width* $h_{\mathcal{G}}$ of a surface mesh $\mathcal{G}$ is given by

$$h_{\mathcal{G}} := \max\{h_\tau : \tau \in \mathcal{G}\}. \tag{4.16}$$

We write h instead of $h_{\mathcal{G}}$ if the mesh $\mathcal{G}$ is clear from the context.

Remark 4.1.11. *For plane panels τ, ρ_τ is the incircle diameter of τ.*
The diameters of τ and τ^{affine} satisfy

$$C_{\text{affine}}^{-1} h_\tau \leq \sup_{\mathbf{x},\mathbf{y} \in \tau^{\text{affine}}} \|\mathbf{x} - \mathbf{y}\| = h_{\tau^{\text{affine}}} \leq c_{\text{affine}}^{-1} h_\tau .$$

Definition 4.1.12. The shape-regularity constant $\kappa_{\mathcal{G}}$ is given by

$$\kappa_{\mathcal{G}} := \max_{\tau \in \mathcal{G}} \frac{h_\tau}{\rho_\tau}. \tag{4.17}$$

For some theorems we will assume, apart from the shape-regularity, that the diameters of all triangles are of the same order of magnitude.

Definition 4.1.13. The constant $q_{\mathcal{G}}$ that describes the quasi-uniformity is given by

$$q_{\mathcal{G}} := h_{\mathcal{G}} / \min\{h_\tau : \tau \in \mathcal{G}\}.$$

Remark 4.1.14. *In order to study the convergence of boundary element methods, we will consider sequences $(\mathcal{G}_\ell)_{\ell \in \mathbb{N}}$ of surface meshes whose mesh width $h_\ell := h_{\mathcal{G}_\ell}$ tends to zero. It is essential that the constant for the shape-regularity $\kappa_\ell := \kappa_{\mathcal{G}_\ell}$ remains uniformly bounded above:*

$$\sup_{\ell \in \mathbb{N}} \kappa_\ell \leq \kappa < \infty. \tag{4.18}$$

In a similar way the constants of quasi-uniformity $q_\ell := q_{\mathcal{G}_\ell}$ have to be bounded above in some theorems:

$$\sup_{\ell \in \mathbb{N}} q_\ell \leq q < \infty. \tag{4.19}$$

We call a mesh family $(\mathcal{G}_\ell)_{\ell \in \mathbb{N}}$ with the property (4.18) shape-regular *and with the property (4.19)* quasi-uniform.

Exercise 4.1.15. *Show the following:*

(a) If the surface mesh $\mathcal{G}_0$ is regular and if finer surface meshes $(\mathcal{G}_\ell)_\ell$ are constructed according to the method described in Remark 4.1.8 then all surface meshes $(\mathcal{G}_\ell)_\ell$ are regular.
(b) The constants concerning shape-regularity and quasi-uniformity are, under the conditions in Part (a), uniformly bounded with respect to ℓ.

4.1.3 Discontinuous Boundary Elements

The boundary element method defines an approximation of the unknown density φ in the boundary integral equation (4.6) which is described by finitely many parameters. This can, for example, be achieved by (piecewise) polynomials on the elements τ of a mesh $\mathcal{G}$.

Example 4.1.16. *(Piecewise Constant Boundary Elements)*
Let $\Gamma = \partial\Omega$ be piecewise smooth and let $\mathcal{G}$ be a – not necessarily regular – surface mesh on Γ. Then $S^0_{\mathcal{G}}$ denotes all piecewise constant functions on the mesh $\mathcal{G}$

$$S^0_{\mathcal{G}} := \{\psi \in L^\infty(\Gamma) \mid \forall \tau \in \mathcal{G} : \psi|_\tau \text{ is constant}\}. \tag{4.20}$$

Since $\psi \in L^\infty(\Gamma)$, we only need to define ψ in the interior of an element, as the boundary $\partial\tau$, i.e., the set of edges and vertices of the panel, is a set of zero measure.

Every function $\psi \in S^0_{\mathcal{G}}$ is defined by its values ψ_τ on the elements $\tau \in \mathcal{G}$ and can be written in the form

$$\psi(\mathbf{x}) = \sum_{\tau \in \mathcal{G}} \psi_\tau b_\tau(\mathbf{x}) \tag{4.21}$$

with the characteristic function $b_\tau : \Gamma \to \mathbb{R}$ of $\tau \in \mathcal{G}$:

$$b_\tau(\mathbf{x}) := \begin{cases} 1 & \mathbf{x} \in \tau, \\ 0 & \text{otherwise}. \end{cases} \tag{4.22}$$

In particular, $S^0_{\mathcal{G}}$ is a vector space of dimension $N = \#\{\tau : \tau \in \mathcal{G}\}$ with basis $\{b_\tau : \tau \in \mathcal{G}\}$.

In many cases the piecewise constant approximation of the unknown density converges too slowly and, instead, one uses polynomials of degree $p \geq 1$. In the same way as in Example 4.1.16 this leads to the boundary element spaces $S^p_{\mathcal{G}}$. For their definition we need polynomials of total degree p on the reference element as well as the convention for multi-indices from (2.67)

$$\mathbb{P}^\Delta_p = \operatorname{span}\left\{\xi^\mu : \mu \in \mathbb{N}^2_0 \wedge |\mu| \leq p\right\}. \tag{4.23}$$

For $p = 1$ and $p = 2$, $\mathbb{P}^\Delta_p$ contains all polynomials of the form

$$\begin{array}{lll} a_{00} + a_{10}\xi_1 + a_{01}\xi_2 & \forall a_{00}, a_{10}, a_{01} \in \mathbb{R} & \text{for } p = 1, \\ a_{00} + a_{10}\xi_1 + a_{01}\xi_2 + a_{20}\xi_1^2 + a_{11}\xi_1\xi_2 + a_{02}\xi_2^2 & \forall a_{00}, a_{10}, a_{01}, a_{20}, a_{11}, a_{02} \in \mathbb{R} & \text{for } p = 2. \end{array}$$

Definition 4.1.17. Let $\Gamma = \partial\Omega$ be piecewise smooth and let $\mathcal{G}$ be a surface mesh of Γ. Then, for $p \in \mathbb{N}_0$,

$$S^p_{\mathcal{G}} := \left\{\psi : \Gamma \to \mathbb{K} \mid \forall \tau \in \mathcal{G} : \psi \circ \chi_\tau \in \mathbb{P}^\Delta_p\right\}. \tag{4.24}$$

We simply write S^p or only S if the reference to the surface mesh $\mathcal{G}$ is obvious.

Remark 4.1.18. *Note that in (4.24) the functions* $\psi \in S^p$ *do not constitute polynomials on the surface* Γ. *Only once they have been "transported back" to the reference element* $\hat{\tau}$ *by means of the element mapping* χ_τ *(see Fig. 4.1) is this the case. The parametrizations* χ_τ *of the elements* $\tau \in \mathcal{G}$ *in Definition 4.1.2 (b,c) are thus part of the set* $S^p_{\mathcal{G}}$. *A change in parametrization* χ_τ *will lead (with the same mesh* $\mathcal{G}$*) to a different* $S^p_{\mathcal{G}}$. *Therefore for a mesh* $\mathcal{G}$ *we summarize the element mappings* χ_τ *in the mapping vector*

$$\chi := \{\chi_\tau : \tau \in \mathcal{G}\} \tag{4.25}$$

and instead of (4.24) we write $S^p_{\mathcal{G},\chi}$.

Remark 4.1.19. *Note that (4.24) also holds for meshes* $\mathcal{G}$ *with quadrilateral elements, i.e., with reference element* $\hat{\tau} = (0,1)^2$. *Since* S^p *does not require continuity across element boundaries, the space of polynomials* $\mathbb{P}^\Delta_p$ *in (4.23) can also be applied to quadrilateral meshes.*

For the realization of the boundary element spaces we need a basis for $\mathbb{P}^\Delta_p$, which we denote by $\widehat{N}_{(i,j)}(\hat{\xi}_1, \hat{\xi}_2)$ and which satisfies

$$\mathbb{P}^\Delta_p = \operatorname{span}\left\{\widehat{N}_{(i,j)} : 0 \le i, j \le p,\ i + j \le p\right\}. \tag{4.26}$$

For example, $\widehat{N}_{(i,j)}(\xi_1, \xi_2) := \hat{\xi}_1^i \hat{\xi}_2^j$, $0 \le i + j \le p$ as in (4.23), would be admissible basis functions.

Remark 4.1.20. *(Nesting of Spaces)*

We have $\mathbb{P}^\Delta_p \subset \mathbb{P}^\Delta_q$ *for all* $p \le q$. *Therefore we can always choose a basis in* $\mathbb{P}^\Delta_q$ *which contains the basis functions from* $\mathbb{P}^\Delta_p$ *as a subset. The basis functions* $\widehat{N}_{(i,j)}$ *in (4.23) have this property.*

Once we have determined a basis $\widehat{N}_{(i,j)}(\hat{\xi})$ on $\hat{\tau}$, every $\psi \in S^p_{\mathcal{G},\chi}$ on a panel $\tau \in \mathcal{G}$ can be written as

$$\psi|_\tau = \sum_{0 \le i+j \le p} \alpha_{i,j} \left(\widehat{N}_{(i,j)} \circ \chi_\tau^{-1}\right)$$

and

$$N^\tau_{(i,j)} := \widehat{N}_{(i,j)} \circ \chi_\tau^{-1} \qquad 0 \le i + j \le p$$

spans the restriction $\{\psi|_\tau : \psi \in S^p(\Gamma, \mathcal{G}, \chi)\}$. In order to give a basis of $S^p_{\mathcal{G},\chi}$ suitable indices, we define

$$\iota_p := \left\{\mu \in \mathbb{N}_0^2 : |\mu| \le p\right\}.$$

Thus we have

$$S^p_{\mathcal{G},\chi} = \operatorname{span}\left\{b_{(\mu,\tau)}(x) : (\mu, \tau) \in \iota_p \times \mathcal{G}\right\}, \tag{4.27}$$

where the global basis functions $b_I(x)$ with the multi-index $I = (\mu, \tau)$ denote the zero extension of the element function N_μ^τ to Γ: For

$$I = (\mu, \tau) \in \iota_p \times \mathcal{G} =: \mathcal{I}(\mathcal{G}, p) =: \mathcal{I} \tag{4.28}$$

we explicitly have

$$b_I(\mathbf{x}) := \begin{cases} N_\mu^\tau(\mathbf{x}), & \mathbf{x} \in \tau, \\ 0 & \text{otherwise.} \end{cases} \tag{4.29}$$

Hence, every ψ can be written as a combination of the basis function $b_I(x)$:

$$\psi(\mathbf{x}) = \sum_{I \in \mathcal{I}} \psi_I \, b_I(\mathbf{x}), \qquad \mathbf{x} \in \tau, \quad \tau \in \mathcal{G}. \tag{4.30}$$

Let $|\mathcal{G}|$ be the number of elements in the mesh $\mathcal{G}$. The dimension of $S_{\mathcal{G},\chi}^p$ or the number of degrees of freedom is then given by

$$N = |\mathcal{G}| \, (p+1)(p+2)/2 = \dim(S_{\mathcal{G},\chi}^p). \tag{4.31}$$

Every function in $\psi \in S_{\mathcal{G},\chi}^p$ is then uniquely characterized by the vector $(\psi_I)_{I \in \mathcal{I}(\mathcal{G},p)} \subset \mathbb{R}^N \cong \mathbb{R}^{\mathcal{I}(\mathcal{G},p)}$ as in (4.30).

4.1.4 Galerkin Boundary Element Method

The simplest boundary element method for Problem (4.6) consists in approximating the unknown density φ in (4.9) by a piecewise constant function $\varphi_S \in S^0(\Gamma, \mathcal{G})$.

Convention 4.1.21. *The boundary element functions depend on the boundary element space $S^p(\Gamma, \mathcal{G}, \chi)$; in particular, they depend on Γ, the surface mesh $\mathcal{G}$ and the polynomial degree p. We will, whenever possible, use the abbreviated notation φ_S instead of $\varphi_{S_{\mathcal{G},\chi}^p}$.*

Inserting (4.30) into (4.6) or into the variational formulation (4.8) leads to a contradiction: since, in general, we have $\varphi_S \neq \varphi$, (4.6) and (4.8) cannot be satisfied with $\varphi = \varphi_S$, which is why the statements have to be weakened. As φ_S is determined by N parameters $\left(\varphi_I^S\right)_{I \in \mathcal{I}}$ [see (4.29)–(4.31)], we are looking for N conditions to determine φ_I^S. In the Galerkin boundary element method we only let the test function η run through a basis of $S_{\mathcal{G}}^p$ in the variational formulation of the boundary integral equation (4.9). The Galerkin approximation of the integral equation (4.9) then reads:

Find $\varphi_S \in S^p_{\mathcal{G},\chi}$ such that

$$b(\varphi_S, \eta_S) = F(\eta_S) \qquad \forall \eta_S \in S^p_{\mathcal{G},\chi}, \tag{4.32}$$

with $b(\cdot,\cdot)$ and $F(\cdot)$ from (4.10) and (4.11) respectively.

Remark 4.1.22. *(i) The Galerkin discretization (4.32) of (4.8) is achieved by restricting the trial and test functions φ, η to the subspace $S^p_{\mathcal{G},\chi} \subset H^{-1/2}(\Gamma)$ in the variational formulation (4.8).*

(ii) The boundary element solution φ_S in (4.32) is independent of the basis chosen for the subspace.

The *computation* of the approximation φ_S requires that we choose a concrete basis for the subspace. Therefore, [see (4.29)–(4.31)] for a fixed $p \in \mathbb{N}_0$, we choose the basis

$$(b_I : I \in \mathcal{I}(\mathcal{G}, p)) \tag{4.33}$$

for $S^p_{\mathcal{G},\chi}$. Then (4.32) is equivalent to the linear system of equations:
Find $\varphi \in \mathbb{R}^N$ such that

$$\mathbf{B}\varphi = \mathbf{F}. \tag{4.34}$$

Here the *system matrix* $\mathbf{B} = (B_{I,J})_{I,J\in\mathcal{I}(\mathcal{G},p)}$ and the right-hand side $\mathbf{F} = (F_J)_{J\in\mathcal{I}(\mathcal{G},p)} \in \mathbb{R}^N$ with $I = (\mu, \tau)$ and $J = (\nu, t)$ are given by

$$\begin{aligned} B_{I,J} &:= b(b_I, b_J) \\ &= \int_\Gamma\int_\Gamma \frac{b_J(\mathbf{x})\, b_I(\mathbf{y})}{4\pi\,\|\mathbf{x}-\mathbf{y}\|} ds_\mathbf{y}\, ds_\mathbf{x} = \int_t\int_\tau \frac{N^t_\nu(\mathbf{x})\, N^\tau_\mu(\mathbf{y})}{4\pi\,\|\mathbf{x}-\mathbf{y}\|} ds_\mathbf{y}\, ds_\mathbf{x} \end{aligned} \tag{4.35}$$

$$F_J := F(b_J) = \int_\Gamma g_D(\mathbf{x}) b_J(\mathbf{x})\, ds_\mathbf{x} = \int_t g_D(\mathbf{x}) N^t_\nu(\mathbf{x})\, ds_\mathbf{x}. \tag{4.36}$$

Remark 4.1.23. *The matrix $\mathbf{B}$ in (4.34) is dense because of (4.35), which means that all entries $B_{I,J}$ are, in general, not equal to zero. Furthermore, the twofold surface integral in (4.35) can very often not be computed exactly, even for polyhedrons, and requires numerical integration methods for its approximation. The influence of this additional approximation will be discussed in Chap. 5. In this chapter we will always assume that the matrix $\mathbf{B}$ can be determined exactly.*

Proposition 4.1.24. *The system matrix $\mathbf{B}$ in (4.34) is symmetric and positive definite.*

Proof. From the symmetry of $b(\varphi, \eta) = b(\eta, \varphi)$ we immediately have

$$B_{I,J} = b(b_I, b_J) = b(b_J, b_I) = B_{J,I},$$

and subsequently $\mathbf{B} = \mathbf{B}^\intercal$. Now let $\varphi \in \mathbb{R}^N$ be arbitrary. Then we have

$$\begin{aligned}\varphi^\mathsf{T}\mathbf{B}\varphi &= \sum_{I,J\in\mathcal{I}(\mathcal{G},p)} \varphi_J\varphi_I B_{I,J} = \sum_{I,J}\varphi_J\varphi_I b(b_I,b_J) = b\left(\sum_I \varphi_I b_I, \sum_J \varphi_J b_J\right)\\ &= b(\varphi_S,\varphi_S) \ge \gamma\|\varphi_S\|^2_{H^{-1/2}(\Gamma)} > 0\end{aligned}$$

if and only if $\varphi_S \neq 0$. Since $\{b_I : I \in \mathcal{I}\}$ is a basis of S^p, we have $\varphi_S \neq 0$ if and only if $\varphi \neq \mathbf{0} \in \mathbb{R}^N$. Therefore $\mathbf{B}$ is positive definite. □

Thus the discrete problem (4.32) or (4.34) has a unique solution $\varphi_S \in S^p_{\mathcal{G}}$. The following proposition supplies us with an estimate for the error $\varphi - \varphi_S$.

Proposition 4.1.25. *Let φ be the exact solution of (4.9). The Galerkin solution φ_S of (4.32) converges quasi-optimally*

$$\|\varphi - \varphi_S\|_{H^{-1/2}(\Gamma)} \le \frac{\|b\|}{\gamma} \min_{\eta_S\in S^p} \|\varphi - \eta_S\|_{H^{-1/2}(\Gamma)}. \tag{4.37}$$

The error satisfies the Galerkin orthogonality

$$b(\varphi - \varphi_S, \eta_S) = 0 \qquad \forall \eta_S \in S^p. \tag{4.38}$$

Proof. We will first prove the statement in (4.38). If we only consider (4.10) for test functions from S^p we can subtract (4.32) and obtain

$$b(\varphi - \varphi_S, \eta_S) = b(\varphi, \eta_S) - b(\varphi_S, \eta_S) = F(\eta_S) - F(\eta_S) = 0 \qquad \forall \eta_S \in S^p.$$

Next we prove (4.37). For the error $e_S = \varphi - \varphi_S$ we have by the ellipticity and the continuity of the boundary integral operator V and (4.38)

$$\begin{aligned}\gamma\|\varphi - \varphi_S\|^2_{H^{-1/2}(\Gamma)} &\le b(e_S,e_S) = b(e_S,\varphi-\varphi_S)\\ &= b(e_S,\varphi) - b(e_S,\varphi_S) = b(e_S,\varphi) - b(e_S,\eta_S) = b(e_S,\varphi-\eta_S)\\ &\le \|b\|\,\|e_S\|_{H^{-1/2}(\Gamma)}\|\varphi-\eta_S\|_{H^{-1/2}(\Gamma)}\end{aligned}$$

for all $\eta_S \in S^p$.

If we cancel $\|e_S\|_{H^{-1/2}(\Gamma)}$ and minimize over $\eta_S \in S^p$ we obtain the assertion (4.37). □

The inequality in (4.37) shows that the Galerkin error $\|\varphi - \varphi_S\|_{H^{-1/2}(\Gamma)}$ coincides with the error of the best approximation of φ in S^p up to a multiplicative constant. This is where the term *quasi-optimality* for the a priori error estimate (4.37) originates.

Remark 4.1.26 (Collocation). *We obtained the Galerkin discretization (4.32) from (4.8) by restricting the trial and test functions φ, η to the subspace $S^p \subset S$. Alternatively, one can insert φ_S into (4.6) and impose the equation*

$$(V\varphi_S)(\mathbf{x}_J) = g_D(\mathbf{x}_J) \qquad J \in \mathcal{I}(\mathcal{G}, p) \tag{4.39}$$

only in N collocation points $\{\mathbf{x}_J : J \in \mathcal{I}\}$. *The solvability of (4.39) depends strongly on the choice of collocation points* $\{\mathbf{x}_J : J \in \mathcal{I}\}$. *Equation (4.39) is also equivalent to a linear system of equations, where the entries of the system matrix* $\mathbf{B}^{coll}$ *are defined by*

$$B_{I,J}^{coll} := \int_{\tau} \frac{b_J(\mathbf{y})}{4\pi \, \|\mathbf{x}_I - \mathbf{y}\|} ds_{\mathbf{y}}. \tag{4.40}$$

Note that $\mathbf{B}^{coll}$ *is again dense, but not symmetric.*

The collocation method (4.39) is widespread in the field of engineering, because the computation of the matrix entries (4.40) only requires the evaluation of one integral over the surface Γ, *instead of, as with the Galerkin method, a twofold integration over* Γ. *However, the stability and convergence of collocation methods on polyhedral surfaces is still an open question, especially with integral equations of the first kind. For integral operators of zero order or equations of the second kind we only have stability results in some special cases. For a detailed discussion on collocation methods we refer to, e.g., [6, 8, 87, 187, 207, 215] and the references contained therein.*

We now return to the Galerkin method.

Remark 4.1.27 (Stability of the Galerkin Projection). *The Galerkin method (4.32) defines a mapping*

$$\Pi_S^p : H^{-1/2}(\Gamma) \to S_{\mathcal{G},\chi}^p : \qquad \Pi_S^p \varphi := \varphi_S,$$

which is called the Galerkin projection. Clearly, Π_S^p *is linear and because of the ellipticity of the boundary integral operator* V *we have*

$$\begin{aligned} \gamma \|\Pi_S^p \varphi\|_{H^{-1/2}(\Gamma)}^2 &= \gamma \|\varphi_S\|_{H^{-1/2}(\Gamma)}^2 \le b(\varphi_S, \varphi_S) = b(\varphi, \varphi_S) \\ &\le \|b\| \|\varphi\|_{H^{-1/2}(\Gamma)} \|\Pi_S^p \varphi\|_{H^{-1/2}(\Gamma)}, \end{aligned}$$

from which we have, after canceling, the boundedness of the Galerkin projection $\Pi_S^p : H^{-\frac{1}{2}}(\Gamma) \to H^{-\frac{1}{2}}(\Gamma)$ *independent of the mesh* $\mathcal{G}$:

$$\|\Pi_S^p \varphi\|_{H^{-1/2}(\Gamma)} \le \frac{\|b\|}{\gamma} \|\varphi\|_{H^{-1/2}(\Gamma)}. \tag{4.41}$$

The quasi-optimality (4.37) and the boundedness of the Galerkin projection combined with the following corollary give us the convergence of the Galerkin BEM.

Corollary 4.1.28. *Let* $(\mathcal{G}_\ell)_{\ell \in \mathbb{N}}$ *be a sequence of meshes on* Γ *with a mesh width* $h_\ell = h_{\mathcal{G}_\ell}$ *and let* $h_\ell \to 0$ *for* $\ell \to \infty$. *Then the sequence* $(\varphi_\ell)_{\ell \in \mathbb{N}}$ *of boundary element solutions (4.32) in* $S_\ell = S_{\mathcal{G}_\ell}^p$ *converges to* φ *for every fixed* $p \in \mathbb{N}_0$.

Proof. Since $S_\ell^0 \subseteq S_\ell^p$ for all $p \in \mathbb{N}_0$, we will only consider the case $p = 0$. S_ℓ^0 are step functions on meshes whose mesh width converges to zero. The density follows from the construction of the Lebesgue spaces

$$\overline{\bigcup_{\ell \in \mathbb{N}} S_\ell^0}^{\|\cdot\|_{L^2(\Gamma)}} = L^2(\Gamma)$$

and from Proposition 2.5.2 we have the dense embedding $L^2(\Gamma) \subset H^{-1/2}(\Gamma)$.

For $\varphi \in H^{-1/2}(\Gamma)$ and an arbitrary $\varepsilon > 0$ we can therefore choose a $\tilde{\varphi}$ from $L^2(\Gamma)$ and an $\ell \in \mathbb{N}$, combined so that $\tilde{\varphi}_\ell \in S_\ell^0$, such that

$$\|\varphi - \tilde{\varphi}\|_{H^{-1/2}(\Gamma)} \leq \varepsilon/2 \quad \text{and} \quad \|\tilde{\varphi} - \tilde{\varphi}_\ell\|_{L^2(\Gamma)} \leq \varepsilon/2.$$

From this we have

$$\|\varphi - \tilde{\varphi}_\ell\|_{H^{-1/2}(\Gamma)} \leq \|\varphi - \tilde{\varphi}\|_{H^{-1/2}(\Gamma)} + \|\tilde{\varphi} - \tilde{\varphi}_\ell\|_{H^{-1/2}(\Gamma)} \leq \frac{\varepsilon}{2} + \frac{\varepsilon}{2} \leq \varepsilon.$$

The quasi-optimality of the Galerkin method gives us

$$\|\varphi - \varphi_\ell\|_{H^{-1/2}(\Gamma)} \leq \frac{\|b\|}{\gamma} \|\varphi - \tilde{\varphi}_\ell\|_{H^{-1/2}(\Gamma)} \leq \varepsilon \frac{\|b\|}{\gamma}.$$

As $\varepsilon > 0$ is arbitrary, we have the assertion for $\ell \to \infty$. □

4.1.5 Convergence Rate of Discontinuous Boundary Elements

We have seen in Proposition 4.1.25 that the approximations $\varphi_S \in S$ from the Galerkin boundary element method approximate the exact solution φ of the equation of the first kind (4.9) quasi-optimally: the error $\varphi - \varphi_S$, which is measured in the "natural" $H^{-1/2}(\Gamma)$-norm, is – up to a multiplicative constant – just as large as

$$\min\left\{\|\varphi - \psi_S\|_{H^{-1/2}(\Gamma)} : \psi_S \in S\right\} \tag{4.42}$$

which is the error of the best approximation in the space S. The convergence rate of the BEM indicates how fast the error converges to zero in relation to an increase in the degrees of freedom N. Here we will only prove the convergence rate for $p = 0$, while the general case will be treated in Sect. 4.3. We begin with the second Poincaré inequality on the reference element $\hat{\tau}$.

Convention 4.1.29. *Variables on the reference element are always marked by a "^". If the variables $\mathbf{x} \in \tau$ and $\hat{\mathbf{x}} \in \hat{\tau}$ appear in the same context this should always be understood in terms of the relation $\mathbf{x} = \chi_\tau(\hat{\mathbf{x}})$. Derivatives with respect to variables in the reference element are also marked by a "^". We will write, for example, $\widehat{\nabla}$ as an abbreviation for $\nabla_{\hat{\mathbf{x}}}$. Should the functions $u : \tau \to \mathbb{K}$ and $\hat{u} : \hat{\tau} \to \mathbb{K}$ appear in the same context, they are connected by the relation $u \circ \chi_\tau = \hat{u}$.*

Proposition 4.1.30. *Let $\hat{\tau} \subset \mathbb{R}^2$ be the reference element, $\hat{\varphi} \in H^1(\hat{\tau})$ and $\hat{\varphi}_0 := \frac{1}{|\hat{\tau}|}\int_{\hat{\tau}} \hat{\varphi}\, d\hat{\mathbf{x}}$. Then there exists some $\hat{c} > 0$ such that*

$$\|\hat{\varphi} - \hat{\varphi}_0\|_{L^2(\hat{\tau})} \leq \hat{c}\|\widehat{\nabla}\hat{\varphi}\|_{L^2(\hat{\tau})}, \tag{4.43}$$

where $\hat{c}$ depends only on $\hat{\tau}$.

Proof. The assertion follows directly from the proof of Corollary 2.5.10. □

In the following we will derive error estimates for a simplified situation. We will discuss the general case in Sect. 4.3. Here we let Γ be a plane manifold in $\mathbb{R}^3$ with a polygonal boundary. As integrals are invariant under rotation and translation, we assume without loss of generality that

$$\Gamma \text{ is a two-dimensional polygonal domain,} \tag{4.44}$$

i.e., we restrict ourselves to the two-dimensional approximation problem in the plane.

Furthermore, let $\mathcal{G} = \{\tau_i : 1 \leq i \leq N\}$ be a surface mesh on Γ of shape-regular triangles with straight edges and with mesh width $h > 0$. Then the triangles $\tau \in \mathcal{G}$ are affinely equivalent to the reference element $\hat{\tau}$ via the transformation (4.14):

$$\tau \ni \mathbf{x} = \chi_\tau(\hat{\mathbf{x}}) = \mathbf{P}_0 + \mathbf{J}\hat{\mathbf{x}}, \qquad \hat{\mathbf{x}} \in \hat{\tau}, \tag{4.45}$$

where $\mathbf{J}$ is the matrix with the columns $\mathbf{P}_1 - \mathbf{P}_0$ and $\mathbf{P}_2 - \mathbf{P}_1$ (see Fig. 4.1). With (4.45) and the chain rule

$$\frac{\partial}{\partial x_\alpha} = \frac{\partial}{\partial \hat{x}_1}\frac{\partial \hat{x}_1}{\partial x_\alpha} + \frac{\partial}{\partial \hat{x}_2}\frac{\partial \hat{x}_2}{\partial x_\alpha} \qquad \alpha = 1, 2,$$

the relation

$$\nabla = \left(\mathbf{J}^{-1}\right)^\intercal \widehat{\nabla}, \qquad d\mathbf{x} = (\det \mathbf{J})\, d\hat{\mathbf{x}} = 2\,|\tau|\, d\hat{\mathbf{x}} \tag{4.46}$$

follows. This leads to the transformation formula for Sobolev norms

$$\begin{aligned} \|\widehat{\nabla}\hat{\varphi}\|^2_{L^2(\hat{\tau})} &= \int_{\hat{\tau}} |\widehat{\nabla}\hat{\varphi}|^2\, d\hat{\mathbf{x}} = \frac{|\hat{\tau}|}{|\tau|}\int_\tau (\nabla\varphi)^\intercal \mathbf{J}\mathbf{J}^\intercal(\nabla\varphi) d\mathbf{x} \\ &\leq \frac{|\hat{\tau}|}{|\tau|}\lambda_\tau \int_\tau \|\nabla\varphi\|^2\, d\mathbf{x}, \end{aligned} \tag{4.47}$$

where λ_τ denotes the largest eigenvalue of $\mathbf{J}\mathbf{J}^\intercal \in \mathbb{R}^{2\times 2}$. Furthermore, we have for the left-hand side of (4.43)

$$\|\hat{\varphi} - \hat{\varphi}_0\|^2_{L^2(\hat{\tau})} = \frac{|\hat{\tau}|}{|\tau|}\|\varphi - \varphi_0\|^2_{L^2(\tau)} \tag{4.48}$$

with $\varphi_0 := \frac{1}{|\tau|}\int_\tau \varphi d\mathbf{x}$. If we combine (4.48) with (4.43) and (4.47) we obtain

$$\|\varphi - \varphi_0\|^2_{L^2(\tau)} = \frac{|\tau|}{|\hat{\tau}|}\|\hat{\varphi} - \hat{\varphi}_0\|^2_{L^2(\hat{\tau})} \le \hat{c}^2\frac{|\tau|}{|\hat{\tau}|}\|\widehat{\nabla}\hat{\varphi}\|^2_{L^2(\hat{\tau})} \le \hat{c}^2\lambda_\tau\,\|\nabla\varphi\|^2_{L^2(\tau)} \quad \forall\tau \in \mathcal{G}. \tag{4.49}$$

Exercise 4.1.32 shows that

$$\lambda_\tau \le \|\mathbf{P}_1 - \mathbf{P}_0\|^2 + \|\mathbf{P}_2 - \mathbf{P}_1\|^2 \le 2h_\tau^2. \tag{4.50}$$

From this we have

$$\|\varphi - \varphi_0\|_{L^2(\tau)} \le \sqrt{2}\hat{c}h_\tau|\varphi|_{H^1(\tau)}. \tag{4.51}$$

Squaring and then summing over all $\tau \in \mathcal{G}$ leads to the following error estimate.

Proposition 4.1.31. *Let (4.44) hold. Let $\mathcal{G}$ be a surface mesh of Γ. Let $\varphi \in L^2(\Gamma)$ with $\varphi|_\tau \in H^1(\tau)$ for all $\tau \in \mathcal{G}$. Then we have the error estimate*

$$\min_{\psi\in S^0_{\mathcal{G}}} \|\varphi - \psi\|_{L^2(\Gamma)} \le \sqrt{2}\hat{c}\left(\sum_{\tau\in\mathcal{G}} h_\tau^2|\varphi|^2_{H^1(\tau)}\right)^{1/2}. \tag{4.52}$$

For $\varphi \in H^1(\Gamma)$ the error estimate can be simplified to

$$\min_{\psi\in S^0_{\mathcal{G}}} \|\varphi - \psi\|_{L^2(\Gamma)} \le \sqrt{2}\hat{c}h_{\mathcal{G}}|\varphi|_{H^1(\Gamma)}. \tag{4.53}$$

Exercise 4.1.32. *Let τ be a plane triangle with straight edges in $\mathbb{R}^2$ with vertices $\mathbf{P}_0, \mathbf{P}_1, \mathbf{P}_2$. Let the matrix $\mathbf{J}$ and the eigenvalue λ_τ be defined as in (4.45) and (4.47) respectively. Show that*

$$\lambda_\tau \le \|\mathbf{P}_1 - \mathbf{P}_0\|^2 + \|\mathbf{P}_2 - \mathbf{P}_1\|^2.$$

From the approximation property we will now derive an error estimate for the Galerkin solution.

Theorem 4.1.33. *Let Γ be the surface of a polyhedron. Let the surface mesh $\mathcal{G}$ consist of triangles with straight edges.*

For the solution φ of the integral equation of the first kind (4.6) we assume that for an $0 \le s \le 1$ we have

$$\varphi \in H^s(\Gamma). \tag{4.54}$$

Then the Galerkin approximation $\varphi_S \in S^0_{\mathcal{G}}$ satisfies the error estimate

$$\|\varphi - \varphi_S\|_{H^{-1/2}(\Gamma)} \le C\,h^{s+1/2}\|\varphi\|_{H^s(\Gamma)} \tag{4.55}$$

Proof. The conditions of the theorem allow us to apply Proposition 4.1.31. With (4.37) we obtain for the Galerkin solution φ_S the error estimate

$$\|\varphi - \varphi_S\|_{H^{-1/2}(\Gamma)} = \|\varphi - \Pi_S^0 \varphi\|_{H^{-1/2}(\Gamma)} \le \frac{\|b\|}{\gamma} \min_{\psi_S \in S_{\mathcal{G}}^0} \|\varphi - \psi_S\|_{H^{-1/2}(\Gamma)}.$$

The definition of the $H^{-1/2}(\Gamma)$-norm gives us

$$\|\varphi - \psi_S\|_{H^{-1/2}(\Gamma)} = \sup_{\eta \in H^{1/2}(\Gamma)\backslash\{0\}} \frac{(\varphi - \psi_S, \eta)_{L^2(\Gamma)}}{\|\eta\|_{H^{1/2}(\Gamma)}}. \tag{4.56}$$

We will first consider the case $\varphi \in H^1(\Gamma)$ and choose ψ_S elementwise as the mean value of φ

$$P\varphi := \psi_S \quad \text{with} \quad \psi_S|_\tau := \frac{1}{|\tau|} \int_\tau \varphi \, d\mathbf{x}, \qquad \tau \in \mathcal{G},$$

i.e., P is the L^2-orthogonal projection onto $S_{\mathcal{G}}^0$. Hence it follows from Proposition 4.1.31 that

$$\|\psi_S\|_{L^2(\Gamma)} \le \|\varphi\|_{L^2(\Gamma)}, \ \|\varphi - \psi_S\|_{L^2(\Gamma)} \le 2\|\varphi\|_{L^2(\Gamma)}, \ \|\varphi - \psi_S\|_{L^2(\Gamma)} \le ch\|\varphi\|_{H^1(\Gamma)}. \tag{4.57}$$

If in Proposition 2.1.62 we choose $T = I - P$ we have $T : L^2(\Gamma) \to L^2(\Gamma)$ and $T : H^1(\Gamma) \to L^2(\Gamma)$. For the norms we have, by (4.57), the estimates

$$\|T\|_{L^2(\Gamma) \leftarrow L^2(\Gamma)} \le 2 \quad \text{and} \quad \|T\|_{L^2(\Gamma) \leftarrow H^1(\Gamma)} \le ch.$$

Proposition 2.1.62 implies that $T : H^s(\Gamma) \to L^2(\Gamma)$ for all $0 \le s \le 1$ and that

$$\|T\|_{L^2(\Gamma) \leftarrow H^s(\Gamma)} \le ch^s.$$

This is equivalent to the error estimate

$$\|\varphi - \psi_S\|_{L^2(\Gamma)} \le c\, h^s \|\varphi\|_{H^s(\Gamma)}. \tag{4.58}$$

In order to derive an error estimate for the $H^{-1/2}(\Gamma)$-norm, we use (4.56) and note that the equality

$$|(\varphi - \psi_S, \eta)_{L^2(\Gamma)}| = |(\varphi - \psi_S, \eta - \eta_S)_{L^2(\Gamma)}|$$

holds for an arbitrary $\eta_S \in S_{\mathcal{G}}^0$. By using $\varphi \in H^s(\Gamma)$, $\eta \in H^{1/2}(\Gamma)$ and (4.58) and by choosing η_S elementwise as the integral mean value of η, we obtain the estimate

$$\begin{aligned} \left|(\varphi - \psi_S, \eta)_{L^2(\Gamma)}\right| &= \left|(\varphi - \psi_S, \eta - \eta_S)_{L^2(\Gamma)}\right| \le \|\varphi - \psi_S\|_{L^2(\Gamma)} \|\eta - \eta_S\|_{L^2(\Gamma)} \\ &\le ch^s \|\varphi\|_{H^s(\Gamma)} h^{1/2} \|\eta\|_{H^{1/2}(\Gamma)}. \end{aligned}$$

□

The error estimate (4.55) shows that the convergence rate $h^{s+1/2}$ of the BEM depends on the regularity of the solution φ. In Sect. 3.2 we stated the regularity – the maximal $s > 0$ such that $\varphi \in H^{-1/2+s}(\Gamma)$ – without knowing the exact solution φ explicitly. Ideally, φ is smooth on the entire surface ($s = \infty$) or at least on every panel. The convergence rate would then be bounded by the polynomial order p of the boundary elements, due to the fact that the following generalization of Theorem 4.1.33 holds.

Corollary 4.1.34. *Let the exact solution of (4.9) satisfy $\varphi \in H^s(\Gamma)$ for an $s \geq 0$. Then the boundary element solution $\varphi_S \in S_{\mathcal{G}}^p$ satisfies the error estimate*

$$\|\varphi - \varphi_S\|_{H^{-1/2}(\Gamma)} \leq c h_{\mathcal{G}}^{1/2+\min(s,p+1)} \|\varphi\|_{H^s(\Gamma)}, \tag{4.59}$$

for a surface mesh $\mathcal{G}$ of the boundary Γ, which consists of triangles with straight edges. Here the constant c depends on p and the shape-regularity of the surface mesh.

The proof of Corollary 4.1.34 will be completed in Sect. 4.3.4 (see Remark 4.3.21).

4.1.6 Model Problem 2: Neumann Problem

Let $\Omega^- \subset \mathbb{R}^3$ be a bounded interior domain with boundary Γ and $\Omega^+ := \mathbb{R}^3 \backslash \overline{\Omega^-}$. For $g_N \in H^{-1/2}(\Gamma)$ we consider the Neumann problem

$$\Delta u = 0 \qquad \text{in } \Omega^+, \tag{4.60}$$

$$\gamma_1 u = g_N \qquad \text{on } \Gamma, \tag{4.61}$$

$$|u(\mathbf{x})| \leq C \|\mathbf{x}\|^{-1} \qquad \text{for } \|\mathbf{x}\| \to \infty. \tag{4.62}$$

The exterior problem (4.60)–(4.62) has a unique solution u, which can be represented as a double layer potential

$$u(\mathbf{x}) = \frac{1}{4\pi} \int_\Gamma \varphi(\mathbf{y}) \frac{\partial}{\partial \mathbf{n}_\mathbf{y}} \frac{1}{\|\mathbf{x} - \mathbf{y}\|} ds_\mathbf{y}, \qquad \mathbf{x} \in \Omega^+. \tag{4.63}$$

Thanks to the jump relations (see Corollary 3.3.12)

$$\frac{1}{4\pi} \int_\Gamma \frac{\partial}{\partial \mathbf{n}_\mathbf{y}} \frac{1}{\|\mathbf{x} - \mathbf{y}\|} ds_\mathbf{y} = \begin{cases} -1 & \mathbf{x} \in \Omega^-, \\ -\frac{1}{2} & \mathbf{x} \in \Gamma \text{ and } \Gamma \text{ is smooth in } \mathbf{x} \\ 0 & \mathbf{x} \in \Omega^+ \end{cases}$$

$u(\mathbf{x})$ in (4.63) does not change if a constant is added to φ. If we put (4.63) into the boundary condition (4.61) we obtain the equation

$$-W\varphi = \frac{\partial}{\partial \mathbf{n_x}} \left(\frac{1}{4\pi} \int_\Gamma \varphi(\mathbf{y}) \frac{\partial}{\partial \mathbf{n_y}} \frac{1}{\|\mathbf{x}-\mathbf{y}\|} ds_{\mathbf{y}} \right) = g_N(\mathbf{x}), \qquad \mathbf{x} \in \Gamma. \tag{4.64}$$

The following remark shows that the derivative $\partial/\partial \mathbf{n_x}$ and the integral do not commute.

Remark 4.1.35. *The normal derivative $\partial/\partial \mathbf{n_x}$, applied to the kernel in (4.64), yields*

$$\frac{\partial^2}{\partial \mathbf{n_x} \partial \mathbf{n_y}} \frac{1}{\|\mathbf{x}-\mathbf{y}\|} = \frac{\langle \mathbf{n_x}, \mathbf{n_y} \rangle}{\|\mathbf{x}-\mathbf{y}\|^3} - 3 \frac{\langle \mathbf{n_x}, \mathbf{x}-\mathbf{y} \rangle \langle \mathbf{n_y}, \mathbf{x}-\mathbf{y} \rangle}{\|\mathbf{x}-\mathbf{y}\|^5}.$$

Therefore the kernel of the associated hypersingular integral operator is not integrable.

There are three possibilities of representing the integral operator $W\varphi$ on the surface: (a) by extending the definition of an integral to strongly singular kernel functions (see [201, 211]), (b) by integration by parts (see Sect. 3.3.4) and (c) by introducing suitable differences of test and trial functions (see [117, Sect. 8.3]). In this section we will consider option (b). The notation and theorems from Sect. 3.3.4 can be simplified for the Laplace problem, so that they read

$$\operatorname{curl}_\Gamma \varphi := \gamma_0 \left(\operatorname{grad} Z_- \varphi \right) \times \mathbf{n},$$
$$b(\varphi, \eta) = \int_\Gamma \int_\Gamma \frac{\langle \operatorname{curl}_\Gamma \varphi \,(\mathbf{y}), \operatorname{curl}_\Gamma \eta \,(\mathbf{x}) \rangle}{4\pi \|\mathbf{x}-\mathbf{y}\|} ds_{\mathbf{y}} ds_{\mathbf{x}},$$

where $Z_- : H^{1/2}(\Gamma) \to H^1(\Omega^-)$ is an arbitrary extension operator (see Theorem 2.6.11 and Exercise 3.3.25).

The variational formulation of the boundary integral equation is given by (see Theorem 3.3.22): Find $\varphi \in H^{1/2}(\Gamma)/\mathbb{K}$ such that

$$b(\varphi, \eta) = -(g_N, \eta)_{L^2(\Gamma)} \qquad \forall \eta \in H^{1/2}(\Gamma)/\mathbb{K}. \tag{4.65}$$

In Theorem 3.5.3 we have already shown that the density φ in (4.63) is the unique solution of the boundary integral equation (4.65). The proof was based on the fact that the bilinear form $b(\cdot, \cdot)$ is symmetric, continuous and $H^{1/2}(\Gamma)/\mathbb{K}$-elliptic.

4.1.7 Continuous Boundary Elements

The Galerkin method is based on the concept of replacing the infinite-dimensional Hilbert space by a finite-dimensional *sub*space. The bilinear form that is associated with the hypersingular integral operator is defined on the Sobolev space

$H^{1/2}(\Gamma)/\mathbb{K}$. As the discontinuous boundary element functions from Example 4.1.16 and Definition 4.1.17 are not contained in $H^{1/2}(\Gamma)/\mathbb{K}$ (see Exercise 2.4.4), we will introduce *continuous* boundary element spaces for the Neumann problem.

We again start with a mesh $\mathcal{G}$ on the boundary Γ. In order to define continuous boundary elements, we assume (see Definition 4.1.4):

$$\text{The surface mesh } \mathcal{G} \text{ is regular.} \tag{4.66}$$

This means that the intersection $\overline{\tau} \cap \overline{\tau}'$ of two different panels is either empty, a vertex or an entire edge. Furthermore, the boundary elements are either triangles or quadrilaterals and are images of the reference triangle or quadrilateral $\hat{\tau}$ respectively (see Fig. 4.1). Note that the boundary edges of the panels "have the same parametrization on both sides" in the case of continuous boundary elements (see Definition 4.1.4).

We assume that the boundary Γ is piecewise smooth (see Definition 2.2.10 and Fig. 4.1) so that the reference mappings $\chi_\tau : \hat{\tau} \to \tau$ can be chosen as smooth diffeomorphisms. As in the case for discontinuous boundary elements, the continuous boundary elements are also piecewise polynomials on the surface Γ. When using discontinuous elements, a boundary element function φ_S is locally a polynomial of degree p in *each* element $\tau \in \mathcal{G}$:

$$\forall \tau \in \mathcal{G}: \qquad \varphi_S \circ \chi_\tau \in \mathbb{P}_p^\Delta(\hat{\tau}).$$

With continuous elements we have for $\tau \in \mathcal{G}$:

$$\varphi_S \circ \chi_\tau \in \mathbb{P}_p^\tau := \begin{cases} \mathbb{P}_p^\Delta & \text{if } \tau \text{ is a triangular element,} \\ \mathbb{P}_p^\square & \text{if } \tau \text{ is a quadrilateral element,} \end{cases} \tag{4.67}$$

where for $p \geq 1$ the polynomial space $\mathbb{P}_p^\Delta$ is defined as in (4.23) and

$$\mathbb{P}_p^\square := \operatorname{span}\{\hat{\xi}_1^i \hat{\xi}_2^j : 0 \leq i, j \leq p\}.$$

Now we come to the definition of continuous boundary element functions of degree $p \geq 1$.

Definition 4.1.36. Let Γ be a piecewise smooth surface, $\mathcal{G}$ a regular surface mesh of Γ and $\chi = \{\chi_\tau : \tau \in \mathcal{G}\}$ the mapping vector. Then the space of continuous boundary elements of degree $p \geq 1$ is given by

$$S_{\mathcal{G},\chi}^{p,0} := \{\varphi \in C^0(\Gamma) \mid \forall \tau \in \mathcal{G} : \varphi|_\tau \circ \chi_\tau \in \mathbb{P}_p^\tau\}.$$

In order to make the distinction between continuous and discontinuous boundary elements of degree p we will from now on denote discontinuous elements by $S_{\mathcal{G},\chi}^{p,-1}$.

Just like the space $S^{p,-1}$ of discontinuous boundary elements, the space $S^{p,0}$ is also finite-dimensional. In the following we will introduce a basis $\{\varphi_I : I \in \mathcal{I}\}$ of $S^{p,0}$. In contrast to $S^{p,-1}$, the support of the basis functions in general consists of more than one panel and the basis functions are defined piecewise on those panels. We begin with the simplest case, $p = 1$.

Example 4.1.37. *(Linear and Bilinear, Continuous Boundary Elements)*
The shape functions $\widehat{N}(\hat{\mathbf{x}})$, $\hat{\mathbf{x}} = (\hat{x}_1, \hat{x}_2)$ *on the reference element* $\hat{\tau}$ *are:*

- *In the case of the unit triangle with vertices* $\mathbf{P}_0 = (0,0)^\mathsf{T}$, $\mathbf{P}_1 = (1,0)^\mathsf{T}$, $\mathbf{P}_2 = (1,1)^\mathsf{T}$ *[see (4.13)], given by*

$$\begin{aligned} \widehat{N}_0(\hat{\mathbf{x}}) &= 1 - \hat{x}_1, \\ \widehat{N}_1(\hat{\mathbf{x}}) &= \hat{x}_1 - \hat{x}_2, \\ \widehat{N}_2(\hat{\mathbf{x}}) &= \hat{x}_2 \end{aligned} \tag{4.68}$$

and

- *In the case of the unit square with vertices* $\mathbf{P}_0 = (0,0)^\mathsf{T}$, $\mathbf{P}_1 = (1,0)^\mathsf{T}$, $\mathbf{P}_2 = (1,1)^\mathsf{T}$, $\mathbf{P}_3 = (0,1)^\mathsf{T}$, *given by*

$$\begin{aligned} \widehat{N}_0(\hat{\mathbf{x}}) &= (1 - \hat{x}_1)(1 - \hat{x}_2), \\ \widehat{N}_1(\hat{\mathbf{x}}) &= \hat{x}_1(1 - \hat{x}_2), \\ \widehat{N}_2(\hat{\mathbf{x}}) &= (1 - \hat{x}_1)\,\hat{x}_2, \\ \widehat{N}_3(\hat{\mathbf{x}}) &= \hat{x}_1\hat{x}_2. \end{aligned} \tag{4.69}$$

We notice that the shape function $\widehat{N}_i$ *is equal to* 1 *at the vertex* $\mathbf{P}_i$ *of the reference element* 1 *and vanishes at all other vertices (see Fig. 4.3).*
It holds $\mathbb{P}_1^{\Delta}(\hat{\tau}) = \operatorname{span}\{\widehat{N}_i : i = 0,1,2\}$ *and* $\mathbb{P}_1^{\square}(\hat{\tau}) = \operatorname{span}\{\widehat{N}_i : i = 0,\ldots 3\}$.

For the definition of the boundary element spaces of polynomial degree p we have to distinguish between quadrilateral elements and triangular elements. For the reference element $\hat{\tau} \in \mathcal{G}$ and $p \in \mathbb{N}_0$ we define the index set

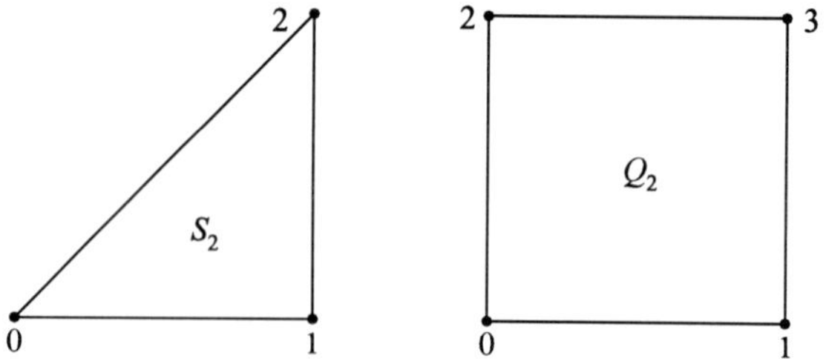

Fig. 4.3 Reference elements $\hat{\tau} = S_2$ *(left)* and $\hat{\tau} = Q_2$ *(right)* and nodal points for $\mathbb{P}_1^{\hat{\tau}}$

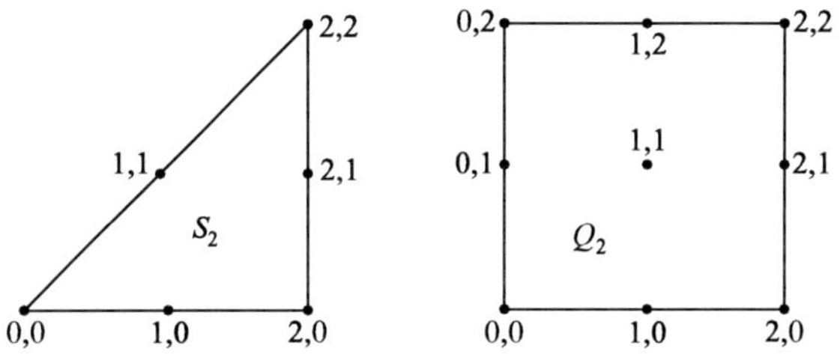

Fig. 4.4 Nodal points $\hat{\mathbf{P}}^{(2)}_{i,j}$ for the reference triangle (*left*) and for the unit square (*right*)

$$\iota_p^{\hat{\tau}} := \begin{cases} \{(i,j) \in \mathbb{N}_0^2 : 0 \le j \le i \le p\} & \text{in the case of the unit triangle,} \\ \{(i,j) \in \mathbb{N}_0^2 : 0 \le i, j \le p\} & \text{in the case of the unit square.} \end{cases} \tag{4.70}$$

We will omit the index $\hat{\tau}$ in $\iota_p^{\hat{\tau}}$ if the reference element is clear from the context.

Example 4.1.38 (Boundary elements of degree $p > 1$). *The trial spaces $\mathbb{P}_p^{\Delta}$, $\mathbb{P}_p^{\square}$ in (4.67) are spanned by the functions $\widehat{N}^{(p)}_{(i,j)} \in \mathbb{P}_p^{\hat{\tau}}$ which will be defined next. The nodal points for the reference element $\hat{\tau}$ are given by*

$$\widehat{P}^{(p)}_{(i,j)} := \left(\frac{i}{p}, \frac{j}{p}\right)^{\mathsf{T}}, \qquad \forall\, (i,j) \in \iota_p^{\hat{\tau}} \tag{4.71}$$

(see Fig. 4.4).

For $(i,j) \in \iota_p^{\hat{\tau}}$ the shape function $\widehat{N}^{(p)}_{(i,j)}$ is characterized by

$$\widehat{N}^{(p)}_{(i,j)} \in \mathbb{P}_p^{\hat{\tau}} \quad \text{and} \quad \widehat{N}^{(p)}_{(i,j)}(\widehat{P}^{(p)}_{(k,\ell)}) = \begin{cases} 1 & (k,\ell) = (i,j), \\ 0 & (k,\ell) \in \iota_p^{\hat{\tau}} \backslash \{(i,j)\} \end{cases}$$

(see Theorem 4.1.39).

Theorem 4.1.39. *Let $k \in \mathbb{N}$. Then every $q \in \mathbb{P}_k^{\hat{\tau}}$ is uniquely determined by its values in $\Sigma_k := \left\{(i/k, j/k) : (i,j) \in \iota_k^{\hat{\tau}}\right\}$.*

The set Σ_k is called *unisolvent* for the polynomial space $\mathbb{P}_k^{\hat{\tau}}$ because of this property.

Proof. A simple calculation shows that

$$\dim \mathbb{P}_k^{\hat{\tau}} = \sharp \Sigma_k .$$

Therefore it suffices to prove either one of the following statements (a) or (b):

(a) For every vector $(b_{\mathbf{z}})_{\mathbf{z}\in\Sigma_k}$ there exists a $q \in \mathbb{P}_k^{\hat{\tau}}$ such that $q(\mathbf{z}) = b_{\mathbf{z}}$ for all $\mathbf{z} \in \Sigma_k$.
(b) If $q \in \mathbb{P}_k^{\hat{\tau}}$ and $q(\mathbf{z}) = 0$ for all $\mathbf{z} \in \Sigma_k$ then $q \equiv 0$.

Case 1: $\hat{\tau} = (0,1)^2$: For $\mu \in \iota_k^{\hat{\tau}}$ we define the function $\widehat{N}_\mu$ by

$$\widehat{N}_\mu(\mathbf{x}) := \prod_{j=1}^{2} \prod_{\substack{i_j=0 \\ i_j \neq \mu_j}}^{k} \frac{kx_j - i_j}{\mu_j - i_j}.$$

Then $\widehat{N}_\mu \in \mathbb{P}_k^{\hat{\tau}}$ with $\widehat{N}_\mu(\mu/k) = 1$ and $\widehat{N}_\mu\left(\frac{i_1}{k}, \frac{i_2}{k}\right) = 0$ for all $(i_1, i_2) \in \iota_k^{\hat{\tau}} \backslash \{\mu\}$. Now let $(b_\mu)_{\mu \in \iota_k^{\hat{\tau}}}$ be arbitrary. Then the polynomial $q \in \mathbb{P}_k^{\hat{\tau}}$

$$q(\mathbf{x}) = \sum_{\mu \in \iota_p^{\hat{\tau}}} b_\mu \widehat{N}_\mu(\mathbf{x})$$

satisfies property (a).
Case 2: $\hat{\tau}$ is the reference triangle. As in Example 4.1.37 we set

$$\hat{\lambda}_1(\mathbf{x}) := 1 - \hat{x}_1, \qquad \hat{\lambda}_2(\mathbf{x}) := \hat{x}_1 - \hat{x}_2, \qquad \hat{\lambda}_3(\mathbf{x}) := \hat{x}_2.$$

Clearly, these functions are in $\mathbb{P}_1^{\hat{\tau}}$ and have the Lagrange property

$$\forall 1 \le i, j \le 3 : \hat{\lambda}_i(\mathbf{A}_j) = \delta_{i,j} \quad \text{with} \quad \mathbf{A}_1 = (0,0)^\mathsf{T}, \mathbf{A}_2 = (1,0)^\mathsf{T}, \mathbf{A}_3 = (1,1)^\mathsf{T}.$$

1. $k = 1$: For a given $(b_i)_{i=1}^3 \in \mathbb{R}^3$, $q \in \mathbb{P}_1$:

$$q(\mathbf{x}) = \sum_{i=1}^{3} b_i \hat{\lambda}_i(\mathbf{x})$$

clearly has the property (a).

2. $k = 2$: For $1 \le i < j \le 3$, $\mathbf{A}_{(i,j)} := (\mathbf{A}_i + \mathbf{A}_j)/2$ denote the midpoints of the edges of $\hat{\tau}$. We define

$$\begin{aligned} \widehat{N}_i &:= \hat{\lambda}_i \left(2\hat{\lambda}_i - 1\right) \quad 1 \le i \le 3, \\ \widehat{N}_{(i,j)} &:= 4\hat{\lambda}_i \lambda_j \qquad 1 \le i < j \le 3. \end{aligned}$$

Then we clearly have $\widehat{N}_k, \widehat{N}_{(i,j)} \in \mathbb{P}_2^{\hat{\tau}}$ and

$$\begin{aligned} &\widehat{N}_i(\mathbf{A}_j) = \delta_{i,j} \;\; \widehat{N}_i(\mathbf{A}_{(k,\ell)}) = 0 && \forall i, k, \ell, \\ &\widehat{N}_{(i,j)}(\mathbf{A}_k) = 0 \;\; \widehat{N}_{(i,j)}(\mathbf{A}_{(k,\ell)}) = \delta_{i,k}\delta_{j,\ell} && \forall i, j, k, \ell. \end{aligned}$$

For a given $\{b_\mathbf{z} : \mathbf{z} \in \Sigma_2\} = \{b_i, b_{(k,\ell)}\}$, the polynomial $q \in \mathbb{P}_2^{\hat{\tau}}$ defined by

$$q(\mathbf{x}) := \sum_{i=1}^{3} b_i \widehat{N}_i(\mathbf{x}) + \sum_{1 \le k < \ell \le 3} b_{(k,\ell)} \widehat{N}_{(k,\ell)}(\mathbf{x})$$

has the property (a).

3. $k = 3$: This case will be treated in Exercise 4.1.40.
4. $k \geq 4$: Let $q \in \mathbb{P}_k^{\Delta}$ with $q(\mathbf{z}) = 0$ for all $\mathbf{z} \in \Sigma_k$. Then q vanishes on all edges of $\hat{\tau}$. Therefore there exists a $\psi \in \mathbb{P}_{k-3}^{\Delta}$ such that

$$q = \hat{\lambda}_1 \hat{\lambda}_2 \hat{\lambda}_3 \psi \quad \text{and} \quad \forall \mathbf{z} \in \Sigma_k \cap \hat{\tau} : \psi(\mathbf{z}) = 0.$$

(Note that $\hat{\tau}$ is open.) The problem can thus be reduced to

$$\left(\psi \in \mathbb{P}_{k-3}^{\Delta}\right) \wedge (\forall \mathbf{z} \in \Sigma_k \cap \hat{\tau} : \psi(\mathbf{z}) = 0) \Longrightarrow \psi \equiv 0. \tag{4.72}$$

Property (b) follows by induction over k as follows.
Let $\hat{\tau}'$ be the triangle with vertices $\mathbf{A} = \left(\frac{2}{k+1}, \frac{1}{k+1}\right)^{\mathsf{T}}$, $\mathbf{B} = \left(\frac{k}{k+1}, \frac{1}{k+1}\right)^{\mathsf{T}}$, $\mathbf{C} = \left(\frac{k}{k+1}, \frac{k-1}{k+1}\right)^{\mathsf{T}}$. Then we have $\Sigma_k \cap \hat{\tau} =: \Sigma_k' \subset \hat{\tau}'$. The transformation

$$T : \hat{\tau} \to \hat{\tau}' : T\xi = \mathbf{A} + \left(1 - \frac{3}{k+1}\right)\xi$$

is affine and therefore $\tilde{\psi} = \psi \circ T \in \mathbb{P}_{k-3}^{\Delta}$. Furthermore, we have $T^{-1}\Sigma_k' = \Sigma_{k-3}$. Hence (4.72) is equivalent to

$$\left(\tilde{\psi} \in \mathbb{P}_{k-3}^{\Delta}\right) \wedge \left(\forall \mathbf{z} \in \Sigma_{k-3} : \tilde{\psi}(\mathbf{z}) = 0\right) \Longrightarrow \tilde{\psi} \equiv 0.$$

This, however, is statement (b) for $k \leftarrow k - 3$. Since the induction hypothesis for $k = 1, 2, 3$ is given by steps 1–3 in the proof, the assertion follows by virtue of the equivalence of the two statements (a) and (b). □

Exercise 4.1.40. *Let $\hat{\tau}$ be the unit triangle. For $\mathbb{P}_3^{\hat{\tau}}$ construct a Lagrange basis for the set of mesh points Σ_3 (see Theorem 4.1.39).*

In combination with the polynomial space $\mathbb{P}_p^{\hat{\tau}}$ on $\hat{\tau}$ we define an interpolation operator $\widehat{I}^p$ for the set of nodal points $\Sigma_p = \left(\widehat{P}_{(i,j)}^{(p)}\right)_{(i,j)\in\iota_p}$ for continuous functions $\varphi \in C^0\left(\overline{\hat{\tau}}\right)$ by

$$\widehat{I}^p \varphi := \sum_{(i,j)\in\iota_p^{\hat{\tau}}} \varphi\left(\widehat{P}_{(i,j)}^{(p)}\right) \widehat{N}_{(i,j)}^{(p)}. \tag{4.73}$$

The Sobolev embedding theorem (Theorem 2.5.4) proves the continuity of the embedding $H^t(\hat{\tau}) \hookrightarrow C^0\left(\overline{\hat{\tau}}\right)$ thanks to $\hat{\tau} \subset \mathbb{R}^2$ for $t > 1$ and therefore $\widehat{I}^p$ is defined on $H^t(\hat{\tau})$, thus

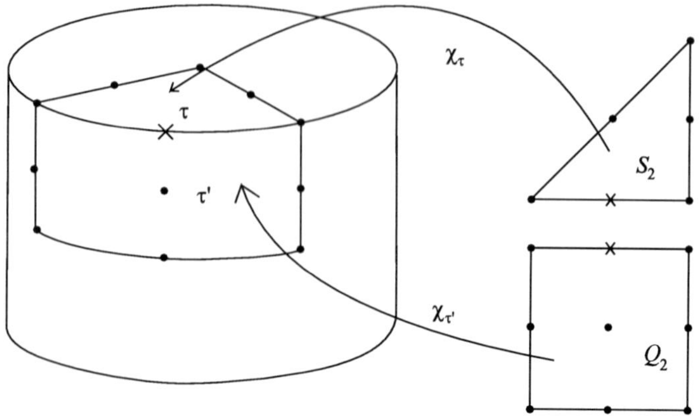

Fig. 4.5 Quadratic triangular and quadrilateral elements which share a common edge. The compatibility of the parametrizations ensures that the midpoints (*cross marks*) of the pullbacks of the common edge in the reference elements are mapped to the same surface points

$$\widehat{I}^p : H^t(\hat{\tau}) \to \mathbb{P}_p^{\hat{\tau}} \quad \text{and continuous:} \quad \left\| \widehat{I}^p \right\|_{C^0(\widehat{\tau}) \leftarrow H^t(\hat{\tau})} < \infty.$$

One obtains the set of nodal points on the surface by lifting the set of nodes on the reference element by means of the element parametrization

$$\mathcal{I} := \left\{ \chi_\tau \left(\widehat{P}_{(i,j)} \right) : \forall \tau \in \mathcal{G}, \quad \forall (i,j) \in \iota_p^{\hat{\tau}} \right\}. \tag{4.74}$$

Clearly, in a mesh $\mathcal{G}$ on Γ there will be nodal points that lie in more than one element, more precisely, that lie in their closures. As an example, consider Fig. 4.5 with two panels that have a common edge.

If the parameter representation $\chi_\tau, \chi_{\tau'}$ of the panels $\tau, \tau' \in \mathcal{G}$ is not compatible, the edge midpoint "×" on the common edge will be mapped to different points in $\hat{\tau}$, $\widehat{\tau'}$, depending on whether it is associated with τ or τ'. Thus, regular element mappings (see Definition 4.1.4) must parametrize edges $e = \overline{\tau} \cap \overline{\tau}'$ "identically from both sides". In the following we will always assume in the definition of continuous boundary elements $S_{\mathcal{G},\chi}^{p,0}$ that $\mathcal{G}$ and χ are regular.

Example 4.1.41 (p-Parametric Boundary Elements). *Let $\mathcal{G}$ be a regular mesh on Γ and let $q \geq 1$ be given and fixed. Then we can approximate a regular, generally non-linear, parametrization $\chi_\tau : \hat{\tau} \longrightarrow \tau \in \mathcal{G}$ by means of a p-parametric element mapping*

$$\widetilde{\chi}_\tau(\hat{\mathbf{x}}) := \sum_{(i,j) \in \iota_q^{\hat{\tau}}} \mathbf{P}_{(i,j)}^{(q)}(\tau) \widehat{N}_{(i,j)}^{(q)}(\hat{\mathbf{x}}), \qquad \hat{\mathbf{x}} \in \hat{\tau}, \tag{4.75}$$

where $\mathbf{P}_{(i,j)}^{(q)}(\tau) := \chi_\tau \left(\widehat{P}_{(i,j)}^{(q)} \right)$ *denotes the lifted nodes of the reference element.*

Remark 4.1.42. *In practical applications the construction (4.75) is used for $p = 1$ and $p = 2$ with the shape functions $\widehat{N}^{(p)}_{(i,j)}$ for the set of points $\widehat{P}^{(p)}_{(i,j)}$ in (4.71). In every case the approximation panel $\tilde{\tau} := \widetilde{\chi}_\tau(\hat{\tau})$ interpolates the exact panel τ at the points $\mathbf{P}^{(p)}_{(i,j)}$. It is known from interpolation theory (see Sect. 7.1.3.1) that, for the quality of the approximation, the choice of interpolation points becomes essential for high orders of approximation such as $p \geq 3$. For $p \geq 3$ the images of the Gauss–Lobatto points for the unit square represent a better choice for the set of nodes $\mathbf{P}^{(p)}_{(i,j)}$. Similar sets of points are known for the unit triangle (see [16, 130]).*

In the following we will always assume that the χ describe the surface Γ exactly. The influence of the approximation of the domain on the accuracy of the boundary element solution is discussed in Chap. 8.

We define the space of the continuous, piecewise polynomial boundary elements of degree $p \geq 1$ by a basis b_I. For this, let $\mathcal{I}$ be, as in (4.74), the set of all nodal points in the mesh $\mathcal{G}$. The basis function $b_\mathbf{P}$ for the nodal point $\mathbf{P} \in \mathcal{I}$ is characterized by the conditions

$$b_\mathbf{P} \in S^{p,0}_{\mathcal{G}} \quad \text{and} \quad b_\mathbf{P}(\mathbf{P}') := \begin{cases} 1 & \text{for } \mathbf{P}' = \mathbf{P}, \\ 0 & \text{for } \mathbf{P}' \neq \mathbf{P}, \quad \mathbf{P}' \in \mathcal{I}. \end{cases} \tag{4.76}$$

For a nodal point $\mathbf{P} \in \mathcal{I}$ we define a local neighborhood of triangles by $\Gamma_\mathbf{P} := \bigcup \{\overline{\tau} : \tau \in \mathcal{G},\ \mathbf{P} \in \overline{\tau}\}$. Then we have

$$\operatorname{supp}(b_\mathbf{P}) = \Gamma_\mathbf{P}. \tag{4.77}$$

In order to derive a local representation of the basis functions by element shape functions, we need a relation between global indices $\mathbf{P} \in \mathcal{I}$ and local indices $(i,j) \in \iota_p^{\hat{\tau}}$. For $\tau \in \mathcal{G}$ and $I = (i,j) \in \iota_p^{\hat{\tau}}$ we define a mapping $\operatorname{ind} : \mathcal{G} \times \iota_p^{\hat{\tau}} \to \mathcal{I}$ by

$$\operatorname{ind}(\tau, I) := \chi_\tau\left(\widehat{P}_{(i,j)}\right) \in \mathcal{I}. \tag{4.78}$$

With this we have, for $\tau \in \mathcal{G}$, $I = (i,j) \in \iota_p^{\hat{\tau}}$ and $\mathbf{P} = \operatorname{ind}(\tau, I) \in \mathcal{I}$, the relation

$$b_\mathbf{P}|_\tau = N^\tau_{(i,j)} := \widehat{N}_I \circ \chi_\tau^{-1}. \tag{4.79}$$

In the following we will show that the functions in $S^{p,0}_{\mathcal{G},\chi}$ are Lipschitz continuous and are thus contained in $H^1(\Gamma)$. In order to compare the Euclidian distance with the surface distance, we introduce the geodesic distance

$$\operatorname{dist}_\Gamma(\mathbf{x}, \mathbf{y}) := \inf\left\{\operatorname{length}\left(\gamma_{\mathbf{x},\mathbf{y}}\right) : \gamma_{\mathbf{x},\mathbf{y}} \text{ is a path in } \Gamma \text{ that connects } \mathbf{x} \text{ and } \mathbf{y}\right\}$$

and the constant g_Γ

$$g_\Gamma := \sup_{\mathbf{x},\mathbf{y}\in\Gamma} \left\{ \frac{\operatorname{dist}_\Gamma(\mathbf{x},\mathbf{y})}{\|\mathbf{x}-\mathbf{y}\|} \right\}. \tag{4.80}$$

Remark 4.1.43. *The functions $\varphi_S \in S^{p,0}_{\mathcal{G},\chi}$ are Lipschitz continuous*

$$|\varphi_S(\mathbf{x}) - \varphi_S(\mathbf{y})| \le C\,\|\mathbf{x}-\mathbf{y}\| \qquad \forall \mathbf{x},\mathbf{y}\in\Gamma,$$

where C depends on Γ, $\mathcal{G}$, χ and g_Γ.

Proof. The continuity of $\varphi_S \in S^{p,0}_{\mathcal{G},\chi}$ follows directly from the definition so that we only need to prove the Lipschitz continuity. Let $\mathbf{x},\mathbf{y}\in\Gamma$ and let $\gamma_{\mathbf{x},\mathbf{y}}$ be a connecting path with minimal length on Γ. Let $(\tau_j)_{j=0}^q \subset \mathcal{G}$ be a minimal subset of $\mathcal{G}$ with the property:

$$\mathbf{x}\in\overline{\tau_0}, \quad \mathbf{y}\in\overline{\tau_q}, \quad \gamma_{\mathbf{x},\mathbf{y}} \subset \bigcup_{j=1}^{q} \overline{\tau_j}$$
$$\forall 1\le j\le q : \overline{\tau_{j-1}} \cap \overline{\tau_j} \text{ is a common edge } e_j \text{ and } e_j \cap \gamma_{\mathbf{x},\mathbf{y}} \neq \emptyset.$$

We fix the points $\mathbf{M}_j$ on $e_j \cap \gamma_{\mathbf{x},\mathbf{y}}$, $1\le j\le q$ and set $\mathbf{M}_0 = \mathbf{x}$ and $\mathbf{M}_{q+1} = \mathbf{y}$. Without loss of generality we assume that all $(\mathbf{M}_j)_{j=0}^{q+1}$ are distinct; otherwise we simply eliminate points that appear in the sequence more than once. Then, by the continuity of φ_S, we have

$$\varphi_S(\mathbf{y}) - \varphi_S(\mathbf{x}) = \varphi_S(\mathbf{M}_{q+1}) - \varphi_S(\mathbf{M}_0) = \sum_{j=0}^{q} \left(\varphi_S(\mathbf{M}_{j+1}) - \varphi_S(\mathbf{M}_j)\right).$$

The points $\mathbf{M}_{j+1}$, $\mathbf{M}_j$ are in the panel τ_j. Since $\varphi_S|_\tau$ is the composition of a polynomial with a diffeomorphism, these restrictions are Lipschitz continuous. With

$$c_\tau := \sup_{\mathbf{x},\mathbf{y}\in\tau} \frac{|\varphi_S(\mathbf{x}) - \varphi_S(\mathbf{y})|}{\|\mathbf{x}-\mathbf{y}\|}$$

we have

$$\left|\varphi_S(\mathbf{M}_{j+1}) - \varphi_S(\mathbf{M}_j)\right| \le c_\tau \left\|\mathbf{M}_{j+1} - \mathbf{M}_j\right\| \le c_\tau L\left(\gamma_{\mathbf{M}_j,\mathbf{M}_{j+1}}\right),$$

where $L\left(\gamma_{\mathbf{M}_j,\mathbf{M}_{j+1}}\right)$ denotes the length of the shortest connecting path in Γ that connects $\mathbf{M}_j$ with $\mathbf{M}_{j+1}$. Finally, with (4.80) we have

$$|\varphi_S(\mathbf{y}) - \varphi_S(\mathbf{x})| \le \left(\max_{1\le j\le q} c_{\tau_j}\right) L(\gamma_{\mathbf{x},\mathbf{y}}) \le g_\Gamma \left(\max_{1\le j\le q} c_{\tau_j}\right) \|\mathbf{x}-\mathbf{y}\|,$$

which is the Lipschitz continuity of φ_S. □

4.1.8 Galerkin BEM with Continuous Boundary Elements

The inclusion $S_{\mathcal{G},\chi}^{p,0} \subset H^{1/2}(\Gamma)$ of the continuous boundary elements permits the Galerkin discretization of the hypersingular boundary integral equation:
Find $\varphi_S \in S_{\mathcal{G}}^{p,0}/\mathbb{K}$ such that

$$b(\varphi_S, \eta_S) = (g_N, \eta_S)_{L^2(\Gamma)} \qquad \forall \eta_S \in S_{\mathcal{G}}^{p,0}/\mathbb{K}. \tag{4.81}$$

The ellipticity (Theorem 3.5.3) implies the existence of a unique solution of Problem (4.81). The system matrix of the hypersingular integral equation has similar properties to the matrix of the single layer potential (see Proposition 4.1.24).

Proposition 4.1.44. *The system matrix* $\mathbf{W}$ *of the bilinear form* $b : S_{\mathcal{G}}^{p,0}/\mathbb{R} \times S_{\mathcal{G}}^{p,0}/\mathbb{R} \to \mathbb{R}$ *in (4.65) is symmetric and positive definite. The entries* $W_{I,J}$, $I, J \in \mathcal{I}$ *have the explicit form*

$$W_{I,J} = \int_\Gamma \int_\Gamma \frac{\langle \operatorname{curl}_\Gamma b_I(\mathbf{x}), \operatorname{curl}_\Gamma b_J(\mathbf{y})\rangle}{4\pi \|\mathbf{x}-\mathbf{y}\|} ds_{\mathbf{y}} ds_{\mathbf{x}} = W_{J,I}. \tag{4.82}$$

The integrals in (4.82) are, according to Remark 4.1.43, weakly singular and therefore the matrix entries are well defined. We can write the actual generation of the matrix by means of integrals over single panels, with the help of the index allocation (4.78). In the following we will give an algorithmic description in the form of a pseudo programming language.

procedure generate_system_matrix;
for all $\tau, t \in \mathcal{G}$ **do begin**
for all $I = (i, i') \in \iota_p^{\hat{\tau}}$, $J = (j, j') \in \iota_{\hat{t}}^{p}$ **do begin**

$$W_{\tau,t}^{I,J} := \int_\tau \int_t G(\mathbf{x}-\mathbf{y}) \left\langle \operatorname{curl}_\Gamma \left(\widehat{N}_{(i,i')} \circ \chi_\tau^{-1}(\mathbf{x}) \right), \operatorname{curl}_\Gamma \left(\widehat{N}_{(j,j')} \circ \chi_t^{-1}(\mathbf{y}) \right) \right\rangle ds_{\mathbf{y}} ds_{\mathbf{x}};$$

$$K := \operatorname{ind}(\tau, I); \quad L := \operatorname{ind}(t, J); \quad W_{K,L} := W_{K,L} + W_{\tau,t}^{I,J}; \tag{4.83}$$

end;end;

Exercise 4.1.45. *Let* $\tau, t \in \mathcal{G}$ *be panels with reference elements* $\hat{\tau}$, $\hat{t}$ *and reference mappings* χ_τ, χ_t. *The Jacobian of the transformation is denoted by* $\mathbf{J}_\tau := \left[\hat{\partial}_1 \chi_\tau, \hat{\partial}_2 \chi_\tau\right]$ *and we set* $\widehat{\nabla}^\perp := \left(\hat{\partial}_2, -\hat{\partial}_1\right)$. *For sufficiently smooth functions* $u : \tau \to \mathbb{R}$ *prove the relation*

$$g_\tau \operatorname{curl}_\Gamma u \circ \chi_\tau = \mathbf{J}_\tau \widehat{\nabla}^\perp \hat{u},$$

where $g_\tau := \sqrt{\det\left(\mathbf{J}_\tau^\intercal \mathbf{J}_\tau\right)}$ *and* $\hat{u} := u \circ \chi_\tau$.

For the local system matrix $W_{\tau,t}^{I,J}$ *in (4.83) we have the representation*

$$\int_{\hat{\tau}}\int_{\hat{t}} \frac{\left\langle \left(\mathbf{J}_\tau \widehat{\nabla}^\perp \widehat{N}_{(i,i')}\right)(\hat{\mathbf{x}}), \left(\mathbf{J}_t \widehat{\nabla}^\perp \widehat{N}_{(j,j')}\right)(\hat{\mathbf{y}})\right\rangle}{4\pi \left\|\chi_\tau(\hat{\mathbf{x}}) - \chi_t(\hat{\mathbf{y}})\right\|} d\hat{\mathbf{y}} d\hat{\mathbf{x}}.$$

(Hint: Use Exercise 3.3.25.)

In the same way as in Proposition 4.1.25 we obtain a quasi-optimal estimate for the Galerkin error for continuous boundary elements on a regular mesh $\mathcal{G}$.

Proposition 4.1.46. *The Galerkin approximation* $\varphi_S \in S_{\mathcal{G}}^{p,0}$ *of the solution* φ *of the hypersingular boundary integral equation converges quasi-optimally:*

$$\|\varphi - \varphi_S\|_{H^{1/2}(\Gamma)/\mathbb{K}} \leq \frac{\|b\|}{\gamma} \min_{\psi_S \in S_{\mathcal{G}}^{p,0}} \|\varphi - \psi_S\|_{H^{1/2}(\Gamma)/\mathbb{K}}. \tag{4.84}$$

The Galerkin projection $\Pi_{\mathcal{G}}^{(p)} : H^{1/2}(\Gamma)/\mathbb{K} \to S_{\mathcal{G}}^{p,0}/\mathbb{K}$, *given by* $\Pi_{\mathcal{G}}^{(p)}\varphi = \varphi_S$, *is stable:*

$$\|\Pi_{\mathcal{G}}^{(p)}\|_{H^{1/2}(\Gamma)/\mathbb{K} \leftarrow H^{1/2}(\Gamma)/\mathbb{K}} \leq \|b\|/\gamma, \tag{4.85}$$

where the norm of the bilinear form $b(\cdot,\cdot)$ *is given by*

$$\|b\| := \sup_{\varphi \in H^{1/2}(\Gamma)\backslash\{0\}} \sup_{\eta \in H^{1/2}(\Gamma)\backslash\{0\}} \frac{b(\varphi,\eta)}{\|\varphi\|_{H^{1/2}(\Gamma)/\mathbb{K}} \|\eta\|_{H^{1/2}(\Gamma)/\mathbb{K}}}$$

[see (2.29)].

Thanks to the stability result (4.85), the search for convergence rates of the Galerkin BEM is again reduced to the study of the approximation properties of the spaces $S_{\mathcal{G}}^{p,0}$.

4.1.9 Convergence Rates with Continuous Boundary Elements

In order to find convergence rates for the boundary element approximation φ_S in (4.81) of the hypersingular equation (4.65), we need approximation properties of the continuous boundary element spaces, which we will now specify. For this, let the boundary Γ be bounded and piecewise smooth in the sense of Definition 2.2.10.

Remark 4.1.47. *The partitioning of* Γ *which is employed in Definition 2.2.10 of piecewise smoothness is denoted here by* $\mathcal{C} = \{\Gamma_i : 1 \leq i \leq q\}$ *instead of* $\mathcal{G}$ *in order to distinguish the notation from the boundary element mesh* $\mathcal{G}$ *and its panels* $\tau \in \mathcal{G}$ *(cf. Definition 4.1.2). In this light, the cardinality* q *of* $\mathcal{C}$ *depends only on* Γ *and is, in particular, independent of the discretization parameters. However, we always assume that the boundary element mesh is compatible with* $\mathcal{C}$ *in the sense that, for any* $\tau \in \mathcal{G}$, *there exists a* $\Gamma_i \in \mathcal{C}$ *with* $\tau \subset \Gamma_i$.

We will prove the approximation property and the convergence rates for the Galerkin solution under the assumption that the exact solution belongs to the space $H^t_{\text{pw}}(\Gamma)$ which we will define next.

Definition 4.1.48. Let Γ be piecewise smooth with partitioning $\mathcal{C} := \{\Gamma_i : 1 \leq i \leq q\}$:

(a) For $t > 1$, the space $H^t_{\text{pw}}(\Gamma)$ contains all functions $\psi \in H^1(\Gamma)$ which satisfy

$$\forall \Gamma_i \in \mathcal{C} : \quad \psi|_{\Gamma_i} \in H^t(\Gamma_i)$$

and is furnished with the graph norm

$$\|\psi\|_{H^t_{\text{pw}}(\Gamma)} := \left(\sum_{\Gamma_i \in \mathcal{C}} \|\psi\|^2_{H^t(\Gamma_i)} \right)^{1/2}. \tag{4.86}$$

(b) For $0 \leq t \leq 1$, the space $H^t_{\text{pw}}(\Gamma)$ equals $H^t(\Gamma)$ and the norm $\|\cdot\|_{H^t_{\text{pw}}(\Gamma)}$ is the usual $H^t(\Gamma)$-norm.

Some properties of the $H^t_{\text{pw}}(\Gamma)$- and the $H^t(\Gamma)$-norms are stated in the next lemma.

Lemma 4.1.49. *(a) Let $t \geq 1$. For any $\psi \in H^t(\Gamma)$, we have*

$$\|\psi\|_{H^t_{\text{pw}}(\Gamma)} \leq \|\psi\|_{H^t(\Gamma)}.$$

(b) Let $s \geq 0$. Let ι denote a finite index set and let $\{v_i : i \in \iota\}$ be a set of functions in $H^s(\Gamma)$. If the supports $\omega_i := \operatorname{supp} v_i$ satisfy

$$|\omega_i \cap \omega_j| = 0 \qquad \forall i, j \in \iota \text{ with } i \neq j,$$

then

$$\left\| \sum_{i \in \iota} v_i \right\|^2_{H^s(\Gamma)} \leq \frac{5}{2} \sum_{i \in \iota} \|v_i\|^2_{H^s(\Gamma)}.$$

Proof. Part a: Let $t \in \mathbb{N}_0$. Then

$$\|\psi\|^2_{H^t(\Gamma)} = \sum_{\Gamma_i \in \mathcal{C}} \|\psi\|^2_{H^t(\Gamma_i)} = \|\psi\|^2_{H^t_{\text{pw}}(\Gamma)}.$$

For $t \in \mathbb{R}_{\geq 0} \backslash \mathbb{N}_0$, let $t = \lfloor t \rfloor + \lambda$ with $\lambda \in]0, 1[$. We employ (2.85) to obtain

$$\|\psi\|^2_{H^t(\Gamma)} = \sum_{|\alpha| \leq \lfloor t \rfloor} |\psi_\alpha|^2_{L^2(\Gamma)} + \sum_{|\alpha| \leq \lfloor t \rfloor} \int_{\Gamma \times \Gamma} \frac{|\psi_\alpha(\mathbf{x}) - \psi_\alpha(\mathbf{y})|^2}{\|\mathbf{x} - \mathbf{y}\|^{2+2\lambda}} ds_{\mathbf{x}} ds_{\mathbf{y}}$$

$$\geq \sum_{\Gamma_i \in \mathcal{C}} \left\{ \sum_{|\alpha| \leq \lfloor t \rfloor} \|\psi_\alpha\|^2_{L^2(\Gamma_i)} + \sum_{|\alpha| \leq \lfloor t \rfloor} \int_{\Gamma_i \times \Gamma_i} \frac{|\psi_\alpha(\mathbf{x}) - \psi_\alpha(\mathbf{y})|^2}{\|\mathbf{x}-\mathbf{y}\|^{2+2\lambda}} ds_{\mathbf{x}} ds_{\mathbf{y}} \right\}$$
$$= \sum_{\Gamma_i \in \mathcal{C}} \|\psi\|^2_{H^t(\Gamma_i)} .$$

Part b: The proof of Part b is as in [91, Satz 3.26]. First, we will consider the case $s \in]0,1[$. We write

$$v = \sum_{i \in \iota} v_i, \qquad D_i := \operatorname{supp} v_i, \qquad D := \bigcup_{i \in \iota} D_i = \operatorname{supp} v$$

and introduce the shorthand

$$\int_{\Gamma'} \int_{\Gamma''} [w]_s^2 := \int_{\Gamma'} \int_{\Gamma''} \frac{|w(\mathbf{x}) - w(\mathbf{y})|^2}{\|\mathbf{x}-\mathbf{y}\|^{2+2s}} ds_{\mathbf{x}} ds_{\mathbf{y}}$$

for any measurable subsets $\Gamma', \Gamma'' \subset \Gamma$ and $w \in H^s(\Gamma)$.

For any $i \in \iota$, we get

$$\int_\Gamma \int_\Gamma [v_i]_s^2 = \int_{D_i} \int_{D_i} [v_i]_s^2 + 2 \int_{D_i} \int_{\Gamma \backslash D_i} [v_i]_s^2 + \underbrace{\int_{\Gamma \backslash D_i} \int_{\Gamma \backslash D_i} [v_i]_s^2}_{=0}$$
$$= \int_{D_i} \int_{D_i} [v_i]_s^2 + 2 \int_{D_i} |v_i(\mathbf{x})|^2 \int_{\Gamma \backslash D_i} \|\mathbf{x}-\mathbf{y}\|^{-2-2s} ds_{\mathbf{y}} ds_{\mathbf{x}}. \quad (4.87)$$

On the other hand,

$$\int_\Gamma \int_\Gamma [v]_s^2 = \int_D \int_\Gamma [v]_s^2 + \int_{\Gamma \backslash D} \int_D [v]_s^2 + \underbrace{\int_{\Gamma \backslash D} \int_{\Gamma \backslash D} [v]_s^2}_{=0}$$
$$= \sum_{i \in \iota} \int_{D_i} \int_{D_i} \underbrace{[v]_s^2}_{=[v_i]_s^2} + \sum_{i \in \iota} \int_{D_i} \int_{\Gamma \backslash D_i} [v]_s^2 + \int_D \int_{\Gamma \backslash D} [v]_s^2 \quad (4.88)$$

and

$$\int_{D_i} \int_{\Gamma \backslash D_i} [v]_s^2 = \int_{D_i} \int_{\Gamma \backslash D_i} \frac{|v(\mathbf{x}) - v(\mathbf{y})|^2}{\|\mathbf{x}-\mathbf{y}\|^{2+2s}} ds_{\mathbf{x}} ds_{\mathbf{y}}$$
$$\leq \underbrace{2 \int_{\Gamma \backslash D_i} |v(\mathbf{x})|^2 \left(\int_{D_i} \frac{1}{\|\mathbf{x}-\mathbf{y}\|^{2+2s}} ds_{\mathbf{y}} \right) ds_{\mathbf{x}}}_{=:J_i}$$

$$+ 2\int_{D_i} \underbrace{|v(\mathbf{y})|^2}_{=|v_i(\mathbf{y})|^2} \int_{\Gamma\backslash D_i} \frac{1}{\|\mathbf{x}-\mathbf{y}\|^{2+2s}} ds_{\mathbf{x}} ds_{\mathbf{y}}$$

$$\stackrel{(4.87)}{=} \int_\Gamma\int_\Gamma [v_i]_s^2 - \int_{D_i}\int_{D_i} [v_i]_s^2 + 2J_i.$$

Inserting this into (4.88) results in

$$\int_\Gamma\int_\Gamma [v]_s^2 \le \sum_{i\in\iota}\left(\int_\Gamma\int_\Gamma [v_i]_s^2 + 2J_i\right) + \int_D\int_{\Gamma\backslash D} [v]_s^2. \tag{4.89}$$

Next, we will investigate the sum over the quantities J_i. Let χ_i denote the characteristic function for $\Gamma\backslash D_i$. Then

$$\begin{aligned}\sum_{i\in\iota} J_i &= \sum_{i\in\iota}\int_{\Gamma\backslash D_i} |v(\mathbf{x})|^2 \left(\int_{D_i} \frac{1}{\|\mathbf{x}-\mathbf{y}\|^{2+2s}} ds_{\mathbf{y}}\right) ds_{\mathbf{x}} \\ &= \sum_{i\in\iota}\int_\Gamma \chi_i(\mathbf{x})\, |v(\mathbf{x})|^2 \left(\int_{D_i} \frac{1}{\|\mathbf{x}-\mathbf{y}\|^{2+2s}} ds_{\mathbf{y}}\right) ds_{\mathbf{x}} \\ &= \int_\Gamma |v(\mathbf{x})|^2 \underbrace{\left(\sum_{i\in\iota}\chi_i(\mathbf{x})\int_{D_i} \frac{1}{\|\mathbf{x}-\mathbf{y}\|^{2+2s}} ds_{\mathbf{y}}\right)}_{=:f(\mathbf{x})} ds_{\mathbf{x}}.\end{aligned} \tag{4.90}$$

Let $j\in\iota$ and let $\mathbf{x}$ be an interior point of D_j, i.e., $\mathbf{x}\in\overset{\circ}{D}_j$. For any $i\in\iota$, we have

$$\chi_i(\mathbf{x}) := \left\{\begin{matrix} 1 \text{ if } \mathbf{x}\in\Gamma\backslash D_i \\ 0 \;\; \text{if } \mathbf{x}\in D_i \end{matrix}\right\} = \left(1-\delta_{i,j}\right).$$

For $\mathbf{x}\in\overset{\circ}{D}_j$ we have

$$f(\mathbf{x}) = \sum_{i\in\iota\backslash\{j\}}\int_{D_i} \frac{1}{\|\mathbf{x}-\mathbf{y}\|^{2+2s}} ds_{\mathbf{y}} = \int_{D\backslash D_j} \frac{1}{\|\mathbf{x}-\mathbf{y}\|^{2+2s}} ds_{\mathbf{y}}.$$

Inserting this into (4.90) results in

$$\begin{aligned}2\sum_{i\in\iota} J_i &= \sum_{j\in\iota} 2\int_{D_j} \underbrace{|v(\mathbf{x})|^2}_{|v_j(\mathbf{x})|^2} \left(\int_{D\backslash D_j} \frac{1}{\|\mathbf{x}-\mathbf{y}\|^{2+2s}} ds_{\mathbf{y}}\right) ds_{\mathbf{x}} \\ &\stackrel{(4.87)}{\le} \sum_{j\in\iota}\int_\Gamma\int_\Gamma \left[v_j\right]_s^2.\end{aligned} \tag{4.91}$$

It remains to estimate the second term in (4.89). We have

$$\begin{aligned}\int_D \int_{\Gamma\backslash D} [v]_s^2 &= \int_D |v(\mathbf{x})|^2 \left(\int_{\Gamma\backslash D} \frac{1}{\|\mathbf{x}-\mathbf{y}\|^{2+2s}} ds_{\mathbf{y}} \right) ds_{\mathbf{x}} \\ &= \sum_{i\in\iota} \int_{D_i} \underbrace{|v(\mathbf{x})|^2}_{=|v_i(\mathbf{x})|^2} \left(\int_{\Gamma\backslash D} \frac{1}{\|\mathbf{x}-\mathbf{y}\|^{2+2s}} ds_{\mathbf{y}} \right) ds_{\mathbf{x}} \\ &\leq \sum_{i\in\iota} \int_{D_i} |v_i(\mathbf{x})|^2 \left(\int_{\Gamma\backslash D_i} \frac{1}{\|\mathbf{x}-\mathbf{y}\|^{2+2s}} ds_{\mathbf{y}} \right) ds_{\mathbf{x}} \\ &\overset{(4.87)}{\leq} \frac{1}{2} \sum_{i\in\iota} \int_\Gamma \int_\Gamma [v_i]_s^2 . \end{aligned} \tag{4.92}$$

The combination of (4.89), (4.91), and (4.92) leads to

$$\int_\Gamma \int_\Gamma [v]_s^2 \leq \frac{5}{2} \sum_{i\in\iota} \int_\Gamma \int_\Gamma [v_i]_s^2 .$$

Because the $L^2(\Gamma)$-norm is additive we obtain

$$\begin{aligned}\left\| \sum_{i\in\iota} v_i \right\|_{H^s(\Gamma)}^2 &= \|v\|_{L^2(\Gamma)}^2 + \int_\Gamma \int_\Gamma [v]_s^2 \leq \sum_{i\in\iota} \|v_i\|_{L^2(\Gamma)}^2 + \frac{5}{2} \sum_{i\in\iota} \int_\Gamma \int_\Gamma [v_i]_s^2 \\ &\leq \frac{5}{2} \sum_{i\in\iota} \|v_i\|_{H^s(\Gamma)}^2 .\end{aligned}$$

The proof for $s \in \mathbb{R}_{>1}\backslash\mathbb{N}$ can be carried out in the same way. Note that the expression $[v]_s$ has to be replaced by $[v_\alpha]_s$, where v_α is defined as in (2.86). □

Proposition 4.1.50. *Let Γ be piecewise smooth and let $\mathcal{G}$ be a surface mesh of Γ:*

(a) Let[1] $\varphi \in H_{\mathrm{pw}}^t(\Gamma)$ for some $t > 1$. Then there exists a continuous interpolation $I_{\mathcal{G}}^p \varphi \in S_{\mathcal{G}}^{p,0}$ with

$$\|\varphi - I_{\mathcal{G}}^p \varphi\|_{H^s(\Gamma)} \leq C\, h_{\mathcal{G}}^{\min\{t,p+1\}-s} \|\varphi\|_{H_{\mathrm{pw}}^t(\Gamma)}, \qquad s \in \{0,1\}, \tag{4.93}$$

where the constant C depends only on p and on the constant $\kappa_{\mathcal{G}}$ from Definition 4.1.12, which describes the shape-regularity of the mesh.

(b) Let $0 \leq s \leq t \leq 1$. Then there exists a continuous operator $Q_{\mathcal{G}} : H^t(\Gamma) \to S_{\mathcal{G}}^{p,0}$ such that, for every $\varphi \in H^t(\Gamma)$, we have

---

[1] In Sect. 4.3.3, we will prove the continuous embedding $H_{pw}^t(\Gamma) \hookrightarrow C^0(\Gamma)$ for $t > 1$ and piecewise smooth Lipschitz surfaces.

$$\|\varphi - Q_{\mathcal{G}}\varphi\|_{H^s(\Gamma)} \leq C h_{\mathcal{G}}^{t-s} \|\varphi\|_{H^t(\Gamma)}.$$

The operator $Q_{\mathcal{G}}$ is stable for $0 \leq s \leq 1$

$$\|Q_{\mathcal{G}}\|_{H^s(\Gamma) \leftarrow H^s(\Gamma)} \leq C.$$

The proof of Proposition 4.1.50 is postponed to Sect. 4.3.5.

With Proposition 4.1.50 we can now derive quantitative error estimates from the quasi-optimality (4.84) of the Galerkin solution φ_S.

Theorem 4.1.51. *Let Γ be a piecewise smooth Lipschitz surface. Furthermore, let $\mathcal{G}$ be a regular surface mesh on Γ. Let $\varphi \in H^t_{\mathrm{pw}}(\Gamma)$ with $t \geq 1/2$. Then we have for the Galerkin approximation $\varphi_S \in S^{p,0}_{\mathcal{G}}$ of (4.65) the error estimate*

$$\|\varphi - \varphi_S\|_{H^{1/2}(\Gamma)/\mathbb{K}} \leq C h^{\min(t,p+1)-1/2} \|\varphi\|_{H^t_{\mathrm{pw}}(\Gamma)}, \tag{4.94}$$

where the constant C depends only on p and, via the constant $\kappa_{\mathcal{G}}$ from Definition 4.1.12, on the shape-regularity of the mesh.

Proof.

Case 1: $t = 1/2$.

For $\varphi \in H^{1/2}(\Gamma)/\mathbb{K}$ it follows from (4.84) that by choosing $\psi_S = 0$ we obtain the boundedness of the error $\|\varphi - \varphi_S\|_{H^{1/2}(\Gamma)/\mathbb{K}}$ by $(\|b\|/\gamma)\|\varphi\|_{H^{1/2}(\Gamma)/\mathbb{K}}$. This yields (4.94) for $t = 1/2$.

Case 2: $t > 1$.

Now let $\varphi \in H^t_{\mathrm{pw}}(\Gamma)$ with $t > 1$. Let $T^p_{\mathcal{G}} : H^t_{\mathrm{pw}}(\Gamma) \to S^{p,0}_{\mathcal{G}}$ be defined by

$$T^p_{\mathcal{G}} := \begin{cases} Q_{\mathcal{G}} & \text{if } t = 1, \\ I^p_{\mathcal{G}} & \text{if } t > 1. \end{cases}$$

Proposition 4.1.50 implies that $T^p_{\mathcal{G}}$ is continuous. The estimate

$$\|\varphi - \varphi_S\|_{H^{1/2}(\Gamma)/\mathbb{K}} \leq \frac{\|b\|}{\gamma}\|\varphi - T^p_{\mathcal{G}}\varphi\|_{H^{1/2}(\Gamma)/\mathbb{K}} \leq \frac{\|b\|}{\gamma}\|\varphi - T^p_{\mathcal{G}}\varphi\|_{H^{1/2}(\Gamma)}$$

follows from the quasi-optimality (4.84), and we have used $\|\varphi\|_{H^{1/2}(\Gamma)/\mathbb{K}} = \min_{c\in\mathbb{R}} \|\varphi - c\|_{H^{1/2}(\Gamma)} \leq \|\varphi\|_{H^{1/2}(\Gamma)}$.

If we apply Proposition 2.1.65 with $X_0 = L^2(\Gamma)$, $X_1 = H^1(\Gamma)$ and $\theta = 1/2$ we obtain the interpolation inequality

$$\|\varphi\|^2_{H^{1/2}(\Gamma)} \leq \|\varphi\|_{L^2(\Gamma)} \|\varphi\|_{H^1(\Gamma)}.$$

With this and with Proposition 4.1.50 it follows for $t \geq 1$ that

$$\|\varphi - T_{\mathcal{G}}^{p}\,\varphi\|^2_{H^{1/2}(\Gamma)} \leq C\,\|\varphi - T_{\mathcal{G}}^{p}\varphi\|_{L^2(\Gamma)}\|\varphi - T_{\mathcal{G}}^{p}\,\varphi\|_{H^1(\Gamma)}$$
$$\leq C\,h^{2\min(t,p+1)-1}\|\varphi\|^2_{H^t_{\mathrm{pw}}(\Gamma)} \tag{4.95}$$

and therefore we have (4.94) for $t > 1$.
Case 3: $1/2 < t \leq 1$.
In this case we prove (4.94) by interpolation. We have for the operator $I - Q_{\mathcal{G}}$ the estimate [cf. Proposition 4.1.50(b)]

$$\|I - Q_{\mathcal{G}}\|_{H^{1/2}(\Gamma)\leftarrow H^{1/2}(\Gamma)} \leq C, \qquad \|I - Q_{\mathcal{G}}\|_{H^{1/2}(\Gamma)\leftarrow H^1(\Gamma)} \leq C\,h^{1/2}.$$

As in the proof of Theorem 4.1.33, the estimate

$$\|(I - Q_{\mathcal{G}})\varphi\|_{H^{1/2}(\Gamma)} \leq Ch^{t-\frac{1}{2}}\|\varphi\|_{H^t(\Gamma)}.$$

follows for $1/2 \leq t \leq 1$ by interpolation of the linear operator $I - Q_{\mathcal{G}}$: $H^t(\Gamma) \to H^{\frac{1}{2}}(\Gamma)$ (see Proposition 2.1.62). □

4.1.10 Model Problem 3: Mixed Boundary Value Problem*

We consider the mixed boundary value problem for the Laplace operator:

$$\Delta u = 0 \quad \text{in } \Omega^-, \qquad u = g_D \quad \text{on } \Gamma_D, \qquad \partial u/\partial \mathbf{n} = g_N \quad \text{on } \Gamma_N \tag{4.96}$$

for given boundary data $g_D \in H^{1/2}(\Gamma_D)$, $g_N \in H^{-1/2}(\Gamma_2)$. For the associated variational formulation we refer to Sect. 2.9.2.3. The approach that allows the discretization of mixed boundary value problems by means of the Galerkin boundary element method is due to [220, 239]. For the treatment of problems with more general transmission conditions we refer to [233].

The problem can be reduced to an integral equation for the pair of densities $(\varphi, \sigma) \in \mathbf{H} = \widetilde{H}^{-1/2}(\Gamma_D) \times \widetilde{H}^{1/2}(\Gamma_N)$. The solution of (4.96) can be represented with the help of Green's representation formula

$$u(\mathbf{x}) = (S\sigma)(\mathbf{x}) - (D\varphi)(\mathbf{x}), \qquad \mathbf{x} \in \Omega^-.$$

The variational formulation of the boundary integral equation reads [see (3.89)]:
Find $(\varphi, \sigma) \in \mathbf{H}$ such that

$$b_{mixed}\left(\begin{pmatrix}\varphi\\ \sigma\end{pmatrix}, \begin{pmatrix}\eta\\ \kappa\end{pmatrix}\right) = (g_D, \eta)_{L^2(\Gamma_D)} + (g_N, \kappa)_{L^2(\Gamma_N)} \qquad \forall\,(\eta, \kappa) \in \mathbf{H} \tag{4.97}$$

* This section should be read as a complement to the core material of this book.

with

$$b_{mixed}\left(\begin{pmatrix}\varphi\\ \sigma\end{pmatrix},\begin{pmatrix}\eta\\ \kappa\end{pmatrix}\right) = (V_{DD}\varphi,\eta)_{L^2(\Gamma_D)} - (K_{DN}\sigma,\eta)_{L^2(\Gamma_D)} + \left(K'_{ND}\varphi,\kappa\right)_{L^2(\Gamma_N)}$$
$$+ (W_{NN}\sigma,\kappa)_{L^2(\Gamma_N)} .$$

The *boundary element discretization* is achieved by a combination of different boundary element spaces on the pieces Γ_D, Γ_N. For this let $\mathcal{G}_D$, $\mathcal{G}_N$ be surface meshes of Γ_D, Γ_N, while we assume that $\mathcal{G}_N$ is regular (see Definition 4.1.4). We use discontinuous boundary elements of order $p_1 \geq 0$ on Γ_D. The inclusion

$$S_{\mathcal{G}_D}^{p_1,-1} \subset \widetilde{H}^{-1/2}(\Gamma_D), \tag{4.98}$$

results, because the zero extension $\psi^\star$ of every function $\psi \in S_{\mathcal{G}_D}^{p_1,-1}$ satisfies the inclusion $\psi^\star \in L^2(\Gamma) \subset H^{-1/2}(\Gamma)$ and thus we have $\psi \in \widetilde{H}^{-1/2}(\Gamma_D)$.

For the approximation of $\sigma \in \widetilde{H}^{1/2}(\Gamma_N)$ we define for $p_2 \geq 1$

$$S_{\mathcal{G}_N,0}^{p_2,0} = \left\{\eta \in S_{\mathcal{G}_N}^{p_2,0} : \eta|_{\partial\Gamma_N} = 0\right\} \tag{4.99}$$

and therefore the boundary values of the functions $\eta \in S_{\mathcal{G}_N,0}^{p_2,0}$ vanish on $\partial\Gamma_N$.

Remark 4.1.52. *The zero extension $\sigma^\star$ of functions $\sigma \in S_{\mathcal{G}_N,0}^{p,0}$ satisfies $\sigma^\star \in S_{\mathcal{G}}^{p,0} \subset H^{1/2}(\Gamma)$, where we have set $\mathcal{G} := \mathcal{G}_D \cup \mathcal{G}_N$.*

With these spaces we can finally formulate the boundary element discretization of (4.97). In the following we will summarize the polynomial orders $p_1 \geq 0$ and $p_2 \geq 1$ in the vector $\mathbf{p} = (p_1, p_2)$.

Find $(\varphi_S, \sigma_S) \in S^{\mathbf{p}} := S_{\mathcal{G}_D}^{p_1,-1} \times S_{\mathcal{G}_N,0}^{p_2,0}$ such that

$$b_{mixed}\left(\begin{pmatrix}\varphi_S\\ \sigma_S\end{pmatrix},\begin{pmatrix}\eta_S\\ \kappa_S\end{pmatrix}\right) = (g_D,\eta_S)_{L^2(\Gamma_D)} + (g_N,\kappa_S)_{L^2(\Gamma_N)} \quad \forall(\eta_S,\kappa_S) \in S^{\mathbf{p}}. \tag{4.100}$$

The norm for functions $(\varphi,\sigma) \in \mathbf{H}$ is given by $\|(\varphi,\sigma)\|_{\mathbf{H}} := \|\varphi\|_{\tilde{H}^{-1/2}(\Gamma_D)} + \|\sigma\|_{\tilde{H}^{1/2}(\Gamma_N)}$. Once more the unique solvability of the boundary element discretization of the integral equation follows from the $\mathbf{H}$-ellipticity (3.112) of the bilinear form b_{mixed}, and from the Galerkin orthogonality of the error, we have the quasi-optimality.

Theorem 4.1.53. *Let $(\varphi,\sigma) \in \mathbf{H}$ be the exact solution of (4.97). The discretization (4.100) has a unique solution $(\varphi_S,\sigma_S) \in S^{\mathbf{p}}$, $\mathbf{p} = (p_1, p_2)$, which converges quasi-optimally:*

$$\|(\varphi,\sigma) - (\varphi_S,\sigma_S)\|_{\mathbf{H}} \leq C_1 \min_{(\eta,\kappa)\in S^{\mathbf{p}}} \|(\varphi,\sigma) - (\eta,\kappa)\|_{\mathbf{H}} . \tag{4.101a}$$

If the exact solution satisfies $(\varphi, \sigma) \in H^s_{\text{pw}}(\Gamma_D) \times H^t_{\text{pw}}(\Gamma_N)$ *for* $s, t \geq 0$ *we have the quantitative estimate*

$$\|(\varphi, \sigma) - (\varphi_S, \sigma_S)\|_{\mathbf{H}} \leq C_2 \Big(h^{\min\{s, p_1+1\}+\frac{1}{2}} \|\varphi\|_{H^s_{\text{pw}}(\Gamma_D)} + h^{\min\{t, p_2+1\}-\frac{1}{2}} \|\sigma\|_{H^t_{\text{pw}}(\Gamma_N)} \Big). \qquad (4.101\text{b})$$

Here the constant C_2 *depends only on* C_1 *in (4.101a), the shape-regularity (see Definition 4.1.12) of the surface meshes* $\mathcal{G}_D$, $\mathcal{G}_N$ *and the polynomial degrees* p_1 *and* p_2.

Proof. For the proof we only need to show the approximation property on the boundary pieces Γ_D and Γ_N. Here we use (4.59) on Γ_D and (4.93) on Γ_N for a sufficiently large $t > 1$. Hence the interpolation $I^p_{\mathcal{G}} \varphi$ in (4.93) is well defined and we have $\varphi|_{\partial\Gamma_N} = I^p_{\mathcal{G}} \varphi|_{\partial\Gamma_N} = 0$. Therefore the zero extension of the difference function satisfies $\left(\varphi - I^p_{\mathcal{G}}\varphi\right)^\star \in H^{1/2}(\Gamma)$ and from (4.93) with $s = 0, 1$ we have:

$$\begin{aligned}\|\left(\varphi - I^p_{\mathcal{G}}\varphi\right)^\star\|_{L^2(\Gamma)} &= \|\varphi - I^p_{\mathcal{G}}\varphi\|_{L^2(\Gamma_N)} \leq C h^{\min(t, p+1)} \|\varphi\|_{H^t_{\text{pw}}(\Gamma_N)},\\ \|\left(\varphi - I^p_{\mathcal{G}}\varphi\right)^\star\|_{H^1(\Gamma)} &= \|\varphi - I^p_{\mathcal{G}}\varphi\|_{H^1(\Gamma_N)} \leq C h^{\min(t, p+1)-1} \|\varphi\|_{H^t_{\text{pw}}(\Gamma_N)}.\end{aligned} \qquad (4.102)$$

Then, by interpolation as in the proof of Theorem 4.1.51 and by the boundedness of the Galerkin projection (see Remark 4.1.27), (4.101b) follows. □

4.1.11 Model Problem 4: Screen Problems*

In this section we will discuss the Galerkin boundary element method for the screen problem from Sect. 3.5.3, which is due to [219].

Hence we again assume that an open manifold Γ_0 is given, which can be extended to a closed Lipschitz surface Γ in $\mathbb{R}^3$ in such a way that we have for $\Gamma_0^c = \Gamma \backslash \overline{\Gamma}_0$

$$\Gamma = \Gamma_0 \cup \Gamma_0^c.$$

In order to avoid technical difficulties, we require that Γ_0 and Γ_0^c be simply connected. We have already introduced the integral equations for the Dirichlet and Neumann screen problems in Sect. 3.5.3:

Dirichlet Screen Problem: For a given $g_D \in H^{1/2}(\Gamma_0)$ find $\varphi \in \widetilde{H}^{-1/2}(\Gamma_0)$ such that

$$(V\varphi, \eta)_{L^2(\Gamma_0)} = (g_D, \eta)_{L^2(\Gamma_0)} \qquad \forall \eta \in \widetilde{H}^{-1/2}(\Gamma_0). \qquad (4.103)$$

* This section should be read as a complement to the core material of this book.

Neumann Screen Problem: For a given $g_N \in H^{-1/2}(\Gamma_0)$ find $\sigma \in \widetilde{H}^{1/2}(\Gamma_0)$ such that

$$(W\sigma, \kappa)_{L^2(\Gamma_0)} = (g_N, \kappa)_{L^2(\Gamma_0)} \qquad \forall \kappa \in \widetilde{H}^{1/2}(\Gamma_0). \tag{4.104}$$

The Galerkin BEM for (4.103) and (4.104) are based on a regular mesh $\mathcal{G}$ of Γ_0 and a boundary element space of polynomial degree $p_1 \geq 0$ for the Dirichlet problem (4.103) and $p_2 \geq 1$ for the Neumann problem (4.104).
Dirichlet Screen Problem: For a given $g_D \in H^{1/2}(\Gamma_0)$ find $\varphi_S \in S_{\mathcal{G}}^{p_1,-1}$ such that

$$(V\psi_S, \eta_S)_{L^2(\Gamma_0)} = (g_D, \eta_S)_{L^2(\Gamma_0)} \qquad \forall \eta_S \in S_{\mathcal{G}}^{p_1,-1}. \tag{4.105}$$

Neumann Screen Problem: For a given $g_N \in H^{-1/2}(\Gamma_0)$ find $\sigma_S \in S_{\mathcal{G},0}^{p_2,0}$ such that

$$(W\sigma_S, \kappa_S)_{L^2(\Gamma_0)} = (g, \kappa)_{L^2(\Gamma_0)} \qquad \forall \kappa \in S_{\mathcal{G},0}^{p_2,0}. \tag{4.106}$$

Note that in $S_0^{p_2,0}$ the boundary data of σ_S on $\partial\Gamma_0$ is set to zero (see Remark 4.1.52). With the ellipticity from Theorem 3.5.9 we immediately have the quasi-optimality of the discretization.

Theorem 4.1.54. *Equations (3.116), (3.117) as well as (4.105), (4.106) have a unique solution and the Galerkin solutions converge quasi-optimally:*

$$\|\psi - \psi_S\|_{\tilde{H}^{-1/2}(\Gamma_0)} \leq C \min_{\eta_S \in S_{\mathcal{G}}^{p_1,-1}} \|\psi - \eta_S\|_{\tilde{H}^{-1/2}(\Gamma_0)}, \tag{4.107a}$$

$$\|\sigma - \sigma_S\|_{\tilde{H}^{1/2}(\Gamma_0)} \leq C \min_{\kappa_S \in S_{\mathcal{G},0}^{p_2,0}} \|\sigma - \kappa_S\|_{\tilde{H}^{1/2}(\Gamma_0)}. \tag{4.107b}$$

If the exact solution of the Dirichlet problem (3.116) is contained in $H^s_{\mathrm{pw}}(\Gamma_0)$ for an $s \geq 0$ we have

$$\|\psi - \psi_S\|_{\tilde{H}^{-\frac{1}{2}}(\Gamma_0)} \leq C_1\, h^{\min(s,p_1+1)+\frac{1}{2}} \|\psi\|_{H^s_{\mathrm{pw}}(\Gamma_0)}. \tag{4.108a}$$

If the exact solution of the Neumann problem is contained in $H^t_{\mathrm{pw}}(\Gamma_0)$ for a $t > 1/2$ we have

$$\|\sigma - \sigma_S\|_{\tilde{H}^{1/2}(\Gamma_0)} \leq C_2\, h^{\min(t,p_2+1)-\frac{1}{2}} \|\sigma\|_{H^t_{\mathrm{pw}}(\Gamma_0)}. \tag{4.108b}$$

Here the constants C_1, C_2 depend only on the respective constant C in (4.107), the shape-regularity (see Definition 4.1.12) of the mesh and the polynomial degrees p_1 and p_2.

Remark 4.1.55. *In general, the exact solutions of the screen problems have edge singularities and therefore they do not have a very high order of regularity s or t in (4.108). Therefore the convergence rates of the Galerkin solutions in (4.108)* *are low, even for higher order discretizations. This problem can be overcome by an anisotropic mesh refinement near $\partial\Gamma_0$. For details we refer to [221].*

4.2 Convergence of Abstract Galerkin Methods

All boundary integral operators in Chap. 4.1 were elliptic, which allowed the use of the Lax–Milgram lemma to prove existence and uniqueness. As we have already seen with the Helmholtz problem, however, in certain practical cases we encounter indefinite boundary integral operators. Here we will show for very general subspaces and especially for non-symmetric and non-elliptic sesquilinear forms, under which circumstances the Galerkin solution $u_S \in S$ exists and the error converges quasi-optimally. An early study on this subject can be found in [223]. For a study on the convergence of general boundary element methods we refer to [215].

4.2.1 Abstract Variational Problem

We would first like to recall the abstract framework from Sect. 2.1.6 and, again, refer, e.g., to [9, Chap. 5], [151, 166, 174] as standard references and additional material.

Let H_1, H_2 be Hilbert spaces and $a(\cdot,\cdot) : H_1 \times H_2 \to \mathbb{C}$ a continuous sesquilinear form:

$$\|a\| = \sup_{u\in H_1\backslash\{0\}} \sup_{v\in H_2\backslash\{0\}} \frac{|a(u,v)|}{\|u\|_{H_1}\|v\|_{H_2}} < \infty, \tag{4.109}$$

and let the (continuous) inf–sup conditions hold: There exists a constant $\gamma > 0$ such that

$$\inf_{u\in H_1\backslash\{0\}} \sup_{v\in H_2\backslash\{0\}} \frac{|a(u,v)|}{\|u\|_{H_1}\|v\|_{H_2}} \geq \gamma > 0, \tag{4.110a}$$

and we have

$$\forall v \in H_2\backslash\{0\} : \sup_{u\in H_1} |a(u,v)| > 0. \tag{4.110b}$$

Then for every functional $F \in H_2'$ the problem

$$\text{Find } u \in H_1 : \qquad a(u,v) = F(v) \qquad \forall v \in H_2 \tag{4.111}$$

has a unique solution, which satisfies

$$\|u\|_{H_1} \leq \frac{1}{\gamma}\|F\|_{H_2'}. \tag{4.112}$$

4.2.2 Galerkin Approximation

We require the following construction of approximating subspaces for the definition of the Galerkin method, which we use to solve (4.111).

For $i = 1, 2$, let $\left(S_\ell^i\right)_{\ell\in\mathbb{N}}$ be given sequences of finite-dimensional, nested subspaces of H_i whose union is dense in H_i

$$\forall \ell \geq 0 : S_\ell^i \subset S_{\ell+1}^i, \quad \dim S_\ell^i < \infty \quad \text{and} \quad \overline{\bigcup_{\ell\in\mathbb{N}} S_\ell^i}^{\|\cdot\|_{H_i}} = H_i, \qquad i = 1, 2 \tag{4.113}$$

and whose respective dimensions satisfy the conditions

$$\begin{aligned} &N_\ell := \dim S_\ell^1 = \dim S_\ell^2 < \infty, \ \ \forall \ell \in \mathbb{N} : N_\ell < N_{\ell+1}, \\ &N_\ell \to \infty \quad \text{for } \ell \to \infty. \end{aligned} \tag{4.114}$$

Since the dimensions of S_ℓ^1 and S_ℓ^2 are equal, it follows that the system matrix for the boundary element method is square.

The density implies the *approximation property*

$$\forall u_i \in H_i : \qquad \lim_{\ell\to\infty} \min\{\|u_i - v\|_{H_i} : v \in S_\ell^i\} = 0. \tag{4.115}$$

Every u_i in H_i can thus be approximated by a sequence $v_\ell^i \in S_\ell^i$. In Sect. 4.1 we have already encountered the spaces $S_\mathcal{G}^{p,0}$ and $S_\mathcal{G}^{p,-1}$, and one obtains a sequence of boundary element spaces by, for example, successively refining an initially coarse mesh $\mathcal{G}_0$.

With the subspaces $\left(S_\ell^i\right)_{\ell\in\mathbb{N}} \subset H_i$ the Galerkin discretization of (4.111) is given by: Find $u_\ell \in S_\ell^1$ such that

$$a(u_\ell, v_\ell) = F(v_\ell) \qquad \forall v_\ell \in S_\ell^2. \tag{4.116}$$

A solution of (4.116) is called a *Galerkin solution*. The existence and uniqueness of the Galerkin solution is proven in the following theorem.

Theorem 4.2.1. *(i) For every functional $F \in H_2'$, (4.116) has a unique solution $u_\ell \in S_\ell^1$ if the discrete inf–sup condition*

$$\inf_{u\in S_\ell^1\setminus\{0\}} \sup_{v\in S_\ell^2\setminus\{0\}} \frac{|a(u,v)|}{\|u\|_{H_1}\|v\|_{H_2}} \geq \gamma_\ell \tag{4.117}$$

holds with a stability constant $\gamma_\ell > 0$ and if

$$\forall v \in S_\ell^2 \setminus \{0\} : \qquad \sup_{u\in S_\ell^1} |a(u,v)| > 0 \tag{4.118}$$

is satisfied.

(ii) *For all ℓ let (4.118) and (4.117) be satisfied with $\gamma_\ell > 0$. Then the sequence $(u_\ell)_\ell \subset H_1$ of Galerkin solutions satisfies the error estimate*

$$\|u - u_\ell\|_{H_1} \leq \left(1 + \frac{\|a\|}{\gamma_\ell}\right) \min_{v \in S_\ell^1} \|u - v\|_{H_1}. \tag{4.119}$$

Proof. Statement (i) follows from Theorem 2.1.44.

For (ii): The difference between (4.116) and (4.111) with $S_\ell^2 \subset H_2$ yields the *Galerkin orthogonality* of the error:

$$a(u - u_\ell, v) = 0 \quad \forall v \in S_\ell^2. \tag{4.120}$$

Owing to the discrete inf–sup condition (4.117) we have

$$\begin{aligned}
\gamma_\ell \|u_\ell\|_{H_1} &\leq \sup_{v \in S_\ell^2 \backslash \{0\}} \frac{|a(u_\ell, v)|}{\|v\|_{H_2}} = \sup_{v \in S_\ell^2 \backslash \{0\}} \frac{|F(v)|}{\|v\|_{H_2}} \\
&\leq \sup_{v \in H_2 \backslash \{0\}} \frac{|F(v)|}{\|v\|_{H_2}} = \sup_{v \in H_2 \backslash \{0\}} \frac{|a(u, v)|}{\|v\|_{H_2}} \leq \|a\| \, \|u\|_{H_1}.
\end{aligned}$$

This means that the statement $Q_\ell u := u_\ell$ defines a linear mapping $Q_\ell : H_1 \to S_\ell^1$ with $\|Q_\ell\|_{H_1 \leftarrow H_1} \leq \|a\|/\gamma_\ell$. For all $w \in S_\ell^1 \subset H_1$ it follows from (4.117) and (4.120) that we have the estimate

$$\|w - Q_\ell w\|_{H_1} \leq \frac{1}{\gamma_\ell} \sup_{v \in S_\ell^2 \backslash \{0\}} \frac{|a(w - Q_\ell w, v)|}{\|v\|_{H_2}} = 0,$$

from which we have the projection property:

$$\forall w \in S_\ell^1 : \qquad Q_\ell w = w.$$

It then follows for all $w \in S_\ell^1 \subset H_1$, that

$$\begin{aligned}
\|u - u_\ell\|_{H_1} &\leq \|u - w\|_{H_1} + \|w - Q_\ell u\|_{H_1} \\
&= \|u - w\|_{H_1} + \|Q_\ell (u - w)\|_{H_1} \\
&\leq \left(1 + \frac{\|a\|}{\gamma_\ell}\right) \|u - w\|_{H_1}.
\end{aligned}$$

Since $w \in S_\ell^1$ was arbitrary, we have proven (4.119). □

Remark 4.2.2. *(i) The Galerkin method (4.116) is called uniformly stable if there exists a constant $\gamma > 0$ that is independent of ℓ such that $\gamma_\ell \geq \gamma > 0$. In this case (4.119) implies the quasi-optimal convergence of the Galerkin solution.*

(ii) *The subspaces S_ℓ^1 and S_ℓ^2 contain different functions: S_ℓ^1 serves to approximate the solution and guarantees the consistency, while S_ℓ^2 guarantees the stability, because of the discrete inf–sup condition [which is equivalent to (4.117)]*

$$\forall u \in S_\ell^1 : \qquad \sup_{v \in S_\ell^2 \setminus \{0\}} \frac{|a(u,v)|}{\|v\|_{H_2}} \geq \gamma_\ell \|u\|_{H_1}. \tag{4.121}$$

Remark 4.2.3. *In Sect. 4.1 we have seen that for the integral equations for the Laplace problem we can always choose $S_\ell^1 = S_\ell^2$. The same property holds for the integral equation formulation of the Helmholtz equation.*

Remark 4.2.4. *Equations (4.117) and (4.118) are equivalent to the conditions*

$$\inf_{v \in S_\ell^2 \setminus \{0\}} \sup_{u \in S_\ell^1 \setminus \{0\}} \frac{|a(u,v)|}{\|u\|_{H_1} \|v\|_{H_2}} \geq \gamma_\ell^* \tag{4.122}$$

with $\gamma_\ell^* > 0$ and

$$\forall u \in S_\ell^1 \setminus \{0\} : \qquad \sup_{v \in S_\ell^2} |a(u,v)| > 0. \tag{4.123}$$

Remark 4.2.5. *For $H_1 = H_2 = H$ and $S_\ell^1 = S_\ell^2 = S_\ell$, (4.117) implies the condition (4.122) with $\gamma_\ell^* = \gamma_\ell$ and vice-versa.*

The Galerkin method (4.116) is equivalent to a linear system of equations. To see this we need to choose bases $\left(b_j^i\right)_{j=1}^{N_\ell}$ of S_ℓ^i, $i = 1, 2$:

$$S_\ell^1 = \operatorname{span}\{b_j^1 : j = 1, \ldots, N_\ell\}, \qquad S_\ell^2 = \operatorname{span}\{b_j^2 : j = 1, \ldots, N_\ell\}.$$

Therefore every $u \in S_\ell^1$ and $v \in S_\ell^2$ has a unique basis representation

$$u = \sum_{j=1}^{N_\ell} u_j b_j^1, \qquad v_\ell = \sum_{j=1}^{N_\ell} v_j\, b_j^2. \tag{4.124}$$

If we insert (4.124) into (4.116) we obtain:

$$\begin{aligned} &\forall v \in S_\ell^2 : \; a(u,v) - F(v) = 0 \Longrightarrow \\ &\forall \mathbf{v} = \left(v_j\right)_{j=1}^{N_\ell} \in \mathbb{C}^{N_\ell} : \quad \sum_{j=1}^{N_\ell} \overline{v}_j \left(\left\{ \sum_{k=1}^{N_\ell} u_k\, a(b_k^1, b_j^2) \right\} - F(b_j^2) \right) = 0 \Longrightarrow \\ &\mathbf{K}_\ell \mathbf{u} = \mathbf{F}_\ell, \end{aligned} \tag{4.125}$$

where the matrix $\mathbf{K}_\ell$ and the vectors $\mathbf{u}$, $\mathbf{F}_\ell$ are given by $\mathbf{u} = \left(u_j\right)_{j=1}^{N_\ell}$ and

$$\left.\begin{array}{ll}(\mathbf{K}_\ell)_{j,k} & := a(b_k^1, b_j^2) \\ (\mathbf{F}_\ell)_j & := F(b_j^2)\end{array}\right\} \qquad 1 \leq j,k \leq N_\ell.$$

The linear system of equations in (4.125) is the basis representation of (4.116). In engineering literature the system matrix $\mathbf{K}_\ell$ is also called the stiffness matrix of the Galerkin method (4.116) and the vector $\mathbf{F}_\ell$ on the right-hand side is called the load vector.

Proposition 4.2.6. *The stiffness matrix $\mathbf{K}_\ell$ in (4.125) is non-singular if and only if we have (4.121) with $\gamma_\ell > 0$.*

Proof. Let $\mathbf{K}_\ell$ be singular. Then there exists a vector $\mathbf{u} = \left(u_j\right)_{j=1}^{N_\ell} \in \mathbb{C}^{N_\ell} \backslash \{0\}$ with $\mathbf{K}_\ell \mathbf{u} = \mathbf{0}$. Since $\left(b_j^1\right)_{j=1}^{N_\ell}$ is a basis of S_ℓ^1 we have for the associated function $u = \sum_{j=1}^{N_\ell} u_j b_j^1 \neq 0$. It follows from (4.125) that $a(u_\ell, v_\ell) = 0$ for all $v_\ell \in S_\ell^2$. This is a contradiction to (4.121) with $\gamma_\ell > 0$.

The inverse statement is proven in the same way. □

4.2.3 Compact Perturbations

Boundary integral operators often appear in the form

$$(A + T)u = F \tag{4.126}$$

with a principal part $A \in L(H, H')$ for which the associated sesquilinear form $a(\cdot,\cdot) : H \times H \to \mathbb{C}$ satisfies the inf–sup conditions

$$\inf_{u \in H\backslash\{0\}} \sup_{v \in H\backslash\{0\}} \frac{|a(u,v)|}{\|u\|_H \, \|v\|_H} \geq \gamma > 0, \tag{4.127}$$

$$\forall v \in H\backslash\{0\} : \qquad \sup_{u \in H} |a(u,v)| > 0 \tag{4.128}$$

and a compact operator $T \in L(H, H')$. Let $t : H \times H \to \mathbb{C}$ be the sesquilinear form that is associated with T. The variational formulation:
Find $u \in H$ such that

$$a(u,v) + t(u,v) = F(v) \qquad \forall v \in H \tag{4.129}$$

is equivalent to (4.126).

The discretization of the variational problem (4.129) is based on a dense sequence of finite-dimensional subspaces $(S_\ell)_{\ell \in \mathbb{N}}$ in H:

For a given $F \in H'$ find $u_\ell \in S_\ell$ such that

$$a(u_\ell, v_\ell) + t(u_\ell, v_\ell) = F(v_\ell) \qquad \forall v_\ell \in S_\ell. \tag{4.130}$$

The following theorem states that the inf–sup condition for the principal part of the sesquilinear form together with the injectivity of the operator $A + T$ ensure well posedness of the continuous problem. Furthermore, the discrete inf–sup conditions for a dense sequence of subspaces imply (a) the well-posedness of the discrete problem, (b) the unique solvability of the continuous problem, and (c) the convergence of the Galerkin solutions to the continuous solution.

Theorem 4.2.7. *Let (4.127) and (4.128) hold, let $T \in L(H, H')$ be compact and $A + T$ injective,*

$$(A + T)u = 0 \Longrightarrow u = 0. \tag{4.131}$$

Then problem (4.126) has a unique solution $u \in H$ for every $F \in H'$.

Furthermore, let $(S_\ell)_\ell$ be a dense sequence of finite-dimensional subspaces in H and $t(\cdot,\cdot)$ the sesquilinear form associated with the compact operator T. We assume that there exist an $\ell_0 > 0$ and a $\gamma > 0$ such that for all $\ell \geq \ell_0$ the discrete inf–sup conditions

$$\inf_{u_\ell \in S_\ell \backslash \{0\}} \sup_{v_\ell \in S_\ell \backslash \{0\}} \frac{|a(u_\ell, v_\ell) + t(u_\ell, v_\ell)|}{\|u_\ell\|_H \|v_\ell\|_H} \geq \gamma \tag{4.132a}$$

and

$$\inf_{v_\ell \in S_\ell \backslash \{0\}} \sup_{u_\ell \in S_\ell \backslash \{0\}} \frac{|a(u_\ell, v_\ell) + t(u_\ell, v_\ell)|}{\|u_\ell\|_H \|v_\ell\|_H} \geq \gamma \tag{4.132b}$$

are satisfied uniformly with respect to ℓ. Then we have:

(i) For all $F \in H'$ and all $\ell \geq \ell_0$ the Galerkin equations (4.130) have a unique solution u_ℓ.
(ii) The Galerkin solutions u_ℓ converge for $\ell \to \infty$ to the unique solution $u \in H$ of the problem (4.126) and satisfy the quasi-optimal error estimate

$$\|u - u_\ell\|_H \leq C \min\{\|u - v_\ell\|_H : v_\ell \in S_\ell\}, \qquad \ell \geq \ell_0$$

with a constant $C > 0$ which is independent of ℓ.

Proof. As $a(\cdot,\cdot)$ satisfies the inf–sup conditions, the associated operator $A : H \to H'$ is an isomorphism with $\|A\|_{H' \leftarrow H} \leq \gamma^{-1}$ [see (2.38)]. Hence (4.126) is equivalent to the Fredholm equation

$$\left(I + A^{-1}T\right)u = A^{-1}f$$

with the compact operator $A^{-1}T : H \to H$ (see Lemma 2.1.29). By (4.131), -1 is not an eigenvalue of $A^{-1}T$ and, from the Fredholm alternative (Theorem 2.1.36), $I + A^{-1}T$ is an isomorphism $\left\|I + A^{-1}T\right\|_{H \leftarrow H} \leq C$. This yields the unique solvability of (4.126) and the continuous dependence on the data.

of (i): Theorem 2.1.44 implies both (i) and the fact that the Galerkin solution depends continuously on the data:

$$\|u_\ell\|_H \leq \frac{1}{\gamma} \|F\|_{H'}. \tag{4.133}$$

of (ii): Let

$$b(u, v) := a(u, v) + t(u, v).$$

Because of (4.133) the sequence $(u_\ell)_\ell$ of Galerkin solutions is uniformly bounded in H. Theorem 2.1.26 thus guarantees the existence of a subsequence $u_{\ell_i} \rightharpoonup u \in H$ that converges weakly in H (in the following we will again denote this sequence by u_ℓ). We will now show that, with this limit u, $b(u, v) = F(v)$ for all $v \in H$. For an arbitrary $v \in H$, $P_\ell v \in S_\ell$ denotes the orthogonal projection:

$$\forall w_\ell \in S_\ell : (v - P_\ell v, w_\ell)_H = 0.$$

Then we have

$$\begin{aligned}|b(u,v) - F(v)| \leq{}& \underbrace{|b(u,v) - b(u_\ell, v)|}_{T_1} + \underbrace{|b(u_\ell, v) - b(u_\ell, P_\ell v)|}_{T_2} \\ &+ \underbrace{|b(u_\ell, P_\ell v) - F(P_\ell v)|}_{T_3} + \underbrace{|F(P_\ell v) - F(v)|}_{T_4}.\end{aligned}$$

For a fixed $v \in H$

$$b(\cdot, v) : H \to \mathbb{C}$$

defines a continuous functional in H'. The definition of weak convergence then yields the convergence of T_1 to 0 for $\ell \to \infty$.

Since $\bigcup_\ell S_\ell$ is dense in H, according to the conditions, we consequently have the consistency of the discretization sequence

$$\|u - P_\ell u\|_H = \inf_{v_\ell \in S_\ell} \|u - v_\ell\|_H \overset{\ell\to\infty}{\to} 0. \tag{4.134}$$

Thus we have for T_4

$$|T_4| = |F(v - P_\ell v)| \leq \|F\|_{H'} \|v - P_\ell v\|_H \overset{\ell\to\infty}{\to} 0.$$

Since $(u_\ell)_\ell$ is uniformly bounded, we have

$$|T_2| \leq (\|A\|_{H' \leftarrow H} + \|T\|_{H' \leftarrow H}) \|u_\ell\|_H \|v - P_\ell v\|_H,$$

and the consistency again implies that $T_2 \to 0$ for $\ell \to \infty$. Finally, we have $T_3 = 0$ since $b(u_\ell, v_\ell) = F(v_\ell)$ for all $v_\ell \in S_\ell$. Therefore u is a solution of (4.126). By (4.131), u is unique.

We have thus shown the unique solvability of Problem (4.126) in H.

By (4.132), $b(\cdot,\cdot)$ satisfies the conditions of Theorem 4.2.1 for $\ell \geq \ell_0$, from which we obtain the quasi-optimality.

□

Remark 4.2.8. *Theorem 4.2.7 only holds if the discrete inf–sup conditions (4.132) are satisfied. In general, the discrete inf–sup conditions do not follow from the density of $(S_\ell)_\ell$ in H combined with (4.127) and (4.128). Instead, they have to be verified for each specific problem.*

In applications concerning boundary integral equations we often encounter the following special case of Theorem 4.2.7.

Theorem 4.2.9. *Let H be a Hilbert space and $(S_\ell)_\ell$ a dense sequence of finite-dimensional subspaces in H. We assume that for the sesquilinear forms $a(\cdot,\cdot)$ and $t(\cdot,\cdot)$ of the variational problem (4.129) we have*

(i) $a(\cdot,\cdot)$ satisfies the ellipticity condition (2.44), i.e., there exists a constant $\alpha > 0$ such that

$$\forall u \in H : \qquad |a(u,u)| \geq \alpha \|u\|_H^2. \tag{4.135}$$

(ii) The operator $T \in L(H, H')$ that is associated with the sesquilinear form $t(\cdot,\cdot) : H \times H \to \mathbb{C}$ is compact.

(iii) We assume that, for $F = 0$, (4.129) only has the trivial solution:

$$\forall v \in H \backslash \{0\} : \qquad a(u,v) + t(u,v) = 0 \Longrightarrow u = 0. \tag{4.136}$$

Then the variational problem (4.129) has a unique solution $u \in H$ for every $F \in H'$.

There exists a constant $\ell_0 > 0$ such that for all $\ell \geq \ell_0$ the Galerkin equations (4.130) have a unique solution $u_\ell \in S_\ell$. The sequence $(u_\ell)_\ell$ of the Galerkin solutions converges to u and, for $\ell \geq \ell_0$, satisfies the quasi-optimal error estimate

$$\|u - u_\ell\|_H \leq C \min_{v_\ell \in S_\ell} \|u - v_\ell\|_H \tag{4.137}$$

with a constant C which is independent of ℓ.

Proof. The H-ellipticity of $a(\cdot,\cdot)$ implies the inf-sup condition (4.127), (4.128), and therefore the unique solvability of (4.129) follows from Theorem 4.2.7.

Now we will turn our attention to the Galerkin equations and prove the inf-sup condition for a sufficiently large ℓ.

We set $b(\cdot,\cdot) = a(\cdot,\cdot) + t(\cdot,\cdot)$ and define the associated operators $B : H \to H'$ and $B_\ell : S_\ell \to S'_\ell$ by

$$\forall\, u, v \in H : \langle Bu, v\rangle_{H'\times H} := b\,(u, v) \quad \text{and}$$
$$\forall\, u_\ell, v_\ell \in S_\ell : \langle B_\ell u_\ell, v_\ell\rangle_{S'_\ell \times S_\ell} := b\,(u_\ell, v_\ell)\,.$$

The norm of $B_\ell u_\ell \in S'_\ell$ is given by

$$\|B_\ell u_\ell\|_{S'_\ell} = \sup_{v_\ell \in S_\ell \backslash \{0\}} \frac{|b\,(u_\ell, v_\ell)|}{\|v_\ell\|_H}$$

and the discrete inf-sup condition (4.132a) is equivalent to

$$\forall u_\ell \in S_\ell \text{ with } \|u_\ell\|_H = 1 \text{ we have: } \exists \ell_0 > 0 \text{ s.t. } \|B_\ell u_\ell\|_{S'_\ell} \geq \gamma \quad \forall \ell \geq \ell_0.$$

We will prove this statement by contradiction by using the conditions given in the theorem. For this we assume:

$$\exists\, (w_\ell)_{\ell\in\mathbb{N}} \text{ with } w_\ell \in S_\ell \text{ and } \|w_\ell\|_H = 1 \text{ such that: } \|B_\ell w_\ell\|_{S'_\ell} \to 0 \text{ for } \ell \to \infty. \tag{4.138}$$

As $(w_\ell)_\ell$ is bounded in H there exists, according to Theorem 2.1.26, a weakly convergent subsequence (which we again denote by $(w_\ell)_\ell$) such that $w_\ell \rightharpoonup w \in H$.

For all $v \in H$, $b\,(\cdot, v)$ defines a continuous, linear functional on H and so we have

$$\forall v \in H : b\,(w_\ell, v) \to b\,(w, v) \quad \text{for } \ell \to \infty.$$

It follows that

$$\|Bw\|_{H'} = \sup_{v \in H\backslash\{0\}} \frac{|b\,(w, v)|}{\|v\|_H} = \sup_{v \in H\backslash\{0\}} \lim_{\ell\to\infty} \frac{|b\,(w_\ell, v)|}{\|v\|_H}. \tag{4.139}$$

In the following we will estimate the numerator on the right-hand side and for this purpose we use the decomposition

$$b\,(w_\ell, v) = b\,(w_\ell, v_\ell) + b\,(w_\ell, v - v_\ell) \tag{4.140}$$

with the H-orthogonal projection $v_\ell = P_\ell v \in S_\ell$. From assumption (4.138) we have

$$|b\,(w_\ell, v_\ell)| \leq \|B_\ell w_\ell\|_{S'_\ell} \|v_\ell\|_H \leq \|B_\ell w_\ell\|_{S'_\ell} \|v\|_H \overset{\ell\to\infty}{\to} 0.$$

The fact that the spaces S_ℓ are dense in H yields for the second term in (4.140)

$$|b\,(w_\ell, v - v_\ell)| \leq \|b\|\, \|w_\ell\|_H\, \|v - v_\ell\|_H \leq \|b\|\, \|v - v_\ell\|_H \overset{\ell\to\infty}{\to} 0.$$

Hence for all $v \in H$ we have the convergence $\lim_{\ell\to\infty} b\,(w_\ell, v) = 0$ and from (4.139) we have $Bw = 0$, which, combined with the injectivity of (4.136), finally gives us $w = 0$.

We will now show the strong convergence $w_\ell \to w$ and begin with the estimate

$$\alpha \|w - w_\ell\|_H^2 \leq |a(w - w_\ell, w - w_\ell)| = |a(w - w_\ell, w) - a(w, w_\ell) + a(w_\ell, w_\ell)| . \tag{4.141}$$

Since T is compact, there exists a subsequence (which we again denote by $(w_\ell)_{\ell \in \mathbb{N}}$) such that $T w_\ell \to T w$ in H'. This can be written in the form

$$\sup_{\substack{v \in H \\ \|v\|_H = 1}} |t(w_\ell, v) - t(w, v)| =: \delta_\ell \overset{\ell \to \infty}{\to} 0,$$

from which we deduce by using $\|w_\ell\|_H = 1$ that

$$|t(w_\ell, w_\ell) - t(w, w_\ell)| \leq \delta_\ell \|w_\ell\|_H = \delta_\ell \overset{\ell \to \infty}{\to} 0.$$

This result, combined with assumption (4.138), yields

$$0 \overset{\ell \to \infty}{\leftarrow} |b(w_\ell, w_\ell)| = |a(w_\ell, w_\ell) + t(w_\ell, w_\ell)| \leq |a(w_\ell, w_\ell) + t(w, w_\ell)| + \delta_\ell,$$

in other words:

$$a(w_\ell, w_\ell) = -t(w, w_\ell) + \tilde{\delta}_\ell \quad \text{with} \quad \lim_{\ell \to \infty} \tilde{\delta}_\ell = 0. \tag{4.142}$$

If we insert this into (4.141) we obtain

$$\alpha \|w - w_\ell\|_H^2 \leq \left| a(w - w_\ell, w) - b(w, w_\ell) + \tilde{\delta}_\ell \right| .$$

The first two terms on the right-hand side are equal to zero because of $w = 0$. We also determined $\lim_{\ell \to 0} \tilde{\delta}_\ell = 0$ in (4.142) so that we have proven $w_\ell \to w = 0$. This, however, is a contradiction to the assumption that $\|w_\ell\|_H = 1$.

Condition (4.132b) can be proven similarly.

The solvability of the Galerkin equation for $\ell \geq \ell_0$ and the error estimate (4.137) then follow from Theorem 4.2.7. □

4.2.4 *Consistent Perturbations: Strang's Lemma*

In this section we will consider variational formulations of boundary integral equations of abstract form:

Find $u \in H$ such that

$$b(u, v) = F(v) \qquad \forall v \in H \tag{4.143}$$

with $F \in H'$.

In general we assume that the sesquilinear form $b(\cdot,\cdot)$ is continuous and injective and that it satisfies a Gårding inequality.

Continuity:

$$\forall u, v \in H : \ |b(u,v)| \leq C_b \|u\|_H \|v\|_H . \tag{4.144}$$

Gårding Inequality:

$$\forall u \in H : |b(u,u) + (Tu,u)_{H' \times H}| \geq \alpha \|u\|_H^2 \tag{4.145}$$

with $\alpha > 0$ and a compact operator $T \in L(H, H')$.

Injectivity:

$$\forall v \in H \backslash \{0\} : \ b(u,v) = 0 \Longrightarrow u = 0. \tag{4.146}$$

Conditions (4.144)–(4.146) yield the prerequisites (i)–(iii) from Theorem 4.2.9 with $t(\cdot,\cdot) := -\langle T\cdot,\cdot\rangle_{H' \times H}$ and $a := b - t$. From Theorem 4.2.9 we derive the unique solvability of (4.143) as well as the stability (and thus the quasi-optimal convergence) of the Galerkin method as follows. For a dense sequence of finite-dimensional boundary element spaces $(S_\ell)_\ell$ in H there exists some $\ell_0 > 0$ such that for all $\ell \geq \ell_0$ the discrete inf–sup conditions

$$\begin{aligned} \inf_{u \in S_\ell \backslash \{0\}} \sup_{v \in S_\ell \backslash \{0\}} \frac{|b(u,v)|}{\|u\|_H \|v\|_H} \geq \gamma > 0 \\ \inf_{v \in S_\ell \backslash \{0\}} \sup_{u \in S_\ell \backslash \{0\}} \frac{|b(u,v)|}{\|u\|_H \|v\|_H} \geq \gamma > 0 \end{aligned} \tag{4.147}$$

hold, while $\gamma > 0$ is independent of ℓ. The Galerkin equations

$$\text{Find } u_\ell \in S_\ell : \qquad b(u_\ell, v) = F(v) \quad \forall v \in S_\ell \tag{4.148}$$

are, by Theorem 4.2.7, uniquely solvable for $\ell \geq \ell_0$ and we have

$$\|u - u_\ell\|_H \leq C \min_{v \in S_\ell} \|u - v\|_H . \tag{4.149}$$

In practical implementations of the Galerkin boundary element method in the form of a computer program it is usually not possible to realize the exact sesquilinear form $b(\cdot,\cdot)$. Instead, one usually uses an *approximative* sesquilinear form $b_\ell(\cdot,\cdot)$. Reasons for this are:

(a) The approximation of the system matrix by means of numerical integration
(b) The use of compressed, approximative representations of the Galerkin equations with cluster or wavelet methods,
(c) The approximation of the exact boundary Γ by means of, for example a polyhedral surface.

The perturbation of the sesquilinear form $b(\cdot,\cdot)$ as well as the functional F leads to the *perturbed Galerkin method*:

Find $\tilde{u}_\ell \in S_\ell$ such that

$$b_\ell(\tilde{u}_\ell, v) = F_\ell(v) \qquad \forall v \in S_\ell. \tag{4.150}$$

For the algorithmic realization of boundary element methods, one of the essential aims is to define the approximations (4.150) in such a way that the solutions $\tilde{u}_\ell$ exist, converge quasi-optimally and – in comparison with the computation of the exact Galerkin solution – can be calculated reasonably rapidly and with little use of computational memory. A sufficient condition in this respect is that the difference $b_\ell(\cdot,\cdot)-b(\cdot,\cdot)$ is "sufficiently small". We will specify this statement in the following.

For the Galerkin discretization we will generally assume in the following that we have chosen a dense sequence $(S_\ell)_\ell \subset H$ of subspaces of dimension $N_\ell := \dim S_\ell < \infty$ which satisfies (4.114).

Let sesquilinear forms $b_\ell : S_\ell \times S_\ell \to \mathbb{C}$ be defined for all $\ell \in \mathbb{N}$. These are *uniformly continuous* if there exists a constant $\widetilde{C}_b$ which is independent of ℓ such that

$$|b_\ell(u_\ell, v_\ell)| \leq \widetilde{C}_b \, \|u_\ell\|_H \, \|v_\ell\|_H \qquad \forall u_\ell, v_\ell \in S_\ell. \tag{4.151}$$

The forms b_ℓ satisfy the *stability condition* if there exists a null sequence $(c_\ell)_{\ell\in\mathbb{N}}$ such that

$$|b(u_\ell, v_\ell) - b_\ell(u_\ell, v_\ell)| \leq c_\ell \|u_\ell\|_H \, \|v_\ell\|_H \qquad \forall u_\ell, v_\ell \in S_\ell. \tag{4.152}$$

The stability condition will imply the existence of a unique solution of the perturbed Galerkin equations for a sufficiently large ℓ (see Theorem 4.2.11).

For the error estimate of the perturbed Galerkin solution we may measure the function u_ℓ on the right-hand side in (4.152) in a stronger norm (see Theorem 4.2.11). In this context $\|\cdot\|_U : S_\ell \to \mathbb{R}_{\geq 0}$ defines a *stronger* norm on S_ℓ if there exists a constant $C > 0$ independent of ℓ such that

$$\|u\|_H \leq C \, \|u\|_U \qquad \forall u \in S_\ell.$$

The perturbed sesquilinear forms $b_\ell : S_\ell \times S_\ell \to \mathbb{C}$ satisfy the *consistency condition with respect to a stronger norm* $\|\cdot\|_U$ if there exists a zero sequence $(\delta_\ell)_{\ell\in\mathbb{N}}$ such that

$$|b(u_\ell, v_\ell) - b_\ell(u_\ell, v_\ell)| \leq \delta_\ell \|u_\ell\|_U \, \|v_\ell\|_H \qquad \forall u_\ell, v_\ell \in S_\ell. \tag{4.153}$$

Remark 4.2.10. *(a) The stability condition and the continuity of $b(\cdot,\cdot)$ imply the uniform continuity of the sesquilinear form $b_\ell(\cdot,\cdot)$.*

(b) The consistency condition follows from the stability condition with $\delta_\ell = C c_\ell$.

(c) *In many practical applications the use of the stronger norm $\|\cdot\|_U$ in (4.153) permits the use of a zero sequence $(\delta_\ell)_\ell$ which converges more rapidly than in (4.152). The convergence rate of the perturbed Galerkin solution is influenced by $(\delta_\ell)_\ell$ and not by $(c_\ell)_\ell$.*

Theorem 4.2.11. *Let the sesquilinear form $b(\cdot,\cdot)$: $H \times H \to \mathbb{C}$ be continuous, injective and let it satisfy a Gårding inequality [see (4.144)–(4.146)]. Let the stability condition (4.152) be satisfied by the approximations b_ℓ.*

Then the perturbed Galerkin method (4.150) is stable. That is, there exist $\tilde{\gamma} > 0$, $\ell_0 > 0$ such that for all $\ell \geq \ell_0$ the discrete inf–sup conditions

$$\begin{aligned} \inf_{u_\ell \in S_\ell \setminus \{0\}} \sup_{v_\ell \in S_\ell \setminus \{0\}} \frac{|b_\ell(u_\ell, v_\ell)|}{\|u_\ell\|_H \, \|v_\ell\|_H} &\geq \tilde{\gamma}, \\ \inf_{v_\ell \in S_\ell \setminus \{0\}} \sup_{u_\ell \in S_\ell \setminus \{0\}} \frac{|b_\ell(u_\ell, v_\ell)|}{\|u_\ell\|_H \, \|v_\ell\|_H} &\geq \tilde{\gamma} \end{aligned} \tag{4.154}$$

hold. The perturbed Galerkin equations (4.150) have a unique solution for $\ell \geq \ell_0$.

If in addition the approximative sesquilinear forms are uniformly continuous and satisfy the consistency condition (4.153) the solutions $\tilde{u}_\ell$ satisfy the error estimate

$$\|u - \tilde{u}_\ell\|_H \leq C \left\{ \min_{w_\ell \in S_\ell} \left(\|u - w_\ell\|_H + \delta_\ell \|w_\ell\|_U\right) + \sup_{v_\ell \in S_\ell \setminus \{0\}} \frac{|F(v_\ell) - F_\ell(v_\ell)|}{\|v_\ell\|_H} \right\}. \tag{4.155}$$

Proof. According to the assumptions, the exact sesquilinear form $b(\cdot,\cdot)$ satisfies the inf–sup conditions (4.147) as well as the stability condition (4.149). We will verify (4.154). For this let $0 \neq u_\ell \in S_\ell \subset H$ be arbitrary. Then we have

$$\begin{aligned} \sup_{v_\ell \in S_\ell \setminus \{0\}} \frac{|b_\ell(u_\ell, v_\ell)|}{\|v_\ell\|_H} &\geq \sup_{v_\ell \in S_\ell \setminus \{0\}} \left(\frac{|b(u_\ell, v_\ell)|}{\|v_\ell\|_H} - \frac{|b(u_\ell, v_\ell) - b_\ell(u_\ell, v_\ell)|}{\|v_\ell\|_H} \right) \\ &\geq \gamma \|u_\ell\|_H - \sup_{v_\ell \in S_\ell} \frac{|b(u_\ell, v_\ell) - b_\ell(u_\ell, v_\ell)|}{\|v_\ell\|_H} \\ &\geq (\gamma - c_\ell) \, \|u_\ell\|_H . \end{aligned} \tag{4.156}$$

If we choose $\ell_0 > 0$ so that $c_\ell < \gamma$ for all $\ell \geq \ell_0$ we have verified the first condition in (4.154). The second condition can be verified in a similar way.

Combined with (4.154), it follows from Theorem 4.2.1(i) that the perturbed Galerkin equations (4.150) have a unique solution for $\ell \geq \ell_0$.

Next, we will prove the error estimate (4.155). Let $u_\ell \in S_\ell$ be the exact Galerkin solution from (4.148). For $\ell \geq \ell_0$ we have, according to (4.156), the following estimate for the perturbed Galerkin solution $\tilde{u}_\ell \in S_\ell$

$$\begin{aligned}
\|u-\tilde{u}_\ell\|_H &\le \|u-u_\ell\|_H + \|u_\ell-\tilde{u}_\ell\|_H \\
&\le \|u-u_\ell\|_H + (\gamma-c_\ell)^{-1} \sup_{v_\ell\in S_\ell\setminus\{0\}} \frac{|b_\ell(u_\ell-\tilde{u}_\ell, v_\ell)|}{\|v_\ell\|_H} \\
&= \|u-u_\ell\|_H + (\gamma-c_\ell)^{-1} \sup_{v_\ell\in S_\ell\setminus\{0\}} \frac{|b_\ell(u_\ell, v_\ell)-F_\ell(v_\ell)|}{\|v_\ell\|_H} \\
&\le \|u-u_\ell\|_H + (\gamma-c_\ell)^{-1} \sup_{v_\ell\in S_\ell\setminus\{0\}} \frac{|b_\ell(u_\ell, v_\ell)-b(u_\ell,v_\ell)| + |F(v_\ell)-F_\ell(v_\ell)|}{\|v_\ell\|_H}.
\end{aligned}$$

We consider the difference term $|b_\ell(u_\ell, v_\ell) - b(u_\ell, v_\ell)|$ and obtain, by using the continuity of b_ℓ and b as well as the consistency condition, for an arbitrary $w_\ell \in S_\ell$

$$\begin{aligned}
|b_\ell(u_\ell, v_\ell) - b(u_\ell, v_\ell)| &\le |b_\ell\,(u_\ell - w_\ell, v_\ell)| + |b_\ell\,(w_\ell, v_\ell) - b\,(w_\ell, v_\ell)| \\
&\quad + |b\,(w_\ell - u_\ell, v_\ell)| \\
&\le \widetilde{C}_b\,\|u_\ell - w_\ell\|_H\,\|v_\ell\|_H + \delta_\ell\,\|w_\ell\|_U\,\|v_\ell\|_H \\
&\quad + C_b\,\|w_\ell - u_\ell\|_H\,\|v_\ell\|_H.
\end{aligned}$$

From this we have

$$\sup_{v_\ell\in S_\ell\setminus\{0\}} \frac{|b_\ell(u_\ell, v_\ell) - b(u_\ell, v_\ell)|}{\|v_\ell\|_H} \le C \min_{w_\ell\in S_\ell} \left(\|u-w_\ell\|_H + \delta_\ell\,\|w_\ell\|_U\right).$$

With $c_\ell < \gamma$ and the consistency condition (4.153) we finally obtain

$$\|u-\tilde{u}_\ell\|_H \le C \min_{w_\ell\in S_\ell} \left\{ \|u-w_\ell\|_H + \frac{1}{\gamma-c_\ell}\left((\|u-w_\ell\|_H + \delta_\ell\|w_\ell\|_U + \sup_{v_\ell\in S_\ell\setminus\{0\}} \frac{|F(v_\ell)-F_\ell(v_\ell)|}{\|v_\ell\|_H}\right)\right\}. \tag{4.157}$$

□

Remark 4.2.12. *In connection with the boundary integral operator V for the single layer potential we have $H = H^{-1/2}(\Gamma)$. Since all the boundary element spaces we have considered so far are contained in $L^2(\Gamma)$, we can choose $\|\cdot\|_U = \|\cdot\|_{L^2(\Gamma)}$ as a stronger norm on S_ℓ. The term $\|w_\ell\|_{L^2(\Gamma)}$ on the right-hand side in (4.155) can be easily estimated if the boundary integral operator is L^2-regular, more specifically if $V^{-1} : H^1(\Gamma) \to L^2(\Gamma)$ is continuous. Let $u \in L^2(\Gamma)$ be the exact solution and $w_\ell := \Pi_\ell u$ the L^2-orthogonal projection of u onto the boundary element space S_ℓ. Then we have $\|w_\ell\|_{L^2(\Gamma)} \le \|u\|_{L^2(\Gamma)} \le C\,\|F\|_{H^1(\Gamma)}$ and, thus for a sufficiently large $\ell \ge \ell_0$:*

$$\|u-\tilde{u}_\ell\|_{H^{-1/2}(\Gamma)} \le C\left\{ \|u-\Pi_\ell u\|_{H^{-1/2}(\Gamma)} + \delta_\ell\|F\|_{H^1(\Gamma)} + \sup_{v_\ell\in S_\ell\setminus\{0\}} \frac{|F(v_\ell)-F_\ell(v_\ell)|}{\|v_\ell\|_{H^{-1/2}(\Gamma)}}\right\}.$$

From this we can deduce how the null sequence $(\delta_\ell)_\ell$ and the consistency of the approximation affect the right-hand side in the error estimate.

The error $\|u - \Pi_\ell u\|_{H^{-1/2}(\Gamma)}$ can be traced back to the approximation properties of S_ℓ. The choice $w_\ell = \Pi_\ell u$ yields, for an arbitrary $v_\ell \in S_\ell$

$$\|u - \Pi_\ell u\|_{H^{-1/2}(\Gamma)} = \sup_{v \in H^{1/2}(\Gamma)\setminus\{0\}} \frac{\left|(u - \Pi_\ell u, v)_{L^2(\Gamma)}\right|}{\|v\|_{H^{1/2}(\Gamma)}}$$
$$= \sup_{v \in H^{1/2}(\Gamma)\setminus\{0\}} \frac{\left|(u - \Pi_\ell u, v - v_\ell)_{L^2(\Gamma)}\right|}{\|v\|_{H^{1/2}(\Gamma)}}.$$

Now we take the infimum over all $v_\ell \in S_\ell$ and obtain

$$\|u - \Pi_\ell u\|_{H^{-1/2}(\Gamma)} \leq \left(\sup_{v \in H^{1/2}\setminus\{0\}} \inf_{v_\ell \in S_\ell \setminus\{0\}} \frac{\|v - v_\ell\|_{L^2(\Gamma)}}{\|v\|_{H^{1/2}(\Gamma)}} \right) \times \left(\inf_{w_\ell \in S_\ell} \|u - w_\ell\|_{L^2(\Gamma)} \right). \tag{4.158}$$

4.2.5 Aubin–Nitsche Duality Technique

Boundary integral equations were derived with the help of the integral equation method (direct and indirect method) for elliptic boundary value problems. In many cases our goal thus is to find the solution of the original boundary value problem by solving the boundary integral equation. The numerical solution of the boundary integral equation then only represents a part of the entire process. (Note, however, that with the direct method the boundary element method yields a quasi-optimal approximation of the unknown Cauchy data.) More importantly, the aim is to find the solution u of the original elliptic differential equation in the domain Ω. This solution can, as we will show here, be extracted from the Galerkin solution of the boundary integral equations with an increased convergence rate, a fact which stems from the representation formula.

Example 4.2.13 (Dirichlet Problem in the Interior, Ω). *Let $\Omega \subset \mathbb{R}^3$ be a bounded Lipschitz domain with boundary Γ and given Dirichlet data $g_D \in H^{1/2}(\Gamma)$. Find $u \in H^1(\Omega)$ such that*

$$\Delta u = 0 \quad \text{in } \Omega, \qquad u|_\Gamma = g_D. \tag{4.159}$$

The fundamental solution for the Laplace operator is given by $G(\mathbf{z}) := (4\pi\|\mathbf{z}\|)^{-1}$. The single layer potential $u(\mathbf{x}) = \int_\Gamma G(\mathbf{x} - \mathbf{y})\,\sigma(\mathbf{y})ds_y$, $\mathbf{x} \in \Omega$, leads to the boundary integral equation: Find $\sigma \in H^{-1/2}(\Gamma)$ such that

$$(V\sigma, \eta)_{L^2(\Gamma)} = (g_D, \eta)_{L^2(\Gamma)} \qquad \forall \eta \in H^{-1/2}(\Gamma), \tag{4.160}$$

where $(\cdot,\cdot)_{L^2(\Gamma)}$ again denotes the continuous extension of the L^2 inner-product to the dual pairing $\langle\cdot,\cdot\rangle_{H^{1/2}(\Gamma)\times H^{-1/2}(\Gamma)}$.

For a subspace $S_\ell \subset H^{-1/2}(\Gamma)$ the Galerkin approximation $\sigma_\ell \in S_\ell$ is defined by: Find $\sigma_\ell \in S_\ell$ such that

$$(V\sigma_\ell, \eta)_{L^2(\Gamma)} = (g_D, \eta)_{L^2(\Gamma)} \qquad \forall \eta \in S_\ell. \tag{4.161}$$

Equation (4.161) has a unique solution which satisfies the quasi-optimal error estimate

$$\|\sigma - \sigma_\ell\|_{H^{-1/2}(\Gamma)} \leq C \min\left\{\|\sigma - v\|_{H^{-1/2}(\Gamma)}, v \in S_\ell\right\}. \tag{4.162}$$

We obtain the approximation of the solution $u(\mathbf{x})$ of the boundary value problem (4.159) by

$$u_\ell(\mathbf{x}) := \int_\Gamma G\left(\mathbf{x}-\mathbf{y}\right)\sigma_\ell(\mathbf{y})\, ds_{\mathbf{y}}, \qquad \mathbf{x} \in \Omega. \tag{4.163}$$

In this section we will derive error estimates for the pointwise error $|u(\mathbf{x}) - u_\ell(\mathbf{x})|$.

4.2.5.1 Errors in Functionals of the Solution

The Aubin–Nitsche technique allows us to estimate errors in the linear functionals of the Galerkin solution. We will first introduce this method for abstract problems as discussed in Sect. 4.2.1. The abstract variational problem reads: For a given $F(\cdot) \in H'$ find a function $u \in H$ such that

$$b(u, v) = F(v) \qquad \forall v \in H. \tag{4.164}$$

Let $(S_\ell)_\ell \subset H$ be a family of dense subspaces that satisfy the discrete inf–sup conditions (4.117), (4.118). Then the Galerkin discretization of (4.164), i.e., find $u_\ell \in S_\ell$ such that

$$b(u_\ell, v_\ell) = F(v_\ell) \qquad \forall v_\ell \in S_\ell, \tag{4.165}$$

has a unique solution. The error $e_\ell = u - u_\ell$ satisfies the Galerkin orthogonality

$$b(u - u_\ell, v_\ell) = 0 \qquad \forall v_\ell \in S_\ell \tag{4.166}$$

as well as the quasi-optimal error estimate

$$\|u - u_\ell\|_H \leq \frac{C}{\gamma_\ell} \min\{\|u - \varphi_\ell\|_H : \varphi_\ell \subset S_\ell\}. \tag{4.167}$$

The Aubin–Nitsche argument estimates the error in *functionals* of the solution.

Theorem 4.2.14. *Let $\mathfrak{G} \in H'$ be a continuous, linear functional on the set of solutions H of Problem (4.164) which satisfies the assumptions (4.109), (4.110). Let $u_\ell \in S_\ell$ be the Galerkin approximation from (4.165) of the solution u. Furthermore, let the discrete inf–sup conditions (4.117), (4.118) be uniformly satisfied: $\gamma_\ell \geq \gamma > 0$.*

Then we have the error estimate

$$|\mathfrak{G}(u) - \mathfrak{G}(u_\ell)| \leq C \, \|u - \varphi_\ell\|_H \|w_{\mathfrak{G}} - \psi_\ell\|_H \tag{4.168}$$

for an arbitrary $\varphi_\ell \in S_\ell$, $\psi_\ell \in S_\ell$, where $w_{\mathfrak{G}}$ is the solution of the dual problem:

$$\textit{Find } w_{\mathfrak{G}} \in H: \qquad b(w, w_{\mathfrak{G}}) = \mathfrak{G}(w) \quad \forall w \in H. \tag{4.169}$$

Proof. From the continuous inf-sup conditions (4.110) Remark 2.1.45 gives us the inf-sup conditions for the adjoint problem, from which we have the existence of a unique solution.

Remark 4.2.4 shows that the discrete inf-sup conditions for $b(\cdot,\cdot)$ induce the discrete inf-sup conditions for the adjoint form $b^*(u,v) = \overline{b(v,u)}$. Therefore the adjoint problem (4.169) has a unique solution $w_{\mathfrak{G}} \in H$ for every $\mathfrak{G}(\cdot) \in H'$. By virtue of $S_\ell \subset H$ and (4.169), (4.166) it follows that

$$\begin{aligned} |\mathfrak{G}(u) - \mathfrak{G}(u_\ell)| &= |\mathfrak{G}(u - u_\ell)| = |b(u - u_\ell, w_{\mathfrak{G}})| \\ &= |b(u - u_\ell, w_{\mathfrak{G}} - v_\ell)| \qquad \forall v_\ell \in S_\ell. \end{aligned}$$

The continuity (4.109) of the form $b(\cdot,\cdot)$ and the error estimate (4.119) together yield (4.168). □

The error estimate (4.168) states that linear functionals $\mathfrak{G}(u)$ of the solution may under certain circumstances converge more rapidly than the energy error $\|u - u_\ell\|_H$. The convergence rate is superior to the rate in the energy norm by a factor $\inf\{\|w_{\mathfrak{G}} - \psi_\ell\|_H : \psi_\ell \in S_\ell\}$. The following example, for which $\mathfrak{G}(\cdot)$ represents an evaluation of the representation formula (4.163) in the domain point $\mathbf{x} \in \Omega$, makes this fact evident.

Example 4.2.15. *With the terminology used in Example 4.2.13, for the error $|u(\mathbf{x}) - u_\ell(\mathbf{x})|$ we have the estimate*

$$\begin{aligned} |u(\mathbf{x}) - u_\ell(\mathbf{x})| \leq C \, \min\left\{\|\sigma - \varphi_\ell\|_{H^{-1/2}(\Gamma)} : \varphi_\ell \in S_\ell\right\} \\ \times \min\left\{\|v_e - \psi_\ell\|_{H^{-1/2}(\Gamma)} : \psi_\ell \in S_\ell\right\} \end{aligned} \tag{4.170}$$

with the solution $v_e \in H^{-1/2}(\Gamma)$ of the dual problem:

Find $v_e \in H^{-\frac{1}{2}}(\Gamma)$ such that

$$(V v_e, \eta)_{L^2(\Gamma)} = (G(\mathbf{x} - \cdot), \eta)_{L^2(\Gamma)} \qquad \forall \eta \in H^{-1/2}(\Gamma). \tag{4.171}$$

With Corollary 4.1.34 we deduce the convergence rate for $S_\ell = S_{\mathcal{G}_\ell}^{p,-1}$

$$|u(\mathbf{x}) - u_\ell(\mathbf{x})| \le C\, h_\ell^{\min(s,p+1)+\frac{1}{2}+\min(t,p+1)+\frac{1}{2}} \|\sigma\|_{H^s(\Gamma)} \|v_e\|_{H^t(\Gamma)} \qquad (4.172)$$

for $s, t > -\frac{1}{2}$ *if* $\|\sigma\|_{H^s(\Gamma)}$ *and* $\|v_e\|_{H^t}$ *are bounded. If we have maximal regularity, i.e.,* $s = t = p+1$, *the result is a doubling of the convergence rate of the Galerkin method. For example, for piecewise constant boundary elements* $p = 0$ *and (4.172) with* $s = t = 1$ *we obtain the estimate*

$$|u(\mathbf{x}) - u_\ell(\mathbf{x})| \le C\, h_\ell^3 \|\sigma\|_{H^1(\Gamma)} \|v_e\|_{H^1(\Gamma)} \qquad (4.173)$$

and, thus, third order convergence for all $\mathbf{x} \in \Omega$. *Note that the constant* C *tends to infinity for* $\operatorname{dist}(\mathbf{x}, \Gamma) \to 0$.

Remark 4.2.16 (Regularity). *Inequality (4.172) only gives a high convergence rate if the solutions* σ, v_e *are sufficiently regular. For the boundary integral operator* V *on smooth surfaces* Γ, *the property* $g_D \in H^{1/2+s}(\Gamma)$ *with* $s \ge 0$ *is sufficient so that* $\sigma \in H^{-1/2+s}(\Gamma)$, *and the property* $G(\mathbf{x} - \cdot) \in H^{1/2+t}(\Gamma)$ *with* $t \ge 0$ *is sufficient so that* $v_e \in H^{-1/2+t}(\Gamma)$ *(see Sect. 3.2). Then we have the estimates*

$$\|\sigma\|_{H^{-1/2+s}(\Gamma)} \le C(s) \|g_D\|_{H^{1/2+s}(\Gamma)}, \quad \|v_e\|_{H^{-1/2+t}(\Gamma)} \le C(t) \|G(\mathbf{x}, \cdot)\|_{H^{1/2+t}(\Gamma)}, \qquad (4.174)$$

with a constant $C(\cdot)$ *which is independent of* g_D *and* G. *Because of the smoothness of the fundamental solution* $G(\mathbf{x} - \cdot)$ *for* $\mathbf{x} \in \Omega$, $\mathbf{y} \in \Gamma$ *we have* $G(\mathbf{x} - \cdot) \in C^\infty(\Gamma)$. *On smooth surfaces this implies the estimate (4.174) for all* $t \ge 0$. *With this (4.172) becomes*

$$|u(\mathbf{x}) - u_\ell(\mathbf{x})| \le C_1(p)\, C_2(\mathbf{x})\, h_\ell^{2(p+1)+1}, \qquad (4.175)$$

where we have $C_2(\mathbf{x}) = \|v_e\|_{H^{p+1}(\Gamma)} \le C(p) \|G(\mathbf{x} - \cdot)\|_{H^{p+2}(\Gamma)}$.

Note that especially for elements of higher order, $C_2(\mathbf{x})$ can become very large for $\mathbf{x}$ near Γ. Formula (4.163) should therefore only be used for points $\mathbf{x}$ in the domain that are sufficiently far away from Γ. For points $\mathbf{x}$ which are very close to the boundary or even lie on Γ, a bootstrapping algorithm has been developed to extract the potentials and arbitrary Cauchy data and their derivatives near and up to the boundary (see [213]).

If a quantity which has been computed or postprocessed by using the Galerkin method converges with an order that is higher than the order of the Galerkin error in the energy norm one speaks of *superconvergence*. Similar to the superconvergence (4.168) of functionals $\mathfrak{G}(\cdot)$ of the Galerkin solution u_ℓ, one can also study the convergence of u_ℓ in norms below the energy norm.

Now let $H = H^s(\Gamma)$ be the Hilbert space for the boundary integral operator $B : H^s(\Gamma) \to H^{-s}(\Gamma)$ of order $2s$ and let $b(\cdot, \cdot)$ be the $H^s(\Gamma)$-elliptic and injective sesquilinear form associated with B:

$$b(u, v) = (Bu, v)_{L^2(\Gamma)} : H^s(\Gamma) \times H^s(\Gamma) \to \mathbb{C}.$$

Here the continuous extension of the $L^2(\Gamma)$ inner-product for the dual pairing $\langle\cdot,\cdot\rangle_{H^s(\Gamma)\times H^{-s}(\Gamma)}$ is again denoted by $(\cdot,\cdot)_{L^2(\Gamma)}$. Furthermore, let $(S_\ell)_\ell$ be a dense sequence of subspaces in $H^s(\Gamma)$and let the discrete inf–sup conditions (4.117), (4.118) hold. Then we have for $t > 0$

$$\|u - u_\ell\|_{H^{s-t}(\Gamma)} = \sup_{v\in H^{-s+t}(\Gamma)\backslash\{0\}} \frac{(v, u - u_\ell)_{L^2(\Gamma)}}{\|v\|_{H^{-s+t}(\Gamma)}}.$$

Let w_v be a solution of the adjoint problem: Find $w_v \in H^s(\Gamma)$ such that

$$b(w, w_v) = (v, w)_{L^2(\Gamma)} \qquad \forall w \in H^s(\Gamma). \tag{4.176}$$

Then with the Galerkin orthogonality (4.166) we have (transferred to the adjoint problem)

$$\begin{aligned}\|u - u_\ell\|_{H^{s-t}(\Gamma)} &= \sup_{v\in H^{-s+t}(\Gamma)\backslash\{0\}} \frac{b(u - u_\ell, w_v)}{\|v\|_{H^{-s+t}(\Gamma)}}\\ &= \sup_{v\in H^{-s+t}(\Gamma)\backslash\{0\}} \frac{b(u - u_\ell, w_v - w_\ell)}{\|v\|_{H^{-s+t}(\Gamma)}}\\ &\leq C\,\|u - u_\ell\|_{H^s(\Gamma)} \sup_{v\in H^{-s+t}(\Gamma)\backslash\{0\}} \frac{\|w_v - w_\ell\|_{H^s(\Gamma)}}{\|v\|_{H^{-s+t}(\Gamma)}}.\end{aligned}$$

Since $w_\ell \in S_\ell$ was arbitrary, we obtain

$$\|u - u_\ell\|_{H^{s-t}(\Gamma)} \leq C\,\|u - u_\ell\|_{H^s(\Gamma)} \sup_{v\in H^{-s+t}(\Gamma)\backslash\{0\}} \inf_{w_\ell\in S_\ell} \frac{\|w_v - w_\ell\|_{H^s(\Gamma)}}{\|v\|_{H^{-s+t}(\Gamma)}}. \tag{4.177}$$

For $t > 0$ higher convergence rates are therefore possible for u_ℓ than in the H^s-norm, assuming that the adjoint problem (4.176) has the regularity

$$v \in H^{-s+t}(\Gamma) \Longrightarrow w_v \in H^{s+t}(\Gamma), \quad \forall 0 \leq t \leq \bar{t}. \tag{4.178}$$

In order to obtain quantitative error estimates with respect to the mesh width h_ℓ we again consider a dense sequence of boundary element spaces $(S_\ell)_\ell$ of order p on regular meshes $\mathcal{G}_\ell$ of mesh width h_ℓ. Then the approximation property

$$\inf_{w_\ell\in S_\ell} \|w_v - w_\ell\|_{H^s(\Gamma)} \leq C\, h_\ell^{\min(p+1,s+t)-s} \|w_v\|_{H^{s+t}(\Gamma)}$$

holds. These ideas are summarized in the following theorem.

Theorem 4.2.17. *Let the sesquilinear form $b(\cdot,\cdot)$ of problem (4.164) satisfy the conditions (4.109), (4.110). Let the exact solution satisfy $u \in H^r(\Gamma)$ with $r \geq s$. We assume that the adjoint problem (4.176) has the regularity (4.178) with $\bar{t} \geq 0$.*

Furthermore, let $(S_\ell)_\ell$ be a dense sequence of boundary element spaces of order p in $H^s(\Gamma)$ on regular meshes $\mathcal{G}_\ell$ of mesh width h_ℓ.

Then we have for the Galerkin solution $u_\ell \in S_\ell$ and $0 \le t \le \bar{t}$ the error estimate

$$\|u - u_\ell\|_{H^{s-t}(\Gamma)} \le C\, h_\ell^{\min(p+1,r)+\min(p+1),s+t)-2s} \|u\|_{H^r(\Gamma)}. \tag{4.179}$$

In particular, in the case of maximal regularity, i.e., for $r \ge p+1, \bar{t} \ge p+1-s$, it thus follows that we have a doubling of the convergence rate of the Galerkin method:

$$\|u - u_\ell\|_{H^{2s-p-1}} \le C\, h^{2(p+1)-2s} \|u\|_{H^{p+1}(\Gamma)}.$$

4.2.5.2 Perturbations

The efficient numerical realization of the Galerkin BEM (4.165) involves, for example, perturbations of the sesquilinear form $b(\cdot,\cdot)$ by quadrature, surface and cluster approximation of the operator or the functional $\mathfrak{G}(\cdot)$, used for the evaluation of the representation formula at a point $\mathbf{x} \in \Omega$. Instead of (4.165) one implements a perturbed boundary element method:
Find $\tilde{u}_\ell \in S_\ell$ such that

$$b_\ell(\tilde{u}_\ell, v) = F_\ell(v) \quad \forall v \in S_\ell \tag{4.180}$$

and instead of $\mathfrak{G}(u_\ell)$ one implements an approximation $\mathfrak{G}_\ell(\tilde{u}_\ell)$. Here we will study the error

$$\mathfrak{G}(u) - \mathfrak{G}_\ell(\tilde{u}_\ell) \tag{4.181}$$

of a linear functional of the solution, for example of the representation formula (see Example 4.2.13). According to Theorem 4.2.11, (4.180) has a unique solution for a sufficiently large ℓ if the exact form $b(\cdot,\cdot)$ satisfies the discrete inf–sup conditions

$$\begin{aligned} \inf_{u_\ell \in S_\ell\setminus\{0\}} \sup_{v_\ell \in S_\ell\setminus\{0\}} \frac{|b(u_\ell, v_\ell)|}{\|u_\ell\|_H \;\|v_\ell\|_H} &\ge \gamma > 0, \\ \inf_{v_\ell \in S_\ell\setminus\{0\}} \sup_{u_\ell \in S_\ell\setminus\{0\}} \frac{|b(u_\ell, v_\ell)|}{\|u_\ell\|_H \;\|v_\ell\|_H} &\ge \gamma > 0 \end{aligned} \tag{4.182}$$

on $S_\ell \times S_\ell$ and if the perturbed form $b_\ell(\cdot,\cdot)$ is uniformly continuous [see (4.151)] and at the same time satisfies the *stability and consistency conditions* (4.152), (4.153). Then for a sufficiently large ℓ we have the error estimate

$$\|u - \tilde{u}_\ell\|_H \le C \left\{ \min_{w_\ell \in S_\ell} (\|u - w_\ell\|_H + \delta_\ell \|w_\ell\|_U) + \sup_{v_\ell \in S_\ell\setminus\{0\}} \frac{|F(v_\ell) - F_\ell(v_\ell)|}{\|v_\ell\|_H} \right\}. \tag{4.183}$$

The perturbations of the right-hand side F and of the functional $\mathfrak{G}$ define the quantities

$$f_\ell := \sup_{v_\ell \in S_\ell \setminus \{0\}} \frac{|F_\ell(v_\ell) - F(v_\ell)|}{\|v_\ell\|_H} \quad \text{and} \quad g_\ell := \sup_{v_\ell \in S_\ell \setminus \{0\}} \frac{|\mathfrak{G}_\ell(v_\ell) - \mathfrak{G}(v_\ell)|}{\|v_\ell\|_H}. \tag{4.184}$$

Note that in many practical applications the perturbations F_ℓ and $\mathfrak{G}_\ell$ are not defined on H but only on S_ℓ. We assume that $(f_\ell)_\ell$ and $(g_\ell)_\ell$ are null sequences and, thus, that there exist constants C_F and C_G such that

$$\|\mathfrak{G}\|_{H'} =: C_G < \infty \quad \text{and} \quad \|F\|_{H'} + f_\ell \leq C_F \qquad \forall \ell \in \mathbb{N}.$$

Theorem 4.2.18. *Let the form $b(\cdot,\cdot)$ satisfy (4.182) and let the perturbed form $b_\ell(\cdot,\cdot)$ satisfy the conditions (4.151)–(4.153). Then, for a sufficiently large ℓ, the error (4.181) has the estimate*

$$\begin{aligned}|\mathfrak{G}(u) - \mathfrak{G}_\ell(\tilde{u}_\ell)| \leq C\|u - u_\ell\|_H \min_{\psi_\ell \in S_\ell} \|w_{\mathfrak{G}} - \psi_\ell\|_H + \frac{C_G}{\gamma} c_\ell \|\tilde{u}_\ell - u\|_H + \frac{C_G}{\gamma} f_\ell \\ + \frac{C_G}{\gamma} \min_{\varphi_\ell \in S_\ell} (c_\ell \|u - \varphi_\ell\|_H + \delta_\ell \|\varphi_\ell\|_U) + \frac{C_F}{\gamma} g_\ell.\end{aligned} \tag{4.185}$$

Proof. By the definition (4.170) of $w_{\mathfrak{G}}$ and the orthogonality of the Galerkin error we have

$$\begin{aligned}|\mathfrak{G}(u) - \mathfrak{G}(\tilde{u}_\ell)| &= |b(u - \tilde{u}_\ell, w_{\mathfrak{G}})| \\ &= |b(u - u_\ell, w_{\mathfrak{G}} - \psi_\ell)| + |b(u_\ell - \tilde{u}_\ell, w_{\mathfrak{G}})|\end{aligned} \tag{4.186}$$

for an arbitrary $\psi_\ell \in S_\ell$. Furthermore, let $w_\ell^{\mathfrak{G}} \in S_\ell$ be the solution of the Galerkin equations

$$b(w_\ell, w_\ell^{\mathfrak{G}}) = b(w_\ell, w_{\mathfrak{G}}) = \mathfrak{G}(w_\ell) \qquad \forall w_\ell \in S_\ell.$$

Then, taking the Galerkin orthogonality into consideration, we have

$$\begin{aligned}|b(u_\ell - \tilde{u}_\ell, w_{\mathfrak{G}})| &= |b(u_\ell - \tilde{u}_\ell, w_\ell^{\mathfrak{G}})| = |b(u_\ell, w_\ell^{\mathfrak{G}}) - b(\tilde{u}_\ell, w_\ell^{\mathfrak{G}})| \\ &\leq |F(w_\ell^{\mathfrak{G}}) - b_\ell(\tilde{u}_\ell, w_\ell^{\mathfrak{G}})| + |(b_\ell - b)(\tilde{u}_\ell, w_\ell^{\mathfrak{G}})| \\ &= |F(w_\ell^{\mathfrak{G}}) - F_\ell(w_\ell^{\mathfrak{G}})| + |(b - b_\ell)(\tilde{u}_\ell, w_\ell^{\mathfrak{G}})|.\end{aligned}$$

We consider the difference $b - b_\ell$ and with the stability and consistency conditions we obtain for an arbitrary $\varphi_\ell \in S_\ell$ the estimate

$$\begin{aligned}\left|(b - b_\ell)(\tilde{u}_\ell, w_\ell^{\mathfrak{G}})\right| &\leq \left|(b - b_\ell)\,(\tilde{u}_\ell - \varphi_\ell, w_\ell^{\mathfrak{G}}))\right| + \left|b\left(\varphi_\ell, w_\ell^{\mathfrak{G}}\right) - b_\ell\left(\varphi_\ell, w_\ell^{\mathfrak{G}}\right)\right| \\ &\leq c_\ell \|\tilde{u}_\ell - \varphi_\ell\|_H \left\|w_\ell^{\mathfrak{G}}\right\|_H + \delta_\ell \|\varphi_\ell\|_U \left\|w_\ell^{\mathfrak{G}}\right\|_H.\end{aligned} \tag{4.187}$$

With this result and with (4.186) we obtain

$$
\begin{aligned}
|\mathfrak{G}(u) - \mathfrak{G}_\ell(\tilde{u}_\ell)| &\le |\mathfrak{G}(u) - \mathfrak{G}(\tilde{u}_\ell)| + |\mathfrak{G}(\tilde{u}_\ell) - \mathfrak{G}_\ell(\tilde{u}_\ell)| \\
&\le |b(u - u_\ell, w_{\mathfrak{G}} - \psi_\ell)| + |(F - F_\ell)(w_\ell^{\mathfrak{G}})| \\
&\quad + |(b - b_\ell)(\tilde{u}_\ell, w_\ell^{\mathfrak{G}})| + |(\mathfrak{G} - \mathfrak{G}_\ell)(\tilde{u}_\ell)| \\
&\le C \|u - u_\ell\|_H \|w_{\mathfrak{G}} - \psi_\ell\|_H + f_\ell \left\| w_\ell^{\mathfrak{G}} \right\|_H \\
&\quad + \left\| w_\ell^{\mathfrak{G}} \right\|_H (c_\ell \|\tilde{u}_\ell - \varphi_\ell\|_H + \delta_\ell \|\varphi_\ell\|_U) + g_\ell \|\tilde{u}_\ell\|_H .
\end{aligned}
\tag{4.188}
$$

According to Theorem 4.2.11, for a sufficiently large ℓ the sequence $(\tilde{u}_\ell)_\ell$ of the perturbed Galerkin solutions is stable and with ℓ_0 from Theorem 4.2.11 we have

$$
\|\tilde{u}_\ell\|_H \le \frac{1}{\gamma} (\|F\|_{H'} + f_\ell) \le \frac{C_F}{\gamma} \qquad \forall \ell \ge \ell_0. \tag{4.189}
$$

We use the discrete inf–sup conditions (4.182) to find a bound for the term $\left\| w_\ell^{\mathfrak{G}} \right\|_H$:

$$
\gamma \|w_\ell^{\mathfrak{G}}\|_H \le \sup_{w_\ell \in S_\ell \backslash \{0\}} \frac{|b(w_\ell, w_\ell^{\mathfrak{G}})|}{\|w_\ell\|_H} = \sup_{w_\ell \in S_\ell \backslash \{0\}} \frac{|\mathfrak{G}(w_\ell)|}{\|w_\ell\|_H} \le \|\mathfrak{G}\|_{H'} = C_G
$$

for $\ell \ge \ell_0$. This yields

$$
\begin{aligned}
|\mathfrak{G}(u) - \mathfrak{G}_\ell(\tilde{u}_\ell)| \le{}& C \|u - u_\ell\|_H \|w_{\mathfrak{G}} - \psi_\ell\|_H + \frac{C_G}{\gamma} f_\ell \\
&+ \frac{C_G}{\gamma} (c_\ell \|\tilde{u}_\ell - \varphi_\ell\|_H + \delta_\ell \|\varphi_\ell\|_U) + \frac{C_F}{\gamma} g_\ell .
\end{aligned}
$$

The triangle inequality $\|\tilde{u}_\ell - \varphi_\ell\|_H \le \|\tilde{u}_\ell - u\|_H + \|u - \varphi_\ell\|_H$ finally yields the assertion. □

The inequality (4.185) can be used to bound the size of the perturbations c_ℓ, δ_ℓ, f_ℓ and g_ℓ in such a way that the functional $\mathfrak{G}_\ell(\tilde{u}_\ell)$ converges with the same rate as the functional $\mathfrak{G}(u_\ell)$ for the original Galerkin method.

To illustrate this we consider $H = H^s(\Gamma)$ and a discretization with piecewise polynomials of order p. Then the optimal convergence rate of the unperturbed Galerkin method is given by $\|u - u_\ell\|_H \le C h_\ell^{p+1-s}$.

Inequality (4.183) shows that the two conditions $\delta_\ell \le C h_\ell^{p+1-s}$ and $f_\ell \le C h_\ell^{p+1-s}$ imposed on the size of the perturbations guarantee that $\|u - \tilde{u}_\ell\|_H \le C h_\ell^{p+1-s}$ converges with the same rate as the unperturbed Galerkin method. The optimal convergence rate for the dual problem is also $\left\| w_{\mathfrak{G}} - w_\ell^{\mathfrak{G}} \right\|_H \le C h_\ell^{p+1-s}$

and it is our aim to control the size of the perturbation in such a way that the functional $\mathfrak{G}_\ell(\tilde{u}_\ell)$ converges at the rate $Ch_\ell^{2p+2-2s}$.

For this to hold, the perturbed sesquilinear forms, the right-hand sides and functionals in (4.152), (4.153) and (4.184) all have to satisfy the estimates

$$c_\ell \leq Ch_\ell^{p+1-s}, \quad \delta_\ell \leq Ch_\ell^{2p+2-2s}, \quad f_\ell \leq Ch_\ell^{2p+2-2s}, \quad g_\ell \leq Ch_\ell^{2p+2-2s}.$$

In the following theorem we will determine a bound for the effect of perturbations $b - b_\ell$ and $F - F_\ell$ on negative norms of the Galerkin error.

Theorem 4.2.19. *Let the assumptions from Theorem 4.2.18 hold for $H = H^s(\Gamma)$, $b : H^s(\Gamma) \times H^s(\Gamma) \to \mathbb{C}$. Furthermore, let the adjoint problem (4.176) satisfy the regularity assumption (4.178) for a $\bar{t} > 0$. Then for a sufficiently large ℓ we have the error estimate*

$$\|u - \tilde{u}_\ell\|_{H^{s-t}(\Gamma)} \leq C\left\{d_{\ell,s,s+t}\|u - u_\ell\|_{H^s(\Gamma)} + c_\ell\|u - \tilde{u}_\ell\|_{H^s(\Gamma)}\right.$$
$$\left.+ f_\ell + \inf_{\varphi_\ell \in S_\ell}\left(c_\ell\|u - \varphi_\ell\|_{H^s(\Gamma)} + \delta_\ell\|\varphi_\ell\|_U\right)\right\} \tag{4.190}$$

for $0 \leq t \leq \bar{t}$ with

$$d_{\ell,s,s+t} := \sup_{w \in H^{s+t}(\Gamma)\setminus\{0\}}\left(\inf_{\psi_\ell \in S_\ell}\frac{\|w - \psi_\ell\|_{H^s(\Gamma)}}{\|w\|_{H^{s+t}(\Gamma)}}\right).$$

Proof. Let $v \in H^{-s+t}(\Gamma)$ be arbitrary and let w_v be the solution of the adjoint problem (4.176) with the right-hand side v. We then have

$$\begin{aligned}(v, u - \tilde{u}_\ell)_{L^2(\Gamma)} &= b(u - \tilde{u}_\ell, w_v)\\ &= b(u - u_\ell, w_v) + \underbrace{b(u_\ell - \tilde{u}_\ell, w_v)}_{(*)}.\end{aligned} \tag{4.191}$$

We consider $(*)$. Let $w_v^\ell \in S_\ell$ be the Galerkin approximation of w_v^ℓ:

$$b\left(v_\ell, w_v^\ell\right) = (w_v, v_\ell)_{L^2(\Gamma)} \qquad \forall v_\ell \in S_\ell.$$

With $v_\ell = u_\ell - \tilde{u}_\ell \in S_\ell$ it follows from the Galerkin orthogonality $b\left(v_\ell, w_v - w_v^\ell\right) = 0$ that we have the relation

$$\begin{aligned}(*) &= b(u_\ell - \tilde{u}_\ell, w_v) = b\left(u_\ell - \tilde{u}_\ell, w_v^\ell\right)\\ &= (b - b_\ell)\left(u_\ell - \tilde{u}_\ell, w_v^\ell\right) + b_\ell\left(u_\ell - \tilde{u}_\ell, w_v^\ell\right)\end{aligned}$$

$$= (b - b_\ell)\left(u_\ell - \tilde{u}_\ell, w_v^\ell\right) + b_\ell\left(u_\ell, w_v^\ell\right) - F_\ell(w_v^\ell)$$

$$= (b - b_\ell)\left(u_\ell - \tilde{u}_\ell, w_v^\ell\right) + (b_\ell - b)\left(u_\ell, w_v^\ell\right) + b\left(u_\ell, w_v^\ell\right) - F_\ell(w_v^\ell)$$

$$= (b - b_\ell)\left(-\tilde{u}_\ell, w_v^\ell\right) + F(w_v^\ell) - F_\ell(w_v^\ell).$$

With this we will estimate (4.191) by using (4.166) as follows. For every $\psi_\ell \in S_\ell$ we have

$$\begin{aligned} |(v, u - \tilde{u}_\ell)_{L^2(\Gamma)}| \le{}& |b(u - u_\ell, w_v - \psi_\ell)| \\ &+ \left|(b - b_\ell)\left(\tilde{u}_\ell, w_v^\ell\right)\right| + \left|F(w_v^\ell) - F_\ell(w_v^\ell)\right|. \end{aligned} \tag{4.192}$$

As in (4.187), we use the consistency condition to prove for an arbitrary $\varphi_\ell \in S_\ell$ the estimate

$$\left|(b - b_\ell)(\tilde{u}_\ell, w_v^\ell)\right| \le \left(c_\ell \|\tilde{u}_\ell - \varphi_\ell\|_{H^s(\Gamma)} + \delta_\ell \|\varphi_\ell\|_U\right) \left\|w_v^\ell\right\|_{H^s(\Gamma)}, \tag{4.193}$$

where $\|\cdot\|_U$ again denotes a stronger norm than $H^s(\Gamma)$.

The regularity assumption (4.178) and the stability of the Galerkin approximations $\left(w_v^\ell\right)_\ell$ of the adjoint problem yield for all $0 \le t < \bar{t}$ and all $v \in H^{-s+t}(\Gamma)$ the estimate

$$\|w_v^\ell\|_{H^s(\Gamma)} \le C\,\|w_v\|_{H^s(\Gamma)} \le C\,\|v\|_{H^{-s}(\Gamma)} \le C\,\|v\|_{H^{-s+t}(\Gamma)}. \tag{4.194}$$

Therefore it follows from (4.192) and (4.193) with (4.184) that

$$\begin{aligned} \|u - \tilde{u}_\ell\|_{H^{s-t}(\Gamma)} ={}& \sup_{v \in H^{-s+t}(\Gamma)\setminus\{0\}} \frac{|(v, u - \tilde{u}_\ell)_{L^2(\Gamma)}|}{\|v\|_{H^{-s+t}(\Gamma)}} \\ \le{}& C\,\|u - u_\ell\|_{H^s(\Gamma)} \sup_{v \in H^{-s+t}(\Gamma)\setminus\{0\}} \left(\inf_{\psi_\ell \in S_\ell} \frac{\|w_v - \psi_\ell\|_{H^s(\Gamma)}}{\|v\|_{H^{-s+t}(\Gamma)}}\right) \\ &+ C \inf_{\varphi_\ell \in S_\ell} \left(c_\ell \|\tilde{u}_\ell - \varphi_\ell\|_{H^s(\Gamma)} + \delta_\ell \|\varphi_\ell\|_U\right) + C f_\ell. \end{aligned}$$

The regularity assumption imposed upon the adjoint problem yields the estimate $\|v\|_{H^{-s+t}(\Gamma)} \ge C^{-1}\,\|w_v\|_{H^{s+t}(\Gamma)}$. Hence we have

$$\begin{aligned} &\sup_{v \subset H^{-s+t}(\Gamma)\setminus\{0\}} \left(\inf_{\psi_\ell \in S_\ell} \frac{\|w_v - \psi_\ell\|_{H^s(\Gamma)}}{\|v\|_{H^{-s+t}(\Gamma)}}\right) \\ &\le C \sup_{w \in H^{s+t}(\Gamma)\setminus\{0\}} \left(\inf_{\psi_\ell \in S_\ell} \frac{\|w - \psi_\ell\|_{H^s(\Gamma)}}{\|w\|_{H^{s+t}(\Gamma)}}\right) = C d_{\ell,s,s+t}. \end{aligned}$$

Note that $d_{\ell,s,t}$ represents an approximation property of the space S_ℓ. Combining these results we have proved that

$$\|u-\tilde{u}_\ell\|_{H^{s-t}(\Gamma)} \leq C\left\{d_{\ell,s,s+t}\|u-u_\ell\|_{H^s(\Gamma)} + c_\ell\|u-\tilde{u}_\ell\|_{H^s(\Gamma)} + f_\ell + \inf_{\varphi_\ell\in S_\ell}\left(c_\ell\|u-\varphi_\ell\|_{H^s(\Gamma)} + \delta_\ell\|\varphi_\ell\|_U\right)\right\}.$$

□

With the help of inequality (4.190) we can determine sufficient conditions on the admissible magnitude of the perturbations c_ℓ, δ_ℓ, f_ℓ and g_ℓ so that the Galerkin error $\|u-\tilde{u}_\ell\|_{H^{s-t}(\Gamma)}$ converges with the same rate as the unperturbed Galerkin solution.

In order to illustrate this, we consider a discretization with piecewise polynomials of order p and assume that the continuous solution satisfies $u \in H^{p+1}(\Gamma)$. Then the optimal convergence rate of the unperturbed Galerkin method is given by $\|u-u_\ell\|_{H^{s-t}(\Gamma)} \leq Ch_\ell^{p+1-s+\min(p+1-s),t)}\|u\|_{H^{p+1}(\Gamma)}$.

Inequality (4.183) shows that the two conditions $\delta_\ell \leq Ch_\ell^{p+1-s}$ and $f_\ell \leq Ch_\ell^{p+1-s}$ imposed on the size of the perturbations guarantee that $\|u-\tilde{u}_\ell\|_{H^s(\Gamma)} \leq Ch_\ell^{p+1-s}$ converges with the same rate as the unperturbed Galerkin method (with respect to the H^s-norm). The optimal convergence rate of the term $d_{\ell,s,s+t}$ is $d_{\ell,s,s+t} \leq Ch_\ell^{\min\{p+1-s,t\}}$ and it is our goal to control the size of the perturbations in such a way that the term $\|u-\tilde{u}_\ell\|_{H^{s-t}(\Gamma)}$ converges at the rate $Ch_\ell^{p+1-s+\min(p+1-s),t)}$. This leads to the following condition for the quantities c_ℓ, δ_ℓ, f_ℓ

$$C\left(h_\ell^{\min\{p+1-s,t\}+p+1-s} + c_\ell h^{p+1-s} + f_\ell + c_\ell h^{p+1-s} + \delta_\ell\right) \leq Ch_\ell^{p+1-s+\min(p+1-s),t)}.$$

For this the perturbed sesquilinear form, right-hand sides and functionals in (4.152), (4.153) and (4.184) have to satisfy the estimates

$$c_\ell \leq Ch_\ell^{\min\{p+1-s,t\}}, \quad \delta_\ell \leq Ch_\ell^{\min\{p+1-s,t\}+p+1-s}, \quad f_\ell \leq Ch_\ell^{\min\{p+1-s,t\}+p+1-s}.$$

4.3 Proof of the Approximation Property

In Sects. 4.1–4.2.5 we have seen that the Galerkin boundary element method produces approximative solutions of boundary integral equations which converge quasi-optimally. Here we will present the proofs of the convergence rates (4.59) and (4.93) of discontinuous and continuous boundary elements on surface meshes $\mathcal{G}$ with mesh width $h > 0$.

In general we will assume that Assumption 4.1.6 holds, i.e., that the panel parametrizations can be decomposed into a regular, affine mapping $\chi_\tau^{\text{affine}}$ and a diffeomorphism χ_Γ, independent of τ, since $\chi_\tau = \chi_\Gamma \circ \chi_\tau^{\text{affine}}$. For $\chi_\tau^{\text{affine}}$ there exist $\mathbf{b}_\tau \in \mathbb{R}^3$ and $\mathbf{B}_\tau \in \mathbb{R}^{3\times 2}$ such that

$$\chi_\tau^{\text{affine}}(\hat{\mathbf{x}}) = \mathbf{B}_\tau \hat{\mathbf{x}} + \mathbf{b}_\tau.$$

The Gram matrix of this mapping is denoted by $\mathbf{G}_\tau := \mathbf{B}_\tau^\intercal \mathbf{B}_\tau \in \mathbb{R}^{2\times 2}$. It is symmetric and positive definite.

Note: The proof of the approximation property has the same structure as the proofs for the finite element methods (see, for example, [27,33,68,115]) and is also based on concepts such as the pullback to the reference element, the shape-regularity and the Bramble–Hilbert lemma.

4.3.1 Approximation Properties on Plane Panels

We use the same notation as in Sect. 4.1.2. Let $\hat{\Gamma}$ be a polyhedral surface with plane sides and let $\mathcal{G}^{\text{affine}}$ be a surface mesh of $\hat{\Gamma}$ which consists of plane triangles or parallelograms. The panels $\tau \in \mathcal{G}^{\text{affine}}$ are images of the reference element $\widehat{\tau}$ under a regular, affine transformation $\chi_\tau^{\text{affine}} : \hat{\tau} \to \tau$.

As in (4.23), for the reference element $\hat{\tau}$ and $p \geq 0$ we denote the space of all polynomials of total degree p by $\mathbb{P}_p^\Delta(\hat{\tau})$, while $\iota_p^{\hat{\tau}}$ denotes the index set for the associated unisolvent set of nodal points [see (4.70) and Theorem 4.1.39].

In preparation for Proposition 4.3.3 we will first prove a norm equivalence.

Lemma 4.3.1. *Let $k \in \mathbb{N}_{\geq 1}$. Then*

$$[u]_{k+1} := |u|_{k+1} + \sum_{(i,j)\in\iota_p^{\hat{\tau}}} \left| u\left(\frac{i}{p}, \frac{j}{p}\right) \right| \tag{4.195}$$

defines a norm on $H^{k+1}(\hat{\tau})$ which is equivalent to $\|\cdot\|_{k+1}$.

Proof. The continuity of the embedding $H^{k+1}(\hat{\tau}) \hookrightarrow C\left(\overline{\hat{\tau}}\right)$ follows from the Sobolev Embedding Theorem (see Theorem 2.5.4), and thus $[\cdot]_{k+1}$ is well defined. Therefore there exists a constant $c_1 \in \mathbb{R}_{>0}$ such that

$$[u]_{k+1} \leq c_1 \|u\|_{k+1} \qquad \forall u \in H^{k+1}(\hat{\tau}).$$

Therefore it remains to show that there exists a constant $c_2 \in \mathbb{R}_{>0}$ such that

$$\|u\|_{k+1} \leq c_2 [u]_{k+1} \qquad \forall u \in H^{k+1}(\hat{\tau}).$$

We prove this indirectly and for this purpose we assume that there exists a sequence $(u_n)_{n\in\mathbb{N}} \subset H^{k+1}(\hat{\tau})$ such that

$$\forall n \in \mathbb{N} : \|u_n\|_{k+1} = 1 \quad \text{and} \quad \lim_{n\to\infty} [u_n]_{k+1} = 0. \tag{4.196}$$

We deduce from Theorem 2.5.6 by induction over k that there exists a subsequence $\left(u_{n_j}\right)_{j\in\mathbb{N}}$ that converges to some $u \in H^k(\hat{\tau})$:

$$\lim_{j\to\infty} \left\|u_{n_j} - u\right\|_k = 0.$$

The second assumption in (4.196) yields

$$\lim_{j\to\infty} \left|u_{n_j} - u\right|_{k+1} = 0.$$

Hence $u \in H^{k+1}(\hat{\tau})$ with $|u|_{k+1} = 0$ and we have

$$\lim_{j\to\infty} \left\|u_{n_j} - u\right\|_{k+1} = 0.$$

Since $|u|_{k+1} = 0$, we have $u \in \mathbb{P}_k$ and the Sobolev Embedding Theorem implies the convergence in the nodal points

$$u(\mathbf{z}) = \lim_{j\to\infty} u_{n_j}(\mathbf{z}) \qquad \forall \mathbf{z} = \left(\frac{i}{p}, \frac{j}{p}\right), \quad (i,j) \in \iota_p^{\hat{\tau}}.$$

Theorem 4.1.39 therefore yields a contradiction to the first assumption in (4.196). □

Lemma 4.3.2 (Bramble–Hilbert Lemma). *Let $k \in \mathbb{N}_0$. Then*

$$\inf_{p\in\mathbb{P}_k} \|u - p\|_{k+1} \le c_2 |u|_{k+1}$$

for all $u \in H^{k+1}(\hat{\tau})$, with c_2 from the proof of Lemma 4.3.1.

Proof. For $k = 0$ the statement follows from the Poincaré inequality (see Corollary 2.5.10).

In the following let $k \ge 1$ and $u \in H^{k+1}(\hat{\tau})$. Thanks to the Sobolev Embedding Theorem the point evaluation of u is well defined. Let $(b_{\mathbf{z}})_{\mathbf{z}\in\Sigma_k}$ be the vector that contains the values of u at the nodal points: $b_{\mathbf{z}} = u(\mathbf{z})$ for all $\mathbf{z} \in \Sigma_k$. Let $p \in \mathbb{P}_k^{\hat{\tau}}$ be the, according to Theorem 4.1.39, unique polynomial with $b_{\mathbf{z}} = p(\mathbf{z})$ for all $\mathbf{z} \in \Sigma_k$. Then, by Lemma 4.3.1,

$$\inf_{q\in\mathbb{P}_k} \|u - q\|_{k+1} \le \|u - p\|_{k+1} \le c_2 [u - p]_{k+1} = c_2 |u|_{k+1}.$$

□

Proposition 4.3.3. *Let* $\widehat{\Pi} : H^{p+1}(\hat{\tau}) \to H^s(\hat{\tau})$ *be linear and continuous for* $0 \le s \le p+1$ *such that*

$$\forall q \in \mathbb{P}_p^{\Delta}(\hat{\tau}) : \ \widehat{\Pi} q = q. \tag{4.197}$$

Then there exists a constant $c = c(\widehat{\Pi})$ *so that*

$$\forall v \in H^{p+1}(\hat{\tau}) : \ \|v - \widehat{\Pi} v\|_{H^s(\hat{\tau})} \le \hat{c}\, |v|_{H^{p+1}(\hat{\tau})}. \tag{4.198}$$

Proof. Let $v \in H^{p+1}(\hat{\tau})$. Then by (4.197) for all $q \in \mathbb{P}_p^{\Delta}(\hat{\tau})$ we have

$$\begin{aligned} v - \widehat{\Pi} v &= v + q - \widehat{\Pi}(v+q) \\ \|v - \widehat{\Pi} v\|_{H^s(\hat{\tau})} &\le \hat{c}\, \|v+q\|_{H^{p+1}(\hat{\tau})} \\ \hat{c} &:= \|I - \widehat{\Pi}\|_{H^s(\hat{\tau}) \leftarrow H^{p+1}(\hat{\tau})}, \end{aligned}$$

where I denotes the identity. Since $q \in \mathbb{P}_p^{\Delta}(\hat{\tau})$ was arbitrary, with Lemma 4.3.2 we deduce

$$\forall v \in H^{p+1}(\hat{\tau}) : \ \|v - \Pi v\|_{H^s(\widehat{\tau})} \le \hat{c} \inf_{q \in \mathbb{P}_p^{\Delta}(\hat{\tau})} \|v+q\|_{H^{p+1}(\hat{\tau})} = \hat{c}\, |v|_{H^{p+1}(\widehat{\tau})}.$$

□

The estimate of the approximation error is proven by a transformation to the reference element.

First we will need some transformation formulas for Sobolev norms. Let $\tau \subset \mathbb{R}^2$ be a plane panel as before (triangle or parallelogram) with an affine parametrization $\chi_\tau^{\text{affine}} : \hat{\tau} \to \tau$. Tangential vectors on τ are defined by $\mathbf{b}_i := \partial \chi_\tau^{\text{affine}} / \partial \hat{x}_i$ for $i = 1, 2$. The (constant) normal vector $\mathbf{n}_\tau$ is oriented in such a way that $(\mathbf{b}_1, \mathbf{b}_2, \mathbf{n}_\tau)$ forms a right system. For $\varepsilon > 0$ we set $I_\varepsilon = (-\varepsilon, \varepsilon)$ and define a neighborhood $U_\varepsilon \subset \mathbb{R}^3$ of τ by

$$U_\varepsilon := \left\{ \mathbf{z} \in \mathbb{R}^3 : \exists\, (\mathbf{x}, \alpha) \in \tau \times I_\varepsilon : \mathbf{z} = \mathbf{x} + \alpha \mathbf{n}_\tau \right\}. \tag{4.199}$$

A function $u \in H^{k+1}(\tau)$ can be extended as a constant on U_ε:

$$u^\star(\mathbf{x} + \alpha \mathbf{n}_\tau) = u(\mathbf{x}) \qquad \forall\, (\mathbf{x}, \alpha) \in \tau \times I_\varepsilon.$$

The surface gradient $\nabla_S u$ is defined by

$$\nabla_S u = \nabla u^\star \big|_\tau, \tag{4.200}$$

which gives us

$$|u|^2_{H^1(\tau)} = \int_\tau \langle \nabla_S u, \nabla_S u \rangle.$$

The pullback of the function u to the reference element is denoted by $\hat{u} := u \circ \chi_\tau^{\text{affine}}$.

Lemma 4.3.4. *We have*

$$\|u\|^2_{L^2(\tau)} = \frac{|\tau|}{|\hat{\tau}|}\int_{\hat{\tau}} |\hat{u}|^2$$

$$|u|^2_{H^1(\tau)} = \frac{|\tau|}{|\hat{\tau}|}\int_{\hat{\tau}} \left\langle \widehat{\nabla}\hat{u}, \mathbf{G}_\tau^{-1}\widehat{\nabla}\hat{u}\right\rangle,$$

where $\widehat{\nabla}$ denotes the two-dimensional gradient in the coordinates of the reference element.

Proof. The transformation formula for surface integrals yields the first equation

$$\int_\tau |u|^2 = \frac{|\tau|}{|\hat{\tau}|}\int_{\hat{\tau}} |\hat{u}|^2 .$$

We define $\chi : \mathbb{R}^3 \to \mathbb{R}^3$ for $\hat{\mathbf{x}} \in \hat{\tau}$ and $x_3 \in \mathbb{R}$ by

$$\chi\left(\hat{\mathbf{x}}, \hat{x}_3\right) := \chi_\tau^{\text{affine}}\left(\hat{\mathbf{x}}\right) + \hat{x}_3\mathbf{n}_\tau = \mathbf{B}_\tau\hat{\mathbf{x}} + \hat{x}_3\mathbf{n}_\tau + \mathbf{b}_\tau$$

and we set $\widehat{U}_\varepsilon := \chi^{-1}\left(U_\varepsilon\right)$. With this we can define the function $\hat{u}^\star : \widehat{U}_\varepsilon \to \mathbb{K}$ by

$$\hat{u}^\star := u^\star \circ \chi$$

and it satisfies $\hat{u}^\star|_{\hat{\tau}} = \hat{u}$ in the sense of traces. The chain rule then yields

$$\left(\nabla_S u\right) \circ \chi_\tau^{\text{affine}} = \left(\nabla u^\star\right) \circ \chi_\tau^{\text{affine}} = \left(\mathbf{J}_\tau^{-1}\right)^{\mathsf{T}} \nabla\hat{u}^\star. \tag{4.201}$$

with the Jacobian $\mathbf{J}_\tau = [\mathbf{B}_\tau, \mathbf{n}_\tau]$ of the transformation χ. From this we have

$$\left(\nabla_S u\right) \circ \chi_\tau^{\text{affine}} = \left(\mathbf{J}_\tau^{-1}\right)^{\mathsf{T}} \widehat{\nabla}\hat{u}^\star\Big|_{\hat{\tau}} .$$

Elementary properties of the vector product give us

$$\mathbf{J}_\tau^{-1}\left(\mathbf{J}_\tau^{-1}\right)^{\mathsf{T}} = \begin{bmatrix} \mathbf{G}_\tau^{-1} & 0 \\ 0 & 1 \end{bmatrix}$$

and from this it follows that

$$\left\|\left(\nabla_S u\right) \circ \chi_\tau^{\text{affine}}\right\|^2 = \left\langle \widehat{\nabla}\hat{u}, \mathbf{G}_\tau^{-1}\widehat{\nabla}\hat{u}\right\rangle \qquad \text{on } \hat{\tau}.$$

Combined with the transformation formula for surface integrals we obtain the assertion. □

Lemma 4.3.5. *We have*

$$\|\mathbf{G}_\tau\| \leq 2h_\tau^2, \qquad \|\mathbf{G}_\tau^{-1}\| \leq \frac{2}{\pi^2}\left(\frac{h_\tau}{\rho_\tau}\right)^4 h_\tau^{-2}. \tag{4.202}$$

Proof. The Jacobian $\mathbf{B}_\tau$ of the affine transformation $\chi_\tau^{\text{affine}}$ has the column vectors $\mathbf{b}_1, \mathbf{b}_2$. The maximal eigenvalue of the symmetric, positive definite matrix $\mathbf{G}_\tau$ can be bounded by the row sum norm

$$\|\mathbf{G}_\tau\| \leq \max_{i=1,2}\left\{\|\mathbf{b}_i\|^2 + \langle \mathbf{b}_1, \mathbf{b}_2\rangle\right\} \leq 2h_\tau^2,$$

since $\mathbf{b}_i$ are edge vectors of τ (see Definition 4.1.2). For the inverse matrix we have

$$\mathbf{G}_\tau^{-1} = \frac{1}{\det \mathbf{G}_\tau}\begin{bmatrix} \|\mathbf{b}_2\|^2 & -\langle \mathbf{b}_1, \mathbf{b}_2\rangle \\ -\langle \mathbf{b}_1, \mathbf{b}_2\rangle & \|\mathbf{b}_1\|^2 \end{bmatrix} = \left(\frac{|\hat{\tau}|}{|\tau|}\right)^2 \begin{bmatrix} \|\mathbf{b}_2\|^2 & -\langle \mathbf{b}_1, \mathbf{b}_2\rangle \\ -\langle \mathbf{b}_1, \mathbf{b}_2\rangle & \|\mathbf{b}_1\|^2 \end{bmatrix}.$$

From this we have for the largest eigenvalue

$$\|\mathbf{G}_\tau^{-1}\| \leq \left(\frac{|\hat{\tau}|}{\pi\rho_\tau^2}\right)^2 2h_\tau^2 \leq \frac{2}{\pi^2}\left(\frac{h_\tau}{\rho_\tau}\right)^4 h_\tau^{-2}.$$

□

Lemma 4.3.4 can be generalized for derivatives of higher order.

Lemma 4.3.6. *Let $\tau \in \mathcal{G}^{\text{affine}}$ be the affine image of the reference element $\hat{\tau}$*

$$\tau = \chi_\tau^{\text{affine}}(\hat{\tau}) \quad \textit{with} \quad \chi_\tau^{\text{affine}}(\hat{\mathbf{x}}) = \mathbf{B}_\tau \hat{\mathbf{x}} + \mathbf{b}_\tau.$$

Then

$$v \in H^k(\tau) \Longleftrightarrow \hat{v} := v \circ \chi_\tau^{\text{affine}} \in H^k(\hat{\tau}), \tag{4.203}$$

which gives us for all $0 \leq \ell \leq k$

$$|v|_{H^\ell(\tau)} \leq C_1 h_\tau^{1-\ell} |\hat{v}|_{H^\ell(\hat{\tau})}, \tag{4.204a}$$

$$|\hat{v}|_{H^\ell(\hat{\tau})} \leq C_2 h_\tau^{\ell-1} |v|_{H^\ell(\tau)} \tag{4.204b}$$

with constants C_1, C_2 that depend only on k and the constant $\kappa_{\mathcal{G}}$, which describes the shape-regularity (see Definition 4.1.12).

Proof. The equivalence (4.203) follows from the chain rule, as the transformation is affine and therefore all derivatives of $\chi_\tau^{\text{affine}}$ are bounded. We will only prove the first inequality, the second can be treated in the same way.

Since $C^\infty(\tau) \cap H^\ell(\tau)$ is dense in $H^\ell(\tau)$ (see Proposition 2.3.10), it suffices to prove the statement for smooth functions.

Let $v^\star$, $\hat{v}$, U_ε, $\widehat{U}_\varepsilon$, $\hat{v}^\star$, χ, $\mathbf{J}_\tau$ be as in the proof of Lemma 4.3.4. In the following α will always denote a three-dimensional multi-index $\alpha \in \mathbb{N}_0^3$ and $\hat{\partial}$ denotes the derivative in the coordinates of the reference element. Then we have

$$|v|^2_{H^\ell(\tau)} = \sum_{|\alpha|=\ell} \int_\tau |\partial^\alpha v^\star|^2 = \frac{|\tau|}{|\hat{\tau}|} \sum_{|\alpha|=\ell} \int_{\hat{\tau}} |(\partial^\alpha v^\star) \circ \chi|^2 .$$

The chain rule then yields

$$((\partial^\alpha v^\star) \circ \chi) = \left((\mathbf{J}_\tau^{-1})^\intercal \widehat{\nabla}^\star\right)^\alpha \hat{v}^\star,$$

where $\widehat{\nabla}^\star$ denotes the three-dimensional gradient (while, in the following, the two-dimensional gradient will be denoted by $\widehat{\nabla}$, as before). For the (transposed) inverse of the Jacobian of χ we have $\left(\mathbf{J}_\tau^{-1}\right)^\intercal = [\mathbf{A}_\tau, \mathbf{n}_\tau]$ with

$$\mathbf{A}_\tau := [\mathbf{a}_1, \mathbf{a}_2] \in \mathbb{R}^{3\times 2}, \quad \mathbf{a}_1 := \frac{|\hat{\tau}|}{|\tau|} (\mathbf{b}_2 \times \mathbf{n}_\tau), \quad \mathbf{a}_2 := \frac{|\hat{\tau}|}{|\tau|} (\mathbf{n}_\tau \times \mathbf{b}_1) .$$

Since $\hat{\partial}_3 \hat{v}^\star = 0$ we obtain

$$((\partial^\alpha v^\star) \circ \chi)|_{\hat{\tau}} = \left(\mathbf{A}_\tau \widehat{\nabla}\right)^\alpha \hat{v}.$$

We use the convention

$$\sum_{\mu \le \alpha} \cdots := \sum_{\mu_1=0}^{\alpha_1} \sum_{\mu_2=0}^{\alpha_2} \sum_{\mu_3=0}^{\alpha_3} \cdots, \quad \binom{\alpha}{\mu} = \binom{\alpha_1}{\mu_1}\binom{\alpha_2}{\mu_2}\binom{\alpha_3}{\mu_3} \quad \text{and} \quad \mathbf{a}^\mu = \prod_{i=1}^3 \mathbf{a}_i^{\mu_i} .$$

for the multi-indices $\mu, \alpha \in \mathbb{N}_0^3$. With this we have

$$\left(\mathbf{A}\widehat{\nabla}\right)^\alpha \hat{v} = \sum_{\mu \le \alpha} \binom{\alpha}{\mu} \mathbf{a}_1^\mu \mathbf{a}_2^{\alpha-\mu} \hat{\partial}_1^\mu \hat{\partial}_2^{\alpha-\mu} \hat{v}.$$

In order to estimate the absolute value, we use

$$\left|\mathbf{a}_{i,j}\right| \le \|\mathbf{a}_i\| \le \frac{h_\tau}{\pi \rho_\tau^2} \le \frac{\kappa_{\mathcal{G}}^2}{\pi} h_\tau^{-1}$$

and obtain with $|\alpha| = \ell$

$$\left|\left(\mathbf{A}\widehat{\nabla}\right)^\alpha \hat{v}\,(\hat{\mathbf{x}})\right|^2 \le C h_\tau^{-2\ell} \sum_{\mu \le \alpha} \left|\hat{\partial}_1^\mu \hat{\partial}_2^{\alpha-\mu} \hat{v}\,(\hat{\mathbf{x}})\right|^2 \tag{4.205}$$

with a constant C, which depends only on ℓ and the constant $\kappa_{\mathcal{G}}$. By integrating over $\hat{\tau}$ we obtain

$$\left\|\partial^\alpha v^\star\right\|_{L^2(\tau)}^2 = \frac{|\tau|}{|\hat{\tau}|}\left\|\left(\mathbf{A}\widehat{\nabla}\right)^\alpha \hat{v}\right\|_{L^2(\hat{\tau})}^2 \leq C h_\tau^{2-2\ell}\,|\hat{v}|_{H^\ell(\hat{\tau})}^2 .$$

If we sum over all α with $|\alpha| = \ell$ we obtain the assertion. □

The following corollary is a consequence of (4.205).

Corollary 4.3.7. *Let $\tau \in \mathcal{G}^{\text{affine}}$ be the affine image of the reference element $\hat{\tau}$*

$$\tau = \chi_\tau^{\text{affine}}(\hat{\tau}) \quad \text{with} \quad \chi_\tau^{\text{affine}}(\hat{\mathbf{x}}) = \mathbf{B}_\tau \hat{\mathbf{x}} + \mathbf{b}_\tau .$$

Then

$$v \in C^k(\tau) \Longleftrightarrow \hat{v} := v \circ \chi_\tau^{\text{affine}} \in C^k(\hat{\tau}),$$

which gives us for all $0 \leq \ell \leq k$

$$|v|_{C^\ell(\tau)} \leq C_1 h_\tau^{-\ell}\,|\hat{v}|_{C^\ell(\hat{\tau})}, \tag{4.206a}$$

$$|\hat{v}|_{C^\ell(\hat{\tau})} \leq C_2 h_\tau^{\ell}\,|v|_{C^\ell(\tau)} \tag{4.206b}$$

with constants C_1, C_2 that depend only on k and the constant $\kappa_{\mathcal{G}}$, which describes the shape-regularity (see Definition 4.1.12).

Theorem 4.3.8. *Let $\tau \in \mathcal{G}^{\text{affine}}$ be the affine image of the reference element $\tau = \chi_\tau^{\text{affine}}(\hat{\tau})$. Let the interpolation operator $\widehat{\Pi} : H^s(\hat{\tau}) \to H^t(\hat{\tau})$ be continuous for $0 \leq t \leq s \leq k+1$ and let*

$$\forall q \in \mathbb{P}_k^{\hat{\tau}} : \quad \widehat{\Pi} q = q \tag{4.207}$$

hold. Then the operator $\Pi : H^s(\tau) \to H^t(\tau)$, which is defined by:

$$\Pi v := \left(\widehat{\Pi}\hat{v}\right) \circ \left(\chi_\tau^{\text{affine}}\right)^{-1} \quad \text{with} \quad \hat{v} := v \circ \chi_\tau^{\text{affine}}, \tag{4.208}$$

satisfies the error estimate

$$\forall v \in H^{k+1}(\tau) : \; |v - \Pi v|_{H^t(\tau)} \leq C h_\tau^{s-t} |v|_{H^s(\tau)} \tag{4.209}$$

for $0 \leq t \leq s \leq k+1$. The constant C depends only on k and the shape-regularity of the surface mesh, more specifically, it depends on the constant $\kappa_{\mathcal{G}}$ in Definition 4.1.12.

Proof. According to Proposition 4.3.3, on the reference element $\hat{\tau}$ we have

$$\left\|\hat{v} - \widehat{\Pi}\hat{v}\right\|_{H^t(\hat{\tau})} \leq \hat{c}\,|\hat{v}|_{H^s(\hat{\tau})}.$$

We transport this estimate from $\hat{\tau}$ to $\tau = \chi_\tau^{\text{affine}}(\hat{\tau})$. With Lemma 4.3.6 we obtain the error estimate for $s = k + 1$

$$|v - \Pi v|_{H^t(\tau)} \leq C h_\tau^{1-t} \left|\hat{v} - \widehat{\Pi}\hat{v}\right|_{H^t(\hat{\tau})} \leq C h_\tau^{1-t} |\hat{v}|_{H^{k+1}(\hat{\tau})} \leq C h_\tau^{k+1-t} |v|_{H^{k+1}(\tau)}.$$

For $s < k + 1$, (4.209) follows from the continuity of $\widehat{\Pi} : H^s(\hat{\tau}) \to H^t(\widehat{\tau})$ by means of interpolation (see proof of Theorem 4.1.33). □

Remark 4.3.9. *The interpolation operator $\widehat{I}^k$ from (4.73) satisfies the prerequisites of Proposition 4.3.3 with $p \leftarrow k \geq 1$ by virtue of the Sobolev Embedding Theorem.*

For $k = 0$, $\widehat{\Pi}$ can be defined as a mean value:

$$\left(\widehat{\Pi} v\right)(\mathbf{x}) = \frac{1}{|\hat{\tau}|} \int_{\hat{\tau}} v \qquad \forall \mathbf{x} \in \hat{\tau}.$$

4.3.2 Approximation on Curved Panels*

In this section we will prove the approximation properties for curved panels that satisfy the following geometric assumptions (see Assumption 4.1.6 and Fig. 4.6).

For $\mathbf{x} \in \tau \in \mathcal{G}$, $\mathbf{n}_\tau(\mathbf{x}) \in \mathbb{S}_2$ denotes unit normal vector to τ at the point $\mathbf{x}$. The orientation is chosen as explained in Sect. 2.2.3 with respect to the chart χ_τ.

Assumption 4.3.10. *For every $\tau \in \mathcal{G}$ with the associated reference mapping $\chi_\tau : \hat{\tau} \to \tau$:*

- *There exists a regular, affine mapping $\chi_\tau^{\text{affine}} : \mathbb{R}^3 \to \mathbb{R}^3$ of the form*

$$\chi_\tau^{\text{affine}}(\hat{\mathbf{x}}, x_3) = \begin{bmatrix} \mathbf{a} & \mathbf{0} \\ \mathbf{0} & 1 \end{bmatrix} \begin{pmatrix} \hat{\mathbf{x}} \\ \hat{x}_3 \end{pmatrix} + \begin{pmatrix} \mathbf{b}_\tau \\ 0 \end{pmatrix}$$

with $\mathbf{a} \in \mathbb{R}^{2\times 2}$, $(\hat{\mathbf{x}}, \hat{x}_3) \in \mathbb{R}^2 \times \mathbb{R}$, $\mathbf{b}_\tau \in \mathbb{R}^2$ and $\det \mathbf{a} > 0$.

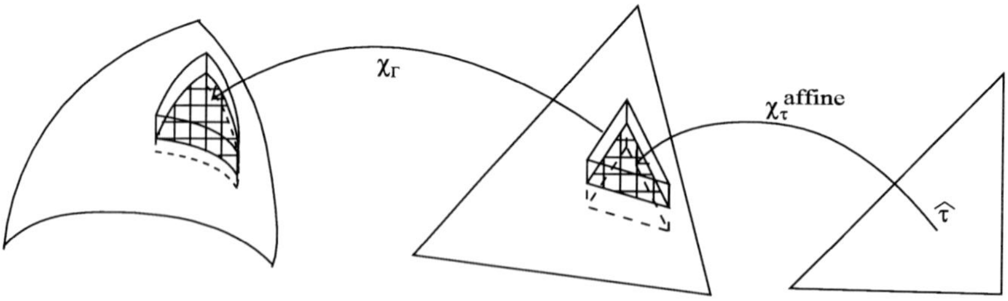

Fig. 4.6 *Left*: curved surface panel τ and three-dimensional neighborhood U_τ. *Middle*: flat surface panel τ^{affine} with neighborhood U_τ^{affine}. *Right*: reference element $\hat{\tau} \subset \mathbb{R}^2$

* This section should be read as a complement to the core material of this book.

- *There exists a C^∞-diffeomorphism $\chi : U \to V$ that is independent of $\mathcal{G}$, with open sets $U, V \subset \mathbb{R}^3$ that satisfy*

$$\tau_\varepsilon^{\text{affine}} \subset U, \; \tau_\varepsilon^{\text{affine}} := \left\{\chi_\tau^{\text{affine}}(\hat{\mathbf{x}}, 0) : \hat{\mathbf{x}} \in \widehat{\tau}\right\} \times (-\varepsilon, \varepsilon),$$
$$\tau_\varepsilon \subset V, \quad \tau_\varepsilon := \{\mathbf{x} + \alpha \mathbf{n}_\tau(\mathbf{x}) : \mathbf{x} \in \tau, \alpha \in (-\varepsilon, \varepsilon)\}$$

for an $\varepsilon > 0$ such that

$$\chi_\tau(\hat{\mathbf{x}}) = \chi \circ \chi_\tau^{\text{affine}}(\hat{\mathbf{x}}, 0).$$

- *For every function $u \in H^k(\tau)$ with a constant extension*

$$u^\star(\mathbf{x} + \alpha \mathbf{n}_\tau(\mathbf{x})) = u(\mathbf{x}) \tag{4.210}$$

we have

$$\partial\left(u^\star \circ \chi \circ \chi_\tau^{\text{affine}}\right)/\partial \hat{x}_3 = 0. \tag{4.211}$$

A situation of this kind was introduced in Example 4.1.7 (also see [170, Chap. 2]). First we will prove a transformation formula for composite functions.

Lemma 4.3.11. *Let $\eta : U \to V$ be a C^∞-diffeomorphism and let $U, V \subset \mathbb{R}^3$ be open sets. For a function $u \in H^k(V)$ we set $\tilde{u} = u \circ \eta$. Then $\tilde{u} \in H^k(U)$ and for all $\alpha \in \mathbb{N}_0^3$, $1 \leq |\alpha| \leq k$, we have*

$$(\partial^\alpha \tilde{u}) \circ \eta^{-1} = \sum_{|\beta|=1}^{|\alpha|} c_\beta \partial^\beta u \tag{4.212}$$

with coefficients c_β that are real linear combinations of products of the form

$$\prod_{r=1}^{|\beta|} \partial^{\mu_r} \eta_{n_r}. \tag{4.213}$$

The relevant indices for $1 \leq r \leq |\beta|$ satisfy the relations $1 \leq n_r \leq 3$, $\mu_r \in \mathbb{N}_0^3$ and $\sum_{r=1}^{|\beta|} |\mu_r| = |\alpha|$.

Proof. For the equivalence $u \in H^k(V) \iff \tilde{u} \in H^k(U)$ it suffices to prove (4.212) for smooth functions. We will prove Formula (4.212) by induction. Let $\mathbf{e}_k$ be the k-th canonical unit vector in $\mathbb{R}^3$.

Initial case: For $|\alpha| = 1$ we obtain explicitly

$$(\partial^\alpha \tilde{u}) \circ \eta^{-1} = \sum_{|\beta|=1} c_\beta \partial^\beta u, \quad \text{where for } \beta = \mathbf{e}_k \text{ we have } c_\beta = \partial^\alpha \eta_k.$$

Hypothesis: Let the statement hold for $|\alpha| \leq i - 1$.

Conclusion: Let $|\alpha| = i$, choose $k = 1, 2$, or 3, and let $\tilde{\alpha} = \alpha - \mathbf{e}_k \in \mathbb{N}_0^3$. Thus we obtain

$$(\partial^\alpha \tilde{u}) \circ \eta^{-1} = \partial_k \left(\partial^{\tilde{\alpha}} \tilde{u}\right) \circ \eta^{-1} = \left(\partial_k \sum_{|\beta|=1}^{i-1} c_\beta \left(\partial^\beta u\right) \circ \eta\right) \circ \eta^{-1}$$

$$= \sum_{|\beta|=1}^{i-1} (\partial_k c_\beta) \, \partial^\beta u + \sum_{|\beta|=1}^{i-1} \sum_{j=1}^{3} (c_\beta \partial_k \eta_j) \left(\partial_j \partial^\beta u\right).$$

This proves the assertion if we show that $\partial_k c_\beta$ and $c_\beta \left(\partial_k \eta_j\right)$ are of the form (4.213). With the Leibniz product rule we obtain

$$\partial_k \prod_{r=1}^{|\beta|} \partial^{\mu_r} \eta_{n_r} = \sum_{j=1}^{|\beta|} (\partial_k \partial^{\mu_j}) \, \eta_{n_j} \prod_{\substack{r=1 \\ r \neq j}}^{|\beta|} \partial^{\mu_r} \eta_{n_r}$$

and the expression on the right-hand side is a linear combination of terms of the form

$$\prod_{r=1}^{|\beta|} \partial^{\tilde{\mu}_r} \eta_{n_r}$$

with $\sum_{r=1}^{|\beta|} |\tilde{\mu}_r| = i$. The assertion follows analogously for the product $c_\beta \left(\partial_k \eta_j\right)$. □

Corollary 4.3.12. *1. Let the conditions of Lemma 4.3.11 be satisfied. Then*

$$C_1^{-1} \|\tilde{u}\|_{L^2(U)} \leq \|u\|_{L^2(V)} \leq C_2 \|\tilde{u}\|_{L^2(U)}$$

and

$$|\tilde{u}|^2_{H^k(U)} \leq C_1 \sum_{i=1}^{k} |u|^2_{H^i(V)} \quad \text{and} \quad |u|^2_{H^k(V)} \leq C_2 \sum_{i=1}^{k} |\tilde{u}|^2_{H^i(U)}.$$

The constants C_1, C_2 depend only on k and the derivatives of η, η^{-1} up to the order $\max\{1, k\}$.

2. *Let Assumption 4.3.10 and the conditions of Lemma 4.3.11 be satisfied with $\eta \leftarrow \chi$. For $\tau \in \mathcal{G}$, $\tau^{\text{affine}} := \chi^{-1}(\tau)$ and $u \in H^k(\tau)$, $\tilde{u}(\hat{\mathbf{x}}) := u \circ \chi(\hat{\mathbf{x}}, 0)$ we have*

$$C_3^{-1} \|\tilde{u}\|^2_{L^2(\tau^{\text{affine}})} \leq \|u\|^2_{L^2(\tau)} \leq C_4 \|\tilde{u}\|^2_{L^2(\tau^{\text{affine}})}$$

and

$$|\tilde{u}|^2_{H^k(\tau^{\text{affine}})} \le C_3 \sum_{i=1}^{k} |u|^2_{H^i(\tau)} \quad \text{and} \quad |u|^2_{H^k(\tau)} \le C_4 \sum_{i=1}^{k} |\tilde{u}|^2_{H^i(\tau^{\text{affine}})} .$$

The constants C_3, C_4 again depend only on k and the derivatives of χ, χ^{-1} up to the order k.

Proof. Statement 1 follows from the transformation formula (4.212).

For the second statement we define a constant extension of u in the direction of the normal as a function $u^\star$, according to (4.210), and note that the normal derivative of $u^\star$ vanishes, i.e., we have $|u^\star|_{H^k(\tau_\varepsilon)} = |u|_{H^k(\tau)}$.

From (4.211) we have $|u^\star \circ \chi|_{H^k(\tau_\varepsilon^{\text{affine}})} = |\tilde{u}|_{H^k(\tau^{\text{affine}})}$ and, thus, we have the assertion in Part 1. □

At the next step we will apply Lemma 4.3.11 to the composite reference mapping and study how far this depends on the panel diameter h_τ.

Lemma 4.3.13. *Let Assumption 4.3.10 and the conditions of Lemma 4.3.11 hold with $\eta \leftarrow \chi$. For $\tau \in \mathcal{G}$ and $u \in H^k(\tau)$, $\tau \subset V$, $\hat{u} := u \circ \chi_\tau$ we have*

$$v \in H^k(\tau) \Longleftrightarrow \hat{v} := v \circ \chi_\tau \in H^k(\hat{\tau}) \tag{4.214}$$

and

$$|u|^2_{H^k(\tau)} \le C_1 h_\tau^{2-2k} \sum_{i=1}^{k} |\hat{u}|^2_{H^i(\hat{\tau})} , \tag{4.215a}$$

$$|\hat{u}|^2_{H^k(\hat{\tau})} \le C_2 h_\tau^{2k-2} \sum_{i=1}^{k} |u|^2_{H^i(\tau)} . \tag{4.215b}$$

The constants C_1, C_2 depend only on k, the constant $\kappa_\mathcal{G}$ of the shape-regularity (see Definition 4.1.12) and the derivatives χ, χ^{-1} up to the order k.

Proof. It follows from Corollary 4.3.12 that

$$|u|^2_{H^k(\tau)} \le C \sum_{i=1}^{k} |\tilde{u}|^2_{H^i(\tau^{\text{affine}})} .$$

We can therefore apply the transformation formulas from Lemma 4.3.6, which gives us the estimates

$$|u|^2_{H^k(\tau)} \le C_1 h_\tau^2 \sum_{i=1}^{k} h^{-2i} |\tilde{u}|^2_{H^i(\tau^{\text{affine}})} \le C_2 h_\tau^{2-2k} \sum_{i=1}^{k} |\hat{u}|^2_{H^i(\hat{\tau})},$$

$$|\hat{u}|^2_{H^k(\hat{\tau})} \le C_3 h_\tau^{2k-2} |\tilde{u}|^2_{H^k(\tau^{\text{affine}})} \le C_4 h_\tau^{2k-2} \sum_{i=1}^{k} |u|^2_{H^i(\tau)}.$$

□

With this we obtain the analogy of Theorem 4.3.8 for curved panels.

Theorem 4.3.14. *Let Assumption 4.3.10 and the conditions from Lemma 4.3.11 hold with* $\eta \leftarrow \chi$. *Let* $\tau \in \mathcal{G}$ *be the image of the reference element* $\hat{\tau}$ *as given by* $\tau = \chi \circ \chi_\tau^{\text{affine}}(\hat{\tau})$. *Let the interpolation operator* $\widehat{\Pi} : H^s(\hat{\tau}) \to H^t(\hat{\tau})$ *satisfy the conditions from Theorem 4.3.8 for* $0 \le t \le s \le k+1$.

Then we have for the operator $\Pi : H^s(\tau) \to H^t(\tau)$, *which is defined by*

$$\Pi v := \left(\widehat{\Pi}\hat{v}\right) \circ \chi_\tau^{-1} \quad \text{with} \quad \hat{v} := v \circ \chi_\tau,$$

the error estimate for $0 \le t \le s \le k+1$

$$\forall v \in H^{k+1}(\tau): \ |v - \Pi v|_{H^t(\tau)} \le C h_\tau^{s-t} \|v\|_{H^s(\tau)}. \tag{4.216}$$

The constant C *depends only on* k, *the shape-regularity of the surface mesh via the constant* $\kappa_{\mathcal{G}}$ *in Definition 4.1.12 and the derivatives of* χ, χ^{-1} *up to the order* k.

Theorem 4.3.8 and Theorem 4.3.14 contain the central, local approximation properties that are combined in Sects. 4.3.4 and 4.3.5 to form error estimates for boundary elements. The easiest way of constructing a global approximation for continuous boundary elements and sufficiently smooth functions is by means of interpolation. For this the functions $u \in H^s_{\text{pw}}(\Gamma)$ need to be continuous. In the following section we will show that this is the case for $s > 1$.

4.3.3 Continuity of Functions in $H^s_{\text{pw}}(\Gamma)$ for $s > 1$

In order to avoid technical difficulties, we will generally assume in this section that we are dealing with the geometric situation from Example 4.1.7(1).

Assumption 4.3.15. Γ *is a piecewise smooth Lipschitz surface that can be parametrized bi-Lipschitz continuously over a polyhedral surface* $\hat{\Gamma}$: $\chi_\Gamma : \hat{\Gamma} \to \Gamma$.

Then the Sobolev spaces $H^s(\Gamma)$ on Γ are defined invariantly for $|s| \le 1$, which means that they do not depend on the chosen parametrization of Γ (see Proposition 2.4.2). For a higher differentiation index $s > 1$, $H^s_{\text{pw}}(\Gamma)$ is defined as in (4.86). These spaces form a scale with

$$L^2(\Gamma) = H^0_{\text{pw}}(\Gamma) \supset H^s_{\text{pw}}(\Gamma) \supset H^t_{\text{pw}}(\Gamma),\ 0 < s < t. \tag{4.217}$$

Lemma 4.3.16. *For $s > 1$ every $u \in H^s_{\text{pw}}(\Gamma)$ is continuous on Γ, i.e., $H^s_{\text{pw}}(\Gamma) \subset C^0(\Gamma)$.*

Proof. Γ is the bi-Lipschitz continuous image of a polyhedral surface: $\Gamma = \chi_\Gamma\left(\hat{\Gamma}\right)$ and therefore it suffices to prove the statement for polyhedral surfaces. Let $\hat{\Gamma}^j$, $1 \leq j \leq J$, be the plane, relatively closed polygonal faces of the polyhedron.

Let $u \in H^s_{\text{pw}}\left(\hat{\Gamma}\right)$ for $s > 1$. The Sobolev Embedding Theorem implies that $u \in C^0\left(\hat{\Gamma}^j\right)$ for all $1 \leq j \leq J$ and, thus, it suffices to prove the continuity across the common edges of the surface pieces $\hat{\Gamma}_j$. For this we consider two pieces $\hat{\Gamma}_i$ and $\hat{\Gamma}_j$ with a common edge $\widehat{E}$. Then there exists an (open) polygonal domain $U \subset \mathbb{R}^2$ and a bi-Lipschitz continuous mapping $\chi : \overline{U} \to \hat{\Gamma}_i \cup \hat{\Gamma}_j$ with the properties

$$\overline{U}_1 := \chi^{-1}\left(\hat{\Gamma}_i\right), \quad \overline{U}_2 := \chi^{-1}\left(\hat{\Gamma}_j\right), \quad \text{and} \quad \chi|_{U_k} \text{ is affine for } k = 1, 2.$$

$$U_1, U_2 \text{ are disjoint and } \overline{U} = \overline{U_1 \cup U_2}.$$

$$e := \chi^{-1}\left(\widehat{E}\right) = \overline{U_1} \cap \overline{U_2}.$$

We only need to show that $w := u \circ \chi$ is continuous over e. Clearly, we have $w_k := w \circ \chi_k \in H^s(U_k)$, $k = 1, 2$, and $w \in H^1(U)$. If we combine this result with the statements from Theorem 2.6.8 and Remark 2.6.10 we obtain the assertion. □

4.3.4 Approximation Properties of $S_{\mathcal{G}}^{p,-1}$

We will now prove the error estimate (4.59) for the following two geometric situations.

Assumption 4.3.17 (Polyhedral Surface). *Γ is the surface of a polyhedron. The mesh $\mathcal{G}$ on Γ consists of plane panels with straight edges with mesh width $h > 0$.*

Assumption 4.3.18 (Curved Surface). *Assumption 4.3.10 holds and the conditions from Lemma 4.3.11 are satisfied with $\eta \leftarrow \chi$.*

Theorem 4.3.19. *Let either Assumption 4.3.17 or Assumption 4.3.18 hold. Let $s \geq 0$. Then there exists an operator $I_{\mathcal{G}}^{p,-1} : H^s_{\text{pw}}(\Gamma) \to S_{\mathcal{G}}^{p,-1}$ such that*

$$\left\| u - I_{\mathcal{G}}^{p,-1} u \right\|_{L^2(\Gamma)} \leq C\, h^{\min(p+1,s)} \|u\|_{H^s(\Gamma)}. \tag{4.218}$$

For a polyhedral surface the constant C depends only on p and the shape-regularity of the mesh $\mathcal{G}$ via the constant $\kappa_{\mathcal{G}}$ from Definition 4.1.12. In the case of a curved *surface it also depends on the derivatives of the global transformations χ, χ^{-1} up to the order k.*

Proof. Let $\widehat{\Pi}_{\hat{\tau}}^{p} : H^s(\hat{\tau}) \to \mathbb{P}_p^{\Delta}$ be the L^2-projection:

$$\left(\widehat{\Pi}_{\hat{\tau}}^{p} u, q\right)_{L^2(\hat{\tau})} = (u, q)_{L^2(\hat{\tau})} \qquad \forall q \in \mathbb{P}_p^{\Delta}. \tag{4.219}$$

We lift this projection to the panels $\tau \in \mathcal{G}$ by means of

$$\left(\Pi_\tau^p u_\tau\right)(\mathbf{x}) := \left(\widehat{\Pi}_{\hat{\tau}}^{p} \hat{u}_\tau\right) \circ \chi_\tau^{-1}(\mathbf{x}) \qquad \forall \mathbf{x} \in \tau,$$

where $u_\tau := u|_\tau$ and $\hat{u}_\tau := u_\tau \circ \chi_\tau$. The operator $I_{\mathcal{G}}^{p,-1}$ then consists of the panelwise composition of Π_τ^p:

$$I_{\mathcal{G}}^{p,-1} u\Big|_\tau := \Pi_\tau^p u \qquad \forall \tau \in \mathcal{G}.$$

Obviously, this defines a mapping from $H_{\mathrm{pw}}^s(\Gamma)$ to $S_{\mathcal{G}}^{p,-1}$. The operator $\widehat{\Pi}_{\hat{\tau}}^{p}$ satisfies the prerequisites of Theorem 4.3.8, because we have for the orthogonal projection:

1. $$\left\|\widehat{\Pi}_{\hat{\tau}}^{p} \hat{v}\right\|_0 \leq \|\hat{v}\|_0 \qquad \forall \hat{v} \in L^2(\hat{\tau}).$$
   Since $\widehat{\Pi}_{\hat{\tau}}^{p} \hat{v}$ is a polynomial in a finite-dimensional space $\mathbb{P}_p^{\Delta}$, all norms are equivalent and there exists a constant $C_p > 0$ such that for all $0 \leq t \leq s \leq p+1$ we have
   $$\left\|\widehat{\Pi}_{\hat{\tau}}^{p} \hat{v}\right\|_s \leq C_p \left\|\widehat{\Pi}_{\hat{\tau}}^{p} \hat{v}\right\|_0 \leq C_p \|\hat{v}\|_0 \leq C_p \|\hat{v}\|_t \qquad \forall \hat{v} \in H^s(\hat{\tau}).$$
2. It follows immediately from the characterization (4.219) that
   $$\widehat{\Pi} q = q \qquad \forall q \in \mathbb{P}_p^{\Delta}.$$

Therefore we can apply (4.209) or (4.216) with $t = 0$ and obtain the error estimate

$$\left|v - I_{\mathcal{G}}^{p,-1} v\right|_{L^2(\tau)} \leq C h_\tau^s \|v\|_{H^s(\tau)} \tag{4.220}$$

for all $v \in H^s(\Gamma)$ with $0 \leq s \leq p+1$. If we then square and sum over all $\tau \in \mathcal{G}$ we obtain the assertion. □

Theorem 4.3.19 gives us error estimates in negative norms by means of the same duality argument as in the proof of Theorem 4.1.33. This is the subject of the following theorem.

Theorem 4.3.20. *Let the assumption from Theorem 4.3.19 be satisfied. Then we have for the interpolation $I_{\mathcal{G}}^{p,-1}$ and $0 \leq t \leq s \leq p+1$ and all $u \in H_{\mathrm{pw}}^s(\Gamma)$ the estimate*

$$\|u - I_{\mathcal{G}}^{p,-1} u\|_{H^{-t}(\Gamma)} \leq C h^{s+t} \|u\|_{H^s(\Gamma)}. \tag{4.221}$$

Proof. The continuous extension of the L^2 inner-product to $H_{\mathrm{pw}}^{-t}(\Gamma) \times H_{\mathrm{pw}}^{t}(\Gamma)$ is again denoted by $(\cdot,\cdot)_0$. Since $I_{\mathcal{G}}^{p,-1}$ consists locally of L^2-orthogonal projections, we have for an arbitrary $\varphi_{\mathcal{G}} \in S_{\mathcal{G}}^{p,-1}$

$$\left\| u - I_{\mathcal{G}}^{p,-1} u \right\|_{H^{-t}(\Gamma)} = \sup_{\varphi \in H^t(\Gamma)\backslash\{0\}} \frac{\left| \left(u - I_{\mathcal{G}}^{p,-1} u, \varphi \right)_0 \right|}{\|\varphi\|_{H^t(\Gamma)}} = \sup_{\varphi \in H^t(\Gamma)\backslash\{0\}} \frac{\left| \left(u - I_{\mathcal{G}}^{p,-1} u, \varphi - \varphi_{\mathcal{G}} \right)_0 \right|}{\|\varphi\|_{H^t(\Gamma)}} \tag{4.222}$$

(see proof of Theorem 4.1.33). If we choose $\varphi_{\mathcal{G}} = I_{\mathcal{G}}^{p,-1}\varphi \in S_{\mathcal{G}}^{p,-1}$, (4.221) follows by means of a twofold application of (4.218). □

Remark 4.3.21. *Corollary 4.1.34 follows from (4.221) with* $t = \frac{1}{2}$.

4.3.5 Approximation Properties of $S_{\mathcal{G}}^{p,0}$

Here we will prove approximation properties of continuous boundary elements that have already been introduced in Proposition 4.1.50.

Theorem 4.3.22. *Let Assumption 4.3.17 or Assumption 4.3.18 hold:*

(a) Then there exists an interpolation operator $I_{\mathcal{G}}^{p,0} : H_{\mathrm{pw}}^{s}(\Gamma) \to S_{\mathcal{G}}^{p,0}$ *such that*

$$\left\| u - I_{\mathcal{G}}^{p,0} u \right\|_{H^t(\Gamma)} \le C h^{s-t} \|u\|_{H_{\mathrm{pw}}^s(\Gamma)} \tag{4.223}$$

for $t = 0, 1$, $1 < s \le p+1$ *and all* $u \in H_{\mathrm{pw}}^{s}(\Gamma)$. *For a polyhedral surface the constant* C *depends only on* p *and on the shape-regularity of the mesh* $\mathcal{G}$ *via the constant* $\kappa_{\mathcal{G}}$ *from Definition 4.1.12. In the case of a curved surface it also depends on the derivatives of the global transformations* χ, χ^{-1} *up to the order* k.

(b) Let $u \in H^s(\Gamma)$ *for some* $1 < s \le p+1$. *Then, for any* $0 \le t \le 1$, *we have*

$$\left\| u - I_{\mathcal{G}}^{p,0} u \right\|_{H^t(\Gamma)} \le C h^{s-t} \|u\|_{H^s(\Gamma)}.$$

Proof. Part a: Lemma 4.3.16 implies that $u \in H_{\mathrm{pw}}^{s}(\Gamma) \subset C^0(\Gamma)$ for $s > 1$. We define $I_{\mathcal{G}}^{p,0} u$ on $\tau \in \mathcal{G}$ by

$$\left(I_{\mathcal{G}}^{p,0} u_\tau \right)(\mathbf{x}) := \left(\widehat{I}^p \hat{u}_\tau \right) \circ \chi_\tau^{-1}(\mathbf{x}) \qquad \forall \mathbf{x} \in \tau \tag{4.224}$$

with $u_\tau := u|_\tau$, $\hat{u}_\tau := u_\tau \circ \chi_\tau$ and the interpolation operator $\widehat{I}^p$ from (4.73) for the set of nodal points Σ_p from Theorem 4.1.39. By Theorem 4.1.39 this operator is well defined and satisfies

$$\begin{aligned}\left(\widehat{I}^p \hat{u}_\tau\right)(\mathbf{z}) &= \hat{u}_\tau(\mathbf{z}) \ \forall \mathbf{z} \in \Sigma_p, \\ \widehat{I}^p q = q & \qquad \forall q \in \mathbb{P}_p^{\hat{\tau}}.\end{aligned}$$

By Lemma 4.3.1 we have on the reference element

$$\begin{aligned}\left\|\widehat{I}^p \hat{u}_\tau\right\|_{H^t(\hat{\tau})} &\leq \left\|\widehat{I}^p \hat{u}_\tau\right\|_{H^{p+1}(\hat{\tau})} \leq c_2 \left(\left|\widehat{I}^p \hat{u}_\tau\right|_{H^{p+1}(\hat{\tau})} + \sum_{\mathbf{z}\in\Sigma^p} \left|\left(\widehat{I}^p \hat{u}_\tau\right)(\mathbf{z})\right|\right) \\ &= c_2 \sum_{\mathbf{z}\in\Sigma^p} |\hat{u}(\mathbf{z})| \\ &\leq c_2 \|\hat{u}\|_{C^0(\overline{\hat{\tau}})} \leq C c_2 \|\hat{u}\|_{H^s(\hat{\tau})}.\end{aligned}$$

Therefore Theorem 4.3.8 or Theorem 4.3.14 is applicable and for $1 < s \leq p+1$ and $t \in \{0, 1\}$ we obtain the estimate

$$\forall u \in H^s(\tau): \left|u_\tau - I_{\mathcal{G}}^{p,0} u_\tau\right|_{H^t(\tau)} \leq C h_\tau^{s-t} \|u_\tau\|_{H^s(\tau)}. \tag{4.225}$$

If we square (4.225) and sum over all $\tau \in \mathcal{G}^{\text{affine}}$ we obtain (4.223).

Part b: By using Lemma 4.1.49 we derive from Part a the estimate

$$\left\|u - I_{\mathcal{G}}^{p,0} u\right\|_{H^t(\Gamma)} \leq C h^{s-t} \|u\|_{H^s(\Gamma)} \tag{4.226}$$

for $t \in \{0, 1\}$. We apply Proposition 2.1.62 with $T = I - I_{\mathcal{G}}^{p,0}$, $Y_0 = Y_1 = H^s(\Gamma)$, $X_0 = L^2(\Gamma)$, $X_1 = H^1(\Gamma)$, and $\theta = t \in (0, 1)$ to interpolate the inequality (4.226). The result is

$$\begin{aligned}\|T\|_{H^t(\Gamma)\leftarrow H^s(\Gamma)} &\leq \|T\|_{L^2(\Gamma)\leftarrow H^s(\Gamma)}^{1-t} \|T\|_{H^1(\Gamma)\leftarrow H^s(\Gamma)}^{t} \\ &\leq \left(C h_{\mathcal{G}}^s\right)^{1-t} \left(C h_{\mathcal{G}}^{s-1}\right)^t = C h_{\mathcal{G}}^{s-t}\end{aligned}$$

and this implies the assertion of Part b. □

Next we investigate the approximation property for functions in $H_{\text{pw}}^s(\Gamma)$ for $0 \leq s \leq 1$. Recall that $H_{\text{pw}}^s(\Gamma) = H^s(\Gamma)$ in this case. In general, functions in $H^s(\Gamma)$ are not continuous and the application of the pointwise interpolation $I_{\mathcal{G}}^{p,0}$ is not defined. We will introduce the *Clément interpolation operator* $Q_{\mathcal{G}} : L^1(\Gamma) \to S_{\mathcal{G}}^{1,0}$ for the approximation of functions in $H^s(\Gamma)$ if $0 \leq s \leq 1$ (cf. [69]). To avoid technicalities, we consider only the case that all panels are (possibly curved) surface triangles. Let $\mathcal{I}$ denote the set of panel vertices with corresponding continuous, piecewise linear nodal basis $(b_{\mathbf{z}})_{\mathbf{z}\in\mathcal{I}}$. For $\mathbf{z} \in \mathcal{I}$ and $\tau \in \mathcal{G}$, we introduce local meshes $\mathcal{G}_{\mathbf{z}}$ and $\mathcal{G}_\tau$ by

$$\mathcal{G}_{\mathbf{z}} := \{\tau \in \mathcal{G} \mid \tau \subset \operatorname{supp} b_{\mathbf{z}}\}, \quad \mathcal{G}_\tau := \left\{t \in \mathcal{G} \mid \bar{t} \cap \bar{\tau} \neq \emptyset\right\}.$$

The corresponding surface patches on Γ are denoted by

$$\omega_{\mathbf{z}} := \bigcup_{\tau\in\mathcal{G}_{\mathbf{z}}} \bar{\tau}, \quad \omega_\tau := \bigcup_{t\in\mathcal{G}_\tau} \bar{t}.$$

For functions $f \in L^1(\Gamma)$ and $\mathbf{z} \in \mathcal{I}$, the functional $\pi_{\mathbf{z}} : L^1(\Gamma) \to \mathbb{C}$ is defined by

$$\pi_{\mathbf{z}}(f) := \frac{1}{|\omega_{\mathbf{z}}|} \int_{\omega_{\mathbf{z}}} f(\mathbf{y})\, ds_{\mathbf{y}}.$$

Remark 4.3.23. *For $\mathbf{z} \in \mathcal{I}$, we set $h_{\mathbf{z}} := \max_{\tau\in\mathcal{G}_{\mathbf{z}}} h_\tau$. There exists a constant C_0 which depends only on the shape-regularity constant $\kappa_{\mathcal{G}}$ such that*

$$h_{\mathbf{z}} \leq C_0 h_\tau \qquad \forall \tau \in \mathcal{G}_{\mathbf{z}}.$$

Definition 4.3.24 (Clément interpolation). The Clément interpolation operator $Q_{\mathcal{G}} : L^1(\Gamma) \to S_{\mathcal{G}}^{1,0}$ is given by

$$Q_{\mathcal{G}} f := \sum_{\mathbf{z}\in\mathcal{I}} \pi_{\mathbf{z}}(f)\, b_{\mathbf{z}}.$$

The proof of the stability and the approximation property of the Clément interpolation employs local pullbacks to two-dimensional polygonal parameter domains and then follows the classical convergence proof in the two-dimensional parameter plane. The next assumption is illustrated in Fig. 4.7.

Assumption 4.3.25. *(a) For any $\mathbf{z} \in \mathcal{I}$, there is a two-dimensional convex and polygonal parameter domain $\widetilde{\omega}_{\mathbf{z}} \subset \mathbb{R}^2$ along with a bi-Lipschitz continuous mapping $\chi_{\mathbf{z}} : \widetilde{\omega}_{\mathbf{z}} \to \omega_{\mathbf{z}}$ which satisfies: For any $\tau \in \mathcal{G}_{\mathbf{z}}$, the pullback $\tilde{\tau} :=$*

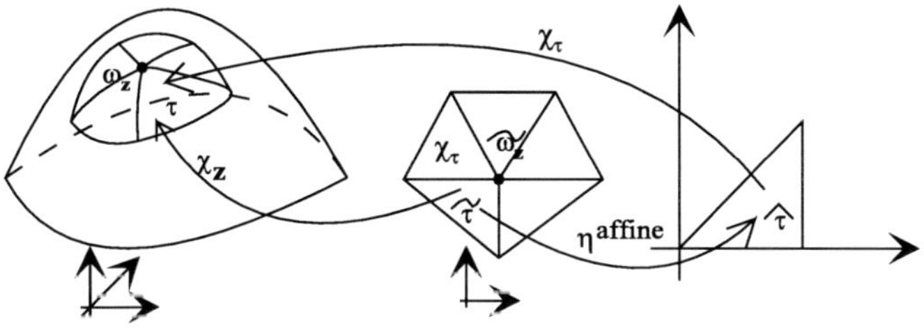

Fig. 4.7 Pullback of a surface patch to a two-dimensional parameter domain

$\chi_{\mathbf{z}}^{-1}(\tau)$ *is a plane panel with straight edges. The pullback* $\tilde{\tau}$ *can be transformed to the reference element* $\hat{\tau}$ *by a regular, affine mapping which is denoted by* η^{affine}.

(b) *The reference mapping (see Definition 4.1.2) is denoted by* $\chi_\tau : \hat{\tau} \to \tau$, *where the reference element is always the unit triangle* $\hat{\tau} = \widehat{S}_2$ *because we only consider triangular panels. For curved panels, Assumption 4.3.18 holds so that*

$$\chi_\tau = \chi \circ \chi_\tau^{\text{affine}}$$

where $\chi_\tau^{\text{affine}}$ *is affine and* $\chi : U \to V$ *is independent of* $\mathcal{G}$.

(c) *For any* $\tau \in \mathcal{G}$, *the image* $\chi_\tau^{\text{affine}}(\hat{\tau})$ *is the plane triangle with straight edges which has the same vertices as* τ, *i.e.,* $\chi_\tau^{\text{affine}}$ *is the componentwise affine interpolation of* χ_τ.

Notation 4.3.26. *If* τ, $\tilde{\tau}$, $\hat{\tau}$, χ_τ, $\chi_\tau^{\text{affine}}$, *etc., appear in the same context their relationships are always as in Assumption 4.3.25.*

Let $g_{\mathbf{z}} \in L^\infty(\widetilde{\omega_{\mathbf{z}}})$ denote the surface element

$$g_{\mathbf{z}}(\mathbf{x}) := \sqrt{\det\left(\mathbf{J}_{\mathbf{z}}^{\intercal}(\mathbf{x})\,\mathbf{J}_{\mathbf{z}}(\mathbf{x})\right)} \quad \forall \mathbf{x} \in \widetilde{\omega_{\mathbf{z}}} \quad \text{a.e.},$$

where $\mathbf{J}_{\mathbf{z}}$ denotes the Jacobian of $\chi_{\mathbf{z}}$. Let the constants θ, Θ be defined by

$$\left\|g_{\mathbf{z}}^{-1}\right\|_{L^\infty(\widetilde{\omega_{\mathbf{z}}})} =: \theta \frac{|\widetilde{\omega_{\mathbf{z}}}|}{|\omega_{\mathbf{z}}|} \quad \text{and} \quad \|g_{\mathbf{z}}\|_{L^\infty(\widetilde{\omega_{\mathbf{z}}})} =: \Theta \frac{|\omega_{\mathbf{z}}|}{|\widetilde{\omega_{\mathbf{z}}}|}. \tag{4.227}$$

Lemma 4.3.27. *Let Assumption 4.3.17 or Assumption 4.3.18 hold:*

(a) *Then,* θ, *respectively* Θ, *in (4.227) can be bounded from below, respectively from above, by constants which depend only on the shape-regularity of the mesh, the ratios*

$$c_1 := \max_{\mathbf{z}\in\mathcal{I}} \max_{\tau\in\mathcal{G}_{\mathbf{z}}} \left\{\frac{|\omega_{\mathbf{z}}|}{|\tau|}\right\} \quad \textit{and} \quad C_1 := \max_{\mathbf{z}\in\mathcal{I}} \max_{\tau\in\mathcal{G}_{\mathbf{z}}} \left\{\frac{|\widetilde{\omega_{\mathbf{z}}}|}{|\tilde{\tau}|}\right\}, \tag{4.228}$$

and, for curved panels, on the global mapping χ *(cf. Assumption 4.3.10).*

(b) *There exists a constant* C_2 *so that, for* $i \in \{1,2\}$ *and any* $\tilde{\mathbf{x}} \in \tilde{\tau} \subset \widetilde{\omega_{\mathbf{z}}}$, *we have*

$$\left\|\partial_i \left\{\left(\chi_\tau - \chi_\tau^{\text{affine}}\right) \circ \eta^{\text{affine}}(\tilde{\mathbf{x}})\right\}\right\| \le C_2 \frac{\operatorname{diam}\tilde{\tau}}{2|\tilde{\tau}|} h_\tau^2. \tag{4.229}$$

Proof. Proof of part a. Let $\tau \in \mathcal{G}_{\mathbf{z}}$ and $\tilde{\tau} := \chi_{\mathbf{z}}^{-1}(\tau)$. The restriction $\chi_{\mathbf{z},\tau} := \chi_{\mathbf{z}}|_{\tilde{\tau}}$ can be written as

$$\chi_{\mathbf{z},\tau} = \chi_\tau \circ \eta^{\text{affine}},$$

where $\chi_\tau : \hat{\tau} \to \tau$ is the reference mapping as in Definition 4.1.2 and $\hat{\tau}$ is the unit triangle as in (4.13). Further, $\eta^{\text{affine}} : \tilde{\tau} \to \hat{\tau}$ is some affine map. For $\mathbf{x} \in \tau$, let

$\hat{\mathbf{x}} := \chi_\tau^{-1}(\mathbf{x})$ and $\tilde{\mathbf{x}} := \left(\eta^{\text{affine}}\right)^{-1}(\hat{\mathbf{x}})$. Then

$$\mathbf{J}_{\mathbf{z},\tau}(\tilde{\mathbf{x}}) = \mathbf{J}_\tau(\hat{\mathbf{x}})\,\mathbf{J}_{\text{affine}}(\tilde{\mathbf{x}}), \tag{4.230}$$

where $\mathbf{J}_{\mathbf{z},\tau}$, $\mathbf{J}_\tau$, $\mathbf{J}_{\text{affine}}$ are the Jacobi matrices of $\chi_{\mathbf{z},\tau}$, χ_τ, η^{affine}, and

$$g_{\mathbf{z}}(\tilde{\mathbf{x}}) = \sqrt{\det\left(\mathbf{J}_{\text{affine}}^{\mathsf{T}}(\tilde{\mathbf{x}})\,\mathbf{G}_\tau(\hat{\mathbf{x}})\,\mathbf{J}_{\text{affine}}(\tilde{\mathbf{x}})\right)} \quad \text{with} \quad \mathbf{G}_\tau(\hat{\mathbf{x}}) := \mathbf{J}_\tau^{\mathsf{T}}(\hat{\mathbf{x}})\,\mathbf{J}_\tau(\hat{\mathbf{x}}).$$

We introduce $\mathbf{G}_{\text{affine}} := \mathbf{J}_{\text{affine}}^{\mathsf{T}}\mathbf{J}_{\text{affine}}$ and employ the multiplication theorem for determinants to obtain

$$g_{\mathbf{z}}(\tilde{\mathbf{x}}) = \left|\det \mathbf{J}_{\text{affine}}\right| \sqrt{\det \mathbf{G}_\tau(\hat{\mathbf{x}})} = \frac{\sqrt{\det \mathbf{G}_\tau(\hat{\mathbf{x}})}}{2\,|\tilde{\tau}|}. \tag{4.231}$$

If τ is a plane triangle with straight edges then

$$\sqrt{\det \mathbf{G}_\tau(\hat{\mathbf{x}})} = 2\,|\tau|.$$

For curved panels, we have $\chi_\tau = \chi \circ \chi_\tau^{\text{affine}}$ (cf. Assumption 4.3.10) and obtain by arguing as in (4.231)

$$c_\chi 2\left|\tau^{\text{affine}}\right| \leq \sqrt{\det \mathbf{G}_\tau(\hat{\mathbf{x}})} \leq C_\chi 2\left|\tau^{\text{affine}}\right| \quad \text{with } \tau^{\text{affine}} := \chi_\tau^{\text{affine}}(\hat{\tau}),$$

where the constants $0 < c_\chi \leq C_\chi$ depend only on χ, i.e., are independent of the discretization parameters. From this we derive, by using the bi-Lipschitz continuity of χ and the shape-regularity of the surface mesh, the estimate

$$2cc_\chi\,|\tau| \leq 2c_\chi h_\tau^2 \leq \sqrt{\det \mathbf{G}_\tau(\hat{\mathbf{x}})} \leq 2C_\chi h_\tau^2 \leq 2CC_\chi\,|\tau|,$$

where c, C depend only on the shape-regularity constant $\kappa_{\mathcal{G}}$. Thus

$$\frac{cc_\chi}{c_1}\frac{|\omega_{\mathbf{z}}|}{|\widetilde{\omega_{\mathbf{z}}}|} \leq cc_\chi\frac{|\tau|}{|\tilde{\tau}|} \leq |g_{\mathbf{z}}(\tilde{\mathbf{x}})| \leq CC_\chi\frac{|\tau|}{|\tilde{\tau}|} \leq CC_\chi C_1\frac{|\omega_{\mathbf{z}}|}{|\widetilde{\omega_{\mathbf{z}}}|}.$$

Proof of part b. The statement is trivial for plane triangles with straight edges because the left-hand side in (4.229) is zero.

Let $\mathbf{z} \in \mathcal{I}$ and assume that $\tau \in \mathcal{G}_{\mathbf{z}}$ is a curved panel. For any $\tilde{\mathbf{x}} \in \tilde{\tau} \subset \widetilde{\omega_{\mathbf{z}}}$, we have

$$\left\|\partial_i\left\{\left(\chi_\tau - \chi_\tau^{\text{affine}}\right)\circ\eta^{\text{affine}}(\tilde{\mathbf{x}})\right\}\right\| \leq \sum_{j=1}^{2}\left\|\partial_j\left(\chi_\tau - \chi_\tau^{\text{affine}}\right)(\hat{\mathbf{x}})\right\|\left|\partial_i\eta_j^{\text{affine}}(\tilde{\mathbf{x}})\right|, \tag{4.232}$$

where $\hat{\mathbf{x}} = \eta^{\text{affine}}(\tilde{\mathbf{x}}) \in \hat{\tau}$.

Let $\widehat{T_p}\chi_\tau$ denote the p-th order Taylor expansion of χ_τ about the barycenter $\widehat{\mathbf{M}}$ of $\hat{\tau}$ and let $\chi_\tau^{\text{affine}} = \widehat{I_1}\chi_\tau$ be the affine interpolation at the vertices of $\hat{\tau}$. Then

$$\chi_\tau - \chi_\tau^{\text{affine}} = \left(\chi_\tau - \widehat{T_1}\chi_\tau\right) + \left(\widehat{T_1}\chi_\tau - \chi_\tau^{\text{affine}}\right) = \left(\chi_\tau - \widehat{T_1}\chi_\tau\right) + \widehat{I_1}\left(\widehat{T_1}\chi_\tau - \chi_\tau\right).$$

For $k = 0, 1$, this splitting leads to the estimate

$$\|\chi_\tau - \chi_\tau^{\text{affine}}\|_{C^k(\hat{\tau})} \leq \left(1 + \left\|\widehat{I_1}\right\|_{C^k(\hat{\tau})\leftarrow C^k(\hat{\tau})}\right) \left\|\chi_\tau - \widehat{T_1}\chi_\tau\right\|_{C^k(\hat{\tau})}.$$

Standard error estimates for two-dimensional Taylor expansions result in

$$\left\|\chi_\tau - \widehat{T_1}\chi_\tau\right\|_{C^0(\hat{\tau})} \leq \frac{1}{2} \max_{0\leq j\leq 2} \left\|\partial_1^j \partial_2^{2-j} \chi_\tau\right\|_{C^0(\hat{\tau})}.$$

Because $\partial_j \widehat{T_1}\chi_\tau = \widehat{T_0}\left(\partial_j \chi_\tau\right)$ we obtain

$$\left\|\partial_j\chi_\tau - \partial_j\widehat{T_1}\chi_\tau\right\|_{C^0(\hat{\tau})} = \left\|\partial_j\chi_\tau - \widehat{T_0}\partial_j\chi_\tau\right\|_{C^0(\hat{\tau})} \leq \max_{0\leq i\leq 1} \left\|\partial_1^i\partial_2^{1-i}\partial_j\chi_\tau\right\|_{C^0(\hat{\tau})}.$$

Thus

$$\|\chi_\tau - \chi_\tau^{\text{affine}}\|_{C^k(\hat{\tau})} \leq \left(1 + \left\|\widehat{I_1}\right\|_{C^k(\hat{\tau})\leftarrow C^k(\hat{\tau})}\right) \max_{0\leq j\leq 2} \left\|\partial_1^j\partial_2^{2-j}\chi_\tau\right\|_{C^0(\hat{\tau})}. \tag{4.233}$$

Next, we will estimate the first factor in (4.233). For any $w \in C^0\left(\overline{\hat{\tau}}\right)$, we have

$$\left\|\widehat{I_1}w\right\|_{C^0(\hat{\tau})} = \max_{\hat{\mathbf{x}} \text{ is a vertex of } \hat{\tau}} |w(\hat{\mathbf{x}})| \leq \|w\|_{C^0(\hat{\tau})}.$$

We denote the vertices of $\hat{\tau}$ by $\widehat{P}_1 = (0,0)$, $\widehat{P}_2 = (1,0)$, $\widehat{P}_3 = (1,1)$ and the values of a continuous function w at $\widehat{P}_j$ by w_j, $1 \leq j \leq 3$. It is easy to see that

$$\left\|\partial_1\widehat{I_1}w\right\|_{C^0(\hat{\tau})} = |w_2 - w_1| \leq \frac{|w_2 - w_1|}{\left\|\widehat{P}_2 - \widehat{P}_1\right\|} \leq \sup_{\hat{\mathbf{x}},\hat{\mathbf{y}}\in\hat{\tau}} \frac{|w(\hat{\mathbf{x}}) - w(\hat{\mathbf{y}})|}{\|\hat{\mathbf{x}} - \hat{\mathbf{y}}\|} \leq \|w\|_{C^1(\hat{\tau})}$$

and, similarly, we obtain the stability of the derivative ∂_2. Hence we have proved that the first factor in (4.233) is bounded from above by 2.

To estimate the second derivative of χ_τ in (4.233) we write the mapping $\chi_\tau^{\text{affine}}$ in the form

$$\chi_\tau^{\text{affine}}(\hat{\mathbf{x}}) = \mathbf{B}_\tau\hat{\mathbf{x}} + \mathbf{b}_\tau$$

with the (constant) Jacobi matrix $\mathbf{B}_\tau \in \mathbb{R}^{3\times 2}$ and $\mathbf{b}_\tau \in \mathbb{R}^3$. The columns of $\mathbf{B}_\tau$ are denoted by $\mathbf{a}_1, \mathbf{a}_2 \in \mathbb{R}^3$. As in the proof of Lemma 4.3.6, we use

$$\partial^{\mu}\left(\chi \circ \chi_{\tau}^{\text{affine}}\right) = \sum_{\substack{\beta \in \mathbb{N}_0^3 \\ |\beta|=\mu_1}} \sum_{\substack{\nu \in \mathbb{N}_0^3 \\ |\nu|=\mu_2}} \frac{\mu!}{\beta!\nu!} \mathbf{a}_1^{\beta} \mathbf{a}_2^{\nu} \left(\partial^{\beta+\nu}\chi\right) \circ \chi_{\tau}^{\text{affine}}.$$

Next, we employ $\left|(\mathbf{B}_{\tau})_{i,j}\right| \leq h_{\tau}$ and obtain for any $\mu \in \mathbb{N}_0^2$ with $|\mu| = 2$

$$\sup_{\hat{\mathbf{x}} \in \hat{\tau}} \left|\partial^{\mu}\left(\chi \circ \chi_{\tau}^{\text{affine}}\right)(\hat{\mathbf{x}})\right| \leq C_3 h_{\tau}^2,$$

where C_3 depends only on the derivatives of χ which, by Assumption 4.3.10, are independent of $\mathcal{G}$. Thus we have proved that

$$\left\|\chi_{\tau} - \chi_{\tau}^{\text{affine}}\right\|_{C^k(\widehat{\tau})} \leq 2C_3 h_{\tau}^2 \tag{4.234}$$

and it remains to estimate the last factor in (4.232). Because η^{affine} is affine, it is straightforward to show that $\mathbf{J}_{\text{affine}}^{-1} \in \mathbb{R}^{2\times 2}$ [cf. (4.230)] has column vectors $\mathbf{B} - \mathbf{A}$ and $\mathbf{C} - \mathbf{B}$, where $\mathbf{A}, \mathbf{B}, \mathbf{C}$ denote the vertices of $\tilde{\tau}$. Hence

$$\mathbf{J}_{\text{affine}} = \frac{1}{2|\tilde{\tau}|} \begin{bmatrix} (\mathbf{C}-\mathbf{B})_2 & -(\mathbf{C}-\mathbf{B})_1 \\ -(\mathbf{B}-\mathbf{A})_2 & (\mathbf{B}-\mathbf{A})_1 \end{bmatrix}.$$

Consequently

$$\left|\partial_i \eta_j^{\text{affine}}\right| \leq \frac{\operatorname{diam} \tilde{\tau}}{2|\tilde{\tau}|}. \tag{4.235}$$

□

As a measure for the distortion of the local patches $\omega_{\mathbf{z}}$ by the pullback, we introduce the constant C_{d} by

$$C_{\text{d}} := \max_{\mathbf{z} \in \mathcal{I}} \left\{|\widetilde{\omega_{\mathbf{z}}}|^{-1/2} \operatorname{diam} \widetilde{\omega_{\mathbf{z}}}\right\}. \tag{4.236}$$

Theorem 4.3.28. *Let Assumption 4.3.25 be satisfied.*

There exist two constants c_1, c_2 depending only on the shape-regularity constant $\kappa_{\mathcal{G}}$ [cf. (4.17)], the constants C_{d} and C_1 [as in (4.228)], and, for curved panels, on the global chart χ so that

$$\|v - Q_{\mathcal{G}}v\|_{L^2(\tau)} \leq c_1 h_{\tau} \|v\|_{H^1(\omega_{\tau})} \quad \text{and} \quad \|Q_{\mathcal{G}}v\|_{H^1(\tau)} \leq \tilde{c}_1 \|v\|_{H^1(\omega_{\tau})} \tag{4.237a}$$

for all $v \in H^1(\Gamma)$ and all triangles $\tau \in \mathcal{G}$. Also,

$$\|v - Q_{\mathcal{G}}v\|_{H^{\sigma}(\Gamma)} \leq c_2 h_{\mathcal{G}}^{s-\sigma} \|v\|_{H^s(\Gamma)} \quad \text{and} \quad \|Q_{\mathcal{G}}\|_{H^s(\Gamma)\leftarrow H^s(\Gamma)} \leq \tilde{c}_2 \tag{4.237b}$$

for any $0 \leq \sigma \leq s \leq 1$ and $v \in H^s(\Gamma)$.

Proof. We present the proof in eight steps (a)–(h).

(a) For $\mathbf{z} \in \mathcal{I}$, let $\chi_{\mathbf{z}} : \widetilde{\omega_{\mathbf{z}}} \to \omega_{\mathbf{z}}$ be the mapping as in Assumption 4.3.25. For $\varphi \in H^1(\omega_{\mathbf{z}})$, the pullback to $\widetilde{\omega_{\mathbf{z}}}$ is denoted by $\tilde{\varphi} := \varphi \circ \chi_{\mathbf{z}}$. The Lipschitz continuity of $\chi_{\mathbf{z}}$ implies that $\tilde{\varphi} \in H^1(\widetilde{\omega_{\mathbf{z}}})$.

We consider $\pi_{\mathbf{z}}(\varphi) \in \mathbb{C}$ as a constant function and obtain

$$\|\varphi - \pi_{\mathbf{z}}(\varphi)\|^2_{L^2(\omega_{\mathbf{z}})} = \int_{\widetilde{\omega_{\mathbf{z}}}} g_{\mathbf{z}}(\tilde{\mathbf{x}}) \left|\tilde{\varphi}(\tilde{\mathbf{x}}) - \pi_{\mathbf{z}}(\varphi)\right|^2 d\tilde{\mathbf{x}}. \tag{4.238}$$

Case 1: First, we consider the case of flat panels with straight edges. Note that, for any $t \in \mathcal{G}_{\mathbf{z}}$ and $\tilde{t} := \chi_{\mathbf{z}}^{-1}(t)$, we have $g_{\mathbf{z}}|_{\tilde{t}} = |t| / |\widetilde{T}|$.

Let $\tau \in \mathcal{G}_{\mathbf{z}}$. Then for any $\tilde{\mathbf{x}} \in \tilde{\tau} = \chi_{\mathbf{z}}^{-1}(\tau)$

$$\begin{aligned}
\tilde{\varphi}(\tilde{\mathbf{x}}) - \pi_{\mathbf{z}}\varphi &= \tilde{\varphi}(\tilde{\mathbf{x}}) - \frac{1}{|\omega_{\mathbf{z}}|}\int_{\omega_{\mathbf{z}}} \varphi = \tilde{\varphi}(\tilde{\mathbf{x}}) - \frac{1}{|\omega_{\mathbf{z}}|}\int_{\widetilde{\omega_{\mathbf{z}}}} g_{\mathbf{z}}\tilde{\varphi} \\
&= \tilde{\varphi}(\tilde{\mathbf{x}}) - \frac{1}{|\omega_{\mathbf{z}}|}\sum_{t\in\mathcal{G}_{\mathbf{z}}}\int_{\tilde{t}} g_{\mathbf{z}}\tilde{\varphi} = \tilde{\varphi}(\tilde{\mathbf{x}}) - \frac{1}{|\omega_{\mathbf{z}}|}\sum_{t\in\mathcal{G}_{\mathbf{z}}}\frac{|t|}{|\widetilde{T}|}\int_{\tilde{t}}\tilde{\varphi} \\
&= \sum_{t\in\mathcal{G}_{\mathbf{z}}}\frac{|t|}{|\omega_{\mathbf{z}}|}\left(\tilde{\varphi}(\tilde{\mathbf{x}}) - \pi_{\tilde{t}}\tilde{\varphi}\right)
\end{aligned} \tag{4.239}$$

with $\pi_{\tilde{t}}\tilde{\varphi} := \frac{1}{|\tilde{t}|}\int_{\tilde{t}}\tilde{\varphi}$. Applying the L^2-norm to both sides yields

$$\|\tilde{\varphi} - \pi_{\mathbf{z}}\varphi\|_{L^2(\tilde{\tau})} \le \sum_{t\in\mathcal{G}_{\mathbf{z}}}\frac{|t|}{|\omega_{\mathbf{z}}|}\|\tilde{\varphi} - \pi_{\tilde{t}}\tilde{\varphi}\|_{L^2(\tilde{\tau})}. \tag{4.240}$$

Because $\widetilde{T} \subset \widetilde{\omega_{\mathbf{z}}}$ are both convex we may apply Corollary 2.5.12 to obtain

$$\begin{aligned}
\|\tilde{\varphi} - \pi_{\tilde{t}}\tilde{\varphi}\|_{L^2(\tilde{\tau})} &\le \|\tilde{\varphi} - \pi_{\tilde{t}}\tilde{\varphi}\|_{L^2(\widetilde{\omega_{\mathbf{z}}})} \le \left(1 + \sqrt{\frac{|\widetilde{\omega_{\mathbf{z}}}|}{|\tilde{t}|}}\right)\frac{\operatorname{diam}\widetilde{\omega_{\mathbf{z}}}}{\pi}|\tilde{\varphi}|_{H^1(\widetilde{\omega_{\mathbf{z}}})} \\
&\le \left(1 + \sqrt{C_1}\right)\frac{\operatorname{diam}\widetilde{\omega_{\mathbf{z}}}}{\pi}|\tilde{\varphi}|_{H^1(\widetilde{\omega_{\mathbf{z}}})},
\end{aligned} \tag{4.241}$$

where C_1 is as in (4.228). Inserting this into (4.240) yields

$$\|\tilde{\varphi} - \pi_{\mathbf{z}}\varphi\|_{L^2(\tilde{\tau})} \le \left(1 + \sqrt{C_1}\right)\frac{\operatorname{diam}\widetilde{\omega_{\mathbf{z}}}}{\pi}|\tilde{\varphi}|_{H^1(\widetilde{\omega_{\mathbf{z}}})}.$$

We sum over all $\tilde{\tau} \subset \widetilde{\omega_{\mathbf{z}}}$ and apply a Cauchy–Schwarz inequality to derive the estimate

$$\|\tilde{\varphi} - \pi_{\mathbf{z}}\varphi\|_{L^2(\widetilde{\omega_{\mathbf{z}}})} \le \sqrt{\operatorname{card}\mathcal{G}_{\mathbf{z}}}\left(1 + \sqrt{C_1}\right)\frac{\operatorname{diam}\widetilde{\omega_{\mathbf{z}}}}{\pi}|\tilde{\varphi}|_{H^1(\widetilde{\omega_{\mathbf{z}}})},$$

where the number of panels ($\operatorname{card}\mathcal{G}_{\mathbf{z}}$) is bounded by a constant which depends only on the shape-regularity of the surface mesh.

The combination with (4.238) leads to

$$\begin{aligned}\|\varphi - \pi_{\mathbf{z}}(\varphi)\|_{L^2(\omega_{\mathbf{z}})} &\leq \sqrt{\|g_{\mathbf{z}}\|_{L^\infty(\tilde{\omega}_{\mathbf{z}})}}\, \|\tilde{\varphi} - \pi_{\mathbf{z}}\varphi\|_{L^2(\tilde{\omega}_{\mathbf{z}})} \\ &\overset{(4.227)}{\leq} C_4\sqrt{|\omega_{\mathbf{z}}|}\frac{\operatorname{diam}\widetilde{\omega_{\mathbf{z}}}}{\sqrt{|\widetilde{\omega_{\mathbf{z}}}|}}\, |\tilde{\varphi}|_{H^1(\tilde{\omega}_{\mathbf{z}})} \\ &\leq C_4 C_{\mathrm{d}} h_{\mathbf{z}}\, |\tilde{\varphi}|_{H^1(\tilde{\omega}_{\mathbf{z}})}\end{aligned}$$

with $C_4 := \sqrt{\Theta \operatorname{card}\mathcal{G}_{\mathbf{z}}}\left(1 + \sqrt{C_1}\right)/\pi$. From Lemma 4.3.6 resp. Lemma 4.3.13 we obtain

$$|\tilde{\varphi}|^2_{H^1(\tilde{\omega}_{\mathbf{z}})} = \sum_{\tilde{\tau}\subset\widetilde{\omega_{\mathbf{z}}}} |\tilde{\varphi}|^2_{H^1(\tilde{\tau})} \leq C_5 \sum_{\tau\in\mathcal{G}_{\mathbf{z}}} |\varphi|^2_{H^1(\tau)} \tag{4.242}$$

and, finally, for any $\tau \subset \omega_{\mathbf{z}}$

$$\|\varphi-\pi_{\mathbf{z}}(\varphi)\|_{L^2(\omega_{\mathbf{z}})} \leq \widetilde{C}_6 h_{\mathbf{z}}\, |\varphi|_{H^1(\omega_{\mathbf{z}})} \leq C_0\widetilde{C}_6 h_\tau\, |\varphi|_{H^1(\omega_{\mathbf{z}})} \quad \text{with } \widetilde{C}_6 = C_{\mathrm{d}} C_4 \sqrt{C_5}. \tag{4.243}$$

Case 2: Next, we consider the general case of curved panels. As in (4.239) we derive

$$\begin{aligned}\tilde{\varphi} - \pi_{\mathbf{z}}\varphi &= \sum_{t\in\mathcal{G}_{\mathbf{z}}} \frac{|t|}{|\omega_{\mathbf{z}}|}\left(\tilde{\varphi} - \frac{1}{|\widetilde{T}|}\int_{\tilde{t}} \frac{|\widetilde{T}|}{|t|} g_{\mathbf{z}}\tilde{\varphi}\right) \\ &= \sum_{t\in\mathcal{G}_{\mathbf{z}}} \frac{|t|}{|\omega_{\mathbf{z}}|}\left\{(\tilde{\varphi} - \pi_{\tilde{t}}\tilde{\varphi}) + \frac{1}{|\widetilde{T}|}\int_{\tilde{t}} d_t\tilde{\varphi}\right\}\end{aligned} \tag{4.244}$$

with $d_t := 1 - \frac{|\widetilde{T}|}{|t|}\, g_{\mathbf{z}}|_{\tilde{t}}$. The first difference in (4.244) can be estimated as in the case of flat panels while, for the second one, we will derive an estimate of d_t. We use the notation as in Assumption 4.3.25 and employ the splitting

$$d_t = \left(1 - \frac{|t^{\mathrm{affine}}|}{|t|}\right) + \frac{1}{2|t|}\left(2\left|t^{\mathrm{affine}}\right| - 2\left|\widetilde{T}\right|(g_{\mathbf{z}}|_{\tilde{t}})\right), \tag{4.245}$$

where $t^{\mathrm{affine}} = \chi_t^{\mathrm{affine}}(\hat{\tau})$ is the plane triangle with straight edges which interpolates τ at its vertices.

We start by estimating the second term in (4.245). We employ the representation (4.231) for Gram's determinant to obtain

$$2\left|\widetilde{T}\right|(g_{\mathbf{z}}|_{\tilde{t}}) = g_t,$$

where g_t is Gram's determinant of the reference map $\chi_t : \hat{\tau} \to t$, i.e.

$$g_t = \|\partial_1\chi_t \times \partial_2\chi_t\|\,.$$

The area $2\left|t^{\text{affine}}\right|$ can be expressed by

$$2\left|t^{\text{affine}}\right| = \left\|\partial_1\chi_t^{\text{affine}} \times \partial_2\chi_t^{\text{affine}}\right\| =: g_t^{\text{affine}} \tag{4.246}$$

Hence

$$\begin{aligned}\left|2\left|t^{\text{affine}}\right| - 2\left|\widetilde{T}\right|(g_{\mathbf{z}}|_{\tilde{t}})\right| &= \left|g_t^{\text{affine}} - g_t\right| = \left|\left\|\partial_1\chi_t^{\text{affine}} \times \partial_2\chi_t^{\text{affine}}\right\| - \|\partial_1\chi_t \times \partial_2\chi_t\|\right| \qquad (4.247)\\ &\le \left\|\partial_1\chi_t^{\text{affine}} \times \partial_2\chi_t^{\text{affine}} - (\partial_1\chi_t \times \partial_2\chi_t)\right\|\\ &\le \left\|\partial_1\left(\chi_t - \chi_t^{\text{affine}}\right) \times \partial_2\chi_t\right\|\\ &\quad + \left\|\partial_1\chi_t^{\text{affine}} \times \partial_2\left(\chi_t - \chi_t^{\text{affine}}\right)\right\|.\end{aligned}$$

We employ (4.234) to obtain

$$\left|2\left|t^{\text{affine}}\right| - 2\left|\widetilde{T}\right|(g_{\mathbf{z}}|_{\tilde{t}})\right| \le 2C_3h_t^2\left(\|\partial_2\chi_t\|_{\mathbf{L}^\infty(\hat{\tau})} + \left\|\partial_1\chi_t^{\text{affine}}\right\|_{\mathbf{L}^\infty(\hat{\tau})}\right).$$

The estimate $\left\|\partial_1\chi_t^{\text{affine}}\right\|_{\mathbf{L}^\infty(\hat{\tau})} \le h_t$ is obvious because t^{affine} interpolates t in its vertices. For the other term, we use

$$\|\partial_2\chi_t\|_{\mathbf{L}^\infty(\hat{\tau})} = \left\|\sum_{j=1}^{3}\left(\partial_j\chi \circ \chi_t^{\text{affine}}\right)\partial_2\left(\chi_t^{\text{affine}}\right)_j\right\|_{\mathbf{L}^\infty(\hat{\tau})} \le Ch_t,$$

where C depends only on the global chart χ but not on the discretization parameters. In summary we have proved that

$$\frac{\left|\left|t^{\text{affine}}\right| - \left|\widetilde{T}\right|(g_{\mathbf{z}}|_{\tilde{t}})\right|}{|t|} \le \frac{C_3Ch_t}{c},$$

where c depends only on the shape-regularity of the mesh and the global chart χ. The first term of the sum in (4.245) can be estimated by using (4.247)

$$\begin{aligned}\left|1 - \frac{\left|t^{\text{affine}}\right|}{|t|}\right| &= \left|\frac{|t| - \left|t^{\text{affine}}\right|}{|t|}\right| \le |t|^{-1}\int_{\hat{\tau}}\left|g_t - g_t^{\text{affine}}\right| d\mathbf{x}\\ &\le \frac{C_3Ch_t^3}{|t|} \le \frac{C_3Ch_t}{c}.\end{aligned}$$

This finishes the estimate of d_t

$$|d_t| \leq 2\frac{C_3 C}{c} h_t.$$

Inserting this into (4.244) and proceeding as in the case of flat panels yields

$$\begin{aligned}
\|\tilde{\varphi} - \pi_{\mathbf{z}}\varphi\|_{L^2(\tilde{\tau})} &\leq \sum_{t \in \mathcal{G}_{\mathbf{z}}} \frac{|t|}{|\omega_{\mathbf{z}}|} \|\tilde{\varphi} - \pi_{\tilde{t}}\tilde{\varphi}\|_{L^2(\tilde{\tau})} + \sum_{t \in \mathcal{G}_{\mathbf{z}}} \frac{|t|}{|\omega_{\mathbf{z}}|} 2\frac{C_3 C}{c} h_t \sqrt{\frac{|\tilde{\tau}|}{|\widetilde{T}|}} \|\tilde{\varphi}\|_{L^2(\tilde{T})} \\
&\overset{(4.241)}{\leq} \left(1 + \sqrt{C_1}\right) \frac{\operatorname{diam} \widetilde{\omega}_{\mathbf{z}}}{\pi} |\tilde{\varphi}|_{H^1(\tilde{\omega}_{\mathbf{z}})} + 2\frac{C_3 C \sqrt{C_1}}{c} h_{\mathbf{z}} \|\tilde{\varphi}\|_{L^2(\tilde{\omega}_{\mathbf{z}})}.
\end{aligned}$$

We sum over all $\tilde{\tau} \subset \widetilde{\omega}_{\mathbf{z}}$ and apply a Cauchy–Schwarz inequality to derive the estimate

$$\begin{aligned}
\|\tilde{\varphi} - \pi_{\mathbf{z}}\varphi\|_{L^2(\tilde{\omega}_{\mathbf{z}})} \leq \sqrt{\operatorname{card} \mathcal{G}_{\mathbf{z}}} \Bigg\{ & \left(1 + \sqrt{C_1}\right) \frac{\operatorname{diam} \widetilde{\omega}_{\mathbf{z}}}{\pi} |\tilde{\varphi}|_{H^1(\tilde{\omega}_{\mathbf{z}})} \\
&+ 2\frac{C_3 C \sqrt{C_1}}{c} h_{\mathbf{z}} \|\tilde{\varphi}\|_{L^2(\tilde{\omega}_{\mathbf{z}})} \Bigg\}.
\end{aligned}$$

From Lemma 4.3.6 resp. Lemma 4.3.13 we obtain the scaling relations

$$|\tilde{\varphi}|^2_{H^1(\tilde{\omega}_{\mathbf{z}})} \leq C_5 |\varphi|^2_{H^1(\omega_{\mathbf{z}})} \text{ and } \tilde{c}_5 \frac{|\widetilde{\omega}_z|}{|\omega_{\mathbf{z}}|} \|\varphi\|^2_{L^2(\omega_{\mathbf{z}})} \leq \|\tilde{\varphi}\|^2_{L^2(\tilde{\omega}_{\mathbf{z}})} \leq \widetilde{C}_5 \frac{|\widetilde{\omega}_z|}{|\omega_{\mathbf{z}}|} \|\varphi\|^2_{L^2(\omega_{\mathbf{z}})}$$

and, finally, for any $\tau \subset \omega_{\mathbf{z}}$

$$\|\varphi - \pi_{\mathbf{z}}\varphi\|_{L^2(\omega_{\mathbf{z}})} \leq \widehat{C}_6 h_{\mathbf{z}} \|\varphi\|_{H^1(\omega_{\mathbf{z}})} \leq C_0 \widehat{C}_6 h_\tau \|\varphi\|_{H^1(\omega_{\mathbf{z}})}, \tag{4.248}$$

where $\widehat{C}_6$ depends on C_1, C_{d}, C_5, $\tilde{c}_5$, $\widetilde{C}_5$, and $\operatorname{card} \mathcal{G}_{\mathbf{z}}$. Let $C_6 := \max\left\{\widetilde{C}_6, \widehat{C}_6\right\}$ [cf. (4.243)].

(b) Let $\tau \in \mathcal{G}$. The set of vertices of τ is denoted by $\mathcal{I}_\tau$. Then

$$\sum_{\mathbf{x} \in \mathcal{I}_\tau} b_{\mathbf{x}} = 1 \qquad \text{on } \tau.$$

By using Step a, we derive

$$\begin{aligned}
\|\varphi - Q_{\mathcal{G}}\varphi\|_{L^2(\tau)} &= \left\| \sum_{\mathbf{z} \in \mathcal{I}_\tau} b_{\mathbf{z}} \left(\varphi - \pi_{\mathbf{z}}(\varphi)\right) \right\|_{L^2(\tau)} \leq \sum_{\mathbf{z} \in \mathcal{I}_\tau} \|b_{\mathbf{z}} \left(\varphi - \pi_{\mathbf{z}}(\varphi)\right)\|_{L^2(\tau)} \\
&\leq \sum_{\mathbf{z} \in \mathcal{I}_\tau} \|\varphi - \pi_{\mathbf{z}}(\varphi)\|_{L^2(\tau)} \leq \sum_{\mathbf{z} \in \mathcal{I}_\tau} \|\varphi - \pi_{\mathbf{z}}(\varphi)\|_{L^2(\omega_{\mathbf{z}})}
\end{aligned}$$

$$\leq C_0 C_6 h_\tau \sum_{\mathbf{z}\in\mathcal{I}_\tau} \|\varphi\|_{H^1(\omega_\mathbf{z})}$$

$$\leq \sqrt{3} C_0 C_6 h_\tau \sqrt{\sum_{\mathbf{z}\in\mathcal{I}_\tau} \|\varphi\|^2_{H^1(\omega_\mathbf{z})}} \tag{4.249}$$

$$\leq \sqrt{3} C_0 C_6 h_\tau \sqrt{\sum_{t\in\omega_\tau} \sum_{\mathbf{z}\in\mathcal{I}_\tau : t\subset\omega_\mathbf{z}} \|\varphi\|^2_{H^1(t)}}$$

$$\leq C_7 h_\tau \|\varphi\|_{H^1(\omega_\tau)} \tag{4.250}$$

with $C_7 := 3C_0C_6$.

(c) By summing over all panels we obtain

$$\|\varphi - Q_\mathcal{G}\varphi\|^2_{L^2(\Gamma)} = \sum_{\tau\in\mathcal{G}} \|\varphi - Q_\mathcal{G}\varphi\|^2_{L^2(\tau)} \leq C_7^2 h_\mathcal{G}^2 \sum_{\tau\in\mathcal{G}} \|\varphi\|^2_{H^1(\omega_\tau)}$$
$$= C_7^2 h_\mathcal{G}^2 \sum_{t\in\mathcal{G}} \sum_{\tau\in\mathcal{G}:t\subset\omega_\tau} \|\varphi\|^2_{H^1(t)} \leq C_8^2 h_\mathcal{G}^2 \|\varphi\|^2_{H^1(\Gamma)},$$

where $C_8 = C_7 C_\sharp^{1/2}$ and

$$C_\sharp := \max_{t\in\mathcal{G}} \operatorname{card}\{\tau\in\mathcal{G} : t\subset\omega_\tau\}$$

depends only on the shape-regularity constant.

(d) For the $L^2(\Gamma)$-stability we repeat the first steps of (4.249) to obtain

$$\|Q_\mathcal{G}\varphi\|_{L^2(\tau)} \leq \sum_{\mathbf{z}\in\mathcal{I}_\tau} \|\pi_\mathbf{z}(\varphi)\|_{L^2(\omega_\mathbf{z})}.$$

The Cauchy-Schwarz inequality yields

$$|\pi_\mathbf{z}(\varphi)| \leq |\omega_\mathbf{z}|^{-1/2} \|\varphi\|_{L^2(\omega_\mathbf{z})}$$

and as in (4.250) we derive

$$\|Q_\mathcal{G}\varphi\|_{L^2(\tau)} \leq \sum_{\mathbf{z}\in\mathcal{I}_\tau} \|\varphi\|_{L^2(\omega_\mathbf{z})} \leq \sqrt{3}\,\|\varphi\|_{L^2(\omega_\tau)}. \tag{4.251}$$

A summation as in Step c results in the $L^2(\Gamma)$-stability of the Clément interpolation operator

$$\|Q_\mathcal{G}\varphi\|_{L^2(\Gamma)} \leq \sqrt{3C_\sharp}\,\|\varphi\|_{L^2(\Gamma)}. \tag{4.252}$$

(e) From Step c and Step d we conclude that

$$\|\varphi - Q_{\mathcal{G}}\varphi\|_{L^2(\Gamma)} \leq C_9 \|\varphi\|_{L^2(\Gamma)} \quad \text{and} \quad \|\varphi - Q_{\mathcal{G}}\varphi\|_{L^2(\Gamma)} \leq C_8 h_{\mathcal{G}} \|\varphi\|_{H^1(\Gamma)}$$

hold with $C_9 := 1 + \sqrt{3C_\sharp}$. Hence the approximation result for the intermediate Sobolev spaces $H^s(\Gamma)$, $s \in]0,1[$, follows by interpolation as in the proof of Theorem 4.1.33.

(f) For the local H^1-stability we proceed as in Step d, respectively as in (4.249). Recall the definition of the surface gradient as in (4.200) and (4.201) to derive

$$|Q_{\mathcal{G}}\varphi|_{H^1(\tau)} = \left\| \sum_{\mathbf{z}\in\mathcal{I}_\tau} \pi_{\mathbf{z}}(\varphi) \nabla_S b_{\mathbf{z}} \right\|_{L^2(\tau)} = \left\| \sum_{\mathbf{z}\in\mathcal{I}_\tau} \left(\pi_{\mathbf{z}}(\varphi) - \pi_{\mathbf{z}_0}(\varphi)\right) \nabla_S b_{\mathbf{z}} \right\|_{L^2(\tau)} \tag{4.253}$$

for any fixed $\mathbf{z}_0 \in \mathcal{I}_\tau$. Let $\bar{\varphi}_\tau := \frac{1}{|\omega_\tau|}\int_{\omega_\tau} \varphi$. Then $\pi_{\mathbf{z}}(\bar{\varphi}_\tau) = \pi_{\mathbf{z}_0}(\bar{\varphi}_\tau) = \bar{\varphi}_\tau$ and

$$\begin{aligned} \left|\pi_{\mathbf{z}}(\varphi) - \pi_{\mathbf{z}_0}(\varphi)\right| &\leq |\pi_{\mathbf{z}}(\varphi) - \pi_{\mathbf{z}}(\bar{\varphi}_\tau)| + \left|\pi_{\mathbf{z}_0}(\bar{\varphi}_\tau) - \pi_{\mathbf{z}_0}(\varphi)\right| \\ &\leq \left|\frac{1}{|\omega_{\mathbf{z}}|}\int_{\omega_{\mathbf{z}}} (\varphi - \bar{\varphi}_\tau)\right| + \left|\frac{1}{|\omega_{\mathbf{z}_0}|}\int_{\omega_{\mathbf{z}_0}} (\bar{\varphi}_\tau - \varphi)\right| \\ &\leq \frac{\|\varphi - \bar{\varphi}_\tau\|_{L^2(\omega_{\mathbf{z}})}}{|\omega_{\mathbf{z}}|^{1/2}} + \frac{\|\varphi - \bar{\varphi}_\tau\|_{L^2(\omega_{\mathbf{z}_0})}}{|\omega_{\mathbf{z}_0}|^{1/2}}. \end{aligned}$$

In a similar fashion to (4.243) and (4.248) one derives for $D \in \{\omega_{\mathbf{z}}, \omega_{\mathbf{z}_0}\}$

$$\|\varphi - \bar{\varphi}_\tau\|_{L^2(D)} \leq \|\varphi - \bar{\varphi}_\tau\|_{L^2(\omega_\tau)} \leq \widetilde{C}_7 (\operatorname{diam}\omega_\tau) \|\varphi\|_{H^1(\omega_\tau)}.$$

Hence

$$\left|\pi_{\mathbf{z}}(\varphi) - \pi_{\mathbf{z}_0}(\varphi)\right| \leq C_{10} \|\varphi\|_{H^1(\omega_\tau)}, \tag{4.254}$$

where C_{10} depends only on the shape-regularity constant and the global parametrization χ.

In Theorem 4.4.2 (with $\ell = 1$ and $m = 0$), we will prove the *inverse inequality* and, thus, obtain the estimate

$$\|\nabla_S b_{\mathbf{z}}\|_{L^2(\tau)} \leq C h_\tau^{-1} \|b_{\mathbf{z}}\|_{L^2(\tau)} \leq C h_\tau^{-1} |\tau|^{1/2} \leq C_{11}, \tag{4.255}$$

where C_{11} depends only on the shape-regularity of the mesh and the global parametrization χ.

By inserting (4.254) and (4.255) into (4.253) we derive

$$|Q_{\mathcal{G}}\varphi|_{H^1(\tau)} \leq 3C_{10}C_{11} \|\varphi\|_{H^1(\omega_\tau)}.$$

The combination with (4.251) leads to the local stability with respect to the $\|\cdot\|_{H^1(\tau)}$-norm and a summation over all panels as in Step c results in the global H^1-stability

$$\|Q_{\mathcal{G}}\varphi\|_{H^1(\Gamma)} \leq C_{12} \|\varphi\|_{H^1(\Gamma)}.$$

(g) Applying Proposition 2.1.62 with $X_0 = Y_0 = L^2(\Gamma)$ and $X_1 = Y_1 = H^1(\Gamma)$ we obtain by interpolation of (4.252)

$$\|Q_{\mathcal{G}}\|_{H^s(\Gamma)\leftarrow H^s(\Gamma)} \leq \|Q_{\mathcal{G}}\|^{1-s}_{L^2(\Gamma)\leftarrow L^2(\Gamma)} \|Q_{\mathcal{G}}\|^{s}_{H^1(\Gamma)\leftarrow H^1(\Gamma)} \leq C_{13}$$

with $C_{13} := \left(3C_{\sharp}\right)^{\frac{1-s}{2}} c_{12}^s$.

(h) Part e and g imply that

$$\|\varphi - Q_{\mathcal{G}}\varphi\|_{L^2(\Gamma)} \leq C_8 h_{\mathcal{G}}^s \|\varphi\|_{H^s(\Gamma)} \text{ and } \|\varphi - Q_{\mathcal{G}}\varphi\|_{H^s(\Gamma)} \leq (1 + C_{13})\|\varphi\|_{H^s(\Gamma)}.$$

We apply Proposition 2.1.62 with $T = I - Q_{\mathcal{G}}$, $Y_0 = Y_1 = H^s(\Gamma)$, $X_0 = L^2(\Gamma)$, $X_1 = H^s(\Gamma)$, and $\theta = \sigma/s \in [0, 1]$ to interpolate these two inequalities. The result is

$$\begin{aligned} \|T\|_{H^\sigma(\Gamma)\leftarrow H^s(\Gamma)} &\leq \|T\|^{1-\theta}_{L^2(\Gamma)\leftarrow H^s(\Gamma)} \|T\|^{\theta}_{H^s(\Gamma)\leftarrow H^s(\Gamma)} \leq \left(C_8 h_{\mathcal{G}}^s\right)^{1-\theta} (1 + C_{13})^{\theta} \\ &= C_{14} h_{\mathcal{G}}^{s-\sigma} \end{aligned}$$

with $C_{14} := C_8^{1-\sigma/s} (1 + C_{13})^{\sigma/s}$ and this implies the first estimate in (4.237b). □

In Sect. 9 we will need an estimate of the surface metric on $\omega_{\mathbf{z}}$ compared with the two-dimensional Euclidean metric on $\widetilde{\omega}_{\mathbf{z}}$. Since $\omega_{\mathbf{z}}$ may consist of several panels, the local Assumptions 4.3.17 and 4.3.18 have to be supplemented by the following, more global Assumption 4.3.29 which states that Γ has to satisfy a cone-type condition and that the minimal angle of the surface mesh has to be bounded below by a positive constant (see Fig. 4.8).

Assumption 4.3.29. *1. For all* $\tau \in \mathcal{G}$, $\mathbf{x} \in \Gamma\backslash\tau$ *and* $\mathbf{y} \in \tau$, *there exist* $c > 0$ *and an* $\mathbf{x}_0 \in \overline{\tau}$ *such that*

$$\|\mathbf{x} - \mathbf{x}_0\| = \operatorname{dist}(\mathbf{x}, \tau) \quad \textit{and} \quad \|\mathbf{x} - \mathbf{y}\|^2 \geq c\left(\|\mathbf{x} - \mathbf{x}_0\|^2 + \|\mathbf{x}_0 - \mathbf{y}\|^2\right).$$

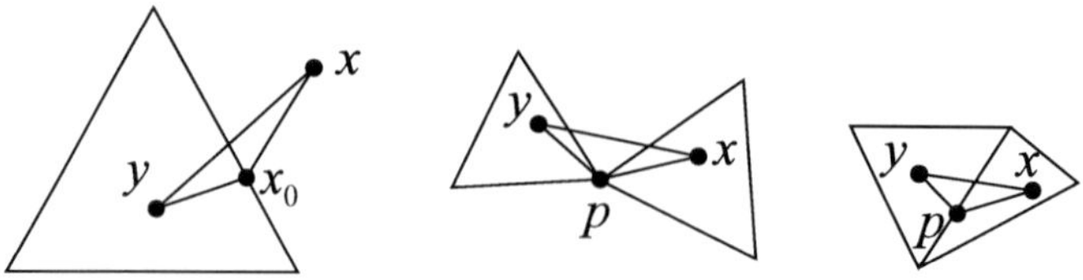

Fig. 4.8 Illustration of the cone and the angle condition for the surface mesh

2. *For all $\tau, t \in \mathcal{G}$ whose intersection consists of at most one point, there exists a point* $\mathbf{p}$ *of* t *such that*

$$\|\mathbf{x}-\mathbf{y}\| \geq c\left(\|\mathbf{x}-\mathbf{p}\|+\|\mathbf{p}-\mathbf{y}\|\right) \qquad \forall \mathbf{x} \in \tau, \forall \mathbf{y} \in t.$$

3. *For all $\tau, t \in \mathcal{G}$ with exactly one common edge $\overline{\tau} \cap \overline{t} = E$ and for all $\mathbf{x} \in \tau$, $\mathbf{y} \in t$ there exists a point $\mathbf{p} \in E$ such that*

$$\|\mathbf{y}-\mathbf{x}\| \geq c\left(\|\mathbf{y}-\mathbf{p}\|+\|\mathbf{p}-\mathbf{x}\|\right).$$

Lemma 4.3.30. *Let Assumption 4.3.29 be satisfied and let Assumption 4.3.17 or Assumption 4.3.18 hold. Then*

$$c\,\|\tilde{\mathbf{x}}-\tilde{\mathbf{y}}\| \leq \frac{\operatorname{diam} \widetilde{\omega_{\mathbf{z}}}}{h_{\mathbf{z}}}\left\|\chi_{\mathbf{z}}(\tilde{\mathbf{x}})-\chi_{\mathbf{z}}(\tilde{\mathbf{y}})\right\| \leq C\,\|\tilde{\mathbf{x}}-\tilde{\mathbf{y}}\| \quad \forall \tilde{\mathbf{x}}, \tilde{\mathbf{y}} \in \widetilde{\omega_{\mathbf{z}}},$$

where C depends only on the global chart χ but is independent of the surface mesh.

Proof. (a) Let $\tau \in \mathcal{G}_{\mathbf{z}}$ be a surface triangle with vertices $\mathbf{A}, \mathbf{B}, \mathbf{C}$. First, we will prove the statement for $\tilde{\mathbf{x}}, \tilde{\mathbf{y}} \in \tilde{\tau} = \chi_{\mathbf{z}}^{-1}(\tau)$.

Let $\tau^{\text{affine}} := \chi_\tau^{\text{affine}}(\hat{\tau})$ be the plane triangle with straight edges which interpolates τ in its vertices. Note that $\chi\left(\tau^{\text{affine}}\right) = \tau$. Hence

$$\chi(\mathbf{x})-\chi(\mathbf{y}) = \mathbf{J}_\chi(\mathbf{w})(\mathbf{x}-\mathbf{y}) \quad \forall \mathbf{x}, \mathbf{y} \in \tau^{\text{affine}}$$

where $\mathbf{J}_\chi \in \mathbb{R}^{3\times 3}$ is the Jacobi matrix of the global chart χ and $\mathbf{w}$ is some point in $\overline{\mathbf{xy}}$. Note that the largest and the smallest eigenvalues $\lambda_{\max}$ and $\lambda_{\min}$ of the positive definite Gram matrix $\mathbf{G}_\chi$ depend only on the global chart χ and are, in particular, independent of the discretization parameters. Thus

$$\sqrt{\lambda_{\min}}\,\|\mathbf{x}-\mathbf{y}\| \leq \|\chi(\mathbf{x})-\chi(\mathbf{y})\| \leq \sqrt{\lambda_{\max}}\,\|\mathbf{x}-\mathbf{y}\| \qquad \forall \mathbf{x}, \mathbf{y} \in \tau^{\text{affine}}.$$

Let $\mathbf{G}_\tau^{\text{affine}} \in \mathbb{R}^{2\times 2}$ denote the (constant) Gram matrix of $\chi_\tau^{\text{affine}}$. From Lemma 4.3.5 we conclude that

$$\left\|\chi_\tau^{\text{affine}}(\hat{\mathbf{x}})-\chi_\tau^{\text{affine}}(\hat{\mathbf{y}})\right\| = \left\langle \mathbf{G}_\tau^{\text{affine}}(\hat{\mathbf{x}}-\hat{\mathbf{y}}), (\hat{\mathbf{x}}-\hat{\mathbf{y}})\right\rangle^{1/2} \leq \sqrt{2}h_\tau\,\|\hat{\mathbf{x}}-\hat{\mathbf{y}}\|$$

for all $\hat{\mathbf{x}}, \hat{\mathbf{y}} \in \hat{\tau}$. Because the matrix $\mathbf{G}_\tau^{\text{affine}}$ is symmetric and positive definite, its minimal eigenvalue $\lambda_{\min}^{\text{affine}}$ can be expressed by

$$\lambda_{\min}^{\text{affine}} = \left\|\left(\mathbf{G}_\tau^{\text{affine}}\right)^{-1}\right\|^{-1}.$$

We employ Lemma 4.3.5 to obtain

$$\left\|\chi_\tau^{\text{affine}}(\hat{\mathbf{x}})-\chi_\tau^{\text{affine}}(\hat{\mathbf{y}})\right\| \geq c h_\tau\,\|\hat{\mathbf{x}}-\hat{\mathbf{y}}\|$$

for all $\hat{\mathbf{x}}, \hat{\mathbf{y}} \in \hat{\tau}$, where C depends only on the shape-regularity of the mesh. Thus we have proved that

$$c\sqrt{\lambda_{\min}}h_\tau \|\hat{\mathbf{x}} - \hat{\mathbf{y}}\| \le \|\chi_\tau(\hat{\mathbf{x}}) - \chi(\hat{\mathbf{y}})\| \le \sqrt{2\lambda_{\max}}h_\tau \|\hat{\mathbf{x}} - \hat{\mathbf{y}}\|$$

for all $\hat{\mathbf{x}}, \hat{\mathbf{y}} \in \hat{\tau}$. Finally, we replace $\hat{\mathbf{x}}$ and $\hat{\mathbf{y}}$ by $\eta^{\text{affine}}(\tilde{\mathbf{x}})$ and $\eta^{\text{affine}}(\tilde{\mathbf{y}})$. From (4.235) we derive the estimate for the largest eigenvalue $\lambda_\eta^{\max}$ of the Gram matrix $\mathbf{G}_\eta$ of η^{affine}

$$\sqrt{\lambda_\eta^{\max}} \overset{(4.235)}{\le} \sqrt{2}\frac{\operatorname{diam}\tilde{\tau}}{2|\tilde{\tau}|} \le C \operatorname{diam}^{-1}\tilde{\tau},$$

where C depends only on the shape-regularity constant and the global chart χ.

For the smallest eigenvalue we use

$$\mathbf{G}_\eta^{-1} = \begin{bmatrix} \|\tilde{\mathbf{e}}_1\|^2 & \langle \tilde{\mathbf{e}}_1, \tilde{\mathbf{e}}_2 \rangle \\ \langle \tilde{\mathbf{e}}_1, \tilde{\mathbf{e}}_2 \rangle & \|\tilde{\mathbf{e}}_2\|^2 \end{bmatrix},$$

where $\widetilde{\mathbf{A}}, \widetilde{\mathbf{B}}, \widetilde{\mathbf{C}}$ denote the vertices of $\tilde{\tau}$ and $\tilde{\mathbf{e}}_1 = \widetilde{\mathbf{B}} - \widetilde{\mathbf{A}}$, $\mathbf{e}_2 = \widetilde{\mathbf{C}} - \widetilde{\mathbf{B}}$. Thus

$$\left\|\mathbf{G}_\eta^{-1}\right\| \le 2 \operatorname{diam}\tilde{\tau}$$

and the minimal eigenvalue $\lambda_\eta^{\min}$ satisfies

$$\sqrt{\lambda_\eta^{\min}} = \left\|\mathbf{G}_\eta^{-1}\right\|^{-1/2} \ge \frac{1}{\sqrt{2}\operatorname{diam}\tilde{\tau}}.$$

The combination of these estimates leads to

$$c\frac{h_\tau}{\operatorname{diam}\tilde{\tau}} \|\tilde{\mathbf{x}} - \tilde{\mathbf{y}}\| \le \|\chi_{\mathbf{z}}(\tilde{\mathbf{x}}) - \chi_{\mathbf{z}}(\tilde{\mathbf{y}})\| \le C\frac{h_\tau}{\operatorname{diam}\tilde{\tau}} \|\tilde{\mathbf{x}} - \tilde{\mathbf{y}}\| \qquad \forall \tilde{\mathbf{x}}, \tilde{\mathbf{y}} \in \tilde{\tau}. \tag{4.256}$$

(b) We assume that $\mathcal{G}_{\mathbf{z}}$ contains more than one panel and consider the case that $\mathbf{x}$ and $\mathbf{y}$ belong to different panels $\tau, t \in \mathcal{G}_{\mathbf{z}}$. Note that

$$c\frac{h_{\mathbf{z}}}{\operatorname{diam}\widetilde{\omega}_{\mathbf{z}}} \le \frac{h_\tau}{\operatorname{diam}\tilde{\tau}} \le C\frac{h_{\mathbf{z}}}{\operatorname{diam}\widetilde{\omega}_{\mathbf{z}}} \qquad \forall \tau \in \mathcal{G}_{\mathbf{z}},$$

where c and C depend only on the global chart χ and the shape-regularity constant. Assumption 4.3.29 implies that one of the following two cases is satisfied:

(i) The panels τ and t share exactly one common edge $\overline{\tau} \cap \overline{t} = E$. Then there exists a point $\mathbf{p} \in E$ such that

$$\|\mathbf{y} - \mathbf{x}\| \ge c\left(\|\mathbf{y} - \mathbf{p}\| + \|\mathbf{p} - \mathbf{x}\|\right).$$

The combination of (4.256) and a triangle inequality leads to

$$\|\mathbf{y}-\mathbf{x}\| \geq \tilde{c}\frac{h_{\mathbf{z}}}{\operatorname{diam}\widetilde{\omega_{\mathbf{z}}}}\left(\|\tilde{\mathbf{y}}-\tilde{\mathbf{p}}\|+\|\tilde{\mathbf{p}}-\tilde{\mathbf{x}}\|\right) \geq \tilde{c}\frac{h_{\mathbf{z}}}{\operatorname{diam}\widetilde{\omega_{\mathbf{z}}}}\|\tilde{\mathbf{y}}-\tilde{\mathbf{x}}\|$$

with $\tilde{\mathbf{p}} := \chi_{\mathbf{z}}^{-1}(\mathbf{p})$. For the upper estimate we use $\overline{\tilde{\mathbf{x}}\tilde{\mathbf{y}}} \subset \widetilde{\omega_{\mathbf{z}}}$ since $\widetilde{\omega_{\mathbf{z}}}$ is convex. Let $(\tilde{\mathbf{p}}_i)_{i=0}^{q}$ be the minimal number of points lying on $\overline{\tilde{\mathbf{x}}\tilde{\mathbf{y}}}$ such that

$$\tilde{\mathbf{p}}_0 = \tilde{\mathbf{x}}, \quad \tilde{\mathbf{p}}_q = \mathbf{y}, \quad \text{and} \quad \forall 1 \leq i \leq q : \overline{\tilde{\mathbf{p}}_{i-1}\tilde{\mathbf{p}}_i} \text{ is contained in some } \tilde{\tau} \subset \widetilde{\omega_{\mathbf{z}}}.$$

Let $\mathbf{p}_i = \chi_{\mathbf{z}}(\tilde{\mathbf{p}}_i)$, $1 \leq i < q$. Then the upper estimate follows from

$$\|\mathbf{y}-\mathbf{x}\| \leq \sum_{i=1}^{q}\|\mathbf{p}_i-\mathbf{p}_{i-1}\| \leq C\frac{h_{\mathbf{z}}}{\operatorname{diam}\widetilde{\omega_{\mathbf{z}}}}\sum_{i=1}^{q}\|\tilde{\mathbf{p}}_i-\tilde{\mathbf{p}}_{i-1}\| = C\frac{h_{\mathbf{z}}}{\operatorname{diam}\widetilde{\omega_{\mathbf{z}}}}\|\tilde{\mathbf{y}}-\tilde{\mathbf{x}}\|.$$

(ii) τ and t share exactly one common point $\{\mathbf{z}\} = \overline{\tau} \cap \overline{t}$. Then

$$\|\mathbf{x}-\mathbf{y}\| \geq c\left(\|\mathbf{x}-\mathbf{z}\|+\|\mathbf{z}-\mathbf{y}\|\right)$$

and the rest of the proof is just a repetition of the arguments as in Case i. □

Lemma 4.3.31. *Let Assumption 4.3.17 or Assumption 4.3.18 be satisfied. For $\tau \in \mathcal{G}_{\mathbf{z}}$, let $\chi_{\mathbf{z},\tau} := \chi_{\mathbf{z}}|_{\tilde{\tau}}$, where $\tilde{\tau} := \chi_{\mathbf{z}}^{-1}(\tau)$. Then, for any $\mu \in \mathbb{N}_0^2$ with $k := |\mu|$,*

$$\|\partial^{\mu}\chi_{\mathbf{z},\tau}\|_{\mathrm{L}^{\infty}(\tilde{\tau})} \leq C\left(\frac{h_{\mathbf{z}}}{\operatorname{diam}\widetilde{\omega_{\mathbf{z}}}}\right)^{k},$$

where C depends only on k, C_1 as in (4.228), C_{d} as in (4.236), and the global chart χ.

Proof. Recall that $\chi_{\mathbf{z},\tau} = \chi \circ \kappa^{\text{affine}}$, where $\kappa^{\text{affine}} = \chi_{\tau}^{\text{affine}} \circ \eta^{\text{affine}}$ is affine. As in the proof of Lemma 4.3.6, we use

$$\partial^{\mu}\left(\chi \circ \kappa^{\text{affine}}\right) = \sum_{\substack{\beta\in\mathbb{N}_0^2\\|\beta|=\mu_1}}\sum_{\substack{\nu\in\mathbb{N}_0^2\\|\nu|=\mu_2}}\frac{\mu!}{\beta!\nu!}\mathbf{a}_1^{\beta}\mathbf{a}_2^{\nu}\left(\partial^{\beta+\nu}\chi\right)\circ\kappa^{\text{affine}},$$

where $\mathbf{a}_1, \mathbf{a}_2$ are the column vectors of the Jacobi matrix of κ^{affine}, that is,

$$(\mathbf{a}_i)_j = \partial_j\kappa_i^{\text{affine}} = \sum_{k=1}^{2}\partial_k\chi_{\tau,i}^{\text{affine}}\partial_j\eta_j^{\text{affine}}.$$

We have $\left|\partial_k \chi_{\tau,i}^{\text{affine}}\right| \leq h_\tau$ and from (4.235) we conclude that $\left|\partial_i \eta_j^{\text{affine}}\right| \leq \operatorname{diam} \tilde{\tau} / (2 |\tilde{\tau}|)$. This leads to $\left|(\mathbf{a}_i)_j\right| \leq h_\tau \operatorname{diam} \tilde{\tau} / |\tilde{\tau}|$. Thus

$$\left|\partial^\mu \left(\chi \circ \kappa^{\text{affine}}\right)\right| \leq C \left(\frac{h_\tau \operatorname{diam} \tilde{\tau}}{|\tilde{\tau}|}\right)^k \leq C \left(C_1 C_{\mathrm{d}} \frac{h_{\mathbf{z}}}{\operatorname{diam} \widetilde{\omega_{\mathbf{z}}}}\right)^k,$$

where C depends only on k and the global chart χ. □

4.4 Inverse Estimates

The spaces $H^s(\Gamma)$ form a scale:

$$H^s(\Gamma) \subseteq H^t(\Gamma), \quad \text{for } t \leq s \tag{4.257}$$

with a continuous embedding: there exists some $C(s,t) > 0$ such that

$$\|u\|_{H^t(\Gamma)} \leq C(s,t) \|u\|_{H^s(\Gamma)}, \qquad \forall u \in H^s(\Gamma). \tag{4.258}$$

Note that the range of s and t may be bounded by the smoothness of the surface (see Sect. 2.4). In general, the inverse of this inequality is false.

Exercise 4.4.1. *Find a sequence of functions $(u_n)_{n\in\mathbb{N}} \in C^\infty([0,1])$ which contradicts the inverse of (4.258) for $s = 0$ and $t = 1$, i.e., which satisfies*

$$\lim_{n\to\infty} \|u_n\|_{H^1([0,1])} / \|u_n\|_{L^2([0,1])} = \infty.$$

However, for boundary element functions there is a valid inverse of (4.258), a so-called *inverse inequality*, where the constant C depends on the dimension of the boundary element space. In the following we will assume that the maximal mesh width h is bounded above by a global constant h_0. For example, we can choose $h_0 = \operatorname{diam} \Gamma$ or otherwise $h_0 = 1$ for sufficiently fine surface meshes. Recall the definition of $\mathbb{P}_k^\tau$ as in (4.67).

Theorem 4.4.2. *Let either Assumption 4.3.17 or Assumption 4.3.18 hold. We have for $0 \leq m \leq \ell$, all $\tau \in \mathcal{G}$ and all $v \in \mathbb{P}_k^\tau$:*

$$\|v\|_{H^\ell(\tau)} \leq C h_\tau^{m-\ell} \|v\|_{H^m(\tau)}.$$

The constant C depends only on h_0, ℓ, k and, for a polyhedral surface, on the shape-regularity of the mesh $\mathcal{G}$ via the constant $\kappa_{\mathcal{G}}$ from Definition 4.1.12. In the case of a curved surface it also depends on the derivatives of the global transformations χ, χ^{-1} up to the order k.

Proof. Owing to the h-independent equivalence of the norms $\|v\|_{H^\ell(\tau)}$ and $\|\tilde{v}\|_{H^\ell(\tau^{\text{affine}})}$ from Corollary 4.3.12, it suffices to consider the case of a plane polyhedral surface.

Case 1: $m = 0$. Since $\mathbb{P}_k^\tau$ is finite-dimensional, all norms on $\mathbb{P}_k^\tau$ are equivalent: There exists a positive constant C_ℓ such that for $0 \le j \le \ell$

$$\|\hat{v}\|_{H^j(\hat{\tau})} \le C_\ell \|\hat{v}\|_{L^2(\hat{\tau})} \qquad \forall \hat{v} \in \mathbb{P}_k^{\hat{\tau}}.$$

With Lemma 4.3.6 or Lemma 4.3.13 it follows for all $v \in \mathbb{P}_k^\tau$ that

$$|v|_{H^j(\tau)} \le C_1 h_\tau^{1-j} |\hat{v}|_{H^j(\hat{\tau})} \le C_\ell C_1 h_\tau^{1-j} \|\hat{v}\|_{L^2(\hat{\tau})} \le C_\ell C_1 C_2 h_\tau^{-j} \|v\|_{L^2(\tau)}.$$

For the $\|\cdot\|_{H^\ell}$-norm, by summing the squares of the seminorm we obtain

$$\|v\|_{H^\ell(\tau)} \le C h_\tau^{-\ell} \|v\|_{L^2(\tau)}, \tag{4.259}$$

where C depends on ℓ, k and the upper bound of the mesh width h_0.

Case 2: $0 < m \le \ell$. For $\ell - m \le n \le \ell$ and $|\alpha| = n$ we write $\partial^\alpha v = \partial^\beta \partial^{\alpha-\beta}$ with $|\beta| = \ell - m$ and $\beta \le \alpha$ componentwise. Then with Case 1 we have

$$\|\partial^\alpha v\|_{L^2(\tau)} \le \left|\partial^{\alpha-\beta} v\right|_{H^{\ell-m}(\tau)} \le C h_\tau^{m-\ell} \left\|\partial^{\alpha-\beta} v\right\|_{L^2(\tau)} \le C h_\tau^{m-\ell} |v|_{H^{n-\ell+m}(\tau)}.$$

Since $|\alpha| = n$ was arbitrary, this result and $n - \ell + m \le m$ together yield

$$|v|_{H^n(\tau)} \le C h_\tau^{m-\ell} |v|_{H^{n-\ell+m}(\tau)} \le C h_\tau^{m-\ell} \|v\|_{H^m(\tau)} \tag{4.260}$$

for an arbitrary $\ell - m \le n \le \ell$. (Note that the constant C in (4.260) depends on n, m, and ℓ. However, n and m are from the finite set $\{0, 1, \ldots, \ell\}$ and – by taking the maximum over n and m – results in a constant C which does not depend on n and m but on ℓ instead.) Inequality (4.259) for $\ell \leftarrow \ell - m$ as well as Estimate (4.260) finally yield the assertion

$$\begin{aligned}
\|v\|_{H^\ell(\tau)}^2 &= \|v\|_{H^{\ell-m}(\tau)}^2 + \sum_{n=\ell-m+1}^{m} |v|_{H^n(\tau)}^2 \\
&\le C \left\{ h_\tau^{2(m-\ell)} \|v\|_{L^2(\tau)}^2 + \sum_{n=\ell-m+1}^{m} h_\tau^{2(m-\ell)} \|v\|_{H^m(\tau)}^2 \right\} \\
&\le C h_\tau^{2(m-\ell)} \|v\|_{H^m(\tau)}^2.
\end{aligned}$$

□

The global version of Theorem 4.4.2 requires the quasi-uniformity of the surface mesh $\mathcal{G}$.

Theorem 4.4.3. *Let either Assumption 4.3.17 or Assumption 4.3.18 hold. Then we have for all* $t, s \in \{0, 1\}$, $t \leq s$, *the estimate*

$$\forall v \in S_{\mathcal{G}}^{p,0} : \|v\|_{H^s(\Gamma)} \leq C h^{t-s} \|v\|_{H^t(\Gamma)}. \tag{4.261}$$

The constant C *depends only on* h_0, p *and, for a polyhedral surface, on the shape-regularity and quasi-uniformity of the mesh* $\mathcal{G}$ *via the constants* $\kappa_{\mathcal{G}}$ *and* $q_{\mathcal{G}}$ *from Definitions 4.1.12 and 4.1.13. In the case of a curved surface it also depends on the derivatives of the global transformations* χ, χ^{-1} *up to the order* k.

Proof. From Theorem 4.4.2 we have

$$\begin{aligned} \|v\|_{H^s(\Gamma)}^2 &= \sum_{\tau \in \mathcal{G}} \|v\|_{H^s(\tau)}^2 \leq C \sum_{\tau \in \mathcal{G}} h_\tau^{2(t-s)} \|v\|_{H^t(\tau)}^2 \leq C \left(\min_{\tau \in \mathcal{G}} h_\tau \right)^{2(t-s)} \|v\|_{H^t(\Gamma)}^2 \\ &\leq \left(C q_{\mathcal{G}}^{2(s-t)} \right) h^{2(t-s)} \|v\|_{H^t(\Gamma)}^2 . \end{aligned}$$

□

Theorem 4.4.3 can be generalized in various ways. In the following we will cite results from [75].

Remark 4.4.4. *(a) Theorem 4.4.3 holds for all* $t, s \in \mathbb{R}$ *with* $0 \leq t \leq s \leq 1$ *or* $-1 \leq t \leq 0 \wedge s = 0$ *(see [75, Theorems 4.1, 4.6]).*
(b) Theorem 4.4.3 is valid for the space $S_{\mathcal{G}}^{p,-1}$ *for all* $t, s \in \mathbb{R}$ *with* $t = 0 \wedge 0 \leq s < 1/2$ *or* $-1 \leq t \leq 0 \wedge s = 0$ *(see [75, Theorems 4.2, 4.6]).*

We will also require estimates between different L^p-norms and discrete ℓ^p-norms for boundary element functions and, thus, we again start with a local result. Here we will always consider the situation where a Lagrange basis is chosen for $\mathbb{P}_k^{\hat{\tau}}$ on $\hat{\tau}$. $\Sigma_\kappa = \left\{ \widehat{\mathbf{P}}_i : i \in \iota_k^{\hat{\tau}} \right\}$ denotes the set of nodal points on $\hat{\tau}$. The Lagrange basis $\left(\widehat{N}_i \right)_{i \in \iota_k^{\hat{\tau}}}$ of $\mathbb{P}_k^{\hat{\tau}}$ satisfies

$$\widehat{N}_i \left(\widehat{\mathbf{P}}_j \right) = \delta_{i,j} \qquad \forall i, j \in \iota_k^{\hat{\tau}}.$$

A vector of coefficients $\mathbf{w} := (w_i)_{i \in \iota_k^{\hat{\tau}}}$ is put into relation with the associated polynomial $\hat{w} \in \mathbb{P}_k^{\hat{\tau}}$ on the reference element by means of

$$\hat{w} := \widehat{P} \mathbf{w} := \sum_{i \in \iota_k^{\hat{\tau}}} w_i \widehat{N}_i .$$

We define the "lifted" function

$$w := P_\tau \mathbf{w} := \sum_{i \in \iota_k^{\hat{\tau}}} w_i N_i \quad \text{with} \quad N_i = \widehat{N}_i \circ \chi_\tau^{-1}$$

analogously.

Theorem 4.4.5. *Let either Assumption 4.3.17 or Assumption 4.3.18 hold. For all $\tau \in \mathcal{G}$ and all $\mathbf{w} := (w_i)_{i \in \iota_k^{\hat{\tau}}}$ we have*

$$\tilde{c} h_\tau \|\mathbf{w}\|_{\ell^2} \le \|P_\tau \mathbf{w}\|_{L^2(\tau)} \le \widetilde{C} h_\tau \|\mathbf{w}\|_{\ell^2} .$$

The constants $\tilde{c}$ and $\widetilde{C}$ depend on the parameters qualitatively in the same way as does C in Theorem 4.4.3.

Proof. From Lemma 4.3.6 or Lemma 4.3.13 we have

$$c h_\tau \|\hat{w}\|_{L^2(\hat{\tau})} \le \|w\|_{L^2(\tau)} \le C h_\tau \|\hat{w}\|_{L^2(\hat{\tau})} \quad \text{with} \quad \hat{w} := w \circ \chi_\tau .$$

Since all norms are equivalent on $\mathbb{P}_k^{\hat{\tau}}$, we have

$$c_k \|\hat{w}\|_{H^{k+1}(\hat{\tau})} \le \|\hat{w}\|_{L^2(\hat{\tau})} \le C_k \|\hat{w}\|_{H^{k+1}(\hat{\tau})} .$$

The equivalence of the $H^{k+1}(\hat{\tau})$-norm and the $[\cdot]_{k+1}$-norm follows from Lemma 4.3.1. Since $\hat{w} \in \mathbb{P}_k^{\hat{\tau}}$,

$$[\hat{w}]_{k+1} = |\hat{w}|_{H^{k+1}(\hat{\tau})} + \sum_{\mathbf{z} \in \Sigma_k} |\hat{w}(\mathbf{z})| = \sum_{\mathbf{z} \in \Sigma_k} |\hat{w}(\mathbf{z})| = \sum_{i \in \iota_p^{\hat{\tau}}} |w_i| = \|\mathbf{w}\|_{\ell^1} . \quad (4.262)$$

Since $\sharp \Sigma_k$ is finite, there exist positive constants c, C depending only on the cardinality of Σ_k, i.e., on k, such that

$$c \|\mathbf{w}\|_{\ell^2} \le \|\mathbf{w}\|_{\ell^1} \le C \|\mathbf{w}\|_{\ell^2} .$$

Combining all these results, we have thus proved that

$$\tilde{c} h_\tau \|\mathbf{w}\|_{\ell^2} \le \|w\|_{L^2(\tau)} \le \widetilde{C} h_\tau \|\mathbf{w}\|_{\ell^2} .$$

□

Corollary 4.4.6. *Let the conditions from Theorem 4.4.5 be satisfied. Then*

$$\hat{c} h_\tau \|w\|_{L^\infty(\tau)} \le \|w\|_{L^2(\tau)} \le \widehat{C} h_\tau \|w\|_{L^\infty(\tau)}$$

for all $w \in \mathbb{P}_k^\tau$. *The constants* $\hat{c}, \widehat{C}$ *qualitatively depend on the parameters in the same way as do* $\tilde{c}, \widetilde{C}$ *in Theorem 4.4.3.*

Proof. If we combine Theorem 4.4.5 with the norm equivalence on finite-dimensional spaces for $\mathbf{w} = (w_i)_{i \in \iota_p^{\hat{\tau}}}$ and $w = P_\tau \mathbf{w}$ it follows that

$$\|w\|_{L^2(\tau)} \leq C h_\tau \|\mathbf{w}\|_{\ell^2} \leq \widehat{C} h_\tau \|\mathbf{w}\|_{\ell^\infty} \leq \widehat{C} h_\tau \|w\|_{L^\infty(\tau)} .$$

Conversely, with the notation from the proof of Theorem 4.4.5 we have

$$\begin{aligned} \|w\|_{L^\infty(\tau)} &= \|\hat{w}\|_{L^\infty(\hat{\tau})} \leq C \|\hat{w}\|_{H^{k+1}(\hat{\tau})} \leq C' [\hat{w}]_{k+1} \overset{(4.262)}{=} C' \|\mathbf{w}\|_{\ell^1} \\ &\leq C'' \|\mathbf{w}\|_{\ell^\infty} \leq C''' \|\mathbf{w}\|_{\ell^2} . \end{aligned}$$

Note that the constants in this estimate depend on the cardinality of Σ_k, i.e., on k. From Theorem 4.4.5 we thus have the lower bound. □

The global version of Theorem 4.4.5 shows an equivalence between boundary element functions and the associated coefficient vector. Let $(b_i)_{i=1}^N$ be the Lagrange basis of the boundary element space S. We define the operator $P : \mathbb{R}^N \to S$ for $\mathbf{w} = (w_i)_{i=1}^N$ by

$$P\mathbf{w} = \sum_{i=1}^{N} w_i b_i .$$

Theorem 4.4.7. *Let Assumption 4.3.17 or Assumption 4.3.18 hold. Then for all* $\mathbf{w} \in \mathbb{R}^N$

$$\check{c} h \|\mathbf{w}\|_{\ell^2} \leq \|P\mathbf{w}\|_{L^2(\Gamma)} \leq \check{C} h \|\mathbf{w}\|_{\ell^2} .$$

The constants $\check{c}, \check{C}$ *qualitatively depend on the parameters in the same way as* $\tilde{c}, \widetilde{C}$ *do in Theorem 4.4.5.*

Proof. Let $\mathbf{w} \in \mathbb{R}^N$ be the coefficient vector of the boundary element function $w = P\mathbf{w}$. For $\tau \in \mathcal{G}$ we can associate a global index $\operatorname{ind}(m, \tau) \in \{1, 2, \ldots, N\}$ on τ with every local degree of freedom $m \in \iota_k^{\hat{\tau}}$. We set $\mathbf{w}_\tau := (\mathbf{w}_{\tau,m})_{m \in \iota_k^{\hat{\tau}}} := \left(\mathbf{w}_{\operatorname{ind}(m,\tau)}\right)_{m \in \iota_k^{\hat{\tau}}}$. With Theorem 4.4.5 we obtain

$$\|P\mathbf{w}\|_{L^2(\Gamma)}^2 = \sum_{\tau \in \mathcal{G}} \|P_\tau \mathbf{w}\|_{L^2(\tau)}^2 \leq C h^2 \sum_{\tau \in \mathcal{G}} \|\mathbf{w}_\tau\|_{\ell^2}^2 .$$

The constant

$$M := \max_{i \in \{1,2,\ldots,N\}} \sharp \left\{ (m, \tau) \in \iota_p^{\hat{\tau}} \times \mathcal{G} : i = \operatorname{ind}(m, \tau) \right\}$$

depends only on the polynomial degree k and on the shape-regularity of the surface mesh. It thus follows that

$$\|P\mathbf{w}\|_{L^2(\Gamma)}^2 \leq CMh^2 \|\mathbf{w}\|_{\ell^2}^2 .$$

The lower bound can be found in a similar way. □

Corollary 4.4.8. *Let either Assumption 4.3.17 or Assumption 4.3.18 hold and let $(b_i)_{i\in\mathcal{I}}$ denote the nodal basis for the boundary element space S. Then*

$$\|b_i\|_{L^\infty(\Gamma)} \leq \breve{C}_1. \tag{4.263}$$

The constant $\breve{C}_1$ depends only on the shape-regularity of the mesh and the polynomial degree of S.

If $S = S_{\mathcal{G}}^{k,0}$ for some $k \geq 1$ then

$$|b_i|_{W^{1,\infty}(\Gamma)} := \|\nabla_S b_i\|_{L^\infty(\Gamma)} \leq \breve{C}_2 h_\tau^{-1} \quad \text{for any } \tau \subset \operatorname{supp} b_i . \tag{4.264}$$

The full $W^{1,\infty}(\Gamma)$-norm is given by $\|\cdot\|_{W^{1,\infty}(\Gamma)} := \max\left\{\|\cdot\|_{L^\infty(\Gamma)}, |\cdot|_{W^{1,\infty}(\Gamma)}\right\}$ and hence

$$\|b_i\|_{W^{1,\infty}(\Gamma)} \leq \breve{C}_3 h_\tau^{-1} \quad \text{for any } \tau \subset \operatorname{supp} b_i . \tag{4.265}$$

Proof. Let $\mathbf{e}_i \in \mathbb{R}^{\mathcal{I}}$ denote the vector with $(\mathbf{e}_i)_i = 1$ and $(\mathbf{e}_i)_j = 0$ otherwise, i.e., $b_i = P\mathbf{e}_i$, Let $\tau \subset \operatorname{supp} b_i$. The combination of Corollary 4.4.6 and Theorem 4.4.5 leads to

$$\|b_i\|_{L^\infty(\tau)} \leq (\hat{c}h_\tau)^{-1} \|b_i\|_{L^2(\tau)} \leq \widetilde{C}/\hat{c}.$$

Because $b_i|_\tau = 0$ for all $\tau \in \mathcal{G}_i$ with $\tau \not\subset \operatorname{supp} b_i$ we have proved (4.263).

For the proof of the second estimate we observe that – as in the proof of Theorem 4.4.3 – it suffices to consider plane panels with straight edges. Hence $\nabla_S b_i$ is a polynomial on every panel τ so that

$$\begin{aligned} \hat{c}h_\tau \|\nabla_S b_i\|_{L^\infty(\tau)} &\overset{\text{Cor. 4.4.6}}{\leq} \|\nabla_S b_i\|_{L^2(\tau)} \overset{\text{Theo. 4.4.2}}{\leq} Ch_\tau^{-1} \|b_i\|_{L^2(\tau)} \\ &\overset{\text{Theo. 4.4.5}}{\leq} C\widetilde{C} \|\mathbf{e}_i\|_{\ell^2} = C\widetilde{C} \end{aligned}$$

from which the assertion follows. □

We can also analyze how far the constants in the norm equivalences depend on the mesh width h in the case of the ℓ^p and $L^p(\Gamma)$-norms with $1 \leq p \leq \infty$. Here we will only require the cases $p = 2$ and $p = \infty$ and refer to [75] for the more general case.

4.5 Condition of the System Matrices

One of the first applications of the inverse inequalities is the estimation of the condition of the system matrices of the integral operators.

Lemma 4.5.1. *Let Assumption 4.3.17 or Assumption 4.3.18 hold. Let* $\mathbf{K}$ *be the system matrix associated with the Galerkin discretization of the single layer operator* V *for the Laplace problem. Then we have*

$$\operatorname{cond}_2(\mathbf{K}) \leq Ch^{-1}.$$

The constant C *depends only on the polynomial degree* p *and the shape-regularity and the quasi-uniformity of the surface mesh* $\mathcal{G}$*, more specifically on the constants* $\kappa_{\mathcal{G}}$ *and* $q_{\mathcal{G}}$ *from Definitions 4.1.12 and 4.1.13. In the case of curved surfaces it also depends on the derivatives of the global transformations* χ*,* χ^{-1} *up to the order* k*.*

Proof. Since $\mathbf{K}$ is symmetric and positive definite, we have

$$\operatorname{cond}_2(\mathbf{K}) = \frac{\lambda_{\max}(\mathbf{K})}{\lambda_{\min}(\mathbf{K})}.$$

In the following we will thus estimate the eigenvalues of $\mathbf{K}$. It follows from the continuity and the $H^{-1/2}$-ellipticity of the bilinear form $(V\cdot,\cdot)_0 : H^{-1/2}(\Gamma) \times H^{-1/2}(\Gamma) \to \mathbb{K}$ that there exist two positive constants γ and C_c such that

$$\gamma \|u\|^2_{H^{-1/2}(\Gamma)} \leq (Vu,u)_0 \leq C_c \|u\|^2_{H^{-1/2}(\Gamma)} \qquad \forall u \in H^{-1/2}(\Gamma).$$

From this it follows with Theorem 4.4.7 that

$$\begin{aligned}\lambda_{\max}(\mathbf{K}) &= \max_{\mathbf{w}=(w_i)_i \in \mathbb{R}^N\backslash\{0\}} \frac{\langle \mathbf{K}\mathbf{w},\mathbf{w}\rangle}{\|\mathbf{w}\|^2} \leq Ch^2 \max_{w\in S\backslash\{0\}} \frac{(Vw,w)_0}{\|w\|^2_{L^2(\Gamma)}} \\ &\leq Ch^2 C_c \max_{w\in S\backslash\{0\}} \frac{\|w\|^2_{H^{-1/2}(\Gamma)}}{\|w\|^2_{L^2(\Gamma)}} \leq Ch^2 C_c.\end{aligned}$$

By Theorem 4.4.7 and Remark 4.4.4 we have for the smallest eigenvalue

$$\begin{aligned}\lambda_{\min}(\mathbf{K}) &= \min_{\mathbf{w}=(w_i)_i \in \mathbb{R}^N\backslash\{0\}} \frac{\langle \mathbf{K}\mathbf{w},\mathbf{w}\rangle}{\|\mathbf{w}\|^2} \geq Ch^2 \min_{w\in S\backslash\{0\}} \frac{(Vw,w)_0}{\|w\|^2_{L^2(\Gamma)}} \\ &\geq Ch^2 \gamma \min_{w\in S\backslash\{0\}} \frac{\|w\|^2_{H^{-1/2}(\Gamma)}}{\|w\|^2_{L^2(\Gamma)}} \geq C'h^2\gamma h.\end{aligned}$$

Thus

$$\lambda_{\max}(\mathbf{K}) / \lambda_{\min}(\mathbf{K}) \leq Ch^{-1}$$

and the lemma follows. □

Exercise 4.5.2. *Show that the system matrix* $\mathbf{K}$ *associated with the hypersingular operator also satisfies the estimate*

$$\operatorname{cond}_2(\mathbf{K}) \leq C h^{-1}$$

under the conditions of Lemma 4.5.1.

Remark 4.5.3. *For the condition of the* mass matrix $\mathbf{M} := \left((b_i, b_j)_{L^2(\Gamma)} \right)_{i,j=1}^{N}$ *we have*

$$\operatorname{cond}_2(\mathbf{M}) \leq C.$$

Proof. Since

$$\langle \mathbf{w}, \mathbf{M}\mathbf{w} \rangle = (P\mathbf{w}, P\mathbf{w})_{L^2(\Gamma)}$$

we can apply Theorem 4.4.7:

$$\check{c}^2 h^2 \leq \min_{\mathbf{w} \in \mathbb{R}^N \setminus \{0\}} \frac{\langle \mathbf{M}\mathbf{w}, \mathbf{w} \rangle}{\|\mathbf{w}\|^2} \leq \max_{\mathbf{w} \in \mathbb{R}^N \setminus \{0\}} \frac{\langle \mathbf{M}\mathbf{w}, \mathbf{w} \rangle}{\|\mathbf{w}\|^2} \leq \check{C}^2 h^2,$$

from which we have the estimate of the condition with $C = \check{C}^2 / \check{c}^2$. □

Estimating the condition of system matrices for equations of the second kind is more problematic, as the stability of the Galerkin discretization for these equations is in many cases still an open question. If we assume that the h-independent stability of the discrete operators is given, the condition of the system matrices for equations of the second kind can be determined in terms of an h-independent constant in the same way as before.

4.6 Bibliographical Remarks and Further Results

In the present chapter, we introduced spaces of piecewise polynomial functions on the boundary manifold Γ, and established approximation properties of these spaces, as the meshwidth h tends to zero, in several function spaces of Sobolev type on Γ. These boundary element spaces are, in a sense, Finite Element spaces on the boundary surface Γ. We also presented a general framework for the convergence analysis of Galerkin boundary element methods, in particular necessary and sufficient conditions for the quasi-optimality of the Galerkin solutions to hold.

For reasons of space, our presentation does not cover the most general cases. For example, the surface meshes upon which the boundary element spaces are built did not allow for local mesh refinement or, more importantly, for *anisotropic local refinements* for example in the vicinity of edges (see, e.g., [75, 87, 234]).

Most of our results do extend to so-called graded, anisotropic meshes (cf. [104, 107, 108]). In addition, besides mesh refinement, analogs of spectral methods or even a combination of mesh refinement and order increase, the so-called *hp-Version BEM*, is conceivable (cf. [222] and the references therein).

Further, for particular classes of boundary integral equations, special choices of subspaces may yield large gains in accuracy versus number of degrees of freedom. Let us mention, for example, the case of high frequency acoustic scattering. Here, the stability of the boundary integral operators depends, of course, on the problem's wave number but, in addition, also the solutions contain high-frequency components which are smooth, but highly oscillatory at large wave numbers, and therefore poorly captured by standard boundary element spaces, unless the fine scale of the unknown functions on the boundary is resolved by sufficient mesh refinement. This strategy may lead, however, to prohibitively large numbers of degrees of freedom. A better approach may be to augment the standard boundary element spaces by explicitly known, dominant asymptotic components of the unknown solution. In high frequency acoustics and electromagnetics, in particular for BIEs obtained from the direct method (where the unknowns are Cauchy data of the domain unknowns), strong results on the asymptotic structure of the solution are available from geometrical optics. These can be used to build boundary element spaces with no or a reduced preasymptotic convergence regime at high wave numbers. We refer e.g. to [5, 57, 153] for recent work on wave number independent Galerkin BEM for acoustics problems.

In this chapter, and throughout this book, we focused on *Galerkin* BEM. We do emphasize, however, that the alternative *collocation* BEM do constitute a powerful competition; for collocation BEM on polyhedra, however, the theory of stability and quasi-optimality is much less mature that in the Galerkin case. Still, since collocation methods do not require the numerical evaluation of double surface integrals, they offer a substantial gain in accuracy versus CPU time.

For this reason, in recent years substantial work has been devoted to collocation based BIEs for high frequency acoustic and electromagnetic scattering. We mention in particular the work of O. Bruno et al. (e.g. [34,35,161]) which is a collocation type boundary element method which combines incorporation of high frequency asymptotics with a degenerate coordinate transformation of the surface in the presence of edges or vertices and a Nyström type collocation procedure. The mathematical error analysis of this method is in progress.

The a priori asymptotic error bounds for Galerkin BEM developed in Sect. 4.2 show that Galerkin BEM exhibit superconvergence in negative Sobolev norms on Γ. This allows us, in particular, to deduce corresponding results for *postprocessed* Galerkin approximations which can be obtained as smooth functionals of the solution. Importantly, *the insertion of the Galerkin solution into the representation formula* is such a postprocessing operation. Therefore superconvergent pointwise approximations of the solution to the underlying boundary value problem at interior points of the domain result usually from Galerkin boundary element approximations. Note that our analysis in Sect. 4.2 reveals the crucial role of Galerkin orthogonality of the discretization in the derivation of superconvergence estimates in negative order norms (indeed, for other discretization schemes such as collocation or Nyström methods, such superconvergence results either do not hold or only with a much smaller gain in asymptotic convergence order). We finally note that the superconvergence error bounds for the solution at points $\mathbf{x}$ in the interior of both the domain Ω or its complement deteriorate as $\mathbf{x}$ approaches Γ. Nevertheless, this deterioration can be remedied and postprocessing procedures can be designed to recover superconvergent solution values and normal and tangential derivatives (required, for example, in shape optimization or uncertainty quantification) of arbitrary order from the Galerkin solution such that the superconvergence bounds are uniform in the distance of $\mathbf{x}$ to the boundary Γ. For the details, we refer to [213].

Chapter 5
Generating the Matrix Coefficients

In order to implement the Galerkin method for boundary integral equations, the approximation of the coefficients of the system matrix and the right-hand side becomes a primary task. The integrals are of the form

$$\int_\Gamma b_i(\mathbf{x})\, b_j(\mathbf{x})\, ds_{\mathbf{x}}, \qquad \int_\Gamma b_i(\mathbf{x})\, r(\mathbf{x})\, ds_{\mathbf{x}} \tag{5.1}$$

and

$$\int_\Gamma b_i(\mathbf{x}) \int_\Gamma k(\mathbf{x},\mathbf{y},\mathbf{y}-\mathbf{x})\, b_j(\mathbf{y})\, ds_{\mathbf{y}} ds_{\mathbf{x}}, \quad \int_\Gamma b_i(\mathbf{x}) \int_\Gamma k(\mathbf{x},\mathbf{y},\mathbf{y}-\mathbf{x})\, r(\mathbf{y})\, ds_{\mathbf{y}} ds_{\mathbf{x}}, \tag{5.2}$$

where b_i denotes the basis functions of the boundary element space and r is a given function (right-hand side). Note that the basis functions are real and therefore complex conjugation becomes obsolete.

The aim of this chapter is to develop and analyze problem-specific integration techniques in order to approximate these integrals. Note that the integrals in (5.1) do not contain any singularities, assuming that the right-hand side r is sufficiently smooth on every panel. Furthermore, the number of non-zero integrals in (5.1) is proportional to the dimension of the boundary element space (and not the square of the dimension) because the basis functions are only locally defined.

First we will define a class of functions, the kernel functions, and derive their characteristic properties (in local coordinates). All the kernel functions that we have dealt with so far are part of this class as are kernels that appear in connection with linear elasticity. Subsequently, we will introduce suitable variable transformations that render the singular integrands analytic. This, in turn permits the numerical approximation of the integral by means of standard quadrature methods. These coordinate transforms are applicable to any integral operator arising in the boundary reduction of strongly elliptic partial differential equations in $\mathbb{R}^3$ (see [137]). They will not depend on the explicit form of the kernel function. This implies that the numerical integration in the computer program can be realized in an abstract manner.

S.A. Sauter and C. Schwab, *Boundary Element Methods*, Springer Series in Computational Mathematics 39, DOI 10.1007/978-3-540-68093-2_5,

We will conclude this chapter with an error analysis. First we will analyze the local quadrature error depending on the order of the quadrature, after which we will estimate the effect this has on the entire discretization.

If not explicitly stated otherwise we will restrict ourselves to the case $d = 3$ and two-dimensional surfaces Γ in the entire chapter.

Remark 5.0.1. *In certain special cases (plane, right-angled panels, kernel functions for the Laplace operator) the integrals in (5.1) and (5.2) can be evaluated exactly (see [157]). However, we prefer the ansatz via numerical quadrature, as this allows for a much larger class of integral operators and is also easier to implement.*

Note: Readers who are more interested in the quadrature formulas and in the required number of quadrature nodes than in the analysis and derivation will find a compact summary of these in Sect. 5.2.4 and Theorem 5.3.30.

5.1 Kernel Functions and Strongly Singular Integrals

The properties of an integral operator of the form

$$(Ku)(\mathbf{x}) := \int_{\Gamma} k(\mathbf{x}, \mathbf{y}, \mathbf{y} - \mathbf{x})\, u(\mathbf{y})\, ds_{\mathbf{y}}$$

are characterized by the properties of the kernel function k and the smoothness of the surface.

5.1.1 Geometric Conditions

In this section we will summarize a set of conditions imposed on the surface and the boundary element mesh. Most of the applications satisfy either these conditions or some weakening of them. The relevant cases in which the conditions can be weakened will be duly noted.

As before we use the notation $\mathcal{G} := \{\tau_1, \ldots, \tau_n\}$ for the boundary element mesh and $\chi_\tau : \widehat{\tau} \to \tau$ for the parametrization over the reference element (unit square/triangle).

Assumption 5.1.1. *The surface mesh $\mathcal{G}$ is regular in the sense of Definition 4.1.4.*

We assume that the functions χ_τ, χ_t are analytic. The details can be found in the following definition. The differential operator $\langle \mathbf{v}, \nabla \rangle^m$ that appears in the definition is defined for a vector $\mathbf{v} \in \mathbb{R}^2$ by

$$\langle \mathbf{v}, \nabla \rangle^m f := \sum_{k=0}^{m} \binom{m}{k} v_1^k v_2^{m-k} \partial_1^k \partial_2^{m-k} f. \tag{5.3}$$

Definition 5.1.2. Let $\tau \in \mathcal{G}$. The parametrization $\chi_\tau : \hat{\tau} \to \tau$ is analytic if there exists an open, complex neighborhood $\hat{\tau}^\star \subset \mathbb{C} \times \mathbb{C}$ of $\overline{\hat{\tau}}$ in which χ_τ can be extended analytically, i.e., for all $\mathbf{z} = (z_1, z_2) \in \hat{\tau}^\star$ there exists a neighborhood $U(\mathbf{z}) \subset \hat{\tau}^\star$ such that

$$\chi_\tau(\mathbf{w}) = \sum_{i=0}^{\infty} \frac{\langle \mathbf{w} - \mathbf{z}, \nabla \rangle^i \chi_\tau(\mathbf{z})}{i!} \qquad \forall \mathbf{w} \in U(\mathbf{z}).$$

The extension is again denoted by χ_τ.

Assumption 5.1.3. *For all $\tau \in \mathcal{G}$ the parametrizations $\chi_\tau : \widehat{\tau} \to \tau$ are analytic.*

The integrals (5.2) over $\Gamma \times \Gamma$ will be split into a sum of integrals over $\tau \times t$, where $\tau, t \in \mathcal{G}$. The quadrature error will depend on the angles between adjacent panels and the distortion between the surface and the Euclidean metric. We will employ Assumption 4.3.29 and the constants therein to quantify this influence. The discretization parameter for the boundary element method is the diameter h of the largest panel of a surface mesh. The error estimates from Chap. 4 describe quantitatively at what rate the error goes to zero as $h \to 0$. Therefore it is essential for local quadrature error analysis that the behavior of the parametrization with respect to the panel diameter is properly understood. In order to describe this behavior we define the quantities $e_1, e_2, \theta : \hat{\tau} \to \mathbb{R}$ as follows:

$$\begin{aligned} e_1(\hat{\mathbf{x}}) &:= \|\partial_1 \chi_\tau(\hat{\mathbf{x}})\|, \quad e_2(\hat{\mathbf{x}}) := \|\partial_2 \chi_\tau(\hat{\mathbf{x}})\|, \\ \cos\theta(\hat{\mathbf{x}}) &:= \frac{\langle \partial_1 \chi_\tau(\hat{\mathbf{x}}), \partial_2 \chi_\tau(\hat{\mathbf{x}}) \rangle}{\|\partial_1 \chi_\tau(\hat{\mathbf{x}})\| \, \|\partial_2 \chi_\tau(\hat{\mathbf{x}})\|}. \end{aligned} \tag{5.4}$$

For plane triangular elements, e_1, e_2, $\cos\theta$ and the ratio e_1/e_2 can be easily estimated by geometric terms related to the triangle.

Example 5.1.4. *Let $\tau \subset \mathbb{R}^3$ be a plane triangle with vertices $\mathbf{A}$, $\mathbf{B}$, $\mathbf{C}$ and interior angles α, β, γ. The reference triangle is denoted by $\widehat{S}$ and has the vertices $(0,0)^\intercal$, $(1,0)^\intercal$, $(1,1)^\intercal$. Then the mapping $\chi_\tau : \widehat{S} \to \tau$ is given by*

$$\chi_\tau(\hat{\mathbf{x}}) = \mathbf{A} + \mathbf{m}\hat{\mathbf{x}}$$

with the 3×2-matrix

$$\mathbf{m} := [\mathbf{B} - \mathbf{A}, \mathbf{C} - \mathbf{B}].$$

We have

$$e_1 = \|\mathbf{B} - \mathbf{A}\|, \qquad e_2 = \|\mathbf{C} - \mathbf{B}\|, \qquad \cos\theta := \frac{\langle \mathbf{B} - \mathbf{A}, \mathbf{C} - \mathbf{B} \rangle}{\|\mathbf{B} - \mathbf{A}\| \, \|\mathbf{C} - \mathbf{B}\|}.$$

In terms of the smallest interior angle of the triangle τ

$$\theta_0(\tau) := \min\{\alpha, \beta, \gamma\} \tag{5.5}$$

we obtain the estimates

$$\theta_0 \leq \theta \leq \pi - \theta_0 \quad \text{and} \quad \sin\theta_0 \leq \frac{\|\mathbf{B}-\mathbf{A}\|}{\|\mathbf{C}-\mathbf{B}\|} = \frac{\sin\gamma}{\sin\alpha} \leq \frac{1}{\sin\theta_0}.$$

This means that for triangulations that only consist of plane triangles, the quantities e_1/e_2 *and* $\cos\theta$ *depend only on the minimal interior angle but not on the refinement of the triangulation.*

Example 5.1.5. *Let* Γ *be a surface and* $\mathcal{G}_0$ *a triangulation with (curved) triangles whose diameter is of order* 1. *We can construct a refinement of* $\mathcal{G}_0$ *as follows. For every* $t \in \mathcal{G}_0$ *we generate a plane triangulation* $\widehat{\mathcal{G}_t}$ *of the reference element* $\hat{t}$, *which is then mapped by* χ_t *onto* Γ, *i.e.,* $\mathcal{G}_t := \left\{\chi_t(\tilde{\tau}) : \tilde{\tau} \in \widehat{\mathcal{G}_t}\right\}$. *The refinement of* $\mathcal{G}_0$ *is given by*

$$\mathcal{G} := \left\{\chi_t(\tilde{\tau}) : \forall \tilde{\tau} \in \widehat{\mathcal{G}}_t,\ \forall t \in \mathcal{G}_0\right\}$$

(see Fig. 5.1). For a triangle $\tau \in \mathcal{G}$ *with* $\tau \subset t \in \mathcal{G}_0$ *the vertices and interior angles of the triangle* $\tilde{\tau} := \chi_t^{-1}(\tau)$ *are denoted by* $\mathbf{A}, \mathbf{B}, \mathbf{C}$ *and* α, β, γ. *Then the parametrization of* τ *is given by*

$$\chi_\tau(\hat{\mathbf{x}}) := \chi_t(\mathbf{A} + \mathbf{m}\hat{\mathbf{x}})$$

with the 2×2*-matrix* $\mathbf{m} = [\mathbf{B}-\mathbf{A}, \mathbf{C}-\mathbf{B}]$. *It follows that*

$$e_1(\hat{\mathbf{x}}) = \langle \mathbf{B}-\mathbf{A}, \mathbf{G}_t(\hat{\mathbf{x}})(\mathbf{B}-\mathbf{A})\rangle^{1/2}, \qquad e_2(\hat{\mathbf{x}}) = \langle \mathbf{C}-\mathbf{B}, \mathbf{G}_t(\hat{\mathbf{x}})(\mathbf{C}-\mathbf{B})\rangle^{1/2}$$

$$\cos\theta(\hat{\mathbf{x}}) = \frac{\langle \mathbf{B}-\mathbf{A}, \mathbf{G}_t(\hat{\mathbf{x}})(\mathbf{C}-\mathbf{B})\rangle}{e_1(\hat{\mathbf{x}})\, e_2(\hat{\mathbf{x}})}$$

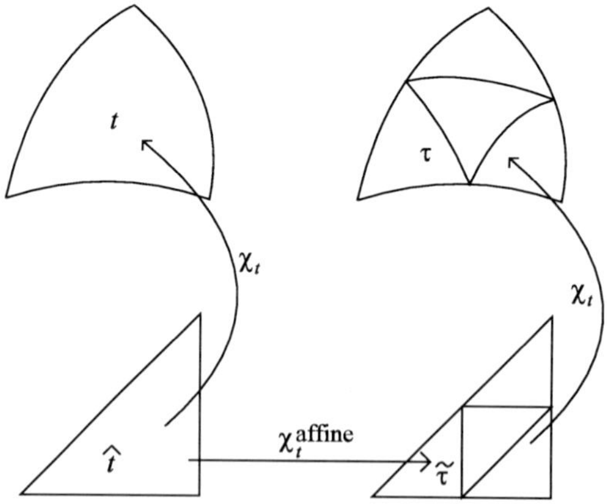

Fig. 5.1 Refinement of the reference element induces a refinement of the surface mesh

with the Gram matrix

$$\mathbf{G}_t(\hat{\mathbf{x}}) := (\mathbf{D}\chi_t(\mathbf{A}+\mathbf{m}\hat{\mathbf{x}}))^\intercal (\mathbf{D}\chi_t(\mathbf{A}+\mathbf{m}\hat{\mathbf{x}}))$$

and the Jacobian $\mathbf{D}\chi_t$. Clearly, the smallest and largest eigenvalues of this matrix can be bounded below and above by $\lambda_{\min}$ and $\lambda_{\max}$ respectively, while these are defined by

$$0 < \lambda_{\min} := \inf_{t\in\mathcal{G}_0} \inf_{\hat{\mathbf{x}}\in\hat{\tau}} \inf_{\xi\in\mathbb{R}^2\setminus\{0\}} \frac{\langle \xi, \mathbf{G}_t(\hat{\mathbf{x}})\,\xi\rangle}{\|\xi\|^2} \leq \sup_{t\in\mathcal{G}_0} \sup_{\hat{\mathbf{x}}\in\hat{\tau}} \sup_{\xi\in\mathbb{R}^2\setminus\{0\}} \frac{\langle \xi, \mathbf{G}_t(\hat{\mathbf{x}})\,\xi\rangle}{\|\xi\|^2}$$
$$=: \lambda_{\max} < \infty.$$

Note that $\lambda_{\min}, \lambda_{\max}$ depend only on the coarsest triangulation $\mathcal{G}_0$ and, in particular, are independent of the refinement of $\mathcal{G}$. Hence it follows that

$$\frac{\lambda_{\min}\sin\theta_0}{\lambda_{\max}} \leq e_1/e_2 \leq \frac{\lambda_{\max}}{\lambda_{\min}\sin\theta_0}$$

with [see (5.5)]

$$\theta_0 := \inf\left\{\theta_0(\tilde{\tau}) : \forall\tilde{\tau}\in\widehat{\mathcal{G}}_t,\ \forall t\in\mathcal{G}_0\right\}.$$

For the term $\cos\theta(\hat{\mathbf{x}})$ we obtain

$$\cos\theta(\hat{\mathbf{x}}) = \frac{\langle \tilde{\mathbf{m}}_1, \mathbf{G}_t(\hat{\mathbf{x}})\,\tilde{\mathbf{m}}_2\rangle}{\langle \tilde{\mathbf{m}}_1, \mathbf{G}_t(\hat{\mathbf{x}})\,\tilde{\mathbf{m}}_1\rangle^{1/2}\langle \tilde{\mathbf{m}}_2, \mathbf{G}_t(\hat{\mathbf{x}})\,\tilde{\mathbf{m}}_2\rangle^{1/2}}$$

with $\tilde{\mathbf{m}}_1 = (\mathbf{B}-\mathbf{A})/\|\mathbf{B}-\mathbf{A}\|$ and $\tilde{\mathbf{m}}_2 = (\mathbf{C}-\mathbf{B})/\|\mathbf{C}-\mathbf{B}\|$. With the function θ_0 from (5.5) we have

$$|\langle \tilde{\mathbf{m}}_1, \tilde{\mathbf{m}}_2\rangle| \leq \cos\theta_0(\tilde{\tau}) < 1.$$

We define the compact set $D := \{(\xi,\zeta)\in\mathbb{S}_2\times\mathbb{S}_2 : \langle\xi,\zeta\rangle \leq \cos\theta_0(\tilde{\tau})\}$. Since $\mathbf{G}_t(\hat{\mathbf{x}})$ is positive definite, the Cholesky decomposition of $\mathbf{G}_t(\hat{\mathbf{x}})$ exists, i.e., there is an upper triangular matrix $\mathbf{R} = \mathbf{R}_t(\hat{\mathbf{x}})$ with $\mathbf{G}_t(\hat{\mathbf{x}}) = \mathbf{R}^\intercal\mathbf{R}$. This results in the estimate

$$|\cos\theta(\hat{\mathbf{x}})| \leq \max_{(\xi,\zeta)\in D} \frac{|\langle \mathbf{R}\xi, \mathbf{R}\zeta\rangle|}{\langle \mathbf{R}\xi, \mathbf{R}\xi\rangle^{1/2}\langle \mathbf{R}\zeta, \mathbf{R}\zeta\rangle^{1/2}} \leq 1. \tag{5.6}$$

The equality on the right-hand side only holds for linearly dependent vectors $\mathbf{R}\xi = c\mathbf{R}\zeta$, i.e., $\xi = c\zeta$ with $c\in\mathbb{R}$. These, however, do not lie in D and it follows that the quotient in (5.6) will always be strictly smaller than one. Since D is compact, it follows that

$$|\cos\theta(\hat{\mathbf{x}})| \leq \left|\cos\theta^\star(\hat{\mathbf{x}})\right| < 1$$

with $0 < \theta^\star(\hat{\mathbf{x}}) < \pi$. This estimate is valid for every $\hat{\mathbf{x}}\in\overline{\hat{S}}$ and from the compactness of $\overline{\hat{S}}$ we again deduce the existence of a $\theta^\star$ $(0 < \theta^\star < \pi)$ depending only on

$\theta_0(\tilde{\tau})$, χ_t and $\lambda_{\min}$, $\lambda_{\max}$, such that

$$\sup_{\hat{\mathbf{x}}\in\widehat{S}} |\cos\theta(\hat{\mathbf{x}})| \le |\cos\theta^\star| < 1.$$

5.1.2 Cauchy-Singular Integrals

We have shown in Sect. 3.3 that all kernel functions $G(\mathbf{x}-\mathbf{y})$, $\gamma_{1,\mathbf{x}}G(\mathbf{x}-\mathbf{y})$ and $\tilde{\gamma}_{1,\mathbf{y}}G(\mathbf{x}-\mathbf{y})$ with G from (3.3) are improperly integrable. For other kernel functions (Example: Elasticity) this is not the case. As all quadrature methods in this chapter can also be applied to *Cauchy-singular* kernel functions without being modified, we will extend the class of kernel functions. The spatial dimension is denoted by $d = 2, 3$.

Definition 5.1.6. The kernel function k is Cauchy-singular if the Cauchy principal value

$$p.v. \int_\Gamma k(\mathbf{x},\mathbf{y},\mathbf{y}-\mathbf{x}) f(\mathbf{y})\, ds_\mathbf{y} := \lim_{\varepsilon\to 0}\int_{\Gamma\backslash B_\varepsilon(\mathbf{x})} k(\mathbf{x},\mathbf{y},\mathbf{y}-\mathbf{x}) f(\mathbf{y})\, ds_\mathbf{y} \quad \forall\mathbf{x}\in\Gamma$$

exists for all functions $f\in L^\infty(\Gamma)$ that are Hölder-continuous with exponent $\lambda > 0$ in a local neighborhood of $\mathbf{x}$.

Remark 5.1.7 is a direct consequence of the definition of improper integrals.

Remark 5.1.7. *For weakly singular kernels k the Cauchy principal value coincides with the improper integral.*

Example 5.1.8. *Let $d = 2$ and $\Gamma = (-1, 2)$. The kernel function $k : \Gamma\times\Gamma \to \mathbb{R}$, $k(x, y) = 1/(x-y)$ is Cauchy-singular for $x\in\Gamma$, but not for $x\in\{-1,2\}$. We have*

$$p.v. \int_\Gamma \frac{f(y)}{x-y}dy = p.v. \int_\Gamma \frac{f(y)-f(x)}{x-y}dy + f(x)\, p.v. \int_\Gamma \frac{1}{x-y}dy.$$

Let f be Hölder-continuous with exponent $\lambda > 0$. Then $|(f(y)-f(x))/(x-y)| \le C|x-y|^{\lambda-1}$ represents an improper upper bound of the first integrand. The second integration yields for $x\in\Gamma$ and a sufficiently small ε

$$\begin{aligned} p.v. \int_\Gamma \frac{1}{x-y}dy &= \lim_{\varepsilon\to 0}\left(\int_{-1}^{x-\varepsilon}\frac{1}{x-y}dy + \int_{x+\varepsilon}^{2}\frac{1}{x-y}dy\right) \\ &= \lim_{\varepsilon\to 0}(\log\varepsilon - \log(1+x) + \log(2-x) - \log\varepsilon) = \log\frac{2-x}{1+x}. \end{aligned} \tag{5.7}$$

This proves the representation

$$p.v. \int_\Gamma \frac{f(y)}{x-y}dy = \int_\Gamma \frac{f(y)-f(x)}{x-y}dy + f(x)\log\frac{2-x}{1+x}. \tag{5.8}$$

If x is a boundary vertex the integral diverges, since in (5.7) there is only one ($\log \varepsilon$) term. Note that the function p.v. $\int_\Gamma f(y) / (\cdot - y)\, dy : \Gamma \to \mathbb{R}$ in (5.8) has logarithmic endpoint singularities for $x = -1, 2$.

Exercise 5.1.9. *Let $\Gamma = (-1,1) \times (-1,1)$. Show that the function $k : \overline{\Gamma} \times \overline{\Gamma} \to \mathbb{R}$, $k(\mathbf{x}, \mathbf{y}) := (x_1 - y_1) / \|\mathbf{x} - \mathbf{y}\|^3$, is Cauchy-singular on $\Gamma \times \Gamma$ but not on $\overline{\Gamma} \times \overline{\Gamma}$.*

Efficient numerical quadrature methods are defined on a reference domain and are then transferred to other domains via pullbacks. For this it is necessary to decompose the twofold integration over Γ into a sum of integrals over pairs of panels and then to transform the integration over a pair of panels into a twofold integral over a reference element. Both steps can be complicated for kernel functions with strong singularities and are derived in the following. The conditions imposed on the kernel function and the associated integral operator will be specified in the following. The general boundary element space on Γ for a surface mesh $\mathcal{G}$ is again denoted by S.

Assumption 5.1.10. *The kernel function k is Cauchy-singular. The associated integral operator*

$$(Ku)(\mathbf{x}) := p.v. \int_\Gamma k(\mathbf{x}, \mathbf{y}, \mathbf{y} - \mathbf{x})\, u(\mathbf{y})\, ds_{\mathbf{y}}$$

is a continuous operator $K : H^\mu(\Gamma) \to H^{-\mu}(\Gamma)$ for some $\mu \in \{-1/2, 0, 1/2\}$ and a continuous mapping $K : S \to L^2(\Gamma)$.

The boundary element space S is embedded in $L^2(\Gamma)$ and $L^\infty(\Gamma)$.

Remark 5.1.11. *The condition $K : S \to L^2(\Gamma)$ allows the decomposition of the outer integral*

$$(Ku, v)_{L^2(\Gamma)} = \sum_{\tau \in \mathcal{G}} (Ku, v)_{L^2(\tau)} \qquad \forall u \in S, \quad \forall v \in L^2(\Gamma), \tag{5.9}$$

where here $(\cdot,\cdot)_{L^2(\Gamma)}$ and $(\cdot,\cdot)_{L^2(\tau)}$ denote the usual L^2 inner-product and not the continuous extension to dual pairings.

Remark 5.1.12. *The integral operators (V, K, K') for the single and double layer potentials from Chap. 3 satisfy Assumption 5.1.10 with $\mu = -1/2, 0, 0$. For $\mu = -1/2$, this follows from Theorem 3.1.16 and the continuous embeddings $S \hookrightarrow H^{-1/2}(\Gamma)$ and $H^{1/2}(\Gamma) \hookrightarrow L^2(\Gamma)$. For $\mu = 0$, this follows from Corollary 3.3.9 and the continuous embedding $S \hookrightarrow L^\infty(\Gamma)$.*

For the hypersingular operator we use the representation from Theorem 3.3.22 as well as the property $\mathbf{curl}_{\Gamma,\mathbf{A},0}\, \varphi, \mathbf{curl}_{\Gamma,\mathbf{A},2\mathbf{b}}\, \psi \in L^\infty(\Gamma)$ for all $\varphi, \psi \in S^{p,0}$ (see Definition 4.1.36). Combining this with previous results we obtain

$$\begin{aligned} b_W(\psi, \varphi) &= \left(V\, \mathbf{curl}_{\Gamma,\mathbf{A},2\mathbf{b}}\, \psi, \mathbf{curl}_{\Gamma,\mathbf{A},0}\, \varphi\right)_{L^2(\Gamma)} + c\left(\widetilde{V}\psi, \varphi\right)_{L^2(\Gamma)} \\ &= \sum_{\tau \in \mathcal{G}} \left\{\left(V\, \mathbf{curl}_{\Gamma,\mathbf{A},2\mathbf{b}}\, \psi, \mathbf{curl}_{\Gamma,\mathbf{A},0}\, \varphi\right)_{L^2(\tau)} + \left(\widetilde{V}\psi, \varphi\right)_{L^2(\tau)}\right\} \end{aligned}$$

with the integral operator V *for the single layer potential and*

$$\left(\widetilde{V}\psi\right)(\mathbf{x}) := c\int_{\Gamma} G(\mathbf{x}-\mathbf{y})\,\psi(\mathbf{y})\left\langle \mathbf{A}^{1/2}\mathbf{n}(\mathbf{x}),\mathbf{A}^{1/2}\mathbf{n}(\mathbf{y})\right\rangle ds_{\mathbf{x}}ds_{\mathbf{y}} \qquad \forall \mathbf{x}\in\Gamma \text{ a.e.}$$

The fact that the integral operator can be localized is, for Cauchy-singular kernel functions, a consequence of the locality of the Cauchy principal value. The restriction of the integration to one panel $\tau \in \mathcal{G}$ leads to the definition

$$(K_\tau u)(\mathbf{x}) := p.v.\int_\tau k(\mathbf{x},\mathbf{y},\mathbf{x}-\mathbf{y})\,u(\mathbf{y})\,ds_{\mathbf{y}}. \tag{5.10}$$

Lemma 5.1.13. *Let Assumption 5.1.10 hold. Let* $u \in L^\infty(\Gamma)$ *and* $u|_\tau \in C^1(\tau)$ *for all* $\tau \in \mathcal{G}$. *Then we have:*

(a) For every $\mathbf{x} \in t \in \mathcal{G}$, *(5.10) is finite.*
(b) For $\mathbf{x} \notin \overline{\tau}$ *the Cauchy principal value coincides with the Riemann integral.*
(c)

$$(Ku)(\mathbf{x}) = \sum_{\tau\in\mathcal{G}} (K_\tau u)(\mathbf{x}) \qquad \forall \mathbf{x}\in t\in\mathcal{G}. \tag{5.11}$$

Let $\mathbf{x} \in \tau \in \mathcal{G}$ and $u \in L^\infty(\Gamma)$ with $u|_\tau \in C^1(\tau)$. We can find an $\varepsilon_0 > 0$ with $\Gamma \cap B_\varepsilon(\mathbf{x}) = \tau \cap B_\varepsilon(\mathbf{x})$ for all $0 < \varepsilon \leq \varepsilon_0$. Since u is differentiable in $\tau \cap B_\varepsilon(\mathbf{x})$, the Cauchy principal value $p.v.\int_{\tau\cap B_\varepsilon(\mathbf{x})} k(\mathbf{x},\mathbf{y},\mathbf{x}-\mathbf{y})\,u(\mathbf{y})\,ds_{\mathbf{y}}$ exists. The local definition of the principal value property and the boundedness of the integrand on $\Gamma\backslash B_{\varepsilon_0}(\mathbf{x})$ together yield that

$$p.v.\int_\tau k(\mathbf{x},\mathbf{y},\mathbf{x}-\mathbf{y})\,u(\mathbf{y})\,ds_{\mathbf{y}}$$

exists for $\mathbf{x} \notin \tau$ as a Riemann integral and for $\mathbf{x} \in \tau$ as a Cauchy principal value. Summing over all $\tau \in \mathcal{G}$ gives us the representation

$$(Ku)(\mathbf{x}) = \sum_{\tau\in\mathcal{G}} p.v.\int_\tau k(\mathbf{x},\mathbf{y},\mathbf{x}-\mathbf{y})\,u(\mathbf{y})\,ds_{\mathbf{y}}. \tag{5.12}$$

□

In Sect. 5.3 we will discuss efficient numerical integration methods for the approximation of integrals of the type $\int_{\tau\times t} k(\mathbf{x},\mathbf{y},\mathbf{y}-\mathbf{x})\,u(\mathbf{y})\,v(\mathbf{x})\,ds_{\mathbf{y}}ds_{\mathbf{x}}$. Lemma 5.1.13 and Assumption 5.1.10 show that this decomposition is unproblematic for Cauchy-singular integrals.

Corollary 5.1.14. *Let the assumptions from Lemma 5.1.13 hold. Then we have for all* $v \in L^2(\Gamma)$

$$(Ku, v)_{L^2(\Gamma)} = \sum_{\tau,t\in\mathcal{G}} (K_t u, v)_{L^2(\tau)}$$

with

$$(K_t u, v)_{L^2(\tau)} = \begin{cases} \int_{\tau\times t} \overline{v}(\mathbf{x})\, k(\mathbf{x},\mathbf{y},\mathbf{y}-\mathbf{x})\, u(\mathbf{y})\, ds_{\mathbf{y}} ds_{\mathbf{x}} & \tau \neq t, \\ \int_{\tau} \overline{v}(\mathbf{x})\, p.v. \int_{\tau} k(\mathbf{x},\mathbf{y},\mathbf{y}-\mathbf{x})\, u(\mathbf{y})\, ds_{\mathbf{y}} ds_{\mathbf{x}} & \tau = t. \end{cases}$$

5.1.3 Explicit Conditions on Cauchy-Singular Kernel Functions

In this subsection we will formulate explicit conditions that describe the analytic properties of the kernel functions (Assumption 5.1.15) and guarantee the existence of the Cauchy principal value (Assumption 5.1.19).

Assumption 5.1.15. *The kernel function has the representation*

$$k(\mathbf{x},\mathbf{y},\mathbf{z}) := \|\mathbf{z}\|^{-s} \sum_{i=0}^{b} \kappa_i(\mathbf{x},\mathbf{y})\, A_i\left(\|\mathbf{z}\|, \frac{\mathbf{z}}{\|\mathbf{z}\|}\right), \qquad \forall \mathbf{x},\mathbf{y} \in \Gamma, \mathbf{z} = \mathbf{y}-\mathbf{x}, \mathbf{x} \neq \mathbf{y} \tag{5.13}$$

with $s \in \mathbb{Z}$ and $b \in \mathbb{N}$. The functions $\kappa_i(\mathbf{x},\mathbf{y})$ and $A_i(r,\xi)$ satisfy the conditions:

1. *For $0 \leq i \leq b$, A_i is analytic on $(0,\rho_0) \times U_0$ where U_0 is a neighborhood of the unit sphere $\mathbb{S}_2$ and $\rho_0 > 0$.*
2. *$\|\mathbf{z}\|^{-s} A_b\left(\|\mathbf{z}\|, \frac{\mathbf{z}}{\|\mathbf{z}\|}\right)$ is improperly integrable in every two-dimensional, bounded neighborhood of the origin.*
3. *The coefficient functions κ_i are contained in $L^\infty(\Gamma\times\Gamma)$ and are uniformly, continuously differentiable on smooth parts $\Gamma_j \times \Gamma_k$ of $\Gamma\times\Gamma$. The order of differentiation depends on the smoothness of the surface pieces $\Gamma_j \times \Gamma_k$. If these are analytic then the coefficients are also analytic.*

We note that practically all kernel functions that are derived by integral equation methods from either scalar elliptic boundary value problems (in $\mathbb{R}^3$) or systems thereof have the form (5.13). For a detailed analysis of fundamental solutions of partial differential equations we refer to [97, 139]. The assumption $b \in \mathbb{N}$ in (5.13) guarantees that *at most* finitely many terms exist that are not improperly integrable. It should be noted that the representation (5.13) is by no means unique.

Example 5.1.16. *The fundamental solution G from (3.3) is of the form (5.13) with $s = 1$, $b = 0$, $\kappa_0 \equiv 1$ and*

$$A_0 : \mathbb{R}_{\geq 0} \times \mathbb{S}_2 \to \mathbb{C}, \qquad A_0(r,\xi) = \frac{1}{4\pi\sqrt{\det \mathbf{A}}} \frac{e^{r(\langle \mathbf{b},\xi\rangle_{\mathbf{A}} - \lambda\|\xi\|_{\mathbf{A}})}}{\|\xi\|_{\mathbf{A}}}.$$

The kernel function of the classical double layer potential also belongs to the introduced class of kernels.

Example 5.1.17. *The kernel of the double layer potential is given by*

$$\widetilde{\gamma_{1,\mathbf{y}}}G\left(\mathbf{x}-\mathbf{y}\right)=\gamma_{1,\mathbf{y}}G\left(\mathbf{x}-\mathbf{y}\right)+2\left\langle \mathbf{n},\mathbf{b}\right\rangle G\left(\mathbf{x}-\mathbf{y}\right).$$

We use (3.30) and obtain

$$\nabla_{\mathbf{z}}G\left(\mathbf{z}\right)=-\frac{1}{4\pi\sqrt{\det\mathbf{A}}}\frac{\mathbf{A}^{-1}\mathbf{z}}{\left\|\mathbf{z}\right\|_{\mathbf{A}}^{3}}+\mathbf{R}\left(\mathbf{z}\right)\quad\text{with}\quad\left\|\mathbf{R}\left(\mathbf{z}\right)\right\|\leq C\left\|\mathbf{z}\right\|^{-1}.$$

It follows that

$$k\left(\mathbf{x},\mathbf{y},\mathbf{z}\right)=\left\|\mathbf{z}\right\|^{-s}\sum_{i=0}^{b}\kappa_{i}\left(\mathbf{x},\mathbf{y}\right)A_{i}\left(\left\|\mathbf{z}\right\|,\frac{\mathbf{z}}{\left\|\mathbf{z}\right\|}\right)$$

with $s=2$, $b=6$,

$$\forall i=0,1,2:\kappa_{i}\left(\mathbf{x},\mathbf{y}\right)=\kappa_{i+3}\left(\mathbf{x},\mathbf{y}\right)=\mathbf{n}_{i+1}\left(\mathbf{y}\right),$$
$$\forall i=0,1,2:A_{i}\left(r,\xi\right)=-\xi_{i}/\left(4\pi\sqrt{\det\mathbf{A}}\left\|\xi\right\|_{\mathbf{A}}^{3}\right),\quad A_{i+3}\left(r,\xi\right)=r^{2}\left(\mathbf{A}\mathbf{R}\left(r\xi\right)\right)_{i+1}$$

and $\kappa_{6}\left(\mathbf{x},\mathbf{y}\right)=2\left\langle \mathbf{n}\left(\mathbf{y}\right),\mathbf{b}\right\rangle$, $A_{6}\left(r,\xi\right)=r^{2}G\left(r\xi\right)$.

The representation can be simplified for the Laplace operator and the parameters can be chosen as $s=2$, $b=3$, $\kappa_{3}=A_{3}\equiv0$,

$$\forall i=0,1,2:\kappa_{i}\left(\mathbf{x},\mathbf{y}\right)=\mathbf{n}_{i+1}\left(\mathbf{y}\right),\quad A_{i}\left(r,\xi\right):=-\xi_{i}/\left(4\pi\right).$$

Exercise 5.1.18. *Let G be the fundamental solution from (3.3). Show that the kernel functions*

$$\gamma_{1,\mathbf{x}}G\left(\mathbf{x}-\mathbf{y}\right),\qquad\gamma_{1,\mathbf{x}}\tilde{\gamma}_{1,\mathbf{y}}G\left(\mathbf{x}-\mathbf{y}\right)$$

belong to the introduced class of kernels.

It is shown in [139] that, for kernel functions of elliptic differential operators of second order in $\mathbb{R}^3$, the order of singularity s in (5.13) is an integer and satisfies $s\leq3$. The case $s=3$ occurs for the kernel function $\gamma_{1,\mathbf{x}}\tilde{\gamma}_{1,\mathbf{y}}G\left(\mathbf{x}-\mathbf{y}\right)$ with G from (3.3). We have, however, shown in (3.3.22) that the (hypersingular) kernel function can be regularized by means of integration by parts. Similar regularization techniques also exist for the integral operators of elasticity (see [127, 169]).

Kernel functions that satisfy Assumption 5.1.15 do not necessarily have a Cauchy principal value. We have to require that the order of singularity s satisfies $s\leq2$ and if $s=2$ it also has to satisfy an antisymmetry condition (see also the parity condition, e.g., in [137, (7.1.74) and Definition 7.1.3]).

Assumption 5.1.19. *The kernel function satisfies Assumption 5.1.15 with* $s\leq2$. *In the case* $s=2$ *let there exist a decomposition*

$$A_i\left(\|\mathbf{z}\|,\frac{\mathbf{z}}{\|\mathbf{z}\|}\right)=A_{i,0}\left(\frac{\mathbf{z}}{\|\mathbf{z}\|}\right)+\|\mathbf{z}\|\,A_{i,1}\left(\|\mathbf{z}\|,\frac{\mathbf{z}}{\|\mathbf{z}\|}\right) \tag{5.14}$$

for $0\le i\le b$ *with functions* $A_{i,0}(\xi):\mathbb{S}_2\to\mathbb{K}$ *and* $A_{i,1}(r,\xi):\mathbb{R}_{\ge 0}\times\mathbb{S}_2\to\mathbb{K}$ *that have the same analytic properties as* A_i *and satisfy*

$$A_{i,0}(\xi)=-A_{i,0}(-\xi)$$

for all $\xi\in\mathbb{S}_2$.

Kernel functions that satisfy Assumption 5.1.19 are called antisymmetric.

Remark 5.1.20. *Improperly integrable kernel functions satisfy Assumption 5.1.19 with* $A_{i,0}\equiv 0$.

5.1.4 Kernel Functions in Local Coordinates

In order to investigate the kernel function in local coordinates, it suffices to analyze an arbitrary term in the sum (5.13). Therefore we assume that the kernel function is given by

$$k(\mathbf{x},\mathbf{y},\mathbf{z}):=\|\mathbf{z}\|^{-s}\,\kappa(\mathbf{x},\mathbf{y})\,A\left(\|\mathbf{z}\|,\frac{\mathbf{z}}{\|\mathbf{z}\|}\right)$$

and satisfies Assumption 5.1.19. We set

$$\hat{k}(\hat{\mathbf{x}},\hat{\mathbf{y}}):=k\left(\chi_\tau(\hat{\mathbf{x}}),\chi_t(\hat{\mathbf{y}}),\chi_t(\hat{\mathbf{y}})-\chi_\tau(\hat{\mathbf{x}})\right),\quad \forall(\hat{\mathbf{x}},\hat{\mathbf{y}})\in\hat{\tau}\times\hat{t}:\hat{\mathbf{x}}\ne\hat{\mathbf{y}} \tag{5.15}$$

and, for the analysis of the behavior of $\hat{k}$ in local coordinates, we distinguish between three cases:

(I) $\overline{\tau},\overline{t}$ touch each other in at most one point.
(II) The intersection $\overline{\tau}\cap\overline{t}$ is a common edge.
(III) The panels are identical $\overline{\tau}=\overline{t}$.

With regular surface meshes one of the above-mentioned cases always applies for two panels $\tau,t\in\mathcal{G}$. Should there be any hanging nodes the situation can be reduced to one of the three above-mentioned cases by appropriate decomposition.

Case I: For $\overline{\tau}\cap\overline{t}=\emptyset$ the kernel function is regular and the analytic properties on the reference domains follow from the analytic properties of the transformations χ_τ and χ_t. We therefore assume in the following that $\overline{\tau}\cap\overline{t}=\mathbf{p}$. In local coordinates we obtain for the difference variable $\mathbf{z}=\mathbf{y}-\mathbf{x}$ the representation

$$\mathbf{z}=\chi_t(\hat{\mathbf{y}})-\chi_\tau(\hat{\mathbf{x}}). \tag{5.16}$$

Without loss of generality we assume that $\chi_\tau(0) = \chi_t(0) = \mathbf{p}$. Clearly, we have $\mathbf{z} = 0$ if and only if $\hat{\mathbf{x}} = \hat{\mathbf{y}} = (0,0)^\mathsf{T}$. Thus, in the case under consideration $\hat{k}$ is analytic outside of $\|\hat{\mathbf{x}}\| + \|\hat{\mathbf{y}}\| \le \varepsilon$ with an arbitrary $\varepsilon > 0$ and it suffices to analyze the singular behavior of $\hat{k}$ in an arbitrarily small neighborhood of 0. If we replace χ_τ and χ_t in (5.16) by their respective Taylor expansions about 0 we obtain

$$\mathbf{z} = \sum_{m=1}^{\infty} \frac{\langle \hat{\mathbf{y}}, \nabla\rangle^m \chi_t(0) - \langle \hat{\mathbf{x}}, \nabla\rangle^m \chi_\tau(0)}{m!}. \tag{5.17}$$

We interpret $(\hat{\mathbf{x}}, \hat{\mathbf{y}})$ as a vector in $\mathbb{R}^4$ and introduce four-dimensional polar coordinates

$$(\hat{\mathbf{x}}, \hat{\mathbf{y}}) = r\xi \tag{5.18}$$

with $\xi = (\hat{\mathbf{x}}, \hat{\mathbf{y}}) / \|(\hat{\mathbf{x}}, \hat{\mathbf{y}})\| \in \mathbb{S}_3$. If we insert this into (5.17) we obtain

$$\mathbf{z} = r \sum_{m=0}^{\infty} r^m l_m(\xi) =: r a_1(r,\xi) \tag{5.19}$$

with

$$l_m(\xi) := \frac{\langle \xi_{34}, \nabla\rangle^{m+1} \chi_t(0) - \langle \xi_{12}, \nabla\rangle^{m+1} \chi_\tau(0)}{(m+1)!} \quad \text{and } \xi_{ij} := (\xi_i, \xi_j), \; 1 \le i, j \le 4.$$

The function a_1 is analytic in $(0, \rho_1) \times U_1$ where U_1 is a neighborhood of the unit sphere $\mathbb{S}_3$ and $\rho_1 > 0$. By Assumption 4.3.29 and the bi-Lipschitz continuity of χ_t, χ_τ there exist $c, \tilde{c}, \hat{c} > 0$ such that

$$\begin{aligned} r\|a_1(r,\xi)\| &= \|r a_1(r,\xi)\| = \|\mathbf{z}\| = \|\mathbf{y} - \mathbf{x}\| \ge c\left(\|\mathbf{y} - \mathbf{p}\| + \|\mathbf{p} - \mathbf{x}\|\right) \\ &\ge c\left(\|\chi_t(\hat{\mathbf{y}}) - \chi_t(0)\| + \|\chi_\tau(\hat{\mathbf{x}}) - \chi_\tau(0)\|\right) \ge \tilde{c}\left(\|\hat{\mathbf{y}}\| + \|\hat{\mathbf{x}}\|\right) \ge \hat{c} r \end{aligned}$$

and therefore a_1 does not have any zeros in $(0, \rho_2) \times U_2$ where U_2 is a neighborhood of $\mathbb{S}_3$ and $\rho_2 > 0$. It follows from this that $a_{2,s}(r,\xi) := \|a_1(r,\xi)\|^s$ is analytic for all $s \in \mathbb{R}$.

Therefore the quotient $\mathbf{z}/\|\mathbf{z}\| = a_1(r,\xi)\, a_{2,-1}(r,\xi)$ is also analytic in $(0, \rho_2) \times U_2$. By combining these representations we obtain

$$\|\mathbf{z}\|^{-s} A\left(\|\mathbf{z}\|, \frac{\mathbf{z}}{\|\mathbf{z}\|}\right) = r^{-s} a_{3,s}(r,\xi) \tag{5.20a}$$

where $a_{3,s}(r,\xi)$ is analytic in $(0, \rho_3) \times U_3$ with $\rho_3 > 0$ and a neighborhood U_3 of $\mathbb{S}_3$.

Case II: The intersection $E := \overline{\tau} \cap \overline{t}$ is a common edge. Without loss of generality we assume that the parametrizations χ_τ, χ_t satisfy the relation

$$\chi_\tau\binom{\xi_1}{0} = \chi_t\binom{\xi_1}{0} \qquad \forall \xi_1 \in [0,1].$$

Hence the difference

$$\mathbf{z} = \mathbf{y} - \mathbf{x} = \chi_t(\hat{\mathbf{y}}) - \chi_\tau(\hat{\mathbf{x}})$$

is zero if and only if the three-dimensional relative coordinates

$$\hat{\mathbf{z}} := (\hat{z}_1, \hat{z}_2, \hat{z}_3)^\intercal := \begin{pmatrix} \hat{y}_1 - \hat{x}_1 \\ \hat{y}_2 \\ \hat{x}_2 \end{pmatrix}$$

are equal to zero. In these coordinates the difference $\mathbf{z}$ has the form

$$\mathbf{z} = \chi_t\binom{\hat{z}_1 + \hat{x}_1}{\hat{z}_2} - \chi_\tau\binom{\hat{x}_1}{\hat{z}_3}.$$

We introduce three-dimensional polar coordinates $\hat{\mathbf{z}} = r\xi$ with $r := \|\hat{\mathbf{z}}\|$ and $\xi = \hat{\mathbf{z}}/\|\hat{\mathbf{z}}\| \in \mathbb{S}_2$ and expand $\mathbf{z}$ with respect to $\hat{\mathbf{z}}$ about zero

$$\begin{aligned} \mathbf{z} = \chi_t\binom{\hat{z}_1 + \hat{x}_1}{\hat{z}_2} - \chi_\tau\binom{\hat{x}_1}{\hat{z}_3} &= \sum_{m=1}^{\infty} r^m \frac{\left\langle \binom{\xi_1}{\xi_2}, \nabla \right\rangle^m \chi_t\binom{\hat{x}_1}{0} - (\xi_3 \partial_2)^m \chi_\tau\binom{\hat{x}_1}{0}}{m!} \\ &= r \sum_{m=0}^{\infty} r^m l_m(\hat{x}_1, \xi) =: r b(\hat{x}_1, r, \xi) \end{aligned} \tag{5.21}$$

with

$$l_m(\hat{x}_1, \xi) := \frac{\left\langle \binom{\xi_1}{\xi_2}, \nabla \right\rangle^{m+1} \chi_t\binom{\hat{x}_1}{0} - (\xi_3 \partial_2)^{m+1} \chi_\tau\binom{\hat{x}_1}{0}}{(m+1)!}.$$

By Assumption 4.3.29 there exists a point on the common edge $\mathbf{p} = \chi_t(\theta, 0) \in E$ with

$$\begin{aligned} r b(\hat{x}_1, r, \xi) &= \|\mathbf{y} - \mathbf{x}\| \geq c\left(\|\mathbf{y} - \mathbf{p}\| + \|\mathbf{p} - \mathbf{x}\|\right) \\ &= c\left(\left\|\chi_t\binom{\hat{y}_1}{\hat{y}_2} - \chi_t\binom{\theta}{0}\right\| + \left\|\chi_\tau\binom{\hat{x}_1}{\hat{x}_2} - \chi_t\binom{\theta}{0}\right\|\right) \\ &\geq c\left((\hat{y}_1 - \theta)^2 + \hat{y}_2^2 + (\hat{x}_1 - \theta)^2 + \hat{x}_2^2\right)^{1/2} \\ &\geq c\left(\frac{(\hat{y}_1 - \hat{x}_1)^2}{2} + \hat{y}_2^2 + \hat{x}_2^2\right)^{1/2} \geq \frac{c}{2}\|\hat{\mathbf{z}}\| = \frac{c}{2} r. \end{aligned}$$

Therefore $b(\hat{x}_1, r, \xi)$ does not have any zeros in $(0, \rho_4) \times U_4$ where U_4 is a neighborhood of $\mathbb{S}_2$ and $\rho_4 > 0$. From this we have that $b_{2,s}(\hat{x}_1, r, \xi) := \|b(\hat{x}_1, r, \xi)\|^s$ is analytic for all $s \in \mathbb{R}$.

As in Case I we deduce the representation

$$\|\mathbf{z}\|^{-s} A\left(\|\mathbf{z}\|, \frac{\mathbf{z}}{\|\mathbf{z}\|}\right) = r^{-s} b_{3,s}\left(x_1, r, \xi\right) \tag{5.20b}$$

with a function $b_{3,s}$ that is analytic in $I_5 \times (0, \rho_5) \times U_5$ with neighborhoods I_5 and U_5 of $[0,1]$ and $\mathbb{S}_2$ respectively and $\rho_5 > 0$.

Case III: The panels τ, t are identical. Since χ_τ is bijective, the difference

$$\mathbf{z} = \mathbf{y} - \mathbf{x} = \chi_\tau\left(\hat{\mathbf{y}}\right) - \chi_\tau\left(\hat{\mathbf{x}}\right)$$

vanishes if and only if the two-dimensional relative coordinates

$$\hat{\mathbf{z}} := \begin{pmatrix} \hat{y}_1 - \hat{x}_1 \\ \hat{y}_2 - \hat{x}_2 \end{pmatrix}$$

are equal to zero. If we expand the difference $\mathbf{z}$ about $\hat{\mathbf{z}} = 0$ we obtain

$$\mathbf{z} = \chi_\tau\left(\hat{\mathbf{z}} + \hat{\mathbf{x}}\right) - \chi_\tau\left(\hat{\mathbf{x}}\right) = \sum_{m=1}^{\infty} \left(\frac{\left\langle \hat{\mathbf{z}}, \nabla \right\rangle^m \chi_\tau}{m!}\right)\left(\hat{\mathbf{x}}\right).$$

We introduce two-dimensional polar coordinates $\hat{\mathbf{z}} = r\xi$ with $\xi = \hat{\mathbf{z}} / \|\hat{\mathbf{z}}\|$ and obtain

$$\mathbf{z} = r \sum_{m=0}^{\infty} r^m l_m\left(\hat{\mathbf{x}}, \xi\right) =: r d\left(\hat{\mathbf{x}}, r, \xi\right) \tag{5.22}$$

with

$$l_m\left(\hat{\mathbf{x}}, \xi\right) = \left(\frac{\left\langle \xi, \nabla \right\rangle^{m+1} \chi_\tau}{(m+1)!}\right)\left(\hat{\mathbf{x}}\right).$$

We deduce from the bi-Lipschitz continuity of χ_τ that

$$r\left\|d\left(\hat{\mathbf{x}}, r, \xi\right)\right\| = \|\mathbf{z}\| = \left\|\chi_\tau\left(\hat{\mathbf{y}}\right) - \chi_\tau\left(\hat{\mathbf{x}}\right)\right\| \geq c\left\|\hat{\mathbf{y}} - \hat{\mathbf{x}}\right\| = c\left\|\hat{\mathbf{z}}\right\| = cr,$$

and therefore $d\left(\hat{\mathbf{x}}, r, \xi\right)$ does not have any zeros in $I_6 \times (0, \rho_6) \times U_6$ with neighborhoods I_6 and U_6 of $\overline{\hat{\tau}}$ and $\mathbb{S}_1$ respectively and $\rho_6 > 0$. From this we have that $d_{2,s}\left(x_1, r, \xi\right) := \left\|b\left(x_1, r, \xi\right)\right\|^s$ is analytic for all $s \in \mathbb{R}$.

As in Case I we deduce the representation

$$k\left(\mathbf{x}, \mathbf{y}, \mathbf{z}\right) = \|\mathbf{z}\|^{-s} A\left(\|\mathbf{z}\|, \frac{\mathbf{z}}{\|\mathbf{z}\|}\right) = r^{-s} d_{3,s}\left(\hat{\mathbf{x}}, r, \xi\right) \tag{5.20c}$$

with a function $d_{3,s}$ that is analytic in $I_7 \times (0, \rho_7) \times U_7$ with neighborhoods I_7 and U_7 of $\overline{\hat{\tau}}$ and $\mathbb{S}_1$ respectively and $\rho_7 > 0$.

For Cauchy-singular kernel functions that satisfy Assumption 5.1.19 we can further improve this result. We use the decomposition from Assumption 5.1.19

$$A\left(\|\mathbf{z}\|,\frac{\mathbf{z}}{\|\mathbf{z}\|}\right)=A_0\left(\frac{\mathbf{z}}{\|\mathbf{z}\|}\right)+\|\mathbf{z}\|\,A_1\left(\|\mathbf{z}\|,\frac{\mathbf{z}}{\|\mathbf{z}\|}\right)$$

with an antisymmetric A_0. By choosing $A=A_0$, $s=2$ and $\kappa\equiv 1$ or $A(r,\xi)=rA_1(r,\xi)$ in (5.20c) we obtain

$$\|\mathbf{z}\|^{-2}\left(A_0\left(\frac{\mathbf{z}}{\|\mathbf{z}\|}\right)+\|\mathbf{z}\|\,A_1\left(\|\mathbf{z}\|,\frac{\mathbf{z}}{\|\mathbf{z}\|}\right)\right)=r^{-2}d_{3,2}(\hat{\mathbf{x}},r,\xi)+r^{-1}d_{3,1}(\hat{\mathbf{x}},r,\xi)\,.$$

Exercise 5.1.21. *From the antisymmetry of A_0 deduce the existence of a function $\tilde{d}_{3,2}$ that is analytic in $I_8\times(0,\rho_8)\times U_8$ with neighborhoods I_8 of $\overline{\tau}$, U_8 of $\mathbb{S}_1$ and $\rho_8>0$, and satisfies*

$$d_{3,2}(\hat{\mathbf{x}},r,\xi)=d_{3,2}(\hat{\mathbf{x}},\xi)+r\tilde{d}_{3,2}(\hat{\mathbf{x}},r,\xi)\,,\qquad d_{3,2}(\hat{\mathbf{x}},\xi)=-d_{3,2}(\hat{\mathbf{x}},-\xi)$$

with $d_{3,2}(\hat{\mathbf{x}},\xi):=d_{3,2}(\hat{\mathbf{x}},0,\xi)$.

Theorem 5.1.22. *Let the kernel function satisfy Assumption 5.1.19:*

(a) Then there exist functions $\tilde{d}_{3,2}$, $\tilde{d}_{3,1}$ that are analytic in $I\times(0,\rho)\times U$, and there exist neighborhoods I of $\overline{\tau}$, U of $\mathbb{S}_1$ and $\rho>0$ such that

$$k(\mathbf{x},\mathbf{y},\mathbf{z})=r^{-2}\tilde{d}_{3,2}(\hat{\mathbf{x}},\xi)+r^{-1}\tilde{d}_{3,1}(\hat{\mathbf{x}},r,\xi)\quad\textit{and}\quad\tilde{d}_{3,2}(\hat{\mathbf{x}},\xi)=-\tilde{d}_{3,2}(\hat{\mathbf{x}},-\xi)$$

with $\mathbf{x}=\chi_\tau(\hat{\mathbf{x}})$, $\mathbf{y}=\chi_\tau(\hat{\mathbf{x}}+r\xi)$, $\mathbf{z}=\mathbf{y}-\mathbf{x}$.

(b) Let $\hat{b}_i$, $\hat{b}_j$ be the basis functions on the reference elements. The integrand $\hat{b}_i(\hat{\mathbf{x}})\,\hat{b}_j(\hat{\mathbf{y}})\,\hat{k}(\hat{\mathbf{x}},\hat{\mathbf{y}})\,g_\tau(\hat{\mathbf{x}})\,g_\tau(\hat{\mathbf{y}})$ [see (5.15)] has the representation

$$r^{-2}\hat{d}_{3,2}(\hat{\mathbf{x}},\xi)+r^{-1}\hat{d}_{3,1}(\hat{\mathbf{x}},r,\xi)$$

with functions $\hat{d}_{3,2}$, $\hat{d}_{3,1}$ that have the same analytic and antisymmetric properties.

(c) There exists a function f that has the same analytic properties as $\tilde{d}_{3,2}$, $\tilde{d}_{3,1}$ from (a) and satisfies

$$\hat{k}(\hat{\mathbf{x}},\hat{\mathbf{x}}-\hat{\mathbf{z}})+\hat{k}(\hat{\mathbf{x}}-\hat{\mathbf{z}},\hat{\mathbf{x}})=r^{-1}f(\hat{\mathbf{x}},r,\xi)\,.\tag{5.23}$$

Proof. The fact that κ is analytic in both variables combined with the fact that χ_τ is analytic passes on this property to

$$\kappa\left(\chi_\tau\left(\hat{\mathbf{x}}\right), \chi_\tau\left(\hat{\mathbf{y}}\right)\right) =: \tilde{\kappa}\left(\hat{\mathbf{x}}, \hat{\mathbf{y}}\right) = \tilde{\kappa}\left(\hat{\mathbf{x}}, \hat{\mathbf{x}} + \hat{\mathbf{z}}\right) = \tilde{\kappa}\left(\hat{\mathbf{x}}, \hat{\mathbf{x}}\right) + \left\langle \hat{\mathbf{z}}, \vec{\kappa}_1\left(\hat{\mathbf{x}}, \hat{\mathbf{z}}\right)\right\rangle$$
$$= \tilde{\kappa}\left(\hat{\mathbf{x}}, \hat{\mathbf{x}}\right) + r\left\langle \xi, \vec{\kappa}_1\left(\hat{\mathbf{x}}, r\xi\right)\right\rangle =: \tilde{\kappa}\left(\hat{\mathbf{x}}, \hat{\mathbf{x}}\right) + r\tilde{\kappa}_1\left(\hat{\mathbf{x}}, r, \xi\right)$$

with analytic functions $\tilde{\kappa}_1$ and a (vector-valued) $\vec{\kappa}_1$. Similar arguments can be applied to the surface element g_τ as well as the basis functions $\hat{b}_i$ and $\hat{b}_j$ and, thus, give us the statements of (a) and (b).

For c: If we substitute $\hat{\mathbf{x}} - \hat{\mathbf{z}} \leftarrow \hat{\mathbf{y}}$ in the second summand of (5.23), apply the already derived representations and use the fact that $\tilde{d}_{3,1}$ is smooth with respect to the first variable, we obtain with $\hat{\mathbf{z}} = r\xi$ for $\hat{k}\left(\hat{\mathbf{x}}, \hat{\mathbf{x}} - \hat{\mathbf{z}}\right) + \hat{k}\left(\hat{\mathbf{y}}, \hat{\mathbf{y}} + \hat{\mathbf{z}}\right)$ the representation

$$r^{-2}\tilde{d}_{3,2}\left(\hat{\mathbf{x}}, -\xi\right) + r^{-1}\tilde{d}_{3,1}\left(\hat{\mathbf{x}}, r, -\xi\right) + r^{-2}\tilde{d}_{3,2}\left(\hat{\mathbf{y}}, \xi\right) + r^{-1}\tilde{d}_{3,1}\left(\hat{\mathbf{y}}, r, \xi\right)$$
$$= r^{-2}\left(\tilde{d}_{3,2}\left(\hat{\mathbf{x}}, -\xi\right) + \tilde{d}_{3,2}\left(\hat{\mathbf{x}} - \hat{\mathbf{z}}, \xi\right)\right) + r^{-1}\left(\tilde{d}_{3,1}\left(\hat{\mathbf{x}}, r, -\xi\right) + \tilde{d}_{3,1}\left(\hat{\mathbf{x}} - \hat{\mathbf{z}}, r, \xi\right)\right)$$
$$=: r^{-2}\left(\tilde{d}_{3,2}\left(\hat{\mathbf{x}}, -\xi\right) + \tilde{d}_{3,2}\left(\hat{\mathbf{x}}, \xi\right) + r\left\langle \xi, \vec{f}\left(\hat{\mathbf{x}}, r, \xi\right)\right\rangle\right) + r^{-1} f_0\left(\hat{\mathbf{x}}, r, \xi\right)$$
$$=: r^{-1} f\left(\hat{\mathbf{x}}, r, \xi\right).$$

□

5.2 Relative Coordinates

The numerical integration is defined on a pair of reference panels and is transferred by means of a transformation to an integration over pairs of panels $\tau \times t$. In general we will assume that the conditions from Sect. 5.1.1 hold. We distinguish between four cases:

1. Identical panels
2. Panels with exactly one common edge
3. Panels with exactly one common vertex
4. Panels with positive distance

Relative coordinates for one-dimensional curves and interval partitionings are due to [134]. In [211] general kernel functions in local coordinates were analyzed and, based on these results, it was shown in [212] that when using simplex coordinates the determinant of the transformation removes the singularity in the integrand, which becomes analytic in a neighborhood of the original singularity.

Relative coordinates for triangular elements were introduced in [197] and [125]. Combining this with the pullback to the reference element was developed in [235]. Erichsen and Sauter [88] contains a compact summary of the required quadrature orders, depending on the underlying operator, the order of approximation and the norm in which the error is measured.

5.2.1 Identical Panels

In order to generate the matrix coefficients the sesquilinear form is evaluated for pairs b_i, b_j of basis functions. In the case of identical panels we consider the integral

$$\int_\tau b_i(\mathbf{x})\, p.v. \int_\tau k(\mathbf{x}, \mathbf{y}, \mathbf{y}-\mathbf{x})\, b_j(\mathbf{y})\, ds_{\mathbf{y}} ds_{\mathbf{x}}. \tag{5.24}$$

For $\mathbf{x} \in \tau$ and a sufficiently small $\varepsilon_0 > 0$ we have $B_\varepsilon(\mathbf{x}) \cap \Gamma \subset \tau$ for all $0 < \varepsilon \leq \varepsilon_0$ and the boundedness of the kernel function on $\tau \times (\tau \backslash B_\varepsilon(\mathbf{x}))$ allows the definition of the integral

$$I_\varepsilon := \int_\tau \int_{\tau \backslash B_\varepsilon(\mathbf{x})} k_1(\mathbf{x}, \mathbf{y}, \mathbf{y}-\mathbf{x})\, ds_{\mathbf{y}} ds_{\mathbf{x}}$$

with

$$k_1(\mathbf{x}, \mathbf{y}, \mathbf{z}) := b_i(\mathbf{x})\, k(\mathbf{x}, \mathbf{y}, \mathbf{z})\, b_j(\mathbf{y}).$$

Let $\chi_\tau : \hat{\tau} \to \tau$ be the transformation of the reference element to τ. We set

$$k_2(\hat{\mathbf{x}}, \hat{\mathbf{y}}) := k_1(\chi_\tau(\hat{\mathbf{x}}), \chi_\tau(\hat{\mathbf{y}}), \chi_\tau(\hat{\mathbf{y}}) - \chi_\tau(\hat{\mathbf{x}}))\, g_\tau(\hat{\mathbf{x}})\, g_\tau(\hat{\mathbf{y}}). \tag{5.25}$$

$\widetilde{B}_\varepsilon(\hat{\mathbf{x}}) := \chi_\tau^{-1}(\Gamma \cap B_\varepsilon(\chi_\tau(\hat{\mathbf{x}})))$ denotes the pullback of the ε-neighborhood of $\mathbf{x}$ (see Fig. 5.2). Note that $\widetilde{B}_\varepsilon(\hat{\mathbf{x}})$ in general does not by any means represent a circular disc in the parameter domain. We will however show that the limit $\varepsilon \to 0$ remains the same if $\widetilde{B}_\varepsilon(\mathbf{x})$ is replaced by the disc $B_\varepsilon(\mathbf{x})$.

Theorem 5.2.1. *Let the kernel function satisfy the assumption (5.1.19). Let $\tau \in \mathcal{G}$ be parametrized by $\chi_\tau \in C^{1+\lambda}(\hat{\tau})$ with $\lambda > 0$. Then*

$$\lim_{\varepsilon \to 0} \int_{\hat{\tau}} \int_{\hat{\tau} \backslash \widetilde{B}_\varepsilon(\hat{\mathbf{x}})} k_2(\hat{\mathbf{x}}, \hat{\mathbf{y}})\, d\hat{\mathbf{y}} d\hat{\mathbf{x}} = \lim_{\varepsilon \to 0} \int_{\hat{\tau}} \int_{\hat{\tau} \backslash B_\varepsilon(\hat{\mathbf{x}})} k_2(\hat{\mathbf{x}}, \hat{\mathbf{y}})\, d\hat{\mathbf{y}} d\hat{\mathbf{x}}.$$

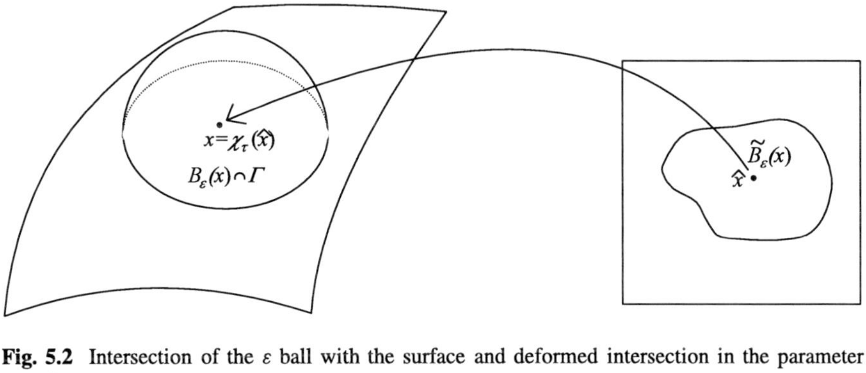

Fig. 5.2 Intersection of the ε ball with the surface and deformed intersection in the parameter domain

Proof. We introduce polar coordinates (r, φ) about $\hat{\mathbf{x}} := \chi_\tau^{-1}(\mathbf{x})$. The parametrization of the boundary $\partial \widetilde{B}_\varepsilon(\hat{\mathbf{x}})$ implicitly defines the function $\rho : [0, \varepsilon_0] \times [-\pi, \pi[\to \mathbb{R}_{\geq 0}$ by

$$\left\| \chi_\tau \left(\hat{\mathbf{x}} + \rho(\varepsilon, \varphi) \binom{\cos\varphi}{\sin\varphi} \right) - \mathbf{x} \right\| = \varepsilon.$$

If we expand with respect to ε (see Exercise 5.2.2) and by using the implicit function theorem we obtain $\rho \in C^{1+\lambda}([0, \varepsilon_0] \times [-\pi, \pi[)$ and

$$\rho(0, \varphi) = 0, \quad \rho(\varepsilon, \varphi) = \varepsilon \rho_\varepsilon(0, \varphi) + O\left(\varepsilon^{1+\lambda}\right) \quad \text{and,} \quad \rho_\varepsilon(0, \varphi) = \left\| D\chi_\tau(\hat{\mathbf{x}}) \binom{\cos\varphi}{\sin\varphi} \right\|^{-1}. \tag{5.26}$$

This leads to the decomposition

$$\int_{\hat{\tau}} \int_{\hat{\tau} \backslash \widetilde{B}_\varepsilon(\hat{\mathbf{x}})} \cdots = \int_{\hat{\tau}} \int_{\hat{\tau} \backslash B_\varepsilon(\hat{\mathbf{x}})} \cdots + \int_{\hat{\tau}} \int_{-\pi}^{\pi} \int_{\rho(\varepsilon,\varphi)}^{\varepsilon} \cdots. \tag{5.27}$$

We will show that the second integral on the right-hand side converges to zero for $\varepsilon \to 0$. The angular integration is split into $[-\pi, 0[$ and $[0, \pi[$ and φ is substituted in the first interval by $\varphi = \tilde{\varphi} - \pi$. For the second integral in (5.27) this yields the representation

$$\int_{\hat{\tau}} \int_0^{\pi} \left(\int_{\rho(\varepsilon,\varphi)}^{\varepsilon} k_2 \left(\hat{\mathbf{x}}, \hat{\mathbf{x}} + r \binom{\cos\varphi}{\sin\varphi} \right) r dr + \int_{\rho(\varepsilon,\varphi-\pi)}^{\varepsilon} k_2 \left(\hat{\mathbf{x}}, \hat{\mathbf{x}} + r \binom{\cos(\varphi-\pi)}{\sin(\varphi-\pi)} \right) r dr \right) d\varphi d\hat{\mathbf{x}}. \tag{5.28}$$

The expansion (5.26) implies that

$$\begin{aligned} \rho(\varepsilon, \varphi - \pi) &= \varepsilon \rho_\varepsilon(0, \varphi - \pi) + O\left(\varepsilon^{1+\lambda}\right) = \varepsilon \rho_\varepsilon(0, \varphi) + O\left(\varepsilon^{1+\lambda}\right) \\ &= \rho(\varepsilon, \varphi) + O\left(\varepsilon^{1+\lambda}\right), \end{aligned}$$

which for the integral in (5.28) yields the representation

$$\begin{aligned} &\int_{\hat{\tau}} \int_0^{\pi} \int_{\rho(\varepsilon,\varphi)}^{\varepsilon} \left(k_2 \left(\hat{\mathbf{x}}, \hat{\mathbf{x}} + r \binom{\cos\varphi}{\sin\varphi} \right) + k_2 \left(\hat{\mathbf{x}}, \hat{\mathbf{x}} - r \binom{\cos(\varphi)}{\sin(\varphi)} \right) \right) r dr d\varphi d\hat{\mathbf{x}} \\ &+ \int_{\hat{\tau}} \int_0^{\pi} \int_{\rho(\varepsilon,\varphi-\pi)}^{\rho(\varepsilon,\varphi)} k_2 \left(\hat{\mathbf{x}}, \hat{\mathbf{x}} - r \binom{\cos(\varphi)}{\sin(\varphi)} \right) r dr d\varphi d\hat{\mathbf{x}}. \end{aligned} \tag{5.29}$$

The kernel function in polar coordinates can be estimated as

$$\left| r k_2 \left(\hat{\mathbf{x}}, \hat{\mathbf{x}} - r \binom{\cos\varphi}{\sin\varphi} \right) \right| \leq C r^{-1}.$$

Since $|\rho(\varepsilon, \varphi) - \rho(\varepsilon, \varphi - \pi)| = O\left(\varepsilon^{1+\lambda}\right)$ the second integral in (5.29) converges to zero and we therefore only consider the first integral. The antisymmetry of the kernel function in local coordinates (see Theorem 5.1.22) implies

$$r\left(k_2\left(\hat{\mathbf{x}}, \hat{\mathbf{x}} + r\binom{\cos\varphi}{\sin\varphi}\right) + k_2\left(\hat{\mathbf{x}}, \hat{\mathbf{x}} - r\binom{\cos(\varphi)}{\sin(\varphi)}\right)\right) = O(1)$$

and therefore the first integral in (5.29) also converges to zero. □

Exercise 5.2.2. *Prove the expansion (5.26).*

This proves that (5.24) coincides with the limit of the integral

$$\widetilde{I}_\varepsilon := \int_0^1 \int_0^{\hat{x}_1} \underbrace{\int_0^1 \int_0^{\hat{y}_1}}_{\|\hat{\mathbf{y}}-\hat{\mathbf{x}}\| \geq \varepsilon} k_2(\hat{\mathbf{x}}, \hat{\mathbf{y}})\, d\hat{\mathbf{y}} d\hat{\mathbf{x}}$$

for $\varepsilon \to 0$. It is our aim to represent the limit $\lim_{\varepsilon \to 0} \widetilde{I}_\varepsilon$ as an integral over a fixed, ε-independent integration domain with an analytic integrand. To achieve this we first introduce relative coordinates $(\hat{\mathbf{x}}, \hat{\mathbf{z}}) = (\hat{\mathbf{x}}, \hat{\mathbf{y}} - \hat{\mathbf{x}})$ that shift the singular behavior of the kernel function into $\hat{\mathbf{z}} = \mathbf{0}$. For now, let $\widehat{\tau}$ be the unit triangle. The results for the unit square will be summarized in Sect. 5.2.4.

Then we have

$$\widetilde{I}_\varepsilon = \int_0^1 \int_0^{\hat{x}_1} \underbrace{\int_{-\hat{x}_1}^{1-\hat{x}_1} \int_{-\hat{x}_2}^{\hat{z}_1+\hat{x}_1-\hat{x}_2}}_{\|\hat{\mathbf{z}}\| \geq \varepsilon} k_2(\hat{\mathbf{x}}, \hat{\mathbf{z}} + \hat{\mathbf{x}})\, d\hat{\mathbf{z}} d\hat{\mathbf{x}}. \tag{5.30}$$

The following two examples serve to illustrate the characteristic behavior of the integrand as well as the strategy for its numerical treatment (see Fig. 5.3).

Example 5.2.3. *Let $k : (-1, 2) \times (-1, 2) \to \mathbb{R}$ be given by $k(x, y) = (y - x)^{-1}$. For smooth functions u, v our aim is to evaluate the integral*

$$I_\varepsilon = \int_{-1}^2 v(x) \int_{\substack{-1 \\ |x-y| \geq \varepsilon}}^2 k(x, y)\, u(y)\, dy dx.$$

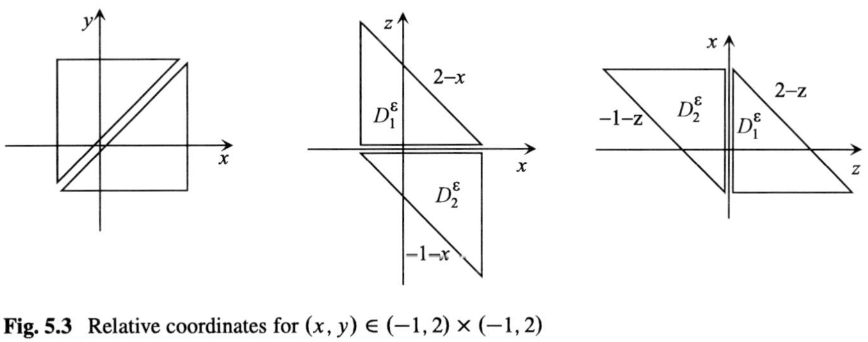

Fig. 5.3 Relative coordinates for $(x, y) \in (-1, 2) \times (-1, 2)$

In this case relative coordinates $z = y - x$ yield

$$I_\varepsilon = \int_{-1}^{2} v(x) \left(\int_{\substack{-1-x \\ |z|\geq\varepsilon}}^{2-x} \frac{u(z+x)}{z} dz \right) dx. \tag{5.31}$$

For simplicity's sake we set $u(z) \equiv 1$ (in general u would have to be expanded about $z = 0$). The inner integration can be evaluated explicitly and yields (for $\varepsilon < \min\{2-x, 1+x\}$)

$$\lim_{\varepsilon\to 0} \int_{\substack{-1-x \\ |z|\geq\varepsilon}}^{2-x} \frac{1}{z} dz = \ln \frac{2-x}{1+x}.$$

This example shows that the result of the inner integration has characteristic (logarithmic) singularities on the boundaries of the panel related to the outer integration and that it therefore is not sufficient to only develop special integration methods for the inner integration. Standard quadrature methods can thus only be used with a significant loss in accuracy. In the case of two-dimensional surfaces it can be shown for Cauchy-singular kernel functions that the result of the inner z-integration behaves as $\sim \log(\operatorname{dist}(\mathbf{x}, \partial\tau))$. The strength of the singular behavior generally depends on the order of singularity of the kernel function and significantly complicates the use of weighted integration formulas. We will show in the following that the integrand can be regularized by simply changing the ordering of the integration variables.

Example 5.2.4. *Taking the auxiliary condition $|z| \geq \varepsilon$ into consideration yields the decomposition of the integration domain in (5.31) as given by*

$$\bigcup_{i=1}^{2} D_i^\varepsilon := \left\{ \begin{matrix} -1 \leq x \leq 2-\varepsilon \\ \varepsilon \leq z \leq 2-x \end{matrix} \right\} \cup \left\{ \begin{matrix} -1+\varepsilon \leq x \leq 2 \\ -1-x \leq z \leq -\varepsilon \end{matrix} \right\}.$$

Changing the ordering of the integration variables in the subdomains $D_{1,2}^\varepsilon$ yields

$$D_1^\varepsilon \cup D_2^\varepsilon = \left\{ \begin{matrix} \varepsilon \leq z \leq 3 \\ -1 \leq x \leq 2-z \end{matrix} \right\} \cup \left\{ \begin{matrix} -3 \leq z \leq -\varepsilon \\ -1-z \leq x \leq 2 \end{matrix} \right\},$$

and we obtain

$$\begin{aligned} \lim_{\varepsilon\to 0} I_\varepsilon &= \lim_{\varepsilon\to 0} \left(\int_\varepsilon^3 \frac{1}{z} \underbrace{\int_{-1}^{2-z} v(x)\, dx}_{=:h^{(1)}(z)} dz + \int_{-3}^{-\varepsilon} \frac{1}{z} \underbrace{\int_{-1-z}^{2} v(x)\, dx}_{h^{(2)}(z)} dz \right) \\ &= \lim_{\varepsilon\to 0} \left(\int_\varepsilon^3 \frac{1}{z} h^{(1)}(z)\, dz + \int_{-3}^{-\varepsilon} \frac{1}{z} h^{(2)}(z)\, dz \right). \end{aligned} \tag{5.32}$$

The substitution $z \leftarrow -z$ in the second integral gives us

$$\lim_{\varepsilon \to 0} I_\varepsilon = \lim_{\varepsilon \to 0} \int_\varepsilon^3 \frac{h^{(1)}(z) - h^{(2)}(-z)}{z} dz.$$

By virtue of $\left|h^{(2)}(z) - h^{(1)}(-z)\right| \leq Cz$ the integrand is bounded and the Cauchy principal value coincides with the Riemann integral

$$\lim_{\varepsilon \to 0} I_\varepsilon = \int_0^3 \frac{h^{(1)}(z) - h^{(2)}(-z)}{z} dz.$$

The integrand has a removable singularity for $z = 0$.

We will apply the one-dimension concept developed in this example to the general situation (5.30) and first change the order of integration.

The integration domain is decomposed as

$$\left\{\begin{array}{c} -1 \leq \hat{z}_1 \leq 0 \\ -1 \leq \hat{z}_2 \leq \hat{z}_1 \\ -\hat{z}_2 \leq \hat{x}_1 \leq 1 \\ -\hat{z}_2 \leq \hat{x}_2 \leq \hat{x}_1 \end{array}\right\} \cup \left\{\begin{array}{c} -1 \leq \hat{z}_1 \leq 0 \\ \hat{z}_1 \leq \hat{z}_2 \leq 0 \\ -\hat{z}_1 \leq \hat{x}_1 \leq 1 \\ -\hat{z}_2 \leq \hat{x}_2 \leq \hat{x}_1 + \hat{z}_1 - \hat{z}_2 \end{array}\right\} \cup \left\{\begin{array}{c} -1 \leq \hat{z}_1 \leq 0 \\ 0 \leq \hat{z}_2 \leq 1 + \hat{z}_1 \\ \hat{z}_2 - \hat{z}_1 \leq \hat{x}_1 \leq 1 \\ 0 \leq \hat{x}_2 \leq \hat{x}_1 + \hat{z}_1 - \hat{z}_2 \end{array}\right\}$$

$$\cup \left\{\begin{array}{c} 0 \leq \hat{z}_1 \leq 1 \\ -1 + \hat{z}_1 \leq \hat{z}_2 \leq 0 \\ -\hat{z}_2 \leq \hat{x}_1 \leq 1 - \hat{z}_1 \\ -\hat{z}_2 \leq \hat{x}_2 \leq \hat{x}_1 \end{array}\right\} \cup \left\{\begin{array}{c} 0 \leq \hat{z}_1 \leq 1 \\ 0 \leq \hat{z}_2 \leq \hat{z}_1 \\ 0 \leq \hat{x}_1 \leq 1 - \hat{z}_1 \\ 0 \leq \hat{x}_2 \leq \hat{x}_1 \end{array}\right\} \cup \left\{\begin{array}{c} 0 \leq \hat{z}_1 \leq 1 \\ \hat{z}_1 \leq \hat{z}_2 \leq 1 \\ \hat{z}_2 - \hat{z}_1 \leq \hat{x}_1 \leq 1 - \hat{z}_1 \\ 0 \leq \hat{x}_2 \leq \hat{z}_1 - \hat{z}_2 + \hat{x}_1 \end{array}\right\}$$

with the auxiliary condition $\|\hat{\mathbf{z}}\| \geq \varepsilon$ into subdomains D_i, $1 \leq i \leq 6$. Note that the $\hat{\mathbf{z}}$-variables describe the outer integration as in (5.32). As in the proof of Theorem 5.2.1 we use the antisymmetry of the principal part of the kernel function to regularize the integrand by means of mirroring. The subdomains D_1, D_2, D_4 are mirrored onto the subdomains D_6, D_5, D_3 by means of the substitution $\hat{\mathbf{z}}^{new} = -\hat{\mathbf{z}}$ (see Fig. 5.4). The $\hat{\mathbf{x}}$-variables in the other integrand are transformed by the substitution $\hat{\mathbf{x}}^{new} = \hat{\mathbf{x}} - \hat{\mathbf{z}}^{new}$. More specifically, on the integration domain D_i we use the linear transformations

$$\begin{pmatrix} \mathbf{x}_{\text{new}}^{(1)} \\ \mathbf{z}_{\text{new}}^{(1)} \end{pmatrix} = \begin{pmatrix} \mathbf{x} \\ -\mathbf{z} \end{pmatrix} \quad \begin{pmatrix} \mathbf{x}_{\text{new}}^{(2)} \\ \mathbf{z}_{\text{new}}^{(2)} \end{pmatrix} = \begin{pmatrix} \mathbf{x} \\ -\mathbf{z} \end{pmatrix} \quad \begin{pmatrix} \mathbf{x}_{\text{new}}^{(3)} \\ \mathbf{z}_{\text{new}}^{(3)} \end{pmatrix} = \begin{pmatrix} \mathbf{x} + \mathbf{z} \\ \mathbf{z} \end{pmatrix}$$

$$\begin{pmatrix} \mathbf{x}_{\text{new}}^{(4)} \\ \mathbf{z}_{\text{new}}^{(4)} \end{pmatrix} = \begin{pmatrix} \mathbf{x} \\ -\mathbf{z} \end{pmatrix} \quad \begin{pmatrix} \mathbf{x}_{\text{new}}^{(4)} \\ \mathbf{z}_{\text{new}}^{(4)} \end{pmatrix} = \begin{pmatrix} \mathbf{x} + \mathbf{z} \\ \mathbf{z} \end{pmatrix} \quad \begin{pmatrix} \mathbf{x}_{\text{new}}^{(6)} \\ \mathbf{z}_{\text{new}}^{(6)} \end{pmatrix} = \begin{pmatrix} \mathbf{x} + \mathbf{z} \\ \mathbf{z} \end{pmatrix}$$

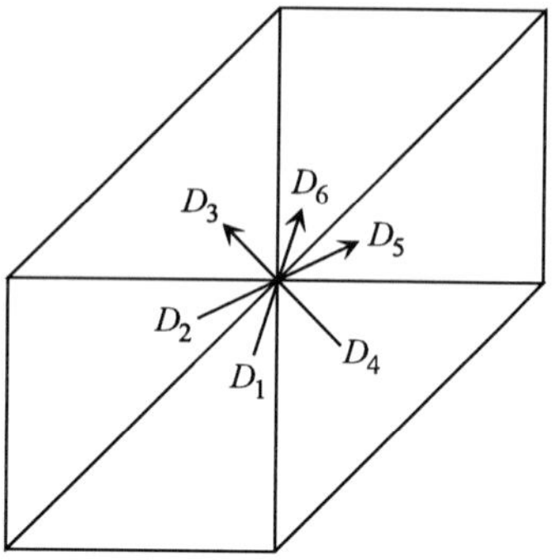

Fig. 5.4 Subdomains D_i, $1 \le i \le 6$, with corresponding directions for reflection

and again denote the new coordinates by $(\mathbf{x}, \mathbf{z})$. This gives us

$$\widetilde{I}_\varepsilon = \underbrace{\int_0^1 \int_{\hat{z}_1}^1}_{\|\hat{\mathbf{z}}\| \ge \varepsilon} \left(\int_{\hat{z}_2}^1 \int_{\hat{z}_2}^{\hat{x}_1} k_2(\hat{\mathbf{x}}, \hat{\mathbf{x}} - \hat{\mathbf{z}}) + k_2(\hat{\mathbf{x}} - \hat{\mathbf{z}}, \hat{\mathbf{x}})\, d\hat{\mathbf{x}} \right) d\hat{\mathbf{z}}$$

$$+ \underbrace{\int_0^1 \int_0^{\hat{z}_1}}_{\|\hat{\mathbf{z}}\| \ge \varepsilon} \left(\int_{\hat{z}_1}^1 \int_{\hat{z}_2}^{\hat{x}_1 - \hat{z}_1 + \hat{z}_2} k_2(\hat{\mathbf{x}}, \hat{\mathbf{x}} - \hat{\mathbf{z}}) + k_2(\hat{\mathbf{x}} - \hat{\mathbf{z}}, \hat{\mathbf{x}})\, d\hat{\mathbf{x}} \right) d\hat{\mathbf{z}}$$

$$+ \underbrace{\int_{-1}^0 \int_0^{1+\hat{z}_1}}_{\|\hat{\mathbf{z}}\| \ge \varepsilon} \left(\int_{\hat{z}_2}^{1+\hat{z}_1} \int_{\hat{z}_2}^{\hat{x}_1} k_2(\hat{\mathbf{x}}, \hat{\mathbf{x}} - \hat{\mathbf{z}}) + k_2(\hat{\mathbf{x}} - \hat{\mathbf{z}}, \hat{\mathbf{x}})\, d\hat{\mathbf{x}} \right) d\hat{\mathbf{z}}.$$

Owing to the smoothness of k with respect to the first argument and, in the case of Cauchy-singular kernels, the antisymmetry of the kernel function [see (5.23)] for $\|\hat{\mathbf{z}}\| \to 0$, the integrands behave as

$$k_2(\hat{\mathbf{x}}, \hat{\mathbf{x}} - \hat{\mathbf{z}}) + k_2(\hat{\mathbf{x}} - \hat{\mathbf{z}}, \hat{\mathbf{x}}) = O\left(\|\hat{\mathbf{z}}\|^{-1}\right)$$

and are therefore improperly integrable. We set $k_2^+(\hat{\mathbf{x}}, \hat{\mathbf{z}}) := k_2(\hat{\mathbf{x}}, \hat{\mathbf{x}} - \hat{\mathbf{z}}) + k_2(\hat{\mathbf{x}} - \hat{\mathbf{z}}, \hat{\mathbf{x}})$. The limit $\varepsilon \to 0$ then yields

$$I = \lim_{\varepsilon \to 0} \widetilde{I}_\varepsilon = \int_0^1 \int_{\hat{z}_1}^1 \left(\int_{\hat{z}_2}^1 \int_{\hat{z}_2}^{\hat{x}_1} k_2^+(\hat{\mathbf{x}}, \hat{\mathbf{z}})\, d\hat{\mathbf{x}} \right) d\hat{\mathbf{z}} \tag{5.33}$$

$$+ \int_0^1 \int_0^{\hat{z}_1} \left(\int_{\hat{z}_1}^1 \int_{\hat{z}_2}^{\hat{x}_1 - \hat{z}_1 + \hat{z}_2} k_2^+(\hat{\mathbf{x}}, \hat{\mathbf{z}})\, d\hat{\mathbf{x}} \right) d\hat{\mathbf{z}}$$

$$+ \int_{-1}^0 \int_0^{1+\hat{z}_1} \left(\int_{\hat{z}_2}^{1+\hat{z}_1} \int_{\hat{z}_2}^{\hat{x}_1} k_2^+(\hat{\mathbf{x}}, \hat{\mathbf{z}})\, d\hat{\mathbf{x}} \right) d\hat{\mathbf{z}}.$$

In the next step these three integrals are transformed onto the four-dimensional reference domain

$$D := \{0 \le w_1 \le 1, 0 \le w_2 \le w_1, 0 \le w_3 \le w_2, 0 \le w_4 \le w_3\}.$$

The associated linear mappings $T_i : D \to D_i$, $(\hat{\mathbf{x}}, \hat{\mathbf{z}}) := T_i \mathbf{w}$ are given for $i = 1, 2, 3$ by the matrices $\mathbf{m}_i$:

$$\mathbf{m}_1 := \begin{bmatrix} 1 & 0 & 0 & 0 \\ 1 & -1 & 1 & 0 \\ 0 & 0 & 0 & 1 \\ 0 & 0 & 1 & 0 \end{bmatrix}, \quad \mathbf{m}_2 := \begin{bmatrix} 1 & 0 & 0 & 0 \\ 0 & 1 & -1 & 1 \\ 0 & 0 & 1 & 0 \\ 0 & 0 & 0 & 1 \end{bmatrix}, \quad \mathbf{m}_3 := \begin{bmatrix} 1 & 0 & 0 & -1 \\ 0 & 1 & 0 & -1 \\ 0 & 0 & 0 & -1 \\ 0 & 0 & 1 & -1 \end{bmatrix}.$$

We have $\det(\mathbf{m}_i) = 1$ for all $1 \le i \le 3$. From this we have for the integral (5.33)

$$I = \sum_{i=1}^{3} \int_0^1 \int_0^{w_1} \int_0^{w_2} \int_0^{w_3} k_2^+ (\mathbf{m}_i \mathbf{w})\, d\mathbf{w}.$$

The simplex coordinates $(\xi, \eta_1, \eta_2, \eta_3)$ transform the unit cube $(0, 1)^4$ onto D:

$$(w_1, w_2, w_3, w_4)^\mathsf{T} = (\xi, \xi\eta_1, \xi\eta_1\eta_2, \xi\eta_1\eta_2\eta_3)^\mathsf{T}.$$

Note that the determinant of the Jacobian of this transformation is equal to $\xi^3\eta_1^2\eta_2$ and that we finally obtain the representation

$$I = \sum_{i=1}^{3} \int_0^1 \int_0^1 \int_0^1 \int_0^1 \xi^3\eta_1^2\eta_2 k_2^+ \left(\xi\mathbf{m}_i \left(1, \eta_1, \eta_1\eta_2, \eta_1\eta_2\eta_3\right)^\mathsf{T}\right) d\eta_1 d\eta_2 d\eta_3 d\xi. \tag{5.34}$$

It is proved in Theorem 5.2.5 that the integrand in this representation is analytic. The integral can therefore be efficiently approximated by means of Gaussian quadrature with respect to each coordinate.

Theorem 5.2.5. *The integrand in (5.34) can be analytically extended with respect to all variables in a complex neighborhood of* $[0, 1]^4$.

Remark 5.2.6. *The size of the complex neighborhood of* $[0, 1]^4$ *in which the integrand in (5.34) can be extended, is estimated in Sect. 5.3.2.3.*

Notation 5.2.7. *A function* $u : \tau \times \mathbb{R}_{\ge 0} \times \mathbb{S}_1$ *has the property (A) if* $\rho > 0$ *and there exist neighborhoods* $\tau^\star$ *of* $\overline{\tau}$ *and* U *of* $\mathbb{S}_1$ *such that* u *can be extended to a function* $u^\star : \tau^\star \times \mathbb{R}_{\ge 0} \times U$ *which is analytic on* $\tau^\star \times [0, \rho] \times U$.

Proof of Theorem 5.2.5: We set $\hat{b}_i := (b_i \circ \chi_\tau)\, g_\tau$ and use $r = \|\hat{\mathbf{z}}\|$ and $\xi = \hat{\mathbf{z}} / \|\hat{\mathbf{z}}\|$. The fact that the basis functions and the surface element g_τ are analytic in local coordinates gives us the representation

$$rk_2^+(\hat{\mathbf{x}},\hat{\mathbf{z}}) = r\left(\hat{b}_i(\hat{\mathbf{x}})\,\hat{b}_j(\hat{\mathbf{x}}-\hat{\mathbf{z}})\,\hat{k}(\hat{\mathbf{x}},\hat{\mathbf{x}}-\hat{\mathbf{z}}) + \hat{b}_i(\hat{\mathbf{x}}-\hat{\mathbf{z}})\,\hat{b}_j(\hat{\mathbf{x}})\,\hat{k}(\hat{\mathbf{x}}-\hat{\mathbf{z}},\hat{\mathbf{x}})\right)$$
$$= \hat{b}_i(\hat{\mathbf{x}})\,\hat{b}_j(\hat{\mathbf{x}})\,r\left(\hat{k}(\hat{\mathbf{x}},\hat{\mathbf{x}}-\hat{\mathbf{z}}) + \hat{k}(\hat{\mathbf{x}}-\hat{\mathbf{z}},\hat{\mathbf{x}})\right) + r^2 R(\hat{\mathbf{x}},r,\xi)$$

with

$$R := \hat{b}_i(\hat{\mathbf{x}})\left(D_r\hat{b}_j\right)(\hat{\mathbf{x}},r,\xi)\,\hat{k}(\hat{\mathbf{x}},\hat{\mathbf{x}}-r\xi) + \hat{b}_j(\hat{\mathbf{x}})\left(D_r\hat{b}_i\right)(\hat{\mathbf{x}},r,\xi)\,\hat{k}(\hat{\mathbf{x}}-r\xi,\hat{\mathbf{x}})$$

and

$$(D_r u)(\hat{\mathbf{x}},r,\xi) := \begin{cases} \dfrac{u(\hat{\mathbf{x}}-r\xi)-u(\hat{\mathbf{x}})}{r} & \text{if } r>0, \\ -\langle \xi, \nabla u(\hat{\mathbf{x}})\rangle & \text{if } r=0. \end{cases}$$

Clearly, the functions $D_r\hat{b}_j$, $D_r\hat{b}_i$ have the property (A). From Theorem 5.1.22(c) we thus also have the property (A) for the function rk_2^+ in $(\hat{\mathbf{x}},r,\xi)$-coordinates. The integrand in (5.34) is therefore analytic with respect to every variable if we can show that the transformation from (ξ,η)-coordinates to $(\hat{\mathbf{x}},r,\xi)$ is analytic. The coordinate systems satisfy the relation

$$(\hat{x}_1,\hat{x}_2,r\cos\varphi,r\sin\varphi)^\mathsf{T} = (\hat{\mathbf{x}},\hat{\mathbf{z}}) = \xi\mathbf{m}_i\,(1,\eta_1,\eta_1\eta_2,\eta_1\eta_2\eta_3)^\mathsf{T}.$$

For $\mathbf{m}_i$, $1\le i\le 3$, we obtain the transformations

$$(\hat{x}_1,\hat{x}_2,r,\varphi)^\mathsf{T} = \begin{cases} \left(\xi,\xi(1-\eta_1+\eta_1\eta_2),\xi\eta_1\eta_2\sqrt{1+\eta_3^2},\operatorname{arccot}\eta_3\right)^\mathsf{T} & i=1 \\ \left(\xi,\xi\eta_1(1-\eta_2+\eta_2\eta_3),\xi\eta_1\eta_2\sqrt{1+\eta_3^2},\arctan\eta_3\right)^\mathsf{T} & i=2 \\ \left(\xi(1-\eta_3),\xi\eta_1(1-\eta_2\eta_3),\xi\eta_1\eta_2\sqrt{\eta_3^2+(1-\eta_3)^2},\arctan\dfrac{1-\eta_3}{\eta_3}\right)^\mathsf{T} & i=3 \end{cases} \tag{5.35}$$

that are clearly analytic with respect to all variables in $(0,1)$. For the determinant of the Jacobian in (5.34) we have in $(\hat{\mathbf{x}},r,\varphi)$-coordinates $\xi^3\eta_1^2\eta_2 = r\times$ (entire function in $\hat{\mathbf{x}},\varphi$) and, thus, Property (A) is transferred from $rk_2^+\left(\hat{\mathbf{x}},r\binom{\cos\varphi}{\sin\varphi}\right)$ to the integrands in (5.34). □

5.2.2 *Common Edge*

We consider two panels $\tau,t\in\mathcal{G}$ with exactly one common edge $E=\overline{\tau}\cap\overline{t}$. In Lemma 5.1.13 it was shown that the integral

$$I_{\tau\times t} := \int_{\tau\times t} b_i(\mathbf{x})\,b_j(\mathbf{y})\,k(\mathbf{x},\mathbf{y},\mathbf{y}-\mathbf{x})\,ds_\mathbf{y}ds_\mathbf{x}$$

exists in this case as an improper Riemann integral. Let the parametrizations $\chi_\tau : \hat{\tau} \to \tau$ and $\chi_t : \hat{t} \to t$ be chosen in such a way that

$$\chi_\tau\binom{s}{0} = \chi_t\binom{s}{0} \qquad \forall s \in [0,1].$$

The integrand is given in local coordinates by

$$k_3(\hat{\mathbf{x}}, \hat{\mathbf{y}}) := \hat{b}_i(\hat{\mathbf{x}})\,\hat{b}_j(\hat{\mathbf{y}})\,k\left(\chi_\tau(\hat{\mathbf{x}}), \chi_t(\hat{\mathbf{y}}), \chi_t(\hat{\mathbf{y}}) - \chi_\tau(\hat{\mathbf{x}})\right) g_\tau(\hat{\mathbf{x}})\, g_t(\hat{\mathbf{y}}), \quad (5.36)$$

i.e.,

$$I_{\tau\times t} = \int_{\hat{\tau}}\int_{\hat{t}} k_3(\hat{\mathbf{x}}, \hat{\mathbf{y}})\, d\hat{\mathbf{y}} d\hat{\mathbf{x}}.$$

We will again first consider the case that $\hat{\tau} = \hat{t}$ is the unit triangle. As before we introduce relative coordinates

$$\hat{\mathbf{z}} = (\hat{y}_1 - \hat{x}_1, \hat{y}_2, \hat{x}_2)^\intercal,$$

in order to fix the location of the singularities of the integrand in $\hat{\mathbf{z}} = \mathbf{0}$. Then

$$I_{\tau\times t} = \int_0^1 \int_{-\hat{x}_1}^{1-\hat{x}_1} \int_0^{\hat{z}_1+\hat{x}_1} \int_0^{\hat{x}_1} k_3(\hat{x}_1, \hat{z}_3, \hat{z}_1 + \hat{x}_1, \hat{z}_2)\, d\hat{\mathbf{z}} d\hat{x}_1. \quad (5.37)$$

In general the result of the $\hat{\mathbf{z}}$-integration, considered as a function of $\hat{x}_1$, has end-point singularities in $(0,1)$. We change the order of integration for the purpose of regularization. The integration domain in (5.37) can be decomposed into five disjoint, four-dimensional polyhedra

$$I_{\tau\times t} = \sum_{i=1}^{5} \int_{D_i} \ldots d\hat{x}_1 d\hat{\mathbf{z}}$$

with

$$\bigcup_{i=1}^{5} D_i := \left\{\begin{array}{c} -1 \le \hat{z}_1 \le 0 \\ 0 \le \hat{z}_2 \le 1 + \hat{z}_1 \\ 0 \le \hat{z}_3 \le \hat{z}_2 - \hat{z}_1 \\ \hat{z}_2 - \hat{z}_1 \le \hat{x}_1 \le 1 \end{array}\right\} \cup \left\{\begin{array}{c} -1 \le \hat{z}_1 \le 0 \\ 0 \le \hat{z}_2 \le 1 + \hat{z}_1 \\ \hat{z}_2 - \hat{z}_1 \le \hat{z}_3 \le 1 \\ \hat{z}_3 \le \hat{x}_1 \le 1 \end{array}\right\} \cup \left\{\begin{array}{c} 0 \le \hat{z}_1 \le 1 \\ 0 \le \hat{z}_2 \le \hat{z}_1 \\ 0 \le \hat{z}_3 \le 1 - \hat{z}_1 \\ \hat{z}_3 \le \hat{x}_1 \le 1 - \hat{z}_1 \end{array}\right\}$$

$$\cup \left\{\begin{array}{c} 0 \le \hat{z}_1 \le 1 \\ \hat{z}_1 \le \hat{z}_2 \le 1 \\ 0 \le \hat{z}_3 \le \hat{z}_2 - \hat{z}_1 \\ \hat{z}_2 - \hat{z}_1 \le \hat{x}_1 \le 1 - \hat{z}_1 \end{array}\right\} \cup \left\{\begin{array}{c} 0 \le \hat{z}_1 \le 1 \\ \hat{z}_1 \le \hat{z}_2 \le 1 \\ \hat{z}_2 - \hat{z}_1 \le \hat{z}_3 \le 1 - \hat{z}_1 \\ \hat{z}_3 \le \hat{x}_1 \le 1 - \hat{z}_1 \end{array}\right\}.$$

The integration domains D_i, $1 \le i \le 5$, are transformed onto the four-dimensional unit cube as in the case of identical panels. For $1 \le i \le 5$ the transformations

$T_i : (0,1)^4 \to D_i$, $(\hat{x}_1, \hat{\mathbf{z}}) := T_i(\xi, \eta)$, have the form

$$T_1\begin{pmatrix}\xi\\ \eta\end{pmatrix} = \xi\begin{pmatrix}1\\ -\eta_1\eta_2\\ \eta_1(1-\eta_2)\\ \eta_1\eta_3\end{pmatrix}, \qquad T_2\begin{pmatrix}\xi\\ \eta\end{pmatrix} = \xi\begin{pmatrix}1\\ -\eta_1\eta_2\eta_3\\ \eta_1\eta_2(1-\eta_3)\\ \eta_1\end{pmatrix},$$

$$T_3\begin{pmatrix}\xi\\ \eta\end{pmatrix} = \xi\begin{pmatrix}1-\eta_1\eta_2\\ \eta_1\eta_2\\ \eta_1\eta_2\eta_3\\ \eta_1(1-\eta_2)\end{pmatrix}, \qquad T_4\begin{pmatrix}\xi\\ \eta\end{pmatrix} = \xi\begin{pmatrix}1-\eta_1\eta_2\eta_3\\ \eta_1\eta_2\eta_3\\ \eta_1\\ \eta_1\eta_2(1-\eta_3)\end{pmatrix}$$

$$T_5\begin{pmatrix}\xi\\ \eta\end{pmatrix} = \xi\begin{pmatrix}(1-\eta_1\eta_2\eta_3)\\ \eta_1\eta_2\eta_3\\ \eta_1\eta_2\\ \eta_1(1-\eta_2\eta_3)\end{pmatrix}.$$

We have for the absolute value of the Jacobian determinant of T_i

$$|\det T_i| = \begin{cases} \xi^3\eta_1^2 & \text{for } i = 1,\\ \xi^3\eta_1^2\eta_2 & \text{for } 2 \le i \le 5.\end{cases}$$

Therefore for the integral $I_{\tau\times t}$ we have derived the representation

$$\begin{aligned} I_{\tau\times t} = \int_{(0,1)^4} &\{\xi^3\eta_1^2 k_3(\xi, \xi\eta_1\eta_3, \xi(1-\eta_1\eta_2), \xi\eta_1(1-\eta_2)) \\ &+\xi^3\eta_1^2\eta_2 [k_3(\xi, \xi\eta_1, \xi(1-\eta_1\eta_2\eta_3), \xi\eta_1\eta_2(1-\eta_3)) \\ &+k_3(\xi(1-\eta_1\eta_2), \xi(\eta_1(1-\eta_2)), \xi, \xi\eta_1\eta_2\eta_3) \\ &+k_3(\xi(1-\eta_1\eta_2\eta_3), \xi\eta_1\eta_2(1-\eta_3), \xi, \xi\eta_1) \\ &+k_3(\xi(1-\eta_1\eta_2\eta_3), \xi\eta_1(1-\eta_2\eta_3), \xi, \xi\eta_1\eta_2)]\}\, d\eta\, d\xi. \end{aligned} \tag{5.38}$$

The following theorem proves that the integrand is analytic with respect to every variable.

Theorem 5.2.8. *The integrand in (5.38) can be extended analytically with respect to every variable in a complex neighborhood of* $[0,1]^4$.

Proof. Using the same arguments as in the proof of Theorem 5.2.5 we only need to show that the transformation from the (ξ, η)-coordinates to the (x_1, r, ξ)-coordinates is analytic. The coordinate systems satisfy the relation

$$(\hat{x}_1, r\cos\varphi\sin\theta, r\sin\varphi\sin\theta, r\cos\theta) = T_i(\xi, \eta).$$

For $i = 1$ we can explicitly determine the transformation

$$\begin{pmatrix} \hat{x}_1 \\ r \\ \varphi \\ \theta \end{pmatrix} = \begin{pmatrix} \xi \\ \xi\eta_1\sqrt{\eta_2^2 + (1-\eta_2)^2 + \eta_3^2} \\ \arctan\frac{\eta_2-1}{\eta_2} \\ \arccos\frac{\eta_3}{\sqrt{\eta_2^2+(1-\eta_2)^2+\eta_3^2}} \end{pmatrix},$$

which is analytic with respect to every variable $\xi, \eta_1, \eta_2, \eta_3$. For $i = 2, 3, 4, 5$ the coordinate transformations can also be given explicitly, which also supplies their analyticity. Here we will refrain from a detailed analysis of the case $2 \leq i \leq 5$. □

5.2.3 Common Vertex

We consider two panels $\tau, t \in \mathcal{G}$ with exactly one common point $\mathbf{p} = \overline{\tau} \cap \overline{t}$. It was shown in Lemma 5.1.13 that the integral

$$I_{\tau\times t} := \int_{\tau\times t} b_i(\mathbf{x})\, b_j(\mathbf{y})\, k(\mathbf{x}, \mathbf{y}, \mathbf{y}-\mathbf{x})\, ds_{\mathbf{y}} ds_{\mathbf{x}}$$

exists in this case as a Riemann integral. Let the parametrizations $\chi_\tau : \hat{\tau} \to \tau$ and $\chi_t : \widehat{t} \to t$ be chosen such that

$$\chi_\tau(\mathbf{0}) = \chi_t(\mathbf{0}) = \mathbf{p}.$$

As in the case of the common edge the integrand in local coordinates is given by

$$k_3(\hat{\mathbf{x}}, \hat{\mathbf{y}}) = \hat{b}_i(\hat{\mathbf{x}})\, \hat{b}_j(\hat{\mathbf{y}})\, k(\chi_\tau(\hat{\mathbf{x}}), \chi_t(\hat{\mathbf{y}}), \chi_t(\hat{\mathbf{y}}) - \chi_\tau(\hat{\mathbf{x}}))\, g_\tau(\hat{\mathbf{x}})\, g_t(\hat{\mathbf{y}}).$$

We introduce four-dimensional relative coordinates

$$\hat{\mathbf{z}} = (\hat{x}_1, \hat{x}_2, \hat{y}_1, \hat{y}_2)^\mathsf{T},$$

in order to fix the location of the singularities of the integrand in $\hat{\mathbf{z}} = \mathbf{0}$.

First, let $\hat{\tau} = \hat{t}$ again be the unit triangle. Then

$$I_{\tau\times t} = \int_0^1 \int_0^{\hat{z}_1} \int_0^1 \int_0^{\hat{z}_3} k_3(\hat{\mathbf{z}})\, d\hat{\mathbf{z}}.$$

In order to again be able to remove the singularity by means of a suitable multilinear transformation, the integration domain has to be decomposed into

$$D_1 \cup D_2 := \begin{Bmatrix} 0 \leq \hat{z}_1 \leq 1 \\ 0 \leq \hat{z}_2 \leq \hat{z}_1 \\ 0 \leq \hat{z}_3 \leq \hat{z}_1 \\ 0 \leq \hat{z}_4 \leq \hat{z}_3 \end{Bmatrix} \cup \begin{Bmatrix} 0 \leq \hat{z}_3 \leq 1 \\ 0 \leq \hat{z}_4 \leq \hat{z}_3 \\ 0 \leq \hat{z}_1 \leq \hat{z}_3 \\ 0 \leq \hat{z}_2 \leq \hat{z}_1 \end{Bmatrix}.$$

For $i = 1, 2$ the transformations $T_i : (0,1)^4 \to D_i, \hat{\mathbf{z}} = T_i(\xi, \eta)$ are given by

$$T_1(\xi,\eta) := \xi\,(1, \eta_1, \eta_2, \eta_2\eta_3)^\mathsf{T}$$
$$T_2(\xi,\eta) := \xi\,(\eta_2, \eta_2\eta_3, 1, \eta_1)^\mathsf{T}.$$

The absolute value of the Jacobian determinant is in both cases equal to $\xi^3\eta_2$. Thus we have derived the representation

$$I_{\tau\times t} = \int_{(0,1)^4} \xi^3\eta_2 \left\{k_3(\xi, \xi\eta_1, \xi\eta_2, \xi\eta_2\eta_3) + k_3(\xi\eta_2, \xi\eta_2\eta_3, \xi, \xi\eta_1)\right\} d\xi d\eta. \tag{5.39}$$

The following theorem shows that the integrand in (5.39) is analytic with respect to every variable.

Theorem 5.2.9. *The integrand in (5.39) can be extended analytically with respect to every variable in a complex neighborhood of* $[0,1]^4$.

Proof. As before, it suffices to show that the transformation from the (ξ, η)-coordinates to four-dimensional polar coordinates is analytic. We recommend the explicit derivation of the transformation formulas as an exercise for the interested reader. □

5.2.4 Overview: Regularizing Coordinate Transformations

In this subsection we will formulate the regularizing coordinate transformations for all occurring cases in a compact form. We assume that the kernel function satisfies the assumption 5.1.19. For $\tau, t \in \mathcal{G}$, $\chi_\tau : \hat{\tau} \to \tau$ and $\chi_t : \hat{t} \to t$ denote the (analytic) parametrizations over the reference elements. In the case of identical panels we assume $\chi_\tau = \chi_t$, in the case of a common edge we assume $\chi_\tau(s,0) = \chi_t(s,0)$ and in the case of a common point we assume $\chi_\tau(\mathbf{0}) = \chi_t(\mathbf{0})$. The integrand in local coordinates defines

$$k_3(\hat{\mathbf{x}}, \hat{\mathbf{y}}) = \hat{b}_i(\hat{\mathbf{x}})\,\hat{b}_j(\hat{\mathbf{y}})\,k\left(\chi_\tau(\hat{\mathbf{x}}), \chi_t(\hat{\mathbf{y}}), \chi_t(\hat{\mathbf{y}}) - \chi_\tau(\hat{\mathbf{x}})\right) g_\tau(\hat{\mathbf{x}})\, g_t(\hat{\mathbf{y}}),$$

and we set

$$I_{\tau\times t} := \int_{\hat{\tau}} p.v. \int_{\hat{t}} k_3(\hat{\mathbf{x}}, \hat{\mathbf{y}})\, d\hat{\mathbf{y}} d\hat{\mathbf{x}}.$$

The unit square is denoted by $\widehat{Q}$ and the unit triangle by $\widehat{S}$.

I. Identical Panels

I.1 Case $\hat{\tau} = \hat{t} = \widehat{S}$

$$
\begin{aligned}
I_{\tau\times t} = \int_{(0,1)^4} \xi^3 \eta_1^2 \eta_2 \Bigg\{ & k_3\left(\xi\begin{pmatrix}1\\1-\eta_1+\eta_1\eta_2\\1-\eta_1\eta_2\eta_3\\1-\eta_1\end{pmatrix}\right) + k_3\left(\xi\begin{pmatrix}1-\eta_1\eta_2\eta_3\\1-\eta_1\\1\\1-\eta_1+\eta_1\eta_2\end{pmatrix}\right)\\
&+ k_3\left(\xi\begin{pmatrix}1\\\eta_1(1-\eta_2+\eta_2\eta_3)\\1-\eta_1\eta_2\\\eta_1(1-\eta_2)\end{pmatrix}\right) + k_3\left(\xi\begin{pmatrix}1-\eta_1\eta_2\\\eta_1(1-\eta_2)\\1\\\eta_1(1-\eta_2+\eta_2\eta_3)\end{pmatrix}\right)\\
&+ k_3\left(\xi\begin{pmatrix}1-\eta_1\eta_2\eta_3\\\eta_1(1-\eta_2\eta_3)\\1\\\eta_1(1-\eta_2)\end{pmatrix}\right) + k_3\left(\xi\begin{pmatrix}1\\\eta_1(1-\eta_2)\\1-\eta_1\eta_2\eta_3\\\eta_1(1-\eta_2\eta_3)\end{pmatrix}\right)\Bigg\} d\eta_1 d\eta_2 d\eta_3 d\xi.
\end{aligned}
$$

I.2 Case $\hat{\tau} = \hat{t} = \widehat{Q}$

$$
\begin{aligned}
I_{\tau\times t} = \int_{(0,1)^4} \xi(1-\xi)(1-\xi\eta_1) \Bigg\{ & k_3\begin{pmatrix}(1-\xi)\eta_3\\(1-\xi\eta_1)\eta_2\\\xi+(1-\xi)\eta_3\\\xi\eta_1+(1-\xi\eta_1)\eta_2\end{pmatrix} + k_3\begin{pmatrix}(1-\xi\eta_1)\eta_2\\(1-\xi)\eta_3\\\xi\eta_1+(1-\xi\eta_1)\eta_2\\\xi+(1-\xi)\eta_3\end{pmatrix}\\
+k_3\begin{pmatrix}(1-\xi)\eta_3\\\xi\eta_1+(1-\xi\eta_1)\eta_2\\\xi+(1-\xi)\eta_3\\(1-\xi\eta_1)\eta_2\end{pmatrix} &+ k_3\begin{pmatrix}(1-\xi\eta_1)\eta_2\\\xi+(1-\xi)\eta_3\\\xi\eta_1+(1-\xi\eta_1)\eta_2\\(1-\xi)\eta_3\end{pmatrix} + k_3\begin{pmatrix}\xi+(1-\xi)\eta_3\\(1-\xi\eta_1)\eta_2\\(1-\xi)\eta_3\\\xi\eta_1+(1-\xi\eta_1)\eta_2\end{pmatrix}\\
+k_3\begin{pmatrix}\xi\eta_1+(1-\xi\eta_1)\eta_2\\(1-\xi)\eta_3\\(1-\xi\eta_1)\eta_2\\\xi+(1-\xi)\eta_3\end{pmatrix} &+ k_3\begin{pmatrix}\xi+(1-\xi)\eta_3\\\xi\eta_1+(1-\xi\eta_1)\eta_2\\(1-\xi)\eta_3\\(1-\xi\eta_1)\eta_2\end{pmatrix} + k_3\begin{pmatrix}\xi\eta_1+(1-\xi\eta_1)\eta_2\\\xi+(1-\xi)\eta_3\\(1-\xi\eta_1)\eta_2\\(1-\xi)\eta_3\end{pmatrix}\Bigg\} d\eta d\xi.
\end{aligned}
$$

II. Common Edge

II.1 Case $\hat{\tau} = \hat{t} = \widehat{S}$

$$
\begin{aligned}
I_{\tau\times t} = \int_{(0,1)^4} \xi^3\eta_1^2 k_3\begin{pmatrix}\xi\\\xi\eta_1\eta_3\\\xi(1-\eta_1\eta_2)\\\xi\eta_1(1-\eta_2)\end{pmatrix} &+ \xi^3\eta_1^2\eta_2 \Bigg\{ k_3\begin{pmatrix}\xi\\\xi\eta_1\\\xi(1-\eta_1\eta_2\eta_3)\\\xi\eta_1\eta_2(1-\eta_3)\end{pmatrix}\\
+k_3\begin{pmatrix}\xi(1-\eta_1\eta_2)\\\xi\eta_1(1-\eta_2)\\\xi\\\xi\eta_1\eta_2\eta_3\end{pmatrix} + k_3\begin{pmatrix}\xi(1-\eta_1\eta_2\eta_3)\\\xi\eta_1\eta_2(1-\eta_3)\\\xi\\\xi\eta_1\end{pmatrix} &+ k_3\begin{pmatrix}\xi(1-\eta_1\eta_2\eta_3)\\\xi\eta_1(1-\eta_2\eta_3)\\\xi\\\xi\eta_1\eta_2\end{pmatrix}\Bigg\} d\eta d\xi.
\end{aligned}
$$

II.2 Case $\hat{\tau} = \widehat{Q}, \quad \hat{t} = \widehat{S}$

$$I_{\tau\times t} = \int_{(0,1)^4} \xi^2(1-\xi)\left\{k_3\begin{pmatrix}\xi(1-\eta_3)+\eta_3\\ \xi\eta_2\\ \xi(1-\eta_1-\eta_3)+\eta_3\\ \xi(1-\eta_1)\end{pmatrix} + k_3\begin{pmatrix}(1-\xi)\eta_3\\ \xi\eta_2\\ \xi(1-\eta_3)+\eta_3\\ \xi\eta_1\end{pmatrix}\right.$$

$$\left. +k_3\begin{pmatrix}\xi(1-\eta_1-\eta_3)+\eta_3\\ \xi\eta_2\\ \xi(1-\eta_3)+\eta_3\\ \xi\end{pmatrix}\right\} + \xi^2\eta_1(1-\xi\eta_1)\left\{k_3\begin{pmatrix}\xi\eta_1(1-\eta_3)+\eta_3\\ \xi\\ \xi\eta_1(1-\eta_2-\eta_3)+\eta_3\\ \xi\eta_1(1-\eta_2)\end{pmatrix}\right.$$

$$\left. +k_3\begin{pmatrix}(1-\xi\eta_1)\eta_3\\ \xi\\ \xi\eta_1+(1-\xi\eta_1)\eta_3\\ \xi\eta_1\eta_2\end{pmatrix} + k_3\begin{pmatrix}\xi\eta_1(1-\eta_2-\eta_3)+\eta_3\\ \xi\\ \xi\eta_1(1-\eta_3)+\eta_3\\ \xi\eta_1\end{pmatrix}\right\} d\eta\, d\xi.$$

II.3 Case $\hat{\tau} = \widehat{S}, \quad \hat{t} = \widehat{Q}$

$$I_{\tau\times t} = \int_{(0,1)^4} \xi^2(1-\xi)\left\{k_3\begin{pmatrix}\xi(1-\eta_1-\eta_3)+\eta_3\\ \xi(1-\eta_1)\\ \xi(1-\eta_3)+\eta_3\\ \xi\eta_2\end{pmatrix} + k_3\begin{pmatrix}\xi(1-\eta_3)+\eta_3\\ \xi\eta_1\\ (1-\xi)\eta_3\\ \xi\eta_2\end{pmatrix}\right.$$

$$\left. +k_3\begin{pmatrix}\xi(1-\eta_3)+\eta_3\\ \xi\\ \xi(1-\eta_1-\eta_3)+\eta_3\\ \xi\eta_2\end{pmatrix}\right\} + \xi^2\eta_1(1-\xi\eta_1)\left\{k_3\begin{pmatrix}\xi\eta_1(1-\eta_2-\eta_3)+\eta_3\\ \xi\eta_1(1-\eta_2)\\ \xi\eta_1(1-\eta_3)+\eta_3\\ \xi\end{pmatrix}\right.$$

$$\left. +k_3\begin{pmatrix}\xi\eta_1+(1-\xi\eta_1)\eta_3\\ \xi\eta_1\eta_2\\ (1-\xi\eta_1)\eta_3\\ \xi\end{pmatrix} + k_3\begin{pmatrix}\xi\eta_1(1-\eta_3)+\eta_3\\ \xi\eta_1\\ \xi\eta_1(1-\eta_2-\eta_3)+\eta_3\\ \xi\end{pmatrix}\right\} d\eta\, d\xi.$$

II.4 Case $\hat{\tau} = \hat{t} = \widehat{Q}$

$$I_{\tau\times t} = \int_{(0,1)^4} \xi^2(1-\xi)\left\{k_3\begin{pmatrix}(1-\xi)\eta_3+\xi\\ \xi\eta_2\\ (1-\xi)\eta_3\\ \xi\eta_1\end{pmatrix} + k_3\begin{pmatrix}(1-\xi)\eta_3\\ \xi\eta_2\\ \xi+(1-\xi)\eta_3\\ \xi\eta_1\end{pmatrix}\right\}$$

$$+\xi^2(1-\xi\eta_1)\left\{k_3\begin{pmatrix}(1-\xi\eta_1)\eta_3+\xi\eta_1\\ \xi\eta_2\\ (1-\xi\eta_1)\eta_3\\ \xi\end{pmatrix}+k_3\begin{pmatrix}(1-\xi\eta_1)\eta_3+\xi\eta_1\\ \xi\\ (1-\xi\eta_1)\eta_3\\ \xi\eta_2\end{pmatrix}\right.$$

$$\left.+k_3\begin{pmatrix}(1-\xi\eta_1)\eta_3\\ \xi\eta_2\\ (1-\xi\eta_1)\eta_3+\xi\eta_1\\ \xi\end{pmatrix}+k_3\begin{pmatrix}(1-\xi\eta_1)\eta_3\\ \xi\\ (1-\xi\eta_1)\eta_3+\xi\eta_1\\ \xi\eta_2\end{pmatrix}\right\}d\eta d\xi.$$

III. Common Vertex

III.1 Case $\hat{\tau}=\hat{t}=\widehat{S}$

$$I_{\tau\times t}=\int_{(0,1)^4}\xi^3\eta_2\left\{k_3(\xi,\xi\eta_1,\xi\eta_2,\xi\eta_2\eta_3)+k_3(\xi\eta_2,\xi\eta_2\eta_3,\xi,\xi\eta_1)\right\}d\eta d\xi.$$

III.2 Case $\hat{\tau}=\widehat{Q},\quad \hat{t}=\widehat{S}$

$$I_{\tau\times t}=\int_{(0,1)^4}\xi^3\eta_2\left\{k_3\begin{pmatrix}\xi\\ \xi\eta_1\\ \xi\eta_2\\ \xi\eta_2\eta_3\end{pmatrix}+k_3\begin{pmatrix}\xi\eta_1\\ \xi\\ \xi\eta_2\\ \xi\eta_2\eta_3\end{pmatrix}\right\}+\xi^3k_3\begin{pmatrix}\xi\eta_1\\ \xi\eta_2\\ \xi\\ \xi\eta_3\end{pmatrix}d\eta d\xi.$$

III.3 Case $\hat{\tau}=\widehat{S},\quad \hat{t}=\widehat{Q}$

$$I_{\tau\times t}=\int_{(0,1)^4}\xi^3\eta_2\left\{k_3\begin{pmatrix}\xi\eta_2\\ \xi\eta_2\eta_3\\ \xi\\ \xi\eta_1\end{pmatrix}+k_3\begin{pmatrix}\xi\eta_2\\ \xi\eta_2\eta_3\\ \xi\eta_1\\ \xi\end{pmatrix}\right\}+\xi^3k_3\begin{pmatrix}\xi\\ \xi\eta_3\\ \xi\eta_1\\ \xi\eta_2\end{pmatrix}d\eta d\xi.$$

III.4 Case $\hat{\tau}=\hat{t}=\widehat{Q}$

$$I_{\tau\times t}=\int_{(0,1)^4}\xi^3\left\{k_3\begin{pmatrix}\xi\\ \xi\eta_1\\ \xi\eta_2\\ \xi\eta_3\end{pmatrix}+k_3\begin{pmatrix}\xi\eta_1\\ \xi\\ \xi\eta_2\\ \xi\eta_3\end{pmatrix}+k_3\begin{pmatrix}\xi\eta_1\\ \xi\eta_2\\ \xi\\ \xi\eta_3\end{pmatrix}+k_3\begin{pmatrix}\xi\eta_1\\ \xi\eta_2\\ \xi\eta_3\\ \xi\end{pmatrix}\right\}d\eta d\xi.$$

5.2.5 Evaluating the Right-Hand Side and the Integral-Free Term

In the following we will briefly deal with the approximation of the integrals

$$\int_\Gamma b_i(\mathbf{x})\, b_j(\mathbf{x})\, ds_{\mathbf{x}}, \qquad \int_\Gamma b_i(\mathbf{x})\, r(\mathbf{x})\, ds_{\mathbf{x}} \tag{5.40}$$

[see (5.1)]. Let $\hat{\tau}$ be the reference element, χ_τ the parametrization and $\hat{b}_i := b_i|_\tau \circ \chi_\tau$, $\hat{r}_\tau := r|_\tau \circ \chi_\tau$. Then

$$\int_\Gamma b_i(\mathbf{x})\, b_j(\mathbf{x})\, ds_{\mathbf{x}} = \sum_{\tau\in\mathcal{G}} \int_{\hat{\tau}} \hat{b}_i(\hat{\mathbf{x}})\, \hat{b}_j(\hat{\mathbf{x}})\, g_\tau(\hat{\mathbf{x}})\, d\hat{\mathbf{x}}.$$

For the unit triangle we transform the integral over $\hat{\tau}$ onto the unit square by means of $\hat{\mathbf{x}} = (\xi, \xi\eta)$ and obtain

$$\int_{\hat{\tau}} \hat{b}_i(\hat{\mathbf{x}})\, \hat{b}_j(\hat{\mathbf{x}})\, g_\tau(\hat{\mathbf{x}})\, d\hat{\mathbf{x}} = \int_0^1 \int_0^1 \xi \hat{b}_i\big(\begin{smallmatrix}\xi\\ \xi\eta\end{smallmatrix}\big) \hat{b}_j\big(\begin{smallmatrix}\xi\\ \xi\eta\end{smallmatrix}\big) g_\tau\big(\begin{smallmatrix}\xi\\ \xi\eta\end{smallmatrix}\big) d\xi d\eta. \tag{5.41}$$

The integrand on the right-hand integral is analytic, since the basis functions in ξ, η-coordinates are polynomials and, thus, analytic. We have already shown in Sect. 5.1, proof of Theorem 5.1.22, that the surface element $g_\tau(\xi, \xi\eta)$ is analytic.

In the same way we have under the condition $r \in L^2(\Gamma)$ the representation

$$\int_\Gamma b_i(\mathbf{x})\, r(\mathbf{x})\, ds_{\mathbf{x}} = \sum_{\tau\in\mathcal{G}} \int_{\hat{\tau}} \hat{b}_i(\hat{\mathbf{x}})\, \hat{r}_\tau(\hat{\mathbf{x}})\, g_\tau(\hat{\mathbf{x}})\, d\hat{\mathbf{x}}$$

for the integral in (5.40) with

$$\int_{\hat{\tau}} \hat{b}_i(\hat{\mathbf{x}})\, \hat{r}_\tau(\hat{\mathbf{x}})\, g_\tau(\hat{\mathbf{x}})\, d\hat{\mathbf{x}} = \int_0^1 \int_0^1 \xi \hat{b}_i\big(\begin{smallmatrix}\xi\\ \xi\eta\end{smallmatrix}\big) \hat{r}_\tau\big(\begin{smallmatrix}\xi\\ \xi\eta\end{smallmatrix}\big) g_\tau\big(\begin{smallmatrix}\xi\\ \xi\eta\end{smallmatrix}\big) d\xi d\eta. \tag{5.42}$$

If $\hat{r}_\tau(\xi, \xi\eta)$ can be analytically extended to a neighborhood of the unit square then, using the same arguments as before, the integrand on the right-hand integral is analytic. However, in practical applications there are also cases in which the right-hand side is not so smooth or even has singularities. In this case one should use adaptive numerical quadrature methods that are able to suitably resolve the singular behavior of the function $\hat{r}_\tau$. Since such methods depend very much on the actual function $\hat{r}_\tau$ under consideration, we will refrain from a general description of these methods and refer to [225] for an introduction to the problem.

Remark 5.2.10. *If the reference element is the unit square the transformation* $\hat{\mathbf{x}} = (\xi, \xi\eta)$ *becomes superfluous. The results concerning analytic properties for the local integrand can be appropriately applied.*

If the direct method is used to formulate the boundary value problem as an integral equation the right-hand side r is usually defined by an integral operator

$$r = \lambda_2 f + K_2 f,$$

and therefore integrals of the form

$$\int_\Gamma b_i(\mathbf{x})\, r(\mathbf{x})\, ds_{\mathbf{x}} = \int_\Gamma \lambda_2(\mathbf{x})\, b_i(\mathbf{x})\, f(\mathbf{x})\, ds_{\mathbf{x}} + \int_\Gamma b_i(\mathbf{x}) \int_\Gamma k_2(\mathbf{x}, \mathbf{y}, \mathbf{x}-\mathbf{y})\, f(\mathbf{y})\, ds_{\mathbf{y}} ds_{\mathbf{x}} \qquad (5.43)$$

have to be evaluated. However, both integrals are of the type (5.40) and (5.2) and can thus be regularized and approximated by the same techniques, assuming that the function f is piecewise analytic. Otherwise adaptive methods have to be employed that take into account the singular behavior of f.

5.3 Numerical Integration

We have shown that the coefficients of the system matrix and the right-hand side can be formulated as integrals over $[0,1]^4$ with an analytic integrand. Such integrals can be efficiently approximated by means of Gaussian quadrature methods (see [225]). In this section we will present the appropriate Tensor-Gauss quadrature for the approximation of these integrals. We will also estimate the minimal number of quadrature points per spatial dimension in order to reach a given approximation tolerance. We will later see that the quadrature order for some integrals has to be chosen proportional to $|\log h|$, i.e., the quadrature order goes to infinity for $h \to 0$. Therefore the error estimates have to depend *explicitly* on not only the mesh width h but also the quadrature order.

The integrals in (5.1) are approximated with quadrature methods of *fixed* order and so allow the use of simple quadrature methods.

Remark 5.3.1. *If the continuous integral operators lead to symmetric bilinear forms or Hermitian sesquilinear forms then the (exact) system matrices* $\mathbf{K}$ *of the Galerkin discretization are, respectively, symmetric or Hermitian. Hence it suffices to approximate and save the upper triangular part of the matrix* $\mathbf{K}$ *by numerical quadrature. Since the symmetry of system matrices is often an essential factor for the convergence of iterative solution methods, this method also automatically guarantees the symmetry of perturbed system matrices.*

5.3.1 Numerical Quadrature Methods

In this subsection we will present simple quadrature methods for squares and triangles, as well as Gaussian quadrature methods of arbitrary order.

5.3.1.1 Simple Quadrature Methods

For $\tau \in \mathcal{G}$, $\hat{\tau}$ denotes the reference element and $\chi_\tau : \hat{\tau} \to \tau$ the local parametrization. The integration of a continuous function $v \in C^0(\overline{\tau})$ is transported back to the reference element as

$$I_\tau(v) := \int_\tau v(\mathbf{x})\, d\mathbf{x} = \int_{\hat{\tau}} \hat{v}(\hat{\mathbf{x}})\, g_\tau(\hat{\mathbf{x}})\, d\hat{\mathbf{x}}.$$

Here $g_\tau(\hat{\mathbf{x}})$ denotes the surface element and $\hat{v} := v|_\tau \circ \chi_\tau$. The numerical quadrature on the *reference element* is given by a mapping $Q : C^0\left(\overline{\hat{\tau}}\right) \to \mathbb{K}$ of the form

$$Q(v) := \sum_{i=1}^{n} \omega_{i,n} v(\xi_{i,n})$$

with weights $\omega_{i,n} \in \mathbb{R}$ and nodes $\xi_{i,n} \in \overline{\hat{\tau}}$. The associated quadrature error $E_\tau : C^0(\overline{\tau}) \to \mathbb{K}$ on the *surface panel* is given by

$$E_\tau(v) := I_\tau(v) - Q(\hat{v} g_\tau). \tag{5.44}$$

The space of all polynomials of maximal degree $m \in \mathbb{N}$ was introduced in (4.23) and denoted by $\mathbb{P}_m^\Delta$. In this section we will use the abbreviation $\mathbb{P}_m = \mathbb{P}_m^\Delta$.

Definition 5.3.2. The numerical quadrature has the degree of exactness $m \in \mathbb{N}_0$ if the quadrature method on the reference element is exact for all polynomials of maximal degree m, i.e.,

$$E_{\hat{\tau}}(v) = 0 \qquad \forall v \in \mathbb{P}_m.$$

The numerical quadrature is stable if

$$\sum_{i=1}^{n} \omega_{i,n} = |\hat{\tau}| \qquad \text{and} \qquad \sum_{i=1}^{n} |\omega_{i,n}| \leq C_Q \sum_{i=1}^{n} \omega_{i,n}.$$

Example 5.3.3. *Let $\hat{\tau}$ be the unit triangle with vertices $(0,0)^\mathsf{T}$, $(1,0)^\mathsf{T}$, $(1,1)^\mathsf{T}$. Then*

$$Q(v) = \frac{v(2/3, 1/3)}{2}$$

defines a quadrature formula with exactness degree 1 and $C_Q = 1$.

A quadrature formula with exactness degree 2 and $C_Q = 1$ is given by

$$Q(v) = \frac{v(1/2,0) + v(1,1/2) + v(1/2,1/2)}{6}.$$

Example 5.3.4. *Let $\hat{\tau}$ be the unit square. Then*

$$Q(v) = v(1/2, 1/2)$$

defines a quadrature formula with exactness degree 1 *and* $C_Q = 1$. *Also*

$$Q(v) = \frac{1}{4}\sum_{i,j=1}^{2} v(\xi_i, \xi_j)$$

with $\xi_1 = \left(1 - 1/\sqrt{3}\right)/2$ *and* $\xi_2 = \left(1 + 1/\sqrt{3}\right)/2$ *defines a quadrature formula with exactness degree* 3 *and* $C_Q = 1$.

Further quadrature formulas for the unit triangle and square can be found in [226].

5.3.1.2 Tensor-Gauss Quadrature

For a continuous function $f : [0,1] \to \mathbb{C}$ we set

$$I(f) = \int_0^1 f dx.$$

Let $(\xi_{i,n}, \omega_{i,n})_{i=1}^{n}$ be the nodes and weights of the Gaussian quadrature of order n with the weight function 1 on the interval $[0,1]$ (see [225]). The associated Gaussian quadrature is given by

$$Q^n(f) = \sum_{i=1}^{n} \omega_{i,n} f(\xi_{i,n}).$$

The quadrature error is denoted by $E^n(f) := I(f) - Q^n(f)$ and satisfies

$$E^n(p) = 0 \qquad \forall p \in \mathbb{P}_{2n-1}.$$

For a function $f : [0,1]^4 \to \mathbb{C}$ we set $\mathbf{I}(f) = \int_{(0,1)^4} f(\mathbf{x})\, d\mathbf{x}$ and for $\mathbf{n} = (n_i)_{i=1}^4 \in \mathbb{N}^4$ we define the Tensor-Gauss quadrature of order $\mathbf{n}$ by

$$\mathbf{Q}^{\mathbf{n}}[f] = \sum_{i=1}^{n_1}\sum_{j=1}^{n_2}\sum_{k=1}^{n_3}\sum_{\ell=1}^{n_4} \omega_{i,n_1}\omega_{j,n_2}\omega_{k,n_3}\omega_{\ell,n_4} f\left(\xi_{i,n_1}, \xi_{j,n_2}, \xi_{k,n_3}, \xi_{\ell,n_4}\right). \tag{5.45}$$

Since all the derived integral representations from the previous section are of the form $\mathbf{I}(f)$ with an analytic integrand $f : [0,1]^4 \to \mathbb{C}$, the Tensor-Gauss quadrature can be used for their approximation. As the numerical integration takes up a significant part of the total computation time used for the numerical solution of boundary integral equations, it is very important that a minimal quadrature order $\mathbf{n} := (n_i)_{i=1}^4$ be found to achieve a prescribed tolerance $\varepsilon > 0$. The condition on $\mathbf{n}$ reads

$$|\mathbf{E}^n(f)| := |\mathbf{I}(f) - \mathbf{Q}^{\mathbf{n}}(f)| \le \varepsilon \|f\|. \tag{5.46}$$

The nodes and weights of the Gaussian formulas up to a high order can be precomputed, for example, with the program GAULEG described in [185, Sect. 4.5]. The output of the procedure "gauleg($a, b, \mathbf{x}, \mathbf{w}, n$)" consists of the arrays $\mathbf{x}(1 : n)$, $\mathbf{w}(1 : n)$ of the nodes and weights of the n-point Gaussian formula Q^n on the interval $[a, b]$.

5.3.2 *Local Quadrature Error Estimates*

We begin with quadrature error estimates for stable quadrature methods with an exactness degree m. We will prove the convergence for $h_\tau \to 0$ and a *fixed* degree of exactness. These can be used for the approximation of the integrals in (5.1).

The integrands in the regularized integral representations (see Sect. 5.2.4) are all analytic. However, they do have poles close to (more specifically: in a complex neighborhood of) the integration domain. In practical applications the occurring kernel functions and their derivatives in local coordinates often have a very complicated form, so that the quadrature error estimates, which often contain high order derivatives of the integrand, are not suitable. We therefore use *derivative free* error representations for analytic integrands for the estimation of the local quadrature error. These are explicit with respect to the order.

5.3.2.1 Local Error Estimates for Simple Quadrature Methods

We will analyze the question whether the parametrization χ_t can be analytically extended, which will form the basis for this discussion. We will restrict ourselves to triangular meshes. The analysis for squares can be performed in a similar way. In order to explicitly analyze the scaling of the size of the triangles we need to impose suitable conditions on the parametrization.

Assumption 5.3.5. *For every $\tau \in \mathcal{G}$ the parametrization χ_τ can be represented as the composition of an affine mapping*

$$\chi_\tau^{\text{affine}} : \mathbb{R}^2 \to \mathbb{R}^2, \qquad \chi_\tau^{\text{affine}}(\hat{\mathbf{x}}) := \mathbf{A}_\tau + \mathbf{m}_\tau \hat{\mathbf{x}}$$

and a mapping $\chi : U \to \Gamma$ with $\overline{\chi_\tau^{\text{affine}}(\hat{\tau})} \subset U$,

$$\chi_\tau = \chi \circ \chi_\tau^{\text{affine}}.$$

The image $\tilde{\tau} := \chi_\tau^{\text{affine}}(\hat{\tau})$ is a plane triangle in $\mathbb{R}^2$. The mapping χ can be extended analytically in a complex neighborhood $U^\star$ with $\overline{\tilde{\tau}} \subset \overline{U} \subset U^\star \subset \mathbb{C} \times \mathbb{C}$ and, in particular, independent of the triangulation $\mathcal{G}$.

There exists a positive constant C_K with the property: For all $\tau, t \in \mathcal{G}$ with $\overline{\tau} \cap \overline{t} \neq \emptyset$ we have

$$h_\tau / h_t \leq C_K \qquad \textit{with} \qquad h_\tau := \operatorname{diam} \tau, \quad h_t := \operatorname{diam} t. \tag{5.47}$$

Remark 5.3.6. *The mapping $\chi : U \to \Gamma$ can be considered as a chart in an atlas $\mathcal{A}$ for Γ. The choice of the chart from $\mathcal{A}$ depends on the panel $\tau \in \mathcal{G}$ through $\tau \subset \chi(U)$. The chart χ itself, however, is independent of τ.*

Remark 5.3.7. *From Condition (5.47) we have the existence of a constant $c_1 > 0$ that depends only on χ so that*

$$c_1 h_\tau \le h_{\tilde{\tau}} \le c_1^{-1} h_\tau \tag{5.48}$$

for the plane panels $\tilde{\tau} := \chi_\tau^{\text{affine}} \circ \chi_\tau^{-1}(\tau)$ and $\tilde{t} := \chi_t^{\text{affine}} \circ \chi_t^{-1}(t)$. Furthermore, it follows that

$$h_{\tilde{\tau}} / h_{\tilde{t}} \le c_1^{-2} h_\tau / h_t \le \widetilde{C}_{\mathrm{K}}$$

with $\widetilde{C}_{\mathrm{K}} = c_1^{-2} C_{\mathrm{K}}$ (see Remark 4.1.11).

The scaling and deforming behavior of the affine mapping $\chi_\tau^{\text{affine}}$ will be characterized in the following by suitable geometric parameters. The vertices of the image triangle $\tilde{\tau} = \chi_\tau^{\text{affine}}(\hat{\tau})$ are denoted by $\mathbf{A}_\tau$, $\mathbf{B}_\tau$, $\mathbf{C}_\tau$ and have a counterclockwise orientation. We call the associated interior angles $\alpha_\tau, \beta_\tau, \gamma_\tau$. Then $\mathbf{m}_\tau = [\mathbf{B}_\tau - \mathbf{A}_\tau, \mathbf{C}_\tau - \mathbf{B}_\tau]$. We set

$$\theta_\tau := \min\{\alpha_\tau, \beta_\tau, \gamma_\tau\}. \tag{5.49}$$

Proposition 5.3.8. *Let $\mathbf{m}_\tau$ be as in Assumption 5.3.5 and let $\lambda_{\max}$ ($\lambda_{\min}$) be the largest (smallest) eigenvalue of $\mathbf{m}_\tau^\intercal \mathbf{m}_\tau$. Then*

$$c h_{\tilde{\tau}}^2 \le \lambda_{\min} \le \lambda_{\max} \le 2 h_{\tilde{\tau}}^2 \quad \text{and} \quad c h_{\tilde{\tau}}^2 \le g_\tau^{\text{affine}} := \sqrt{\det\left(\mathbf{m}_\tau^\intercal \mathbf{m}_\tau\right)} \le C h_{\tilde{\tau}}^2$$

with constants $c, C > 0$ that depend only on θ_τ.

Proof. We set $e_1 := \|\mathbf{B}_\tau - \mathbf{A}_\tau\|$ and $e_2 = \|\mathbf{C}_\tau - \mathbf{B}_\tau\|$ and note that $e_i \le h_{\tilde{\tau}}$, $1 \le i \le 2$. The upper bound for the singular values of $\mathbf{m}_\tau$ results from

$$\begin{aligned}
|\langle \mathbf{m}_\tau \xi, \mathbf{m}_\tau \xi \rangle| &= \xi_1^2 e_1^2 + 2\xi_1 \xi_2 \langle \mathbf{B}_\tau - \mathbf{A}_\tau, \mathbf{C}_\tau - \mathbf{B}_\tau \rangle + \xi_2^2 e_2^2 \\
&\le (|\xi_1| e_1 + |\xi_2| e_2)^2 \le 2 h_{\tilde{\tau}}^2 \|\xi\|^2 .
\end{aligned}$$

Elementary geometric relations on $\tilde{\tau}$ and the binomial formula yield

$$\begin{aligned}
\frac{\langle \mathbf{m}_\tau \xi, \mathbf{m}_\tau \xi \rangle}{e_1 e_2} &= \xi_1^2 \frac{e_1}{e_2} - 2\xi_1 \xi_2 \cos\beta + \xi_2^2 \frac{e_2}{e_1} \\
&\ge (1 - \cos\beta) \left(\xi_1^2 \frac{e_1}{e_2} + \xi_2^2 \frac{e_2}{e_1} \right) \ge 2 \sin^2 \frac{\beta}{2} \min\left\{ \frac{e_1}{e_2}, \frac{e_2}{e_1} \right\} \|\xi\|^2 .
\end{aligned}$$

The estimate for the eigenvalues finally follows from

$$\sin^2\frac{\beta}{2} \geq \sin^2\frac{\theta_\tau}{2} \quad \text{and} \quad \min\left\{\frac{e_1}{e_2}, \frac{e_2}{e_1}\right\} = \min\left\{\frac{\sin\gamma}{\sin\alpha}, \frac{\sin\alpha}{\sin\gamma}\right\} \geq \sin\theta_\tau.$$

The estimate for g_τ^{affine} is a result of the representation $g_\tau^{\text{affine}} = e_1 e_2 |\sin\beta|$. □

We consider stable quadrature methods with a degree of exactness m. The stability implies that for all $f \in C^0\left(\hat{\tau}\right)$

$$|Q(f)| = \left|\sum_{i=1}^{n} \omega_{i,n} f(\xi_{i,n})\right| \leq \max_{1\leq i\leq n} |f(\xi_{i,n})| \sum_{i=1}^{n} |\omega_{i,n}| \leq C_Q \|f\|_{C^0(\hat{\tau})}.$$

If we apply this estimate to the product $g_\tau \hat{v}$ we obtain

$$|Q(g_\tau \hat{v})| \leq C_Q \|g_\tau \hat{v}\|_{C^0(\hat{\tau})} \leq C_Q \|g_\tau\|_{C^0(\hat{\tau})} \|\hat{v}\|_{C^0(\hat{\tau})}.$$

Lemma 5.3.9. *Let Assumption 5.3.5 be satisfied, i.e., $\chi_\tau = \chi \circ \chi_\tau^{\text{affine}}$. Then*

$$\|g_\tau\|_{C^0(\hat{\tau})} \leq Ch_\tau^2,$$

where C depends only on the global parametrization and the quantity θ_τ from (5.49).

Proof. The multiplication theorem for determinants yields

$$\begin{aligned} g_\tau(\hat{\mathbf{x}}) &= \sqrt{\det\left((D\chi_\tau)^{\mathsf{T}}(D\chi_\tau)\right)} = |\det \mathbf{m}_\tau| \sqrt{\det\left(\mathbf{G}_\tau \circ \chi_\tau^{\text{affine}}(\hat{\mathbf{x}})\right)} \\ &= |\det \mathbf{m}_\tau| \left(g_\tau^\chi \circ \chi_\tau^{\text{affine}}\right)(\hat{\mathbf{x}}) \end{aligned} \tag{5.50}$$

with the Gram matrix

$$\mathbf{G}_\tau(\hat{\mathbf{w}}) := (D\chi(\hat{\mathbf{w}}))^{\mathsf{T}} D\chi(\hat{\mathbf{w}}) \in \mathbb{R}^{2\times 2},$$

which depends only on the global chart χ and $g_\tau^\chi := \sqrt{\det \mathbf{G}_\tau}$. From this we have

$$\|g_\tau\|_{C^0(\hat{\tau})} \leq C_\chi |\det \mathbf{m}_\tau| \leq Ch_\tau^2,$$

where C depends only on the global parametrization and the quantity θ_τ from (5.49). □

Corollary 5.3.10. *Let Assumption 5.3.5 be satisfied. Then the stability estimate*

$$|E_\tau(v)| \leq Ch_\tau^2 \|v\|_{C^0(\overline{\tau})} \qquad \forall v \in C^0(\overline{\tau})$$

holds for the error E_τ from (5.44).

Proof. The triangle inequality combined with Lemma 5.3.9 yields

$$\begin{aligned}|E_\tau(v)| &= \left|\int_\tau v(\mathbf{x})\,d\mathbf{x} - Q(\hat{v}g_\tau)\right| \le |\tau|\,\|v\|_{C^0(\overline{\tau})} + \left|\sum_{i=1}^{n} \omega_{i,n}(\hat{v}g_\tau)(\xi_{i,n})\right| \\ &\le C_1 h_\tau^2 \|v\|_{C^0(\overline{\tau})} + C_Q C_2 h_\tau^2 \|\hat{v}\|_{C^0(\overline{\hat{\tau}})} = \left(C_1 + C_Q C_2\right) h_\tau^2 \|v\|_{C^0(\overline{\tau})}.\end{aligned}$$

□

We now turn to the main error estimate for stable quadrature methods with an exactness degree m.

Theorem 5.3.11. *Let Assumption 5.3.5 be satisfied, i.e., $\chi_\tau = \chi \circ \chi_\tau^{\text{affine}}$. Let the quadrature method be stable and let it have the degree of exactness m.*

Then there exists a constant C, which depends on C_Q, m, the global parametrization χ and θ_τ from (5.49), such that for all functions $v \in H^{m^+}(\tau)$ with $m^+ = \max\{2, m+1\}$ the quadrature error satisfies the estimate

$$|E_\tau(v)| \le C h_\tau^{m+2} \|v\|_{H^{m^+}(\tau)}.$$

Proof. Let $\hat{v} := v \circ \chi_\tau$ and observe that due to the degree of exactness we have

$$E_\tau(v) = E_{\hat{\tau}}(\hat{v}g_\tau) = E_{\hat{\tau}}(\hat{v}g_\tau - p) \qquad \forall p \in \mathbb{P}_m.$$

Assumption 5.3.5 and the Sobolev embedding theorem imply $\hat{v}g_\tau \in C^0(\hat{\tau})$.

First let $m \ge 1$. We choose $p := \left(\widehat{I}^m(\hat{v}g_\tau)\right) \circ \chi_\tau^{-1}$ with the interpolation $\widehat{I}^m$ from (4.73) and obtain

$$\begin{aligned}|E_{\hat{\tau}}(\hat{v}g_\tau - p)| &\overset{\text{Corollary 5.3.10}}{\le} C\left\|\hat{v}g_\tau - \widehat{I}^m(\hat{v}g_\tau)\right\|_{C^0(\hat{\tau})} \le C\left\|\hat{v}g_\tau - \widehat{I}^m(\hat{v}g_\tau)\right\|_{H^{m+1}(\hat{\tau})} \\ &\overset{\text{Lemma 4.3.1}}{\le} C\left|\hat{v}g_\tau - \widehat{I}^m(\hat{v}g_\tau)\right|_{H^{m+1}(\hat{\tau})} \le C\,|\hat{v}g_\tau|_{H^{m+1}(\hat{\tau})}.\end{aligned}$$

The Leibniz rule for products yields

$$\begin{aligned}|\hat{v}g_\tau|^2_{H^{m+1}(\hat{\tau})} &= \sum_{|\alpha|=m+1} \|\partial^\alpha(\hat{v}g_\tau)\|^2_{L^2(\hat{\tau})} = \sum_{|\alpha|=m+1}\left\|\sum_{\mu\le\alpha}\binom{\alpha}{\mu}(\partial^\mu\hat{v})(\partial^{\alpha-\mu}g_\tau)\right\|_{L^2(\hat{\tau})} \\ &\le C_m \sum_{r=0}^{m+1} |\hat{v}|_{H^r(\hat{\tau})}\,|g_\tau|_{C^{m+1-r}(\hat{\tau})} \\ &\overset{(4.204b),\,(4.215b)}{\le} C_m \|v\|_{H^{m+1}(\tau)} \sum_{r=0}^{m+1} h_\tau^{r-1}\,|g_\tau|_{C^{m+1-r}(\hat{\tau})},\end{aligned}$$

where

$$\binom{\alpha}{\mu} := \binom{\alpha_1}{\mu_1}\binom{\alpha_2}{\mu_2} \quad \text{and} \quad \sum_{\mu\leq\alpha} \ldots := \sum_{\substack{\mu_1\leq\alpha_1\\ \mu_2\leq\alpha_2}} \ldots .$$

As in formula (5.50) we have

$$g_\tau = |\det \mathbf{m}_\tau|\, g_\tau^\chi \circ \chi_\tau^{\text{affine}}$$

and Corollary 4.3.7 leads to

$$|g_\tau|_{C^{m+1-r}(\hat{\tau})} \leq C h_\tau^{m+3-r},$$

where the constant C depends only on m, χ, and the shape-regularity of the panel. Thus

$$|\hat{v} g_\tau|^2_{H^{m+1}(\hat{\tau})} \leq C h_\tau^{m+2} \|v\|_{H^{m+1}(\tau)}$$

and, all in all we have proved that

$$|E_\tau(v)| \leq C h_\tau^{m+2} \|v\|_{H^{m+1}(\tau)}.$$

For $m = 0$ we replace the interpolation of $\widehat{I}^m(\hat{v} g_\tau)$ by the integral mean of $\hat{v} g_\tau$ and apply Poincaré's inequality (cf. Corollary 2.5.12). □

Corollary 5.3.12. *Let the assumptions from Theorem 5.3.11 be satisfied and let $r \in C^{m^+}(\overline{\tau})$ and $v \in S$ with local polynomial degree p. Then*

$$|E_\tau(vr)| \leq C h_\tau^{m+2} \|v\|_{H^p(\tau)} \|r\|_{C^{m^+}(\tau)}. \tag{5.51a}$$

For two boundary element functions $u, v \in S$ we have

$$|E_\tau(uv)| \leq C h_\tau^{m+1} \|u\|_{H^p(\tau)} \|v\|_{H^p(\tau)}. \tag{5.51b}$$

Proof. By Theorem 5.3.11 it suffices to prove $\|vr\|_{H^{m^+}(\tau)} \leq C \|v\|_{H^p(\tau)} \times \|r\|_{C^{m^+}(\tau)}$. We set $\tilde{v} = v \circ \chi^{-1}$, $\tilde{r} = r \circ \chi^{-1}$ and $\tilde{\tau} := \chi^{-1}(\tau) \subset \mathbb{R}^2$. Note that the mapping χ^{-1} is independent of the mesh $\mathcal{G}$. It follows from Corollary 4.3.12 that

$$\|vr\|^2_{H^{m^+}(\tau)} \leq C \sum_{j=1}^{m^+} |\tilde{v}\tilde{r}|^2_{H^j(\tilde{\tau})}.$$

The Leibniz rule for products yields

$$D^\alpha(\tilde{v}\tilde{r}) = \sum_{\mu\leq\alpha} \binom{\alpha}{\mu} (D^\mu \tilde{v})(D^{\alpha-\mu}\tilde{r}).$$

With $j = |\alpha|$ we then have the estimate

$$\|D^{\alpha}(\tilde{v}\tilde{r})\|_{L^2(\tilde{\tau})} \leq C\,\|\tilde{v}\|_{H^j(\tilde{\tau})}\,\|\tilde{r}\|_{C^j(\tilde{\tau})}\,.$$

Since $\tilde{v}$ is a polynomial of maximal degree p, we have $\|\tilde{v}\|_{H^k(\tilde{\tau})} = 0$ for $k > p$ and hence

$$\|\tilde{v}\|_{H^k(\tilde{\tau})} = \|\tilde{v}\|_{H^p(\tilde{\tau})}\,.$$

All in all, we have shown that

$$\|\tilde{v}\tilde{r}\|_{H^{m+}(\tilde{\tau})} \leq C\,\|\tilde{v}\|_{H^p(\tilde{\tau})}\,\|\tilde{r}\|_{C^{m+}(\tilde{\tau})}\,.$$

Finally, the transformation to the surface element (see Corollary 4.3.12) yields the first assertion.

For the product of boundary element functions we obtain by (5.51a) and by using Corollary 4.4.6 the estimate

$$\begin{aligned}|E_\tau(uv)| &\leq Ch_\tau^{m+2}\,\|u\|_{H^p(\tau)}\,\|\tilde{v}\|_{C^{m+}(\tilde{\tau})} = Ch_\tau^{m+2}\,\|u\|_{H^p(\tau)}\,\|\tilde{v}\|_{C^p(\tilde{\tau})}\\ &\leq Ch_\tau^{m+1}\,\|u\|_{H^p(\tau)}\,\|v\|_{H^p(\tau)}\,.\end{aligned}$$

□

5.3.2.2 Derivative Free Quadrature Error Estimates for Analytic Integrands

In this section we will present the classical, derivative free quadrature error estimates for analytic integrands, which are due to Davis [81, (4.6.1.11)]. Let $\mathcal{E}^{\rho}_{a,b} \subset \mathbb{C}$ be the closed ellipse with the focal points at $z = a, b$, semimajor axis $\bar{a} > (b-a)/2$ and semiminor axis $\bar{b} > 0$ (see Fig. 5.5). The sum of the half-axes is denoted by

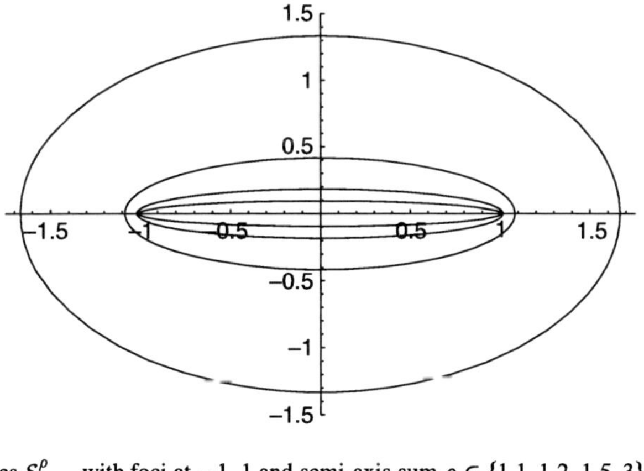

Fig. 5.5 Ellipses $\mathcal{E}^{\rho}_{-1,1}$ with foci at $-1, 1$ and semi-axis sum $\rho \in \{1.1, 1.2, 1.5, 3\}$

$\rho = \bar{a}+\bar{b}$. For $a = 0$ and $b = 1$ we abbreviate $\mathcal{E}_{0,1}^{\rho}$ by $\mathcal{E}^{\rho}$. A classical derivative free error estimate of Gaussian quadrature for analytic integrands can be found in [81].

For $f : [0,1] \to \mathbb{C}$, Q^n denotes the Gaussian quadrature with n nodes, scaled to $[0,1]$, $I(f)$ denotes the integral of f over $[0,1]$ and E^n denotes the associated error.

Theorem 5.3.13. *Let $f : [0,1] \to \mathbb{C}$ be analytic with the analytic extension $f^\star$ on an ellipse $\mathcal{E}^\rho \subset \mathbb{C}$, $\rho > 1/2$.*

Then

$$|I - Q^n| \le C\,(2\rho)^{-2n} \max_{z\in\partial\mathcal{E}^\rho} \left|f^\star(z)\right|.$$

This one-dimensional error estimate can be easily transferred to the error of the Tensor-Gauss quadrature.

Definition 5.3.14. For $1 \le i \le d$ and $-\infty < a_i < b_i < \infty$ let $\omega := \bigotimes_{i=1}^{d} [a_i, b_i] \subset \mathbb{R}^d$. A continuous function $f : \omega \to \mathbb{C}$ is called componentwise analytic if there exists a $(\rho_i)_{i=1}^d \in \mathbb{R}^d$ with $\rho_i > (b_i - a_i)/2$, $1 \le i \le d$ such that for all $1 \le i \le d$ and all $\mathbf{x} \in \omega$ the function

$$f_{i,\mathbf{x}} : [a_i, b_i] \to \mathbb{C}, \qquad f_{i,\mathbf{x}}(t) := f(x_1, \ldots, x_{i-1}, t, x_{i+1}, \ldots x_d)$$

can be extended to an analytic function $f_{i,\mathbf{x}} : \mathcal{E}_{a_i,b_i}^{\rho_i} \to \mathbb{C}$.

Theorem 5.3.15. *Let $f : [0,1]^d \to \mathbb{C}$ be componentwise analytic and let $(\rho_i)_{i=1}^d$ be as in Definition 5.3.14. Then the error for the Gaussian quadrature with n_i nodes per coordinate direction, $1 \le i \le d$, satisfies the estimate*

$$|\mathbf{E}^{\mathbf{n}} f| \le \sum_{i=1}^{d} \max_{\mathbf{x}\in[0,1]^d} |E^{n_i} f_{i,\mathbf{x}}| \le \sum_{i=1}^{d} C_i\,(2\rho_i)^{-2n_i} \max_{\mathbf{x}\in[0,1]^d} \max_{z\in\partial\mathcal{E}^{\rho_i}} |f_{i,\mathbf{x}}(z)|.$$

The constants C_i, $1 \le i \le d$, are as in Theorem 5.3.13.

Proof. It suffices to consider the case $d = 2$, as the statement for $d = 1$ has already been treated in Theorem 5.3.13 and the result for $d > 2$ follows by means of induction. We use a classical tensor product argument. Let $g : [0,1] \to \mathbb{C}$, $f : [0,1]^2 \to \mathbb{C}$ be analytic and $(\rho_i)_{i=1}^2$ as in Definition 5.3.14. We set

$$\begin{aligned} I_1 g &:= \textstyle\int_0^1 g(t)\,dt, & (I_2 f)(t) &:= \textstyle\int_0^1 f(t,s)\,ds, \\ Q_1^{n_1} g &:= \textstyle\sum_{j=1}^{n_1} \omega_{j,n}\, g(\xi_{j,n}), & (Q_2^{n_2} f)(t) &:= \textstyle\sum_{j=1}^{n_2} \omega_{j,n_2} f(t, \xi_{j,n_2}). \end{aligned}$$

This yields the error

$$\begin{aligned} \mathbf{E}^{\mathbf{n}} f &= \left(I_1 I_2 - Q_1^{n_1} Q_2^{n_2}\right) f = \left(I_1 I_2 - I_1 Q_2^{n_2} + I_1 Q_2^{n_2} - Q_1^{n_1} Q_2^{n_2}\right) f \\ &= I_1\left(E_2^{n_2} f\right) + E_1^{n_1}\left(Q_2^{n_2} f\right) \end{aligned}$$

with $E_i^{n_i} := I_i - Q_i^{n_i}$. For $i = 1, 2$ the integral $I_i : C^0([0,1]) \to \mathbb{C}$ is continuous, i.e.,

$$|I_1 g| \leq \max_{t\in[0,1]} |g(t)|, \qquad |(I_2 f)(t)| \leq \max_{s\in[0,1]} |f(t,s)|.$$

The weights of the Gaussian quadrature are positive and add up to the length of the interval $\sum_{j=1}^{n} \omega_{j,n} = 1$ (see [225]). It follows that

$$\begin{aligned} |(\mathbf{I} - \mathbf{Q}^{\mathbf{n}}) f| &\leq \max_{t\in[0,1]} \left|E_2^{n_2} f(t,\cdot)\right| + \sum_{j=1}^{n_2} \omega_{j,n_2} \left|E_1^{n_1} f\left(\cdot, \xi_{j,n_2}\right)\right| \\ &\leq \max_{t\in[0,1]} \left|E_2^{n_2} f(t,\cdot)\right| + \max_{t\in[0,1]} \left|E_1^{n_1} f(\cdot, t)\right|. \end{aligned}$$

□

5.3.2.3 Estimates of the Analyticity Ellipses of the Regularized Integrands

In this subsection we will estimate the size of the analyticity domains of the regularized integrands from Sect. 5.2.4 as well as those that have been extended on these domains. We again distinguish between four cases: identical panels, panels with a common edge, with a common point and with a positive distance to each other. The estimates of the integrands on the analyticity ellipses are always derived according to the same concept: First the integrals are transformed onto the unit triangle or square. By means of suitable expansions in local coordinates the singularity is determined in the transformed, complex coordinates. More specifically, the size of the ellipses on which the integrands can be extended analytically is estimated. Then other properties of the integrands on the ellipses are estimated, such as the Gram determinant, basis functions and kernel functions. We also explicitly determine the dependency on the panel diameter h_τ, h_t so that the constants in the estimates generally depend only on the polynomial degree p of the boundary element space and the shape-regularity of the mesh.

Case 1: Identical Panels

We first consider the case $\hat{\tau} = \hat{t} = \widehat{S}$ and use the representation *I.1* from Sect. 5.2.4. In view of the definition of k_3 [see (5.36)] we analyze the analyticity domains of the basis functions, the surface elements and the kernel functions separately. We begin with the kernel function in local coordinates. We again assume that the panel τ can be written as the composition of a global chart χ and an affine mapping $\chi_\tau^{\text{affine}} : \mathbb{R}^2 \to \mathbb{R}^2$: $\chi_\tau = \chi \circ \chi_\tau^{\text{affine}}$. Let the affine part again be of the type $\chi_\tau^{\text{affine}}(\hat{\mathbf{x}}) = \mathbf{A}_\tau + \mathbf{m}_\tau \hat{\mathbf{x}}$.

For the parametrizations under consideration the difference variable $\mathbf{z} - \mathbf{y} - \mathbf{x}$ has the following representation in two-dimensional polar coordinates [see (5.22)]

$$\begin{aligned}\mathbf{z} &= \chi_\tau(\hat{\mathbf{z}}+\hat{\mathbf{x}}) - \chi_\tau(\hat{\mathbf{x}}) = h_\tau r \sum_{m=0}^{\infty} (h_\tau r)^m\, l_m\left(\chi_\tau^{\text{affine}}(\hat{\mathbf{x}}), \xi\right) \\ &=: (h_\tau r)\, d\left(\chi_\tau^{\text{affine}}(\hat{\mathbf{x}}), h_\tau r, h_\tau^{-1}\mathbf{m}_\tau \xi\right)\end{aligned}$$

with

$$l_m(\hat{\mathbf{w}}, \xi) := \left(\frac{\left\langle h_\tau^{-1}\mathbf{m}_\tau \xi, \nabla\right\rangle^{m+1} \chi}{(m+1)!}\right)(\hat{\mathbf{w}})$$

and a function d that depends only on the global chart χ but not on the triangulation χ_τ. In the same way as in (5.20c) we deduce the representation

$$k(\mathbf{x}, \mathbf{y}, \mathbf{z}) = (h_\tau r)^{-s}\, d_{3,s}\left(\chi_\tau^{\text{affine}}(\hat{\mathbf{x}}), h_\tau r, h_\tau^{-1}\mathbf{m}_\tau \xi\right) \tag{5.52}$$

with a function $d_{3,s}$ that depends only on the global chart χ and the kernel function k.

The basis functions $\hat{b}_i, \hat{b}_j$ on the reference element are polynomials of degree p and have the representation in polar coordinates

$$B_{i,j}(\hat{\mathbf{x}}, r, \xi) := \hat{b}_i(\hat{\mathbf{x}})\, \hat{b}_j(\hat{\mathbf{x}} + r\xi),$$

which is independent of the triangulation $\mathcal{G}$. With this we have for the integrand $k_2(\hat{\mathbf{x}}, \hat{\mathbf{x}} + r\xi)$ from (5.25) the representation [see (5.50)]

$$\begin{aligned} r k_2(\hat{\mathbf{x}}, \hat{\mathbf{x}} + r\xi) &= r\left(\det \mathbf{m}_\tau\right)^2 B_{i,j}(\hat{\mathbf{x}}, r, \xi)\, g_\tau^\chi(\hat{\mathbf{w}})\, g_\tau^\chi(\hat{\mathbf{v}})\, (h_\tau r)^{-s} \times \\ &\quad d_{3,s}\left(\hat{\mathbf{w}}, h_\tau r, h_\tau^{-1}\mathbf{m}_\tau \xi\right), \end{aligned} \tag{5.53}$$

where $\hat{\mathbf{w}} := \chi_\tau^{\text{affine}}(\hat{\mathbf{x}})$ and $\hat{\mathbf{v}} := \chi_\tau^{\text{affine}}(\hat{\mathbf{x}} + r\xi)$. The functions $B_{i,j}, g_\tau^\chi, d_{3,2}$ depend only on the global chart χ, the polynomial degree p and the coefficients of the kernel function k and, in particular, they are independent of the triangulation. They can be extended to suitable complex neighborhoods of the parameter domains, the size of which also does not depend on the triangulation.

The numerical quadrature was not formulated in polar coordinates but with respect to the (ξ, η)-coordinates, which are mapped to $(\hat{\mathbf{x}}, r, \varphi)$-coordinates by means of the transformation from (5.35). The transformation with the index i in (5.35) corresponds to the summand i in (5.34) and it is denoted by $\mathfrak{T}_i$:

$$(\hat{\mathbf{x}}, r, \varphi) = \mathfrak{T}_i(\xi, \eta). \tag{5.54}$$

As these transformations are again analytic and independent of the triangulation, we obtain the analyticity with respect to the coordinates $(\xi, \eta) \in [0,1]^4$.

We use the following notation to describe these neighborhoods. For $\rho > 0$ and $i \in \{1, 2, 3, 4\}$ we set

$$\vec{\mathcal{E}}_{\rho}^{(i)} := \underbrace{(0,1)\times(0,1)\times\ldots\times(0,1)}_{(i-1)\text{ times}}\times\mathcal{E}_{\rho}\times\underbrace{(0,1)\times(0,1)\times\ldots\times(0,1)}_{(4-i)\text{ times}}.$$

Lemma 5.3.16. *Let the kernel function k satisfy Assumption 5.1.19 with $s \in \mathbb{Z}_{\leq 2}$. Let $\sigma := \min\{1, s\}$. There exist constants $\rho_1 > 0$ and $\rho_2 > 1/2$ that depend only on θ_τ from (5.49), the global chart χ, the coefficients of the kernel function and the polynomial degree p such that the integrand from Sect. 5.2.4 (I.1)*

$$k_4 : (0,1)^4 \to \mathbb{C} \qquad k_4(\xi,\eta) = \xi^3\eta_1^2\eta_2\sum_{i=1}^{3} k_3\left(\hat{\mathbf{x}}, \hat{\mathbf{x}} + r\begin{pmatrix}\cos\varphi\\ \sin\varphi\end{pmatrix}\right) \tag{5.55}$$

can be analytically extended to $\vec{\mathcal{E}}^{(1)}_{\rho_1/h_\tau} \cup \bigcup_{j=2}^{4} \vec{\mathcal{E}}^{(j)}_{\rho_2}$ after substituting as in (5.54). We have the estimates

$$\sup_{(\xi,\eta)\in\vec{\mathcal{E}}^{(1)}_{\rho_1/h_\tau}} |k_4(\xi,\eta)| \leq C h_\tau^{1-2p},$$

$$\sup_{(\xi,\eta)\in\vec{\mathcal{E}}^{(i)}_{\rho_2}} |k_4(\xi,\eta)| \leq C h_\tau^{4-\sigma}$$

for $2 \leq i \leq 4$.

Proof. We only need to analyze each summand in (5.55) separately and in the following we will abbreviate $\mathfrak{T}_i$ to $\mathfrak{T}$.

We will first consider weakly singular kernel functions with $s \leq 1$. The size of the analyticity ellipses can be determined from the representation (5.53) and from

$$\left\|h_\tau^{-1}\mathbf{m}_\tau\xi\right\| \leq |\xi_1|\,\frac{\|\mathbf{B}_\tau - \mathbf{A}_\tau\|}{h_\tau} + |\xi_2|\,\frac{\|\mathbf{C}_\tau - \mathbf{B}_\tau\|}{h_\tau} \leq c_1^{-1}\sqrt{2}\,\|\xi\|$$

[with c_1 from (5.48)]. In order to estimate the integrands on the analyticity ellipses we consider the functions in (5.53) separately. To estimate the functions $B_{i,j}$ in (ξ,η)-coordinates, we consider the transformations that appear in Sect. 5.2.4, *I.1*, and obtain

$$B_{i,j}\circ\mathfrak{T}(\xi,\eta) = \hat{b}_i(\xi\Lambda_1(\eta))\,\hat{b}_j(\xi\Lambda_2(\eta))$$

with functions Λ_1, Λ_2, that are affine with respect to every variable. Since $\hat{b}_i$ and $\hat{b}_j$ are polynomials of degree p, we deduce that

$$\sup_{(\xi,\eta)\in\vec{\mathcal{E}}^{(1)}_{\rho_1/h_\tau}} \left|B_{i,j}\circ\mathfrak{T}(\xi,\eta)\right| \leq C h_\tau^{-2p} \text{ and } \sup_{(\xi,\eta)\in\vec{\mathcal{E}}^{(i)}_{\rho_2}} \left|B_{i,j}\circ\mathfrak{T}(\xi,\eta)\right| \leq C,\ 2\leq i\leq 4, \tag{5.56}$$

with a constant C that depends only on the polynomial degree p.

We now consider the roots of the Gram determinants $g_\tau(\hat{\mathbf{x}})$, $g_\tau(\hat{\mathbf{y}})$ and define for $i = 1, 2$

$$U_{\rho_1,i} := \left\{ \mathbf{A}_\tau + \xi \mathbf{m}_\tau \Lambda_i(\eta) : (\xi,\eta) \in \overrightarrow{\mathcal{E}}^{(1)}_{\rho_1/h_\tau} \right\}.$$

Depending on θ_τ and the global chart χ we can choose $\rho_1 > 0$ sufficiently small so that we have

$$U_{\rho_1,i} \subset U^\star,$$

where $U^\star \subset \mathbb{C} \times \mathbb{C}$ is the domain onto which χ can be extended analytically. From this we have for $i = 1, 2$

$$\sup_{(\xi,\eta)\in\overrightarrow{\mathcal{E}}^{(1)}_{\rho_1/h_\tau}} \left| g_\tau^\chi \circ \chi_\tau^{\text{affine}}(\xi\Lambda_i(\eta)) \right| \leq \sup_{\hat{\mathbf{w}}\in U^\star(\tau)} \left| g_\tau^\chi(\hat{\mathbf{w}}) \right| \leq C,$$

where C depends only on the global chart χ. In the same way we can show that

$$\sup_{(\xi,\eta)\in\overrightarrow{\mathcal{E}}^{(j)}_{\rho_2}} \left| g_\tau^\chi \circ \chi_\tau^{\text{affine}}(\xi\Lambda_i(\eta)) \right| \leq C, \qquad i = 1, 2, \quad j = 2, 3, 4.$$

We consider the singular term $(h_\tau r)^{-s}$ in (5.53) together with the factors $\xi^3\eta_1^2\eta_2$. In (ξ,η)-coordinates we have

$$\frac{\xi^3\eta_1^2\eta_2}{(h_\tau r)^s} = \frac{\xi^3\eta_1^2\eta_2}{\left(h_\tau\xi\eta_1\eta_2\sqrt{\Lambda_4(\eta_3)}\right)^s} = \frac{\xi^2\eta_1}{h_\tau^s\Lambda_4^{s/2}(\eta_3)}(\xi\eta_1\eta_2)^{1-s}$$

with a polynomial Λ_4 which satisfies $\Lambda_4(\eta_3) > 0$ for all $\eta_3 \in [0,1]$. It follows that

$$\sup_{(\xi,\eta)\in\overrightarrow{\mathcal{E}}^{(1)}_{\rho_1/h_\tau}} \frac{\xi^2\eta_1}{h_\tau^s\Lambda_4^{s/2}(\eta_3)}(\xi\eta_1\eta_2)^{1-s} \leq C h_\tau^{-3},$$

$$\sup_{(\xi,\eta)\in\overrightarrow{\mathcal{E}}^{(i)}_{\rho_2}} \frac{\xi^2\eta_1}{h_\tau^s\Lambda_4^{s/2}(\eta_3)}(\xi\eta_1\eta_2)^{1-s} \leq C h_\tau^{-s}$$

for $i = 2, 3, 4$. Finally, we need to estimate the factor $d_{3,s}\left(\hat{\mathbf{w}}, h_\tau r, h_\tau^{-1}\mathbf{m}_\tau\xi\right)$ in (ξ,η)-coordinates. We obtain the estimate

$$\sup_{(\xi,\eta)\in\overrightarrow{\mathcal{E}}^{(1)}_{\rho_1/h_\tau}} \left| d_{3,s}\left(\chi_\tau^{\text{affine}}(\hat{\mathbf{x}}), h_\tau r, h_\tau^{-1}\mathbf{m}_\tau\binom{\cos\varphi}{\sin\varphi}\right)\Big|_{(\hat{\mathbf{x}},r,\varphi)=\mathfrak{T}(\xi,\eta)} \right| \leq C,$$

$$\sup_{(\xi,\eta)\in\overrightarrow{\mathcal{E}}^{(i)}_{\rho_2}} \left| d_{3,s}\left(\chi_\tau^{\text{affine}}(\hat{\mathbf{x}}), h_\tau r, h_\tau^{-1}\mathbf{m}_\tau\binom{\cos\varphi}{\sin\varphi}\right)\Big|_{(\hat{\mathbf{x}},r,\varphi)=\mathfrak{T}(\xi,\eta)} \right| \leq C$$

for $i = 2, 3, 4$ with the same arguments as for g_τ. With $(\det \mathbf{m}_\tau)^2 \le C h_\tau^4$ we have proved that

$$\sup_{(\xi,\eta)\in \vec{\mathcal{E}}^{(1)}_{\rho_1/h_\tau}} |k_4(\xi,\eta)| \le C h_\tau^{1-2p} \quad \text{and} \quad \sup_{(\xi,\eta)\in \vec{\mathcal{E}}^{(i)}_{\rho_2}} |k_4(\xi,\eta)| \le C h_\tau^{4-s}, \quad i = 2, 3, 4.$$

We now consider Cauchy-singular kernel functions that satisfy Assumption 5.1.19 with $s = 2$. The integrand from Sect. 5.2.4, *I.1*, is a sum of pairs of kernel functions of the type $k_2(\hat{\mathbf{x}}, \hat{\mathbf{x}} - \hat{\mathbf{z}}) + k_2(\hat{\mathbf{x}} - \hat{\mathbf{z}}, \hat{\mathbf{x}})$. We can therefore apply Theorem 5.1.22(c) and show that the above-mentioned arguments can be applied to the terms $k_2(\hat{\mathbf{x}}, \hat{\mathbf{x}} - \hat{\mathbf{z}}) + k_2(\hat{\mathbf{x}} - \hat{\mathbf{z}}, \hat{\mathbf{x}})$. Thus we can also prove the statement for the kernel function from Assumption 5.1.19. □

Proposition 5.3.17. *The statements from Lemma 5.3.16 can be directly transferred to identical squares, as the variable transformations given in Sect. 5.2.4,* I.2, *are of the same type as for triangles.*

Case 2: Panels with a Common Edge

We apply the same ansatz that we developed for identical panels to panels with exactly one common edge. We first consider the case of two triangles $\tau, t \in \mathcal{G}$ with $\chi_\tau(\xi, 0) = \chi_t(\xi, 0)$ for all $\xi \in (0,1)$ and use the representation *II.1* from Sect. 5.2.4. We again assume that the local charts χ_τ and χ_t can be written as a composition of global charts χ_1, χ_2 and affine transformations

$$\chi_\tau = \chi_1 \circ \chi_\tau^{\text{affine}}, \qquad \chi_t = \chi_2 \circ \chi_t^{\text{affine}}.$$

For the parametrization in three-dimensional polar coordinates $\hat{\mathbf{z}} = r\xi$, $r = \|\hat{\mathbf{z}}\|$, $\xi = \hat{\mathbf{z}} / \|\hat{\mathbf{z}}\|$ [see (5.21)] under consideration, the three-dimensional difference variable $\hat{\mathbf{z}} = (\hat{y}_1 - \hat{x}_1, \hat{y}_2, \hat{x}_2)^{\mathsf{T}}$ has the representation

$$\begin{aligned}\mathbf{z} = \chi_t\left(\begin{smallmatrix}\hat{z}_1+\hat{x}_1\\ \hat{z}_2\end{smallmatrix}\right) - \chi_\tau\left(\begin{smallmatrix}\hat{x}_1\\ \hat{z}_3\end{smallmatrix}\right) &= h_t r \sum_{m=0}^{\infty} (h_t r)^m\, l_m\left(\chi_t^{\text{affine}}\left(\begin{smallmatrix}\hat{x}_1\\ 0\end{smallmatrix}\right), \chi_\tau^{\text{affine}}\left(\begin{smallmatrix}\hat{x}_1\\ 0\end{smallmatrix}\right), \xi\right)\\ &=: (h_t r)\, b\left(\chi_t^{\text{affine}}\left(\begin{smallmatrix}\hat{x}_1\\ 0\end{smallmatrix}\right), \chi_\tau^{\text{affine}}\left(\begin{smallmatrix}\hat{x}_1\\ 0\end{smallmatrix}\right), h_t r, h_t^{-1}\mathbf{m}_t\left(\begin{smallmatrix}\xi_1\\ \xi_2\end{smallmatrix}\right), h_t^{-1}\xi_3\left(\mathbf{C}_\tau - \mathbf{B}_\tau\right)\right)\end{aligned}$$

with

$$l_m(\hat{\mathbf{v}}, \hat{\mathbf{w}}, \xi) := \frac{\left\langle h_t^{-1}\mathbf{m}_t\left(\begin{smallmatrix}\xi_1\\ \xi_2\end{smallmatrix}\right), \nabla\right\rangle^{m+1} \chi_2(\hat{\mathbf{v}}) - \left\langle h_t^{-1}\xi_3\left(\mathbf{C}_\tau - \mathbf{B}_\tau\right), \nabla\right\rangle^{m+1} \chi_1(\hat{\mathbf{w}})}{(m+1)!}$$

and a function b that depends only on the global charts χ_1, χ_2 but not on the triangulation. In the same way as in (5.20b) we deduce the representation

$$k(\mathbf{x}, \mathbf{y}, \mathbf{z}) = (h_t r)^{-s}\, b_{3,s}\left(\chi_t^{\text{affine}}\left(\begin{smallmatrix}\hat{x}_1\\ 0\end{smallmatrix}\right), \chi_\tau^{\text{affine}}\left(\begin{smallmatrix}\hat{x}_1\\ 0\end{smallmatrix}\right), h_t r, h_t^{-1}\mathbf{m}_t\left(\begin{smallmatrix}\xi_1\\ \xi_2\end{smallmatrix}\right), h_t^{-1}\xi_3\left(\mathbf{C}_\tau - \mathbf{B}_\tau\right)\right)$$

with a function $b_{3,s}$ that depends only on the global charts $\chi_{1,2}$ and the kernel function k.

The numerical integration is not carried out in $(\hat{x}_1, r, \xi)$-coordinates but in simplex coordinates (ξ, η). The associated transformation is denoted by $\mathfrak{T}_j$, $(\hat{x}_1, r, \xi) = \mathfrak{T}_j(\xi, \eta)$, where the index j refers to the single summands in the representation from Sect. 5.2.4(*II.1–II.4*). It will be omitted in the following. The transformations are analytic with respect to every variable. It is easy to verify that the 3. – 5. components of $\mathfrak{T}$ are independent of ξ, which is why we introduce the abbreviation

$$\xi_i = \mathfrak{T}_{i+2}(\eta) \qquad 1 \le i \le 3.$$

The size of the analyticity ellipses of the local integrands is estimated in the following lemma.

Lemma 5.3.18. *Let the kernel function satisfy Assumption 5.1.19 with $s \in \mathbb{Z}_{\le 2}$. Let the function $k_5 : (0,1)^4 \to \mathbb{C}$ denote one of the integrands from Sect. 5.2.4,* II.1–II.4. *Then there exist constants $\rho_1 > 0$ and $\rho_2 > 1/2$ that depend only on $\theta_\tau, \theta_t, C_K, c_1$ from (5.49), (5.47), (5.48), the global charts $\chi_{1,2}$, the coefficients of the kernel function and the polynomial degree p so that k_5 can be analytically extended to $\vec{\mathcal{E}}^{(1)}_{\rho_1/h_t} \cup \bigcup_{j=2}^{4} \vec{\mathcal{E}}^{(j)}_{\rho_2}$. The estimates*

$$\sup_{(\xi,\eta)\in\vec{\mathcal{E}}^{(1)}_{\rho_1/h_t}} |k_5(\xi,\eta)| \le C h_t^{1-2p},$$

$$\sup_{(\xi,\eta)\in\vec{\mathcal{E}}^{(i)}_{\rho_2}} |k_5(\xi,\eta)| \le C h_t^{4-s}$$

hold for $i = 2, 3, 4$.

Proof. The components of the integrand are each analyzed separately and we begin with the kernel function in local coordinates.

The analysis of the surface elements and basis functions is done in the same way as in the case for the identical panels and leads to

$$g_\tau(\hat{\mathbf{x}})\, g_t(\hat{\mathbf{y}}) = |\det \mathbf{m}_\tau|\, |\det \mathbf{m}_t| \left(g_1 \circ \chi_\tau^{\text{affine}}\right)(\hat{\mathbf{x}}) \left(g_2 \circ \chi_t^{\text{affine}}\right)(\hat{\mathbf{y}})$$

with the surface elements $g_{1,2}$ for the global charts $\chi_{1,2}$.

The coordinate transformations in the case of a common edge (see Sect. 5.2.4) can be written as

$$\hat{\mathbf{x}} = \Lambda_1(\xi,\eta), \qquad \hat{\mathbf{y}} = \Lambda_2(\xi,\eta), \qquad r = \xi\eta_1^{\ell}\sqrt{\Lambda_3(\eta)} \tag{5.57}$$

with functions Λ_i, $i = 1, 2$ that are affine with respect to every variable and a quadratic polynomial Λ_3 with $\Lambda_3(\eta) > 0$ for all $\eta \in [0,1]^3$. The power ℓ in (5.57) is equal to one for the transformations in Sect. 5.2.4, *II.1* and zero for the transformations *II.2–II.4*. Note that the mappings Λ_i, $1 \le i \le 3$, are independent of the triangulation and the surface parametrization.

The determinant of the Jacobian of the transformation $(\xi, \eta) \to (\hat{\mathbf{x}}, \hat{\mathbf{y}})$ from Sect. 5.2.4 is of the form

$$\det\left(D\binom{\hat{\mathbf{x}}}{\hat{\mathbf{y}}}\right)(\xi, \eta) = \xi^2 \eta_1^{\ell'} p(\xi, \eta), \tag{5.58}$$

with a polynomial p of maximal degree 1 with respect to ξ and maximal degree 2 with respect to the η-variables. For the power ℓ' we have

$$\ell' := \begin{cases} 2 \text{ for the transformation } \mathit{II.1} \\ 0 \text{ for the transformations } \mathit{II.2\text{-}4} \end{cases}$$

Using this, we can deduce a representation for the integrand $k_3(\hat{\mathbf{x}}, \hat{\mathbf{y}})$ in (5.36). Thus, with $\hat{\mathbf{w}} := \chi_\tau^{\text{affine}}(\hat{\mathbf{x}})$, $\hat{\mathbf{v}} := \chi_t^{\text{affine}}(\hat{\mathbf{y}})$ and the substitutions (5.57), we obtain

$$\begin{aligned} k_3(\hat{\mathbf{x}}, \hat{\mathbf{y}}) &= |\det \mathbf{m}_\tau| \, |\det \mathbf{m}_t| \, g_1(\hat{\mathbf{w}}) \, g_2(\hat{\mathbf{v}}) \, \hat{b}_i(\hat{\mathbf{x}}) \, \hat{b}_j(\hat{\mathbf{y}}) \\ &\times (h_t r)^{-s} \, b_{3,s}\left(\chi_t^{\text{affine}}\binom{\hat{x}_1}{0}, \chi_\tau^{\text{affine}}\binom{\hat{x}_1}{0}, h_t r, h_t^{-1}\mathbf{m}_t\binom{\mathfrak{T}_1(\eta)}{\mathfrak{T}_2(\eta)}, h_t^{-1}\mathfrak{T}_3(\eta) \right. \\ &\times \left. (\mathbf{C}_\tau - \mathbf{B}_\tau)\right). \end{aligned} \tag{5.59}$$

The product of r^{-s} with the determinant of the Jacobian is, since $s \in \mathbb{Z}_{\leq 2}$,

$$(h_t r)^{-s} \xi^2 \eta_1^{\ell'} p(\xi, \eta) = h_t^{-s} \xi^{2-s} \tilde{p}(\xi, \eta) \tag{5.60}$$

a polynomial and therefore analytic. More specifically, $\tilde{p}$ is a polynomial with maximal degree 1 with respect to ξ and maximal degree $4 - s$ with respect to the η-variables. The size of the analyticity ellipses can be determined from the representation above, while using $\|\mathbf{C}_\tau - \mathbf{B}_\tau\| / h_t \leq c_1^{-1} C_{\mathrm{K}}$ [see (5.47), (5.48)].

In order to estimate the integrands on the analyticity ellipses we consider the functions in (5.59), (5.60) separately. Since $\hat{b}_i$ and $\hat{b}_j$ are polynomials of degree p we deduce that

$$\begin{aligned} \sup_{(\xi,\eta) \in \vec{\mathcal{E}}^{(1)}_{\rho_1/h_\tau}} \left| \hat{b}_i(\Lambda_1(\xi, \eta)) \, \hat{b}_j(\Lambda_2(\xi, \eta)) \right| &\leq C h_t^{-2p} \\ \sup_{(\xi,\eta) \in \vec{\mathcal{E}}^{(\ell)}_{\rho_2}} \left| \hat{b}_i(\Lambda_1(\xi, \eta)) \, \hat{b}_j(\Lambda_2(\xi, \eta)) \right| &\leq C, \quad 2 \leq \ell \leq 4, \end{aligned} \tag{5.61}$$

with a constant C that depends only on the polynomial degree p.

We now turn our attention to the surface elements g_1, g_2 (for the global charts χ_1, χ_2) and define

$$\begin{aligned} U_{\rho_1,\tau} &:= \left\{ \mathbf{A}_\tau + \mathbf{m}_\tau \Lambda_1(\xi, \eta) : (\xi, \eta) \in \vec{\mathcal{E}}^{(1)}_{\rho_1/h_t} \right\}, \\ U_{\rho_1,t} &:= \left\{ \mathbf{A}_t + \mathbf{m}_t \Lambda_2(\xi, \eta) : (\xi, \eta) \in \vec{\mathcal{E}}^{(1)}_{\rho_1/h_t} \right\}. \end{aligned}$$

At the same time note that $h_t \sim h_\tau$ [see (5.47)]. Depending on θ_τ and the global chart χ_1 we can choose a sufficiently small $\rho_1 > 0$ such that

$$U_{\rho_1,\tau} \subset U_1^\star, \qquad U_{\rho_1,t} \subset U_2^\star,$$

where $U_i^\star \subset \mathbb{C} \times \mathbb{C}$, $i = 1, 2$, denotes the domain on which χ_i can be analytically extended. From this we have

$$\sup_{(\xi,\eta)\in \overrightarrow{\mathcal{E}}^{(1)}_{\rho_1/h_t}} \left| g_1 \circ \chi_\tau^{\text{affine}} \left(\Lambda_1 \left(\xi, \eta \right) \right) \right| \leq \sup_{\hat{\mathbf{w}} \in U^\star(\tau)} \left| g_1 \left(\hat{\mathbf{w}} \right) \right| \leq C,$$

where C depends only on the global chart χ_1. Analogously, we deduce that

$$\sup_{(\xi,\eta)\in \overrightarrow{\mathcal{E}}^{(i)}_{\rho_2}} \left| g_1 \circ \chi_\tau^{\text{affine}} \left(\Lambda_1 \left(\xi, \eta \right) \right) \right| \leq C, \qquad i = 2, 3, 4,$$

and the corresponding estimates for g_2.

We consider the singular term $(h_\tau r)^{-s}$ in (5.59) in combination with the determinant of the Jacobian (5.58), (5.60) and obtain

$$\sup_{(\xi,\eta)\in \overrightarrow{\mathcal{E}}^{(1)}_{\rho_1/h_\tau}} (h_\tau r)^{-s} \xi^2 \eta_1^{\ell'} p \left(\xi, \eta \right) \leq C h_t^{-3}$$

$$\sup_{(\xi,\eta)\in \overrightarrow{\mathcal{E}}^{(i)}_{\rho_2}} (h_\tau r)^{-s} \xi^2 \eta_1^{\ell'} p \left(\xi, \eta \right) \leq C h_\tau^{-s}, \qquad i = 2, 3, 4.$$

Finally, we need to estimate the factor $b_{3,s} (\cdot)$ from (5.59) in (ξ, η)-coordinates. With the same arguments as for g_τ we obtain the estimates

$$\sup_{(\xi,\eta)\in \overrightarrow{\mathcal{E}}^{(1)}_{\rho_1/h_t}} \left| b_{3,s} \left(\chi_t^{\text{affine}} \binom{\hat{x}_1}{0}, \chi_\tau^{\text{affine}} \binom{\hat{x}_1}{0}, h_t r, h_t^{-1} \mathbf{m}_t \binom{\mathfrak{T}_1(\eta)}{\mathfrak{T}_2(\eta)}, h_t^{-1} \mathfrak{T}_3 \left(\eta \right) \left(\mathbf{C}_\tau - \mathbf{B}_\tau \right) \right) \right| \leq C,$$

$$\sup_{(\xi,\eta)\in \overrightarrow{\mathcal{E}}^{(i)}_{\rho_2}} \left| b_{3,s} \left(\chi_t^{\text{affine}} \binom{\hat{x}_1}{0}, \chi_\tau^{\text{affine}} \binom{\hat{x}_1}{0}, h_t r, h_t^{-1} \mathbf{m}_t \binom{\mathfrak{T}_1(\eta)}{\mathfrak{T}_2(\eta)}, h_t^{-1} \mathfrak{T}_3 \left(\eta \right) \left(\mathbf{C}_\tau - \mathbf{B}_\tau \right) \right) \right| \leq C.$$

With $|\det \mathbf{m}_\tau| \, |\det \mathbf{m}_t| \leq C h_t^4$ we have proved that

$$\sup_{(\xi,\eta)\in \overrightarrow{\mathcal{E}}^{(1)}_{\rho_1/h_t}} \left| k_5 \left(\xi, \eta \right) \right| \leq C h_t^{1-2p} \quad \text{and} \quad \sup_{(\xi,\eta)\in \overrightarrow{\mathcal{E}}^{(i)}_{\rho_2}} \left| k_5 \left(\xi, \eta \right) \right| \leq C h_\tau^{4-s}$$

for $i = 2, 3, 4$. □

Case 3: Panels with a Common Vertex

We first consider the case of two triangles $\tau, t \in \mathcal{G}$ with $\chi_\tau (0, 0) = \chi_t (0, 0)$ and use the representation from Sect. 5.2.4, *III*. We assume that the local charts χ_τ and χ_t

can be written as a composition of global charts χ_1, χ_2 with affine transformations

$$\chi_\tau = \chi_1 \circ \chi_\tau^{\text{affine}}, \qquad \chi_t = \chi_2 \circ \chi_t^{\text{affine}}.$$

For the parametrization under consideration the four-dimensional difference variable $\hat{\mathbf{z}} = (\hat{y}_1, \hat{y}_2, \hat{x}_1, \hat{x}_2)^\mathsf{T}$ has the representation in four-dimensional polar coordinates $\hat{\mathbf{z}} = r\xi$, $r = \|\hat{\mathbf{z}}\|$, $\xi = \hat{\mathbf{z}}/\|\hat{\mathbf{z}}\|$ [see (5.18)]

$$\begin{aligned}\mathbf{z} &= \chi_t\binom{\hat{z}_1}{\hat{z}_2} - \chi_\tau\binom{\hat{z}_3}{\hat{z}_4} = h_t r \sum_{m=0}^{\infty} (h_t r)^m\, l_m\left(\chi_t^{\text{affine}}\binom{0}{0}, \chi_\tau^{\text{affine}}\binom{0}{0}, \xi\right) \\ &=: (h_t r)\, b\left(\chi_t^{\text{affine}}\binom{0}{0}, \chi_\tau^{\text{affine}}\binom{0}{0}, h_t r, h_t^{-1}\mathbf{m}_t\binom{\xi_1}{\xi_2}, h_t^{-1}\mathbf{m}_\tau\binom{\xi_3}{\xi_4}\right)\end{aligned}$$

with

$$l_m(\hat{\mathbf{v}}, \hat{\mathbf{w}}, \xi) := \frac{\left\langle h_t^{-1}\mathbf{m}_t\binom{\xi_1}{\xi_2}, \nabla\right\rangle^{m+1} \chi_2(\hat{\mathbf{v}}) - \left\langle h_t^{-1}\mathbf{m}_\tau\binom{\xi_3}{\xi_4}, \nabla\right\rangle^{m+1} \chi_1(\hat{\mathbf{w}})}{(m+1)!}$$

and a function b that depends only on the global charts χ_1, χ_2 but not on the triangulation. Similarly to (5.20a) we deduce the representation

$$k(\mathbf{x}, \mathbf{y}, \mathbf{z}) = (h_t r)^{-s}\, a_{3,s}\left(\chi_t^{\text{affine}}\binom{0}{0}, \chi_\tau^{\text{affine}}\binom{0}{0}, h_t r, h_t^{-1}\mathbf{m}_t\binom{\xi_1}{\xi_2}, h_t^{-1}\mathbf{m}_\tau\binom{\xi_3}{\xi_4}\right) \tag{5.62}$$

with a function $a_{3,s}$ that depends only on the global charts $\chi_{1,2}$ and the kernel function k.

The transformation from simplex coordinates to polar coordinates is again denoted by $\mathfrak{T}$, i.e., $(r, \xi) = \mathfrak{T}(\xi, \eta)$, and it is analytic with respect to every variable. It is easily verified that the 2. – 5. components depend on $\mathfrak{T}$, not on ξ, and we therefore introduce the abbreviation

$$\xi_i = \mathfrak{T}_{i+1}(\eta) \qquad 1 \le i \le 4.$$

The size of the analyticity ellipses of the local integrand is estimated in the following lemma.

Lemma 5.3.19. *Let the kernel function satisfy Assumption 5.1.19 with $s \in \mathbb{Z}_{\le 2}$. Let the function $k_6 : (0,1)^4 \to \mathbb{C}$ denote one of the integrands from Sect. 5.2.4, III.1–III.4. Then there exist constants $\rho_1 > 0$ and $\rho_2 > 1/2$ that depend only on $\theta_\tau, \theta_t, C_{\mathrm{K}}, c_1$ from (5.49), (5.47), (5.48), the global charts $\chi_{1,2}$, the coefficients of the kernel function and the polynomial degree p so that k_6 can be analytically extended to $\vec{\mathcal{E}}^{(1)}_{\rho_1/h_t} \cup \bigcup_{i=2}^4 \vec{\mathcal{E}}^{(i)}_{\rho_2}$. The estimates*

$$\sup_{(\xi,\eta)\in\overrightarrow{\mathcal{E}}^{(1)}_{\rho_1/h_t}} |k_6(\xi,\eta)| \le C h_t^{1-2p},$$
$$\sup_{(\xi,\eta)\in\overrightarrow{\mathcal{E}}^{(i)}_{\rho_2}} |k_6(\xi,\eta)| \le C h_t^{4-s} \tag{5.63}$$

hold for $2 \le i \le 4$.

Proof. The proof of this lemma can be accomplished with the same arguments and analogous estimates as for the case of the common edge. □

Case 4: Panels with a Positive Distance

We will now turn our attention to the case of two panels $\tau, t \in \mathcal{G}$ at a positive distance apart. Let

$$d_{\tau,t} := \operatorname{dist}(\tau,t) := \inf_{(x,y)\in\tau\times t} \|\mathbf{x}-\mathbf{y}\| > 0. \tag{5.64}$$

In local coordinates the integrand is again denoted by

$$k_3(\hat{\mathbf{x}},\hat{\mathbf{y}}) = g_\tau(\hat{\mathbf{x}})\, g_t(\hat{\mathbf{y}})\, \hat{b}_i(\hat{\mathbf{x}})\, \hat{b}_j(\hat{\mathbf{y}})\, k(\chi_\tau(\hat{\mathbf{x}}), \chi_t(\hat{\mathbf{y}}), \chi_t(\hat{\mathbf{y}}) - \chi_\tau(\hat{\mathbf{x}})) \tag{5.65}$$

and is analytic with respect to every coordinate. We again assume that the parametrizations χ_τ, χ_t can be represented by means of global charts χ_1, χ_2 that do not depend on the surface mesh and by affine mappings $\chi_\tau^{\text{affine}}$, χ_t^{affine}

$$\chi_\tau = \chi_1 \circ \chi_\tau^{\text{affine}}, \qquad \chi_t = \chi_2 \circ \chi_t^{\text{affine}}.$$

Lemma 5.3.20. *Let the kernel function satisfy Assumption 5.1.19. Let (5.64) and* $\hat{\tau} = \hat{t} = \widehat{Q}$ *hold for* $\tau, t \in \mathcal{G}$.

Then there exists a positive constant $\rho > 1/2$ *that depends only on* θ_τ, θ_t *from (5.49), the global charts* $\chi_{1,2}$, *the coefficients of the kernel function and the polynomial degree* p *so that* k_3 *can be analytically extended to* $\left(\bigcup_{i=1}^{2} \overrightarrow{\mathcal{E}}^{(i)}_{\tilde{\rho}(\tau,t)}\right) \cup \left(\bigcup_{j=3}^{4} \overrightarrow{\mathcal{E}}^{(j)}_{\tilde{\rho}(t,\tau)}\right)$ *with* $\tilde{\rho}(\tau,t) := \rho \max\{d_{\tau,t}/h_\tau, 1\}$. *We have the estimates*

$$\sup_{(\hat{\mathbf{x}},\hat{\mathbf{y}})\in\overrightarrow{\mathcal{E}}^{(i)}_{\tilde{\rho}(\tau,t)}} |k_3(\hat{\mathbf{x}},\hat{\mathbf{y}})| \le C h_\tau^2 h_t^2 \left(\frac{d_{\tau,t}}{h_\tau}\right)^p d_{\tau,t}^{-s} \qquad i = 1,2,$$
$$\sup_{(\hat{\mathbf{x}},\hat{\mathbf{y}})\in\overrightarrow{\mathcal{E}}^{(j)}_{\tilde{\rho}(t,\tau)}} |k_3(\hat{\mathbf{x}},\hat{\mathbf{y}})| \le C h_\tau^2 h_t^2 \left(\frac{d_{\tau,t}}{h_t}\right)^p d_{\tau,t}^{-s} \qquad j = 3,4. \tag{5.66}$$

Proof. We first consider the statement for the variable $\hat{x}_1$.

The scaling of the affine mapping $\chi_\tau^{\text{affine}}$ combined with the distance condition (5.64) yield the existence of a constant $\rho > 0$ such that

$$U_{\rho_1,\tau} := \left\{ \mathbf{A}_\tau + \hat{x}_1 (\mathbf{B}_\tau - \mathbf{A}_\tau) + \hat{x}_2 (\mathbf{C}_\tau - \mathbf{B}_\tau) : (\hat{\mathbf{x}}, \hat{\mathbf{y}}) \in \overrightarrow{\mathcal{E}}^{(1)}_{\rho d_{\tau,t}/h_\tau} \right\}$$

is contained in $U_1^\star$. Here $U_1^\star \subset \mathbb{C} \times \mathbb{C}$ denotes the domain to which χ_1 can be extended analytically. We have a corresponding result for the remaining variables. As before, for $i = 1, 2$ and $j = 3, 4$, we deduce the estimates

$$\sup_{(\hat{\mathbf{x}},\hat{\mathbf{y}}) \in \overrightarrow{\mathcal{E}}^{(i)}_{\rho/h_\tau}} |k_3 (\hat{\mathbf{x}}, \hat{\mathbf{y}})| \leq C h_\tau^2 h_t^2 \left(\frac{d_{\tau,t}}{h_\tau} \right)^p d_{\tau,t}^{-s},$$

$$\sup_{(\hat{\mathbf{x}},\hat{\mathbf{y}}) \in \overrightarrow{\mathcal{E}}^{(j)}_{\rho/h_\tau}} |k_3 (\hat{\mathbf{x}}, \hat{\mathbf{y}})| \leq C h_\tau^2 h_t^2 \left(\frac{d_{\tau,t}}{h_t} \right)^p d_{\tau,t}^{-s}.$$

□

Proposition 5.3.21. *If either $\hat{\tau}$ or $\hat{t}$ is the unit triangle we precede $\chi_\tau^{\text{affine}}$ or χ_t^{affine} by the mapping $(\zeta_1, \zeta_2) \to (\zeta_1, \zeta_1 \zeta_2)$ (with Jacobian determinant ζ_1), which maps the unit square to the unit triangle. The above-mentioned analysis can be repeated for the composite mapping and one obtains results that are analogous to those in Lemma 5.3.20.*

Remark 5.3.22. *The constants in the quadrature error estimates from this section depend on the shape-regularity of the triangles and the polynomial order. In [203] and [202] numerical quadrature for degenerate (non-shape-regular) panels is introduced and analyzed, which are used for the adaptive hp-version of the boundary element method (see [222]).*

5.3.2.4 Quadrature Orders for Regularized Kernel Functions

The estimate for the analyticity domains for the regularized integrands allows the use of derivative free error estimates from Sect. 5.3.2.2.

The Singular Case

Let $k_\star : (0,1)^4 \to \mathbb{C}$ denote one of the integrands from Sect. 5.2.4.*I–III*. The number of nodes with respect to the ξ-integration is denoted by n_1 and with respect to the η_i-integrations, $1 \leq i \leq 3$, by n_2. We set $\mathbf{n} = (n_1, n_2, n_2, n_2)$, and p again denotes the polynomial degree of the boundary element space.

Theorem 5.3.23. *The approximations of the integrals* I–III *by means of Tensor-Gauss quadrature converge exponentially with respect to the number of nodes:*

$$|\mathbf{E}^{\mathbf{n}} k_\star| \leq C h_\tau^{1-2p} (\rho_1 h_\tau)^{2n_1} + C h_t^{4-s} (2\rho_2)^{-2n_2}$$

with $\rho_1 > 0$ and $\rho_2 > 1/2$.

Proof. Note that for two triangles $\tau, t \in \mathcal{G}$ with $\overline{\tau} \cap \overline{t} \neq \emptyset$ we have $ch_\tau \leq h_t \leq Ch_\tau$ [see (5.47)]. By combining Lemma 5.3.16, Proposition 5.3.17, Lemma 5.3.18, Lemma 5.3.19 and Theorem 5.3.15 we obtain the assertion. □

The Regular Case

Let $\tau, t \in \mathcal{G}$ with a positive distance $d_{\tau,t}$ [see (5.64)] and for $\hat{\tau} = \hat{t} = \widehat{Q}$ let the function k_3 be as in (5.65). If either $\hat{\tau}$ or $\hat{t}$ is the unit triangle we first apply the mapping $(\zeta_1, \zeta_2) \to (\zeta_1, \zeta_1\zeta_2)$ as in Proposition 5.3.21 and the resulting function is again denoted by k_3. The polynomial degree of the boundary element space is again denoted by p.

Theorem 5.3.24. *The approximation of the integral $\int_{(0,1)^4} k_3(\hat{\mathbf{x}}, \hat{\mathbf{y}})\, d\hat{\mathbf{x}} d\hat{\mathbf{y}}$ by means of Tensor-Gauss quadrature converges exponentially with respect to the number of nodes*

$$|\mathbf{E}^{\mathbf{n}} k_3| \leq C\,(h_\tau h_t)^2\, d_{\tau,t}^{-s} \left(\left(\frac{d_{\tau,t}}{h_\tau} \right)^p (2\tilde{\rho}(\tau,t))^{-2n_3} + \left(\frac{d_{\tau,t}}{h_t} \right)^p (2\tilde{\rho}(t,\tau))^{-2n_4} \right)$$

with $\tilde{\rho}(\tau,t) = \rho \max\{d_{\tau,t}/h_\tau, 1\}$ and $\rho > 1/2$.

Proof. The statement follows from Lemma 5.3.20, Proposition 5.3.21 and Theorem 5.3.15. □

5.3.3 The Influence of Quadrature on the Discretization Error

In Chap. 4 we introduced the Galerkin boundary element method for the abstract variational problem: Find $u \in H$ such that

$$a(u,v) = F(v) \qquad \forall v \in H \tag{5.67}$$

with

$$a(u,v) = \int_\Gamma \lambda_1(\mathbf{x})\, u(\mathbf{x})\, \overline{v}(\mathbf{x})\, ds_{\mathbf{x}} + \int_\Gamma p.v. \int_\Gamma k(\mathbf{x},\mathbf{y},\mathbf{y}-\mathbf{x})\, u(\mathbf{y})\, \overline{v}(\mathbf{x})\, ds_{\mathbf{x}} ds_{\mathbf{y}}. \tag{5.68}$$

The boundary element space is given by the abstract notation $S \subset H$ and the local nodal basis is denoted by $(b_i)_{i=1}^N$. With this we can define the linear system of equations

$$\mathbf{A}\mathbf{u} = \mathbf{F}$$

with

$$\mathbf{A}_{i,j} = a(b_j, b_i) \qquad 1 \leq i,j \leq N \quad \text{and} \quad \mathbf{F}_i = F(b_i) \qquad 1 \leq i \leq N.$$

The coefficient vector $\mathbf{u}$ is associated with the Galerkin solution by $u_S = \sum_{i=1}^N \mathbf{u}_i b_i$. The approximation of the matrix entries and also the right-hand side by means of numerical quadrature leads to a "perturbed" linear system of equations

$$\widetilde{A}\tilde{\mathbf{u}} = \widetilde{F},$$

which in turn can be written as a variational problem: Find $\tilde{u}_S \in S$ such that

$$\tilde{a}\left(\tilde{u}_S, v\right) = \widetilde{F}\left(v\right), \qquad \forall v \in S.$$

The error $u - \tilde{u}_S$ was abstractly analyzed in Sect. 4.2.4. In this subsection we will apply these results to the perturbation by numerical quadrature and derive a relation between the convergence rate of the Galerkin discretization and the local quadrature order.

First we will need some notation. The quadrature method for the approximation of the integrals

$$I_{\tau\times t}^{i,j} := \int_{\tau} p.v. \int_{t} k\left(\mathbf{x}, \mathbf{y}, \mathbf{y}-\mathbf{x}\right) b_i\left(\mathbf{x}\right) b_j\left(\mathbf{y}\right) ds_{\mathbf{y}} ds_{\mathbf{x}} \tag{5.69}$$

is denoted by $Q_{\tau\times t}^{i,j}$. The associated error is given by

$$E_{\tau\times t}^{i,j} := I_{\tau\times t}^{i,j} - Q_{\tau\times t}^{i,j}.$$

We use the convention that for boundary element functions u, v the coefficient vector is always denoted by $\mathbf{u}, \mathbf{v} \in \mathbb{C}^N$ in the basis representation.

For boundary element functions $u, v \in S$ we set

$$I_{\tau\times t}\left(u, v\right) := \sum_{i,j=1}^{N} \mathbf{u}_i \bar{\mathbf{v}}_j I_{\tau\times t}^{i,j} \tag{5.70}$$

and, similarly, define $Q_{\tau\times t}\left(u, v\right)$, $E_{\tau\times t}\left(u, v\right)$. Note that the sum in (5.70) can be reduced to a sum over i, j with $|\operatorname{supp} b_i \cap \tau| > 0 \wedge \left|\operatorname{supp} b_j \cap t\right| > 0$. This motivates the definition of the index set $\mathcal{I}_\tau$ by

$$\mathcal{I}_\tau := \left\{i : |\operatorname{supp} b_i \cap \tau| > 0\right\}. \tag{5.71}$$

Finally, we set

$$E_{\tau\times t}^{\max} := \max_{(i,j)\in\mathcal{I}_\tau\times\mathcal{I}_t} \left|E_{\tau\times t}^{i,j}\right|. \tag{5.72}$$

Assumption 5.3.25(a), (b) is satisfied for all boundary element spaces from Chap. 4.

Assumption 5.3.25. *(a) There exists a constant $P > 0$ such that*

$$\max\left\{\sharp\mathcal{I}_\tau : \tau \in \mathcal{G}\right\} \le P.$$

(b) *There exist constants $\lambda_{\min}$, $\lambda_{\max}$ that depend only on the polynomial degree p such that the spectrum $\sigma(\hat{\mathbf{m}})$ of the matrix*

$$\hat{\mathbf{m}} := \left(\int_{\hat{\tau}} \hat{b}_i(\hat{\mathbf{x}})\, \hat{b}_j(\hat{\mathbf{x}})\, d\hat{\mathbf{x}} \right)_{i,j \in \mathcal{I}_{\hat{\tau}}}$$

satisfies the estimate

$$0 < \lambda_{\min} \leq \lambda \leq \lambda_{\max} < \infty \qquad \forall \lambda \in \sigma(\hat{\mathbf{m}}).$$

As $\left(\hat{b}_i\right)_{i \in \mathcal{I}_{\hat{\tau}}}$ is a basis in $\mathbb{P}_p$ and $(\cdot,\cdot)_{L^2(\hat{\tau})}$ defines an inner product on $\mathbb{P}_p$, $\hat{\mathbf{m}}$ is positive definite and the existence of the constants $\lambda_{\min}$, $\lambda_{\max}$ is guaranteed. However, note that these can go to either zero or infinity as the polynomial degree p increases.

Example 5.3.26. *Let $\hat{\tau}$ be the unit triangle. For:*

- *$p = 0$ we have $\hat{\mathbf{m}} = \left(\frac{1}{2}\right)$ and $\lambda_{\min} = \lambda_{\max} = 1/2$.*
- *$p = 1$ we have*

$$\hat{\mathbf{m}} = \left(\int_0^1 \int_0^{\hat{x}_1} \hat{b}_i(\hat{\mathbf{x}})\, b_j(\hat{\mathbf{x}})\, d\hat{\mathbf{x}} \right)_{i,j=1}^{3} = \begin{bmatrix} \frac{1}{12} & \frac{1}{24} & \frac{1}{24} \\ \frac{1}{24} & \frac{1}{12} & \frac{1}{24} \\ \frac{1}{24} & \frac{1}{24} & \frac{1}{12} \end{bmatrix}$$

and $\lambda_{\min} = 1/24$, $\lambda_{\max} = 1/6$.

Let $\hat{\tau}$ be the unit square. For:

- *$p = 0$ we have $\hat{\mathbf{m}} = (1)$ and $\lambda_{\min} = \lambda_{\max} = 1$.*
- *$p = 1$ we have*

$$\hat{\mathbf{m}} = \left(\int_0^1 \int_0^1 \hat{b}_i(\hat{\mathbf{x}})\, b_j(\hat{\mathbf{x}})\, d\hat{\mathbf{x}} \right)_{i,j=1}^{4} = \begin{bmatrix} \frac{1}{9} & \frac{1}{18} & \frac{1}{36} & \frac{1}{18} \\ \frac{1}{18} & \frac{1}{9} & \frac{1}{18} & \frac{1}{36} \\ \frac{1}{36} & \frac{1}{18} & \frac{1}{9} & \frac{1}{18} \\ \frac{1}{18} & \frac{1}{36} & \frac{1}{18} & \frac{1}{9} \end{bmatrix}$$

and $\lambda_{\min} = 1/36$, $\lambda_{\max} = 1/4$.

Lemma 5.3.27. *Let Assumptions 5.3.5 and 5.3.25 be satisfied. Then*

$$\left|E_{\tau\times t}(u,v)\right| \leq C h_\tau^{-1} h_t^{-1} \left|E_{\tau\times t}^{\max}\right| \|u\|_{L^2(\tau)} \|v\|_{L^2(t)}$$

for all $u, v \in S$, where C depends only on $\lambda_{\min}$, $\lambda_{\max}$, $\theta_0(\tau)$, $\theta_0(t)$ [see (5.5)] and P from Assumption 5.3.25.

Proof. We use the Cauchy–Schwarz inequality and obtain

$$\begin{aligned}|E_{\tau\times t}(u,v)| &= \left|\sum_{(i,j)\in\mathcal{I}_\tau\times\mathcal{I}_t} \mathbf{u}_i\bar{\mathbf{v}}_j E^{i,j}_{\tau\times t}\right| \le |E^{\max}_{\tau\times t}| \sum_{i\in\mathcal{I}_\tau}|\mathbf{u}_i| \sum_{j\in\mathcal{I}_t}|\mathbf{v}_j| \\ &\le |E^{\max}_{\tau\times t}|\, P \sqrt{\sum_{i\in\mathcal{I}_\tau}|\mathbf{u}_i|^2}\sqrt{\sum_{j\in\mathcal{I}_t}|\mathbf{v}_j|^2}.\end{aligned}$$

Furthermore, with $\hat{u} := u|_\tau \circ \chi_\tau$ and $g^{\max}_\tau := \max_{\hat{\mathbf{x}}\in\hat{\tau}} |g_\tau(\hat{\mathbf{x}})|$ we have the representation

$$\int_\tau |u|^2\,dx = \int_{\hat{\tau}} (g_\tau(\hat{\mathbf{x}}))\,|\hat{u}(\hat{\mathbf{x}})|^2\,d\hat{\mathbf{x}} \le g^{\max}_\tau \int_{\hat{\tau}} |\hat{u}(\hat{\mathbf{x}})|^2\,d\hat{\mathbf{x}}.$$

With $\mathbf{u}_\tau = (\mathbf{u}_i)_{i\in\mathcal{I}_\tau}$ and the matrix $\hat{\mathbf{m}}$ from Assumption 5.3.25 we obtain

$$\int_\tau |u|^2\,dx \le g^{\max}_\tau \mathbf{u}^\intercal_\tau \hat{\mathbf{m}}\bar{\mathbf{u}}_\tau \le g^{\max}_\tau \lambda_{\max} \sum_{i\in\mathcal{I}_\tau} |\mathbf{u}_i|^2.$$

In a similar way, we can show that

$$\int_\tau |u|^2\,dx \ge g^{\min}_\tau \lambda_{\min} \sum_{i\in\mathcal{I}_\tau} |\mathbf{u}_i|^2.$$

From this we have for the error $E_{\tau\times t}$ the estimate

$$|E_{\tau\times t}(u,v)| \le |E^{\max}_{\tau\times t}| \frac{P}{\sqrt{g^{\min}_\tau g^{\min}_t \lambda_{\min}}} \|u\|_{L^2(\tau)} \|v\|_{L^2(t)}.$$

Assumption 5.3.5 combined with Proposition 5.3.8 implies, as in (5.50) and in the proof of Lemma 5.3.16, the estimate

$$ch^2_\tau \le g^{\min}_\tau \le Ch^2_\tau,$$

from which we have the assertion. □

The following corollary is a by-product of the previous proof.

Corollary 5.3.28. *Under the conditions set out in Lemma 5.3.27 we have*

$$c\,\|u\|_{L^2(\Gamma)} \le h\,\|\mathbf{u}\| \le C\,\|u\|_{L^2(\Gamma)} \qquad \forall u \in S,$$

where $\mathbf{u}$ *denotes the coefficient vector of the boundary element function* u *in basis representation and* $\|\cdot\|$ *is the Euclidean vector norm in* $\mathbb{R}^N$*. The constants* c, C *depend only on* $\lambda_{\min}$, $\lambda_{\max}$, $\theta_0(\tau)$, $\theta_0(t)$ *[see (5.5)] and* P *from Assumption 5.3.25.*

We will now deduce the total error from the local error estimates. We consider the variational problem: Find $u \in S$ such that

$$\int_\Gamma \lambda_1(\mathbf{x})\, u(\mathbf{x})\, \overline{v}(\mathbf{x})\, ds_{\mathbf{x}} + \int_\Gamma \overline{v}(\mathbf{x}) \left(p.v. \int_\Gamma k(\mathbf{x},\mathbf{y},\mathbf{y}-\mathbf{x})\, u(\mathbf{y})\, ds_{\mathbf{y}} \right) ds_{\mathbf{x}}$$
$$= \int_\Gamma r(\mathbf{x})\, \overline{v}(\mathbf{x})\, ds_{\mathbf{x}} \qquad (5.73)$$

for all $v \in S$.

The integrals

$$\int_\Gamma b_i(\mathbf{x}) \int_\Gamma k(\mathbf{x},\mathbf{y},\mathbf{y}-\mathbf{x})\, b_j(\mathbf{y})\, ds_{\mathbf{y}} ds_{\mathbf{x}}$$

are regularized (see Sect. 5.2.4) and then approximated by means of the Tensor-Gauss quadrature method. The integrals

$$\int_\Gamma \lambda_1(\mathbf{x})\, b_i(\mathbf{x})\, b_j(\mathbf{x})\, ds_{\mathbf{x}}$$

are approximated by stable quadrature methods with an exactness degree m_1 and the integrals

$$\int_\Gamma b_i(\mathbf{x})\, r(\mathbf{x})\, ds_{\mathbf{x}}$$

by stable quadrature methods with an exactness degree m_2.

Let $\widetilde{A}$ be the Galerkin matrix determined by numerical quadrature and $\tilde{a}(u,v) := \langle \widetilde{A}\mathbf{u}, \mathbf{v} \rangle$ the associated, perturbed sesquilinear form. Let $\widetilde{F}$ be the right-hand side determined by numerical quadrature. We set

$$E_{\max} := \max_{\tau,t \in \mathcal{G}} h_\tau^{-1} h_t^{-1} E_{\tau\times t}^{\max}.$$

Theorem 5.3.29. *Let Assumptions 5.3.5, 5.3.25 and the conditions from Corollary 5.3.12 be satisfied. Let the sesquilinear form* $a : H^\mu(\Gamma) \times H^\mu(\Gamma) \to \mathbb{C}$ *in (5.68) be continuous, injective and coercive for some* $\mu \in \left\{-\frac{1}{2}, 0, \frac{1}{2}\right\}$*, and let* p *denote the polynomial degree of the boundary element space* S.

Then

$$|a(u,v) - \tilde{a}(u,v)| \le C h^{m_1+1+2\mu-2p} \|u\|_{H^\mu(\Gamma)} \|v\|_{H^\mu(\Gamma)}$$
$$+ E_{\max}(\sharp\mathcal{G}) \|u\|_{L^2(\Gamma)} \|v\|_{L^2(\Gamma)}$$

for all $u, v \in S$. *The constant* C *does not depend on* h *but, in general, it does depend on* $\lambda_{\min}$, $\lambda_{\max}$, $\theta_0(\tau)$, $\theta_0(t)$ *[see (5.5)],* P *from Assumption 5.3.25 and the quasi-uniformity of the mesh (see Definition 4.1.13).*

Proof. Let E_τ be as in (5.44) and $E_{\tau\times t}$ as in (5.70). We define $\lceil\mu\rceil$ as the smallest integer that satisfies $\lceil\mu\rceil \geq \mu$. We observe that every introduced boundary element space satisfies $S \subset H^{\lceil\mu\rceil}(\Gamma)$.

The quadrature error in the sesquilinear form in (5.73) can be estimated by the inverse inequality (see Theorem 4.4.2) according to

$$\begin{aligned}
|a(u,v) - \tilde{a}(u,v)| &\leq \sum_{\tau\in\mathcal{G}} |E_\tau(u,v)| + \sum_{\tau,t\in\mathcal{G}} |E_{\tau\times t}(u,v)| \\
&\leq C \sum_{\tau\in\mathcal{G}} h_\tau^{m_1+1} \|u\|_{H^p(\tau)} \|v\|_{H^p(\tau)} + \sum_{\tau,t\in\mathcal{G}} h_\tau^{-1} h_t^{-1} E_{\tau\times t}^{\max} \|u\|_{L^2(\tau)} \|v\|_{L^2(t)} \\
&\leq Ch^{m_1+1+2\lceil\mu\rceil-2p} \sum_{\tau\in\mathcal{G}} \|u\|_{H^{\lceil\mu\rceil}(\tau)} \|v\|_{H^{\lceil\mu\rceil}(\tau)} + E_{\max} \sum_{\tau\in\mathcal{G}} \|u\|_{L^2(\tau)} \sum_{t\in\mathcal{G}} \|v\|_{L^2(t)} \\
&\leq Ch^{m_1+1+2\lceil\mu\rceil-2p} \|u\|_{H^{\lceil\mu\rceil}(\Gamma)} \|v\|_{H^{\lceil\mu\rceil}(\Gamma)} + E_{\max}(\sharp\mathcal{G}) \|u\|_{L^2(\Gamma)} \|v\|_{L^2(\Gamma)} \\
&\leq Ch^{m_1+1+2\mu-2p} \|u\|_{H^{\mu}(\Gamma)} \|v\|_{H^{\mu}(\Gamma)} + E_{\max}(\sharp\mathcal{G}) \|u\|_{L^2(\Gamma)} \|v\|_{L^2(\Gamma)} .
\end{aligned}$$

□

With respect to Theorems 4.2.11 and 4.2.18 we can now determine the necessary quadrature orders. For a prescribed consistency tolerance $|a(u,v) - \tilde{a}(u,v)| \leq \delta$ the local quadrature order can be determined by Theorem 5.3.29 and, by means of Theorems 5.3.23 and 5.3.24, the number of Gaussian points per coordinate direction can be determined.

We will carry out this process for characteristic examples. It is our goal to choose the number of quadrature points in such a way that the order of convergence of the original Galerkin method is maintained for the Galerkin method that is perturbed by the quadrature. For this we assume that the Galerkin method converges with an optimal order of convergence, i.e., the exact solution is sufficiently regular. In this discussion we will restrict ourselves to shape-regular, quasi-uniform meshes (see Definitions 4.1.12 and 4.1.13). For the general case we refer to [105, 106].

We assume that the sesquilinear form $a : H^\mu(\Gamma) \times H^\mu(\Gamma) \to \mathbb{C}$ in (5.68) is continuous, injective and coercive with a $\mu \in \left\{-\frac{1}{2}, 0, \frac{1}{2}\right\}$. The local polynomial degree of the boundary element spaces is denoted by $p \in \mathbb{N}_0$. If the surface is sufficiently smooth and the solution of Problem (5.67) is sufficiently regular the error estimate for the Galerkin error (see Sect. 4.3)

$$\|u - u_S\|_{H^\mu(\Gamma)} \leq Ch^{p+1-\mu} \|u\|_{H^{p+1}(\Gamma)} \tag{5.74}$$

holds. In order to also obtain this order of convergence for the perturbed problem, we choose the tolerance δ in Theorem 4.2.11 as $\delta = Ch^{p+1-\mu}$. From the estimate in Theorem 5.3.29 and with the inverse estimate we obtain for all $u, v \in S$

$$|a(u,v) - \tilde{a}(u,v)| \leq C\left(h^{m_1-2p+2\mu+1}\|u\|_{H^\mu(\Gamma)} + E_{\max}(\sharp\mathcal{G})h^{\tilde{\mu}}\|u\|_{L^2(\Gamma)}\right)\|v\|_{H^\mu(\Gamma)},$$

$$|a(u,v) - \tilde{a}(u,v)| \leq C\left(h^{m_1-2p+2\mu+1} + E_{\max}(\sharp\mathcal{G})\, h^{2\tilde{\mu}}\right) \|u\|_{H^\mu(\Gamma)} \|v\|_{H^\mu(\Gamma)},$$

with $\tilde{\mu} := \min\{\mu, 0\}$. If the local quadrature error and the local degree of exactness m_1 satisfy the estimates

$$\begin{aligned} \left|E_{\tau\times t}^{i,j}\right| &\le Ch^{p+5-\mu-\tilde{\mu}} \quad \forall \tau, t \in \mathcal{G}, \quad \forall (i,j) \in \mathcal{I}(\tau) \times \mathcal{I}(t), \\ m_1 &\ge 3(p-\mu) \end{aligned} \tag{5.75}$$

we deduce with $\sharp\mathcal{G} \le Ch^{-2}$ that

$$\begin{aligned} |a(u,v) - \tilde{a}(u,v)| &\le Ch^{p+1-\mu} \|u\|_{H^{\max\{\mu,0\}}(\Gamma)} \|v\|_{H^{\mu}(\Gamma)}, \\ |a(u,v) - \tilde{a}(u,v)| &\le C\left(h^{p+1-\mu} + h^{p+1-\mu+\tilde{\mu}}\right) \|u\|_{H^{\mu}(\Gamma)} \|v\|_{H^{\mu}(\Gamma)}. \end{aligned} \tag{5.76}$$

Since $p+1-\mu > 0$ and $p+1-\mu+\tilde{\mu} \ge \min\{1, 1-\mu\} \ge 1/2$ the terms $h^{p+1-\mu}$ and $h^{p+1-\mu+\tilde{\mu}}$ converge to zero for $h \to 0$, and therefore the Galerkin method with quadrature is stable and consistent for $h \le h_0$.

Theorem 5.3.30. *Let the mesh $\mathcal{G}$ be quasi-uniform with $h \le h_0 < 1$ and let the assumptions from Theorem 5.3.29 hold. The order of singularity of the kernel function is denoted by s as in Assumption 5.1.19. Let the quadrature orders be chosen as*

$$n_1 \ge \frac{3p+4-\mu-\tilde{\mu}}{2}, \qquad n_2 \ge \frac{(p+1+s-\mu-\tilde{\mu})\,|\log h|}{2\log(2\rho_2)} \tag{5.77}$$

in the singular case and as

$$n_3 = n_4 \ge \frac{\log\left(h^{-2p-1+\mu+\tilde{\mu}} d_{\tau,t}^{p-s}\right)}{2\log(2\rho\chi_{\tau,t})} \tag{5.78}$$

with $\chi_{\tau,t} := \max\{\operatorname{dist}(\tau,t)/h, 1\}$ in the regular case. We assume that the exactness degrees for the approximation of the integrals in (5.1) satisfy

$$m_1 \ge 3(p-\mu) \quad \text{and} \quad m_2 \ge 2(p-\mu). \tag{5.79}$$

Then the Galerkin method with quadrature is stable [see (4.152)].

If the exact solution u is contained in $H^t(\Gamma)$ for a $t \in [\max\{0,\mu\}, p+1]$ and the right-hand side in (5.73) satisfies $r \in C^{m_2+1}(\Gamma)$ the solution $\tilde{u}_S$ of the perturbed problem satisfies the error estimate

$$\|u - \tilde{u}_S\|_{H^{\mu}(\Gamma)} \le Ch^{t-\mu} \|u\|_{H^t(\Gamma)}.$$

Before we prove this theorem we will make a remark concerning its consequences.

Remark 5.3.31. *The estimates (5.77), (5.78) show that the quadrature orders n_2, n_3, n_4 should grow logarithmically with $h \to 0$ for singular integrals and for the nearly-singular integrals with* dist $(\tau, t) \sim O(h)$. *In the far field, i.e., for $\chi_{\tau,t} \sim h^{-1}$ and $d_{\tau,t} \sim 1$, the quadrature orders are independent of the step size h.*

Theorem 5.3.30 is explicit with respect to the step size h. The quantities ρ and ρ_2 usually depend on the polynomial degree p, which is, however, always fixed for the discretizations under consideration.

Proof of Theorem 5.3.30. Let $k_\star : (0,1)^4 \to \mathbb{C}$ denote one of the integrands from Sect. 5.2.4.*I–III*. If we insert (5.77) in the estimates from Theorem 5.3.23 we obtain

$$|\mathbf{E}^{\mathbf{n}} k_\star| \leq C_1 h_\tau^{5+p-\mu-\tilde{\mu}}$$

with a constant C_1 that depends only on the constants C, ρ_1 from Theorem 5.3.23 as well as $p, \mu, \tilde{\mu}$.

If we combine the estimate (5.78) with the estimate from Theorem 5.3.24, by virtue of $h \leq 1$ and $1 \leq \chi_{\tau,t} = \max\{d_{\tau,t}/h, 1\}$ we obtain

$$|\mathbf{E}^{\mathbf{n}} k_3| \leq C_2 h^{p+5-\mu-\tilde{\mu}},$$

with a constant C_2 that depends on the same parameters as C_1.

Since the choice of m_1 in (5.79) is the same as in (5.75), we have the estimates (5.76) and, thus, the stability and consistency of the sesquilinear form.

The consistency estimate for the approximation of the right-hand side $\int_\Gamma v(\mathbf{x})\, r(\mathbf{x})\, d\mathbf{x}$ still needs to be shown. We then have with E_τ^m from (5.44), Corollary 5.3.12 and the inverse inequality

$$\begin{aligned}
\left|F(\mathbf{v}) - \widetilde{F}(\mathbf{v})\right| &= \sum_{\tau \in \mathcal{G}} |E_\tau^{m_2}(rv)| \leq C \sum_{\tau \in \mathcal{G}} h_\tau^{m_2+2} \|v\|_{H^p(\tau)} \|r\|_{C^{m_2+1}(\tau)} \\
&\leq C h^{m_2+2-p+\mu} (\sharp\mathcal{G})^{1/2} \|v\|_{H^\mu(\Gamma)} \|r\|_{C^{m_2+1}(\Gamma)} \\
&\leq C h^{p+1-\mu} \|v\|_{H^\mu(\Gamma)} \|r\|_{C^{m_2+1}(\Gamma)}.
\end{aligned}$$

We deduce from Theorem 4.2.11 with $\mu_+ := \max\{\mu, 0\}$ that

$$\begin{aligned}
\|u - \tilde{u}_S\|_{H^\mu(\Gamma)} \leq C\Big\{ &\min_{w_\ell \in S_\ell} \left(\|u - w_\ell\|_{H^\mu(\Gamma)} + h^{p+1-\mu} \|w_\ell\|_{H^{\mu_+}(\Gamma)} \right) \\
&+ h^{p+1-\mu} \|r\|_{C^{m_2+1}(\Gamma)} \Big\}.
\end{aligned} \tag{5.80}$$

Let w_ℓ be the best approximation of the solution u with respect to the $H^{\mu_+}(\Gamma)$-norm. It satisfies $\|w_\ell\|_{H^{\mu_+}(\Gamma)} \leq \|u\|_{H^{\mu_+}(\Gamma)}$ and it follows from Theorem 4.2.17 that $\|u - w_\ell\|_{H^\mu(\Gamma)} < C h^{t-\mu} \|u\|_{H^t(\Gamma)}$. The condition $u \in H^t$ for a $t \in [\mu_+, p+1]$ guarantees the existence of $\|u\|_{H^{\mu_+}(\Gamma)}$. □

The error estimates that we have developed thus far can not be directly applied to hypersingular kernel functions that have been regularized by means of integration by parts (see Theorem 3.3.22). The sesquilinear form contains an additive term of the form

$$\int_{\Gamma\times\Gamma} G(\mathbf{x}-\mathbf{y}) \langle Du(\mathbf{x}), Dv(\mathbf{y})\rangle \, ds_{\mathbf{x}} ds_{\mathbf{y}},$$

where D denotes a tangential derivative of first order which satisfies

$$\|Du\|_{L^2(\Gamma)} \le C \|u\|_{H^1(\Gamma)}.$$

These integrals can be regularized with the transformations that were developed in Sect. 5.2 and can then be approximated by Tensor-Gauss quadrature methods. However, the error estimates have to be slightly modified. We will summarize these modifications:

(a) The size of the analyticity ellipses remains qualitatively unchanged in the singular case. In order to estimate the integrand in relative coordinates, the basis functions $B_{i,j} := \langle Db_i, Db_j\rangle$ have to be estimated on the analyticity ellipses. In general, the operator D also contains derivatives of order zero so that the polynomial degree of the basis functions is not necessarily reduced by one order after applying D. Therefore the estimates (5.56), (5.61), (5.63) and (5.66) remain valid unchanged. Note that for the order of singularity of the kernel functions we have $s = 1$ in the representation through integration by parts.
(b) It follows from the arguments above that the error estimates from Theorem 5.3.23 and Theorem 5.3.24 (with $s = 1$) remain valid.
(c) The hypersingular integral equation in Theorem 3.3.22 is an equation of first kind so that we have $\lambda_1 = 0$ in (5.68).
(d) For the error analysis we set

$$I_{\tau\times t}^{i,j} := \int_{\tau\times t} k(\mathbf{x},\mathbf{y},\mathbf{y}-\mathbf{x}) \langle Db_i(\mathbf{x}), Db_j(\mathbf{y})\rangle \, ds_{\mathbf{x}} ds_{\mathbf{y}},$$

and we denote the associated quadrature method by $Q_{\tau\times t}^{i,j}$ and the quadrature error by $E_{\tau\times t}^{i,j}$. The error $E_{\max}$ is now defined by the new quantities $I_{\tau\times t}^{i,j}$, $Q_{\tau\times t}^{i,j}$ as in (5.72). Under the same conditions as in Theorem 5.3.29 we have

$$\begin{aligned} |a(u,v) - \tilde{a}(u,v)| &\le E_{\max}(\sharp\mathcal{G}) \|u\|_{L^2(\Gamma)} \|v\|_{L^2(\Gamma)} \\ &\le E_{\max}(\sharp\mathcal{G}) \|u\|_{H^{1/2}(\Gamma)} \|v\|_{H^{1/2}(\Gamma)}. \end{aligned}$$

(e) The formulas for the quadrature orders in Theorem 5.3.30 are applied with $s = 1$ and the results are carried over as is appropriate.

5.3.4 *Overview of the Quadrature Orders for the Galerkin Method with Quadrature*

5.3.4.1 Integral Equations of Negative Order

We first consider the boundary integral operator $V : H^{-1/2}(\Gamma) \to H^{1/2}(\Gamma)$ for the single layer potential. The following table summarizes the number of quadrature nodes and the degree of exactness for various polynomial degrees. Note that there is no integral of the form (5.1). The quadrature order for regular integrals depends on h and $\chi_{\tau,t} = \max\{d_{\tau,t}/h, 1\}$, i.e., $n_3 = n_4 = n_{reg}(h, \chi_{\tau,t})$. The case $n_{reg}(h, 1)$ corresponds to the almost singular case, i.e., $\operatorname{dist}(\tau, t) \sim h$ and the case $n_{reg}(h, h^{-1})$ corresponds to far field integrals, i.e., $\operatorname{dist}(\tau, t) = O(1)$.

	n_1	n_2	$n_{reg}(h, \chi_{\tau,t})$	$n_{reg}(h, 1)$	$n_{reg}(h, h^{-1})$	m_2
$p = 0$	3	$\lceil 3C_1 \lvert\log h\rvert \rceil$	$\left\lceil \frac{\lvert\log(h^{-2} d_{\tau,t}^{-1})\rvert}{2\log(2\rho\chi_{\tau,t})} \right\rceil$	$\lceil C_2 \frac{3}{2} \lvert\log h\rvert \rceil$	1	1
$p = 1$	4	$\lceil 4C_1 \lvert\log h\rvert \rceil$	$\left\lceil \frac{\lvert\log(h^{-4})\rvert}{2\log(2\rho\chi_{\tau,t})} \right\rceil$	$\lceil C_2 2 \lvert\log h\rvert \rceil$	2	3

The constants C_1, C_2 are independent of p, h and $\operatorname{dist}(\tau, t)$.

5.3.4.2 Equations of Order Zero

We now consider the boundary integral operator for the double layer potential or for the adjoint double layer potential. We consider the mapping as an operator $K : L^2(\Gamma) \to L^2(\Gamma)$. If we apply the formulas from Theorem 5.3.30 to an operator of order zero, we obtain the following expressions for the number of quadrature nodes or for the required exactness degrees. Note that in this case we encounter an integral of the type 5.1.

	n_1	n_2	$n_{reg}(h, \chi_{\tau,t})$	$n_{reg}(h, 1)$	$n_{reg}(h, Ch^{-1})$	m_1	m_2
$p = 0$	2	$\lceil 3C_3 \lvert\log h\rvert \rceil$	$\left\lceil \frac{\lvert\log(h^{-1} d_{\tau,t}^{-2})\rvert}{2\log(2\rho\chi_{\tau,t})} \right\rceil$	$\lceil \frac{3}{2} C_4 \lvert\log h\rvert \rceil$	1	0	0
$p = 1$	4	$\lceil 4C_3 \lvert\log h\rvert \rceil$	$\left\lceil \frac{\lvert\log(h^{-3} d_{\tau,t}^{-1})\rvert}{2\log(2\rho\chi_{\tau,t})} \right\rceil$	$\lceil 2C_4 \lvert\log h\rvert \rceil$	2	3	2

5.3.4.3 Equations of Positive Order

We now consider the boundary integral operator $W : H^{1/2}(\Gamma) \to H^{-1/2}(\Gamma)$ of order 1. The following table gives us the required number of quadrature nodes and exactness degrees. Note that there is no term $\int_\Gamma \lambda_1 uv d\mathbf{x}$, i.e., m_1 need not be considered. The minimal polynomial degree is $p = 1$ and here we consider the ansatz by means of integration by parts.

	n_1	n_2	$n_{reg}(h, \chi_{\tau,t})$	$n_{reg}(h, 1)$	$n_{reg}(h, Ch^{-1})$	m_2
$p = 1$	4	$\lceil 2C_5 \lvert\log h\rvert \rceil$	$\left\lceil \frac{5\lvert\log h\rvert}{4\log(2\rho\chi_{\tau,t})} \right\rceil$	$\lceil C_6 \lvert\log h\rvert \rceil$	2	1

Remark 5.3.32. *Our analysis of the quadrature in the far field can be refined by using the higher regularity of the boundary element functions. More precisely, we often have $S \subset H^{\mu+s}(\Gamma)$ with an $s > 0$, where $H^\mu(\Gamma)$ denotes the energy space [see (5.74)]. Details can be found in [200] and [140].*

5.4 Additional Results and Quadrature Techniques

In the work of O. Bruno et al. (e.g. [34, 35, 161]) on frequency robust boundary integral equation based solvers for electromagnetic and acoustic scattering, a key component is the use of frequency dependent numerical integration to deal efficiently with the highly oscillatory integrand functions (the oscillations stemming from the Helmholtz kernel and its derivatives but also on special, oscillatory non-polynomial shape functions which resolve the high frequency components of the solution). This requires the use of stationary phase asymptotics for the efficient numerical evaluation of the oscillatory surface integrals. We refer to, e.g., [36] for details and applications. An alternative is the use of frequency adapted quadrature methods, such as Filon's quadrature rule which has been used by [163].

Chapter 6
Solution of Linear Systems of Equations

The Galerkin boundary element method transforms the boundary integral equation to the linear system of equations

$$\mathbf{K}\mathbf{u} = \mathbf{f}, \tag{6.1}$$

where $\mathbf{K}$ is the system matrix of the integral operator and $\mathbf{f}$ is the load vector. In this chapter we study the efficient solution of (6.1). If the dimension of the linear system of equations is very large, i.e., $N = \dim \mathbf{K} \sim 10^4 - 10^6$, direct methods such as Gauss elimination become impractical, as their complexity grows proportionally to N^3. Instead, one should use iterative methods to solve the system. As will be explained in Chap. 7, iterative methods for the solution of linear systems of equations do *not* require that the matrix $\mathbf{K}$ be known explicitly. Their complexity is dominated by the cost of a matrix-vector multiplication multiplied by the number of iteration steps for the computation of a sufficiently accurate solution of the linear system. The *cluster method*, which is discussed in Chap. 7, provides an approximative matrix-vector multiplication with a complexity of $O(N \log N)$. This is achieved with the help of an alternative representation of the Galerkin discretization. As opposed to this, the evaluation of $\mathbf{K}\mathbf{u}$ with a dense matrix $\mathbf{K}$ in the usual basis representation has a complexity of $O(N^2)$.

Since (6.1) already contains the discretization error, an exact solution $\mathbf{u} = \mathbf{K}^{-1}\mathbf{f}$ is not necessary. It suffices to solve (6.1) approximately with a precision which is of the same size as the discretization error. Therefore in this chapter we discuss the most important iterative methods for the solution of (6.1): the cg method by Hestenes and Stiefel (see [129]) for a symmetric (*Hermitian* in the complex case) $\mathbf{K}$, as well as certain steepest descent methods that are of the same type as the *minimal residual methods* for a non-symmetric (*non-Hermitian* in the complex case) $\mathbf{K}$. The convergence rate of classical iterative methods is determined by the condition of the matrix $\mathbf{K}$. Equations of the second kind are usually well-conditioned, in which case the convergence rate of iterative methods is independent of the dimension of the matrix. The condition of the matrix $\mathbf{K}$ for equations of the first kind usually grows at a rate of h^{-1} with a decreasing mesh width h (see Lemma 4.5.1). This means that the number of iterations necessary in order to reach a prescribed stopping condition grows as the dimension increases. If the dimension of the matrix $\mathbf{K}$ becomes so large that the computing time for the iterative solver dominates the overall solution

S.A. Sauter and C. Schwab, *Boundary Element Methods*, Springer Series in Computational Mathematics 39, DOI 10.1007/978-3-540-68093-2_6,

process, preconditioning methods should be used to improve the condition of the transformed system. We will discuss an example of this situation in Sect. 6.5.

6.1 cg Method

First we recall the definition of positive definite matrices. Let $\mathbb{K} \in \{\mathbb{R}, \mathbb{C}\}$. The Euclidean inner product on $\mathbb{K}^N$ is defined by

$$\langle \mathbf{u}, \mathbf{v} \rangle = \sum_{i=1}^{N} \mathbf{u}_i \overline{\mathbf{v}_i},$$

where $\alpha \to \overline{\alpha}$ denotes complex conjugation. For a matrix $\mathbf{A} \in \mathbb{K}^{N\times N}$ the adjoint matrix is defined by $\mathbf{A}^{\mathrm{H}} := \left(\overline{\mathbf{A}_{j,i}}\right)_{i,j=1}^{N}$. $\mathbf{A}$ is Hermitian if $\mathbf{A} = \mathbf{A}^{\mathrm{H}}$. $\mathbf{A}$ is positive definite if

$$\langle \mathbf{Au}, \mathbf{u} \rangle > 0 \qquad \forall \mathbf{u} \in \mathbb{K}^N \backslash \{0\}.$$

The cg ("conjugate-gradient") method by Hestenes and Stiefel for the solution of (6.1) with positive definite matrices $\mathbf{K}$ is based on the derivation of (6.1) by minimizing a quadratic functional, i.e., computing $\mathbf{u}^\star \in \mathbb{K}^N$ so that

$$J(\mathbf{u}^\star) = \min\left\{J(\mathbf{u}) : \mathbf{u} \in \mathbb{K}^N\right\} \quad \text{with} \quad J(\mathbf{u}) := \frac{1}{2}\langle \mathbf{K}\mathbf{u}, \mathbf{u}\rangle - \operatorname{Re}\langle \mathbf{f}, \mathbf{u}\rangle . \tag{6.2}$$

For $\mathbf{u} \in \mathbb{K}^N$, $\mathbf{r}(\mathbf{u}) := \mathbf{f} - \mathbf{Ku}$ denotes the *residual* of $\mathbf{u}$.

6.1.1 cg Basic Algorithm

The cg algorithm constructs a sequence $(\mathbf{u}_i)_{i\in\mathbb{N}}$ of vectors $\mathbf{u}_i \in \mathbb{K}^N$ that converges to the exact solution of (6.1). We begin with the cg algorithm in its most basic version.

Algorithm 6.1.1.
Initialization:

$$\begin{aligned}
\mathbf{u}_0 &\in \mathbb{K}^N \textit{ given} \\
\mathbf{r}_0 &:= \mathbf{f} - \mathbf{K}\mathbf{u}_0 \\
\mathbf{s}_0 &\in \mathbb{K}^N \\
\mathbf{p}_0 &:= \mathbf{s}_0 \in \mathbb{K}^N
\end{aligned} \tag{6.3}$$

Iterations: *For* $i = 0, 1, 2, \ldots$ *evaluate*

$$\begin{aligned}
\alpha_{i+1} &\in \mathbb{K}\,,\\
\mathbf{u}_{i+1} &:= \mathbf{u}_i + \alpha_{i+1}\,\mathbf{p}_i\,,\\
\mathbf{r}_{i+1} &:= \mathbf{r}_i - \alpha_{i+1}\,\mathbf{K}\mathbf{p}_i\,,\\
\mathbf{s}_{i+1} &\in \mathbb{K}^N\,,\\
\beta_{i+1} &\in \mathbb{K}\,,\\
\mathbf{p}_{i+1} &:= \mathbf{s}_{i+1} + \beta_{i+1}\,\mathbf{p}_i \in \mathbb{K}^N\,.
\end{aligned} \tag{6.4}$$

We still need to determine α_i, β_i, $\mathbf{s}_i$. For this we note that for the derivative of J in the direction of a vector $\mathbf{z} \in \mathbb{K}^N$ at the point $\mathbf{u}_i$ we have

$$\langle J'(\mathbf{u}_i), \mathbf{z}\rangle = \mathrm{Re}\,\langle \mathbf{K}\mathbf{u}_i - \mathbf{f}, \mathbf{z}\rangle = -\,\mathrm{Re}\,\langle \mathbf{r}_i, \mathbf{z}\rangle\,. \tag{6.5}$$

Therefore $\mathbf{u}_i \in \mathbb{K}^N$ minimizes $J(\cdot)$ on an affine subspace $Z \subset \mathbb{K}^N$ if and only if $\mathrm{Re}\,\langle \mathbf{r}_i, \mathbf{z}\rangle = 0$ for all $\mathbf{z} \in Z$.

In the following Z will be spanned by the directions of descent $\mathbf{p}_j$ for $j < i$. We define α_{i+1} by

$$\alpha_{i+1} := \frac{\langle \mathbf{r}_i, \mathbf{p}_i\rangle}{\langle \mathbf{K}\mathbf{p}_i, \mathbf{p}_i\rangle}\,, \quad \text{if} \quad \langle \mathbf{K}\mathbf{p}_i, \mathbf{p}_i\rangle \neq 0\,. \tag{6.6}$$

We have $\langle \mathbf{r}_0, \mathbf{p}_0\rangle = \langle \mathbf{r}_0, \mathbf{s}_0\rangle$ and from (6.4) we have $\langle \mathbf{r}_i, \mathbf{p}_{i-1}\rangle = 0$ for $i \geq 1$, from which we deduce by using (6.4) that

$$\langle \mathbf{r}_i, \mathbf{p}_i\rangle = \langle \mathbf{r}_i, \mathbf{s}_i + \beta_i\,\mathbf{p}_{i-1}\rangle = \langle \mathbf{r}_i, \mathbf{s}_i\rangle, \qquad i \geq 1. \tag{6.7}$$

With (6.6) this yields

$$\alpha_{i+1} = \frac{\langle \mathbf{r}_i, \mathbf{s}_i\rangle}{\langle \mathbf{K}\mathbf{p}_i, \mathbf{p}_i\rangle}\,. \tag{6.8}$$

We determine β_{i+1} in Algorithm 6.1.1. We have

$$\begin{aligned}
\overline{\alpha}_{i+1}\langle \mathbf{r}_{i+2}, \mathbf{p}_i\rangle &= \overline{\alpha}_{i+1}\langle \mathbf{r}_{i+1} - \alpha_{i+2}\,\mathbf{K}\mathbf{p}_{i+1}, \mathbf{p}_i\rangle = \alpha_{i+2}\langle \mathbf{p}_{i+1}, -\alpha_{i+1}\mathbf{K}\mathbf{p}_i\rangle\\
&= \alpha_{i+2}\langle \mathbf{s}_{i+1} + \beta_{i+1}\,\mathbf{p}_i, \mathbf{r}_{i+1} - \mathbf{r}_i\rangle\,.
\end{aligned}$$

We have already seen that the new residual $\mathbf{r}_{i+2}$ is perpendicular to the direction $\mathbf{p}_{i+1}$. It is now our aim to determine β_{i+1} in such a way that $\mathbf{r}_{i+2}$ is also perpendicular to $\mathbf{p}_i$. We will show in Proposition 6.1.3 that $\mathbf{r}_{i+2}$ is then perpendicular to all $\mathbf{p}_j$, $j \leq i + 1$.

For $\langle \mathbf{r}_i, \mathbf{s}_i \rangle \neq 0$ we thus set $\alpha_{i+2} \langle \mathbf{s}_{i+1} + \beta_{i+1} \mathbf{p}_i, \mathbf{r}_{i+1} - \mathbf{r}_i \rangle = 0$ and obtain [using (6.7)]

$$\beta_{i+1} := \frac{\langle \mathbf{s}_{i+1}, \mathbf{r}_{i+1} - \mathbf{r}_i \rangle}{\langle \mathbf{s}_i, \mathbf{r}_i \rangle} . \tag{6.9}$$

One possible choice of $\mathbf{s}_i$ which ensures that $\langle \mathbf{r}_i, \mathbf{s}_i \rangle \neq 0$ for $\mathbf{r}_i \neq \mathbf{0}$ is

$$\mathbf{s}_i = \mathbf{r}_i, \; i = 0, 1, 2, \ldots . \tag{6.10}$$

Algorithm 6.1.1 combined with (6.8)–(6.10) defines the so-called *Polak–Ribière* version of the cg method [183].

6.1.2 Preconditioning Methods

We will see that the convergence rate of the cg method depends on the condition of the matrix $\mathbf{K}$. If it is large the computational complexity of the iterative solver may start to dominate the entire discretization. A remedy for this is the use of preconditioning, which we will consider on an abstract level in this section. For every regular matrix $\mathbf{C} \in \mathbb{K}^{N \times N}$ the solution of the preconditioned system

$$\mathbf{CKu} = \mathbf{Cf} \tag{6.11}$$

solves (6.1). By choosing $\mathbf{C}$ in a suitable way, we aim to decrease the condition of the system matrix $\mathbf{CK}$ in (6.11) considerably compared to the condition of $\mathbf{K}$. Note that we obtain Algorithm 6.1.1 if we choose $\mathbf{C} = \mathbf{I}$. In this section we will present the cg algorithm for the preconditioned system (6.11).

Let $\mathbf{C} : \mathbb{K}^N \to \mathbb{K}^N$ be positive definite. As an alternative to $\mathbf{s}_i = \mathbf{r}_i$ in (6.10) we may choose

$$\mathbf{s}_i = \mathbf{C}\mathbf{r}_i . \tag{6.12}$$

Then we have $\langle \mathbf{r}_i, \mathbf{s}_{i+1} \rangle = \langle \mathbf{r}_i, \mathbf{C}\mathbf{r}_{i+1} \rangle = \langle \mathbf{C}\mathbf{r}_i, \mathbf{r}_{i+1} \rangle = \langle \mathbf{s}_i, \mathbf{r}_{i+1} \rangle$. We have seen in Sect. 6.1.1 that the given choices for α_{i+1} [see (6.8)] and β_{i+1} [see (6.9)] determine the relation $\langle \mathbf{p}_i, \mathbf{r}_{i+1} \rangle = \langle \mathbf{p}_{i-1}, \mathbf{r}_{i+1} \rangle = 0$. From this and with (6.3) we have

$$\begin{aligned} &\text{for } i = 0 \qquad \langle \mathbf{s}_i, \mathbf{r}_{i+1} \rangle = \langle \mathbf{p}_0, \mathbf{r}_1 \rangle = 0 , \\ &\text{for } i \geq 1 \qquad \langle \mathbf{s}_i, \mathbf{r}_{i+1} \rangle = \langle \mathbf{p}_i - \beta_i \mathbf{p}_{i-1}, \mathbf{r}_{i+1} \rangle = 0 . \end{aligned} \tag{6.13}$$

Inserting this into (6.9) we obtain

$$\beta_{i+1} = \frac{\langle \mathbf{s}_{i+1}, \mathbf{r}_{i+1} \rangle}{\langle \mathbf{s}_i, \mathbf{r}_i \rangle}. \tag{6.14}$$

Algorithm 6.1.1, (6.8), (6.12) and (6.14) give us the *Fletcher–Reeves* version of the cg method (see [94]).

Remark 6.1.2. *1. The Hestenes–Stiefel version of the cg method uses*

$$\beta_{i+1} := \frac{\langle \mathbf{r}_{i+1} - \mathbf{r}_i, \mathbf{s}_{i+1} \rangle}{\langle \mathbf{r}_i - \mathbf{r}_{i-1}, \mathbf{s}_i \rangle}, \quad i \geq 1 . \tag{6.15}$$

In view of (6.13), this variant is equivalent to the Fletcher–Reeves version. However, for problems in nonlinear optimization the resulting methods, in general, are different.

2. *The directions $\mathbf{s}_i$ in (6.12) are opposite to the gradients of $J(\cdot)$ in the inner product $\langle \cdot, \cdot \rangle_{\mathbf{C}} := \langle \cdot, \mathbf{C} \cdot \rangle$ and thus correct the previous directions of descent. More precisely, we have*

$$\langle \nabla J\,(\mathbf{u}_i)\,, \mathbf{s}_i \rangle_{\mathbf{C}} = \mathrm{Re}\,\langle \mathbf{K}\mathbf{u}_i - \mathbf{f}, \mathbf{C}\mathbf{s}_i \rangle = -\,\mathrm{Re}\,\langle \mathbf{r}_i, \mathbf{C}\mathbf{s}_i \rangle = -\,\|\mathbf{s}_i\|^2 < 0.$$

3. *The Fletcher–Reeves and Polak–Ribière versions have their origins in the extension of the cg method to non-linear problems.*

6.1.3 Orthogonality Relations

For $i \geq 0$ we define the Krylov space of order i by

$$\mathcal{K}^i := \{p(\mathbf{C}\mathbf{K})\mathbf{s}_0 : p \in \mathbb{P}_i(\mathbb{K})\}, \tag{6.16}$$

where $\mathbb{P}_i(\mathbb{K})$ denotes polynomials of degree $\leq i$ with coefficients in $\mathbb{K}$.

Proposition 6.1.3. *As long as the cg method does not terminate because of zero division we have*

$$\forall i \geq 0 : \quad \mathrm{span}\{\mathbf{p}_0, \ldots, \mathbf{p}_i\} = \mathrm{span}\{\mathbf{s}_0, \ldots, \mathbf{s}_i\}, \tag{6.17a}$$

$$\forall 0 \leq j < i : \quad \langle \mathbf{r}_i, \mathbf{p}_j \rangle = 0 \quad \textit{and} \quad \langle \mathbf{K}\mathbf{p}_i, \mathbf{p}_j \rangle = 0, \tag{6.17b}$$

$$\forall i \geq 0 : \quad \mathrm{span}\{\mathbf{p}_0, \ldots, \mathbf{p}_i\} = \mathcal{K}^i . \tag{6.17c}$$

Proof. (a) is trivial. We will prove (b). We have $\langle \mathbf{r}_1, \mathbf{p}_0 \rangle = 0$. We assume that (b) holds for an $i \geq 1$, more precisely, we assume that $\langle \mathbf{r}_i, \mathbf{p}_j \rangle = 0$ for $j < i$ and $\langle \mathbf{K}\mathbf{p}_{i-1}, \mathbf{p}_j \rangle = 0$ for $j < i-1$. As we have already seen in Sect. 6.1.1, we have $\langle \mathbf{r}_{i+1}, \mathbf{p}_i \rangle = 0$ and $\langle \mathbf{r}_{i+1}, \mathbf{p}_{i-1} \rangle = 0$. For $j < i-1$ we obtain

$$\begin{aligned}\langle \mathbf{r}_{i+1}, \mathbf{p}_j \rangle &= \langle \mathbf{r}_i - \alpha_{i+1}\,\mathbf{K}\mathbf{p}_i, \mathbf{p}_j \rangle \\ &= -\alpha_{i+1} \langle \mathbf{K}(\mathbf{s}_i + \beta_i\,\mathbf{p}_{i-1}), \mathbf{p}_j \rangle \\ &= -\alpha_{i+1} \langle \mathbf{s}_i, \mathbf{K}\mathbf{p}_j \rangle\end{aligned}$$

$$= \alpha_{i+1}(\overline{\alpha_{j+1}})^{-1}\left\langle \mathbf{s}_i, \mathbf{r}_{j+1} - \mathbf{r}_j \right\rangle$$

$$= \alpha_{i+1}(\overline{\alpha_{j+1}})^{-1}\langle \mathbf{r}_i, \mathbf{s}_{j+1} - \mathbf{s}_j \rangle$$

$$\stackrel{(6.4)}{=} \alpha_{i+1}(\overline{\alpha_{j+1}})^{-1}\langle \mathbf{r}_i, \mathbf{p}_{j+1} - \left(1 + \beta_{j+1}\right)\mathbf{p}_j + \beta_j \mathbf{p}_{j-1} \rangle$$

$$\stackrel{\text{(induction assumption)}}{=} 0,$$

and for $j < i$: $\langle \mathbf{K}\mathbf{p}_i, \mathbf{p}_j \rangle = -\frac{1}{\alpha_{i+1}} \langle \mathbf{r}_{i+1} - \mathbf{r}_i, \mathbf{p}_j \rangle = 0.$

We prove (c) by recursion. We have $\mathbf{s}_0 = \mathbf{p}_0$ and thus $\mathcal{K}^0 = \operatorname{span}\{\mathbf{s}_0\} = \operatorname{span}\{\mathbf{p}_0\}$. Now let $\mathcal{K}^i = \operatorname{span}\{\mathbf{p}_0, \ldots, \mathbf{p}_i\}$ for $i \geq 1$ be already proven. For vectors $\mathbf{a} \in \mathbb{K}^N$ and subspaces $V \subset \mathbb{K}^N$ we use the notation $\mathbf{a} + V = \operatorname{span}\{\alpha\mathbf{a} + \mathbf{v} : \alpha \in \mathbb{K}, \mathbf{v} \in V\}$ in this proof. Then with $\beta_{i+1}\mathbf{p}_i \in \operatorname{span}\{\mathbf{p}_0, \ldots, \mathbf{p}_i\} = \mathcal{K}^i$ and (6.12) we have the equality

$$\begin{aligned} \operatorname{span}\{\mathbf{p}_0, \ldots, \mathbf{p}_{i+1}\} &= \operatorname{span}\{\mathbf{p}_0, \ldots, \mathbf{p}_i\} + \mathbf{p}_{i+1} \\ &= \mathcal{K}^i + (\mathbf{s}_{i+1} + \beta_{i+1}\mathbf{p}_i) \\ &= \mathcal{K}^i + \mathbf{C}\mathbf{r}_{i+1}. \end{aligned}$$

Since $\mathbf{C}\mathbf{r}_i = \mathbf{s}_i \in \operatorname{span}\{\mathbf{p}_0, \ldots, \mathbf{p}_i\}$ it follows from (a) that

$$\begin{aligned} \operatorname{span}\{\mathbf{p}_0, \ldots, \mathbf{p}_{i+1}\} &= \mathcal{K}^i + \mathbf{C}(\mathbf{r}_i - \alpha_{i+1}\mathbf{K}\mathbf{p}_i) \\ &= \mathcal{K}^i + \mathbf{C}\mathbf{K}\mathbf{p}_i. \end{aligned}$$

It then follows that $\operatorname{span}\{\mathbf{p}_0, \ldots, \mathbf{p}_{i+1}\} = \mathcal{K}^i + \mathbf{C}\mathbf{K}\mathbf{p}_i \subseteq \mathcal{K}^{i+1}$. We then obtain the equality by considering the dimensions. □

6.1.4 *Convergence Rate of the cg Method*

In case there are no rounding errors, Proposition 6.1.3 implies that the cg method terminates at the latest after N steps with the solution $\mathbf{u}^*$ of $\mathbf{K}\mathbf{u} = \mathbf{f}$. A far more important property of the cg method lies in the fact that it already offers very good approximations of $\mathbf{u}^*$ after only $i \ll N$ steps, as we will show in this section.

Let $\mathbf{u}_* = \mathbf{K}^{-1}\mathbf{f}$ be the solution of (6.2). Then we have

$$J(\mathbf{u}) = \frac{1}{2}\left(\|\mathbf{u} - \mathbf{u}_*\|_{\mathbf{K}}^2 - \|\mathbf{u}_*\|_{\mathbf{K}}^2\right), \tag{6.18}$$

where $\|\mathbf{u}\|_{\mathbf{K}}^2 := \langle \mathbf{u}, \mathbf{K}\mathbf{u} \rangle = \langle \mathbf{K}\mathbf{u}, \mathbf{u} \rangle$ denotes the *energy norm* for $\mathbf{K}$. Since $J'(\mathbf{u}_{i+1}) = -\operatorname{Re}\langle \mathbf{r}_{i+1}, \cdot \rangle$ vanishes on $\mathcal{K}^i = \operatorname{span}\{\mathbf{p}_0, \ldots, \mathbf{p}_i\}$, $J(\cdot)$ assumes its minimum over $\mathbf{u}_0 + \mathcal{K}^i$ in $\mathbf{u}_{i+1}$. Therefore we have for all real polynomials $p \in \mathbb{P}_i$:

$$\|\mathbf{u}_{i+1} - \mathbf{u}_*\|_{\mathbf{K}} \leq \|p(\mathbf{C}\mathbf{K})\,\mathbf{s}_0 + \mathbf{u}_0 - \mathbf{u}_*\|_{\mathbf{K}}. \tag{6.19}$$

Since $\mathbf{s}_0 = \mathbf{CK}(\mathbf{u}_* - \mathbf{u}_0)$ it follows for all $p \in \mathbb{P}_{i+1}$ with $p(0) = 1$ that we have the estimate

$$\|\mathbf{u}_{i+1} - \mathbf{u}_*\|_{\mathbf{K}} \leq \|p(\mathbf{CK})\,(\mathbf{u}_0 - \mathbf{u}_*)\|_{\mathbf{K}}\,. \tag{6.20}$$

Exercise 6.1.4. *Show for positive definite matrices* $\mathbf{K}, \mathbf{C}$ *and for an arbitrary polynomial* p *with real coefficients that*

$$\|p\,(\mathbf{CK})\,\mathbf{w}\|_{\mathbf{K}} \leq \|\mathbf{w}\|_{\mathbf{K}} \max\{|p(\lambda)| : \lambda \in \sigma\} \qquad \forall \mathbf{w} \in \mathbb{K}^N,$$

where σ *denotes the spectrum of the matrix* $\mathbf{CK}$.

One of the consequences of (6.20) is an estimate of the error after i iterations.

Proposition 6.1.5. *For all* $i \geq 0$ *and all real polynomials* $p \in \mathbb{P}_i$ *with* $p(0) = 1$ *we have*

$$\|\mathbf{u}_i - \mathbf{u}_*\|_{\mathbf{K}} \leq \max\{|p(\lambda)| : \lambda \in \sigma\}\ \|\mathbf{u}_0 - \mathbf{u}_*\|_{\mathbf{K}}\,. \tag{6.21}$$

We obtain a convergence estimate from (6.21). However, we first need a preparatory lemma.

Lemma 6.1.6. *Let* $0 < a < b$. *The problem*

$$\min\{\max\{|p(\lambda)| : \lambda \in [a,b]\} : p \in P_i \wedge p(0) = 1\} \tag{6.22}$$

has a unique solution

$$p(z) = \frac{T_i\left(\frac{b+a-2z}{b-a}\right)}{T_i\left(\frac{b+a}{b-a}\right)}\,,$$

where $T_k(x)$ *is the Čebyšev polynomial of order* k *on* $(-1,1)$ *which satisfies* $|T_k(x)| \leq 1$ *for* $|x| \leq 1$ *and* $T_k(1) = 1$.

If we combine this result with (6.21) we obtain the statement concerning convergence.

Theorem 6.1.7. *Let*

$$\sigma(\mathbf{CK}) \subseteq [a,b], \quad \kappa := b/a \geq 1\,.$$

Then for all $\mathbf{u}_0 \in \mathbb{K}^N$ *and* $i \geq 1$

$$\|\mathbf{u}_i - \mathbf{u}_*\|_{\mathbf{K}} \leq \frac{1}{T_i\left(\frac{\kappa+1}{\kappa-1}\right)}\|\mathbf{u}_0 - \mathbf{u}_*\|_{\mathbf{K}} \leq 2\left(\frac{\sqrt{\kappa}-1}{\sqrt{\kappa}+1}\right)^i \|\mathbf{u}_0 - \mathbf{u}_*\|_{\mathbf{K}}\,. \tag{6.23}$$

Proof. Proposition 6.1.5 and Lemma 6.1.6 give us the first inequality. Furthermore, we have $T_i(x) = \cos(i \arccos x)$ or

$$T_i(x) = \frac{1}{2}\left((x + \sqrt{x^2-1})^i + (x - \sqrt{x^2-1})^i\right), \tag{6.24}$$

while there are only even powers of $\sqrt{x^2-1}$ and thus (6.24) is defined for all $x \in \mathbb{R}$. We then have

$$T_i(x) \geq \frac{1}{2}\left(x + \sqrt{x^2-1}\right)^i \qquad \text{for } x \geq 1$$

and from

$$(x + \sqrt{x^2-1})|_{x=\frac{\kappa+1}{\kappa-1}>1} = \frac{\sqrt{\kappa}+1}{\sqrt{\kappa}-1}$$

we obtain the assertion. □

6.1.5 *Generalizations**

We will briefly discuss generalizations of Theorem 6.1.7. Firstly, the restriction to finite-dimensional spaces $\mathbb{K}^N$ is not important. The cg method can just as well be defined for infinite-dimensional Hilbert spaces, in which case (6.23) still remains valid if X, Y are Hilbert spaces over $\mathbb{K}$, $\langle\cdot,\cdot\rangle_{Y\times X} : Y \times X \to \mathbb{K}$ is a sesquilinear form and $A \in \mathcal{L}(X, Y)$ satisfies

$$\forall u, v \in X: \quad \langle Au, v\rangle_{Y\times X} = \overline{\langle Av, u\rangle}_{Y\times X}. \tag{6.25}$$

If $Y = X$ and $A, C \in \mathcal{L}(X, Y)$ with $CA = I + K$ and a compact $K : X \to X$ we can again determine a convergence rate from (6.21). In this respect we note that the spectrum $\sigma(K)$ of K is discrete and only has an accumulation point at zero [see Theorem 2.1.36(ii)]. Now let $\varepsilon > 0$ be given and let $q(z)$ be a polynomial such that $q(0) = 1$ and

$$q(\lambda) = 0 \qquad \forall \lambda \in \sigma(I+K); |\lambda - 1| > \varepsilon. \tag{6.26}$$

Since there are only finitely many $\lambda \in \sigma(I+K)$ that satisfy the condition $|\lambda - 1| > \varepsilon$ for every $\varepsilon > 0$, q in (6.26) exists. With $r(z) = (1-z)^j$ and $p_j(z) = q(z)r(z)$ we have $p_j(0) = 1$, degree of $p_j = j+$ degree of q and

$$\max\{|p_j(z)| : z \in \sigma(I+K)\} \leq \varepsilon^j \max\{|q(z)| : z \in \sigma(I+K)\}.$$

* This section should be read as a complement to the core material of this book.

In the following $\mathbf{u}_i$ again denotes the i-th iterate of the cg method. For all $i \geq$ degree q from (6.21) we deduce the estimate

$$\|\mathbf{u}_i - \mathbf{u}_*\|_\mathbf{K} \leq \max\{|q(z)| : z \in \sigma(I + K)\}\varepsilon^{(i-\text{degree}\, q)}\|\mathbf{u}_0 - \mathbf{u}_*\|_\mathbf{K}\,.$$

Proposition 6.1.8. *Let X be a Hilbert space over $\mathbb{K}$ and $CA = I + K$ with a compact $K : X \to X$. Let $\mathbf{u}_i$ be the i-th iterate of the cg method.*

Then for all $\varepsilon > 0$ there exists a constant $C(\varepsilon)$ such that for all i we have

$$\|\mathbf{u}_i - \mathbf{u}_*\|_\mathbf{K} \leq C\,\varepsilon^i\,\|\mathbf{u}_0 - \mathbf{u}_*\|_\mathbf{K}\,. \tag{6.27}$$

6.2 Descent Methods for Non-symmetric Systems

We consider

$$\mathbf{Ku} = \mathbf{f} \tag{6.28}$$

in $\mathbb{R}^N$ with a non-symmetric $\mathbf{K}$. The symmetric or skew symmetric parts of $\mathbf{K}$ are given by

$$\mathbf{M} = \frac{1}{2}(\mathbf{K} + \mathbf{K}^\mathrm{T}), \qquad \mathbf{R} = \frac{1}{2}(\mathbf{K} - \mathbf{K}^\mathrm{T})\,. \tag{6.29}$$

We then have $\mathbf{K} = \mathbf{M} + \mathbf{R}$. For an arbitrary matrix $\mathbf{X}$ let $\lambda_{\min}(\mathbf{X})$ and $\lambda_{\max}(\mathbf{X})$ be the eigenvalues with the smallest and largest absolute values respectively and let $\rho(\mathbf{X}) := |\lambda_{\max}(\mathbf{X})|$ be the spectral radius of $\mathbf{X}$. $\sigma(\mathbf{X})$ denotes the spectrum of $\mathbf{X}$. For a non-singular $\mathbf{X}$ the condition number with respect to the Euclidean norm $\|\cdot\|$ is given by $\kappa(\mathbf{X}) = \|\mathbf{X}\|\,\|\mathbf{X}^{-1}\|$.

6.2.1 Descent Methods

The general form of descent methods for the solution of (6.28) is described in Algorithm 6.2.1. Here we will restrict ourselves to the case $\mathbb{K} = \mathbb{R}$.

Algorithm 6.2.1 (Descent Method).

$$\mathbf{u}_0 \in \mathbb{R}^N \quad \textit{(Initial Vector)} \tag{6.30a}$$

$$\mathbf{r}_0 = \mathbf{f} - \mathbf{Ku}_0 \tag{6.30b}$$

$$\alpha_i := \frac{\langle \mathbf{r}_i, \mathbf{Kp}_i\rangle}{\langle \mathbf{Kp}_i, \mathbf{Kp}_i\rangle} \tag{6.30c}$$

$$\mathbf{u}_{i+1} := \mathbf{u}_i + \alpha_i\,\mathbf{p}_i \tag{6.30d}$$

$$\mathbf{r}_{i+1} := \mathbf{r}_i - \alpha_i\,\mathbf{Kp}_i \tag{6.30e}$$

$$\textit{Compute } \mathbf{p}_{i+1}\,. \tag{6.30f}$$

The choice (6.30c) minimizes the residual

$$\|\mathbf{r}_{i+1}\| = \|\mathbf{f} - \mathbf{K}(\mathbf{u}_i + \alpha \mathbf{p}_i)\| ,$$

with respect to α and therefore we have $\|\mathbf{r}_{i+1}\| \leq \|\mathbf{r}_i\|$ in every step. We still need to determine the algorithm that provides us with $\mathbf{p}_i$. We set

$$\mathbf{p}_{i+1} := \mathbf{r}_{i+1} + \sum_{j=0}^{i} \beta_j^{(i)} \mathbf{p}_j \qquad (6.31a)$$

$$\beta_j^{(i)} := -\langle \mathbf{K}\,\mathbf{r}_{i+1}, \mathbf{K}\mathbf{p}_j\rangle \,/\, \langle \mathbf{K}\mathbf{p}_j, \mathbf{K}\mathbf{p}_j\rangle . \qquad (6.31b)$$

The vector $\mathbf{p}_{i+1}$ from (6.30) and (6.31) minimizes $E(\mathbf{w}) := \|\mathbf{f} - \mathbf{K}\mathbf{w}\|$ over the affine space $\mathbf{u}_0 + \mathrm{span}\{\mathbf{p}_0, \ldots, \mathbf{p}_i\}$. (6.30) and (6.31) together define a generalized conjugate residual method. Ignoring round-off errors it produces the solution of (6.28) in at most N steps, just like the cg method.

Storing all the $\mathbf{p}_j$ in (6.31a) requires too much memory for large N. Therefore we replace (6.31a) by

$$\mathbf{p}_{i+1} = \mathbf{r}_{i+1} + \sum_{j=i-k+1}^{i} \beta_j^{(i)} \mathbf{p}_j \qquad (6.32)$$

for a $k \geq 0$ with $\beta_j^{(i)}$ as in (6.31b). Note that for $k = 0$ we have

$$\mathbf{p}_{i+1} = \mathbf{r}_{i+1} . \qquad (6.33)$$

In this case (6.30) is also called the "minimal residual method" (MR) or otherwise Orthomin(k).

6.2.2 *Convergence Rate of MR and Orthomin(k)*

Lemma 6.2.2. *The vectors* $(\mathbf{u}_i)$, $(\mathbf{r}_i)$ *and* $(\mathbf{p}_i)$ *for the MR or Orthomin* (k) *satisfy*

$$\langle \mathbf{K}\mathbf{p}_i, \mathbf{K}\mathbf{p}_j\rangle = 0 \quad j = i-k, \ldots, i-1,\ i \geq k , \qquad (6.34a)$$

$$\langle \mathbf{r}_i, \mathbf{K}\mathbf{p}_j\rangle = 0 \quad j = i-k-1, \ldots, i-1,\ i \geq k+1 , \qquad (6.34b)$$

$$\langle \mathbf{r}_i, \mathbf{K}\mathbf{p}_i\rangle = \langle \mathbf{r}_i, \mathbf{K}\mathbf{r}_i\rangle . \qquad (6.34c)$$

Proof. (6.34a) follows from (6.31), (6.32) by induction over i. We recommend the proof as an exercise.

The relation in (6.34b) follows by induction over i. Thus we let (6.34b) hold for all $i \le I$. Then with (6.30e) we have

$$\langle \mathbf{r}_{i+1}, \mathbf{K}\mathbf{p}_j \rangle = \langle \mathbf{r}_i, \mathbf{K}\mathbf{p}_j \rangle - \alpha_i \langle \mathbf{K}\mathbf{p}_i, \mathbf{K}\mathbf{p}_j \rangle .$$

For $i - k \le j < i \le I$ all terms on the right-hand side vanish, according to the induction hypothesis and (6.34a). For $j = i = I$ the right-hand side vanishes because of the definition (6.30c) of α_i. This shows that (6.34b) also holds in the case $i = I + 1$. For (6.34c) we multiply (6.31a) and (6.32) by $\mathbf{K}$ and then form the inner product with $\mathbf{r}_i$: it follows that

$$\langle \mathbf{r}_i, \mathbf{K}\mathbf{p}_i \rangle = \langle \mathbf{r}_i, \mathbf{K}\mathbf{r}_i \rangle + \sum_{j=i-k}^{i-1} \beta_j^{(i-1)} \langle \mathbf{r}_i, \mathbf{K}\mathbf{p}_j \rangle = \langle \mathbf{r}_i, \mathbf{K}\mathbf{r}_i \rangle ,$$

because of (6.34b). □

Theorem 6.2.3. *Let the symmetric part* $\mathbf{M}$ *of* $\mathbf{K}$ *be positive semidefinite. The sequence* $(\mathbf{r}_i)$ *of the residuals of MR or Orthomin* (k) *satisfies*

$$\|\mathbf{r}_i\| \le \left(1 - \frac{\lambda_{\min}(\mathbf{M})^2}{\lambda_{\max}(\mathbf{K}^T\mathbf{K})}\right)^{\frac{i}{2}} \|\mathbf{r}_0\|. \tag{6.35}$$

We need a preparatory lemma for the proof.

Lemma 6.2.4. *For* $(\mathbf{p}_i)$ *and* $(\mathbf{r}_i)$ *of MR or Orthomin*(k)

$$\langle \mathbf{K}\mathbf{p}_i, \mathbf{K}\mathbf{p}_i \rangle \le \langle \mathbf{K}\mathbf{r}_i, \mathbf{K}\mathbf{r}_i \rangle . \tag{6.36}$$

The proof uses (6.31a), (6.32) and (6.34a) and we recommend it to the reader as an exercise.

Proof of Theorem 6.2.3. From (6.30e) we have

$$\begin{aligned}
\|\mathbf{r}_{i+1}\|^2 &= \langle \mathbf{r}_i, \mathbf{r}_i \rangle - 2\alpha_i \langle \mathbf{r}_i, \mathbf{K}\mathbf{p}_i \rangle + \alpha_i^2 \langle \mathbf{K}\mathbf{p}_i, \mathbf{K}\mathbf{p}_i \rangle \\
&= \|\mathbf{r}_i\|^2 - 2\frac{\langle \mathbf{r}_i, \mathbf{K}\mathbf{p}_i \rangle^2}{\langle \mathbf{K}\mathbf{p}_i, \mathbf{K}\mathbf{p}_i \rangle} + \frac{\langle \mathbf{r}_i, \mathbf{K}\mathbf{p}_i \rangle^2}{\langle \mathbf{K}\mathbf{p}_i, \mathbf{K}\mathbf{p}_i \rangle} \\
&= \|\mathbf{r}_i\|^2 - \frac{\langle \mathbf{r}_i, \mathbf{K}\mathbf{p}_i \rangle^2}{\langle \mathbf{K}\mathbf{p}_i, \mathbf{K}\mathbf{p}_i \rangle} .
\end{aligned}$$

With this it follows from (6.34c) and (6.36) that

$$\frac{\|\mathbf{r}_{i+1}\|^2}{\|\mathbf{r}_i\|^2} = 1 - \frac{\langle \mathbf{r}_i, \mathbf{K}\mathbf{p}_i\rangle}{\langle \mathbf{r}_i, \mathbf{r}_i\rangle} \frac{\langle \mathbf{r}_i, \mathbf{K}\mathbf{p}_i\rangle}{\langle \mathbf{K}\mathbf{p}_i, \mathbf{K}\mathbf{p}_i\rangle}$$

$$\leq 1 - \frac{\langle \mathbf{r}_i, \mathbf{K}\mathbf{r}_i\rangle}{\langle \mathbf{r}_i, \mathbf{r}_i\rangle} \frac{\langle \mathbf{r}_i, \mathbf{K}\mathbf{r}_i\rangle}{\langle \mathbf{K}\mathbf{r}_i, \mathbf{K}\mathbf{r}_i\rangle}.$$

Since $\langle \mathbf{v}, \mathbf{R}\mathbf{v}\rangle = 0$ for all $\mathbf{v} \in \mathbb{R}^N$ with $\mathbf{R}$ from (6.29) we have

$$\frac{\langle \mathbf{r}_i, \mathbf{K}\mathbf{r}_i\rangle}{\langle \mathbf{r}_i, \mathbf{r}_i\rangle} = \frac{\langle \mathbf{r}_i, \mathbf{M}\mathbf{r}_i\rangle}{\langle \mathbf{r}_i, \mathbf{r}_i\rangle} \geq \lambda_{\min}(\mathbf{M}).$$

From this and from

$$\frac{\langle \mathbf{r}_i, \mathbf{K}\mathbf{r}_i\rangle}{\langle \mathbf{K}\mathbf{r}_i, \mathbf{K}\mathbf{r}_i\rangle} = \frac{\langle \mathbf{r}_i, \mathbf{r}_i\rangle}{\langle \mathbf{r}_i, \mathbf{K}^\mathrm{T}\mathbf{K}\mathbf{r}_i,\rangle} \frac{\langle \mathbf{r}_i, \mathbf{K}\mathbf{r}_i\rangle}{\langle \mathbf{r}_i, \mathbf{r}_i\rangle} \geq \frac{\lambda_{\min}(\mathbf{M})}{\lambda_{\max}(\mathbf{K}^\mathrm{T}\mathbf{K})}$$

we have the assertion (6.35). □

6.3 Iterative Solvers for Equations of Negative Order

In this section we discuss the iterative solution of the integral equation for the single layer operator V, i.e., an operator of the first kind and negative order.

Numerical approximations of the entries of the system matrix and the right-hand side (see Chap. 5) or the cluster method (see Chap. 7) both lead to a perturbed system of Galerkin equations

$$\widetilde{K}\tilde{\mathbf{u}}_S = \tilde{\mathbf{f}}. \tag{6.37}$$

The size of the perturbations should be chosen in such a way that if the system of equations is solved exactly the associated boundary element solution $\tilde{u}_S \in S$ should converge optimally. As we have already seen in Chap. 4, Lemma 4.5.1, the condition number

$$\kappa(\mathbf{K}) = \lambda_{\max}(\mathbf{K})/\lambda_{\min}(\mathbf{K})$$

of the exact Galerkin matrix $\mathbf{K}$ for a boundary integral operator of the first kind with order ± 1 on a quasi-uniform mesh $\mathcal{G}$ on Γ with mesh width h behaves as

$$\kappa(\mathbf{K}) \leq Ch^{-1} \leq CN^{1/2}. \tag{6.38}$$

If the approximation $\widetilde{K}$ in (6.37) satisfies suitable stability conditions (6.38) also holds for $\widetilde{K}$, as the following exercise will show.

Exercise 6.3.1. *Let* $\mathbf{K}$ *be the exact Galerkin matrix of a boundary element method for an elliptic boundary integral equation of the first kind and of order -1 on a shape-regular, quasi-uniform mesh $\mathcal{G}$ with mesh width h. Furthermore, let $\widetilde{K}$ be a* *stable, consistent, symmetric approximation of* $\mathbf{K}$*, more precisely, we assume that the perturbed Galerkin solution* $\tilde{u}_S \in S$ *from (6.37) converges at an optimal rate. Show that the condition number of the perturbed matrix* $\widetilde{K}$ *satisfies*

$$\kappa(\widetilde{K}) = \lambda_{\max}(\widetilde{K})/\lambda_{\min}(\widetilde{K}) \leq Ch^{-1} = CN^{1/2} \tag{6.39}$$

for a sufficiently small mesh width $h \leq h_0$.

The previous exercise, together with the symmetry of the approximation $\widetilde{K}$ and the convergence estimate (6.23) in Theorem 6.1.7 for the cg method, all combine to give us estimates for the error of the approximation $\tilde{u}_S^j$ after j steps of the cg method.

For this let $\tilde{\mathbf{u}}_S^j$ be the j-th iterate of the cg method applied to (6.37) and let $\tilde{u}_S^j \in S = S_{\mathcal{G}}^{p,-1}$ be the associated boundary element solution. The connection between the Euclidean vector norm $\|\tilde{\mathbf{u}}_S^j - \tilde{\mathbf{u}}_S\|$ of the error in the j-th iterate and the error $\|\tilde{u}_S^j - \tilde{u}_S\|_{H^{-1/2}(\Gamma)}$ in the boundary element solution is established by normalizing the basis functions of S as follows. Since the basis $(b_I)_{I\in\mathcal{I}}$ of $S_{\mathcal{G}}^{p,-1}$ is constructed separately for every panel τ, it can easily be made orthonormal with respect to the $L^2(\Gamma)$ inner-product. In this case we have

$$\tilde{u}_S = \sum_{I\in\mathcal{I}} (\tilde{\mathbf{u}}_S)_I \, b_I \in S \Longrightarrow \|\tilde{u}_S\|_{L^2(\Gamma)} = \|\tilde{\mathbf{u}}_S\|, \tag{6.40}$$

where $\|\cdot\|$ again denotes the Euclidean norm.

Proposition 6.3.2. *Let* $\widetilde{K}$ *be a stable, consistent and symmetric approximation of the Galerkin matrix* $\mathbf{K}$ *of the boundary integral equation of the first kind for the single layer potential. Furthermore, let* $\left(\tilde{\mathbf{u}}_S^j\right)_{j=0}^{\infty}$ *be the sequence of the iterated vectors of the cg method with the initial vector* $\tilde{\mathbf{u}}_S^0 = \mathbf{0}$ *and the approximate solutions* $\tilde{u}_S^j \in S$ *associated with* $\tilde{\mathbf{u}}_S^j$*. Let the basis of* S *be* L^2*-orthonormal and let the inverse inequality from Theorem 4.4.3 hold. Then the error estimate*

$$\|\tilde{u}_S - \tilde{u}_S^j\|_{H^{-1/2}(\Gamma)} \leq C(1 - h^{1/2})^j h^{-1/2} \|f\|_{H^{1/2}(\Gamma)} \tag{6.41}$$

holds.

Proof. We apply the convergence estimate (6.23) with the unit matrix $\mathbf{C} = \mathbf{I}$, which is the cg method without preconditioning, and with the matrix $\widetilde{K}$. According to Exercise 6.3.1 it follows for a sufficiently small $h \leq h_0$ that

$$\kappa(\widetilde{K}) \leq Ch^{-1} \leq CN^{1/2}.$$

With this, Theorem 6.1.7 gives us the estimate

$$\left\| \tilde{u}_S - \tilde{u}_S^j \right\|_{H^{-1/2}(\Gamma)} \leq C \left\| \tilde{u}_S - \tilde{u}_S^j \right\|_{L^2(\Gamma)} = C \left\| \tilde{\mathbf{u}}_S - \tilde{\mathbf{u}}_S^j \right\| \tag{6.42}$$
$$\leq C \left(1 - h^{1/2}\right)^j \|\tilde{\mathbf{u}}_S\| \leq C \left(1 - h^{1/2}\right)^j h^{-1/2} \|\tilde{u}_S\|_{H^{-1/2}(\Gamma)} .$$

The stability of the perturbation and of the Galerkin method combined with the ellipticity of the integral equation together yield

$$\|\tilde{u}_S\|_{H^{-1/2}(\Gamma)} \leq C_1 \|u_S\|_{H^{-1/2}(\Gamma)} \leq C_2 \|u\|_{H^{-1/2}(\Gamma)} \leq C_3 \|f\|_{H^{1/2}(\Gamma)},$$

which gives us the assertion. □

Remark 6.3.3. *The factor $h^{-1/2}$ in (6.41) can eventually be omitted if the Galerkin projection is stable in L^2 so that $\|\tilde{\mathbf{u}}_S\|$ in (6.42) can be estimated in terms of the right-hand side f (with constants independent of h) under appropriate regularity assumptions.*

In order to obtain the order of convergence for the approximation $\tilde{u}_S^j$, we need to set the error bound (6.41) in the same way as for the discretization error of the unperturbed Galerkin method. In the $H^{-1/2}(\Gamma)$-norm the Galerkin method converges with an order of $O(h^{p+3/2})$ if the exact solution u has maximal regularity, more precisely, if we have $u \in H^{p+1}(\Gamma)$. This means that we terminate the cg method after j_0 steps, where j_0 is chosen so that the following condition holds

$$\left(1 - h^{1/2}\right)^{j_0} h^{-1/2} < C h^{p+3/2}. \tag{6.43}$$

Proposition 6.3.4. *For a sufficiently small $h \leq h_0$ the cg method without preconditioning yields an approximation $\tilde{u}_S^{j_0} \in S$ of the Galerkin solution $u_S \in S$ that converges with an optimal order after*

$$j_0 \geq C \, |\log h| \, h^{-1/2} \sim C N^{1/4} \log N \tag{6.44}$$

steps.

We note that a cg step requires a matrix-vector multiplication $\mathbf{u} \longmapsto \widetilde{K}\mathbf{u}$. For dense matrices the evaluation thereof has a complexity of $O\left(N^2\right)$. In this case the total complexity of the entire solution process is proportional to $N^{9/4} \log N$. For a large dimension $N \sim 10^4 - 10^6$ this behavior leads to unacceptable computational costs, even for modern supercomputers.

In Chap. 7 we will introduce the cluster method for a fast matrix-vector multiplication, which only needs $O(N(\log N)^a)$ arithmetic operations with $a \geq 0$. This gives us an almost linear total complexity of $O(N^{5/4}(\log N)^{a+1})$ for the cg method without preconditioning.

Exercise 6.3.5. *In Proposition 6.3.4 we assume that the approximation $\widetilde{K}$ of the matrix $\mathbf{K}$ is symmetric. Study the case in which $\widetilde{K}$ is a stable, consistent and* *non-symmetric approximation of a positive definite matrix* **K** *and in which an MR or Orthomin*(k) *method is used for the iterative solution of (6.37).*

In many cases the complexity of the cg method and its non-symmetric variants is reasonable so that preconditioning is not necessary.

If, however, the dimension of the linear system of equations is very large, i.e., $N \gg 10^5$, preconditioning methods can be applied to reduce the condition number. For example, we can achieve a condition number of $O(|\log h|)$ if we precondition using the *Haar-Multiwavelet basis* of the subspace S [178,205]. Wavelet discretizations of integral equations are discussed in [190, 191, 208, 236]. In [227], [228] and [206] a wavelet construction is described which is also suited to very complex surfaces.

6.4 Iterative Solvers for Equations of Positive Order

Integral equations of positive order are similar to differential equations. *Multi-grid methods* belong to the most efficient solution methods for finite element discretizations. They can also be applied to integral equations without too many modifications. In this chapter we will give an introduction to multi-grid methods that have been adapted to integral equations. For a detailed discussion we refer to [114] and [116].

6.4.1 Integral Equations of Positive Order

In this section we will introduce multi-grid methods for the efficient solution of integral equations of positive order. We consider an integral operator of the form

$$(Ku)(\mathbf{x}) = \int_\Gamma k(\mathbf{x}, \mathbf{y})\, u(\mathbf{y})\, ds_{\mathbf{y}} \qquad \forall \mathbf{x} \in \Gamma.$$

Γ denotes the surface of a three-dimensional domain and $\mathcal{G}$ denotes a surface mesh of Γ.

Assumption 6.4.1. *The operator* $K : H^{1/2}(\Gamma) \to H^{-1/2}(\Gamma)$ *is continuous with the continuity constant* C_c *and elliptic on a closed subspace* $V \subset H^{1/2}(\Gamma)$:

$$b(v, v) := (v, Kv)_{L^2(\Gamma)} \geq \gamma_{\text{ell}} \|v\|^2_{H^{1/2}(\Gamma)} \qquad \forall v \in V. \tag{6.45}$$

For integral equations of positive order the kernel function is usually hypersingular and the integral is defined by means of a suitable regularization.

For a right-hand side $f \in V'$ we are looking for a function $u \in V$ such that

$$b(u,v) = f(v) \qquad \forall v \in V. \tag{6.46}$$

The Galerkin or collocation discretization transforms the integral equation into a coefficient matrix $\mathbf{K} \in \mathbb{C}^{N \times N}$ and into the right-hand side $\mathbf{f} \in \mathbb{C}^N$. The coefficient vector $\mathbf{u}$ in the basis representation of the boundary element solution $u \in S$ is the solution of the linear system of equations

$$\mathbf{Ku} = \mathbf{f}. \tag{6.47}$$

Since it is difficult to define point evaluations for integral equations with a hypersingular kernel function, we will restrict ourselves to Galerkin discretizations. The conformity $S \subset V \subset H^{1/2}(\Gamma)$ implies the Lipschitz continuity of the boundary element function. Some proofs require inverse inequalities and we denote the minimal constant in

$$\|u\|_{H^{1/2}(\Gamma)} \leq Ch^{-1/2} \|u\|_{L^2(\Gamma)} \qquad \forall u \in S \tag{6.48}$$

by C_{inv}. It depends on the regularity of the mesh $\mathcal{G}$ [see Chap. 4, Theorem 4.4.3 and Remark 4.4.4(a)]. The minimal and maximal constant C and c in

$$|\operatorname{supp} b_i| \leq Ch^2 \quad \text{and} \quad |\operatorname{supp} b_i| \geq ch^2 \tag{6.49}$$

are denoted by C_{supp} and c_{supp} respectively.

We will now provide an estimate for the diagonal elements of the matrix $\mathbf{K}$ for later applications.

Lemma 6.4.2. *There exist positive constants c_{d} and C_{d} such that*

$$c_{\mathrm{d}} h_i \leq |\mathbf{K}_{i,i}| \leq C_{\mathrm{d}} h_i \qquad \forall 1 \leq i \leq N \tag{6.50}$$

with $h_i := \max\{h_\tau : \tau \in \mathcal{G} \wedge \tau \subset \operatorname{supp} b_i\}$. Here the constants c_{d}, C_{d} depend only on Γ, the minimal angle of the surface mesh, the constants from local inverse inequalities and the coefficients of the kernel function.

Proof. The continuity and ellipticity (cf. Assumption 6.4.1) implies that

$$\gamma_{\mathrm{ell}} \|b_i\|^2_{H^{1/2}(\Gamma)} \leq \mathbf{K}_{i,i} \leq C_{\mathrm{c}} \|b_i\|^2_{H^{1/2}(\Gamma)}.$$

From Remark 4.4.4a with $s = 1/2$ and $t = 0$ we deduce that

$$\|b_i\|_{H^{1/2}(\Gamma)} \leq Ch^{-1/2} \|b_i\|_{L^2(\Gamma)} \overset{\text{Th. 4.4.5}}{\leq} Ch^{1/2}.$$

The lower estimate is proved in two steps. Recall that (b_i) is a Lagrange basis. Owing to the h-independent equivalence of the norms $\|v\|_{H^\ell(\tau)}$ and $\|\tilde{v}\|_{H^\ell(\tau^{\text{affine}})}$ from Corollary 4.3.12, it suffices to consider the case of a plane polyhedral surface.

Step a: Estimate with respect to the H^1-seminorm.

For a basis function b_i we select some $\tau \in \mathcal{G}$ with $\tau \subset \operatorname{supp} b_i$ and denote the reference map by $\chi_\tau : \hat{\tau} \to \tau$. Let $\widehat{N}_\mu = b_i|_\tau \circ \chi_\tau$ denote the local shape function for some $\mu \in \iota_p^{\hat{\tau}}$ [cf. Example 4.1.37 and (4.70)]. Then we obtain

$$\begin{aligned} |b_i|^2_{H^1(\Gamma)} \geq |b_i|^2_{H^1(\tau)} &= \int_{\hat{\tau}} g_\tau(\hat{\mathbf{x}}) \left\langle \mathbf{G}_\tau^{-1} \nabla \widehat{N}_\mu(\hat{\mathbf{x}}), \nabla \widehat{N}_\mu(\hat{\mathbf{x}}) \right\rangle d\hat{\mathbf{x}} \\ &\overset{\text{Lemma 4.3.6}}{\geq} c h_\tau^2 \int_{\hat{\tau}} \left\langle \mathbf{G}_\tau^{-1} \nabla \widehat{N}_\mu(\hat{\mathbf{x}}), \nabla \widehat{N}_\mu(\hat{\mathbf{x}}) \right\rangle d\hat{\mathbf{x}} \end{aligned} \tag{6.51}$$

with the inverse of Gram's matrix $\mathbf{G}_\tau^{-1} := \mathbf{J}_\tau^{-1} \left(\mathbf{J}_\tau^{-1}\right)^\intercal$ and the Jacobian $\mathbf{J}_\tau$ of χ_τ. Note that the minimal eigenvalue of $\mathbf{G}_\tau^{-1}$ is the reciprocal of the maximal eigenvalue of $\mathbf{G}_\tau$ so that from Lemma 4.3.5 and (6.51) we derive

$$|b_i|_{H^1(\Gamma)} \geq c \left|\widehat{N}_\mu\right|_{H^1(\hat{\tau})}.$$

There exists some $\hat{\mathbf{z}} \in \overline{\hat{\tau}}$ such that

$$\left\|\nabla \widehat{N}_\mu(\hat{\mathbf{z}})\right\| = \sup_{\hat{\mathbf{x}} \in \hat{\tau}} \left\|\nabla \widehat{N}_\mu(\hat{\mathbf{x}})\right\| \geq 1.$$

The smoothness of $\widehat{N}_\mu$ implies that there exist some $\hat{t} \subset \hat{\tau}$ and some constants $\alpha, \beta \in]0,1[$ which depend only on the polynomial degree p such that

$$\forall \hat{\mathbf{x}} \in \hat{t} \qquad \alpha \left\|\nabla N_\mu(\hat{\mathbf{z}})\right\| \leq \left\|\nabla \widehat{N}_\mu(\hat{\mathbf{x}})\right\| \quad \text{and} \quad \left|\hat{t}\right| \geq \beta \left|\hat{\tau}\right|.$$

Thus

$$|b_i|_{H^1(\Gamma)} \geq c \left|\widehat{N}_\mu\right|_{H^1(\hat{\tau})} \geq c \left|\widehat{N}_\mu\right|_{H^1(\hat{t})} \geq c\alpha \left\|\nabla N_\mu(\hat{\mathbf{z}})\right\| \sqrt{\left|\hat{t}\right|} \geq c\alpha \sqrt{\beta \left|\hat{\tau}\right|}.$$

Step b: $H^{1/2}$-estimate.

From Remark 4.4.4a with $s = 1$ and $t = 1/2$ we deduce the second inequality in

$$c\alpha \sqrt{\beta \left|\hat{\tau}\right|} \leq |b_i|_{H^1(\Gamma)} \leq C h_i^{-1/2} \|b_i\|_{H^{1/2}(\Gamma)}.$$

□

6.4.2 Iterative Methods

In this subsection we will summarize some of the properties of iterative methods. For a more elaborate treatment we recommend [116]. The proofs that we do not provide here can be found in [116].

Since the large dimension of the matrix $\mathbf{K}$ does not allow the use of direct solution methods (complexity $O\left(N^3\right)$), we should use iterative methods to solve (6.47). In the following we will consider linear iterative methods of the type:

$$\mathbf{u}^{(i+1)} := \mathbf{u}^{(i)} - \mathbf{W}\left(\mathbf{K}\mathbf{u}^{(i)} - \mathbf{f}\right). \tag{6.52}$$

Clearly, the iteration process is defined entirely by the choice of the matrix $\mathbf{W}$ and the initial vector $\mathbf{u}^{(0)}$.

Some of the common iterative methods include the Jacobi method, the Gauss–Seidel method and the SOR method (Successive OverRelaxation method). They are based on the decomposition

$$\mathbf{K} = \mathbf{D} - \mathbf{L} - \mathbf{R} \tag{6.53}$$

with the diagonal matrix $\mathbf{D} := \operatorname{diag}\mathbf{K}$, the strictly lower triangular matrix $\mathbf{L}$ and the strictly upper triangular matrix $\mathbf{R}$.

Example 6.4.3. *The Jacobi method is defined by* $\mathbf{W} = \mathbf{D}^{-1}$. *Componentwise the iterative procedure reads*

$$\mathbf{u}_j^{(i+1)} := \mathbf{u}_j^{(i)} - \frac{1}{\mathbf{K}_{j,j}}\left(\sum_{k=1}^{N}\mathbf{K}_{j,k}\mathbf{u}_k^{(i)} - \mathbf{f}_j\right) \qquad 1 \le j \le N.$$

In many applications the method has to be damped by a parameter $\omega > 0$. *The damped Jacobi method is defined by* $\mathbf{W} = \omega\mathbf{D}^{-1}$.

Example 6.4.4. *The Gauss–Seidel method is characterized by the choice* $\mathbf{W} = (\mathbf{L} - \mathbf{D})^{-1}$. *Note that for this method the triangular matrix* $\mathbf{L} - \mathbf{D}$ *does not have to be inverted. One only needs to solve a system of equations with a triangular matrix. Componentwise this reads*

$$\mathbf{u}_j^{(i+1)} = \mathbf{u}_j^{(i)} - \frac{1}{\mathbf{K}_{j,j}}\left(\sum_{k=1}^{j-1}\mathbf{K}_{j,k}\mathbf{u}_k^{(i+1)} + \sum_{k=j}^{N}\mathbf{K}_{j,k}\mathbf{u}_k^{(i)} - \mathbf{f}_j\right) \qquad 1 \le j \le N. \tag{6.54}$$

One obtains the damped version by introducing a positive factor ω *so that* $\mathbf{W} := \omega\left(\mathbf{L} - \mathbf{D}\right)^{-1}$.

Example 6.4.5. *The SOR method is obtained by introducing a parameter* $\omega > 0$ *in (6.54) which yields*

$$\mathbf{u}_j^{(i+1)} = \mathbf{u}_j^{(i)} - \frac{\omega}{\mathbf{K}_{j,j}} \left(\sum_{k=1}^{j-1} \mathbf{K}_{j,k} \mathbf{u}_k^{(i+1)} + \sum_{k=j}^{N} \mathbf{K}_{j,k} \mathbf{u}_k^{(i)} - \mathbf{f}_j \right) \qquad 1 \leq j \leq N.$$

In the following we will summarize the basic convergence results for iterative methods of the form (6.52). We begin with the definition of convergence.

Definition 6.4.6. An iterative method of the form (6.52) is called convergent if for all $\mathbf{f} \in \mathbb{C}^N$ there exists a limit $\mathbf{u}$ of the iteration that does not depend on the initial value $\mathbf{u}^{(0)} \in \mathbb{C}^N$.

The convergence of a linear iterative method can be similarly defined by means of the *spectral radius* of the *iteration matrix*.

Definition 6.4.7. The spectral radius of a matrix $\mathbf{A} \in \mathbb{C}^{N \times N}$ is given by

$$\rho(\mathbf{A}) := \max\{|\lambda| : \lambda \text{ is an eigenvalue of } \mathbf{A}\}.$$

The iteration matrix of a linear iterative method of the form (6.52) is given by

$$\mathbf{T} := \mathbf{I} - \mathbf{W}\mathbf{K}.$$

Remark 6.4.8. *A linear iterative method of the form (6.52) with a regular matrix* $\mathbf{K}$ *has the representation*

$$\mathbf{u}^{(i+1)} = \mathbf{T}\mathbf{u}^{(i)} + (\mathbf{I} - \mathbf{T})\,\mathbf{K}^{-1}\mathbf{f}.$$

Theorem 6.4.9. *The iterative method (6.52) converges if and only if* $\rho(\mathbf{T}) < 1$. *The limit of the iteration is the solution of the linear system of equations.*

Proof. Theorem 3.2.7 and Corollary 3.2.8 from [116]. □

In order to *assess* an iterative method for a special application, quantitative convergence results play a decisive role. For a vector norm $\|\cdot\|$ on $\mathbb{C}^N$ the associated matrix norm is also denoted by $\|\cdot\|$.

$$\|\mathbf{A}\| := \sup_{\mathbf{v} \in \mathbb{C}^N \setminus \{0\}} \frac{\|\mathbf{A}\mathbf{v}\|}{\|\mathbf{v}\|}.$$

Theorem 6.4.10. *Let* $\|\cdot\|$ *be an associated matrix norm. A sufficient condition for the convergence of an iteration (6.52) with an iteration matrix* $\mathbf{T}$ *is given by the estimate* $\|\mathbf{T}\| < 1$. *The error* $\mathbf{e}^{(i)} := \mathbf{u}^{(i)} - \mathbf{u}$ *satisfies the estimate*

$$\left\|\mathbf{e}^{(i+1)}\right\| \leq \|\mathbf{T}\| \left\|\mathbf{e}^{(i)}\right\|, \qquad \left\|\mathbf{e}^{(i)}\right\| \leq \|\mathbf{T}\|^i \left\|\mathbf{e}^{(0)}\right\|.$$

Proof. Theorem 3.2.10 in [116]. □

The condition $\|\mathbf{T}\| < 1$ is only sufficient for the convergence of the iterative method.

Remark 6.4.11. *(a) Let $\|\cdot\|$ be an associated matrix norm. Then for every matrix $\mathbf{A} \in \mathbb{R}^{N\times N}$ we have the estimate*

$$\rho(\mathbf{A}) \leq \|\mathbf{A}\|.$$

(b) For every matrix $\mathbf{A}$ and every $\varepsilon > 0$ there exists an associated matrix norm $\|\cdot\| = \|\cdot\|_{\mathbf{A},\varepsilon}$ with

$$\|\mathbf{A}\| \leq \rho(\mathbf{A}) + \varepsilon.$$

Proof. Part a: Let $\mathbf{w}$ be an eigenvector associated with the eigenvalue λ of $\mathbf{A}$ with the largest absolute value. Then

$$\|\mathbf{A}\| = \sup_{\mathbf{v}\in\mathbb{C}^N\setminus\{0\}} \frac{\|\mathbf{A}\mathbf{v}\|}{\|\mathbf{v}\|} \geq \frac{\|\mathbf{A}\mathbf{w}\|}{\|\mathbf{w}\|} = |\lambda| = \rho(\mathbf{A}).$$

The proof of Part b uses the Jordan canonical form of the matrix $\mathbf{A}$ and can be found in, e.g., [224, Theorem 6.9.2]. □

The remark shows that for a convergent iterative method there always exists an associated matrix norm which satisfies the conditions from Theorem 6.4.10. For linear systems of equations with a positive definite coefficient matrix $\mathbf{A} \in \mathbb{C}^{N\times N}$, choosing the norm

$$\|\mathbf{x}\|_{\mathbf{A}} := \langle \mathbf{x}, \mathbf{A}\mathbf{x}\rangle^{1/2} \quad \text{with} \quad \langle \mathbf{y}, \mathbf{A}\mathbf{x}\rangle := \sum_{i,j=1}^{N} \mathbf{y}_i \mathbf{A}_{i,j} \overline{\mathbf{x}}_j$$

is often a good option to prove that $\|\mathbf{T}\|_{\mathbf{A}} < 1$.

As an example, we study the convergence behavior of the Jacobi method in dependence on the condition number of the matrix. For a detailed discussion on iterative methods we refer to [116, 225].

Theorem 6.4.12 (Jacobi Method). *Let $\mathbf{K}$ in (6.47) be positive definite. The largest (smallest) eigenvalue of $\widetilde{K} := \mathbf{D}^{-1/2}\mathbf{K}\mathbf{D}^{-1/2}$ is denoted by Λ (λ). The optimal damping parameter for the Jacobi method is given by $\omega_{opt} := 2/(\Lambda + \lambda)$ and the norm of the associated iteration matrix satisfies*

$$\rho(\mathbf{T}) = \|\mathbf{T}\|_{\mathbf{K}} = \|\mathbf{T}\|_{\mathbf{A}} = \frac{\kappa - 1}{\kappa + 1} \tag{6.55}$$

with the condition number κ of $\widetilde{K}$.

The following theorem will elaborate on the condition κ of $\widetilde{K}$.

Theorem 6.4.13. *Let Assumptions 5.3.5, 5.3.25 and let (6.50) be satisfied. Let the Galerkin matrix in (6.47) be positive definite. Then the iteration matrix of the damped Jacobi method with an optimal damping parameter satisfies*

$$\rho(\mathbf{T}) \leq 1 - ch,$$

where the positive constant c depends on C_{inv} *[see* (6.48)*)],* γ_{ell} *[see (6.45)],* C_{c}, c_{d}, C_{d} *(see Lemma 6.4.2) and the constants from Corollary 5.3.28.*

Proof. We begin by estimating the largest and smallest eigenvalues of $\mathbf{K}$ (see Chap. 4, proof of Lemma 4.5.1 and Exercise 4.5.2).

For a coefficient vector $\mathbf{u} \in \mathbb{C}^N$, u denotes the associated boundary element function. Since $\mathbf{K}$ is positive definite, we have for the largest and smallest eigenvalues of $\mathbf{K}$

$$\lambda_{\max} = \sup_{\mathbf{u} \in \mathbb{C}^N \setminus \{\mathbf{0}\}} \frac{\langle \mathbf{u}, \mathbf{K}\mathbf{u} \rangle}{\langle \mathbf{u}, \mathbf{u} \rangle} \quad \text{and} \quad \lambda_{\min} = \inf_{\mathbf{u} \in \mathbb{C}^N \setminus \{\mathbf{0}\}} \frac{\langle \mathbf{u}, \mathbf{K}\mathbf{u} \rangle}{\langle \mathbf{u}, \mathbf{u} \rangle}$$

respectively. The estimate for the largest eigenvalue stems from the continuity of K, the inverse inequality (6.48) and Corollary 5.3.28. Thus

$$\begin{aligned} \langle \mathbf{u}, \mathbf{K}\mathbf{u} \rangle &= (u, Ku)_{L^2(\Gamma)} \leq \|u\|_{H^{1/2}(\Gamma)} \|Ku\|_{H^{-1/2}(\Gamma)} \leq C_{\text{c}} \|u\|^2_{H^{1/2}(\Gamma)} \\ &\leq C^2_{\text{inv}} C_{\text{c}} h^{-1} \|u\|^2_{L^2(\Gamma)} \leq Ch \|\mathbf{u}\|^2 \qquad \text{for } \lambda_{\max} \leq Ch. \end{aligned} \tag{6.56}$$

The lower bound for the smallest eigenvalue

$$\lambda_{\min} \geq ch^2$$

can be obtained from the ellipticity of $\mathbf{K}$ by

$$\langle \mathbf{u}, \mathbf{K}\mathbf{u} \rangle = (u, Ku)_{L^2(\Gamma)} \geq \gamma_{\text{ell}} \|u\|^2_{H^{1/2}(\Gamma)} \geq \gamma_{\text{ell}} \|u\|^2_{L^2(\Gamma)} \geq ch^2 \|\mathbf{u}\|^2 .$$

For $\mathbf{v} \in \mathbb{C}^N$ we set $\mathbf{u} = \mathbf{D}^{-1/2}\mathbf{v}$ and obtain the estimates

$$\tilde{\lambda}_{\max} \leq C \quad \text{and} \quad \tilde{\lambda}_{\min} \geq ch$$

for the largest and smallest eigenvalues $\tilde{\lambda}_{\max}$, $\tilde{\lambda}_{\min}$ of $\widetilde{K}$ from the inequalities

$$\begin{aligned} \left\langle \mathbf{v}, \widetilde{K}\mathbf{v} \right\rangle &= \langle \mathbf{u}, \mathbf{K}\mathbf{u} \rangle \leq Ch \|\mathbf{u}\|^2 = Ch \left\| \mathbf{D}^{-1/2}\mathbf{v} \right\|^2 \leq \widetilde{C} \|\mathbf{v}\|^2 , \\ \left\langle \mathbf{v}, \widetilde{K}\mathbf{v} \right\rangle &= \langle \mathbf{u}, \mathbf{K}\mathbf{u} \rangle \geq ch^2 \|\mathbf{u}\|^2 = ch^2 \left\| \mathbf{D}^{-1/2}\mathbf{v} \right\|^2 \geq \tilde{c}h \|\mathbf{v}\|^2 \end{aligned} \tag{6.57}$$

by using Lemma 6.4.2.

This proves the estimate $\kappa \leq Ch^{-1}$ for the condition number of $\widetilde{K}$. If we insert this into (6.55) we obtain the assertion. □

Corollary 6.4.14. *Under the conditions of Theorem 6.4.13 the spectrum of the Galerkin discretization of* $\mathbf{K}$ *in (6.47) satisfies*

$$\sigma(\mathbf{K}) \subset \left[ch^2, Ch\right]$$

with positive constants c and C.

The cg method from Sect. 6.1 is an example of a non-linear iterative method. If the continuous boundary integral operator is symmetric and elliptic the Galerkin discretization leads to a positive definite system matrix and the cg method becomes applicable.

Theorem 6.4.15. *Let the conditions from Theorem 6.4.13 be satisfied. The i-th iteration in the cg method (without preconditioning) is denoted by $\mathbf{u}^{(i)}$. Then the iteration error $\mathbf{e}^{(i)} := \mathbf{u}^{(i)} - \mathbf{u}$ satisfies the estimate*

$$\left\|\mathbf{e}^{(i)}\right\|_{\mathbf{K}} \leq 2\left(1 - C\sqrt{h}\right)^{i} \left\|\mathbf{e}^{(0)}\right\|_{\mathbf{K}}.$$

Proof. We apply Theorem 6.1.7. From Corollary 6.4.14 we have for $\kappa = \lambda_{\max}/\lambda_{\min}$ the estimate $\kappa \leq Ch^{-1}$. For the estimate (6.23) we use

$$\frac{\sqrt{\kappa}-1}{\sqrt{\kappa}+1} \leq 1 - C\sqrt{h}$$

and thus obtain the assertion. □

Remark 6.4.16. *Theorem 6.4.15 implies that the number of iterations necessary to solve the linear system of equations grows as the dimension of* $\mathbf{K}$ *increases.*

Under similar conditions to those in Proposition 6.3.4 we obtain an asymptotic complexity of $O\left(\sharp It \times \sharp MVM\right)$ for the solution of the linear system of equations (6.47), where $\sharp It := \left(N^{1/4}\log N\right)$ denotes the number of iterations and $\sharp MVM$ the complexity of a matrix-vector multiplication. For the cluster method we have $\sharp MVM \leq CN\log^{a} N$ with $a \geq 0$.

Remark 6.4.16 shows that the complexity for the solution of the linear system of equations has a growth which is faster than linear-logarithmic. For very large problems ($N \sim 10^5 - 10^6$) the computational costs for the solution of the linear system of equations will dominate the total costs. In the following subsection we will introduce a method which can solve the linear system of equations (6.47) at a linear-logarithmic rate.

6.4.3 Multi-grid Methods*

Multi-grid methods are well suited to solve linear systems of equations that result from the discretization of differential operators or integral operators of positive order. In this subsection we will introduce multi-grid methods for the problem (6.47)

* This section should be read as a complement to the core material of this book.

and refer to the monograph [114] and the textbook [116] for a detailed treatment. Here we will only give a brief presentation, which is adapted to boundary integral equations of positive order.

6.4.3.1 Motivation

The elliptic integral operator $K : H^{1/2}(\Gamma) \to H^{-1/2}(\Gamma)$ acts on a function $u \to Ku$ as a differentiation. Multi-grid methods can be applied to general, elliptic problems of positive order. The basic idea behind multi-grid methods can best be explained by using an elliptic differential operator which has been discretized by finite differences. Transferring this method to elliptic boundary integral equations of positive order does not require any modification and is presented afterwards.

We consider the Poisson model problem on the unit interval $I = (0, 1)$. Find a function u such that

$$\begin{aligned} -u'' &= f \qquad \text{in } I, \\ u(0) &= u(1) = 0. \end{aligned} \tag{6.58}$$

For the discretization we use an equidistant mesh $\Theta := \{ih : 1 \le i \le N\}$ with $h := 1/(N+1)$. It is our aim to determine an approximation of the solution at the mesh points $x \in \Theta$. This mesh function is denoted by $\mathbf{u} \in \mathbb{R}^N$. In order to determine a system of equations for $\mathbf{u}$, the second derivative is replaced by the difference approximation

$$-u''(x) \approx \frac{-u(x-h) + 2u(x) - u(x+h)}{h^2}.$$

If we use this at every mesh point $x \in \Theta$ and then take the zero boundary conditions into consideration we obtain the linear system of equations

$$\mathbf{L}\mathbf{u} = \mathbf{f} \tag{6.59}$$

with a tridiagonal coefficient matrix $\mathbf{L}$ and the vector $\mathbf{f} \in \mathbb{R}^N$

$$\mathbf{L} := h^{-2} \begin{bmatrix} 2 & -1 & 0 & \dots & 0 \\ -1 & \ddots & \ddots & \ddots & \vdots \\ 0 & \ddots & & & 0 \\ \vdots & \ddots & & & -1 \\ 0 & \dots & 0 & -1 & 2 \end{bmatrix}, \qquad \mathbf{f}_i := f\left(\frac{i}{h}\right) \qquad \forall 1 \le i \le N.$$

The solution $\mathbf{u}$ is the required approximative solution at the mesh points ih, $1 \le i \le N$. For this problem the eigensystem can be given explicitly.

Proposition 6.4.17. *The eigenvalues and eigenvectors* $\left(\lambda^{(k)}, \alpha^{(k)}\right)$, $1 \leq k \leq N$, *of the matrix* $\mathbf{L}$ *are given by*

$$\alpha^{(k)} := \sqrt{2h}\,(\sin(kjh\pi))_{j=1}^{N} \qquad \lambda^{(k)} := 4h^{-2}\sin^2\frac{k\pi h}{2}.$$

The damped Jacobi iteration applied to (6.59) can be written in the form

$$\mathbf{u}^{(i+1)} = \mathbf{u}^{(i)} - \frac{\omega h^2}{2}\left(\mathbf{L}\mathbf{u}^{(i)} - \mathbf{f}\right)$$

and we limit the damping parameter to

$$\omega \in \left]0, 1\right[. \tag{6.60}$$

The iteration matrix is then given by

$$\mathbf{T}_{\omega}^{\mathrm{Jac}} := \mathbf{I} - \frac{\omega h^2}{2}\mathbf{L}$$

and the eigenvectors coincide with those of the matrix $\mathbf{L}$. The eigenvalues read

$$\Lambda^{(k)} := 1 - 2\omega\sin^2\frac{k\pi h}{2}. \tag{6.61}$$

Lemma 6.4.18. *Let (6.60) hold. Then the spectral radius of the iteration matrix* $\mathbf{T}_{\omega}^{\mathrm{Jac}}$ *satisfies*

$$\rho\left(\mathbf{T}_{\omega}^{\mathrm{Jac}}\right) = 1 - 2\omega\sin^2\frac{\pi h}{2} = 1 - \frac{1}{2}\pi^2 h^2\omega + O\left(h^4\right).$$

Proof. It is easy to see that $\Lambda^{(k)}$ in (6.61) assumes its maximal value for $k = 1$. □

This lemma explains the slow convergence of the Jacobi method for the linear system of equations (6.59). The iteration error satisfies the recursion

$$\mathbf{e}^{(i+1)} = \mathbf{T}_{\omega}^{\mathrm{Jac}}\mathbf{e}^{(i)}. \tag{6.62}$$

We expand the vector $\mathbf{e}^{(i)}$ with respect to the eigenvectors $\left(\alpha^{(k)}\right)_{k=1}^{N}$

$$\mathbf{e}^{(i)} = \sum_{k=1}^{N} c_k^{(i)}\alpha^{(k)}$$

and insert this into the representation (6.62) and thus obtain

$$\mathbf{e}^{(i+1)} = \sum_{k=1}^{N} c_k^{(i+1)} \alpha^{(k)} \quad \text{with} \quad c_k^{(i+1)} := c_k^{(i)} \Lambda^{(k)} \quad \text{and} \quad \Lambda^{(k)} \text{ as in (6.61).}$$

As the index k increases, the oscillations of the eigenfunctions $\alpha^{(k)}$ grow. The following lemma shows that the parts of the error $\mathbf{e}^{(i)}$ that have high oscillations are reduced by a fixed factor with every Jacobi iteration.

Lemma 6.4.19. *Let the damping parameter satisfy* $\omega = 1/2$. *For* $k \geq \frac{N+1}{2}$ *we have*

$$\left|\Lambda^{(k)}\right| \leq 1/2$$

and the associated coefficients $c_k^{(i+1)}$ *of the error* $\mathbf{e}^{(i+1)}$ *in the eigenvector representation satisfy*

$$\left|c_k^{(i+1)}\right| \leq 1/2 \left|c_k^{(i)}\right|.$$

Proof. Choosing $\omega = 1/2$ gives us

$$\Lambda^{(k)} = 1 - \sin^2 \frac{k\pi}{2(N+1)},$$

and the right-hand side, considered as a function of k, $1 \leq k \leq N$, is monotonically decreasing. For $k \geq \frac{N+1}{2}$ we thus have

$$\Lambda^{(k)} \leq 1 - \sin^2 \frac{\pi}{4} = \frac{1}{2},$$

and the assertion follows from this. □

The slow convergence of the Jacobi method can thus be explained by the slow reduction of the low frequency parts of the error. On the other hand, the error $\mathbf{e}^{(i)}$ is already smooth after very few iteration steps, as opposed to the solution $\mathbf{u}$. The error satisfies the equation

$$\mathbf{L}\mathbf{e}^{(i)} = \mathbf{L}\mathbf{u}^{(i)} - \mathbf{L}\mathbf{u} = \mathbf{L}\mathbf{u}^{(i)} - \mathbf{f} =: \mathbf{d}. \tag{6.63}$$

The right-hand side is the negative residual of the iteration $\mathbf{u}^{(i)}$ and in connection with multi-grid methods it is called the *defect*. The vector $\mathbf{d}$ can be easily computed by using the previous iteration and (6.63) represents the equation for the error $\mathbf{e}^{(i)}$. If one were to solve (6.63) for $\mathbf{e}^{(i)}$ one would have solved the original system of equations $\mathbf{u} = \mathbf{u}^{(i)} - \mathbf{e}^{(i)}$. Since (6.63) is of the same type as (6.59), the iterative solution, e.g., with the Jacobi method, would have the same complexity as the solution of (6.59). The essential difference between (6.59) and (6.63) lies in the differing smoothness of the solutions $\mathbf{u}$ and $\mathbf{e}^{(i)}$. We expect that the solution of (6.63) will be smooth after very few Jacobi steps, as opposed to the solution $\mathbf{u}$.

The basic idea of a *two-grid method* consists in approximating the solution $\mathbf{e}^{(i)}$ from (6.63) by means of a coarser discretization. For this let $\mathbf{L}_{\text{coarse}}$ be the discretization of the problem (6.58) on a coarser mesh $\Theta_{\text{coarse}} := \{ih_{\text{coarse}} : 1 \leq i \leq N_{\text{coarse}}\}$

with $h_{\text{coarse}} = (N_{\text{coarse}} + 1)^{-1}$ and $N_{\text{coarse}} \sim N/2$. A coefficient vector $\mathbf{v} \in \mathbb{R}^N$ is transported to the coarse mesh by means of a *restriction* $\mathbf{R} : \mathbb{R}^N \to \mathbb{R}^{N_{\text{coarse}}}$ and a vector $\mathbf{w} \in \mathbb{R}^{N_{\text{coarse}}}$ is prolonged to the finer mesh by means of a *prolongation* $\mathbf{P} : \mathbb{R}^{N_{\text{coarse}}} \to \mathbb{R}^N$. Both mappings $\mathbf{R}$, $\mathbf{P}$ will be concretely introduced in Definition 6.4.21.

Formally one obtains the approximation $\tilde{\mathbf{e}}^{(i)}$ of (6.63) in three steps, as follows:

1. Restriction of the defect in (6.63):

$$\mathbf{d}_{\text{coarse}} := \mathbf{R}\mathbf{d}$$

2. Solving the equation for the coarse mesh

$$\mathbf{e}_{\text{coarse}} := \mathbf{L}_{\text{coarse}}^{-1}\mathbf{d}_{\text{coarse}}$$

3. Transfer to the fine mesh

$$\tilde{\mathbf{e}}^{(i)} := \mathbf{P}\mathbf{e}_{\text{coarse}}$$

The new iteration is then obtained by the correction $\mathbf{u}^{(i+1)} = \mathbf{u}^{(i)} - \tilde{\mathbf{e}}^{(i)}$.

The *multi-grid method* is obtained by again applying the two-grid method to the equation

$$\mathbf{L}_{\text{coarse}}\mathbf{e}_{\text{coarse}} = \mathbf{d}_{\text{coarse}}$$

and then iterating this process for increasingly coarse meshes.

The multi-grid method described in this subsection can be applied to general, elliptic differential and integral equations of positive order. It is based on the property that classical iterative methods rapidly reduce the high-frequency parts of the error. The more large-scale parts can then be approximated by coarser discretizations. It is because of this reason that the iterative method within the multi-grid algorithm is also called the smoothing method.

In the next subsection we will apply this concept to integral equations of positive order and at the same time define the multi-grid method.

6.4.3.2 Multi-grid Method for Integral Equations of Positive Order

We have already explained in the motivation for this subsection that the efficiency of multi-grid methods is based on a *hierarchy* of discretizations. Therefore we assume that a sequence of surface meshes $(\mathcal{G}_\ell)_{\ell=0}^{\ell_{\max}}$ is given and that the linear system of equations has to be solved on the level $\ell_{\max}$. In the simplest case such meshes can be obtained by refining a coarse mesh.

Example 6.4.20. *Let Γ be the surface of a polyhedron and let $\mathcal{G}_0$ be a (coarse) triangulation of Γ. A family of fine triangulations $(\mathcal{G}_\ell)_{\ell=0}^{\ell_{\max}}$ is obtained by recursively connecting the midpoints of the sides of every triangle $\tau \in \mathcal{G}_{\ell-1}$ and thus subdividing τ into four congruent triangles (see Remark 4.1.8). The mesh points of the mesh Θ_ℓ are denoted by $\left(\mathbf{x}_{i,\ell}\right)_{i=1}^{N_\ell}$.*

On an abstract level we assume that the surface mesh $\mathcal{G}$ is the finest mesh of a mesh family $(\mathcal{G}_\ell)_{\ell=0}^{\ell_{\max}}$, i.e., $\mathcal{G} = \mathcal{G}_{\ell_{\max}}$, with a strictly decreasing mesh width h_ℓ. The Galerkin discretization of the integral equation (6.46) with boundary elements on the meshes $\mathcal{G}_\ell$, $0 \leq \ell \leq \ell_{\max}$ leads to a family of linear systems of equations

$$\mathbf{K}_\ell \mathbf{u}_\ell = \mathbf{f}_\ell \qquad 0 \leq \ell \leq \ell_{\max}. \tag{6.64}$$

It is our goal to efficiently solve the equation on a fine mesh

$$\mathbf{K}_{\ell_{\max}} \mathbf{u}_{\ell_{\max}} = \mathbf{f}_{\ell_{\max}}. \tag{6.65}$$

To combine simple iterative methods with a coarse grid correction we need to define transfer operators from coarse to fine meshes and vice-versa. For $0 \leq \ell \leq \ell_{\max}$ we introduce the notation $X_\ell := \mathbb{C}^{N_\ell}$ for the space of mesh functions.

The canonical choice for the prolongation $\mathbf{P}_{\ell,\ell-1} : X_{\ell-1} \to X_\ell$ can be obtained by interpreting a coefficient vector $\mathbf{u} \in X_{\ell-1}$ as a boundary element function. For this we assume that the boundary element spaces S_ℓ are nested

$$S_0 \subset S_1 \subset \ldots \subset S_{\ell_{\max}} \subset H^s(\Gamma). \tag{6.66}$$

The boundary element basis for S_ℓ is denoted by $\left(b_{i,\ell}\right)_{i=1}^{N_\ell}$.

Every coefficient vector $\mathbf{u} \in X_{\ell-1}$ is uniquely associated with a boundary element function $u \in S_{\ell-1}$ by

$$u = P_{\ell-1}\mathbf{u} := \sum_{i=1}^{N_{\ell-1}} \mathbf{u}_i b_{i,\ell-1}, \tag{6.67}$$

which, owing to (6.66), also satisfies $u \in S_\ell$. In S_ℓ it again has a unique basis representation

$$u = \sum_{i=1}^{N_\ell} \alpha_i b_{i,\ell} \tag{6.68}$$

and the operation $S_\ell \ni u \to \alpha \in X_\ell$ defines the operator $R_\ell : S_\ell \to X_\ell$. The composite mapping $X_{\ell-1} \ni \mathbf{u} \to u \to \alpha \in X_\ell$ defines the prolongation $\mathbf{P}_{\ell,\ell-1} : X_{\ell-1} \to X_\ell$.

Definition 6.4.21. Let (6.66) hold. Let the operators $P_\ell : X_\ell \to S_\ell$ and $R_\ell : S_\ell \to X_\ell$ be given by (6.67) and (6.68). The canonical prolongation $\mathbf{P}_{\ell,\ell-1} : X_{\ell-1} \to X_\ell$ is the composite mapping

$$\mathbf{P}_{\ell,\ell-1} = R_\ell P_{\ell-1}.$$

The restriction $\mathbf{R}_{\ell-1,\ell} : X_\ell \to X_{\ell-1}$ is the adjoint of $\mathbf{P}_{\ell,\ell-1}$ and it is characterized by

$$\left\langle \mathbf{R}_{\ell-1,\ell}\mathbf{v}, \mathbf{u} \right\rangle = \left\langle \mathbf{v}, \mathbf{P}_{\ell,\ell-1}\mathbf{u} \right\rangle \qquad \forall \mathbf{v} \in X_\ell, \quad \mathbf{u} \in X_{\ell-1}.$$

Remark 6.4.22. *The canonical prolongation* $\mathbf{P}_{\ell,\ell-1}$ *is represented by a rectangular matrix of the dimension* $N_\ell \times N_{\ell-1}$. *For the restriction we have* $\mathbf{R}_{\ell-1,\ell} = \mathbf{P}^{\mathsf{T}}_{\ell,\ell-1}$.

Example 6.4.23. *The mesh hierarchy described in Example 6.4.20 has the following property: For every mesh point* $\mathbf{x} \in \Theta_\ell$ *we either have* $\mathbf{x} \in \Theta_{\ell-1}$ *or there exist two coarse mesh points* $\mathbf{y}, \mathbf{z} \in \Theta_{\ell-1}$ *that are connected by a panel edge and satisfy* $\mathbf{x} = (\mathbf{y} + \mathbf{z})/2$. *In the case of continuous, piecewise linear boundary elements we then have for the prolongation*

$$\left(\mathbf{P}_{\ell,\ell-1}\right)_{i,j} = b_{j,\ell-1}\left(x_{i,\ell}\right) = \begin{cases} 1 & \text{if } \mathbf{x}_{i,\ell} = \mathbf{y}_{j,\ell-1} \\ \frac{1}{2} & \text{if } \overline{\mathbf{x}_{i,\ell}, \mathbf{y}_{j,\ell-1}} \text{ form a panel edge in } \mathcal{G}_\ell, \\ 0 & \text{otherwise.} \end{cases}$$

We have now defined all the components of the two-grid method and can describe it in an algorithmic form. The application of one step of a simple iterative method, for example, the Jacobi iteration, to a mesh function $\mathbf{u}_\ell$ defines the mapping $\mathbf{S}_\ell(\mathbf{u}_\ell, \mathbf{f}_\ell)$, where $\mathbf{f}_\ell$ denotes the right-hand side of the associated system of equations $\mathbf{K}_\ell \mathbf{u}_\ell = \mathbf{f}_\ell$. In the following algorithm we will apply this *smoothing iteration* ν times. The required number of smoothing iterations is estimated in Theorem 6.4.37 according to $\nu \geq \bar{\nu}$ with $\bar{\nu} = O(1)$. In many applications $\nu = 2, 3$ proves to be a suitable choice.

Algorithm 6.4.24 (Two-Grid Method). *The iteration step of the two-grid method for the solution of (6.65) is called by* $\mathbf{TGM}\left(\mathbf{u}_{\ell_{\max}}, \mathbf{f}_{\ell_{\max}}\right)$ *and is defined as follows.*

procedure *TGM*$(\mathbf{u}_\ell, \mathbf{f}_\ell)$;
begin

$$\begin{aligned} &\textbf{for } i := 1 \textbf{ to } \nu \textbf{ do } \mathbf{u}_\ell := \mathbf{S}_\ell(\mathbf{u}_\ell, \mathbf{f}_\ell); \\ &\mathbf{d}_{\ell-1} := \mathbf{R}_{\ell-1,\ell}(\mathbf{K}_\ell \mathbf{u}_\ell - \mathbf{f}_\ell); \\ &\mathbf{c}_{\ell-1} := \mathbf{K}_{\ell-1}^{-1} \mathbf{d}_{\ell-1}; \\ &\mathbf{u}_\ell := \mathbf{u}_\ell - \mathbf{P}_{\ell,\ell-1} \mathbf{c}_{\ell-1}; \end{aligned} \tag{6.69}$$

end;

Remark 6.4.25. *The two-grid method defines a linear iterative method with the iteration matrix*

$$\mathbf{T}^{TGM} := \left(\mathbf{I}_\ell - \mathbf{P}_{\ell,\ell-1}\mathbf{K}_{\ell-1}^{-1}\mathbf{R}_{\ell-1,\ell}\mathbf{K}_\ell\right)\left(\mathbf{T}^{OGM}\right)^\nu,$$

where $\mathbf{T}^{OGM}$ *denotes the iteration matrix of the simple iterative method (smoothing method). For the damped Jacobi method we have, for example,* $\mathbf{T}^{OGM} := \mathbf{I}_\ell - \omega \mathbf{D}_\ell^{-1}\mathbf{K}_\ell$.

For practical applications the two-grid method can not yet be recommended, as on the level $\ell - 1$ there is still one system of linear equations that has to be solved per iteration. The idea of the multi-grid method is to again replace the linear system of equations

$$\mathbf{K}_{\ell-1}\mathbf{c}_{\ell-1} = \mathbf{d}_{\ell-1}$$

by a two-grid method and to repeat this procedure until we reach the coarsest level $\ell = 0$. The method includes a control parameter $\gamma \in \{1, 2\}$ which will be explained at a later stage.

Algorithm 6.4.26 (Multi-grid Method). *An iteration of the multi-grid method for the solution of (6.65) is called by* $\mathbf{MGM}\left(\ell_{\max}, \mathbf{u}_{\ell_{\max}}, \mathbf{f}_{\ell_{\max}}\right)$ *and is defined as follows.*

procedure *MGM*$(\ell, \mathbf{u}_\ell, \mathbf{f}_\ell)$;
begin
  if $\ell = 0$ ***then*** $\mathbf{u}_0 := \mathbf{K}_0^{-1}\mathbf{f}_0$
  else begin

$$\begin{aligned}
&\textbf{for } i := 1 \textbf{ to } \nu \textbf{ do } \mathbf{u}_\ell := \mathbf{S}_\ell\left(\mathbf{u}_\ell, \mathbf{f}_\ell\right);\\
&\mathbf{d}_{\ell-1} := \mathbf{R}_{\ell-1,\ell}\left(\mathbf{K}_\ell \mathbf{u}_\ell - \mathbf{f}_\ell\right);\\
&\mathbf{c}_{\ell-1} := \mathbf{0};\\
&\textit{for } i := 1 \textit{ to } \gamma \textit{ do MGM}\left(\ell - 1, \mathbf{c}_{\ell-1}, \mathbf{d}_{\ell-1}\right);\\
&\mathbf{u}_\ell := \mathbf{u}_\ell - \mathbf{P}_{\ell,\ell-1}\mathbf{c}_{\ell-1};
\end{aligned} \tag{6.70}$$

  end;
end;

Remark 6.4.27. *It is easy to verify that the recursion in the multi-grid algorithm reaches the level* $\ell = 0$ *after* $\ell_{\max}$ *steps and then terminates, which means that the algorithm is well defined.*

Figure 6.1 illustrates the succession of the single recursion steps of a multi-grid iteration for the cases $\gamma = 1, 2$. The names V-cycle ($\gamma = 1$) and W-cycle ($\gamma = 2$) stem from the shape of the mesh transitions in Fig. 6.1.

6.4.3.3 Nested Iterations

The multi-grid iteration starts on the finest mesh, descends to the coarsest mesh level and then prolongs the corrections over the different levels up to the finest level. In practical applications the following situation often occurs. The finest discretization level $\ell_{\max}$ is not known a priori. We start with very coarse discretizations and the associated Galerkin solutions are computed thereon. It is then decided whether the current mesh needs to be refined to improve the precision. Thus the object is to solve a *sequence* of linear systems of equations on the levels $\ell = 0, 1, 2, \ldots$. This situation can be used to the advantage of the multi-grid iteration by prolonging the previously computed solution $\mathbf{u}_\ell$ to the refined mesh $\mathcal{G}_{\ell+1}$ where it in turn defines a suitable initial value for the multi-grid iteration on the level $\ell + 1$. This procedure is called *nested iteration* and can be described in an algorithmic form as follows.

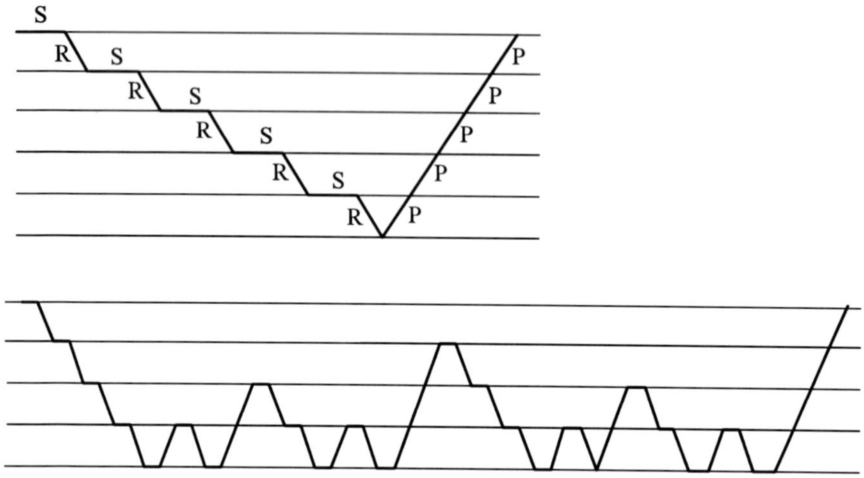

Fig. 6.1 Schematic representation of a multi-grid iteration with a V-cycle (*top*) for $\gamma = 1$ and $\ell = 5$ and with a W-cycle (*bottom*) for $\gamma = 2$ and $\ell = 4$

Algorithm 6.4.28 (Nested Iteration). *The nested iteration calls the multi-grid procedure* **MGM**, *starting with the coarsest level. Let* $\mathbf{u}_0 := \mathbf{K}_0^{-1}\mathbf{f}_0$ *be the solution of the equation for the coarse mesh.*

procedure Nested_Iteration;
for $\ell := 1$ ***to*** $\ell_{\max}$ ***do***
begin
  $\mathbf{u}_\ell := \mathbf{P}_{\ell,\ell-1}\mathbf{u}_{\ell-1}$;
  for $i := 1$ ***to*** m_ℓ ***do*** **MGM** $(\mathbf{u}_\ell, \mathbf{f}_\ell)$;
end;

In connection with the convergence analysis for nested iterations we will prove that $m_\ell = m = O(1)$ can be chosen independently of ℓ. It turns out in practical applications that the algorithm already converges for $m = 1, 2$.

6.4.3.4 Convergence Analysis for Multi-grid Methods

The proof of convergence for the multi-grid method is subdivided into an analysis of the smoothing property and an analysis of the coarse grid correction. For the sake of clarity we will first consider the convergence of the two-grid method.

Convergence of the Two-Grid Method

In Sect. 6.4.3.1 we called a mesh function smooth if coefficients in the expansion with respect to the eigenvectors of the system matrix have small absolute value for the eigenvectors of high frequency. Since we are considering integral operators of positive order, the smoothness can also be described by the measure $\left\|\mathbf{K}_\ell \mathbf{T}_\ell^\nu \mathbf{e}\right\|$, where $\mathbf{e}$ denotes the current iteration error and $\mathbf{S}_\ell^{(\nu)}$ denotes the ν-fold application of the smoothing iteration. Here and in the following $\|\cdot\|$ denotes the Euclidean norm.

Definition 6.4.29 (Smoothing Property). A smoothing iteration with iteration matrix $\mathbf{T}_\ell$, $\ell \geq 0$, satisfies the smoothing property with the exponent $s \in \mathbb{R}$ if there exist functions $\eta(\nu)$ and $\bar{\nu}(h)$ that are independent of ℓ such that:

1. $\left\|\mathbf{K}_\ell \mathbf{T}_\ell^\nu\right\| \leq \eta(\nu)\, h_\ell^{-s} \qquad \forall 0 \leq \nu \leq \bar{\nu}(h_\ell),\ \ell \geq 1.$
2. $\lim_{\nu\to\infty} \eta(\nu) = 0.$
3. $\lim_{h\to 0} \bar{\nu}(h) = \infty \qquad$ or $\qquad \bar{\nu}(h) = \infty.$

Theorem 6.4.30. *Let the conditions from Theorem 6.4.13 be satisfied for the matrices* $\mathbf{K}_\ell$, $0 \leq \ell \leq \ell_{\max}$, *from (6.64). Then there exist* $0 < \underline{\omega} < \overline{\omega} < 1$ *independent of the refinement of the discretization such that the smoothing property holds for the damped Jacobi method for all* $\omega \in [\underline{\omega}, \overline{\omega}]$ *with exponent* $s = -1$ *and* $\bar{\nu}(h) = \infty$.

Proof. For the sake of simplicity we will omit the index ℓ in the proof. We have

$$\mathbf{K}\mathbf{T}^\nu = \mathbf{K}\left(\mathbf{I} - \omega \mathbf{D}^{-1}\mathbf{K}\right)^\nu = \frac{1}{\omega}\mathbf{D}^{1/2}\mathbf{X}\left(\mathbf{I}-\mathbf{X}\right)^\nu \mathbf{D}^{1/2} \tag{6.71}$$

with the positive definite matrix $\mathbf{X} = \omega \mathbf{D}^{-1/2}\mathbf{K}\mathbf{D}^{-1/2}$. From (6.57) we have

$$\sigma\left(\mathbf{D}^{-1/2}\mathbf{K}\mathbf{D}^{-1/2}\right) \subset [0, C]$$

with a constant C which is independent of the refinement of the discretization. If we choose $\omega \in \left[c, C^{-1}\right]$ with $0 < c < C^{-1}$ the spectrum of $\mathbf{X}$ is contained in the interval $[0, 1]$. It is shown in Lemma 6.4.31 that this yields

$$\left\|\mathbf{X}\left(\mathbf{I}-\mathbf{X}\right)^\nu\right\| \leq \eta_0(\nu)$$

with $\eta_0(\nu)$ from (6.72). Lemma 6.4.2 implies that

$$\left\|\mathbf{D}^{1/2}\right\|^2 \leq Ch,$$

from which we have the assertion by using the Cauchy–Schwarz inequality. □

Lemma 6.4.31. *(a) Let* $\mathbf{X}$ *be a positive definite matrix for which* $\mathbf{I} - \mathbf{X}$ *is also positive definite. Then*

$$\left\|\mathbf{X}\left(\mathbf{I}-\mathbf{X}\right)^\nu\right\| \leq \eta_0(\nu)$$

for all $\nu \geq 0$, *where the function* $\eta_0(\nu)$ *is defined by*

$$\eta_0(\nu) := \nu^\nu / (\nu+1)^{\nu+1}. \tag{6.72}$$

(b) The asymptotic behavior of $\eta_0(\nu)$ as $\nu \to \infty$ is given by

$$\eta_0(\nu) = \frac{1}{e\nu} + O\left(\nu^{-2}\right).$$

Proof. It is well known from linear algebra (see [116, Theorem 2.8.1]) that for every matrix $\mathbf{A} \in \mathbb{C}^{N\times N}$ there exist a unitary matrix $\mathbf{Q}$ and an upper triangular matrix $\mathbf{U}$ such that

$$\mathbf{A} = \mathbf{QUQ}^H.$$

By means of induction one can show that for a polynomial p the representation

$$p(\mathbf{A}) = \mathbf{Q}\, p(\mathbf{U})\, \mathbf{Q}^H$$

is valid. It follows from the theorem on determinant multiplication that the characteristic polynomials of $p(\mathbf{A})$ and $p(\mathbf{U})$ coincide. Since the product of two upper triangular matrices is again an upper triangular matrix, it follows by induction that $p(\mathbf{U})$ is an upper triangular matrix with the diagonal elements $(p(\mathbf{U}))_{i,i} = p(\mathbf{U}_{i,i})$. As $\{\mathbf{U}_{i,i} : 1 \leq i \leq N\}$ is the spectrum $\sigma(\mathbf{A})$ of $\mathbf{A}$ we have proved that

$$\sigma(p(\mathbf{A})) = p(\sigma(\mathbf{A})) := \{p(\lambda) : \lambda \in \sigma(\mathbf{A})\}.$$

Since $\mathbf{X}$ is positive definite, so is $\mathbf{B} := \mathbf{X}(\mathbf{I}-\mathbf{X})^\nu$ and with $p(\xi) = \xi(1-\xi)^\nu$ we thus have

$$\begin{aligned}\left\|\mathbf{X}(\mathbf{I}-\mathbf{X})^\nu\right\| &= \max\{\lambda : \lambda \in \sigma(\mathbf{B})\} = \max\{p(\lambda) : \lambda \in \sigma(\mathbf{X})\}\\ &\leq \max\{p(\xi) : \xi \in [0,1]\}.\end{aligned}$$

Simple analysis provides the maximum $\xi_0 = (\nu+1)^{-1}$ as well as the equality

$$p(\xi_0) = \frac{1}{1+\nu}\left(1 - \frac{1}{1+\nu}\right)^\nu = \eta_0(\nu).$$

Part (b) follows by analyzing η_0. □

The iteration matrix of the two-grid method has the representation (see Remark 6.4.25)

$$\mathbf{T}^{TGM} := \left(\mathbf{K}_\ell^{-1} - \mathbf{P}_{\ell,\ell-1}\mathbf{K}_{\ell-1}^{-1}\mathbf{R}_{\ell-1,\ell}\right)\mathbf{K}_\ell\mathbf{T}_\ell^\nu \tag{6.73}$$

and can be estimated in the Euclidean norm as

$$\left\|\mathbf{T}^{TGM}\right\| \leq \left\|\mathbf{K}_\ell^{-1} - \mathbf{P}_{\ell,\ell-1}\mathbf{K}_{\ell-1}^{-1}\mathbf{R}_{\ell-1,\ell}\right\| \left\|\mathbf{K}_\ell\mathbf{T}_\ell^\nu\right\|. \tag{6.74}$$

We have already estimated the second factor on the right-hand side in connection with the smoothing property in Theorem 6.4.30. The first factor compares the

Galerkin solution on different mesh levels and is therefore called the approximation property.

Definition 6.4.32 (Approximation Property). The two-grid method has the approximation property with the exponent $s \in \mathbb{R}$ if

$$\left\| \mathbf{K}_\ell^{-1} - \mathbf{P}_{\ell,\ell-1}\mathbf{K}_{\ell-1}^{-1}\mathbf{R}_{\ell-1,\ell} \right\| \leq C h_\ell^s \tag{6.75}$$

with a constant C which is independent of ℓ, $\ell_{\max}$ and h_ℓ.

In order to prove the approximation property for the two-grid method we need the following assumption.

Assumption 6.4.33. *There exists a constant $C_G > 0$ such that for all right-hand sides $f \in L^2(\Gamma)$ the Galerkin solution $u_\ell \in S_\ell$ of (6.46) for the mesh $\mathcal{G}_\ell$ satisfies the estimate*

$$\|u - u_\ell\|_{L^2(\Gamma)} \leq C_G h_\ell \|f\|_{L^2(\Gamma)},$$

where u denotes the continuous solution of (6.46).

Lemma 6.4.34. *Let Assumption 6.4.1 be satisfied and let the inverse operator*

$$K^{-1} : L^2(\Gamma) \to H^1(\Gamma)$$

be continuous. Let the mesh family $\mathcal{G}_\ell$ be quasi-uniform and let the boundary element space $S_\ell^{(1)}$ satisfy $S_\ell^{(1)} \subset S_\ell$, where $S_\ell^{(1)}$ denotes the continuous, piecewise linear boundary element space. Let the conditions from Theorem 4.2.17 be satisfied.

Then Assumption 6.4.33 holds.

Proof. For $f \in L^2(\Gamma)$ the continuous solution satisfies $u \in H^1(\Gamma)$. With Proposition 4.1.46 and the quasi-optimality of the Galerkin discretization we have

$$\|u - u_\ell\|_{H^{1/2}(\Gamma)} \leq C \inf_{v \in S_\ell} \|u - v\|_{H^{1/2}(\Gamma)} \leq C \inf_{v \in S_\ell^{(1)}} \|u - v\|_{H^{1/2}(\Gamma)}.$$

The approximation property (see Proposition 4.1.50) and the regularity of the integral operator K yield

$$\inf_{v \in S_\ell^{(1)}} \|u - v\|_{H^{1/2}(\Gamma)} \leq C h_\ell^{1/2} \|u\|_{H^1(\Gamma)} \leq C h_\ell^{1/2} \|f\|_{L^2(\Gamma)}.$$

The estimate

$$\|u - u_\ell\|_{L^2(\Gamma)} \leq C h_\ell \|f\|_{L^2(\Gamma)}$$

follows by the Aubin–Nitsche duality argument (see Theorem 4.2.17). □

In order to prove the approximation property, first for the two-grid and then for the multi-grid method, we need a weak condition on the relation between consecutive mesh widths.

Assumption 6.4.35. *The mesh widths h_ℓ, $0 \le \ell \le \ell_{\max}$, satisfy*

$$c_1 < h_\ell / h_{\ell-1} \le C_1 \tag{6.76}$$

with constants c_1, C_1 that do not depend on ℓ.

Theorem 6.4.36. *Let Assumptions 5.3.5, 5.3.25, 6.4.33 and 6.4.35 be satisfied. Let the Galerkin matrix in (6.47) be positive definite.*

Then the two-grid method has the approximation property with exponent $s = -1$, where the positive constant C in (6.75) does not depend on the mesh width h_ℓ but does depend on the shape-regularity of the mesh.

Proof. Let $\mathbf{f}_\ell \in \mathbb{C}^{N_\ell}$ be arbitrary. We set $\mathbf{A}_\ell := \mathbf{K}_\ell^{-1} - \mathbf{P}_{\ell,\ell-1}\mathbf{K}_{\ell-1}^{-1}\mathbf{R}_{\ell-1,\ell}$ and prove that $\|\mathbf{A}_\ell \mathbf{f}_\ell\| \le C h_\ell^{-1} \|\mathbf{f}_\ell\|$.

(a) In the first step we construct a continuous function $f_\ell \in S_\ell$ with the property that $\mathbf{f}_\ell$ is the right-hand side of the Galerkin discretization. We use the ansatz

$$f_\ell = \sum_{i=1}^{N_\ell} \beta_{\ell,i} b_{\ell,i}$$

and from the condition $\left(f_\ell, b_{\ell,i}\right)_{L^2(\Gamma)} = (\mathbf{f}_\ell)_i$, $1 \le i \le N_\ell$, we determine the coefficient vector β_ℓ as the solution of the linear system of equations

$$\sum_{j=1}^{N_\ell} \beta_{\ell,j} \left(b_{\ell,j}, b_{\ell,i}\right)_{L^2(\Gamma)} = (\mathbf{f}_\ell)_i \qquad 1 \le i \le N_\ell.$$

By using the matrix $\mathbf{M}_\ell := \left(\left(b_{\ell,j}, b_{\ell,i}\right)_{L^2(\Gamma)}\right)_{i,j=1}^{N_\ell}$ we obtain the compact representation

$$\beta_\ell = \mathbf{M}_\ell^{-1} \mathbf{f}_\ell. \tag{6.77}$$

(b) In the second step the vectors $\mathbf{u}_\ell := \mathbf{K}_\ell^{-1}\mathbf{f}_\ell$ and $\mathbf{u}_{\ell,\ell-1} := \mathbf{P}_{\ell,\ell-1}\mathbf{K}_{\ell-1}^{-1}\mathbf{R}_{\ell-1,\ell}\mathbf{f}_\ell$ are interpreted as Galerkin solutions of auxiliary problems.

The vector $\mathbf{u}_\ell$ is the coefficient vector of the Galerkin solution of problem (6.46): Find $u_\ell \in S_\ell$ such that

$$b\left(u_\ell, v\right) = \left(f_\ell, v\right)_{L^2(\Gamma)} \qquad \forall v \in S_\ell.$$

We now consider $\mathbf{u}_{\ell,\ell-1}$ and set $\mathbf{f}_{\ell-1} := \mathbf{R}_{\ell-1,\ell}\mathbf{f}_\ell$. The corresponding right-hand side $f_{\ell-1}$ is defined by

$$f_{\ell-1} = \sum_{i=1}^{N_{\ell-1}} \beta_{\ell-1,i} b_{\ell-1,i}$$

with

$$\beta_{\ell-1} := \mathbf{M}_{\ell-1}^{-1}\mathbf{f}_{\ell-1}.$$

Therefore the vector $\mathbf{u}_{\ell-1} := \mathbf{K}_{\ell-1}^{-1}\mathbf{f}_{\ell-1}$ is the coefficient vector of the Galerkin solution: Find $u_{\ell-1} \in S_{\ell-1}$ such that

$$b\left(u_{\ell-1}, v\right) = \left(f_{\ell-1}, v\right)_{L^2(\Gamma)} \qquad \forall v \in S_{\ell-1}.$$

Then we have $\mathbf{u}_{\ell,\ell-1} := \mathbf{P}_{\ell,\ell-1}\mathbf{u}_{\ell-1}$.

(c) Splitting the term $\mathbf{A}_\ell \mathbf{f}_\ell$.

By Definition 6.4.21 we obtain

$$\mathbf{A}_\ell \mathbf{f}_\ell = \mathbf{u}_\ell - \mathbf{u}_{\ell,\ell-1} = R_\ell u_\ell - \mathbf{P}_{\ell,\ell-1} R_{\ell-1} u_{\ell-1}.$$

If we use $\mathbf{P}_{\ell,\ell-1} = R_\ell P_{\ell-1}$ and $P_{\ell-1} R_{\ell-1} = I_{\ell-1}$ on $S_{\ell-1}$ we obtain

$$\mathbf{u}_\ell - \mathbf{u}_{\ell,\ell-1} = R_\ell \left(u_\ell - u_{\ell-1}\right).$$

(d) By using Corollary 5.3.28 we have for the norms of R_ℓ, P_ℓ

$$\begin{aligned} c_P h_\ell \left\|\mathbf{u}\right\| &\le \left\|P_\ell \mathbf{u}\right\|_{L^2(\Gamma)} \le C_P h_\ell \left\|\mathbf{u}\right\|, \\ C_P^{-1} h_\ell^{-1} \left\|u\right\|_{L^2(\Gamma)} &\le \left\|R_\ell u\right\| \le c_P^{-1} h_\ell^{-1} \left\|u\right\|_{L^2(\Gamma)} \end{aligned} \tag{6.78}$$

for all $u \in S_\ell$ and $\mathbf{u} \in \mathbb{C}^N$. Hence we have proved that

$$\left\|\mathbf{u}_\ell - \mathbf{u}_{\ell,\ell-1}\right\| \le C h_\ell^{-1} \left\|u_\ell - u_{\ell-1}\right\|_{L^2(\Gamma)}. \tag{6.79}$$

If we insert the continuous solution u of (6.46) with $f = f_\ell$ into the right-hand side of (6.79) we obtain

$$\left\|\mathbf{u}_\ell - \mathbf{u}_{\ell,\ell-1}\right\| \le C h_\ell^{-1} \left(\left\|u_\ell - u\right\|_{L^2(\Gamma)} + \left\|u - u_{\ell-1}\right\|_{L^2(\Gamma)}\right). \tag{6.80}$$

(e) Estimating the differences $u_\ell - u$ and $u - u_{\ell-1}$
Assumption 6.4.33 yields

$$\left\|u_\ell - u\right\|_{L^2(\Gamma)} \le C h_\ell \left\|f_\ell\right\|_{L^2(\Gamma)}.$$

and

$$\left\|u_{\ell-1} - u\right\|_{L^2(\Gamma)} \le C h_{\ell-1} \left\|f_{\ell-1}\right\|_{L^2(\Gamma)}.$$

(f) Estimating the functions f_ℓ, $f_{\ell-1}$

In the following the norms $\|f_\ell\|_{L^2(\Gamma)}$ and $\|f_{\ell-1}\|_{L^2(\Gamma)}$ are expressed in terms of the vector $\mathbf{f}_\ell$. We have

$$\|f_\ell\|_{L^2(\Gamma)} \le Ch_\ell \|\beta_\ell\| \tag{6.81}$$

and similarly

$$\|f_{\ell-1}\|_{L^2(\Gamma)} \le Ch_{\ell-1} \|\beta_{\ell-1}\| . \tag{6.82}$$

We first estimate the norm of β_ℓ. We have

$$\|\beta_\ell\| = \left\|\mathbf{M}_\ell^{-1}\mathbf{f}_\ell\right\| \le \lambda_{\min}^{-1} \|\mathbf{f}_\ell\|$$

with the smallest eigenvalue $\lambda_{\min}$ of the positive definite matrix $\mathbf{M}_\ell$. We can estimate this value by

$$\lambda_{\min} = \inf_{\mathbf{v}\in\mathbb{C}^{N_\ell}\setminus\{\mathbf{0}\}} \frac{\langle \mathbf{v}, \mathbf{M}_\ell \mathbf{v}\rangle}{\langle \mathbf{v}, \mathbf{v}\rangle} = \inf_{\mathbf{v}\in\mathbb{C}^{N_\ell}\setminus\{\mathbf{0}\}} \frac{(P_\ell \mathbf{v}, P_\ell \mathbf{v})_{L^2(\Gamma)}}{\langle \mathbf{v}, \mathbf{v}\rangle} = \inf_{\mathbf{v}\in\mathbb{C}^{N_\ell}\setminus\{\mathbf{0}\}} \frac{\|P_\ell \mathbf{v}\|^2_{L^2(\Gamma)}}{\|\mathbf{v}\|^2} \ge ch_\ell^2. \tag{6.83}$$

With (6.81) it follows that

$$\|f_\ell\| \le Ch_\ell^{-1} \|\mathbf{f}_\ell\| .$$

We now turn our attention to (6.82). The definitions of $\beta_{\ell-1}$ and $\mathbf{f}_{\ell-1}$ combined with (6.83) and (6.76) yield

$$\|\beta_{\ell-1}\| = \left\|\mathbf{M}_{\ell-1}^{-1}\mathbf{f}_{\ell-1}\right\| \le ch_\ell^{-2} \|\mathbf{f}_{\ell-1}\| = ch_\ell^{-2} \left\|\mathbf{R}_{\ell-1,\ell}\mathbf{f}_\ell\right\| \le ch_\ell^{-2} \left\|\mathbf{R}_{\ell-1,\ell}\right\| \|\mathbf{f}_\ell\|.$$

We still need to estimate the norm of the operator $\mathbf{R}_{\ell-1,\ell}$. With (6.78) we have

$$\left\|\mathbf{R}_{\ell-1,\ell}\right\| = \left\|\mathbf{P}_{\ell,\ell-1}^\intercal\right\| = \left\|\mathbf{P}_{\ell,\ell-1}\right\| \le \|R_\ell\| \, \|P_{\ell-1}\| \le C, \tag{6.84}$$

so that in all we have proved that

$$\|f_\ell\| \le Ch_\ell^{-1} \|\mathbf{f}_\ell\| \quad \text{and} \quad \|f_{\ell-1}\|_{L^2(\Gamma)} \le Ch_{\ell-1}^{-1} \|\mathbf{f}_\ell\| .$$

(g) Estimating the operator $\mathbf{A}_\ell$

The approximation property follows from

$$\begin{aligned} \|\mathbf{A}_\ell \mathbf{f}_\ell\| = \left\|\mathbf{u}_\ell - \mathbf{u}_{\ell,\ell-1}\right\| &\overset{(6.80)}{\le} Ch_\ell^{-1} \left(\|u_\ell - u\|_{L^2(\Gamma)} + \|u - u_{\ell-1}\|_{L^2(\Gamma)}\right) \\ &\overset{(e)}{\le} Ch_\ell^{-1} \left(h_\ell \|f_\ell\|_{L^2(\Gamma)} + h_{\ell-1} \|f_{\ell-1}\|_{L^2(\Gamma)}\right) \\ &\overset{(f)}{\le} Ch_\ell^{-1} \|\mathbf{f}_\ell\| . \end{aligned} \tag{6.85}$$

□

Theorem 6.4.37. *Let the conditions from Theorems 6.4.30 and 6.4.36 be satisfied. Then there exist $0 < \underline{\omega} < \overline{\omega} < 1$ and $\bar{\nu} > 0$ independent of the refinement of the discretization so that the norm of the two-grid method converges with respect to the Euclidean norm with $\nu \geq \bar{\nu}$ smoothing steps of the Jacobi method. The damping parameter $\omega \in [\underline{\omega}, \overline{\omega}]$ is independent of the mesh width h and the discretization level ℓ. For the iteration matrix the estimate*

$$\left\| \mathbf{T}_\ell^{TGM} \right\| < 1$$

holds.

Proof. Combine the decomposition (6.74) with Theorems 6.4.30 and 6.4.36. □

W-Cycle Convergence

In the next step we will prove the convergence of the multi-grid method. The convergence proofs for the W and V-cycles are very different. We begin with the simpler W-cycle multi-grid algorithm. The proof makes use of the fact that the multi-grid method (W-cycle) can be regarded as a small perturbation of the two-grid algorithm. We require an additional condition which will be stated next.

Assumption 6.4.38. *The iteration matrix $\mathbf{T}_\ell$ of the smoothing iteration satisfies*

$$\left\| \mathbf{T}_\ell^\nu \right\| \leq C_{\mathrm{it}}$$

for all $\ell \geq 1$ and $0 < \nu \leq \bar{\nu} = \min_{\ell \geq 1} \bar{\nu}(h_\ell)$ with $\bar{\nu}$ from Definition 6.4.29.

This assumption is satisfied for the Jacobi method.

Lemma 6.4.39. *Let the conditions from Theorem 6.4.13 be satisfied for the matrices $\mathbf{K}_\ell$, $0 \leq \ell \leq \ell_{\max}$, from (6.64) and let $0 < \underline{\omega} < \overline{\omega}$ be as in Theorem 6.4.30. Then the damped Jacobi method satisfies Assumption 6.4.38 for all $\omega \in [\underline{\omega}, \overline{\omega}]$.*

Proof. For the Jacobi method we have

$$\mathbf{T}_\ell^\nu := \left(\mathbf{I}_\ell - \omega \mathbf{D}_\ell^{-1} \mathbf{K}_\ell\right)^\nu = \mathbf{D}_\ell^{-1/2} \left(\mathbf{I}_\ell - \mathbf{X}_\ell\right)^\nu \mathbf{D}_\ell^{1/2} \tag{6.86}$$

with $\mathbf{X}_\ell$ from (6.71).

For diagonal matrices $\mathbf{D}_\ell$ we have

$$\left\| \mathbf{D}_\ell^{-1/2} \right\| \left\| \mathbf{D}_\ell^{1/2} \right\| \leq C.$$

As was already shown in the proof of Theorem 6.4.30, the matrix in the brackets on the right-hand side of (6.86) satisfies $\|\mathbf{I}_\ell - \mathbf{X}_\ell\| \leq 1$. The Cauchy–Schwarz inequality then yields the assertion. □

Lemma 6.4.40. *Let Assumptions 5.3.5, 5.3.25, 6.4.35 and 6.4.38 be satisfied. Let* $\mathbf{T}_\ell$ *denote the iteration matrix for the smoothing method and* C_{TGM} *the norm of the iteration matrix of the two-grid method. Then*

$$\left\| \mathbf{K}_{\ell-1}^{-1} \mathbf{R}_{\ell-1,\ell} \mathbf{K}_\ell \mathbf{T}_\ell^\nu \right\| \le c \left(C_{\mathrm{it}} + C_{\mathrm{TGM}} \right)$$

Proof. We use the decomposition

$$\mathbf{P}_{\ell,\ell-1} \mathbf{K}_{\ell-1}^{-1} \mathbf{R}_{\ell-1,\ell} \mathbf{K}_\ell \mathbf{T}_\ell^\nu = \mathbf{T}_\ell^\nu - \left(\mathbf{K}_\ell^{-1} - \mathbf{P}_{\ell,\ell-1} \mathbf{K}_{\ell-1}^{-1} \mathbf{R}_{\ell-1,\ell} \right) \mathbf{K}_\ell \mathbf{T}_\ell^\nu .$$

By Assumption 6.4.38, the first summand on the right-hand side satisfies the estimate $\left\| \mathbf{T}_\ell^\nu \right\| \le C_{\mathrm{it}}$ and the second summand is the iteration matrix of the two-grid method. It then follows that

$$\left\| \mathbf{P}_{\ell,\ell-1} \mathbf{K}_{\ell-1}^{-1} \mathbf{R}_{\ell-1,\ell} \mathbf{K}_\ell \mathbf{T}_\ell^\nu \right\| \le C_{\mathrm{it}} + C_{\mathrm{TGM}} .$$

Corollary 5.3.28 [see (6.78)] and Assumption 6.4.35 yield

$$\left\| \mathbf{P}_{\ell,\ell-1} \mathbf{v} \right\| = \left\| R_\ell P_{\ell-1} \mathbf{v} \right\| \ge c_1 h_\ell^{-1} \left\| P_{\ell-1} \mathbf{v} \right\|_{L^2(\Gamma)} \ge c_2 h_\ell^{-1} h_{\ell-1} \left\| \mathbf{v} \right\| \ge c_3 \left\| \mathbf{v} \right\|$$

from which we have the assertion. □

In the next theorem we derive a recursive representation for the iteration matrix of the multi-grid method.

Theorem 6.4.41. *Let* $\mathbf{T}_\ell$ *be the iteration matrix for the smoothing method and let* $\mathbf{T}_\ell^{TGM}$ *be the iteration matrix for the two-grid method with respect to the levels* ℓ, $\ell - 1$. *The iteration matrix* $\mathbf{T}_\ell^{MGM}$ *of the multi-grid method can be recursively represented as*

$$\mathbf{T}_0^{MGM} = \mathbf{0}, \qquad \mathbf{T}_1^{MGM} = \mathbf{T}_1^{TGM},$$
$$\mathbf{T}_\ell^{MGM} = \mathbf{T}_\ell^{TGM} + \mathbf{P}_{\ell,\ell-1} \left(\mathbf{T}_{\ell-1}^{MGM} \right)^\gamma \mathbf{K}_{\ell-1}^{-1} \mathbf{R}_{\ell-1,\ell} \mathbf{K}_\ell \mathbf{T}_\ell^\nu .$$

Proof. The proof is achieved by induction over the levels $\ell = 0, 1, \ldots$. For $\ell = 0, 1$ the statement is clear, since the solution is exact on the level $\ell = 0$ and since on the level $\ell = 1$ the multi-grid method and the two-grid method are identical.

In order to identify the iteration matrix $\mathbf{T}_\ell^{CGC}$ of the coarse grid correction, we use (6.70) and set $\mathbf{f}_\ell = \mathbf{0}$. Then

$$\mathbf{d}_{\ell-1} = \mathbf{R}_{\ell-1,\ell} \mathbf{K}_\ell \mathbf{u}_\ell .$$

If we denote the first iterate $\mathbf{c}_{\ell-1} = \mathbf{0}$ in (6.70) by $\mathbf{c}_{\ell-1}^{(0)}$ and the iterate belonging to the index i in (6.70) by $\mathbf{c}_{\ell-1}^{(i)}$ we obtain

$$\mathbf{c}_{\ell-1}^{(i)} = \mathbf{T}_{\ell-1}^{MGM} \mathbf{c}_{\ell-1}^{(i-1)} + \mathbf{N}_{\ell-1}^{MGM} \mathbf{d}_{\ell-1}$$

with the matrix $\mathbf{N}_{\ell-1}^{MGM} := \left(\mathbf{I}_{\ell-1} - \mathbf{T}_{\ell-1}^{MGM}\right)\mathbf{K}_{\ell-1}^{-1}$ (see Remark 6.4.8). Since $\mathbf{c}_{\ell-1}^{(0)} = \mathbf{0}$ it follows that

$$\mathbf{c}_{\ell-1}^{(\gamma)} = \left(\sum_{i=0}^{\gamma-1} \left(\mathbf{T}_{\ell-1}^{MGM}\right)^i\right) \mathbf{N}_{\ell-1}^{MGM} \mathbf{d}_{\ell-1} = \left(\sum_{i=0}^{\gamma-1} \left(\mathbf{T}_{\ell-1}^{MGM}\right)^i\right) \left(\mathbf{I}_{\ell-1} - \mathbf{T}_{\ell-1}^{MGM}\right) \mathbf{K}_{\ell-1}^{-1} \mathbf{d}_{\ell-1}$$
$$= \left(\mathbf{I}_{\ell-1} - \left(\mathbf{T}_{\ell-1}^{MGM}\right)^{\gamma}\right) \mathbf{K}_{\ell-1}^{-1} \mathbf{R}_{\ell-1,\ell} \mathbf{K}_{\ell} \mathbf{u}_{\ell}.$$

The iteration matrix of the coarse grid correction (6.70) is then given by

$$\begin{aligned}\mathbf{T}_{\ell}^{CGC} &= \mathbf{I}_{\ell} - \mathbf{P}_{\ell,\ell-1}\left(\mathbf{I}_{\ell-1} - \left(\mathbf{T}_{\ell-1}^{MGM}\right)^{\gamma}\right)\mathbf{K}_{\ell-1}^{-1}\mathbf{R}_{\ell-1,\ell}\mathbf{K}_{\ell} \\ &= \left(\mathbf{K}_{\ell}^{-1} - \mathbf{P}_{\ell,\ell-1}\mathbf{K}_{\ell-1}^{-1}\mathbf{R}_{\ell-1,\ell}\right)\mathbf{K}_{\ell} + \mathbf{P}_{\ell,\ell-1}\left(\mathbf{T}_{\ell-1}^{MGM}\right)^{\gamma}\mathbf{K}_{\ell-1}^{-1}\mathbf{R}_{\ell-1,\ell}\mathbf{K}_{\ell}.\end{aligned} \tag{6.87}$$

Combined with the preceding smoothing step $\mathbf{T}_{\ell}^{\nu}$ and with (6.73) we obtain the assertion

$$\mathbf{T}_{\ell}^{MGM} = \mathbf{T}_{\ell}^{TGM} + \mathbf{P}_{\ell,\ell-1}\left(\mathbf{T}_{\ell-1}^{MGM}\right)^{\gamma}\mathbf{K}_{\ell-1}^{-1}\mathbf{R}_{\ell-1,\ell}\mathbf{K}_{\ell}\mathbf{T}_{\ell}^{\nu}.$$

□

As an abbreviation we set $\zeta_{\ell} := \left\|\mathbf{T}_{\ell}^{MGM}\right\|$. In the following we will derive a recursive estimate for ζ_{ℓ}.

Lemma 6.4.42. *Let Assumptions 5.3.5, 5.3.25, 6.4.35 and 6.4.38 be satisfied. For the numbers ζ_{ℓ} the recursive estimate*

$$\zeta_0 = 0 \qquad \text{and for } 1 \le \ell \le \ell_{\max} : \zeta_{\ell} \le \left\|\mathbf{T}_{\ell}^{TGM}\right\| + C\zeta_{\ell-1}^{\gamma}$$

holds, where C depends only on $\left\|\mathbf{T}_{\ell}^{TGM}\right\|$, c_P, C_P from (6.78), c_1, C_1 from (6.76) and C_{it} from (6.4.38).

Proof. The recursive representation of the iteration matrix of the multi-grid method from Theorem 6.4.41 can be estimated by using the triangle and Cauchy–Schwarz inequalities as follows:

$$\zeta_0 = 0, \qquad \zeta_1 = \left\|\mathbf{T}_{\ell}^{TGM}\right\|,$$
$$\zeta_{\ell} \le \left\|\mathbf{T}_{\ell}^{TGM}\right\| + \zeta_{\ell-1}^{\gamma}\left\|\mathbf{P}_{\ell,\ell-1}\right\|\left\|\mathbf{K}_{\ell-1}^{-1}\mathbf{R}_{\ell-1,\ell}\mathbf{K}_{\ell}\mathbf{T}_{\ell}^{\nu}\right\|.$$

Lemma 6.4.40 and $\mathbf{P}_{\ell,\ell-1} = R_{\ell}P_{\ell-1}$ with (6.78) then yield

$$\zeta_{\ell} \le \left\|\mathbf{T}_{\ell}^{TGM}\right\| + C_{\star}\zeta_{\ell-1}^{\gamma} \tag{6.88}$$

with a constant $C_\star$ which depends only on $\left\|\mathbf{T}_\ell^{TGM}\right\|$, c_P, C_P from (6.78), c_1, C_1 from (6.76) and C_{it} from (6.4.38). □

Theorem 6.4.43. *Let the conditions from Lemma 6.4.42 be satisfied.*

Then there exists a constant $\bar{\nu} > 0$ such that the iteration matrix for the multi-grid method with W-cycle and $\nu \geq \bar{\nu}$ smoothing steps satisfies the estimate

$$\left\|\mathbf{T}_\ell^{MGM}\right\| \leq C < 1$$

with a constant C which is independent of ℓ and the mesh width h_ℓ.

Proof. The W-cycle is defined by choosing $\gamma = 2$ in Algorithm 6.4.26. Without loss of generality we assume in (6.88) that

$$C_\star > 1. \tag{6.89}$$

We define an auxiliary sequence $(x_\ell)_{\ell=0}^{\ell_{\max}}$ by $x_0 := 0$ and for $\ell = 1, 2, \ldots, \ell_{\max}$ by $x_\ell := 1 + \eta x_{\ell-1}^2$ with $\eta = C_{\text{TGM}} C_\star$. We clearly have $\zeta_\ell \leq C_{\text{TGM}} x_\ell$ for all $0 \leq \ell \leq \ell_{\max}$ and $(x_\ell)_{\ell=0}^{\ell_{\max}}$ is strictly increasing. In the case $\eta \leq 1/4$ the auxiliary sequence is bounded (for example, by 2) and the limit $x_\star$ is given by

$$x_\star = \frac{1 + \sqrt{1 - 4\eta}}{2\eta}.$$

Combining these results we have shown that under the condition $C_{\text{TGM}} C_\star \leq 1/4$ the norm of the iteration matrix of the multi-grid method (W-cycle) satisfies the estimate [see (6.89)]

$$\zeta_\ell \leq C_{\text{TGM}} \frac{1 + \sqrt{1 - 4\eta}}{2\eta} = \frac{1 + \sqrt{1 - 4 C_{\text{TGM}} C_\star}}{2 C_\star} \leq C_\star^{-1} < 1.$$

By (6.72) the minimal number of smoothing steps can always be chosen such that the associated two-grid method satisfies $C_{\text{TGM}} C_\star \leq 1/4$. □

V-Cycle Convergence

The above argument cannot be applied to the V-cycle. Since the multi-grid iteration with the V-cycle requires far less computational time than the W-cycle, we will also present the more complicated convergence analysis associated with it.

The essential differences compared to the W-cycle convergence consist in restricting the method to symmetric smoothers and using convergence results with respect to the *energy* norm $\|\cdot\|_{\mathbf{K}_\ell}$ instead of the Euclidean norm. We will specify these assumptions in the following.

Assumption 6.4.44. *The matrices* $\mathbf{K}_\ell$ *are positive definite. For the restriction we have* $\mathbf{R}_{\ell-1,\ell} = \mathbf{P}_{\ell,\ell-1}^{\mathsf{T}}$ *for all* ℓ.

In the next step the multi-grid (and two-grid) method is generalized to form a symmetric method.

Definition 6.4.45. Let $\mathbf{K}$ be positive definite and let a smoothing method $\mathbf{S}_\ell\left(\mathbf{u}^{(i)},\mathbf{f}\right)$ of the form (6.52) be given. The adjoint smoothing method $\mathbf{S}^H\left(\mathbf{u}^{(i)},\mathbf{f}\right)$ is given by

$$\mathbf{u}^{(i+1)} := \mathbf{u}^{(i)} - \mathbf{W}^H\left(\mathbf{K}\mathbf{u}^{(i)} - \mathbf{f}\right).$$

Definition 6.4.46. The symmetric multi-grid (and two-grid) method results by adding ν post-smoothing steps with the adjoint smoothing iteration. The following lines

$$\textbf{for } i := 1 \textbf{ to } \nu \textbf{ do } \mathbf{u}_\ell := \mathbf{S}_\ell^H\left(\mathbf{u}_\ell,\mathbf{f}_\ell\right);$$

have to be added to the end of the program in (6.70) [and (6.69)].

For the V-cycle we assume that the coarse mesh matrices are defined by the Galerkin product

$$\mathbf{K}_{\ell-1} := \mathbf{R}_{\ell-1,\ell}\mathbf{K}_\ell\mathbf{P}_{\ell,\ell-1}. \tag{6.90}$$

Remark 6.4.47. *By Assumption 6.4.44 we have for the Galerkin product*

$$\mathbf{K}_{\ell-1} = \mathbf{P}_{\ell,\ell-1}^{\mathsf{T}}\mathbf{K}_\ell\mathbf{P}_{\ell,\ell-1}.$$

Assumption 6.4.48. *The smoothing method is Hermitian:* $\mathbf{W}_\ell^H = \mathbf{W}_\ell$ *for all* $0 \le \ell \le \ell_{\max}$.

For positive definite system matrices $\mathbf{K}_\ell$ *the matrices* $\mathbf{W}_\ell^{-1} - \mathbf{K}_\ell$ *are also positive definite.*

Assuming these conditions, the iteration matrix for the symmetric two-grid method is given by

$$\mathbf{T}_\ell^{TGM} := \mathbf{T}_\ell^\nu\left(\mathbf{K}_\ell^{-1} - \mathbf{P}_{\ell,\ell-1}\mathbf{K}_{\ell-1}^{-1}\mathbf{P}_{\ell,\ell-1}^{\mathsf{T}}\right)\mathbf{K}_\ell\mathbf{T}_\ell^\nu.$$

Lemma 6.4.49. *Let Assumptions 5.3.5 and 5.3.25 be satisfied. Let the Galerkin matrix in (6.47) be positive definite. Then, for a sufficiently small parameter domain* $0 < \underline{\omega} < \overline{\omega}$, *the Jacobi method satisfies Assumption 6.4.48 for all damping parameters* $\omega \in [\underline{\omega}, \overline{\omega}]$.

Proof. For the damped Jacobi method we have $\mathbf{W}_\ell^{-1} = \omega^{-1}\mathbf{D}_\ell$. For positive definite system matrices, $\mathbf{W}_\ell$ is therefore also Hermitian and the first part from Assumption 6.4.48 is satisfied.

From Lemma 6.4.2 we have

$$\left\|\mathbf{W}_\ell^{-1}\right\| \geq c\omega^{-1} h_\ell$$

and, by (6.56), the system matrix $\mathbf{K}_\ell$ satisfies the estimate

$$\|\mathbf{K}_\ell\| \leq C h_\ell.$$

It then follows that the smallest eigenvalue $\lambda_{\min}$ of $\mathbf{W}_\ell^{-1} - \mathbf{K}_\ell$ satisfies

$$\lambda_{\min} \geq \left(c\omega^{-1} - C\right) h_\ell.$$

Therefore $\mathbf{W}_\ell^{-1} - \mathbf{K}_\ell$ is positive definite for all damping parameters $0 < \omega < c/C$. □

The V-cycle convergence is proved in terms of the energy norm $\|\cdot\|_{\mathbf{K}_\ell}$. For this we need the approximation property of multi-grid methods in terms of the energy norm.

Assumption 6.4.50. *For all* $1 \leq \ell \leq \ell_{\max}$ *we have*

$$\left\|\mathbf{W}_\ell^{-1/2}\left(\mathbf{K}_\ell^{-1} - \mathbf{P}_{\ell,\ell-1}\mathbf{K}_{\ell-1}^{-1}\mathbf{R}_{\ell-1,\ell}\right)\mathbf{W}_\ell^{-1/2}\right\| \leq C_{\mathrm{A}}$$

with a constant C_{A} *which is independent of* ℓ, $\ell_{\max}$ *and the mesh width* h_ℓ.

Lemma 6.4.51. *Let the conditions from Theorem 6.4.36 hold. Then Assumption 6.4.50 is satisfied for the Jacobi method.*

Proof. We combine the statement

$$\left\|\mathbf{K}_\ell^{-1} - \mathbf{P}_{\ell,\ell-1}\mathbf{K}_{\ell-1}^{-1}\mathbf{R}_{\ell-1,\ell}\right\| \leq C_1 h_\ell^{-1}$$

from Theorem 6.4.36 with Lemma 6.4.2

$$\left\|\mathbf{W}_\ell^{-1/2}\right\|\left\|\mathbf{W}_\ell^{-1/2}\right\| \leq \omega^{-2}\left\|\mathbf{D}_\ell^{1/2}\right\|^2 \leq C_2\omega^{-2} h_\ell$$

and obtain

$$\left\|\mathbf{W}_\ell^{-1/2}\left(\mathbf{K}_\ell^{-1} - \mathbf{P}_{\ell,\ell-1}\mathbf{K}_{\ell-1}^{-1}\mathbf{R}_{\ell-1,\ell}\right)\mathbf{W}_\ell^{-1/2}\right\| \leq C_1 C_2 \omega^{-2}.$$

□

We have now gathered all the necessary conditions for the convergence of the V-cycle multi-grid method and have used the damped Jacobi method as a smoothing iteration to check its validity. The iteration matrix for the multi-grid method with V-cycle is denoted by $\mathbf{T}_\ell^V$.

Theorem 6.4.52. *Let Assumptions 6.4.44, 6.4.48, 6.4.50 and (6.90) be satisfied. Then the symmetric multi-grid method as a V-cycle converges in the energy norm at the following rate*

$$\rho\left(\mathbf{T}_\ell^V\right) = \left\|\mathbf{T}_\ell^V\right\|_{\mathbf{K}_\ell} \leq \frac{C_{\mathrm{A}}}{C_{\mathrm{A}} + 2\nu} < 1. \tag{6.91}$$

Proof. (see [116, Theorem 10.7.15]). The recursion from Theorem 6.4.41 can be simplified for the V-cycle ($\gamma = 1$) to

$$\mathbf{T}_0^V = 0, \qquad \forall \ell \geq 1 : \mathbf{T}_\ell^V = \mathbf{T}_\ell^{TGM} + \mathbf{T}_\ell^\nu \mathbf{P}_{\ell,\ell-1}\mathbf{T}_{\ell-1}^V \mathbf{K}_{\ell-1}^{-1}\mathbf{R}_{\ell-1,\ell}\mathbf{K}_\ell \mathbf{T}_\ell^\nu. \tag{6.92}$$

To make use of the symmetric structure of the V-cycle, we insert the transformations

$$\begin{aligned}
\check{\mathbf{T}}_\ell^{TGM} &:= \mathbf{K}_\ell^{1/2}\mathbf{T}_\ell^{TGM}\mathbf{K}_\ell^{-1/2}, & \check{\mathbf{T}}_\ell^{(\nu)} &:= \mathbf{K}_\ell^{1/2}\mathbf{T}_\ell^{\nu}\mathbf{K}_\ell^{-1/2}, \\
\check{\mathbf{P}}_{\ell,\ell-1} &:= \mathbf{K}_\ell^{1/2}\mathbf{P}_{\ell,\ell-1}\mathbf{K}_{\ell-1}^{-1/2}, & \check{\mathbf{R}}_{\ell-1,\ell} &= \mathbf{K}_{\ell-1}^{-1/2}\mathbf{R}_{\ell-1,\ell}\mathbf{K}_\ell^{1/2},
\end{aligned}$$

into (6.92) and use (6.73) to obtain the representation

$$\begin{aligned}
\check{\mathbf{T}}_\ell^V &:= \mathbf{K}_\ell^{1/2}\mathbf{T}_\ell^V\mathbf{K}_\ell^{-1/2} = \check{\mathbf{T}}_\ell^{TGM} + \check{\mathbf{T}}_\ell^{(\nu)}\check{\mathbf{P}}_{\ell,\ell-1}\check{\mathbf{T}}_{\ell-1}^V\check{\mathbf{R}}_{\ell-1,\ell}\check{\mathbf{T}}_\ell^{(\nu)} \\
&= \check{\mathbf{T}}_\ell^{(\nu)}\left(\mathbf{I}_\ell - \check{\mathbf{P}}_{\ell,\ell-1}\left(\mathbf{I}_\ell - \check{\mathbf{T}}_{\ell-1}^V\right)\check{\mathbf{R}}_{\ell-1,\ell}\right)\check{\mathbf{T}}_\ell^{(\nu)} \qquad (6.93) \\
&= \check{\mathbf{T}}_\ell^{(\nu)}\left(\mathbf{Q}_\ell + \check{\mathbf{P}}_{\ell,\ell-1}\check{\mathbf{T}}_{\ell-1}^V\check{\mathbf{R}}_{\ell-1,\ell}\right)\check{\mathbf{T}}_\ell^{(\nu)} \qquad (6.94)
\end{aligned}$$

with

$$\mathbf{Q}_\ell := \mathbf{I}_\ell - \check{\mathbf{P}}_{\ell,\ell-1}\check{\mathbf{R}}_{\ell-1,\ell}. \tag{6.95}$$

It follows from Assumption 6.4.44, 6.4.48 and (6.90) that $\mathbf{Q}_\ell^2 = \mathbf{Q}_\ell = \mathbf{Q}_\ell^H$. Therefore the mapping $\mathbf{Q}_\ell$ is an orthogonal projection (see Exercise 6.4.53) and is thus positive semidefinite. By induction and with (6.94) the property "$\check{\mathbf{T}}_0^V$ is positive semidefinite" is transferred to $\check{\mathbf{T}}_\ell^V$ for all $\ell \geq 0$.

Then the Euclidean norm of $\check{\mathbf{T}}_\ell^V$ is given by the largest eigenvalue of $\check{\mathbf{T}}_\ell^V$ and we have the equivalence of the two statements (6.96), (6.97)

$$\left\|\mathbf{T}_\ell^V\right\|_{\mathbf{K}_\ell} = \left\|\check{\mathbf{T}}_\ell^V\right\| \leq \zeta_\ell \tag{6.96}$$

$$\sigma\left(\check{\mathbf{T}}_\ell^V\right) \subset [0, \zeta_\ell]. \tag{6.97}$$

We assume by induction that $\sigma\left(\check{\mathbf{T}}_{\ell-1}^V\right) \subset [0, \zeta_{\ell-1}]$ with $\zeta_{\ell-1} = C_{\mathrm{A}}/\left(C_{\mathrm{A}} + 2\nu\right)$. Since $\check{\mathbf{T}}_0^V = \mathbf{0}$, this assumption is clearly satisfied for $\ell = 0$.

The spectral radius $\rho(\cdot)$ (see Definition 6.4.7) coincides with the largest eigenvalue for positive semidefinite matrices. From (6.93) and Exercise 6.4.53 we obtain

$$\rho\left(\check{\mathbf{T}}_\ell^V\right) \leq \rho\left(\check{\mathbf{T}}_\ell^{(\nu)}\left(\mathbf{I}_\ell - (1-\zeta_{\ell-1})\check{\mathbf{P}}_{\ell,\ell-1}\check{\mathbf{R}}_{\ell-1,\ell}\right)\check{\mathbf{T}}_\ell^{(\nu)}\right)$$
$$= \rho\left(\check{\mathbf{T}}_\ell^{(\nu)}\left((1-\zeta_{\ell-1})\mathbf{Q}_\ell + \zeta_{\ell-1}\mathbf{I}_\ell\right)\check{\mathbf{T}}_\ell^{(\nu)}\right). \tag{6.98}$$

It follows from the approximation property (see Assumption 6.4.50) that

$$\rho\left(\mathbf{K}_\ell^{-1} - \mathbf{P}_{\ell,\ell-1}\mathbf{K}_{\ell-1}^{-1}\mathbf{R}_{\ell-1,\ell}\right) \leq C_{\mathrm{A}}\rho\left(\mathbf{W}_\ell\right)$$

(see Exercise 6.4.53.c). The eigenvalue inequality (see Exercise 6.4.53.d) is inserted into the transformation $\mathbf{K}_\ell^{1/2}(\cdot)\mathbf{K}_\ell^{1/2}$, which yields the following spectral estimate for $\mathbf{Q}_\ell$

$$0 \leq \rho\left(\mathbf{Q}_\ell\right) = \rho\left(\mathbf{I}_\ell - \mathbf{K}_\ell^{1/2}\mathbf{P}_{\ell,\ell-1}\mathbf{K}_{\ell-1}^{-1}\mathbf{R}_{\ell-1,\ell}\mathbf{K}_\ell^{1/2}\right)$$
$$\leq C_{\mathrm{A}}\rho\left(\check{\mathbf{X}}_\ell\right) \quad \text{with} \quad \check{\mathbf{X}}_\ell := \mathbf{K}_\ell^{1/2}\mathbf{W}_\ell\mathbf{K}_\ell^{1/2}.$$

The projection property of $\mathbf{Q}_\ell$ implies the alternative estimate $\sigma\left(\mathbf{Q}_\ell\right) \subset [0,1]$. Therefore we have for all $\alpha \in [0,1]$

$$0 \leq \rho\left(\mathbf{Q}_\ell\right) \leq \alpha C_{\mathrm{A}}\rho\left(\check{\mathbf{X}}_\ell\right) + (1-\alpha).$$

This inequality, substituted in (6.98), yields

$$\rho\left(\check{\mathbf{T}}_\ell^V\right) \leq \rho\left(\check{\mathbf{T}}_\ell^{(\nu)}\left((1-\zeta_{\ell-1})\left(\alpha C_{\mathrm{A}}\check{\mathbf{X}}_\ell + (1-\alpha)\mathbf{I}_\ell\right) + \zeta_{\ell-1}\mathbf{I}_\ell\right)\check{\mathbf{T}}_\ell^{(\nu)}\right). \tag{6.99}$$

We set $\beta := (1-\zeta_{\ell-1})(1-\alpha) + \zeta_{\ell-1}$ and note that for all $\alpha \in [0,1]$ we have the inclusion $\beta \in [\zeta_{\ell-1}, 1]$. By using this relation to express α in terms of β we obtain the estimate for (6.99)

$$\rho\left(\check{\mathbf{T}}_\ell^V\right) \leq \rho\left(\check{\mathbf{T}}_\ell^{(\nu)}\left((1-\beta)C_{\mathrm{A}}\check{\mathbf{X}}_\ell + \beta\mathbf{I}_\ell\right)\check{\mathbf{T}}_\ell^{(\nu)}\right) \qquad \forall \zeta_{\ell-1} \leq \beta \leq 1. \tag{6.100}$$

The matrix $\check{\mathbf{T}}_\ell^{(\nu)}$ has the representation

$$\check{\mathbf{T}}_\ell^{(\nu)} = \mathbf{K}_\ell^{1/2}\mathbf{T}_\ell^\nu\mathbf{K}_\ell^{-1/2} = \left(\mathbf{I}_\ell - \mathbf{K}_\ell^{1/2}\mathbf{W}_\ell\mathbf{K}_\ell^{1/2}\right)^\nu = \left(\mathbf{I}_\ell - \check{\mathbf{X}}_\ell\right)^\nu.$$

Therefore the matrix on the right-hand side of (6.100) in the argument of ρ is the polynomial

$$f(\xi,\beta) := (1-\xi)^{2\nu}\left((1-\beta)C_{\mathrm{A}}\xi + \beta\right)$$

with $\xi = \check{\mathbf{X}}_\ell$. Therefore the expression

$$m(\beta) := \max\left\{f(\xi,\beta) : 0 \leq \xi \leq 1\right\}$$

is an upper bound for the spectral radius. We choose $\beta = \zeta_{\ell-1} = C_A / (C_A + 2\nu)$. An analysis of the function $f(\cdot, \zeta_{\ell-1})$ shows that it is strictly decreasing with respect to the first argument within the parameter domain under consideration. It follows that

$$\rho\left(\check{\mathbf{T}}_\ell^V\right) \leq f(0, \zeta_{\ell-1}) = \zeta_{\ell-1}$$

and by induction we have $\zeta_\ell = C_A / (C_A + 2\nu)$. The fact that (6.96) and (6.97) are equivalent finally yields the stated estimate in (6.91).

The left-hand equality in (6.91) follows from the similarity of the matrices $\mathbf{T}_\ell^V$ and $\check{\mathbf{T}}_\ell^V$ as given by

$$\left\|\mathbf{T}_\ell^V\right\|_{\mathbf{K}_\ell} = \left\|\mathbf{K}_\ell^{1/2}\mathbf{T}_\ell^V\mathbf{K}_\ell^{-1/2}\right\| = \left\|\check{\mathbf{T}}_\ell^V\right\| = \rho\left(\check{\mathbf{T}}_\ell^V\right) = \rho\left(\mathbf{T}_\ell^V\right).$$

□

Exercise 6.4.53.
Prove the following:

(a) Let Assumptions 6.4.44, 6.4.48 and (6.90) be satisfied. Let the matrices $\mathbf{K}_\ell$ *be positive definite for all* ℓ. *Then* $\mathbf{Q}_\ell$ *from (6.95) is an orthogonal projection.*

(b) Let $\mathbf{A}$, $\mathbf{B}$ *be two positive definite matrices. Then*

$$\rho(\mathbf{A} + \mathbf{B}) \leq \rho(\mathbf{A}) + \rho(\mathbf{B}).$$

(c) For positive definite matrices $\mathbf{A}$, $\mathbf{B}$ *the spectral inequality*

$$\rho(\mathbf{B}) \leq C\rho\left(\mathbf{A}^{-1}\right)$$

follows from $\left\|\mathbf{A}^{1/2}\mathbf{B}\mathbf{A}^{1/2}\right\| \leq C$.

(d) Let $\mathbf{A}$, $\mathbf{B}$, $\mathbf{C}$ *be positive definite. Then* $\rho(\mathbf{A}) \leq \rho(\mathbf{B})$ *yields the inequality*

$$\rho\left(\mathbf{C}^{1/2}\mathbf{A}\mathbf{C}^{1/2}\right) \leq \rho\left(\mathbf{C}^{1/2}\mathbf{B}\mathbf{C}^{1/2}\right).$$

Convergence of the Nested Iteration

In this subsection we will present the convergence analysis for the nested iteration (Algorithm 6.4.28). To do this we begin with suitable assumptions concerning the prolongation and the multi-grid method which is called in every iteration.

Assumption 6.4.54. *Let* $\mathbf{u}_\ell$, $\mathbf{u}_{\ell-1}$ *be solutions of (6.64) for successive mesh levels and let* $\mathbf{P}_{\ell,\ell-1}$ *be the prolongation in Algorithm 6.4.28. Then*

$$\left\|\mathbf{u}_\ell - \mathbf{P}_{\ell,\ell-1}\mathbf{u}_{\ell-1}\right\| \leq C_1 h_\ell^{-1} \|\mathbf{f}_\ell\| \quad \text{and} \quad \left\|\mathbf{P}_{\ell,\ell-1}\right\| \leq C_2.$$

Assumption 6.4.55. *The multi-grid method which is called in Algorithm 6.4.28 has a mesh-independent convergence rate:*

$$\left\| \mathbf{T}_\ell^{MGM} \right\| \leq \zeta < 1 \qquad \forall \ell \geq 1.$$

Remark 6.4.56. *The first estimate from Assumption 6.4.54 follows from (6.85) if the conditions of Theorem 6.4.36 are given:*

$$\left\| \mathbf{u}_\ell - \mathbf{P}_{\ell,\ell-1} \mathbf{u}_{\ell-1} \right\| \leq C h_\ell^{-1} \left\| \mathbf{f}_\ell \right\|.$$

The second estimate follows with (6.84) if the conditions of Theorem 6.4.36 are given.

Assumption 6.4.55 corresponds to the statements from Theorems 6.4.43 and 6.4.52. We will only consider the convergence analysis for the L^2 and Euclidean norm. We remind the reader that the V-cycle convergence was only proved in the energy norm. The convergence of the nested iteration in the energy norm can be derived similarly and is recommended as an exercise.

Let f be the continuous right-hand side in (6.46). Then the right-hand side $\mathbf{f}_\ell$ in (6.64) can be estimated by using

$$\left|(\mathbf{f}_\ell)_i\right| := \left|(f, b_{\ell,i})_{L^2(\Gamma)}\right| \leq \|f\|_{L^2(\operatorname{supp} b_{\ell,i})} \|b_{\ell,i}\|_{L^2(\operatorname{supp} b_{\ell,i})} \leq C h_\ell \|f\|_{L^2(\operatorname{supp} b_{\ell,i})}$$

and the finite intersection of the supports. The estimate is given by

$$\|\mathbf{f}_\ell\| \leq C_3 h_\ell \|f\|_{L^2(\Gamma)} < \infty. \tag{6.101}$$

The convergence of the nested iteration can be shown under these conditions. If a vector $\mathbf{a} \in \mathbb{R}^{N_\ell}$ and a function $a \in S_\ell$ appear in the same context their relation is given by $a = \sum_{i=1}^{N_\ell} \mathbf{a}_i b_{\ell,i}$.

Theorem 6.4.57. *Let Assumptions 6.4.54, 6.4.55 hold and let (6.78), (6.101) be satisfied. Let the iteration number $m_\ell = m$ in Algorithm 6.4.28 satisfy*

$$C_2 c_P^{-1} \zeta^m < 1$$

with c_P from (6.78). Then the nested iteration yields approximations $\tilde{\mathbf{u}}_\ell$ of the exact solution $\mathbf{u}_\ell$ of (6.64) which satisfy the error estimate

$$\|u_\ell - \tilde{u}_\ell\|_{L^2(\Gamma)} \leq g\left(\zeta^m\right) C_1 C_P h_\ell \|f\|_{L^2(\Gamma)} \tag{6.102}$$

with

$$g(x) := \frac{C_3 x}{1 - C_2 c_P^{-1} x}$$

if the initial value $\tilde{u}_0$ satisfies inequality (6.102).

Proof. According to the conditions, we have (6.102) for $\ell = 0$. We make the recursive assumption that (6.102) is satisfied for the levels $\leq \ell - 1$. The initial error $\mathbf{u}_\ell^0 - \mathbf{u}_\ell$ with $\mathbf{u}_\ell^0 := \mathbf{P}_{\ell,\ell-1}\tilde{\mathbf{u}}_{\ell-1}$ can be estimated by

$$\begin{aligned}
\left\|\mathbf{u}_\ell^0 - \mathbf{u}_\ell\right\| &\leq \left\|\mathbf{P}_{\ell,\ell-1}\mathbf{u}_{\ell-1} - \mathbf{u}_\ell\right\| + \left\|\mathbf{P}_{\ell,\ell-1}\left(\mathbf{u}_{\ell-1} - \tilde{\mathbf{u}}_{\ell-1}\right)\right\| \\
&\leq \left\|\mathbf{P}_{\ell,\ell-1}\mathbf{u}_{\ell-1} - \mathbf{u}_\ell\right\| + \left\|\mathbf{P}_{\ell,\ell-1}\right\| \left\|\mathbf{u}_{\ell-1} - \tilde{\mathbf{u}}_{\ell-1}\right\| \\
&\leq C_1 h_\ell^{-1} \left\|\mathbf{f}_\ell\right\| + C_2 \left\|\mathbf{u}_{\ell-1} - \tilde{\mathbf{u}}_{\ell-1}\right\| \\
&\leq C_1 C_3 \left\|f\right\|_{L^2(\Gamma)} + C_2 c_P^{-1} h_{\ell-1}^{-1} \left\|u_{\ell-1} - \tilde{u}_{\ell-1}\right\|_{L^2(\Gamma)} \\
&\leq C_1 \left(C_3 + C_2 c_P^{-1} g\left(\zeta^m\right)\right) \left\|f\right\|_{L^2(\Gamma)}.
\end{aligned}$$

Estimate (6.78) implies for the associated boundary element functions that

$$\left\|u_\ell^0 - u_\ell\right\|_{L^2(\Gamma)} \leq C_1 C_P \left(C_3 + C_2 c_P^{-1} g\left(\zeta^m\right)\right) h_\ell \left\|f\right\|_{L^2(\Gamma)}.$$

After m iterations the initial error $u_\ell^0 - u_\ell$ on the level ℓ is reduced as described by the multi-grid convergence properties

$$\left\|u_\ell^m - u_\ell\right\|_{L^2(\Gamma)} \leq \zeta^m \left\|u_\ell^0 - u_\ell\right\|_{L^2(\Gamma)} \leq \left\{\zeta^m \left(C_3 + C_2 c_P^{-1} g\left(\zeta^m\right)\right)\right\} C_1 C_P h_\ell \left\|f\right\|_{L^2(\Gamma)}.$$

The definition of g implies that $\{\ldots\} = g\left(\zeta^m\right)$ and from this we have the assertion. □

Exercise 6.4.58. *Prove the convergence of the nested iteration for the multi-grid method with V-cycle under suitable conditions.*

6.5 Multi-grid Methods for Equations of Negative Order*

The efficiency of multi-grid methods consists in the combination of the smoothing properties of the operator and the smoothing method as well as the approximation property of the coarse grid correction. The smoothing property is closely connected with the mapping properties of the operator $K : H^s(\Gamma) \to H^{-s}(\Gamma)$ with $s > 0$. For operators of negative order (example: single layer potential) the operator and the associated simple iterative method are no longer smoothing. High-frequency eigenfunctions correspond to small eigenvalues and vice-versa.

---

* This section should be regarded as a complement to the actual focus of the book.

In this subsection we present an approach with which the mapping properties of the integral operators of negative order are reversed. This in turn allows the application of multi-grid methods. This approach is due to Bramble, Leyk and Pasciak (see [29, 30]).

As a model problem we consider the Galerkin discretization of the boundary integral equation for the single layer potential $V : H^{-1/2}(\Gamma) \to H^{1/2}(\Gamma)$. Let $f \in H^{1/2}(\Gamma)$ be given and let the bilinear form $b : H^{-1/2}(\Gamma) \times H^{-1/2}(\Gamma) \to \mathbb{R}$ be defined as

$$b(u, v) := (Vu, v)_{L^2(\Gamma)} = \int_{\Gamma\times\Gamma} \frac{v(\mathbf{x})\, u(\mathbf{y})}{4\pi \|\mathbf{x} - \mathbf{y}\|} ds_{\mathbf{y}} ds_{\mathbf{x}}. \tag{6.103}$$

$S \subset H^{-1/2}(\Gamma)$ denotes the boundary element space and the Galerkin solution is characterized by

$$\forall v \in S : \quad b(u, v) = f(v).$$

The basis representation $u = \sum_{i=1}^{N} \mathbf{u}_i b_i$ transfers the problem to the linear system of equations

$$\mathbf{Vu} = \mathbf{f} \tag{6.104}$$

and our aim is its efficient solution. For simplicity we assume here that $\mathbf{V}$ is symmetric and positive definite (see Proposition 4.1.24). Perturbations, for example, due to numerical quadrature, can be treated as in Exercise 6.3.1 or 6.3.5.

Definition 6.5.1. The surface gradient of a function $u \in H^1(\Gamma)$ is given by

$$\nabla_\Gamma u := \gamma_0 \nabla Z u$$

with the trace extension Z from Theorem 2.6.11 and the trace operator γ_0 from Theorem 2.6.8.

With this the bilinear form $w : H^1(\Gamma) \times H^1(\Gamma) \to \mathbb{R}$ can be defined by

$$w(u, v) = \int_\Gamma (\langle \nabla_\Gamma u, \nabla_\Gamma v\rangle + uv)\, dx \qquad \forall u, v \in H^1(\Gamma). \tag{6.105}$$

In order to avoid technical difficulties we assume that the boundary element space satisfies the inclusion

$$S \subset H^1(\Gamma). \tag{6.106}$$

The Galerkin discretization of the bilinear form w yields the sparse matrix

$$\mathbf{W}_{j,i} := w(b_i, b_j) \qquad \forall 1 \le i, j \le N.$$

Proposition 6.5.2. *The bilinear form w in (6.105) is $H^1(\Gamma)$-elliptic.*

Proof. The statement follows directly from $w(u, u) = \|u\|^2_{H^1(\Gamma)}$. □

A direct consequence is the fact that the matrix $\mathbf{W}$ is positive definite. The operator $W : H^1(\Gamma) \to H^{-1}(\Gamma)$ can be associated with the bilinear form as given by

$$\langle Wu, v\rangle_{H^{-1}(\Gamma)\times H^1(\Gamma)} = w(u,v) \qquad \forall u, v \in H^1(\Gamma).$$

The connection between the continuous operators W and V and the matrices $\mathbf{W}$ and $\mathbf{V}$ will be studied in the following.

First let P and R be as in Definition 6.4.21, while the index ℓ is omitted here. The mapping $T : H^{-1}(\Gamma) \to \mathbb{R}^N$ is defined by

$$Tf = \left((f, b_i)_{L^2(\Gamma)}\right)_{i=1}^N \qquad \forall f \in H^{-1}(\Gamma).$$

Finally, the *mass* or *Gram matrix* is given as in (6.77) by

$$\mathbf{M}_{j,i} := \left(b_i, b_j\right)_{L^2(\Gamma)} \qquad 1 \leq i, j \leq N.$$

Therefore the L^2-orthogonal projection $Q : H^s(\Gamma) \to S$ for $s \in [0,1]$ has the representation $Q = P\mathbf{M}^{-1}T$.

Proposition 6.5.3. *Let $u \in S$ and let $\mathbf{u} = \left(u_j\right)_{j=1}^N$ be the associated coefficient vector. Then*

$$\left(\mathbf{W}\mathbf{M}^{-1}\mathbf{V}\mathbf{u}\right)_i = (WQVu, b_i)_{L^2(\Gamma)}.$$

Proof. The assertion follows from

$$\begin{aligned}(WQVu, b_i)_{L^2(\Gamma)} &= \left(WP\mathbf{M}^{-1}TVu, b_i\right)_{L^2(\Gamma)} = \left(WP\mathbf{M}^{-1}\left((Vu, b_k)_{L^2(\Gamma)}\right)_{k=1}^N, b_i\right)_{L^2(\Gamma)} \\ &= \left(W\sum_{m=1}^N \left(\mathbf{M}^{-1}\mathbf{V}\mathbf{u}\right)_m b_m, b_i\right)_{L^2(\Gamma)} = \sum_{m=1}^N \left(\mathbf{M}^{-1}\mathbf{V}\mathbf{u}\right)_m \mathbf{W}_{i,m} = \left(\mathbf{W}\mathbf{M}^{-1}\mathbf{V}\mathbf{u}\right)_i.\end{aligned}$$

□

The matrix-matrix multiplication $\mathbf{W}\mathbf{M}^{-1}\mathbf{V}$ corresponds to the Galerkin discretization of the composition WQV. This composition approximates the operator WV. For smooth surfaces, $\Gamma \in C^\infty$, it is possible to show the mapping properties $WV : H^s(\Gamma) \to H^{s-1}(\Gamma)$ and the continuous invertibility $(WV)^{-1} : H^{s-1}(\Gamma) \to H^s(\Gamma)$ (see [29]).

Since the matrix $\mathbf{W}\mathbf{M}^{-1}\mathbf{V}$ is the product of positive definite matrices, it is regular and the eigenvalues are real and positive. Therefore the solution of the linear system of equations

$$\mathbf{W}\mathbf{M}^{-1}\mathbf{V}\mathbf{u} = \mathbf{W}\mathbf{M}^{-1}\mathbf{f} \qquad (6.107)$$

solves (6.104). Since $\mathbf{W}\mathbf{M}^{-1}\mathbf{V}$ can be considered as the discretization of the continuous, regular operator WV of order $+1$, the multi-grid method from Sect. 6.4.3 can be dirctly used for the solution of (6.107). It is essential for the efficiency of the method that $\mathbf{W}$ be sparse so that a matrix-vector multiplication with a complexity of $O(N)$ can be implemented. For the convergence analysis we refer to [29–31].

6.6 Further Remarks and Results on Iterative Solvers of BIEs

From the convergence bounds for the basic algorithms for the iterative numerical solution of the large, densely populated linear systems of equations, it is clear that preconditioning is an important issue. This is particularly so on complicated geometries, and for parameter dependent boundary integral operators which arise in acoustic and electromagnetic scattering. In particular, in Galerkin BEM for BIEs arising in electromagnetic scattering, the large, nontrivial null space of the electric field boundary integral equation stalls most "black-box" iterative solvers.

Here, construction of a wave number dependent preconditioner is essential and highly nontrivial. One approach towards such preconditioners is the use of the Calderón projector for electromagnetic scattering. This was proposed and analyzed in [65–67]. The preconditioners thus obtained are based on suitable discretizations of, in the Calderón projector, conjugate boundary integral operators and require therefore the discretization of these additional boundary integral operators.

The question of how to preserve the Calderón identities under discretization was answered recently, in the case of electromagnetic scattering, with tools from algebraic geometry in [4, 38].

Chapter 7
Cluster Methods

Partial differential equations can be directly discretized by means of difference methods or finite element methods (domain methods). For this, a mesh has to be generated over the d-dimensional domain Ω, which is a difficult task for complicated geometries and exterior domains – if the domain is unbounded the generation of a mesh with finitely many mesh cells is impossible, leading to further complications. In comparison, the generation of a $(d-1)$-dimensional surface mesh for the boundary element method is a much simpler task. The latter method has the additional advantage that the degrees of freedom only occur on the surface and the dimension of the system of equations is decidedly smaller than for domain methods. If, for example, we generate a mesh of the unit cube in $\mathbb{R}^3$ with a uniform Cartesian mesh and use the number of mesh points N as degrees of freedom then only $O\left(N^{2/3}\right)$ of these actually lie on the surface, which means that in the BEM the dimension of the system matrix is significantly reduced (in proportion to the number of degrees of freedom). On the other hand, difference methods and finite elements lead to sparse system matrices. This means that the memory requirements only grow linearly as a function of the number of unknowns. At first glance this seems to be a disadvantage of the matrix representation of boundary integral operators, where the system matrices are dense. Storing these would repeal the advantage gained by reducing the dimension of the computational domain through the boundary element method. In the example considered above, one would need $O(N)$ memory units to store the system matrix for the domain discretization and $O\left(N^{4/3}\right)$ flops for the boundary element method. The *cluster methods* use an alternative (approximative) representation of the discrete integral operator and allows the storage of the operator with $O\left(N \log^{\kappa} N\right)$ memory units, where N denotes the number of degrees of freedom on the surface and $\kappa \approx 4-6$ depends on the problem.

Direct elimination methods such as the Gauss or Cholesky decompositions are not suited to higher-dimensional problems as their complexity grows cubically with respect to the dimension. Iterative methods prove to be much more efficient. The cluster representation of the integral operator allows matrix-vector multiplications, which are the elementary operations in iterative methods for linear systems of equations, to be performed in $O\left(N \log^{\kappa} N\right)$ arithmetic operations. The representation does, however, not permit the use of direct elimination methods, as the matrix elements are usually not evaluated. The cluster method was first developed for

S.A. Sauter and C. Schwab, *Boundary Element Methods*, Springer Series in Computational Mathematics 39, DOI 10.1007/978-3-540-68093-2_7,

collocation methods and is due to W. Hackbusch and Z.P. Nowak (see [123, 124]). The extension to Galerkin methods was introduced in [125, 197]. Closely related to the panel clustering method is the multipole method which has been developed independently (see [193]).

We will first introduce and describe the cluster method on an abstract level. Then we will present the algorithmic description of the method. We will conclude this chapter with an analysis of the error in the Galerkin solution introduced by the panel clustering approximation and of the overall complexity of the algorithm.

Note: According to the reader's specific interests, there are different ways of reading this chapter. Here are a few guidelines.

For readers that are interested in the concrete, algorithmic realization of the cluster method we recommend studying the Sects. 7.1.1, 7.1.2, 7.1.3.1, 7.1.4.4 and 7.2.1 in detail. The required orders of expansion can be found in Table (7.73).

If the main point of interest is the abstract error analysis of the Galerkin method perturbed by the cluster method we recommend Sects. 7.1.1, 7.1.3.3, the introduction to Sect. 7.1.4 and Sect. 7.3.2.

Readers that are interested in the derivation of local error estimates for the Čebyšev interpolation should focus on Sects. 7.1.3.1 and 7.3.1.1.

Finally, understanding the complexity estimates in Sect. 7.4 requires a certain knowledge of Sects. 7.1.2, 7.1.3.3 and 7.1.4.4.

7.1 The Cluster Algorithm

7.1.1 Conditions on the Integral Operator

We will first define the cluster method for Galerkin discretizations. In Sect. 7.5 we will present the necessary modifications for collocation methods.

Let $\mathcal{G}$ be a given boundary element mesh on a surface Γ and let K be an abstract boundary integral operator of the form

$$K[u](\mathbf{x}) := \int_\Gamma k(\mathbf{x},\mathbf{y})\, u(\mathbf{y})\, ds_{\mathbf{y}} \qquad \text{for all } \mathbf{x} \in \Gamma. \tag{7.1}$$

The following assumption describes the class of kernel functions that will be considered in this section.

Assumption 7.1.1. *The kernel function $k : \Gamma \times \Gamma \to \mathbb{C}$ is the directional derivative of a global kernel function $G : \mathbb{R}^d \times \mathbb{R}^d \to \mathbb{C}$ that is smooth for $\mathbf{x} \neq \mathbf{y}$:*

$$k(\mathbf{x},\mathbf{y}) = D_{\mathbf{x}} D_{\mathbf{y}} G(\mathbf{x},\mathbf{y}), \tag{7.2}$$

with the differential operators $D_{\mathbf{x}}$, $D_{\mathbf{y}}$ of order 0 or 1.

The cluster method will be formulated for this class of integral operators. We will impose further conditions on the global kernel function G once we deal with results on convergence and complexity.

Example 7.1.2. *For the kernel function of the single layer potential, G is given by the fundamental solution (3.3) and $k(\mathbf{x}, \mathbf{y}) := G(\mathbf{x} - \mathbf{y})$.*

For the kernel function of the double layer potential we have $k(\mathbf{x}, \mathbf{y}) = \tilde{\gamma}_{1,\mathbf{y}} G(\mathbf{x} - \mathbf{y})$ with $\tilde{\gamma}_1 = \langle \mathbf{n}\gamma_0, \mathbf{A}\,\mathrm{grad}\cdot + 2\mathbf{b}\cdot\rangle$ [see (2.107)].

For the kernel function of the adjoint double layer potential we have $k(\mathbf{x}, \mathbf{y}) = \gamma_{1,\mathbf{x}} G(\mathbf{x} - \mathbf{y})$ with $\gamma_1 = \langle \mathbf{n}\gamma_0, \mathbf{A}\,\mathrm{grad}\cdot\rangle$.

The kernel function of the hypersingular operator (without integration by parts, see Sect. 3.3.4) satisfies $k(\mathbf{x}, \mathbf{y}) = \gamma_{1,\mathbf{x}} \tilde{\gamma}_{1,\mathbf{y}} G(\mathbf{x} - \mathbf{y})$.

7.1.2 Cluster Tree and Admissible Covering

The cluster method is based on an approximation of the kernel function on $\Gamma \times \Gamma$. More precisely, first the *global kernel function* $G : \mathbb{R}^d \times \mathbb{R}^d \to \mathbb{C}$ is approximated in suitable domains, from which the approximation of the actual *kernel function* is constructed. Our approach consists in using the cluster algorithm to represent and evaluate the Galerkin discretization of boundary integral equations in a memory efficient way. We begin by clustering the *degrees of freedom* of the Galerkin discretization (as opposed to directly clustering the panels). The advantage of this approach lies in the fact that the algorithmic realization is simpler than the alternative approach, i.e., defining the clusters as the union of panels. This factor is especially relevant to the realization of data structures. The set of degrees of freedom is denoted by $\mathcal{I}$ [see, for example, (4.28)].

Definition 7.1.3. A cluster is the union of one or more indices from $\mathcal{I}$.

The efficiency of the cluster method is based on the organization of the index set in a hierarchical cluster tree.

Definition 7.1.4. The nodes of the cluster tree $\mathcal{T}$ are clusters. The set of all nodes is denoted by T and satisfies:

1. $\mathcal{I}$ is a node of T.
2. The set of leaves $\mathrm{Leaves}(\mathcal{T}) \subset T$ of $\mathcal{T}$ corresponds to the degrees of freedom $i \in \mathcal{I}$ and is given by
$$\mathrm{Leaves}(\mathcal{T}) := \{\{i\} : i \in \mathcal{I}\}.$$
3. For every node σ from $T \backslash \mathrm{Leaves}(\mathcal{T})$ there exists a minimal set $\Sigma(\sigma)$ of nodes in $T \backslash \{\sigma\}$ that satisfies
$$\sigma = \bigcup_{\tilde{\sigma} \in \Sigma(\sigma)} \tilde{\sigma}. \tag{7.3}$$

The set $\Sigma(\sigma)$ is called the sons of σ. The father father $(\tilde{\sigma})$ of a cluster $\tilde{\sigma} \in T \backslash \{\mathcal{I}\}$ is that cluster $\sigma = \text{father}(\tilde{\sigma}) \in T$ with $\tilde{\sigma} \in \Sigma(\sigma)$.
The edges of the cluster tree $\mathcal{T}$ are those pairs of nodes $(\sigma, s) \in T \times T$ that satisfy either $\sigma \in \Sigma(s)$ or $s \in \Sigma(\sigma)$.

If the cluster tree $\mathcal{T}$ is clear from the context we will write "Leaves" instead of "Leaves $(\mathcal{T})$".

We emphasize that the clustering concept is not restricted to boundary element methods but is an abstract concept for organizing large sets as hierarchical trees (see [118, 119]). The following example illustrates the clustering of a set of panels.

Example 7.1.5. *Let $\ell_{\max} \in \mathbb{N}$ be chosen and fixed. We set $N = 2^{\ell_{\max}}$ and $h = (N-1)^{-1}$. The mesh points $x_i := (i-1)h$, $1 \le i \le N$, form the set $\mathcal{I}$. A mesh of the interval $(0,1)$ is given by the panels $\tau_i = (x_{i-1}, x_i)$, $2 \le i \le N$. The set of nodes T of the cluster tree consists of the clusters*

$$\sigma_{i,\ell} := \left\{ \frac{i-1}{2^\ell} N + 1, \frac{i-1}{2^\ell} N + 2, \ldots, \frac{i}{2^\ell} N \right\}, \quad \forall 0 \le \ell \le \ell_{\max}, \quad 1 \le i \le 2^\ell.$$

The set of sons of a node $\sigma_{i,\ell} \in T \backslash$ Leaves is given by

$$\Sigma(\sigma_{i,\ell}) = \{\sigma_{2i-1,\ell+1}, \sigma_{2i,\ell+1}\}.$$

For $\ell_{\max} = 3$ the cluster tree is illustrated in the following figure.

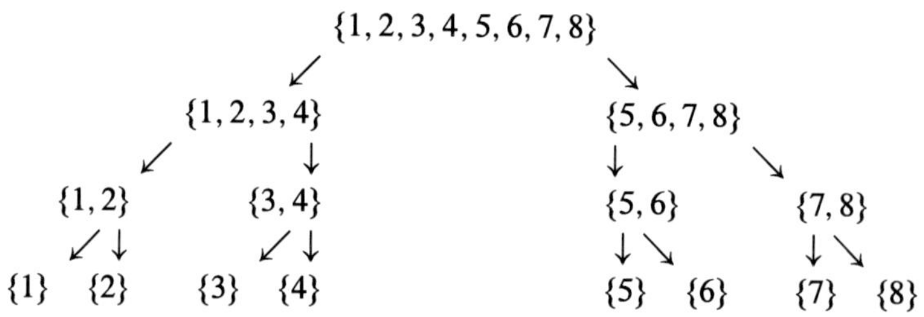

In the next step each cluster $\sigma \in T$ is assigned a *geometric cluster* and a *diameter*. The basis functions of the boundary element space are again denoted by b_i, $i \in \mathcal{I}$.

Definition 7.1.6. Every cluster $\sigma \in T$ is associated with a geometric cluster $\Gamma_\sigma \subset \Gamma$:

$$\Gamma_\sigma := \bigcup_{i \in \sigma} \operatorname{supp} b_i.$$

The cluster box Q_σ of a cluster $\sigma \in T$ is the minimal axiparallel cuboid which contains Γ_σ. The cluster center $\mathbf{M}_c$ is the center of mass of the cluster box.

The cluster diameter is given by

$$\operatorname{diam}\sigma = \sup_{x,y\in Q_\sigma} \|x-y\| .$$

In the original version (see [124]) of the cluster method the size of a cluster was determined by using the minimal ball that contains Γ_σ. Using cuboids, the sides of which are parallel to the coordinate axes (which we will henceforth call axiparallel cuboids), is advantageous from an algorithmic point of view and we have chosen that approach here.

Note that neither the cluster tree nor its generation from the set $\mathcal{I}$ is unique. In the following we will present a construction which only requires the index set $\mathcal{I}$ and the associated surface pieces $\Gamma_{\{i\}}$ as its input. With the notation from Definition 7.1.6, $Q_{\mathcal{I}}$ denotes the minimal axiparallel cuboid that contains Γ. For an arbitrary cuboid Q we define the set of sons by

$$\begin{aligned} \Sigma(Q): \ & \text{set of the 8 congruent subcuboids} \\ & \text{that result by bisecting the edges of } Q. \end{aligned} \tag{7.4}$$

Remark 7.1.7. *For a set $\omega \subset \mathbb{R}^3$, $Box(\omega)$ denotes the minimal axiparallel cuboid that contains ω. Note that $Box(\tau)$ can easily be determined for panels $\tau \in \mathcal{G}$ as is $Box(Box(\omega_1) \cup Box(\omega_2))$ if $Box(\omega_i)$, $i = 1, 2$, is known.*

The cluster tree is generated "in steps". The statements

$$\ell := 0; \qquad \mathcal{L} := \mathcal{I}; \qquad T := \mathcal{I}; \qquad E := \emptyset; \qquad \textbf{generate\_cluster\_tree}(\mathcal{L});$$

generates the cluster tree, where the subroutine **generate_cluster_tree** is defined as follows.

Algorithm 7.1.8 (Cluster Tree).
Comment: Generating the tree structure:

```
procedure generate_cluster_tree(L);
begin
    for all σ ∈ L do begin
        initialize Σ(σ) := ∅;
        generate (temporarily) the set Σ(Q_σ);
        for all Q ∈ Σ(Q_σ) do begin
            initialize a (temporary) node s := ∅;
            for all i ∈ σ do if M_{i} ∈ Q do s := s ∪ {i};
            if s ≠ ∅ and s ≠ σ then Σ(σ) := Σ(σ) ∪ {s};
        end;
        L := Σ(σ) ∪ L\{σ}; T := T ∪ Σ(σ);
    end;
    if L ≠ ∅ then generate_cluster_tree(L);
end;
```

Comment: Generating the cluster boxes.
Comment: Initialization:

for all $i \in \mathcal{I}$ ***do*** $Q_{\{i\}} := \emptyset$;
for all $\tau \in \mathcal{G}$ ***and all*** $(b_i : \operatorname{supp} b_i \cap \tau \neq \emptyset)$ ***do*** $Q_{\{i\}} := Box\left(Q_{\{i\}} \cup Box(\tau)\right)$;

Comment: For the remaining clusters:

$T_{temp} := T \backslash$ Leaves; ***for all*** $c \in T_{temp}$ ***do generate_clusterbox***(c, T_{temp}, Q_c);

Comment: The procedure **generate_clusterbox** is defined as follows:

procedure generate_clusterbox(c, T_{temp}, Q_c);
begin
for all $\tilde{c} \in \Sigma(c)$ ***do begin***
if $Q_{\tilde{c}}$ *has not yet been generated* ***then generate_clusterbox***$(\tilde{c}, T_{temp}, Q_{\tilde{c}})$;
$Q_c := Box(Q_c \cup Q_{\tilde{c}})$;
end;
$T_{temp} := T_{temp} \backslash \{c\}$;
end;

Comment: Computation of the cluster diameter:

for all $c \in T$ ***do*** $\operatorname{diam} c := \operatorname{diam} Q_c$;

Remark 7.1.9. *The construction ensures that a cluster c is either a leaf, i.e., $c = \{i\}$ for some $i \in \mathcal{I}$, or has a non-empty set of sons that are all different from σ.*

Remark 7.1.10. *The subdivision of a cluster is controlled by the geometric subdivision of the associated axiparallel box. Alternatively, the subdivision can also be controlled by the cardinality of the sons. Here the index set σ is divided in such a way that the cardinality of the sons is as large as possible. However, for the purpose of error estimates the geometric subdivision is more advantageous.*

A pair of clusters $(\sigma, s) \in T \times T$ is uniquely associated with a submatrix by $\mathbf{K}^{(\sigma,s)} := (\mathbf{K}_{i,j})_{\substack{i \in \sigma \\ j \in s}}$. The idea of the cluster method consists in finding an approximate representation of such submatrices with significantly reduced memory requirements. This entails a reduced complexity for arithmetic operations such as multiplying a matrix block by a vector. The approximation can be applied to submatrices $\mathbf{K}^{(\sigma,s)}$ for which the cluster pair (σ, s) is sufficiently well separated. The details are given in the following definition.

Definition 7.1.11. (a) The distance between two clusters $\sigma, s \in T$ is given by

$$\operatorname{dist}(\sigma, s) := \inf_{(x,y) \in Q_\sigma \times Q_s} \|x - y\| . \tag{7.5}$$

(b) For $\eta \in \mathbb{R}_{>0}$ two clusters $\sigma, s \in T$ are called admissible if

$$\eta \operatorname{dist}(\sigma, s) \geq \max\{\operatorname{diam} \sigma, \operatorname{diam} s\} .$$

The admissibility condition allows the matrix, or rather the index set $\mathcal{I} \times \mathcal{I}$, to be decomposed into admissible cluster pairs and non-admissible pairs of single indices. On the admissible cluster pairs the associated submatrices are replaced by the cluster method representation and on the non-admissible index pairs the associated matrix elements are computed and stored in the conventional manner. The set of admissible cluster pairs in this decomposition is denoted by P^{far} and the set of non-admissible index pairs by P^{near}. The union $P^{far} \cup P^{near}$ yields the decomposition P. It is essential for the efficiency that this (non-unique) decomposition consist of as few elements as possible. For the algorithmic realization we need a structure of descendants and predecessors on the set of cluster pairs. For $(\sigma, s) \in T \times T$ we define

$$\Sigma(\sigma, s) := \begin{cases} \Sigma(\sigma) \times \Sigma(s) & (\sigma, s) \in (T \backslash \text{Leaves}) \times (T \backslash \text{Leaves}), \\ \Sigma(\sigma) \times \{s\} & (\sigma, s) \in (T \backslash \text{Leaves}) \times \text{Leaves}, \\ \{\sigma\} \times \Sigma(s) & (\sigma, s) \in \text{Leaves} \times (T \backslash \text{Leaves}), \\ \emptyset & (\sigma, s) \in \text{Leaves} \times \text{Leaves}. \end{cases}$$

The program line

$$P^{near} := P^{far} := \emptyset; \quad \textbf{divide}\left(\mathcal{I} \times \mathcal{I}, P^{near}, P^{far}\right);$$

generates this decomposition, where the recursive subroutine **divide** is defined as follows.

Algorithm 7.1.12.

procedure divide$\left((c, s), P^{near}, P^{far}\right)$;
begin
  if (c, s) is admissible **then** $P^{far} := P^{far} \cup \{(c, s)\}$
  else if $(c, s) \in \text{Leaves} \times \text{Leaves}$ **then** $P^{near} := P^{near} \cup \{(c, s)\}$
  else for all $(\tilde{c}, \tilde{s}) \in \sum(\sigma, s)$ **do divide**$\left((\tilde{c}, \tilde{s}), P^{near}, P^{far}\right)$;
end;

The data structures for the algorithmic realization of the cluster method should be chosen in such a way that, for every degree of freedom, the set of the associated near-field degrees of freedom and, for every cluster, the set of associated far-field clusters are stored. For $i \in \mathcal{I}$ and $c \in T$ we define

$$P^{near}(\{i\}) := \{\{j\} \in \text{Leaves} : (\{i\}, \{j\}) \in P^{near}\}, \tag{7.6}$$

$$P^{far}(c) := \left\{\sigma \in T : (c, \sigma) \in P^{far}\right\}. \tag{7.7}$$

7.1.3 Approximation of the Kernel Function

The kernel function is approximated on admissible pairs of geometric clusters. In the first step the global kernel function G is approximated on three-dimensional domains and then the approximation of the kernel function k is defined as the directional derivative of this expansion. Here we consider the approximation by interpolation in detail and briefly summarize possible alternatives.

7.1.3.1 Čebyšev Interpolation

Interpolation is well suited for the kernel approximation because of its easy algorithmic implementation and because of the fact that it can be applied to a large class of kernel functions.

In this section we will introduce the Čebyšev interpolation algorithmically. The error analysis can be found in Sect. 7.3. For a detailed introduction to Lagrange interpolation we refer to [225] and [99].

The space of all three-dimensional polynomials of maximal degree m in every component is denoted by $\mathbb{Q}_m$

$$\mathbb{Q}_m := \left\{ \sum_{i,j,k=0}^{m} \alpha_{i,j,k} x_1^i x_2^j x_3^k : a_{i,j,k} \in \mathbb{C} \right\}. \tag{7.8}$$

Let $D \subset \mathbb{R}^3$ and let $f \in C^0(D)$ be a continuous function. The *interpolation problem* reads: For a given set of nodes $\mathcal{Z} = \left\{\xi^{(i)} : 1 \leq i \leq q\right\} \subset D$ find a function $p \in \mathbb{Q}_{m-1}$ such that

$$p(\xi) = f(\xi) \qquad \forall \xi \in \mathcal{Z}. \tag{7.9}$$

This problem does not have a solution in general. On axiparallel cuboids and Cartesian interpolation nodes the solution of the interpolation problem can, however, be easily formulated. To do this we introduce, for $m \in \mathbb{N}_0$, the sets

$$\mathcal{J}_m := \left\{\mu \in \mathbb{N}^3 \mid \forall 1 \leq i \leq 3 : 1 \leq \mu_i \leq m\right\}. \tag{7.10}$$

Convention 7.1.13. *For* $\mathbf{a} = (a_i)_{i=1}^3$, $\mathbf{b} = (b_i)_{i=1}^3 \in \mathbb{R}^3$ *we will always assume in the following that* $b_i > a_i$, $1 \leq i \leq 3$, *and consider axiparallel cuboids of the form* $Q_{\mathbf{a},\mathbf{b}} := [a_1, b_1] \times [a_2, b_2] \times [a_3, b_3]$. *The associated coordinate intervals are denoted by* $Q_{\mathbf{a},\mathbf{b}}^{(i)} = [a_i, b_i]$, $1 \leq i \leq 3$.

Definition 7.1.14 (Tensorized Lagrange basis). Let $I := [a, b]$ be a real interval with $b > a$. Let $\left\{\xi^{(1)}, \xi^{(2)}, \ldots, \xi^{(m)}\right\} \subset I$ be a set of points with

$$a = \xi^{(1)} < \xi^{(2)} < \ldots < \xi^{(m-1)} < \xi^{(m)} = b.$$

For $i = 1, 2, \ldots, m$ the Lagrange polynomial $L^{(i,m)} : I \to \mathbb{R}$ for the interpolation node $\xi^{(i)}$ is given by

$$L^{(i,m)}(x) := \prod_{\substack{j=1 \\ j \neq i}}^{m} \frac{x - \xi^{(j)}}{\xi^{(i)} - \xi^{(j)}}.$$

Let Convention 7.1.13 hold. Let the mesh $\mathcal{Z} = \mathcal{Z}_1 \times \mathcal{Z}_2 \times \mathcal{Z}_3$ with $\mathcal{Z}_j = \{\xi^{(1,j)}, \xi^{(2,j)}, \ldots, \xi^{(m,j)}\}$, $1 \leq j \leq 3$, be given. For $\mu \in \mathcal{J}_m$ we set $\xi^{(\mu)} := \left(\xi^{(\mu_i,i)}\right)_{i=1}^{3} \in \mathcal{Z}$. With this the tensorized Lagrange basis is given by

$$L^{(\mu,m)}(\mathbf{x}) := L^{(\mu_1,m)}(x_1)\, L^{(\mu_2,m)}(x_2)\, L^{(\mu_3,m)}(x_3).$$

The nodes $\xi^{(j)}$ and the associated Lagrange functions generally depend on the interval I and the order m. The index m in $\xi^{(i,m)}$ and $L^{(\mu,m)}$ will always be omitted in the following, assuming that it is clear from the context.

The Lagrange basis functions $L^{(\mu)}$ are polynomials of maximal degree $m-1$ in every component and have the property

$$L^{(\mu)}(\mathbf{x}) = \begin{cases} 1 & \text{for } \mathbf{x} = \xi^{(\mu)}, \\ 0 & \text{for } \mathbf{x} \in \mathcal{Z} \backslash \{\xi^{\mu}\}. \end{cases}$$

Let $\mathbf{a}$ and $\mathbf{b}$ be as in Convention 7.1.13 and let the mesh $\mathcal{Z}$ on the cuboid $Q_{\mathbf{a},\mathbf{b}}$ be as in Definition 7.1.14. Then the solution of the interpolation problem (7.9) can be given explicitly as

$$p_m(\mathbf{x}) = \sum_{\mu \in \mathcal{J}_m} f\left(\xi^{(\mu)}\right) L^{(\mu)}(\mathbf{x}). \tag{7.11}$$

The choice of the Čebyšev nodes as interpolation nodes has stability advantages compared to the equidistant division of the intervals. The following classical error representation illustrates to what extend the interpolation error depends on the choice of interpolation nodes.

Theorem 7.1.15. *Let $b > a$, $I = [a,b]$ and $f \in C^m(I)$. For a set of interpolation nodes $\mathcal{Z} = \left(\xi^{(i)}\right)_{i=1}^{m}$ we set $p_m := \sum_{i=1}^{m} f\left(\xi^{(i)}\right) L^{(i)}$. Then for all $x \in I$ there exists some $\theta_x \in I$ such that*

$$f(x) - p_m(x) = \frac{f^{(m)}(\theta_x)}{m!} \prod_{i=1}^{m} \left(x - \xi^{(i)}\right). \tag{7.12}$$

Choosing the *Čebyšev nodes* minimizes the product $\prod_{i=1}^{m}\left(x - \xi^{(i)}\right)$ in the error term. We will first introduce the Čebyšev nodes for the interval $[-1,1]$ and summarize some of the properties of the Čebyšev polynomials.

The Čebyšev polynomials can be recursively defined by $T_0(x) := 1$, $T_1(x) := x$ and for $k = 2, 3, \ldots$ by

$$T_{k+1}(x) := 2x T_k(x) - T_{k-1}(x). \tag{7.13}$$

We obtain, for example,

$$T_2(x) = 2x^2 - 1, \qquad T_3(x) = 4x^3 - 3x, \qquad T_4(x) = 8x^4 - 8x^2 + 1.$$

Their roots are real, pairwise different and lie in the interval $[-1, 1]$. More precisely, for the Čebyšev polynomial T_n, $n \geq 1$, they have the form

$$\xi^{(i,n)} := \cos \frac{2i-1}{2n}\pi \qquad 1 \leq i \leq n.$$

Definition 7.1.16 (Čebyšev Interpolation on $[-1, 1]$). The Čebyšev interpolation of a continuous function $f \in C^0([-1, 1])$ is given by

$$\Pi^{(m)}[f] := \sum_{i=1}^{m} f\left(\xi^{(i)}\right) L^{(i)}$$

with respect to the nodes $\left(\xi^{(i)}\right)_{i=1}^{m}$.

For $\mu \in \mathbb{N}_0^3$ and $\mathbf{x} = (x_i)_{i=1}^3$ the tensorized Čebyšev polynomials are given by

$$T_\mu(\mathbf{x}) := T_{\mu_1}(x_1) T_{\mu_2}(x_2) T_{\mu_3}(x_3). \tag{7.14}$$

For the three-dimensional unit cube $Q = I \times I \times I$ with $I = [-1, 1]$ the interpolation nodes are given by the tensorized one-dimensional Čebyšev nodes

$$\xi^{(\mu)} := \left(\xi_i^{(\mu_i)}\right)_{i=1}^{3} \qquad \mu \in \mathcal{J}_m.$$

Definition 7.1.17 (Čebyšev Interpolation on Q). The Čebyšev interpolation of a continuous function $f \in C^0(Q)$ is given by

$$\vec{\Pi}^{(m)}[f] := \sum_{\mu \in \mathcal{J}_m} f\left(\xi^{(\mu)}\right) L^{(\mu)}.$$

For an axiparallel cuboid $Q_{\mathbf{a},\mathbf{b}}$ we define the Čebyšev interpolation by means of the affine transformation $\chi : Q \to Q_{\mathbf{a},\mathbf{b}}$ given by

$$\chi(\hat{\mathbf{x}}) := \left(a_i + (b_i - a_i)\frac{x_i + 1}{2}\right)_{i=1}^{3}.$$

The transformed Čebyšev nodes form the set

$$\Theta_{\mathbf{a},\mathbf{b}}^{(m)} := \left\{\chi\left(\xi^{(\mu)}\right) : \mu \in \mathcal{J}_m\right\}.$$

Definition 7.1.18 (Čebyšev Interpolation on $Q_{\mathbf{a},\mathbf{b}}$). The Čebyšev interpolation of a continuous function $f \in Q_{\mathbf{a},\mathbf{b}}$ is given by

$$\vec{\Pi}^{(m)}_{\mathbf{a},\mathbf{b}}[f] := \sum_{\xi^{(\mu)} \in \Theta^{(m)}_{\mathbf{a},\mathbf{b}}} f\left(\xi^{(\mu)}\right) L^{(\mu)},$$

where the Lagrange functions correspond to the set $\Theta^{(m)}_{\mathbf{a},\mathbf{b}}$.

The approximation of the kernel function starts with the approximation of the global kernel function on a pair of axiparallel cuboids $Q_{\mathbf{a},\mathbf{b}} \times Q_{\mathbf{c},\mathbf{d}}$.

Definition 7.1.19 (Čebyšev Interpolation on $Q_{\mathbf{a},\mathbf{b}} \times Q_{\mathbf{c},\mathbf{d}}$). For $f \in C^0(Q_{\mathbf{a},\mathbf{b}} \times Q_{\mathbf{c},\mathbf{d}})$ the tensorized Čebyšev interpolation is given by

$$\vec{\Pi}^{(m)}_{[\mathbf{a},\mathbf{b}],[\mathbf{c},\mathbf{d}]}[f](\mathbf{x},\mathbf{y}) := \sum_{\xi^{(\mu)} \in \Theta^{(m)}_{\mathbf{a},\mathbf{b}}} \sum_{\zeta^{(\nu)} \in \Theta^{(m)}_{\mathbf{c},\mathbf{d}}} f\left(\xi^{(\mu)}, \zeta^{(\nu)}\right) L^{(\mu)}(\mathbf{x}) L^{(\nu)}(\mathbf{y}).$$

In Assumption 7.1.1 we restricted ourselves to kernel functions that were either the global kernel function or a derivative thereof. We define the approximation of the kernel function by applying the derivatives in (7.2) to the global kernel function.

Definition 7.1.20 (Čebyšev Approximation of the Kernel Function). Let $b = (\sigma, s) \in P^{far}$ be a pair of admissible clusters and let Γ_σ, Γ_s be the associated geometric clusters. The cluster boxes are denoted by $Q_\sigma =: Q_{\mathbf{a},\mathbf{b}}$ and $Q_s =: Q_{\mathbf{c},\mathbf{d}}$. Let the kernel function have the representation

$$k(\mathbf{x},\mathbf{y}) = D_\mathbf{x} D_\mathbf{y} G(\mathbf{x},\mathbf{y})$$

and let it satisfy Assumption 7.1.1. Then the Čebyšev approximation $k_m : \Gamma_\sigma \times \Gamma_s \to \mathbb{C}$ of the kernel function is given by

$$k_b := D_\mathbf{x} D_\mathbf{y} \vec{\Pi}^{(m)}_{[\mathbf{a},\mathbf{b}],[\mathbf{c},\mathbf{d}]} G.$$

The representation of the kernel approximation in separated coordinates plays the key role for the efficiency of the cluster method. For $b = (\sigma, s) \in P^{far}$ we obtain the abstract representation of the kernel approximation

$$k^{(m)}_b(\mathbf{x},\mathbf{y}) = \sum_{\mu,\nu \in \iota_m} \kappa_{\mu,\nu}(b)\, \Phi^{(\mu)}_\sigma(\mathbf{x})\, \Psi^{(\nu)}_s(\mathbf{y}) \tag{7.15}$$

with

$$\iota_m := \mathcal{J}_m, \quad \kappa_{\mu,\nu}(b) := G\left(\xi^{(\mu)}, \zeta^{(\nu)}\right), \quad \Phi^{(\mu)}_\sigma(\mathbf{x}) := D_\mathbf{x} L^{(\mu)}(\mathbf{x}),$$
$$\Psi^{(\nu)}_s(\mathbf{y}) := D_\mathbf{y} L^{(\nu)}(\mathbf{y}). \tag{7.16}$$

Definition 7.1.21. Let $b \in P^{far}$. An approximation k_b of the kernel function is semi-separable if the $\mathbf{x}$ and $\mathbf{y}$ dependence is factored as in (7.15).

7.1.3.2 Multipole Expansion

The multipole expansion was originally developed for the Laplace problem. In contrast to the six-dimensional expansion for the Čebyšev interpolation, the multipole expansion is four-dimensional and thus more efficient. However, the expansion has to be developed for each kernel function separately in contrast to the more general interpolation-based approach for the panel-clustering method. So far, multipole expansions are developed for the Laplace problem, linear elasticity, and the Helmholtz problem. In the following we will give the expansion of the boundary integral operator for the single layer potential of the Laplace problem, more specifically, for the kernel function of the Coulomb potential in $\mathbb{R}^3$

$$k(\mathbf{x}, \mathbf{y}) := \|\mathbf{x} - \mathbf{y}\|^{-1},$$

which is the fundamental solution of the Laplace operator up to a factor $(4\pi)^{-1}$. It can be shown that the multipole expansion and the Taylor expansion are identical if the Cartesian coordinates are replaced by spherical coordinates. For a detailed study we refer to [77, 78, 110, 111, 193, 194].

Definition 7.1.22. Let $b = (\sigma, s) \in P^{far}$ be an admissible block and let Q_σ, Q_s be the associated axiparallel cuboid. The kernel expansion for the block b is about the point $\mathbf{M}_b := \mathbf{M}_\sigma - \mathbf{M}_s$ with the centers of mass $\mathbf{M}_\sigma$, $\mathbf{M}_s$ of Q_σ, Q_s (see Definition 7.1.6).

The multipole expansion (cf. [111]) uses the spherical harmonics Y_ℓ^m and the associated Legendre functions P_ℓ^m as expansion functions. For $\ell, m \in \mathbb{N}_0$ with $m \leq \ell$ we have (cf. [1, 8.6.6])

$$P_\ell^m(x) := (-1)^m \left(1 - x^2\right)^{m/2} \left(\frac{d}{dx}\right)^m P_\ell(x)$$

with the Legendre polynomials

$$P_\ell(x) := \frac{1}{2^\ell \ell!} \left(\frac{d}{dx}\right)^\ell \left(x^2 - 1\right)^\ell, \qquad \ell \in \mathbb{N}_0.$$

For $\ell \in \mathbb{N}_0$ and $m \in \mathbb{Z}$ with $|m| \leq \ell$ the spherical harmonics have the representation

$$Y_\ell^m(\mathbf{x}) := c_{\ell,m} P_\ell^{|m|}(\cos\theta)\, e^{im\phi} \qquad \mathbf{x} \in \mathbb{S}_2$$

with

$$c_{\ell,m} := \sqrt{\frac{(2\ell+1)\,(\ell-|m|)!}{4\pi\,(\ell+|m|)!}},$$

where $(\theta,\phi) \in [0,\pi] \times [0,2\pi[$ denote the spherical coordinates of the point $\mathbf{x}$ on the unit sphere $\mathbb{S}_2$. For an admissible block $b = (\sigma,s) \in P^{far}$ we define expansion coefficients

$$\kappa_{\mu,\nu}(b) := \frac{1}{C_{\mu_1+\nu_1}^{\mu_2+\nu_2}\,\|\mathbf{M}_s - \mathbf{M}_\sigma\|^{\mu_1+\nu_1+1}} Y_{\mu_1+\nu_1}^{\mu_2+\nu_2}\left(\frac{\mathbf{M}_s - \mathbf{M}_\sigma}{\|\mathbf{M}_s - \mathbf{M}_\sigma\|}\right)$$

with

$$C_\ell^m := \frac{i^{|m|}}{\sqrt{(\ell-m)!\,(\ell+m)!}}$$

and expansion functions in (7.15) given by

$$\Phi_\sigma^{(\mu)}(\mathbf{x}) := \Psi_\sigma^{(\mu)}(\mathbf{x}) := C_{\mu_1}^{\mu_2}\,\|\mathbf{x} - \mathbf{M}_\sigma\|^{\mu_1}\,Y_{\mu_1}^{-\mu_2}\left(\frac{\mathbf{x} - \mathbf{M}_\sigma}{\|\mathbf{x} - \mathbf{M}_\sigma\|}\right).$$

The kernel approximation on $\Gamma_\sigma \times \Gamma_s$ of the order $m = m_b$ is thus given by

$$k_b(\mathbf{x},\mathbf{y}) = \sum_{(\mu,\nu)\in\iota_m} \kappa_{\mu,\nu}(b)\,\Phi_\sigma^{(\mu)}(\mathbf{x})\,\Psi_s^{(\nu)}(\mathbf{y})$$

with

$$\iota_m := \left\{(\mu,\nu) \in \mathbb{Z}^2 \times \mathbb{Z}^2 : \begin{pmatrix} 0 \le \mu_1 < m \\ -\mu_1 \le \mu_2 \le \mu_1 \\ 0 \le \nu_1 < m - \mu_1 \\ -\nu_1 \le \nu_2 \le \nu_1 \end{pmatrix}\right\}. \tag{7.17}$$

7.1.3.3 Abstract Cluster Approximation

The Čebyšev interpolation and the multipole expansion are two examples for approximating the kernel function by a *semi-separable* expansion. For special kernel functions other expansions may be more suitable. The following assumption summarizes the abstract conditions for the approximation by the cluster method.

Assumption 7.1.23. *There exist an admissibility condition on the set of all cluster pairs and constants* $0 < \gamma < 1$, $0 < C < \infty$ *and* $s \in \mathbb{R}$ *with the following property: For all admissible blocks* $b = (c,\sigma) \in P^{far}$ *there exists a family of semi-separable approximations* $\left\{k_{\mathbf{b}}^{(m)} : \Gamma_c \times \Gamma_\sigma \to \mathbb{C}\right\}_{m\in\mathbb{N}}$ *of the form*

$$k_b^{(m)}(\mathbf{x},\mathbf{y}) = \sum_{(\nu,\mu)\in\iota_m} \kappa_{\nu,\mu}(b)\,\Phi_c^{(\mu)}(\mathbf{x})\,\Psi_\sigma^{(\nu)}(\mathbf{y}) \tag{7.18}$$

with

$$\left|k(\mathbf{x},\mathbf{y}) - k_b^{(m)}(\mathbf{x},\mathbf{y})\right| \leq C\gamma^m \operatorname{dist}^{-s}(c,\sigma) \quad \forall (\mathbf{x},\mathbf{y}) \in \Gamma_c \times \Gamma_\sigma. \tag{7.19}$$

Furthermore, there exist constants $d_1, d_2 \in \mathbb{N}$ *and* $0 < C < \infty$ *such that the index sets* ι_m *have the property* $\iota_m \subset \mathbb{Z}^{d_1} \times \mathbb{Z}^{d_2}$ *for all* $m \in \mathbb{N}$ *and such that the one-sided restrictions*

$$\mathcal{L}_m := \left\{\nu \mid \exists \mu \in \mathbb{Z}^{d_2} : (\nu,\mu) \in \iota_m\right\} \quad \text{and} \quad \mathcal{R}_m := \left\{\mu \mid \exists \nu \in \mathbb{Z}^{d_1} : (\nu,\mu) \in \iota_m\right\} \tag{7.20}$$

satisfy the conditions

$$\begin{aligned} \sharp\mathcal{L}_m \leq C\,(m+1)^{d_1} & \text{ and } \quad \sharp\mathcal{R}_m \leq C\,(m+1)^{d_2}, \\ |\nu| < C\,(m+1) & \quad \text{for all } \nu \in \mathcal{L}_m \cup \mathcal{R}_m, \\ \left.\begin{array}{r} \mathcal{L}_m \subset \mathcal{L}_M \\ \mathcal{R}_m \subset \mathcal{R}_M \end{array}\right\} & \quad \text{for all } 0 \leq m \leq M. \end{aligned}$$

Remark 7.1.24. *The functions* $\Phi_c^{(\mu)}$, $\Psi_\sigma^{(\nu)}$ *and the coefficients* $\kappa_{\mu,\nu}(b)$ *in the expansion (7.18) usually depend on the order of expansion* m.

Exercise 7.1.25. *Show that the index set* ι_m *from (7.16) and (7.17) satisfies the conditions from Assumption 7.1.23 imposed on the index sets* $\mathcal{L}_m$ *and* $\mathcal{R}_m$.

7.1.4 The Matrix-Vector Multiplication in the Cluster Format

The representation of the integral operator by the cluster method can now be formulated with the help of the near and far-fields P^{near} and P^{far} and the abstract approximations $k_b(\mathbf{x},\mathbf{y})$. In the following note that the boundary elements $\tau \in \mathcal{G}$ are images of the *open* reference element $\hat{\tau}$.

The Figs. 7.1 and 7.2 illustrate the memory organization for the cluster method and the algorithm for the matrix-vector multiplication.

For a boundary element function $u \in S$ we denote the coefficient vector in the basis representation by $\mathbf{u} \in \mathbb{R}^{\mathcal{I}}$, where again the set of degrees of freedom $\mathcal{I}$ is used to index the basis functions:

$$u = \sum_{i \in \mathcal{I}} \mathbf{u}_i b_i.$$

The decomposition of the index set $\mathcal{I} \times \mathcal{I}$ in $P^{near} \cup P^{far}$ induces the representation

$$\begin{aligned} (Ku, v)_{L^2(\Gamma)} = & \sum_{(\{i\},\{j\}) \in P^{near}} \overline{\mathbf{v}}_i \mathbf{u}_j \int_\Gamma b_i(\mathbf{x}) \int_\Gamma k(\mathbf{x},\mathbf{y})\, b_j(\mathbf{y})\, ds_{\mathbf{y}} ds_{\mathbf{x}} \\ & + \sum_{b=(c,\sigma) \in P^{far}} \sum_{(i,j) \in (c,\sigma)} \overline{\mathbf{v}}_i \mathbf{u}_j \int_{\Gamma_c \times \Gamma_\sigma} b_i(\mathbf{x})\, k(\mathbf{x},\mathbf{y})\, b_j(\mathbf{y})\, ds_{\mathbf{x}} ds_{\mathbf{y}} \end{aligned}$$

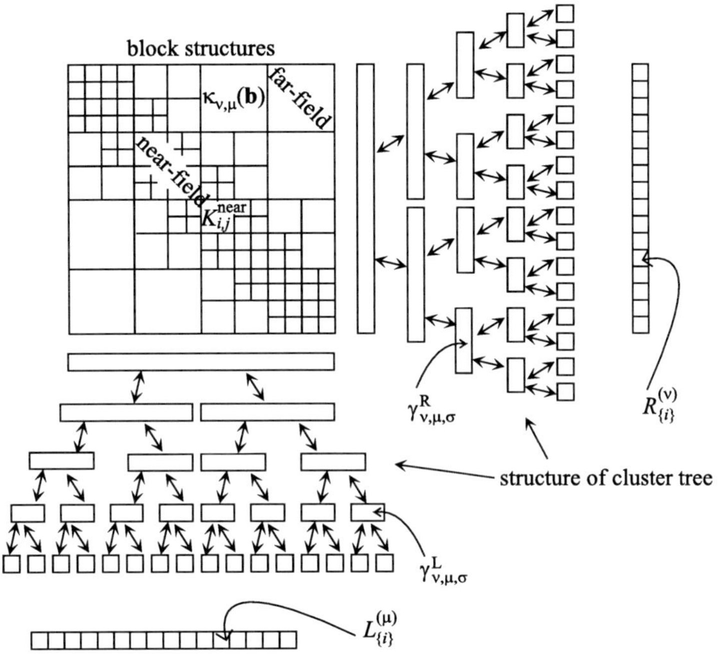

Fig. 7.1 Memory scheme for the cluster algorithm. The near-field matrix $\mathbf{K}^{near}$ and expansion coefficients $\kappa_{\nu,\mu}(\mathbf{b})$ are stored in the block structure P, the shift coefficients $\gamma^{L}_{\nu,\mu,\sigma}$, $\gamma^{R}_{\nu,\mu,\sigma}$ in the cluster tree structure and the basis expansion coefficients $L^{(\nu)}_{\{i\}}$, $R^{(\nu)}_{\{i\}}$ in the vector structure

with Γ_c as in Definition 7.1.6. If we replace the kernel function on the block $\Gamma_c \times \Gamma_\sigma$ by the abstract cluster approximation we obtain the cluster approximation of the integral operator.

Definition 7.1.26 (Cluster Method Approximation of the Bilinear Form). Let an order of expansion $m \in \mathbb{N}_0$ and a family of local kernel approximations be given as in Assumption 7.1.23. Then the cluster approximation of the sesquilinear form $(Ku, v)_{L^2(\Gamma)}$ is given by

$$(K_{PC}u, v)_{L^2(\Gamma)} := \langle \mathbf{K}_{near}\mathbf{u}, \mathbf{v}\rangle + \sum_{b=(c,\sigma)\in P^{far}} \sum_{(\nu,\mu)\in\iota_m} \kappa_{\mu,\nu}(b)\, L_c^{(\nu)}(v)\, R_\sigma^{(\mu)}(u) \tag{7.21}$$

with the sparse near-field matrix

$$(\mathbf{K}_{near})_{i,j} = \begin{cases} \int_\Gamma b_i(\mathbf{x}) \int_\Gamma k(\mathbf{x},\mathbf{y})\, b_j(\mathbf{y})\, ds_{\mathbf{y}} ds_{\mathbf{x}} & \text{if } (\{i\},\{j\}) \in P^{near}, \\ 0 & \text{otherwise} \end{cases}$$

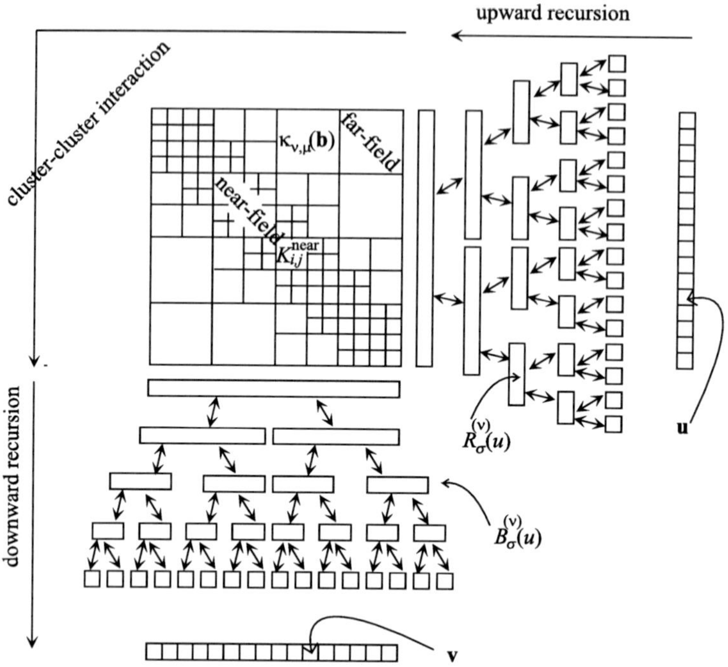

Fig. 7.2 Computation scheme for the cluster method: upward recursion, cluster–cluster interaction and downward recursion

and the far-field coefficients

$$L_c^{(\nu)}(v) := \sum_{i \in c} \overline{\mathbf{v}}_i \int_{\Gamma_c} b_i(\mathbf{x})\, \Phi_c^{(\nu)}(\mathbf{x})\, ds_{\mathbf{x}} \quad \text{and}$$

$$R_\sigma^{(\mu)}(u) := \sum_{i \in \sigma} \mathbf{u}_i \int_{\Gamma_\sigma} b_i(\mathbf{x})\, \Psi_\sigma^{(\mu)}(\mathbf{x})\, ds_{\mathbf{x}} \tag{7.22}$$

for all $c, \sigma \in T$ and $\nu \in \mathcal{L}_m$, $\mu \in \mathcal{R}_m$.

Note that the integrals in (7.22) can be reduced to integrals over $\Gamma_c \cap \operatorname{supp} b_i$ and $\Gamma_\sigma \cap \operatorname{supp} b_i$, which means that they are each the sum of a small number (bounded independent of the mesh width) of integrals over single panels.

In the following we will deal with the efficient evaluation of the representation (7.21) by means of the cluster hierarchy.

The explicit choice of the order of expansion m depends on the precision required from the cluster approximation. The cluster pairs $(c, \sigma) \in P^{far}$ correspond to the non-local character of the integral operators. In order to achieve an efficient algorithm, the coupling between the clusters is broken up wherever possible so as to be able to perform the computations separately on the single clusters. This will be done by a one-sided restriction of the index sets and by inheritance of block sizes on clusters.

It is essential for the efficient realization of the cluster algorithm that the function oriented representation (7.21) be translated into an algebraic coefficient representation. We begin with the near-field and recall the definition of the index set [see (7.6)]

$$P^{near}(\{i\}) = \{\{j\} \in \text{Leaves} : (\{i\}, \{j\}) \in P^{near}\}.$$

Remark 7.1.27. *For the elements of the near-field matrix we have*

$$\mathbf{K}_{i,j}^{near} = 0 \qquad \forall i \in \mathcal{I}, \forall \{j\} \notin P^{near}(\{i\}).$$

The matrix $\mathbf{K}_{i,j}^{near}$ is sparse, which here means that per line i we only need to store the coefficients for the indices $\{j\} \in P^{near}(\{i\})$. We can then implement a matrix-vector multiplication with the near-field matrix as given by

$$\sum_{j \in \mathcal{I}} \mathbf{K}_{i,j}^{near} \mathbf{u}_j = \sum_{\{j\} \in P^{near}(\{i\})} \mathbf{K}_{i,j}^{near} \mathbf{u}_j. \tag{7.23}$$

For the far-field, first, the index set ι_m is separated [see (7.20)]: we define

$$\mathcal{R}_m(\nu) := \{\mu \in \mathcal{R}_m : (\nu, \mu) \in \iota_m\} \qquad \forall \nu \in \mathcal{L}_m.$$

Then we have

$$\iota_m = \{(\nu, \mu) \mid \nu \in \mathcal{L}_m \text{ and } \mu \in \mathcal{R}_m(\nu)\}. \tag{7.24}$$

By means of the one-sided restrictions $P^{far}(c)$ [see (7.7)] we obtain for the second summand in (7.21) the representation (see Exercise 7.1.28)

$$\sum_{c \in T} \sum_{\nu \in \mathcal{L}_m} L_c^{(\nu)}(v)\, B_c^{(\nu)}(u) \tag{7.25}$$

with

$$B_c^{(\nu)}(u) := \sum_{\sigma \in P^{far}(c)} \sum_{\mu \in \mathcal{R}_m(\nu)} \kappa_{\nu,\mu}(b)\, R_\sigma^{(\mu)}(u). \tag{7.26}$$

Exercise 7.1.28. *Prove the representations (7.25) and (7.26).*

The sum in (7.26) corresponds to the evaluation of the cluster–cluster coupling: The indices ν on one cluster c are coupled with the indices μ on another cluster σ via the expansion coefficients $\kappa_{\nu,\mu}(b)$.

We still need to define the efficient evaluation of the sum in (7.25) and the far-field coefficients. For both tasks we use the hierarchical structure of the cluster tree to evaluate recursively the approximation by the cluster method.

7.1.4.1 Computation the Far-Field Coefficients

For the recursive calculation of the far-field coefficients by means of the cluster tree it does not suffice that the index set $\mathcal{I}$ is organized hierarchically in the cluster tree. We also require the expansion functions $\Phi_c^{(\nu)}$ and $\Psi_c^{(\mu)}$ to have a hierarchical structure formalized in the abstract Assumption 7.1.30 below.

We would like to motivate the abstract Assumption 7.1.30 by a concrete example. If the monomials (centered in the cluster centers $\mathbf{M}_c$) are used as a basis of the expansion system, i.e.,

$$\Phi_c^{(\nu)}(\mathbf{x}) := (\mathbf{x} - \mathbf{M}_c)^{\nu} / \nu! \qquad \text{for all } c \in T \text{ and } \nu \in \mathcal{L}_m,$$

we have for the restrictions to the sons

$$\Phi_c^{(\nu)}|_{\Gamma_{\tilde{c}}} = \sum_{\mu \in \mathcal{L}_m} \gamma_{\nu,\mu,\tilde{c}} \Phi_{\tilde{c}}^{(\mu)} \qquad \text{for all } c \in T, \tilde{c} \in \Sigma(c) \text{ and all } \nu \in \mathcal{L}_m \tag{7.27}$$

with the coefficients

$$\gamma_{\nu,\mu,\tilde{c}} := \begin{cases} \dfrac{(\mathbf{M}_{\tilde{c}} - \mathbf{M}_c)^{\nu-\mu}}{(\nu - \mu)!} & \mu \leq \nu, \\ 0 & \text{otherwise.} \end{cases} \tag{7.28}$$

Exercise 7.1.29. *Prove the representation (7.27).*

The hierarchy of the approximation system is abstractly formulated in Assumption 7.1.30.

Assumption 7.1.30. *For all $c, \sigma \in T$, all sons $\tilde{c} \in \Sigma(c)$, $\tilde{\sigma} \in \Sigma(\sigma)$ and expansion functions $\left\{\Phi_c^{(\nu)}\right\}_{\nu \in \mathcal{L}_m}$ and $\left\{\Psi_\sigma^{(\nu)}\right\}_{\nu \in \mathcal{R}_m}$, the refinement relations*

$$\Phi_c^{(\nu)}|_{\Gamma_{\tilde{c}}} = \sum_{\mu \in \mathcal{L}_m} \gamma_{\nu,\mu,\tilde{c}}^{L} \Phi_{\tilde{c}}^{(\mu)} \qquad \text{and } \Psi_\sigma^{(\nu)}|_{\Gamma_{\tilde{\sigma}}} = \sum_{\mu \in \mathcal{R}_m} \gamma_{\nu,\mu,\tilde{\sigma}}^{R} \Psi_{\tilde{\sigma}}^{(\mu)} \tag{7.29}$$

hold with suitable shift coefficients $\gamma_{\nu,\mu,\tilde{c}}^{L}, \gamma_{\nu,\mu,\tilde{\sigma}}^{R} \in \mathbb{C}$.

The computation of the far-field coefficients $R_\sigma^{(\nu)}(u)$ begins with the computing and storing of the *basis far-field coefficients* for every index (leaf) $i \in \mathcal{I}$. The definition of the far-field coefficients implies that

$$R_{\{j\}}^{(\nu)}(b_i) = \begin{cases} \displaystyle\int_{\operatorname{supp} b_j} b_i(\mathbf{x}) \Psi_{\{i\}}^{(\nu)}(\mathbf{x}) \, ds_{\mathbf{x}} & \text{if } i = j, \\ 0 & \text{otherwise,} \end{cases}$$

and we define the basis far-field coefficients by

$$R_{\{i\}}^{(\nu)} := R_{\{i\}}^{(\nu)}(b_i) \qquad \forall i \in \mathcal{I}, \quad \forall \nu \in \mathcal{R}_m.$$

By using these coefficients we obtain a recursion for the calculation of the coefficients $R_\sigma^{(\nu)}(u)$:

1. For all leaves $i \in \mathcal{I}$ compute

$$R_{\{i\}}^{(\nu)}(u) = \mathbf{u}_i R_{\{i\}}^{(\nu)}. \tag{7.30}$$

2. For all clusters $\sigma \in T \setminus$ Leaves recursively calculate from the leaves to the roots

$$R_\sigma^{(\nu)}(u) = \sum_{\tilde{\sigma} \in \Sigma(\sigma)} \sum_{\mu \in \mathcal{R}_m} \gamma_{\nu,\mu,\tilde{\sigma}}^R R_{\tilde{\sigma}}^{(\mu)}(u) \qquad \text{for all } \nu \in \mathcal{R}_m. \tag{7.31}$$

The representation (7.30) follows directly from the linearity of the functional $R_\sigma^{(\nu)}$ and from the representation (7.22). For the representation (7.31) we used the additivity of the integral, more specifically, we used the decomposition

$$\begin{aligned}
R_\sigma^{(\nu)}(u) &= \sum_{i \in \sigma} \mathbf{u}_i \int_{\Gamma_\sigma} b_i(\mathbf{x}) \Psi_\sigma^{(\nu)}(\mathbf{x})\, ds_{\mathbf{x}} \\
&= \sum_{\tilde{\sigma} \in \Sigma(\sigma)} \sum_{i \in \tilde{\sigma}} \mathbf{u}_i \int_{\Gamma_{\tilde{\sigma}}} b_i(\mathbf{x}) \Psi_\sigma^{(\nu)}(\mathbf{x})\, ds_{\mathbf{x}} \\
&= \sum_{\mu \in \mathcal{R}_m} \sum_{\tilde{\sigma} \in \Sigma(\sigma)} \sum_{i \in \tilde{\sigma}} \gamma_{\nu,\mu,\tilde{\sigma}}^R \mathbf{u}_i \int_{\Gamma_{\tilde{\sigma}}} b_i(\mathbf{x}) \Psi_{\tilde{\sigma}}^{(\mu)}(\mathbf{x})\, ds_{\mathbf{x}} \\
&= \sum_{\tilde{\sigma} \in \Sigma(\sigma)} \sum_{\mu \in \mathcal{R}_m} \gamma_{\nu,\mu,\tilde{\sigma}}^R R_{\tilde{\sigma}}^{(\mu)}(u),
\end{aligned}$$

which in turn uses the geometric hierarchy (7.3) of the cluster tree and the refinement relation (7.29).

7.1.4.2 Cluster–Cluster Interaction

In order to compute the coefficients $B_c^{(\nu)}(u)$ we use the representation (7.26). The algorithmic evaluation of the sum is then a recursive procedure.

7.1.4.3 Evaluating the Cluster Approximation of a Matrix-Vector Multiplication

The evaluation of the sum (7.25) can be described as a transposition of the upwards recursion [see (7.31)]. The cluster parts $B_c^{(\nu)}(u)$ that were already computed in

(7.26) are distributed to the sons. For this we consider the local situation of a cluster $c \in T \backslash$ Leaves with sons $\Sigma(c)$. The sum in (7.25) includes all clusters from T. Therefore the sum over $\{c\} \cup \Sigma(c)$ appears as a partial sum

$$\sum_{\nu \in \mathcal{L}_m} L_c^{(\nu)}(b_i) B_c^{(\nu)}(u) + \sum_{\tilde{c} \in \Sigma(c)} \sum_{\mu \in \mathcal{L}_m} L_{\tilde{c}}^{(\mu)}(b_i) B_{\tilde{c}}^{(\mu)}(u). \tag{7.32}$$

If we replace $L_c^{(\nu)}(b_i)$ in the first term by the refinement relation (7.29) [see (7.31)]

$$L_c^{(\nu)}(b_i) = \sum_{\tilde{c} \in \Sigma(c)} \sum_{\mu \in \mathcal{L}_m} \gamma_{\nu,\mu,\tilde{c}}^L L_{\tilde{c}}^{(\mu)}(b_i)$$

we obtain the representation (7.33) for (7.32)

$$\sum_{\tilde{c} \in \Sigma(c)} \sum_{\mu \in \mathcal{L}_m} L_{\tilde{c}}^{(\mu)}(b_i) \left\{ B_{\tilde{c}}^{(\mu)}(u) + \sum_{\nu \in \mathcal{L}_m} \gamma_{\nu,\mu,\tilde{c}}^L B_c^{(\nu)}(u) \right\}. \tag{7.33}$$

The recursion for the evaluation of the sum (7.25) is based on an update of the multipliers $B_c^{(\nu)}(u)$ in (7.25) as given by the brackets $\{\ldots\}$ in (7.33). The modified multipliers are recursively calculated from the root to the leaves and then only have to be multiplied by the basis far-field coefficients. These only have to be calculated once and can then be stored. For all $i \in \mathcal{I}$ these are defined by

$$L_{\{i\}}^{(\nu)} := \int_{\Gamma_{\{i\}}} \Phi_{\{i\}}^{(\nu)} b_i \, ds \qquad \forall \nu \in \mathcal{L}_m. \tag{7.34}$$

With these coefficients, the recursion for the evaluation of the matrix-vector multiplication can be formulated by means of the known quantities:

1. For $c = \mathcal{I}$: For all $\mu \in \mathcal{L}_m$ we define

$$\widetilde{B}_c^{(\mu)}(u) := B_c^{(\mu)}(u). \tag{7.35}$$

2. For all clusters $c \in T \backslash \{\mathcal{I}\}$, calculate recursively from the roots to the leaves:

$$\widetilde{B}_{\tilde{c}}^{(\mu)}(u) := B_{\tilde{c}}^{(\mu)}(u) + \sum_{\nu \in \mathcal{L}_m} \gamma_{\nu,\mu,\tilde{c}}^L \widetilde{B}_c^{(\nu)}(u) \qquad \forall \tilde{c} \in \Sigma(c), \ \forall \mu \in \mathcal{L}_m. \tag{7.36}$$

3. The cluster approximation of the component $(\mathbf{Ku})_i$ is then given by

$$(\widetilde{\mathbf{Ku}})_i = (\mathbf{K}_{near}\mathbf{u})_i + \sum_{\mu \in \mathcal{L}_m} L_{\{i\}}^{(\mu)} \widetilde{B}_{\{i\}}^{(\mu)}(u). \tag{7.37}$$

7.1.4.4 Algorithmic Description of the Cluster Method

As an overview, we will summarize all steps of the cluster method in this subsection.

Algorithm 7.1.31 (Cluster Method). *We assume that a surface mesh $\mathcal{G}$ of Γ and a kernel function $k : \Gamma \times \Gamma \to \mathbb{C}$ are given. Furthermore, an admissibility condition with an associated expansion system $\Phi_c^{(\nu)}$, $\Psi_c^{(\nu)}$ has to be given. This is done in the form of a rule for calculating the expansion coefficients $\kappa_{\nu,\mu}(b)$ in (7.18) as well as the shift coefficients $\gamma_{\nu,\mu,\tilde{c}}^L$ and $\gamma_{\nu,\mu,\tilde{c}}^R$ in (7.29). The chosen order of expansion is denoted by m.*

(I) Preparatory Phase

1. *Generate the cluster tree T from $\mathcal{G}$ according to Algorithm 7.1.8.*
2. *Generate the near and far-field P^{near} and P^{far} according to the procedure **divide** (see Algorithm 7.1.12).*
3. *Compute and store the shift coefficients $\gamma_{\nu,\mu,\tilde{c}}^L$ and $\gamma_{\nu,\mu,\tilde{\sigma}}^R$ for all clusters $c, \sigma \in T$.*
4. *Compute and store the expansion coefficients $\kappa_{\nu,\mu}(b)$ for all blocks $b \in P^{far}$.*
5. *For all $i \in \mathcal{I}$ and $\mu \in \mathcal{L}_m$, $\nu \in \mathcal{R}_m$ evaluate the integrals*

$$L_{\{i\}}^{(\mu)} := \int_{\operatorname{supp} b_i} b_i(\mathbf{x})\, \Phi_{\{i\}}^{(\mu)}(\mathbf{x})\, ds_{\mathbf{x}} \qquad R_{\{i\}}^{(\nu)} := \int_{\operatorname{supp} b_i} b_i(\mathbf{x})\, \Psi_{\{i\}}^{(\nu)}(\mathbf{x})\, ds_{\mathbf{x}}$$

   and store them.
6. *For all $i \in \mathcal{I}$, $\{j\} \in P^{near}(\{i\})$ evaluate the integrals*

$$(\mathbf{K}_{near})_{i,j} := \int_{\operatorname{supp} b_i} b_i(\mathbf{x}) \int_{\operatorname{supp} b_j} k(\mathbf{x}, \mathbf{y})\, b_j(\mathbf{y})\, ds_{\mathbf{y}} ds_{\mathbf{x}}$$

   and store the near-field matrix in a compressed form (see Remark 7.1.27).

(II) Matrix-Vector Multiplication

1. ***Upwards Recursion:***

   The computation of the coefficients $\widetilde{R}_c^{(\nu)} := R_c^{(\nu)}(u)$ for all $c \in T$ is done with the program line

$$T_{temp} := T;\ \textbf{upward\_pass}\left(T_{temp}, \mathcal{I}, \left(\widetilde{R}_c^{(\nu)}\right)_{\nu\in\mathcal{R}_m}\right);$$

   *where the recursive procedure **upward_pass** is defined as follows.*

   procedure upward_pass $\left(T_{temp}, c, \left(\widetilde{R}_c^{(\nu)}\right)_{\nu\in\mathcal{R}_m}\right)$;
   begin
   if $c \in$ Leaves ***then begin*** $\left(\widetilde{R}_c^{(\nu)}\right)_{\nu\in\mathcal{R}_m} := \left(R_c^{(\nu)}\right)_{\nu\in\mathcal{R}_m}$;$T_{temp}$:=$T_{temp}\backslash\{c\}$;***end***

else for all $\tilde{c} \in \Sigma(c)$ *do begin*

if $\left(\widetilde{R}_{\tilde{c}}^{(\tilde{\nu})}\right)_{\tilde{\nu}\in\mathcal{R}_m}$ *is not yet computed* ***then begin***

upward_pass $\left(T_{temp}, \tilde{c}, \left(\widetilde{R}_{\tilde{c}}^{(\tilde{\nu})}\right)_{\tilde{\nu}\in\mathcal{R}_m}\right)$; $T_{temp} := T_{temp} \backslash \{\tilde{c}\}$;

end;

for all $\nu \in \mathcal{R}_m$ ***do*** $\widetilde{R}_c^{(\nu)} := \sum_{\tilde{\nu}\in\mathcal{R}_m} \gamma_{\nu,\tilde{\nu},\tilde{c}}^{R} \widetilde{R}_{\tilde{c}}^{(\tilde{\nu})}$;

end;

end;

2. ***Evaluation of the Cluster–Cluster Coupling:***

   For the cluster–cluster coupling we use the sum representation in (7.26). The algorithmic representation is straightforward.

3. ***Downwards Recursion:***

   The evaluation of the matrix-vector multiplication is based on the recursion (7.36) and is achieved with the program line

$$T_{temp} := T;\ \textbf{for all } c \in T_{temp}\ \textbf{do downward\_pass}\left(T_{temp}, c, \left(\widetilde{B}_c^{(\mu)}\right)_{\mu\in\mathcal{L}_m}\right);$$

where the recursive procedure ***downward_pass*** *is defined as follows. The father* $F(c)$ *(see Definition 7.1.4) of a cluster* $c \in T\backslash\{\mathcal{I}\}$ *is characterized by* $c \in \Sigma(F(c))$.

procedure downward_pass $\left(T_{temp}, c, \left(\widetilde{B}_c^{(\mu)}\right)_{\mu\in\mathcal{L}_m}\right)$;

begin

if $c = \mathcal{I}$ ***then begin*** $\left(\widetilde{B}_c^{(\mu)}\right)_{\mu\in\mathcal{L}_m} := \left(B_c^{(\mu)}\right)_{\mu\in\mathcal{L}_m}$; $T_{temp} := T_{temp}\backslash\{c\}$;***end***

else

if $\left(\widetilde{B}_{F(c)}^{(\tilde{\mu})}\right)_{\tilde{\mu}\in\mathcal{L}_m}$ *is not yet computed* ***then begin***

downward_pass $\left(T_{temp}, F(c), \left(\widetilde{B}_{F(c)}^{(\tilde{\mu})}\right)_{\tilde{\mu}\in\mathcal{L}_m}\right)$;

$T_{temp} := T_{temp}\backslash\{F(c)\}$;

end;

for all $\mu \in \mathcal{L}_m$ ***do*** $\widetilde{B}_c^{(\mu)} := B_c^{(\mu)} + \sum_{\tilde{\mu}\in\mathcal{L}_m} \gamma_{\tilde{\mu},\mu,c}^{L} \widetilde{B}_{F(c)}^{(\tilde{\mu})}$;

end;

end;

4. ***Approximation of the Matrix-Vector Multiplication:***

   The evaluation of $\mathbf{v} := \widetilde{\mathbf{K}}\mathbf{u}$ *is computed according to*

$$\mathbf{v}_i := \sum_{\{j\}\in P^{near}(\{i\})} (\mathbf{K}_{near})_{i,j}\,\mathbf{u}_j + \sum_{\mu\in\mathcal{L}_m} L_{\{i\}}^{(\mu)} \widetilde{B}_{\{i\}}^{(\mu)}. \tag{7.38}$$

7.2 Realization of the Subalgorithms

The abstract formulation of the cluster algorithm 7.1.31 will now be concretely formulated for the class of kernel functions from Assumption 7.1.1. We will consider the algorithmic realization of the Čebyšev interpolation in detail. This method is suitable for the cluster approximation of a large class of kernel functions. It is also easy to implement, as we do not use any analytic properties of the specific kernel functions. Therefore, in order to modify the approximation for a specific kernel function, only the global kernel function has to be provided as a subroutine in the computer program. If in Assumption 7.1.1 the *kernel* function $k(\mathbf{x}, \mathbf{y})$ is defined as the derivative of the global kernel function $G(\mathbf{x}, \mathbf{y})$, the expansion functions of the global kernel functions also have to be replaced by their derivatives.

The expansion is based on the Čebyšev interpolation of the kernel function on pairs of axiparallel cubes $Q_1 \times Q_2 \in \mathbb{R}^3 \times \mathbb{R}^3$ and thus is six-dimensional.

For some kernel functions certain specific expansions may be more efficient. For example, the multipole expansion of the fundamental solution of the Laplace problem is only four-dimensional. Four-dimensional expansions can also be derived for the fundamental solution of the Helmholtz equation. For details we refer to [77, 78, 111, 194].

7.2.1 Algorithmic Realization of the Čebyšev Approximation

The essential steps for the algorithmic realization of the cluster approximation by means of Čebyšev interpolation consist in the computation of the expansion coefficients, the basis far-field coefficients and the shift coefficients.

Computation of the Expansion Coefficients

The efficient and stable evaluation of the Čebyšev interpolation is first defined for the one-dimensional case. Let Γ_σ be a real interval, $f \in C^0(\Gamma_\sigma)$ and let $\left(\xi^{(i,m)}\right)_{i=1}^m$ be the Čebyšev interpolation nodes scaled to Γ_σ. The index m is omitted if the order of expansion is clear from the context. We use the Lagrange representation of the interpolation polynomial

$$f_m(x) := \sum_{i=1}^{m} f_i L^{(i)}(x) \tag{7.39}$$

with $f_i := f\left(\xi^{(i)}\right)$ and $L^{(i)}$ as in Definition 7.1.14. The Lagrange functions $L^{(i)}$ and coefficients f_i depend on the interval Γ_σ and we write $L^{(\sigma,i)}$, $f_{\sigma,i}$ to make this dependency evident.

In the next step this algorithm is generalized so that it applies to the global kernel function $G : Q_c \times Q_\sigma \to \mathbb{C}$ with $b := (c, \sigma) \in P^{far}$. We set $Q_c =: \iota_1 \times \iota_2 \times \iota_3$ and $Q_\sigma := \lambda_1 \times \lambda_2 \times \lambda_3$ with bounded intervals $\iota_k \subset \mathbb{R}$ and $\lambda_k \subset \mathbb{R}$, $k = 1, 2, 3$.

This leads to

$$G(\mathbf{x},\mathbf{y}) \approx G_m(\mathbf{x},\mathbf{y}) = \sum_{\mu,\nu \in \mathcal{J}_m} \kappa_{\mu,\nu}(b)\, L^{(c,\mu)}(\mathbf{x})\, L^{(\sigma,\nu)}(\mathbf{y}) \tag{7.40}$$

with $\mathcal{J}_m$ as in (7.10), the expansion coefficients $\kappa_{\mu,\nu}(b) := G\left(\xi^{(c,\mu)}, \xi^{(\sigma,\nu)}\right)$ and

$$L^{(c,\mu)}(\mathbf{x}) := L^{(\iota_1,\mu_1)}(x_1)\, L^{(\iota_2,\mu_2)}(x_2)\, L^{(\iota_3,\mu_3)}(x_3)\,.$$

We assume that the kernel function $k(\mathbf{x},\mathbf{y})$ of the boundary integral equation is the global kernel function or a suitable derivative of it:

$$k(\mathbf{x},\mathbf{y}) = D_1 D_2 G(\mathbf{x},\mathbf{y})$$

with derivatives D_1 (with respect to $\mathbf{x}$) and D_2 (with respect to $\mathbf{y}$) of order at most 1. We obtain the approximation of the kernel function by applying D_1, D_2 to the Čebyšev interpolation of the global kernel function. From an algorithmic point of view the question arises how to compute the coefficients in the derivative of the Lagrange representation. As an approximation we use [see (7.40)]

$$D_1 D_2 G(\mathbf{x},\mathbf{y}) \approx D_1 D_2 G_m(\mathbf{x},\mathbf{y}) = \sum_{\mu,\nu \in \mathcal{J}_m} \kappa_{\mu,\nu}(b)\, D_1 L^{(c,\mu)}(\mathbf{x})\, D_2 L^{(\sigma,\nu)}(\mathbf{y})\,.$$

This means that the expansion coefficients of the kernel function coincide with those of the global kernel function. Therefore the expansion functions are the derivatives of the original expansion functions.

Computation of the Shift Coefficients

By integrating the expansion coefficients $D_1 L^{(\sigma,\nu)}$, $D_2 L^{(\sigma,\nu)}$ over Γ_σ we define the far-field coefficients for the algorithms **upward** and **downward_pass** (see Algorithm 7.1.31). We need the basis expansion coefficients $R_\sigma^{(\nu)}(b_i)$, $L_c^{(\mu)}(b_i)$ for the initialization of the recursion. We need the shift coefficients $\gamma^L_{\mu,\nu,c}$ and $\gamma^R_{\mu,\nu,c}$ to evaluate the recursion step.

We begin with the algorithm to compute the shift coefficients and thus begin with the one-dimensional expansion functions $L^{(c,i)}$ (see Definition 7.1.14) on an interval Γ_c. Let $\tilde{c} \in \Sigma(c)$ be a son of c in the cluster tree. The Čebyšev nodes with respect to $\tilde{c}$ are denoted by $\left(\xi^{(\tilde{c},j)}\right)_{j=1}^m$. Then we have on $\Gamma_{\tilde{c}}$

$$L^{(c,i)}\Big|_{\Gamma_{\tilde{c}}} = \sum_{j=1}^{i} a_{i,j,\tilde{c}} L^{(\tilde{c},j)} \qquad \text{with } a_{i,j,\tilde{c}} := L^{(c,i)}\left(\xi^{(\tilde{c},j)}\right).$$

In the following we will present an algorithm to compute the one-dimensional shift coefficients. We use the representation

$$L^{(c,i)}\left(\xi^{(\tilde{c},j)}\right) = \prod_{\substack{k=1\\k\neq i}}^{m} \frac{\xi^{(\tilde{c},j)} - \xi^{(c,k)}}{\xi^{(c,i)} - \xi^{(c,k)}}$$

and for $1 \leq j \leq m$ define the numbers

$$\omega_j := \prod_{k=1}^{m} \left(\xi^{(\tilde{c},j)} - \xi^{(c,k)}\right) \quad \text{and} \quad \beta_j := \prod_{\substack{k=1\\k\neq j}}^{m} \left(\xi^{(c,j)} - \xi^{(c,k)}\right). \tag{7.41}$$

The shift coefficients $L^{(c,i)}\left(\xi^{(\tilde{c},j)}\right)$ can be computed with these quantities as given by

$$a_{i,j,\tilde{c}} = L^{(c,i)}\left(\xi^{(\tilde{c},j)}\right) = \begin{cases} \dfrac{\omega_j}{\left(\xi^{(\tilde{c},j)} - \xi^{(c,i)}\right)\beta_i} & \text{if } \xi^{(\tilde{c},j)} \neq \xi^{(c,i)}, \\ 1 & \text{otherwise.} \end{cases} \tag{7.42}$$

We will now consider the multi-dimensional case. Let c be a cluster and $\left(L^{(c,\mu)}\right)_{\mu\in\mathcal{J}_m}$ the set of the associated Lagrange functions. Let $\tilde{c} \in \Sigma(c)$ be a son of c in the cluster tree with a minimal axiparallel box $Q_{\tilde{c}} = \iota_1 \times \iota_2 \times \iota_3$. Owing to the uniqueness of the interpolation the restriction of the expansion functions $L^{(c,\mu)}$ to the geometric cluster $\Gamma_{\tilde{c}}$ has the representation

$$L^{(c,\mu)}\Big|_{\Gamma_{\tilde{c}}} = \sum_{\nu\in\mathcal{J}_m} \gamma_{\mu,\nu,\tilde{c}} L^{(\tilde{c},\nu)}$$

with the shift coefficients

$$\gamma_{\mu,\nu,\tilde{c}} := a_{\mu_1,\nu_1,\iota_1} a_{\mu_2,\nu_2,\iota_2} a_{\mu_3,\nu_3,\iota_3},. \tag{7.43}$$

By virtue of the linearity of the differentiation we have

$$D_2 L^{(c,\mu)}\Big|_{\Gamma_{\tilde{c}}} = \sum_{\nu\in\mathcal{J}_m} \gamma_{\mu,\nu,\tilde{c}} D_2 L^{(\tilde{c},\nu)}$$

with the same shift coefficients as for the original functions $L^{(c,\mu)}$. This yields the definition

$$\gamma^{L}_{\mu,\nu,\tilde{c}} := \gamma^{R}_{\mu,\nu,\tilde{c}} := \gamma_{\mu,\nu,\tilde{c}}. \tag{7.44}$$

Computing the Basis Far-Field Coefficients

In the next step we will present a method to compute the basis far-field coefficients. For this we have to evaluate integrals of the form

$$\int_\tau b_i(\mathbf{x}) D_1 L^{(c,\nu)}(\mathbf{x})\, ds_{\mathbf{x}} \quad \text{and} \quad \int_\tau b_i(\mathbf{x}) D_2 L^{(\sigma,\nu)}(\mathbf{x})\, ds_{\mathbf{x}}. \tag{7.45}$$

If the expansion functions are polynomial on every panel τ and the mesh $\mathcal{G}$ consists of plane panels, the integrals can be evaluated exactly (cf. [112, 117, 157]). We can then derive recursion formulas by means of integration by parts. However, we recommend using Gaussian quadrature because of its increased stability, simple implementation and flexibility, which also allows for the efficient approximation of general (analytic) expansion functions and curved panels τ. Note that for plane triangles and polynomial expansion functions Gaussian quadrature already yields the exact integral value with relatively few interpolation nodes.

Example 7.2.1. *Let τ be a plane triangle with vertices $\mathbf{A}, \mathbf{B}, \mathbf{C} \in \mathbb{R}^3$. Let the expansion function $\Phi_\tau^{(\nu)}(\mathbf{x})$ be a polynomial of degree ν_i with respect to the variables x_i and let the basis function b_i be of degree p. Then we have*

$$\int_\tau b_i(\mathbf{x})\, \Phi_\tau^{(\nu)}(\mathbf{x})\, ds_{\mathbf{x}} = \int_0^1 \int_0^{\xi_1} q(\xi)\, d\xi_1 d\xi_2 = \int_0^1 \int_0^1 t_1 q(t_1, t_1 t_2)\, dt_1 dt_2 \tag{7.46}$$

with

$$q(\xi) = 2|\tau|\, \hat{B}_i(\xi) \left(\Phi_\tau^{(\nu)} \circ \chi_\tau\right)(\xi) \quad \textit{and} \quad \chi_\tau(\xi) = \mathbf{A} + [\mathbf{B}-\mathbf{A}, \mathbf{C}-\mathbf{B}]\, \xi,$$

where the basis function $\hat{B}_i(\xi)$ on the reference element is a polynomial of degree p. Therefore q is a polynomial of total degree $|\nu| + p$ and the integrand in the integral on the right-hand side of (7.46) is a polynomial of degree $|\nu| + p + 1$ in t_1 and $|\nu| + p$ in t_2. Let n be the smallest integer such that

$$2n - 1 \geq |\nu| + p + 1 \tag{7.47}$$

and let $\left(\omega_{k,n}, \xi_{k,n}\right)_{k=1}^n$ be the scaled weights and interpolation nodes on the interval $(0,1)$ of the associated Gaussian quadrature formula. If we apply the tensor version of this Gaussian quadrature on the right-hand side of (7.46) we obtain

$$\int_\tau b_i(\mathbf{x})\, \Phi_\tau^{(\nu)}(\mathbf{x})\, ds_{\mathbf{x}} = 2|\tau| \sum_{k,\ell=1}^n \omega_{k,n}\omega_{\ell,n}\xi_{k,n} \hat{B}_i\left(\xi_{k,n}, \xi_{k,n}\xi_{\ell,n}\right) \Phi_\tau^{(\nu)}\left(\xi_{k,\ell,n}\right) \tag{7.48}$$

with the transformed Gaussian points $\xi_{k,\ell,n} := \chi_\tau\left(\xi_{k,n}, \xi_{k,n}\xi_{\ell,n}\right)$.

Remark 7.2.2. *For a curved triangle the transformation $\chi_\tau : \widehat{\tau} \to \tau$ is non-linear. In this case the integrand in (7.45) on the reference element is, in general, not a polynomial. Gaussian quadrature should again be used for the approximation. The quadrature order n with respect to each coordinate direction should be chosen according to (7.47).*

Computing the basis far-field coefficients requires an efficient method to evaluate the expansion functions $D_1 L^{(c,\mu)}$ and $D_2 L^{(\sigma,\nu)}$ at the quadrature points.

If either D_1 or D_2 represents the identity, $L_{c,\mu}$ is evaluated according to the recursion (7.42).

In the following let $D_1 := \langle \mathbf{w}(\mathbf{x}), \nabla \rangle$ for a vector $\mathbf{w}(\mathbf{x}) = (w_k(\mathbf{x}))_{k=1}^3 \in \mathbb{C}^3$. The minimal axiparallel cuboid for a cluster c is denoted by $Q_c = \iota_1 \times \iota_2 \times \iota_3$. Then we have

$$\begin{aligned} D_1 L^{(c,\mu)}(\mathbf{x}) &= \langle \mathbf{w}(\mathbf{x}), \nabla \rangle L^{(c,\mu)}(\mathbf{x}) \\ &= \sum_{k=1}^{3} w_k(\mathbf{x}) \left(\prod_{\substack{\ell=1 \\ \ell \neq k}}^{3} L^{(\iota_\ell,\mu_\ell)}(x_\ell) \right) \partial_k L^{(\iota_k,\mu_k)}(x_k). \end{aligned} \tag{7.49}$$

Therefore we need an efficient algorithm to evaluate the derivative of the one-dimensional Lagrange functions $\left(\left(L^{(i)}\right)'(\zeta) \right)_{i=1}^{m}$. We distinguish between two cases:

1. Let $\zeta \notin \{\xi^{(i)} : 1 \le i \le m\}$. Then for $1 \le i \le m$ we have

$$\left(L^{(i)}\right)'(\zeta) = L^{(i)}(\zeta) \sum_{\substack{j=1 \\ j \neq i}}^{m} \frac{1}{\zeta - \xi^{(j)}}.$$

2. Let $\zeta = \xi^{(k)}$ for a $1 \le k \le m$. Then for $1 \le i \le m$ we have

$$\left(L^{(i)}\right)'(\zeta) = \begin{cases} L^{(k)}(\zeta) \sum\limits_{\substack{j=1 \\ j \neq k}}^{m} \dfrac{1}{\zeta - \xi^{(j)}} & i = k, \\ \widetilde{L}^{(i)}(\zeta) \dfrac{1}{\xi^{(i)} - \xi^{(k)}} & i \neq k, \end{cases} \tag{7.50}$$

where

$$\widetilde{L}^{(i)}(\zeta) := \prod_{\substack{j=1 \\ j \neq i,k}}^{m} \frac{\zeta - \xi^{(j)}}{\xi^{(i)} - \xi^{(j)}}. \tag{7.51}$$

denotes the Lagrange polynomial for the interpolation node $\xi^{(i)}$, $i \neq k$, of order $m-2$ for the reduced set of interpolation nodes $\{\xi^{(i)} : 1 \le i \le m\} \setminus \{\xi^{(k)}\}$.

In the following we will present the algorithm to evaluate (7.49). Let $c \in T$ and let $Q_c = \iota_1 \times \iota_2 \times \iota_3$ be the associated cluster box. The Čebyšev nodes in Q_c are denoted by $\xi^{(\mu)} = \left(\xi_k^{(\mu_k)}\right)_{k=1}^{3}$, $\mu \in \mathcal{J}_m$.

Algorithm 7.2.3. *The subroutine* ***evaluate*** **_*D*** $L_c(\zeta)$ *generates the evaluation of the Lagrange polynomials* $\left(D_1 L^{(c,\nu)}\right)_{\nu \in \mathcal{J}_m}$ *at a point* $\zeta = (\zeta_i)_{i=1}^3$.

procedure evaluate **_*DL*** $_c(\zeta)$;
begin
for $k := 1$ ***to*** 3 ***do begin***

$$\omega_k := \prod_{i=1}^{m} \left(\zeta_k - \xi_k^{(i)}\right);$$

for $j := 1$ ***to*** m ***do begin***

$$\beta_k^{(j)} := \prod_{\substack{i=1 \\ i \neq j}}^{m} \left(\xi_k^{(j)} - \xi_k^{(i)}\right);$$

$$L_k^{(j)} := \begin{cases} \dfrac{\omega_k}{\left(\zeta_k - \xi_k^{(j)}\right)\beta_k^{(j)}} & \textit{if } \zeta_k \neq \xi_k^{(j)}, \\ 1 & \textit{otherwise} \end{cases}$$

end;

if $\zeta_k \notin \left\{\xi_k^{(i)} : 1 \leq i \leq m\right\}$ ***then***
for $j := 1$ ***to*** m ***do begin***

$$\lambda_k^{(j)} := \sum_{\substack{i=1 \\ i \neq j}}^{m} \frac{1}{\zeta - \xi_k^{(i)}}; \qquad (DL)_k^{(j)} := L_k^{(j)} \lambda_k^{(j)};$$

end;
else begin
choose i *with* $\zeta_k = \xi_k^{(i)}$;

$$\lambda_k^{(i)} := \sum_{\substack{j=1 \\ j \neq i}}^{m} \frac{1}{\zeta_k - \xi_k^{(j)}}; \qquad (DL)_k^{(i)} := L_k^{(i)} \lambda_k^{(i)};$$

for $j \in \{1, \ldots, m\} \setminus \{i\}$ ***do*** $(DL)_k^{(j)} := \widetilde{L}_k^{(j)} / \left(\xi_k^{(j)} - \xi_k^{(i)}\right)$;
(see Remark 7.2.4)
end;
end;
for all $\mu \in \mathcal{J}_m$ ***do***

$$D_1 L^{c,\mu}(\zeta) = \sum_{k=1}^{3} w_k(\zeta) (DL)_k^{(\mu_k)} \prod_{\substack{\ell=1 \\ \ell \neq k}}^{3} L_\ell^{\mu_\ell};$$

end;

Remark 7.2.4. *(a) Evaluating the quantities* $\widetilde{L}_{k,j}$ *defined in (7.51) is realized from an algorithmic point of view in the same way as for the Lagrange polynomials. For the sake of clarity this step was not explicitly formulated in Algorithm 7.2.3.*

(b) The query $\zeta_k \notin \left\{\xi_k^{(i)} : 1 \leq i \leq m\right\}$ *in Algorithm 7.2.3 is numerically unstable because of roundoff errors. If* ζ_k *coincides with an interpolation node* $\xi_k^{(i)}$ *up to machine accuracy, the second, numerically more stable case in (7.50) should be chosen.*

7.2.2 Expansion with Variable Order

By imposing moderate conditions on the surface mesh and the integral operator it can be shown that the complexity of the cluster method is proportional to $(\sharp P) \times m^{\lambda}$, where m denotes the order of expansion, $\lambda \approx 4 - 7$ and $\sharp P$ denotes the number of blocks in the decomposition P. It is shown in Sect. 7.3 that the order of expansion m should be chosen proportional to $\log N$ to maintain the order of convergence of the overall discretization. For the number of blocks in P the estimate $\sharp P \leq CN$ can be shown for shape-regular and quasi-uniform surface meshes. This yields the asymptotic complexity bound $O\left(N \log^{\lambda} N\right)$ for the cluster method on trees with $O(N)$ leaves for some $\lambda > 0$, as $N \to \infty$. The logarithmic terms do have a negative impact on the computational complexity for large, practical applications and are also the reason that the breakeven point (compared to the standard matrix-oriented representation) is quite large and, typically, lies between $N \sim 10^3$ to 2×10^4. In this section we will explain briefly how this logarithmic term can be avoided without any additional algorithmic cost for certain classes of boundary integral operators. A detailed description of the cluster method with variable order can be found in [26, 199].

First we will combine clusters with the same level within a cluster tree to a *cluster level* by $T_0 := \{\mathcal{I}\}$ and recursively for $\ell > 0$ by

$$T_{\ell} := \{c \in T : \text{father}(c) \in T_{\ell-1}\}. \tag{7.52}$$

The maximal cluster level is denoted by $\ell_{\max}$. The level of a cluster $c \in T_{\ell}$ is defined by $\text{level}(c) := \ell$.

The following assumptions serve to simplify the representation and can be generalized.

Assumption 7.2.5. *(i) The cluster tree is balanced:* $\forall \sigma \in \text{Leaves} : \sigma \in T_{\ell_{\max}}$.
(ii) All blocks $b = (c, \sigma) \in P$ *consist of clusters of the same level:* $c, \sigma \in T_{\ell}$ *for a* $0 \leq \ell \leq \ell_{\max}$.

The level hierarchy of the clusters is inherited by the blocks $b = (c, \sigma) \in P$ as given by

$$\text{level}(b) := \text{level}(c), \tag{7.53}$$

and the sets P_ℓ contain all the blocks of the level ℓ. The cluster method with variable expansion order takes advantage of the fact that a *high* expansion order is only necessary on *large* blocks $b \in P_\ell$ (with a *small* index ℓ). These levels, however, only contain *few* blocks. Conversely, it can be shown under suitable conditions on the boundary integral operator that on small blocks $b \in P_\ell$ (i.e., ℓ *close* to $\ell_{\max}$) an expansion order $m = O(1)$ is sufficient for the approximation and that the required precision is achieved through the small size of the blocks. For example, we have for the number of small blocks $\sharp P_{\ell_{\max}} \sim CN$.

For the parameters $\alpha, \beta \geq 0$ we define the distribution function for the expansion order m_ℓ on the blocks $b \in P_\ell$ by

$$m_\ell := \lceil \alpha (\ell_{\max} - \ell) + \beta \rceil , \tag{7.54}$$

where $\lceil x \rceil$ denotes the smallest integer y such that $y \geq x$.

The formal changes to the cluster algorithm with variable order are marginal compared to the original version. We summarize them below. We use the algorithmic description from Sect. 7.1.4.4 to indicate the changes:

1. In the procedures **generate_cluster_tree** and **divide** every cluster and block is recursively assigned its appropriate level.
2. The definition of the shift coefficients $\gamma^L_{\nu,\tilde{\nu},\tilde{c}}$ and $\gamma^R_{\mu,\tilde{\mu},\tilde{\sigma}}$ for $\tilde{c}, \tilde{\sigma} \in T_\ell$ remains unchanged. However, the index sets are reduced: $\nu \in \mathcal{L}_{m_{\ell-1}}$, $\tilde{\nu} \in \mathcal{L}_{m_\ell}$, $\mu \in \mathcal{R}_{m_{\ell-1}}$, $\tilde{\mu} \in \mathcal{R}_{m_\ell}$. We have a similar result for the expansion coefficients $\kappa_{\mu,\nu}(b)$, which, for $b \in P^{far} \cap P_\ell$, only have to be computed and stored for the indices $(\mu, \nu) \in \iota_{m_\ell}$.
3. The basis far-field coefficients $L^{(\mu)}_{\{i\}}$, $R^{(\nu)}_{\{i\}}$ have to be computed for the indices $\mu \in \mathcal{L}_{m_{\lceil \beta \rceil}}$ [see (7.54)] and $\nu \in \mathcal{R}_{m_{\lceil \beta \rceil}}$.
4. In the procedure **upward_pass**, for $c \in T_\ell$ the expressions $\nu \in \mathcal{R}_m$ have to be replaced by $\nu \in \mathcal{R}_{m_\ell}$ and the expressions $\tilde{\nu} \in \mathcal{R}_m$ by $\tilde{\nu} \in \mathcal{R}_{m_{\ell+1}}$.
   In the evaluation of the cluster–cluster coupling the expansion order m has to be replaced by m_ℓ for $b \in P_\ell$.
   In the procedure **downward_pass**, for all $c \in T_\ell$ the expressions $\mu \in \mathcal{L}_m$ have to be replaced by $\mu \in \mathcal{L}_{m_\ell}$ and the expressions $\tilde{\mu} \in \mathcal{L}_m$ by $\tilde{\mu} \in \mathcal{L}_{m_{\ell-1}}$.
   In (7.38), m has to be replaced by $m_{\lceil \beta \rceil}$.

Remark 7.2.6. *Modification (2) implies that the expansion functions Φ^ν_c, Ψ^ν_c on T_ℓ, $\ell < \ell_{\max}$, are approximated by means of the expansion functions on $T_{\ell_{\max}}$.*

We will illustrate the reduction in complexity by using a uniform mesh with

$$N = 4^{\ell_{\max}}, \quad \sharp P_\ell := 4^\ell .$$

The number of all expansion coefficients $\kappa_{\nu,\mu}(b)$ is a measure for the complexity of the method. For the cluster method with variable order it grows linearly with the dimension of the problem

$$\sum_{\ell=0}^{\ell_{\max}} \sharp P_\ell \times (m_\ell)^\lambda = \sum_{\ell=0}^{\ell_{\max}} 4^\ell \left\lceil \alpha \left(\ell_{\max} - \ell\right) + \beta \right\rceil^\lambda$$
$$\leq N \sum_{\ell=0}^{\ell_{\max}} 4^{-\ell} \left(\alpha \ell + \beta\right)^\lambda \leq \left(\alpha + \beta\right)^\lambda N \sum_{\ell=0}^{\infty} 4^{-\ell} \ell^\lambda =: C_\lambda \left(\alpha + \beta\right)^\lambda N.$$

Proposition 7.2.7. *The cluster method with variable expansion order (7.54) requires a storage capacity of $O(N)$* `real` *numbers. The evaluation of a matrix-vector multiplication requires $O(N)$ arithmetic operations.*

A detailed study of this approach can be found in [26, 199].

7.3 Error Analysis for the Cluster Method

The error of the cluster method approximation originates when the kernel function is replaced by the kernel expansion on an admissible pair of clusters. In this section we will estimate this *local* error and analyze its influence on the overall discretization. The *global* error estimate will be based on the abstract assumption given in 7.1.23. We will show in the next section that the Čebyšev interpolation satisfies this assumption in particular for the global kernel function $G(\mathbf{x}, \mathbf{y})$ of the differential operator L as well as for the kernel functions that are derived from it.

7.3.1 *Local Error Estimates*

In this section we will analyze the error for the approximation of general kernel functions by Čebyšev interpolation. Error estimates for the Taylor and multipole expansions for the Laplace and Helmholtz problem can be found in, for example, [110, 111, 122, 125, 194, 199].

7.3.1.1 Local Error Estimates for the Čebyšev Interpolation

We begin with error estimates for the three-dimensional Čebyšev interpolation of a function $f : [-1,1]^3 \to \mathbb{C}$ and transfer this to general axiparallel cuboids by means of an affine pullback. Error estimates with respect to the L^∞-norm for functions $f : Q_1 \times Q_2 \to \mathbb{C}$ on axiparallel cuboids Q_1, Q_2 can be obtained by means of a tensor argument. Since, in general, the kernel function is defined as the derivative of the global kernel function, we will also derive error estimates with respect to the $W^{1,\infty}$-norm at the end of this section.

First we will summarize some of the properties of Čebyšev polynomials. We refer to [192] for the proofs. The one-dimensional Čebyšev polynomials are again denoted by T_m [see (7.13)] and their n-th derivative by $T_m^{(n)}$. For the tensorized

Čebyšev polynomials we use the multi-index $\mu \in \mathbb{N}_0^3$ and write $T_\mu(\mathbf{x}) = \prod_{i=1}^3 T_{\mu_i}(x_i)$ [see (7.14)].

Corollary 7.3.1. *(a) For all* $x \in [-1,1]$ *and* $n, m \in \mathbb{N}_0$

$$\left|T_m^{(n)}(x)\right| \leq \tau_m^{(n)} \quad \textit{with} \quad \tau_m^{(n)} := \prod_{i=0}^{n-1} \frac{m^2 - i^2}{2i+1}. \tag{7.55}$$

(b) For all $\mathbf{x} \in [-1,1]^3$

$$\left|T_\mu(\mathbf{x}) - \vec{\Pi}^{(m)}\left[T_\mu\right](\mathbf{x})\right| = 0 \qquad \forall \mu \in \mathbb{N}_0^3 : \max_{1\leq i\leq 3} \mu_i \leq m-1,$$
$$\left|T_\mu(\mathbf{x}) - \vec{\Pi}^{(m)}\left[T_\mu\right](\mathbf{x})\right| \leq 2 \qquad \forall \mu \in \mathbb{N}_0^3 : \max_{1\leq i\leq 3} \mu_i > m-1.$$

Proof. Part (a) follows from [192, Theorem 2.24].

For (b) The first part of the statement is trivial, owing to the uniqueness of the Čebyšev interpolation polynomial. First we prove the second statement for the one-dimensional case. The univariate Čebyšev interpolation can alternatively be written in the form

$$\Pi^{(m)}(f) = \sum_{k=-m+1}^{m-1} f_k T_k(x) \quad \text{with} \quad f_k := \frac{1}{m}\sum_{i=1}^{m} f\left(\xi^{(i,m)}\right) T_k\left(\xi^{(i,m)}\right).$$

By using the orthogonality of the Čebyšev polynomials with respect to the set of interpolation nodes

$$\frac{2}{m}\sum_{i=1}^{m} T_\ell\left(\xi^{(i,m)}\right) T_k\left(\xi^{(i,m)}\right) = \begin{cases} 2(-1)^s & \text{if } \ell/(2m) = s \in \mathbb{Z} \text{ and } k = 0, \\ (-1)^s & \text{if } k \neq 0 \text{ and } \frac{|k+\ell|}{2m} = s \in \mathbb{Z} \text{ or } \frac{|k-\ell|}{2m} = s \in \mathbb{Z}, \\ 0 & \text{otherwise.} \end{cases}$$

for $|k| < m$ and $\ell \in \mathbb{Z}$ (see [192, p. 49]) we obtain

$$\Pi^{(m)} T_\ell = \gamma_{\ell,m} T_{\eta_{\ell,m}} \tag{7.56}$$

with

$$\begin{array}{lll} \gamma_{\ell,m} := (-1)^{\frac{\ell-\ell \bmod (2m)}{2m}} & \eta_{\ell,m} := \ell \bmod (2m) & \text{if } \ell \bmod (2m) < m, \\ \gamma_{\ell,m} := -(-1)^{\frac{\ell-\ell \bmod (2m)}{2m}} & \eta_{\ell,m} := 2m - \ell \bmod (2m) & \text{if } \ell \bmod (2m) > m, \\ \gamma_{\ell,m} := 0 & \eta_{\ell,m} := \ell \bmod (2m) & \text{if } \ell \bmod (2m) = m. \end{array} \tag{7.57}$$

This and part (a) imply that $\left|\Pi^{(m)}(T_\ell)(x)\right| \leq 1$ for all $x \in [-1,1]$.

The second part of the corollary follows from the triangle inequality and by applying the one-dimensional argument to every factor of the tensor $T_\mu(\mathbf{x}) = T_{\mu_1}(x_1)\, T_{\mu_2}(x_2)\, T_{\mu_3}(x_3)$.

□

Remark 7.3.2. *The first coefficients $\tau_m^{(n)}$ in (7.55) are given by*

$$\tau_m^{(0)} = 1, \quad \tau_m^{(1)} = m^2, \quad \tau_m^{(2)} = \frac{m^2\left(m^2-1\right)}{3}, \quad \tau_m^{(3)} = \frac{m^2\left(m^2-1\right)\left(m^2-4\right)}{15}$$

and are monotonically increasing for $m = 0, 1, 2, \ldots$ and a fixed n.

Since the complexity of the cluster method strongly depends on the required polynomial degree m it is important to derive as accurately as possible an error estimate for the interpolation error. The error representation (7.12) is not suitable for this. In the same way as we did for numerical quadrature, we will apply the interpolation to functions that can be analytically extended to complex neighborhoods of the coordinate intervals under consideration.

In the following we will recall the classical derivative free interpolation error estimates for analytic functions, which are due to Davis [81, (4.6.1.11)]. Let $\mathcal{E}_{a,b}^{\rho} \subset \mathbb{C}$ again be the closed ellipse with focal points at a and b, $a < b$, with the semimajor axis $\bar{a} > (b-a)/2$ and the semiminor axis $\bar{b} > 0$ (see Sect. 5.3.2.2). The sum of the semi-axes is denoted by $\rho = \bar{a} + \bar{b}$. For the three-dimensional version we consider an axiparallel cuboid $Q_{\mathbf{a},\mathbf{b}}$ as in Convention 7.1.13. The ellipses $\mathcal{E}_{a_i,b_i}^{\rho_i}$, $1 \leq i \leq 3$, now refer to the coordinate intervals $Q_{\mathbf{a},\mathbf{b}}^{(i)}$ and, once tensorized, yield the domain $\overrightarrow{\mathcal{E}}_{\mathbf{a},\mathbf{b}}^{\rho} := \bigotimes_{i=1}^{3} \mathcal{E}_{a_i,b_i}^{\rho_i}$. For the cuboid $Q_{\mathbf{a},\mathbf{b}}$ we define the index $\iota \in \{1,2,3\}$ by $\frac{2\rho_\iota}{b_\iota - a_\iota} = \operatorname{argmin}_{i=1,2,3}\left\{\frac{2\rho_i}{b_i - a_i}\right\}$ and denote

$$\rho_{\min} := \rho_\iota \quad \text{and} \quad L := (b_\iota - a_\iota)/2. \tag{7.58}$$

In the case $\mathbf{b} = (1,1,1)^{\mathsf{T}} =: \mathbf{1}$ and $\mathbf{a} = -\mathbf{1}$ we omit the indices $\mathbf{a}, \mathbf{b}$ for the quantities $\mathcal{E}$ and Q.

A classical error estimate for Čebyšev interpolation of analytic functions can be found in [81].

Lemma 7.3.3. *Let $d = 3$, $Q = [-1,1]^d$ and let a function $f \in C^0(Q)$ be given that can be extended to an analytic function $f^\star$ on $\overrightarrow{\mathcal{E}}^{\rho}$ with $\rho_i > 1$, $1 \leq i \leq 3$. Then for the Čebyšev interpolation $p_m = \overrightarrow{\Pi}^{(m)}[f]$ the error estimate*

$$\|f - p_m\|_{C^0(Q)} \leq \sqrt{d}\, 2^{d/2+1} \rho_{\min}^{-m} \left(1 - \rho_{\min}^{-2}\right)^{-d/2} M_\rho(f)$$

holds with

$$M_\rho(f) := \max_{z \in \overrightarrow{\mathcal{E}}^{\rho}} \left|f^\star(z)\right|.$$

Proof. We will only sketch the proof, by using the interpolation error estimates for analytic functions taken from [81]. For this we introduce the inner product

$$(f,g)_\rho := \int_{\overrightarrow{\mathcal{E}}^{\rho}_{\mathbf{a},\mathbf{b}}} \frac{f(\mathbf{z})\overline{g(\mathbf{z})}}{\prod_{i=1}^{d}\sqrt{|1-z_i^2|}} d\mathbf{z}$$

and the Hilbert space

$$L^2\left(\overrightarrow{\mathcal{E}}^{\rho}_{\mathbf{a},\mathbf{b}}\right) := \left\{f : f \text{ is analytic in } \overrightarrow{\mathcal{E}}^{\rho}_{\mathbf{a},\mathbf{b}} \text{ and } \|f\|_\rho := (f,f)_\rho^{1/2} < \infty\right\}.$$

The fact that this is indeed a Hilbert space is proved in [81, Chap. 9.2, Sect. III] (it is, in fact, a separable, closed subspace of the Lebesgue space $L^2\left(\overrightarrow{\mathcal{E}}^{\rho}_{\mathbf{a},\mathbf{b}}\right)$). This space has two properties that are essential for our application: (a) Evaluation at points on $L^2\left(\overrightarrow{\mathcal{E}}^{\rho}_{\mathbf{a},\mathbf{b}}\right)$ is well defined and the associated operator is continuous. More precisely, there exists a constant C such that

$$\sup_{\mathbf{z}\in\overrightarrow{\mathcal{E}}^{\rho}_{\mathbf{a},\mathbf{b}}} |f(\mathbf{z})| \le C\|f\|_\rho \qquad \forall f \in L^2\left(\overrightarrow{\mathcal{E}}^{\rho}_{\mathbf{a},\mathbf{b}}\right).$$

This estimate is essential for the application under consideration, as the interpolation is based on point evaluations. (b) $L^2\left(\overrightarrow{\mathcal{E}}^{\rho}_{\mathbf{a},\mathbf{b}}\right)$ is a Hilbert space and thus permits the application of strong tools from functional analysis of Hilbert spaces.

In the following let $\mathbf{a} = -\mathbf{1}$ and $\mathbf{b} = \mathbf{1}$.

The scaled Čebyšev polynomials

$$\widetilde{T}_\mu(\mathbf{z}) := c_\mu T_\mu(\mathbf{z}) \quad \text{with} \quad c_\mu := \left(\frac{2}{\pi}\right)^{d/2} \prod_{i=1}^{d}\left(\rho_i^{2\mu_i} + \rho_i^{-2\mu_i}\right)^{-1/2} \qquad \forall \mu \in \mathbb{N}_0^3 \tag{7.59}$$

define a complete orthonormal system for $L^2\left(\overrightarrow{\mathcal{E}}^{\rho}_{\mathbf{a},\mathbf{b}}\right)$ with respect to the inner product $(\cdot,\cdot)_\rho$ (see [81]). For an arbitrary, bounded functional E on $L^2\left(\overrightarrow{\mathcal{E}}^{\rho}_{\mathbf{a},\mathbf{b}}\right)$ we have

$$|E(f)| \le \|E\|_\rho \|f\|_\rho, \tag{7.60}$$

where $\|E\|_\rho$ denotes the operator norm, which satisfies

$$\|E\|_\rho = \sup_{f\in L^2\left(\overrightarrow{\mathcal{E}}^{\rho}_{\mathbf{a},\mathbf{b}}\right)\setminus\{0\}} \frac{|E(f)|}{\|f\|_\rho} = \sqrt{\sum_{\mu\in\mathbb{N}_0^d} \left|E\left(\widetilde{T}_\mu\right)\right|^2}$$

owing to the orthonormality of $\left(\widetilde{T}_\mu\right)_{\mu\in\mathbb{N}_0^d}$. Let E be the error of the Čebyšev interpolation at a point $\mathbf{x} \in Q$, i.e.,

$$E(f) = f(\mathbf{x}) - \overrightarrow{\Pi}^{(m)}[f](\mathbf{x}).$$

From $E(p) = 0$ for all $p \in \mathbb{Q}_{m-1}$ and Corollary 7.3.1 we have

$$\begin{aligned}
\sum_{\mu\in\mathbb{N}_0^d} \left|E\left(\widetilde{T}_\mu\right)\right|^2 &= \sum_{\mu\in\mathbb{N}_0^d} c_\mu^2 \left|E(T_\mu)\right|^2 = \sum_{\substack{\mu\in\mathbb{N}_0^d\\ |\mu|_\infty\geq m}} c_\mu^2 \left|E(T_\mu)\right|^2 \\
&\leq 4 \sum_{\substack{\mu\in\mathbb{N}_0^d\\ |\mu|_\infty\geq m}} c_\mu^2 \leq 4\left(\frac{2}{\pi}\right)^d \sum_{i=1}^d \left(\sum_{\substack{\mu\in\mathbb{N}_0^d\\ \mu_i\geq m}} \prod_{j=1}^d \rho_j^{-2\mu_j} \right) \\
&\leq 4\left(\frac{2}{\pi}\right)^d \sum_{i=1}^d \rho_i^{-2m} \left(\sum_{\substack{\mu\in\mathbb{N}_0^d\\ \mu_i\geq m}} \rho_i^{-2(\mu_i-m)} \prod_{\substack{j=1\\ j\neq i}}^d \rho_j^{-2\mu_j} \right) \\
&\leq 4\left(\frac{2}{\pi}\right)^d \sum_{i=1}^d \rho_i^{-2m} \left(\sum_{\mu\in\mathbb{N}_0^d} \prod_{j=1}^d \rho_j^{-2\mu_j} \right) \\
&\leq 4\left(\frac{2}{\pi}\right)^d \rho_{\min}^{-2m} d \sum_{\mu\in\mathbb{N}_0^d} \rho_{\min}^{-2|\mu|} = 4\left(\frac{2}{\pi}\right)^d \rho_{\min}^{-2m} d \left(1-\rho_{\min}^{-2}\right)^{-d}.
\end{aligned}$$

In view of (7.60) we still need to estimate the norm $\|f\|_\rho$. We have

$$\|f\|_\rho^2 = \int_{\overrightarrow{\mathcal{E}}^\rho_{\mathbf{a},\mathbf{b}}} \frac{f(\mathbf{z})\overline{f(\mathbf{z})}}{\prod_{i=1}^d \sqrt{\left|1-z_i^2\right|}} d\mathbf{z} \leq \left(\sup_{\mathbf{z}\in\overrightarrow{\mathcal{E}}^\rho} |f(\mathbf{z})|\right)^2 \|1\|_\rho^2.$$

It follows from $\pi^{d/2}\widetilde{T}_{\mathbf{0}} = 1$ and the orthonormality of the system $\widetilde{T}_\mu$ that

$$\|f\|_\rho^2 \leq \pi^d M_\rho^2(f). \tag{7.61}$$

□

Transferring this error estimate to general, axiparallel cuboids can be achieved by means of an affine transformation. For this let $\mathbf{a}, \mathbf{b}$ and let $Q_{\mathbf{a},\mathbf{b}}$ be as in Convention 7.1.13 and $Q = [-1,1]^3$. The transformation

$$\chi: Q \to Q_{\mathbf{a},\mathbf{b}}; \qquad \chi(\hat{\mathbf{x}}) = \left(a_i + (b_i - a_i)\frac{x_i+1}{2}\right)_{i=1}^3$$

is affine. The Čebyšev interpolation nodes on $Q_{\mathbf{a},\mathbf{b}}$ can be obtained by transforming the one-dimensional interpolation nodes $\xi^{(i)}$ to $[a_i, b_i]$

$$\forall \mu \in \mathcal{J}_m: \quad \xi^{(\mu)} := \chi\left(\xi^{(\mu_1)}, \xi^{(\mu_2)}, \xi^{(\mu_3)}\right), \qquad \Theta^{(m)}_{\mathbf{a},\mathbf{b}} := \left\{\xi^{(\mu)} : \mu \in \mathcal{J}_m\right\}.$$

Then the Čebyšev interpolant on $Q_{\mathbf{a},\mathbf{b}}$ is given by

$$\vec{\Pi}^{(m)}_{\mathbf{a},\mathbf{b}}(f) = \sum_{\mu\in\mathcal{J}_m} f\left(\xi^{(\mu)}\right) L^{(\mu)},$$

where the Lagrange functions $L^{(\mu)}$ refer to the set of interpolation nodes $\Theta^{(m)}_{\mathbf{a},\mathbf{b}}$.

Theorem 7.3.4. *Let $d = 3$ and $Q_{\mathbf{a},\mathbf{b}}$ be as in Convention 7.1.13. We assume that the function $f \in C^0(Q_{\mathbf{a},\mathbf{b}})$ can be extended to an analytic function $f^\star$ on $\vec{\mathcal{E}}^{\rho}_{\mathbf{a},\mathbf{b}}$ with $\rho_i > (b_i - a_i)/2$, $1 \le i \le 3$. Then the Čebyšev interpolant $p_m = \vec{\Pi}^{(m)}[f]$ satisfies the error estimate*

$$\|f - p_m\|_{C^0(Q)} \le \sqrt{d}\,2^{d/2+1}\left(\frac{L}{\rho_{\min}}\right)^m \left(1 - \left(\frac{L}{\rho_{\min}}\right)^2\right)^{-d/2} M_\rho(f)$$

with

$$M_\rho(f) := \max_{\mathbf{z}\in\vec{\mathcal{E}}^{\rho}_{\mathbf{a},\mathbf{b}}} \left|f^\star(\mathbf{z})\right|$$

and $\rho_{\min}$, L as in (7.58).

Proof. Let $f \in C^{m+1}(Q_{\mathbf{a},\mathbf{b}})$ and let the affine transformation $\chi : Q \to Q_{\mathbf{a},\mathbf{b}}$ with $Q = (-1,1)^d$ be defined as before. We set $\hat{f} = f \circ \chi$ and denote the Čebyšev interpolant of $\hat{f}$ on Q by $\hat{p}_m$. Then we have $\vec{\Pi}^{(m)}_{\mathbf{a},\mathbf{b}}[f] = \hat{p}_m \circ \chi^{-1}$ and obtain

$$f - \vec{\Pi}^{(m)}_{\mathbf{a},\mathbf{b}}[f] = \left(\hat{f} - \hat{p}_m\right) \circ \chi^{-1}.$$

The transformed ellipse $\vec{\mathcal{E}}^{\hat{\rho}} := \chi^{-1}\vec{\mathcal{E}}^{\rho}_{\mathbf{a},\mathbf{b}}$ satisfies $\hat{\rho} = (2\rho_i/(b_i - a_i))_{i=1}^3$, and we set

$$\hat{\rho}_{\min} := \min\{2\rho_i/(b_i - a_i) : 1 \le i \le 3\}.$$

Now we can apply the error estimate from the previous theorem and obtain

$$\begin{aligned}\left\|f - \vec{\Pi}_{\mathbf{a},\mathbf{b}}[f]\right\|_{C^0(Q_{\mathbf{a},\mathbf{b}})} = \left\|\hat{f} - \hat{p}_m\right\|_{C^0(Q)} &\le 2^{d/2+1}\sqrt{d}\,\hat{\rho}_{\min}^{-m}\left(1 - \hat{\rho}_{\min}^{-2}\right)^{-d/2} M_{\hat{\rho}}\left(\hat{f}\right) \\ &\le 2^{d/2+1}\sqrt{d}\left(\frac{L}{\rho_{\min}}\right)^m\left(1 - \left(\frac{L}{\rho_{\min}}\right)^2\right)^{-d/2} M_\rho(f).\end{aligned}$$

□

Remark 7.3.5. *Since $\rho_{\min} > L$, it follows from the estimate above that the Čebyšev interpolation converges exponentially with respect to the order m.*

The approximation of the kernel function of the boundary integral operator is based on the (local) approximation of the global kernel function $G : Q_{\mathbf{a},\mathbf{b}} \times Q_{\mathbf{c},\mathbf{d}} \rightarrow \mathbb{C}$, where $Q_{\mathbf{a},\mathbf{b}}, Q_{\mathbf{c},\mathbf{d}} \subset \mathbb{R}^3$ denote axiparallel cuboids as in Convention 7.1.13. The Čebyšev interpolant of G of order m is defined by

$$\vec{\Pi}^{(m)}_{[\mathbf{a},\mathbf{b}],[\mathbf{c},\mathbf{d}]}[G](\mathbf{x},\mathbf{y}) := \sum_{\mu,\nu\in\mathcal{J}_m} G\left(\xi^{(\mu)},\xi^{(\nu)}\right) L^{(\mu)}(\mathbf{x})\, L^{(\nu)}(\mathbf{y})$$
$$\forall\,(\mathbf{x},\mathbf{y}) \in [\mathbf{a},\mathbf{b}] \times [\mathbf{c},\mathbf{d}]\,.$$

In the following we will transfer the statement from Theorem 7.3.4 to this situation. Let $9\rho^{(1)}_{\min}, L^{(1)}$ (or $\rho^{(2)}_{\min}, L^{(2)}$) be the constants from (7.58) for the cuboid $Q_{\mathbf{a},\mathbf{b}}$ (or $Q_{\mathbf{c},\mathbf{d}}$). We fix $(\rho_{\min}, L) \in \left\{\left(\rho^{(1)}_{\min}, L^{(1)}\right), \left(\rho^{(2)}_{\min}, L^{(2)}\right)\right\}$ by

$$\rho_{\min}/L = \min\left\{\rho^{(1)}_{\min}/L^{(1)}, \rho^{(2)}_{\min}/L^{(2)}\right\}.$$

Theorem 7.3.6. *Let $Q_{\mathbf{a},\mathbf{b}}, Q_{\mathbf{c},\mathbf{d}}$ be axiparallel cuboids as in Convention 7.1.13. We assume that the function $f \in C^0(Q_{\mathbf{a},\mathbf{b}} \times Q_{\mathbf{c},\mathbf{d}})$ can be extended on $\vec{\mathcal{E}}^{\rho_1}_{\mathbf{a},\mathbf{b}} \times \vec{\mathcal{E}}^{\rho_2}_{\mathbf{c},\mathbf{d}}$ with $(\rho_1)_i > (b_i - a_i)/2$ and $(\rho_2)_i > (d_i - c_i)/2$, $1 \le i \le 3$. Then the Čebyšev interpolant $p_m = \vec{\Pi}^{(m)}_{[\mathbf{a},\mathbf{b}],[\mathbf{c},\mathbf{d}]}[f]$ satisfies the error estimate*

$$\|f - p_m\|_{C^0(Q_{\mathbf{a},\mathbf{b}}\times Q_{\mathbf{c},\mathbf{d}})} \le C^{(0)}_{\rho_{\min}/L}\left(\frac{L}{\rho_{\min}}\right)^m M_{\rho_1\times\rho_2}(f)$$

with

$$M_{\rho_1\times\rho_2}(f) := \max_{(\mathbf{v},\mathbf{w})\in\vec{\mathcal{E}}^{\rho_1}_{\mathbf{a},\mathbf{b}}\times\vec{\mathcal{E}}^{\rho_2}_{\mathbf{c},\mathbf{d}}} \left|f^\star(\mathbf{v},\mathbf{w})\right|$$

and

$$C^{(0)}_\rho := \sqrt{d}\,2^{d+3/2}\left(1-\rho^{-2}\right)^{-d}.$$

Proof. First let $\mathbf{b} := \mathbf{d} := \mathbf{1}$ and $\mathbf{a} := \mathbf{c} := -\mathbf{1}$. We adapt the arguments from Lemma 7.3.3 to the tensorized case. To this end, we introduce the inner product

$$(f,g)_{\rho_1\times\rho_2} := \int_{\vec{\mathcal{E}}^{\rho_1}}\int_{\vec{\mathcal{E}}^{\rho_2}} \frac{f(\mathbf{v},\mathbf{w})\,\overline{g(\mathbf{v},\mathbf{w})}}{\prod_{i=1}^d\sqrt{|1-v_i^2|}\prod_{i=1}^d\sqrt{|1-w_i^2|}}\,d\mathbf{w}\,d\mathbf{v}$$

for analytic functions $f, g \in \vec{\mathcal{E}}^{\rho_1} \times \vec{\mathcal{E}}^{\rho_2}$ as well as the Hilbert space

$$L^2\left(\vec{\mathcal{E}}^{\rho_1}\times\vec{\mathcal{E}}^{\rho_2}\right) := \Big\{f : f \text{ is analytic in } \vec{\mathcal{E}}^{\rho_1}\times\vec{\mathcal{E}}^{\rho_2} \text{ and}$$
$$\|f\|_{\rho_1\times\rho_2} := (f,f)^{1/2}_{\rho_1\times\rho_2} < \infty\Big\}.$$

The statements from the proof of Lemma 7.3.3 can be applied to $L^2\left(\vec{\mathcal{E}}^{\rho_1}\times\vec{\mathcal{E}}^{\rho_2}\right)$ analogously.

The polynomials $\widetilde{T}_{\mu,\nu}(\mathbf{v},\mathbf{w}) := c_\mu T_\mu(\mathbf{v})\, c_\nu T_\nu(\mathbf{w})$ with c_μ from (7.59) constitute for $\nu,\mu\in\mathbb{N}_0^3$ a complete orthonormal system for $L^2\left(\vec{\mathcal{E}}^{\rho_1}\times\vec{\mathcal{E}}^{\rho_2}\right)$ with respect to the inner product $(\cdot,\cdot)_{\rho_1\times\rho_2}$. Let E be the error of Čebyšev interpolation at a point $(\mathbf{x},\mathbf{y})\in Q\times Q$, i.e.,

$$E(f) = f(\mathbf{x},\mathbf{y}) - \vec{\Pi}^{(m)}_{[-1,1],[-1,1]}[f](\mathbf{x},\mathbf{y}).$$

It follows from $E(p)=0$ for all $p\in\mathbb{Q}_{m-1}\times\mathbb{Q}_{m-1}$ and the second estimate in Corollary 7.3.1(b) that

$$\begin{aligned}\left|E\left(T_{\mu,\nu}\right)\right| &= \left|T_{\mu,\nu}(\mathbf{x},\mathbf{y}) - \vec{\Pi}^{(m)}_{[-1,1],[-1,1]}\left(T_{\mu,\nu}\right)(\mathbf{x},\mathbf{y})\right| \\ &\le \left|T_\mu(\mathbf{x})\right|\left|T_\nu(\mathbf{y})\right| + \left|\vec{\Pi}^{(m)}\left(T_\mu\right)(\mathbf{x})\right|\left|\vec{\Pi}^{(m)}\left(T_\nu\right)(\mathbf{y})\right| \le 2\end{aligned}$$

and thus

$$\begin{aligned}\sum_{\mu,\nu\in\mathbb{N}_0^d}\left|E\left(\widetilde{T}_{\mu,\nu}\right)\right|^2 &= \sum_{\mu,\nu\in\mathbb{N}_0^d} c_\mu^2 c_\nu^2\left|E\left(T_{\mu,\nu}\right)\right|^2 \\ &= \sum_{\substack{(\mu,\nu)\in\mathbb{N}_0^d\times\mathbb{N}_0^d\\ |(\mu,\nu)|_\infty\ge m}} c_\mu^2 c_\nu^2\left|E\left(T_{\mu,\nu}\right)\right|^2 \le 4\sum_{\substack{(\mu,\nu)\in\mathbb{N}_0^d\times\mathbb{N}_0^d\\ |(\mu,\nu)|_\infty\ge m}} c_\mu^2 c_\nu^2 \\ &\le 4\left(\frac{2}{\pi}\right)^{2d}\sum_{q=1}^{2}\sum_{i=1}^{d}\sum_{\substack{(\mu^{(1)},\mu^{(2)})\in\mathbb{N}_0^d\times\mathbb{N}_0^d\\ \mu_i^{(q)}\ge m}}\left(\rho_i^{(q)}\right)^{-2\mu_i^{(q)}} \\ &\quad\times\prod_{r=1}^{2}\prod_{\substack{j=1\\(r,j)\ne(q,i)}}^{d}\left(\rho_j^{(r)}\right)^{-2\mu_j^{(r)}} \\ &\le 4\left(\frac{2}{\pi}\right)^{2d}\rho_{\min}^{-2m}(2d)\sum_{(\mu,\nu)\in\mathbb{N}_0^d\times\mathbb{N}_0^d}\prod_{j=1}^{d}\left(\rho_j^{(1)}\right)^{-2\mu_j}\left(\rho_j^{(2)}\right)^{-2\nu_j} \\ &\le 4\left(\frac{2}{\pi}\right)^{2d}\rho_{\min}^{-2m}(2d)\left(1-\rho_{\min}^2\right)^{-2d}.\end{aligned}$$

The norm $\|f\|_{\rho_1\times\rho_2}$ is estimated in the same way as in (7.61). Thus,

$$\|f\|^2_{\rho_1\times\rho_2} \leq \pi^{2d} M^2_{\rho_1\times\rho_2}(f).$$

The results for the general axiparallel cuboids follow by an affine transformation. □

The kernel function of the boundary integral operator under consideration is either the global kernel function or a directional derivative thereof. In the latter case we define the approximation of the kernel function by the directional derivative of the approximation of the global kernel function. The corresponding error estimate can be found in the following two theorems.

Theorem 7.3.7. *Let $Q_{\mathbf{a},\mathbf{b}}$, $Q_{\mathbf{c},\mathbf{d}}$ be axiparallel cuboids as in Convention 7.1.13. Let $f \in C^1(Q_{\mathbf{a},\mathbf{b}} \times Q_{\mathbf{c},\mathbf{d}})$ and $D \in \left\{\frac{\partial}{\partial x_i}, \frac{\partial}{\partial y_i} : 1 \leq i \leq d\right\}$. We assume that the function Df can be analytically extended to $\vec{\mathcal{E}}^{\rho_1}_{\mathbf{a},\mathbf{b}} \times \vec{\mathcal{E}}^{\rho_2}_{\mathbf{c},\mathbf{d}}$ with $(\rho_1)_i > (b_i - a_i)/2$ and $(\rho_2)_i > (d_i - c_i)/2$, $1 \leq i \leq 3$. Then the Čebyšev interpolant $p_m = \vec{\Pi}^{(m)}_{[\mathbf{a},\mathbf{b}],[\mathbf{c},\mathbf{d}]}[f]$, $m \geq 2$, satisfies the error estimate*

$$\|D(f - p_m)\|_{C^0(Q_{\mathbf{a},\mathbf{b}}\times Q_{\mathbf{c},\mathbf{d}})} \leq C^{(1)}_{\rho_{\min}/L} \left(\frac{\rho_{\min}/L + 1}{2\rho_{\min}/L}\right)^{m-1} M_{\rho_1\times\rho_2}(Df). \tag{7.62}$$

with

$$C^{(1)}_\rho := \sqrt{dc_\rho} 2^{d+3/2} \left(\frac{\rho^2}{\rho^2 - 1}\right)^{d-1/2}$$

and c_ρ from (7.63).

Corollary 7.3.8. *The constant $C^{(1)}_{\rho_{\min}}$ tends to zero for $\rho_{\min} \to \infty$ at the rate $(\log \rho_{\min})^{-2}$ while for $\rho_{\min} \to 1$ it grows as $(\rho_{\min} - 1)^{-d-2}$.*

For the proof of Theorem 7.3.7 we need a lemma.

Lemma 7.3.9. *For $m \in \mathbb{N}_0$ and $\rho > 1$*

$$\sum_{k=m}^{\infty} \rho^{-2k} k^4 \leq \widetilde{C}_\rho \left(\frac{\rho + 1}{2\rho}\right)^{2m}$$

with

$$\widetilde{C}_\rho := \left(\frac{2}{e \ln\left(\frac{\rho+1}{2}\right)}\right)^4 \frac{4\rho^2}{(3\rho + 1)(\rho - 1)} \tag{7.63}$$

Proof. An elementary analysis yields

$$\frac{k^4}{\left(\frac{\rho+1}{2}\right)^{2k}} \leq \left(\frac{2}{e \ln\left(\frac{\rho+1}{2}\right)}\right)^4.$$

From this we obtain

$$\sum_{k=m}^{\infty} \rho^{-2k} k^4 \leq \left(\frac{2}{e \ln \left(\frac{\rho+1}{2} \right)} \right)^4 \sum_{k=m}^{\infty} \left(\frac{\rho+1}{2\rho} \right)^{2k} = \widetilde{C}_\rho \left(\frac{\rho+1}{2\rho} \right)^{2m}.$$

□

Proof of Theorem 7.3.7. We first consider the non-tensorized, one-dimensional case $d = 1$ and $f : C^1([-1,1], \mathbb{C})$. We set $g := f'$ with the analytic extension $g^\star : \mathcal{E}^\rho \to \mathbb{C}$. The approximation of g can be represented by $g_m := \left(\Pi^{(m)} g^{(-1)}\right)'$ with $g^{(-1)}(x) := \int_{-1}^{x} g(s)\, ds$. The error functional E is defined in this case by

$$E(g) := (g - g_m)(\mathbf{x}, \mathbf{y}).$$

The norm $\|E\|_\rho$ again has to be computed and we use the previously developed method. The definition of g_m implies that $E(p) = 0$ for all $p \in \mathbb{P}_{m-2}$.

To estimate $\left|E\left(\widetilde{T}_k\right)\right|$ we need to find an upper bound for $\left(\Pi^{(m)} T_k^{(-1)}\right)'$ and use (see [192, Exercise 1.1.4])

$$T_k^{(-1)} = \beta_{1+k} T_{k+1} + \beta_{1-k} T_{k-1} + \alpha_k \quad \text{with} \quad \beta_s := \begin{cases} 1/(2s) & s \neq 0 \\ 0 & \text{otherwise} \end{cases}$$

and a constant α_k. With (7.56) and $\gamma_{\ell,m}$ and $\eta_{\ell,m}$ from (7.57) we obtain

$$\Pi^{(m)} T_k^{(-1)} = \gamma_{k+1,m} \beta_{1+k} T_{\eta_{k+1,m}} + \gamma_{k-1,m} \beta_{1-k} T_{\eta_{k-1,m}} + \alpha_k$$

and by differentiation

$$\left(\Pi^{(m)} T_k^{(-1)}\right)' = \gamma_{k+1,m} \beta_{1+k} T'_{\eta_{k+1,m}} + \gamma_{k-1,m} \beta_{1-k} T'_{\eta_{k-1,m}}.$$

Corollary 7.3.1 implies for all $x \in [-1,1]$ and $\tau_k^{(1)}$ as in (7.55) that

$$\left|T_k'(x)\right| \leq k^2.$$

In all we have shown that

$$\left|\left(\Pi^{(m)} T_k^{(-1)}\right)'(x)\right| \leq \beta_{1+k} \eta_{k+1,m}^2 + |\beta_{1-k}|\, \eta_{k-1,m}^2 \leq (m-1)^2.$$

The norm of E can therefore be estimated in the same way as in the proof of Theorem 7.3.3. With Lemma 7.3.9 we obtain

$$\sum_{k=0}^{\infty} \left|E\left(\widetilde{T}_k\right)\right|^2 = \sum_{k=m-1}^{\infty} c_k^2 \left|E\left(T_k\right)\right|^2 \le \sum_{k=m-1}^{\infty} c_k^2 \left(k^2 + (m-1)^2\right)^2$$
$$\le 4 \sum_{k=m-1}^{\infty} c_k^2 k^4 \le \frac{8}{\pi} \sum_{k=m-1}^{\infty} \rho^{-2k} k^4 \le \frac{8}{\pi} \widetilde{C}_\rho \left(\frac{\rho+1}{2\rho}\right)^{2m-2}.$$

Now let $\mathbf{b} := \mathbf{d} := \mathbf{1}$ and $\mathbf{a} := \mathbf{c} := -\mathbf{1}$. Without loss of generality we choose $D = \frac{\partial}{\partial x_1}$. Let E be the derivative D of the error of the Čebyšev interpolant at one point $(\mathbf{x}, \mathbf{y}) \in Q \times Q$

$$E(f) = Df(\mathbf{x}, \mathbf{y}) - D\overrightarrow{\Pi}^{(m)}_{[-\mathbf{1},\mathbf{1}],[-\mathbf{1},\mathbf{1}]} \left[f^{(-1)}\right](\mathbf{x}, \mathbf{y}),$$

where $f^{(-1)}$ is an antiderivative of f with respect to the first variable. Let $\mathbb{Q}_{m-1}^-$ be the set of all polynomials $p \in \mathbb{Q}_{m-1}$ with $(x_1 p) \in \mathbb{Q}_{m-1}$. It follows from $E(p) = 0$ for all $p \in \mathbb{Q}_{m-1}^- \times \mathbb{Q}_{m-1}$ combined with the previous results for $\mu_1 \ge m-1$ that we have the estimate

$$\left|E\left(T_{\mu,\nu}\right)\right| = \left|DT_{\mu,\nu}(\mathbf{x}, \mathbf{y}) - D\overrightarrow{\Pi}^{(m)}_{[-\mathbf{1},\mathbf{1}],[-\mathbf{1},\mathbf{1}]} \left(T_{\mu,\nu}^{(-1)}\right)(\mathbf{x}, \mathbf{y})\right|$$
$$\le \left|DT_\mu(\mathbf{x})\right| \left|T_\nu(\mathbf{y})\right| + \left|D\overrightarrow{\Pi}^{(m)} \left(T_\mu^{(-1)}\right)(\mathbf{x})\right| \left|\overrightarrow{\Pi}^{(m)}\left(T_\nu\right)(\mathbf{y})\right| \le 2\mu_1^2.$$

For $\mu_1 < m-1$ we use $\left(T_{\mu_1} - \Pi^{(m)} T_{\mu_1}^{(-1)}\right)' = 0$ and also obtain

$$\left|E\left(T_{\mu,\nu}\right)\right| = \left|T'_{\mu_1}(x_1)\right| \left|\left(\prod_{i=2}^3 T_{\mu_i}(x_i)\right) T_\nu(\mathbf{y})\right.$$
$$- \left(\prod_{i=2}^3 \Pi^{(m)} T_{\mu_i}(x_i)\right)$$
$$\left.\Pi^{(m)}_{[-1,1]} T_\nu\left(x_j\right)\right| \le \mu_1^2 (1+1) = 2\mu_1^2.$$

We denote the Kronecker symbol by $\delta_{(q,i),(r,j)}$ and for $\mu \in \mathbb{N}_0^d$ we set $\mu_+ = \mu + e_1$ with $e_1 = (1, 0, 0, \dots, 0)^\mathsf{T}$. Then we have

$$\sum_{\mu,\nu \in \mathbb{N}_0^d} \left|E\left(\widetilde{T}_{\mu,\nu}\right)\right|^2 = \sum_{\mu,\nu \in \mathbb{N}_0^d} c_\mu^2 c_\nu^2 \left|E\left(T_{\mu,\nu}\right)\right|^2$$
$$= \sum_{\substack{(\mu,\nu) \in \mathbb{N}_0^d \times \mathbb{N}_0^d \\ \left|(\mu_+,\nu)\right|_\infty \ge m}} c_\mu^2 c_\nu^2 \left|E\left(T_{\mu,\nu}\right)\right|^2 \le 4 \sum_{\substack{(\mu,\nu) \in \mathbb{N}_0^d \times \mathbb{N}_0^d \\ \left|(\mu_+,\nu)\right|_\infty \ge m}} c_\mu^2 c_\nu^2 \mu_1^4$$

$$\leq 4\left(\frac{2}{\pi}\right)^{2d}\sum_{q=1}^{2}\sum_{i=1}^{d}\sum_{\substack{(\mu^{(1)},\mu^{(2)})\in\mathbb{N}_0^d\times\mathbb{N}_0^d\\ \mu_i^{(q)}+\delta_{(q,i),(1,1)}\geq m}}\left(\mu_1^{(1)}\right)^4\left(\rho_i^{(q)}\right)^{-2\mu_i^{(q)}}$$

$$\times\prod_{r=1}^{2}\prod_{\substack{j=1\\ (r,j)\neq(q,i)}}^{d}\left(\rho_j^{(r)}\right)^{-2\mu_j^{(r)}}$$

$$\leq 4\left(\frac{2}{\pi}\right)^{2d}\sum_{q=1}^{2}\sum_{i=1}^{d}\sum_{\substack{(\mu^{(1)},\mu^{(2)})\in\mathbb{N}_0^d\times\mathbb{N}_0^d\\ \mu_i^{(q)}+\delta_{(q,i),(1,1)}\geq m}}\left(\mu_1^{(1)}\right)^4\rho_{\min}^{-2|(\mu^{(1)}+\mu^{(2)})|}$$

$$\overset{\text{Lem. 7.3.9}}{\leq} 8\widetilde{C}_{\rho_{\min}}d\left(\frac{2}{\pi}\right)^{2d}\left(\frac{\rho_{\min}^2}{\rho_{\min}^2-1}\right)^{2d-1}\left(\frac{\rho_{\min}+1}{2\rho_{\min}}\right)^{2m-2}.$$

The norm $\|Df\|_{\rho_1\times\rho_2}$ is estimated in the same way as in (7.61). Thus

$$\|f\|_{\rho_1\times\rho_2}^2\leq\pi^{2d}M_{\rho_1\times\rho_2}^2(Df).$$

In all we have shown that

$$\|D(f-p_m)\|_{C^0(Q_{-1,1}\times Q_{-1,1})}\leq\widetilde{C}_{\rho_{\min}}\left(\frac{\rho_{\min}+1}{2\rho_{\min}}\right)^{m-1}M_{\rho_1\times\rho_2}(Df).$$

The result for general axiparallel cuboids can again be achieved by an affine transformation. We again first consider the one-dimensional, non-tensorized case that $g:[a,b]\to\mathbb{C}$ and note that

$$g_m(\chi(\hat{x})):=\left(\frac{d}{dx}\Pi^{(m)}g^{(-1)}\right)(\chi(\hat{x}))=\left(\frac{d}{d\hat{x}}\Pi_{[-1,1]}^{(m)}\hat{g}^{(-1)}\right)(\hat{x})=:\hat{g}_m(\hat{x})$$

with $\hat{g}:=g\circ\chi$. It follows that

$$\|g-g_m\|_{L^\infty(a,b)}=\|\hat{g}-\hat{g}_m\|_{L^\infty(-1,1)}.$$

Now we can apply the error estimates for the unit interval.

From the previous results we deduce for the tensorized case

$$\|D(f-p_m)\|_{C^0(Q_{\mathbf{a},\mathbf{b}}\times Q_{\mathbf{c},\mathbf{d}})}\leq\widetilde{C}_{\rho_{\min}/L}\left(\frac{\rho_{\min}/L+1}{2\rho_{\min}/L}\right)^{m-1}M_{\hat{\rho}_1\times\hat{\rho}_2}\left(\widehat{Df}\right).$$

The assertion finally follows by means of a transform to the original coordinates

$$M_{\hat{\rho}_1\times\hat{\rho}_2}\left(\widehat{Df}\right)=M_{\rho_1\times\rho_2}(Df).$$

□

In the following theorem we consider the error of the approximation of the second derivative of the global kernel function.

Theorem 7.3.10. *Let $Q_{\mathbf{a},\mathbf{b}}$, $Q_{\mathbf{c},\mathbf{d}}$ be axiparallel cuboids as in Convention 7.1.13. Let $f \in C^2(Q_{\mathbf{a},\mathbf{b}} \times Q_{\mathbf{c},\mathbf{d}})$ and $D_1 \in \left\{\frac{\partial}{\partial x_i} : 1 \le i \le d\right\}$, $D_2 \in \left\{\frac{\partial}{\partial y_i} : 1 \le i \le d\right\}$. We assume that the function $D_1 D_2 f$ can be extended to $\overrightarrow{\mathcal{E}}^{\rho_1}_{\mathbf{a},\mathbf{b}} \times \overrightarrow{\mathcal{E}}^{\rho_2}_{\mathbf{c},\mathbf{d}}$ with $(\rho_1)_i > (b_i - a_i)/2$ and $(\rho_2)_i > (d_i - c_i)/2$, $1 \le i \le 3$. Then the Čebyšev interpolant $p_m = \overrightarrow{\Pi}^{(m)}_{[\mathbf{a},\mathbf{b}],[\mathbf{c},\mathbf{d}]}[f]$, $m \ge 2$, satisfies the error estimate*

$$\|D_1 D_2 (f - p_m)\|_{C^0(Q_{\mathbf{a},\mathbf{b}} \times Q_{\mathbf{c},\mathbf{d}})} \le C^{(2)}_\rho \left(\frac{\rho_{\min}/L + 1}{2\rho_{\min}/L}\right)^{m-1} M_{\rho_1 \times \rho_2}(D_1 D_2 f).$$

with

$$C^{(2)}_\rho := \widetilde{C}_\rho \sqrt{d}\, 2^{d+3/2} \left(\frac{\rho^2}{\rho^2 - 1}\right)^{d-1}.$$

The proof is completely analogous to that of the previous theorem and will therefore not be detailed.

For the error analysis of the cluster method we apply the Čebyšev interpolation to the global kernel function for separated cluster boxes Q_1, Q_2. To be able to use the estimate for the interpolation error we need to estimate the modulus $M_{\rho_1 \times \rho_2}$ for the (derivative of the) global kernel function.

Here we will restrict ourselves to the fundamental solution $G : \mathbb{R}^3 \backslash \{0\} \to \mathbb{C}$

$$G(\mathbf{z}) = \frac{1}{4\pi\sqrt{\det \mathbf{A}}} \frac{e^{\langle \mathbf{b}, \mathbf{z}\rangle_{\mathbf{A}} - \lambda \|\mathbf{z}\|_{\mathbf{A}}}}{\|\mathbf{z}\|_{\mathbf{A}}}, \qquad \lambda^2 := c + \|\mathbf{b}\|^2_{\mathbf{A}} \tag{7.64}$$

from (3.3) and the kernel functions derived from it

$$\begin{aligned} k_1(\mathbf{x}, \mathbf{y}) &= G(\mathbf{x} - \mathbf{y}), & k_2(\mathbf{x}, \mathbf{y}) &= \tilde{\gamma}_{1,\mathbf{y}} G(\mathbf{x} - \mathbf{y}), \\ k_3(\mathbf{x}, \mathbf{y}) &= \gamma_{1,\mathbf{x}} G(\mathbf{x} - \mathbf{y}), & k_4(\mathbf{x}, \mathbf{y}) &= \gamma_{1,\mathbf{x}} \tilde{\gamma}_{1,\mathbf{y}} G(\mathbf{x} - \mathbf{y}) \end{aligned} \tag{7.65}$$

with the conormal derivative γ_1 and the modified conormal derivative $\tilde{\gamma}_1$ [see (2.103), (2.107)]. Let η be as in Definition 7.1.11.

Lemma 7.3.11. *There exist constants C_1, C_2 that depend only on the coefficients $\mathbf{A}$, $\mathbf{b}$, c in (7.64) and Γ with the following properties.*

Let $(\sigma, s) \in P^{far}$ and $Q_\sigma =: Q_{\mathbf{a},\mathbf{b}}$, $Q_s =: Q_{\mathbf{c},\mathbf{d}}$. With

$$\rho_{1,i} := \frac{|b_i - a_i|}{2}\left(1 + \frac{2}{C_1 \eta}\right) \quad \text{and} \quad \rho_{2,i} := \frac{|d_i - c_i|}{2}\left(1 + \frac{2}{C_1 \eta}\right)$$

for $1 \le i \le 3$, $\partial_{\mathbf{x}}^{\mu}\partial_{\mathbf{y}}^{\nu}G(\mathbf{x}-\mathbf{y})$, $|\mu+\nu| \le 2$, *can be analytically extended to* $\vec{\mathcal{E}}_{\mathbf{a,b}}^{\rho_1} \times \vec{\mathcal{E}}_{\mathbf{c,d}}^{\rho_2}$ *and satisfies the estimate*

$$\sup_{(\mathbf{x},\mathbf{y})\in\vec{\mathcal{E}}_{\mathbf{a,b}}^{\rho_1}\times\vec{\mathcal{E}}_{\mathbf{c,d}}^{\rho_2}} \left|\partial_{\mathbf{x}}^{\mu}\partial_{\mathbf{y}}^{\nu}G(\mathbf{x}-\mathbf{y})\right| \le \left(\frac{C_2}{\operatorname{dist}(Q_\sigma,Q_s)}\right)^{1+|\mu+\nu|}.$$

Proof. Let $(\sigma,s) \in P^{far}$ be an admissible far-field block with associated cluster boxes $Q_\sigma =: Q_{\mathbf{a,b}}$, $Q_s =: Q_{\mathbf{c,d}}$. The singular behavior of the function $\partial_{\mathbf{x}}^{\mu}\partial_{\mathbf{y}}^{\nu}G(\mathbf{x}-\mathbf{y})$ is characterized by the function

$$g_\lambda : Q_\sigma \times Q_s \to \mathbb{R}, \qquad g_\lambda(\mathbf{x},\mathbf{y}) := \|\mathbf{x}-\mathbf{y}\|^{-\lambda},$$

which we will consider first. It follows from the admissibility condition

$$\eta \operatorname{dist}(Q_\sigma,Q_s) \ge \max\{\operatorname{diam} Q_\sigma, \operatorname{diam} Q_s\} \tag{7.66}$$

that g_λ can be extended with respect to every variable to ellipses $\mathcal{E}_{a_i,b_i}^{\rho_{1,i}}$ or $\mathcal{E}_{c_i,d_i}^{\rho_{2,i}}$, $1 \le i \le 3$, where

$$\rho_{1,i} = \frac{|b_i-a_i|}{2}\left(1+\frac{1}{3\sqrt{3}}\frac{\operatorname{dist}(Q_\sigma,Q_s)}{|b_i-a_i|/2}\right) \ge \frac{|b_i-a_i|}{2}\left(1+\frac{2}{C\eta}\right),$$

$C := 3\sqrt{3}$ and $\rho_{2,i} \ge \frac{|d_i-c_i|}{2}(1+2/(C\eta))$. The function g_λ can be estimated on these ellipses by

$$\sup_{(\mathbf{x},\mathbf{y})\in\vec{\mathcal{E}}_{[\mathbf{a,b}]}^{\rho_1}\times\vec{\mathcal{E}}_{[\mathbf{c,d}]}^{\rho_2}} g_\lambda(\mathbf{x},\mathbf{y}) \le \left(\frac{C}{\operatorname{dist}(Q_\sigma,Q_s)}\right)^{\lambda}.$$

This result can be directly transferred to the function $\|\mathbf{x}-\mathbf{y}\|_{\mathbf{A}}^{-\lambda}$ for the metric induced by $\mathbf{A}$. This is achieved by replacing the constant C by a constant that depends on $\mathbf{A}$.

For the fundamental solution $G(\mathbf{x}-\mathbf{y})$ or its derivatives $\partial_{\mathbf{x}}^{\mu}\partial_{\mathbf{y}}^{\nu}G(\mathbf{x}-\mathbf{y})$ the size of the analyticity ellipses remains unchanged. For the estimate of the function on the analyticity ellipses we obtain

$$\sup_{(\mathbf{x},\mathbf{y})\in\vec{\mathcal{E}}_{\mathbf{a,b}}^{\rho_1}\times\vec{\mathcal{E}}_{\mathbf{c,d}}^{\rho_2}} \left|\partial_{\mathbf{x}}^{\mu}\partial_{\mathbf{y}}^{\nu}G(\mathbf{x}-\mathbf{y})\right| \le \left(\frac{C}{\operatorname{dist}(Q_\sigma,Q_s)}\right)^{1+|\mu+\nu|},$$

where C depends on the coefficients $\mathbf{A}$, $\mathbf{b}$, c of the differential operator and on Γ.

□

By combining Lemma 7.3.11 and Theorems 7.3.6, 7.3.7 and 7.3.10 we obtain the estimates for the blockwise approximations:

$$\tilde{k}_1(\mathbf{x},\mathbf{y}) := \overrightarrow{\Pi}^{(m)}_{[\mathbf{a},\mathbf{b}],[\mathbf{c},\mathbf{d}]} G(\mathbf{x}-\mathbf{y}), \qquad \tilde{k}_2(\mathbf{x},\mathbf{y}) := \tilde{\gamma}_{1,\mathbf{y}} \overrightarrow{\Pi}^{(m)}_{[\mathbf{a},\mathbf{b}],[\mathbf{c},\mathbf{d}]} G(\mathbf{x}-\mathbf{y}),$$
$$\tilde{k}_3(\mathbf{x},\mathbf{y}) := \gamma_{1,\mathbf{x}} \overrightarrow{\Pi}^{(m)}_{[\mathbf{a},\mathbf{b}],[\mathbf{c},\mathbf{d}]} G(\mathbf{x}-\mathbf{y}), \; \tilde{k}_4(\mathbf{x},\mathbf{y}) := \gamma_{1,\mathbf{x}} \tilde{\gamma}_{1,\mathbf{y}} \overrightarrow{\Pi}^{(m)}_{[\mathbf{a},\mathbf{b}],[\mathbf{c},\mathbf{d}]} G(\mathbf{x}-\mathbf{y}) \tag{7.67}$$

of the kernel functions from (7.65) on admissible far-field blocks $(\sigma, s) \in P^{far}$ with associated cluster boxes $Q_\sigma =: Q_{\mathbf{a},\mathbf{b}}$ and $Q_s =: Q_{\mathbf{c},\mathbf{d}}$.

Theorem 7.3.12. *There exist a sufficiently small $\eta_0 \in (0,1)$ and constants $0 < c_1 < 1$, $C_0 > 0$ that depend only on the coefficients $\mathbf{A}$, $\mathbf{b}$, c and Γ with the following properties.*

For all $0 < \eta \le \eta_0$ in Definition 7.1.11 the approximations (7.67) of the kernel functions from (7.65) satisfy on all admissible far-field blocks $(\sigma, s) \in P^{far}$ the error estimate

$$\left\| k_j - \tilde{k}_j \right\|_{C^0(Q_{\mathbf{a},\mathbf{b}} \times Q_{\mathbf{c},\mathbf{d}})} \le C_0 c_1^m \left(\operatorname{dist}(Q_\sigma, Q_s)\right)^{-1-\nu_j}$$

with

$$\nu_1 := 0, \quad \nu_2 := \nu_3 := 1 \quad \text{and} \quad \nu_4 := 2.$$

Proof. Lemma 7.3.11 combined with Definition (7.58) yields

$$\rho_{\min}/L = 1 + \frac{2}{C\eta}. \tag{7.68}$$

If we insert this into the error estimate from Theorem 7.3.6 we obtain

$$\left\| k_1 - \tilde{k}_1 \right\|_{C^0(Q_{\mathbf{a},\mathbf{b}} \times Q_{\mathbf{c},\mathbf{d}})} \le C_1 \left(\frac{C\eta}{2 + C\eta} \right)^m \frac{1}{\operatorname{dist}(Q_\sigma, Q_s)}$$

with a constant C_1 that depends only on Γ, η_0 and the coefficients $\mathbf{A}$, $\mathbf{b}$, c.

For the first derivatives $D \in \{\partial/\partial x_i, \partial/\partial y_i : 1 \le i \le 3\}$ we combine Theorem 7.3.7 with (7.68), which gives us

$$\left\| D\left(k_1 - \tilde{k}_1\right) \right\|_{C^0(Q_{\mathbf{a},\mathbf{b}} \times Q_{\mathbf{c},\mathbf{d}})} \le C_2 \left(\frac{1 + C\eta}{2 + C\eta} \right)^{m-1} \left(\frac{1}{\operatorname{dist}(Q_\sigma, Q_s)} \right)^2$$

with a constant C_2 that again depends only on Γ, η_0 and the coefficients $\mathbf{A}$, $\mathbf{b}$, c. If we choose $c_1 := \frac{1+C\eta}{2+C\eta}$ and $C_0 := C_2/c_1$ we obtain the asserted estimate.

For the mixed derivatives $D_1 \in \{\partial/\partial x_i : 1 \le i \le 3\}$ and $D_2 \in \{\partial/\partial y_i : 1 \le i \le 3\}$ we use Theorem 7.3.10 and obtain

$$\left\| D_1 D_2 \left(k_1 - \tilde{k}_1\right) \right\|_{C^0(Q_{\mathbf{a},\mathbf{b}} \times Q_{\mathbf{c},\mathbf{d}})} \le C_3 \left(\frac{1 + C\eta}{2 + C\eta} \right)^{m-1} \left(\frac{1}{\operatorname{dist}(Q_\sigma, Q_s)} \right)^3.$$

The derivatives $\tilde{\gamma}_1$, γ_1 can be written as a linear combination of the above-mentioned differential operators with L^∞-coefficients. This yields the estimates from the assertion for $k_j - \tilde{k}_j$, $j = 2, 3, 4$. □

Remark 7.3.13. *The error estimates prove the exponential convergence of the kernel approximations with respect to the expansion order m. Note that by employing the classical error estimate*

$$\left\| u - \Pi^{(m)} u \right\|_{C^0([-1,1])} \leq (1 + c_m) \inf_{v \in \mathbb{P}_m} \| u - v \|_{C^0([-1,1])}$$

with the Lebesgue constant

$$c_m := \sup_{u \in C^0([-1,1]) \backslash \{0\}} \left(\| \Pi u \|_{C^0([-1,1])} / \| u \|_{C^0([-1,1])} \right)$$

and the estimate $c_m \leq Cm$ we would obtain the far more pessimistic estimate

$$\left\| k_1 - \tilde{k}_1 \right\| \leq C_0 m^6 c_1^{-m} \left(\operatorname{dist}\left(Q_\sigma, Q_s \right) \right)^{-1}.$$

Since the previous, derivative-free error estimates due to [81] take advantage of the analyticity of the kernel function, we were able to avoid the factor m^6.

Remark 7.3.14. *The size of the constants C_0 and c_1 for special kernel functions can be found, for example, in [125] and [122]. For the Taylor approximation of the fundamental solution of the Laplace operator we obtain $C_0 = 1$ and $c_1 = \eta$.*

Remark 7.3.15. *The explicit dependency with respect to the coefficients* $\mathbf{A}$, $\mathbf{b}$, c *in the fundamental solution has to be analyzed from case to case. For the Helmholtz problem with a large wave number $c \ll -1$ the quantity $M_{\rho_1 \times \rho_2}(k_1)$ grows exponentially with respect to $\sqrt{|c|}$ and the expansion order m has to be chosen proportional to $\sqrt{|c|}$.*

7.3.2 Global Error Estimates

Replacing the kernel function locally by the cluster approximation defines an approximation $\tilde{b}(\cdot,\cdot)$ of the sesquilinear form $b(\cdot,\cdot)$. We have derived local error estimates in Sect. 7.3.1.1 for the Čebyšev interpolation. In this section we will use the abstract assumption 7.1.23 imposed on the local accuracy of the approximation of the kernel function. From this we will use the Strang lemma to derive an estimate for the error $b(\cdot,\cdot) - \tilde{b}(\cdot,\cdot)$.

7.3.2.1 L^2-Estimate for the Clustering Error Without Integration by Parts

We consider the class of kernel functions from Assumption 7.1.1 and use the admissibility condition from Definition 7.1.11 with a fixed $\eta \in (0,1)$. Let $b : S \times S \to \mathbb{C}$ be the sesquilinear form for the integral operator K and let $\tilde{b}$ be the approximation that is defined by the cluster representation.

First we will introduce some surface and mesh dependent constants that will be needed for the error estimates.

Assumption 7.3.16, which is related to the geometry of the surface Γ, excludes strongly folded surfaces. For $0 < r < R \leq \operatorname{diam} \Gamma$ and $\mathbf{x} \in \Gamma$ we define the annular domain $A_{R,r}(\mathbf{x})$ by

$$A_{R,r}(\mathbf{x}) := \left\{ \mathbf{y} \in \mathbb{R}^3 : r < \|\mathbf{x} - \mathbf{y}\| < R \right\}.$$

Assumption 7.3.16. *There exist constants $C_\Gamma, D_\Gamma > 0$ such that the two-dimensional surface measure of the intersection $\Gamma_{R,r}(\mathbf{x}) := \Gamma \cap A_{R,r}(\mathbf{x})$ satisfies the estimate*

$$|\Gamma_{R,r}(\mathbf{x})| \leq C_\Gamma \left(R^2 - r^2 \right).$$

for all $\mathbf{x} \in \Gamma$ and $0 < r < R < D_\Gamma$.

Assumption 7.3.16 implies the estimate

$$|\omega| \leq C_\Gamma (\operatorname{diam} \omega)^2$$

for all subsets $\omega \subset \Gamma$ with the (Euclidean) diameter $\operatorname{diam} \omega$.

For a given surface mesh $\mathcal{G}$ the constant $q_\mathcal{G}$ indicates the quasi-uniformity of the mesh (see Definition 4.1.13) and $\kappa_\mathcal{G}$ describes the shape-regularity of the elements (see Definition 4.1.12). The minimal constant in the inverse estimate (see Corollary 4.4.6)

$$\|u\|_{L^\infty(\tau)} \leq C h_\tau^{-1} \|u\|_{L^2(\tau)} \qquad \forall \tau \in \mathcal{G}, \quad \forall u \in S \tag{7.69}$$

is denoted by C_{inv}. For $s \geq 0$ we need the auxiliary function $C_s : \mathbb{R}_{>0} \to \mathbb{R}_{>0}$

$$C_s(h) := h^{-2} \begin{cases} 1 & 0 \leq s < 2, \\ 1 + |\log h| & s = 2, \\ h^{2-s} & s > 2. \end{cases} \tag{7.70}$$

Convention 7.3.17. *In general, we will assume for the following theorems that either Assumption 4.3.17 or Assumption 4.3.18 is satisfied. The constants in the following theorems usually depend on the polynomial degree in $S_\mathcal{G}^p$ and on the mesh $\mathcal{G}$ via the constants $\kappa_\mathcal{G}$, $q_\mathcal{G}$, C_{inv}. In the case of curved surfaces they also depend on the derivatives of the global transformations χ, χ^{-1}, even if this is not explicitly stated.*

Theorem 7.3.18. *Let Assumption 7.1.23 be satisfied with $s \geq 0$ and $\gamma \in]0,1[$ and let Assumption 7.3.16 hold. For the cluster approximation $\tilde{b}$ of the sesquilinear form b of the integral operator K the following holds with the notation introduced above:*

$$\left| b(u,v) - \tilde{b}(u,v) \right| \leq C\varepsilon \|u\|_{0,\Gamma} \|v\|_{0,\Gamma} \qquad \text{for all } u, v \in S$$

with

$$\varepsilon := \gamma^m C_s(h) \tag{7.71}$$

and a constant C that depends only on s, η, Γ, the constants from Assumption 7.3.16 and the parameters described in Convention 7.3.17.

Proof. We use the notation from Assumption 7.1.23 and Definition 7.1.11 and by applying (7.19) we thus obtain the estimate

$$\begin{aligned} |E(u,v)| := \left| b(u,v) - \tilde{b}(u,v) \right| &\leq \sum_{b=(c,\sigma)\in P^{far}} \int_{\Gamma_c \times \Gamma_\sigma} |v(\mathbf{x})| \, \Big| k(\mathbf{x},\mathbf{y}) \\ &\quad - k_b^{(m)}(\mathbf{x},\mathbf{y}) \Big| \, |u(\mathbf{y})| \, ds_{\mathbf{y}} ds_{\mathbf{x}} \\ &\leq C\gamma^m \sum_{b=(c,\sigma)\in P^{far}} \int_{\Gamma_c \times \Gamma_\sigma} \frac{|v(\mathbf{x})| \, |u(\mathbf{y})|}{\operatorname{dist}^s(c,\sigma)} ds_{\mathbf{y}} ds_{\mathbf{x}}. \end{aligned}$$

For all cluster pairs $(c,\sigma) \in P^{far}$ the admissibility condition implies the estimate

$$\operatorname{dist}(c,\sigma) = \inf_{(\mathbf{x},\mathbf{y})\in Q_c \times Q_\sigma} \|\mathbf{x} - \mathbf{y}\| \geq \eta^{-1} \max\{\operatorname{diam} c, \operatorname{diam} \sigma\} \geq (q_{\mathcal{G}}\eta)^{-1} h$$

with mesh width $h = h(\mathcal{G})$. The geometric far-field blocks are thus contained in

$$(\Gamma \times \Gamma)^{far} := \left\{ \mathbf{x}, \mathbf{y} \in \Gamma : \|\mathbf{x} - \mathbf{y}\| \geq (q_{\mathcal{G}}\eta)^{-1} h \right\}.$$

Since

$$\begin{aligned} \operatorname{dist}(c,\sigma) = \operatorname{dist}(Q_c, Q_\sigma) &\geq \|\mathbf{x} - \mathbf{y}\| - \operatorname{diam} Q_c - \operatorname{diam} Q_\sigma \\ &\geq \|\mathbf{x} - \mathbf{y}\| - 2\eta \operatorname{dist}(c,\sigma) \end{aligned}$$

for all $(\mathbf{x},\mathbf{y}) \in Q_c \times Q_\sigma$ we obtain the estimate

$$\operatorname{dist}(c,\sigma) \geq \frac{1}{1+2\eta} \|\mathbf{x} - \mathbf{y}\|.$$

Therefore $s \geq 0$ yields the estimate

$$|E(u,v)| \leq C(1+2\eta)^s \gamma^m \|u\|_{L^\infty(\Gamma)} \|v\|_{L^\infty(\Gamma)} \int_{(\Gamma\times\Gamma)^{far}} \frac{1}{\|\mathbf{x} - \mathbf{y}\|^s} ds_{\mathbf{x}} ds_{\mathbf{y}}.$$

In Corollary 7.3.19 we show that

$$\int_{(\Gamma\times\Gamma)^{far}} \frac{1}{\|\mathbf{x}-\mathbf{y}\|^s} ds_{\mathbf{x}} ds_{\mathbf{y}} \leq c_s h^2 C_s(h)$$

with $C_s(h)$ as in (7.70) and a constant c_s that depends only on s, Γ and η.

For a boundary element function u there exists a panel $\tau \in \mathcal{G}$ such that

$$\|u\|_{L^\infty(\Gamma)} = \|u\|_{L^\infty(\tau)} \leq C_{\text{inv}} h_\tau^{-1} \|u\|_{L^2(\tau)} \leq C_{\text{inv}} q_{\mathcal{G}} h^{-1} \|u\|_{L^2(\Gamma)}.$$

Combining this we obtain

$$|E(u,v)| \leq \widehat{C} C_s(h) \gamma^m \|u\|_{L^2(\Gamma)} \|v\|_{L^2(\Gamma)}$$

with

$$\widehat{C} := c_s C_{\text{inv}}^2 q_{\mathcal{G}}^2 (1+2\eta)^s.$$

□

Corollary 7.3.19. *Let Assumption 7.3.16 be satisfied. Then we have*

$$\int_{(\Gamma\times\Gamma)^{far}} \frac{1}{\|\mathbf{x}-\mathbf{y}\|^s} ds_{\mathbf{x}} ds_{\mathbf{y}} \leq c_s h^2 C_s(h)$$

with $C_s(h)$ as in (7.70) and a constant c_s that depends only on s, Γ and η as well as the constants from Assumption 7.3.16.

Proof. (a) Let $0 \leq s < 2$ For this case the statement follows from the fact that $\|\mathbf{x}-\mathbf{y}\|^{-s}$ is improperly integrable:

$$\int_{(\Gamma\times\Gamma)^{far}} \frac{1}{\|\mathbf{x}-\mathbf{y}\|^s} ds_{\mathbf{y}} ds_{\mathbf{x}} \leq \int_{\Gamma\times\Gamma} \frac{1}{\|\mathbf{x}-\mathbf{y}\|^s} ds_{\mathbf{y}} ds_{\mathbf{x}} < \infty.$$

(b) Let $s \geq 2$ and C_Γ, D_Γ as in Assumption 7.3.16.

We set $\delta := (q_{\mathcal{G}}\eta)^{-1} h$ and $n := \lceil D_\Gamma/\delta \rceil$, where $\lceil a \rceil$ denotes the smallest integer larger than or equal to a. For $\mathbf{x} \in \Gamma$ we set $\Gamma_i(\mathbf{x}) := \Gamma \cap A_{(i+1)\delta, i\delta}(\mathbf{x})$, $1 \leq i \leq n-1$, and $\Gamma_n(\mathbf{x}) := \Gamma \cap A_{\operatorname{diam}\Gamma, n\delta}(\mathbf{x})$.

Then with Assumption 7.3.16 we have

$$\begin{aligned}
I_s &:= \int_{(\Gamma\times\Gamma)^{far}} \|\mathbf{x}-\mathbf{y}\|^{-s} ds_{\mathbf{y}} = \int_\Gamma \sum_{i=1}^{n-1} \int_{\Gamma_i(\mathbf{x})} \|\mathbf{x}-\mathbf{y}\|^{-s} ds_{\mathbf{y}} ds_{\mathbf{x}} \\
&\quad + \int_\Gamma \int_{\Gamma_n(\mathbf{x})} \|\mathbf{x}-\mathbf{y}\|^{-s} ds_{\mathbf{y}} ds_{\mathbf{x}} \\
&\leq \int_\Gamma \sum_{i=1}^{n-1} |\Gamma_i(\mathbf{x})| (i\delta)^{-s} ds_{\mathbf{x}} + D_\Gamma^{-s} |\Gamma|^2
\end{aligned}$$

$$\leq C_\Gamma \int_\Gamma \sum_{i=1}^{n-1} \frac{((i+1)\,\delta)^2 - (i\delta)^2}{(i\delta)^s} ds_{\mathbf{x}} + D_\Gamma^{-s} |\Gamma|^2$$

$$= C_\Gamma \delta^{2-s} |\Gamma| \sum_{i=1}^{n-1} \frac{2i+1}{i^s} + D_\Gamma^{-s} |\Gamma|^2 .$$

For $s = 2$ we have

$$I_2 \leq C_\Gamma |\Gamma| \left(2 \log(n-1) + 2 + \frac{\pi^2}{6} \right) + D_\Gamma^{-2} |\Gamma|^2 .$$

Since D_Γ depends only on Γ, it follows for a sufficiently large $n = \lceil q_{\mathcal{G}} \eta D_\Gamma / h \rceil$ or a sufficiently small h that the statement

$$I_2 \leq C\,(1 + \log n) \leq \widetilde{C}\,(1 + |\log h|)$$

is satisfied. For $s > 2$ we have $\sum_{i=1}^{n-1} \frac{2i+1}{i^s} < \sum_{i=1}^{\infty} \frac{2i+1}{i^s} < \infty$, and the assertion follows from $\delta^{2-s} \leq C h^{2-s}$. □

7.3.2.2 L^2-Estimates for the Cluster Method with Integration by Parts

The estimate in Corollary 7.3.19 is too pessimistic for hypersingular kernel functions ($s = 3$) if we regularize by means of integration by parts (see Theorem 3.3.22). Applying integration by parts to the hypersingular integral operator for the general elliptic boundary value problem yields the representation

$$b(u,v) = b_0(u,v) + b_1(u,v),$$

where $b_1(u,v)$ contains a kernel function that satisfies

$$|k_1(\mathbf{x},\mathbf{y})| \leq C \|\mathbf{x}-\mathbf{y}\|^{-\alpha} \qquad \text{for all } \mathbf{x},\mathbf{y} \in \Gamma \text{ almost everywhere}$$

with $\alpha \leq 1$ and where $b_0(u,v)$ has the form

$$\int_{\Gamma\times\Gamma} k_0(\mathbf{x},\mathbf{y})\, D_1[u](\mathbf{y})\, D_2[\overline{v}](\mathbf{x})\, ds_{\mathbf{y}} ds_{\mathbf{x}}.$$

D_1 and D_2 denote differential operators in the tangent plane of Γ with an order equal to or smaller than one, i.e.,

$$\|D_1 u\|_{L^2(\Gamma)} \leq C \|u\|_{H^1(\Gamma)} \quad \text{and} \quad \|D_2 v\|_{L^2(\Gamma)} \leq C \|v\|_{H^1(\Gamma)} .$$

The kernel function of b_0 is weakly singular

$$|k_0(\mathbf{x},\mathbf{y})| \leq C \|\mathbf{x}-\mathbf{y}\|^{-1} \qquad \text{for all } \mathbf{x},\mathbf{y} \in \Gamma \text{ with } \mathbf{x} \neq \mathbf{y}.$$

The cluster method can now be applied separately to each of the sesquilinear forms b_0 and b_1 so that Theorem 7.3.18 can be applied with slight modifications. In the error estimate for the approximation of the bilinear form b_1, s in (7.71) has to be replaced by α. The error estimate for the cluster approximation of b_0 reads:

$$\left|b_0(u,v) - \tilde{b}_0(u,v)\right| \leq C\varepsilon \|u\|_{H^1(\Gamma)} \|v\|_{H^1(\Gamma)} \qquad \text{for all } u,v \in H^1(\Gamma),$$

where $s = 1$ has to be chosen in the definition (7.71) of ε.

7.3.2.3 Stability and Consistency of the Cluster Method

We assume that the sesquilinear form $b : H^s(\Gamma) \times H^s(\Gamma) \to \mathbb{C}$ in (5.68) is continuous, injective and coercive with an $s \in \left\{-\frac{1}{2}, 0, \frac{1}{2}\right\}$.

For a given right-hand side $F \in H^{-s}(\Gamma)$ we are seeking $u \in H^s(\Gamma)$ such that

$$b(u,v) = F(v) \qquad \forall v \in H^s(\Gamma).$$

The conforming boundary element space $S \subset H^s(\Gamma)$ is defined on a surface mesh of Γ with a local polynomial degree $p \in \mathbb{N}_0$ (see Chap. 4). The Galerkin solution $u_S \in S$ satisfies

$$b(u_S,v) = F(v) \qquad \forall v \in S.$$

Since the influence of the quadrature error was already studied in Chap. 5, we assume in this analysis of the error introduced by the cluster method that the near-field integrals are computed exactly. Then the cluster method defines a perturbed sesquilinear form $\tilde{b} : S \times S \to \mathbb{C}$. This leads to the perturbed Galerkin solution $\tilde{u}_S \in S$

$$\tilde{b}(\tilde{u}_S,v) = F(v) \qquad \forall v \in S. \tag{7.72}$$

In this section we will estimate the error $u - \tilde{u}_S$ which is introduced by the cluster method.

We will assume that the right-hand side in (7.72) is computed without any numerical errors, i.e., the components of the vectors $F(b_j)$ in the right-hand side of the linear system of equations are evaluated exactly. If this is not the case the influence of this additional error can be analyzed as in Sect. 5.3.3.

If the surface is sufficiently smooth and the solution of the problem (5.67) is sufficiently regular the following asymptotic estimate for the unperturbed Galerkin discretization error holds (see Sect. 4.3):

$$\|u - u_S\|_{H^s(\Gamma)} \leq Ch^{p+1-s} \|u\|_{H^{p+1}(\Gamma)}.$$

In order to achieve this order of convergence for the perturbed approximation as well, we will use Theorem 4.2.11.

Let $s_- := \min\{s, 0\}$ and $s_+ := \max\{s, 0\}$. The stability and consistency conditions (4.151) and (4.153) follow from Theorem 7.3.18 and the inverse estimate, and they are given by

$$\left|b(u,v) - \tilde{b}(u,v)\right| \leq C C_s(h)\, \gamma^m h^{s_-} \|u\|_{H^{s_+}(\Gamma)} \|v\|_{H^s(\Gamma)},$$
$$\left|b(u,v) - \tilde{b}(u,v)\right| \leq C C_s(h)\, \gamma^m h^{2s_-} \|u\|_{H^s(\Gamma)} \|v\|_{H^s(\Gamma)}$$

for all $u, v \in S$. Therefore the conditions

$$C_s(h)\, \gamma^m h^{2s_-} \overset{h\to 0}{\to} 0 \qquad \text{and} \qquad C_s(h)\, \gamma^m h^{s_-} \leq C h^{p+1-s}$$

are sufficient for the stability and consistency of the cluster method. The following table gives the required orders of expansion for $s = \{-1/2, 0, 1/2\}$, corresponding to weakly singular, Cauchy singular, and hypersingular boundary integral equations.

	$s = -\frac{1}{2}$	$s = 0$	$s = 1/2$
m	$(p+4)\left\|\frac{\log h}{\log\gamma}\right\|$	$\frac{(p+3)\|\log h\| + \log(1+\|\log h\|)}{\|\log\gamma\|}$	$(p+7/2)\left\|\frac{\log h}{\log\gamma}\right\|$

(7.73)

Theorem 7.3.20. *Let the conditions from Theorem 7.3.18 be satisfied. Let the sesquilinear form $b : H^s(\Gamma) \times H^s(\Gamma)$ of the boundary integral operator be continuous, coercive and injective for an $s \in \{-1/2, 0, 1/2\}$. Furthermore, let the order of expansion be chosen as given by Table (7.73). Then the Galerkin solution with the cluster method converges with the same order as the exact Galerkin solution.*

For hypersingular kernel functions the order of expansion refers to the direct representation. If we apply integration by parts for the purpose of regularization, the order of singularity is reduced (see Sect. 7.3.2.2). The order of expansion m is reduced accordingly. We refer to [107] for details.

7.4 The Complexity of the Cluster Method

In this subsection we will show that for all kernel functions $k(\mathbf{x}, \mathbf{y})$ which satisfy Assumption 7.1.23 the cluster method reduces the storage and computing complexity from $O(N^2)$ for the matrix-oriented representation of the Galerkin method to $O(N \log^\kappa N)$. We note that the class of kernel functions that we consider here is considerably more general than the $(4\pi \|\mathbf{x} - \mathbf{y}\|)^{-1}$-kernel for the Laplace equation. For this specific kernel, complexity estimates are proved in, e.g., [111].

The estimate consists of two parts. First we will show that, if we impose suitable conditions on the surface and the surface mesh, the number of blocks in the covering is of the order N. Then the algorithmic complexity of the cluster method is estimated per block and cluster, from which we determine the overall complexity.

7.4.1 Number of Clusters and Blocks

We will estimate the number of clusters and blocks for quasi-uniform meshes and balanced cluster trees. Details can be found in the following assumption. The level of a cluster and the maximal number of levels within a cluster tree are defined as in (7.52).

Assumption 7.4.1. *1. Let* $i \in \mathcal{I} \iff \{i\} \in \text{Leaves}$.
2. *The mesh width* h *of the mesh* $\mathcal{G}$ *and the maximal number of levels* $\ell_{\max}$ *in the cluster tree satisfy*

$$2^{\ell_{\max}} h \geq C_{\mathrm{r}} \tag{7.74}$$

with a constant C_{r} *that does not depend on* h *or* $\ell_{\max}$.
3. *Every cluster* $c \in T \backslash \text{Leaves}$ *has at least four sons.*
4. *Every block* $b = (c, s) \in P$ *satisfies* $\text{level}(c) = \text{level}(s)$.

Remark 7.4.2. *Algorithm 7.1.8 does not necessarily produce cluster trees that satisfy Assumptions 7.4.1.(2) and 7.4.1.(3). This problem can be easily resolved by using a post-processing algorithm to redefine the set of sons for all clusters* $\sigma \in T$ *with* $\sharp\Sigma(\sigma) < 4$ *according to* $\Sigma(\sigma) := \bigcup_{s \in \Sigma(\sigma)} \Sigma(s)$.

Assumptions 7.4.1.(1) and 7.4.1.(2) simplify the complexity estimates for the cluster method. We recommend using Algorithm 7.1.8 in an unchanged form for the numerical computations. An algorithm for the generation of a cluster tree that satisfies Assumptions 7.4.1(1)–(4) is described in [199].

The constants in the following assumption depend on the polynomial degree of the boundary element space while we do not track this dependence explicitly.

Assumption 7.4.3. *There exist constants* c_p *and* C_p *such that*

$$c_p \leq (\sharp\mathcal{G}) / (\sharp\mathcal{I}) \leq C_p.$$

First we will derive estimates for the number of clusters in the cluster tree.

Lemma 7.4.4. *Let* $N := \sharp\mathcal{I}$ *and let Assumption 7.4.1(1) hold. Then*

$$\sharp T \leq \frac{4}{3} N.$$

Proof. Let T_ℓ be the set of all clusters with a tree depth of $\ell \in \mathbb{N}_0$ [see (7.52)]

$$T_\ell := \{c \in T : \text{level}(c) = \ell\}$$

We have $\sharp T_{\ell_{\max}} = \sharp\mathcal{I} = N$ and, thus, recursively

$$\sharp T_{\ell-1} \leq \frac{1}{4} \sharp T_\ell.$$

By summing we obtain

$$N\sum_{i=0}^{\ell_{\max}} 4^{-i} \le N\sum_{i=0}^{\infty} 4^{-i} = \frac{4}{3}N.$$

□

In order to estimate the number of blocks in P we use the admissibility condition and a result concerning the size of the clusters $c \in T_\ell$. The constant that describes the quasi-uniformity from Definition 4.1.13 is again denoted by $q_{\mathcal{G}}$ and the minimal constant in (7.69) is denoted by C_{inv}.

Corollary 7.4.5. *Let Assumptions 7.3.16, 7.4.1 and 7.4.3 hold. Let the cluster tree be generated by the procedure* **generate_cluster_tree** *(see Algorithm 7.1.8). Then for* $\sigma \in T_\ell$

$$\begin{aligned} \operatorname{diam}\sigma &\le 2^{-\ell}\operatorname{diam}\Gamma + 4h, \\ |\Gamma_\sigma| &\ge (q_{\mathcal{G}} C_{\text{inv}})^{-2}\, 4^{\ell_{\max}-\ell-1} h^2, \end{aligned} \tag{7.75}$$

where diam $(\cdot)$ *again denotes the Euclidean diameter.*

Proof. (i) Let $\sigma \in T\backslash$Leaves be a cluster with cluster box Q_σ and let $Q \in \Sigma(Q_\sigma)$ be a congruent sub-cuboid [see (7.4)]. The maximal edge length of Q_σ is denoted by L. Q is associated with the cluster $s = s(Q)$, which contains all the indices $i \in \mathcal{I}$ with $\mathbf{M}_{\{i\}} \in Q$ (see Definition 7.1.6). The support of the basis function corresponding to the degree of freedom i satisfies $\operatorname{diam}(\operatorname{supp} b_i) \le 2h$. Therefore the maximal edge length of the cluster box $Q_{\{i\}}$ is bounded from above by $2h$. The union $\bigcup_{i\in s} Q_{\{i\}}$ is thus contained in a cuboid with the center of mass $\mathbf{M}_s$ and the maximal edge length $L/2 + 2h$. Hence it follows that

$$\operatorname{diam} s \le \frac{1}{2}L + 2h. \tag{7.76}$$

Clearly we have for the root $\mathcal{I} \in T$

$$\operatorname{diam}\mathcal{I} = \operatorname{diam}\Gamma.$$

With (7.76) we thus conclude for $\sigma \in T_\ell$ by induction that

$$\operatorname{diam}\sigma \le 2^{-\ell}\operatorname{diam}\Gamma + 2h\sum_{i=0}^{\ell-1} 2^{-\ell} \le 2^{-\ell}\operatorname{diam}\Gamma + 4h.$$

(ii) By Assumption 7.4.1.(3) every cluster $\sigma \in T\backslash$Leaves has at least four sons. With $\ell := \operatorname{level}(\sigma)$ it thus follows that

$$\sharp\sigma \ge 4^{\ell_{\max}-\ell}. \tag{7.77}$$

Since every panel has at most 4 vertices (for triangles only 3 vertices), Γ_σ contains at least $\sharp\sigma/4$ different panels. We obtain a lower bound for the surface of a panel with inequality (7.69) by choosing $u \equiv 1$. This yields

$$|\tau| = \|1\|_{L^2(\tau)}^2 \geq h_\tau^2/C_{\mathrm{inv}}^2 \|1\|_{L^\infty(\tau)}^2 = (h_\tau/C_{\mathrm{inv}})^2 \geq (q_{\mathcal{G}} C_{\mathrm{inv}})^{-2} h^2 \qquad \forall \tau \in \mathcal{G}.$$

It follows that

$$|\Gamma_\sigma| \geq (q_{\mathcal{G}} C_{\mathrm{inv}})^{-2} 4^{\ell_{\max}-\ell-1} h^2. \tag{7.78}$$

□

Corollary 7.4.6. *Let the conditions from Corollary 7.4.5 be satisfied. Then there exists a constant C that depends only on $q_{\mathcal{G}}$, C_{inv}, C_p and* diam Γ *such that*

$$\operatorname{diam} \sigma \leq C 2^{-\ell} \tag{7.79}$$

for all $\sigma \in T_\ell$.

Proof. Estimate (7.75) with $\sigma = \mathcal{I}$ implies that

$$|\Gamma| \geq (q_{\mathcal{G}} C_{\mathrm{inv}})^{-2} 4^{\ell_{\max}-1} h^2$$

and it follows that

$$h \leq 2 q_{\mathcal{G}} C_{\mathrm{inv}} 2^{-\ell_{\max}} \operatorname{diam} \Gamma. \tag{7.80}$$

Therefore the constant in (7.79) can be chosen as $C = (1 + 8 q_{\mathcal{G}} C_{\mathrm{inv}}) \operatorname{diam} \Gamma$. □

With this corollary the number of blocks in P can be estimated.

Theorem 7.4.7. *Let the assumptions from Corollary 7.4.5 be satisfied. Then*

$$\sharp P \leq CN,$$

where C depends only on η, C_r, C_Γ, $q_{\mathcal{G}}$, C_{inv}, C_p and diam Γ.

Proof. Let $b = (\sigma, s) \in P$. The construction of the covering P of $\mathcal{I} \times \mathcal{I}$ by means of the procedure **divide** implies that the father $\tilde{b} = (\tilde{\sigma}, \tilde{s}) :=$ *father* (b) is not contained in P^{far} and that the admissibility condition for $\tilde{b}$ is violated. Therefore we have

$$\eta \operatorname{dist}(\tilde{\sigma}, \tilde{s}) < \max\{\operatorname{diam} \tilde{\sigma}, \operatorname{diam} \tilde{s}\}. \tag{7.81}$$

It follows from $\sigma \subset \tilde{\sigma}$ that $\Gamma_\sigma \subset \Gamma_{\tilde{\sigma}}$, and combined with (7.81) we obtain

$$\operatorname{dist}(\sigma, s) \leq \operatorname{dist}(\tilde{\sigma}, \tilde{s}) + \operatorname{diam} \tilde{\sigma} + \operatorname{diam} \tilde{s} \leq \left(\frac{1}{\eta} + 2\right) \max\{\operatorname{diam} \tilde{\sigma}, \operatorname{diam} \tilde{s}\}.$$

We set $\ell :=$ level $(\sigma) =$ level (s). Corollary 7.4.6 implies the estimate

$$\operatorname{dist}(\sigma, s) \leq C\,(1/\eta + 2)\, 2^{-\ell}. \tag{7.82}$$

For all $(\mathbf{x},\mathbf{y}) \in (\Gamma_\sigma, \Gamma_s)$ we have

$$\|\mathbf{x}-\mathbf{y}\| \leq \operatorname{dist}(\sigma, s) + \operatorname{diam}\sigma + \operatorname{diam} s,$$

and it follows from (7.82) that

$$\|\mathbf{x}-\mathbf{y}\| \leq C\,(1/\eta + 2)\,2^{-\ell} + 2C2^{-\ell} =: C_1 2^{-\ell}.$$

This estimate means that the subset of $\Gamma \times \Gamma$ which is covered by the blocks $(\sigma, s) \in P_\ell$ (more precisely by $(\Gamma_\sigma \times \Gamma_s)$) is contained in

$$(\Gamma \times \Gamma)_\ell := \left\{(\mathbf{x},\mathbf{y}) : \|\mathbf{x}-\mathbf{y}\| \leq C_1 2^{-\ell}\right\}. \tag{7.83}$$

For $\mathbf{x} \in \Gamma$ we set

$$\Gamma_\ell(\mathbf{x}) := \left\{\mathbf{y} \in \Gamma : \|\mathbf{x}-\mathbf{y}\| \leq C_1 2^{-\ell}\right\}.$$

Assumption 7.3.16 yields

$$|\Gamma_\ell(\mathbf{x})| \leq C\,4^{-\ell}$$

and it follows that

$$|(\Gamma \times \Gamma)_\ell| \leq C\,|\Gamma|\,4^{-\ell}. \tag{7.84}$$

With the help of (7.75) the surface of the block can be estimated by

$$|(\Gamma_\sigma, \Gamma_s)| = |\Gamma_\sigma|\,|\Gamma_s| \geq 16^{\ell_{\max}-\ell-1}\left(h/\left(q_{\mathcal{G}} C_{\mathrm{inv}}\right)\right)^4. \tag{7.85}$$

This means that the number of elements P_ℓ [see (7.53)] is bounded by the quotient of the right-hand sides in (7.84) and (7.85):

$$\sharp P_\ell \leq C \frac{4^\ell}{h^4 16^{\ell_{\max}}}. \tag{7.86}$$

If we use (7.74) and sum over all levels we obtain $\sharp P_\ell \leq C_2 4^\ell$ and

$$\sharp P = \sum_{\ell=0}^{\ell_{\max}} \sharp P_\ell \leq C_2 \sum_{\ell=0}^{\ell_{\max}} 4^\ell \leq \frac{C_2}{3} 4^{\ell_{\max}+1}. \tag{7.87}$$

If we then choose $\sigma = \mathcal{I}$ and $\ell = 0$ in Estimate (7.77) we obtain the assertion

$$\sharp P \leq C_3\,(\sharp \mathcal{I}).$$

□

Keeping in mind (7.26) and (7.38) we still need estimates for the cardinalities of the sets $P^{far}(c)$, $c \in T$, and $P^{near}(c)$, $c \in$ Leaves.

Theorem 7.4.8. *Let the conditions from Corollary 7.4.5 be satisfied. Then for all $\sigma \in T$ and $c \in$ Leaves*

$$\sharp P^{far}(\sigma) + \sharp P^{near}(c) \leq C,$$

where C depends only on η, C_r, C_Γ, C_p, $\operatorname{diam}\Gamma$ and the parameters described in Convention 7.3.17.

Proof. Let $\sigma \in T_\ell$ and $P^{far}(\sigma)$ be defined as in (7.7). It follows from (7.83) that all clusters $s \in P^{far}(\sigma)$ (more precisely, the geometric clusters Γ_s) are contained in

$$\Gamma_\ell(\sigma) := \bigcup_{\mathbf{x} \in \Gamma_\sigma} \left\{ \mathbf{y} \in \Gamma : \|\mathbf{x} - \mathbf{y}\| \leq C_1 2^{-\ell} \right\}.$$

Let $\mathbf{M}_\sigma$ be the center of the smallest ball that contains Γ_σ and let r_σ be its radius. Clearly we have with Corollary 7.4.6 that

$$r_\sigma \leq \operatorname{diam}\sigma \leq C 2^{-\ell},$$

and that $\Gamma_\ell(\sigma)$ is contained in a ball B_σ around $\mathbf{M}_\sigma$ with radius $C_2 2^{-\ell} := (C_1 + C) 2^{-\ell}$. The surface measure of the intersection $B_\sigma \cap \Gamma$ can be estimated with Assumption 7.3.16 according to

$$|B_\sigma \cap \Gamma| \leq C C_2^2 4^{-\ell}. \tag{7.88}$$

Assumption 7.4.1(4) implies $P^{far}(\sigma) \subset T_\ell$ and Corollary 7.4.5 gives us that every cluster $s \in P^{far}(\sigma)$ satisfies the estimate

$$|\Gamma_s| \geq 4^{\ell_{\max} - \ell - 1} (h/c)^2. \tag{7.89}$$

The quotient of the right-hand sides in (7.88) and (7.89) constitutes an upper bound for the number $\sharp P^{far}(\sigma)$

$$\sharp P^{far}(\sigma) \leq \frac{C_3}{h^2 4^{\ell_{\max}}}$$

and combined with Assumption 7.4.1(2) we obtain the assertion.

The proof of the estimate for the near-field clusters $P^{near}(c)$, $c \in$ Leaves, uses the same arguments as above and is therefore omitted. □

7.4.2 *The Algorithmic Complexity of the Cluster Method*

In order to estimate the complexity of the cluster method we will make the following simplifying assumptions. For a more general discussion we refer to [109].

Assumption 7.4.9. *Let Assumptions 7.3.16, 7.4.1 and 7.4.3 hold. The cluster approximation is defined by Čebyšev interpolation. The order of expansion is chosen as* $m = O(\log N)$ *[see (7.73)].*

Generating the Cluster Tree

The cluster tree is generated level-by-level by the procedure **generate_cluster_tree** (Algorithm 7.1.8). The set of sons is generated for all clusters belonging to the current level. This results in a total computational complexity of $O(N)$ per level and $O(N\ell_{\max})$ for the entire cluster tree. If we use (7.77) with $\sigma = \mathcal{I}$ we obtain

$$\ell_{\max} \leq \log N / \log 4$$

and $O(N \log N)$ arithmetic operations for the generation of the cluster tree. This requires a memory capacity of $O(N)$ floating-point numbers.

Generating the Covering

The covering P is generated by the procedure **divide** by a recursion over the tree levels. On every level $\ell \geq 1$ all pairs $(\sigma, s) \in T_\ell \times T_\ell$ whose fathers do *not* satisfy the admissibility condition are checked for admissibility. Therefore the associated geometric blocks $(\Gamma_\sigma, \Gamma_s)$ are contained in $(\Gamma \times \Gamma)_\ell$ [see (7.83)] and their number can be estimated as in (7.86). By summing over all levels we obtain [see (7.87)] a computational and memory complexity which is proportional to $O(N)$.

Generating the Shift Coefficients

For the computation of the shift coefficients in the Lagrange representation we use the recursion (7.42) combined with Definitions (7.43) and (7.44)

$$\gamma_{\mu,\nu,\tilde{c}} := a_{\mu_1,\nu_1,\iota_1} a_{\mu_2,\nu_2,\iota_2} a_{\mu_3,\nu_3,\iota_3}.$$

Computing the coefficients a_{i,j,ι_k} for all $1 \leq i \leq m$, $1 \leq j \leq i$ and $1 \leq k \leq 3$ is performed by recursion (7.42) in $O(m^2)$ operations. It is only necessary to compute and store the coefficients a_{i,j,ι_k} in this phase. These coefficients can be multiplied by each other as is necessary during the upwards and downwards paths, for example, for the computation of

$$\widetilde{R}_c^{(\nu)} := \sum_{\tilde{\nu} \in \mathcal{R}_m} \gamma^R_{\nu,\tilde{\nu},\tilde{c}} \widetilde{R}_{\tilde{c}}^{(\tilde{\nu})}.$$

The computational complexity and memory requirements therefore consist of at most $O\left((\sharp T) \times m^2\right) = O\left(N \log^2 N\right)$ operations.

Computing the Expansion Coefficients

For every block $b = (\sigma, s) \in P^{far}$ there exist $O\left(m^6\right)$ expansion coefficients $\kappa_{\mu,\nu}(b)$ that are defined by

$$G\left(\xi^{(\mu)}, \zeta^{(\nu)}\right)_{\mu,\nu \in \mathcal{J}_m}$$

for the interpolation points $\left(\xi^{(\mu)}, \zeta^{(\nu)}\right)_{\mu,\nu \in \mathcal{J}_m} \subset Q_\sigma \times Q_s$. The overall computational complexity for the computation of all far-field coefficients therefore consists of $O\left(\sharp P^{far} \times m^6\right) = O\left(N \log^6 N\right)$ operations and requires memory $O\left(\sharp P^{far} \times m^6\right) = O\left(N \log^6 N\right)$ floating-point numbers.

Computing the Basis Far-Field Coefficients

We will restrict ourselves here to determining the complexity for the computation of the coefficients

$$L_{\{i\}}^{(\mu)} = \int_{\operatorname{supp} b_i} b_i(\mathbf{x})\, \Phi_{\{i\}}^{(\mu)}(\mathbf{x})\, ds_{\mathbf{x}}.$$

The coefficients $R_{\{i\}}^{(\nu)}$ can be computed with the same computational cost.

We achieve this by decomposing $\operatorname{supp} b_i$ into $O(1)$ many panels and by applying the quadrature described in (7.48). Per panel the integrand has to be evaluated in $O\left(m^2\right)$ interpolation nodes. For any panel $\tau \subset \operatorname{supp} b_i$ we denote the associated set of quadrature points by $\left(\zeta^{(\nu)}\right)_{\nu \in I_m}$. In order to evaluate the integrals $\left(L_{\{i\}}^{(\mu)}\right)_{\mu \in \mathcal{L}_m}$ we need to evaluate $\left(\Phi_{\{i\}}^{(\mu)}\left(\zeta^{(\nu)}\right)\right)_{\substack{\mu \in \mathcal{L}_m \\ \nu \in I_m}}$. We use the Algorithm **evaluate_D**L_c to evaluate the expansion functions $\Phi_{\{i\}}^{(\mu)}$. This requires a computational cost of $O\left(m^3\right)$ operations per point $\zeta^{(\nu)}$ and a cost of $O\left(m^5\right)$ operations for all points $\left(\zeta^{(\nu)}\right)_{\nu \in I_m}$. Analogously, the evaluation of the sum in (7.48) for all $\nu \in \mathcal{L}_m$ requires $O\left(m^5\right)$ arithmetic operations by assuming precomputed values $\left(\Phi_{\{i\}}^{(\mu)}\left(\zeta^{(\nu)}\right)\right)_{\substack{\mu \in \mathcal{L}_m \\ \nu \in I_m}}$. Overall, this results in a computational complexity of $O\left((\sharp \mathcal{I}) \times m^5\right) = O\left(N \log^5 N\right)$ operations.

Computing the Near-Field Matrix

We have for the number of near-field matrix elements

$$\sharp P^{near} \leq \sharp P \leq CN.$$

In Chap. 5 quadrature methods were developed that allow for an approximation of the near-field matrix entries with the required precision in $O\left(\log^4 N\right)$ operations per element. The total cost for the computation of the near-field matrix therefore consists of $O\left(N \log^4 N\right)$ operations and the memory capacity in a sparse matrix representation consists of $O\left(N\right)$ floating-point numbers.

All the computations that we have discussed so far only need to be done once during a preparatory phase. The following steps have to be performed once for every matrix-vector multiplication.

Evaluating the Upwards Recursion

Per cluster the operation

$$\forall \nu \in \mathcal{R}_m: \quad \widetilde{R}_c^{(\nu)} = \sum_{\tilde{\nu} \in \mathcal{R}_m} \gamma^R_{\nu,\tilde{\nu},\tilde{c}} \widetilde{R}_{\tilde{c}}^{(\tilde{\nu})}$$

has to be evaluated. Since $\sharp\mathcal{R}_m = O\left(m^3\right)$ this yields an overall complexity of $O\left(\sharp T \times m^6\right) = O\left(N \log^6 N\right)$. Storing all coefficients $\widetilde{R}_c^{(\nu)}$ requires storage of $O\left(N \log^3 N\right)$ floating-point numbers.

Evaluating the Cluster–Cluster Coupling

Per block the expression

$$B_c^{(\nu)} = \sum_{\sigma \in P^{far}(c)} \sum_{\mu \in \mathcal{R}_m(\nu)} \kappa_{\nu,\mu}\left(b\right) \widetilde{R}_\sigma^{(\mu)}$$

has to be evaluated. For every $c \in T$ Theorem 7.4.8 gives us the estimate $\sharp P^{far}\left(c\right) = O\left(1\right)$. For $\nu \in \mathcal{L}_m$ we have for the Čebyšev interpolation the equality

$$\sharp\mathcal{R}_m\left(\nu\right) = \sharp\mathcal{R}_m = \sharp\mathcal{J}_m = m^3.$$

This yields an overall complexity of $O\left(\left(\sharp T\right) \times 1 \times m^6\right) = O\left(N \log^6 N\right)$ operations. Storing all coefficients $\left(B_c^{(\nu)}\right)_{\substack{c \in T\\ \nu \in \mathcal{L}_m}}$ requires $O\left(N \log^3 N\right)$ floating-point numbers.

Evaluating the Downwards Recursion

The complexity for the downwards recursion is the same as for the upwards recursion and therefore consists of $O\left(N\log^6 N\right)$ operations and requires a memory capacity of $O\left(N\log^3 N\right)$ floating-point numbers.

Evaluating the Matrix-Vector Product

The evaluation of the matrix-vector multiplication by means of the near-field matrix and the precalculated coefficients $L_{\{i\}}^{(\mu)}$ and $\widetilde{B}_{\{i\}}^{(\mu)}$ is given by

$$\forall i \in \mathcal{I}: \quad \mathbf{v}_i := \sum_{\{j\}\in P^{near}(\{i\})} (\mathbf{K}_{near})_{i,j}\,\mathbf{u}_j + \sum_{\mu\in\mathcal{L}_m} L_{\{i\}}^{(\mu)}\widetilde{B}_{\{i\}}^{(\mu)}.$$

For every $i \in \mathcal{I}$ we have $\sharp P^{near}(\{i\}) = O(1)$ (see Theorem 7.4.8). This yields an overall complexity of $O\left(\sharp\mathcal{I} \times \left(1+m^3\right)\right) = O\left(N\log^3 N\right)$ operations for this phase. Storing the resulting vector $\mathbf{v}$ requires space for N floating-point numbers.

Theorem 7.4.10. *Let Assumption 7.4.9 be satisfied. Then the computational cost of the cluster method for the approximation of a matrix-vector multiplication is given by $O\left(N\log^6 N\right)$ operations and the memory required is given by $O\left(N\log^6 N\right)$ floating-point numbers.*

For special kernel functions (Laplace/Helmholtz) the asymptotic complexity can be reduced by choosing special approximation systems (see Sect. 7.1.3.2). For a detailed description of these modifications we refer to [77, 78, 110, 111, 194].

Remark 7.4.11. *The estimates that we have presented here illustrate the reduction in asymptotic complexity of the cluster method in comparison with the usual matrix oriented representations. Whether this asymptotic behavior becomes evident for problems with a size that is more common in practical applications, i.e., $N = 10^3 - 2\times 10^4$, strongly depends on the efficiency of the implementation. Numerical experiments have shown that the reductions in computational time and memory capacity become evident from about $N = 500$ for standard problems (Laplace) (see [122, 148, 152, 198, 205]).*

7.5 Cluster Method for Collocation Methods

The discretization of a boundary integral operator by means of the collocation method requires a boundary element space S of dimension $N = \dim S$ and a set of collocation points $\mathcal{I} \subset \Gamma$ with $N = \sharp\mathcal{I}$ (see Remark 4.1.26). We assume that the set $\mathcal{I}$ is chosen in such a way that the interpolation problem:

$$\text{Find } u \in S \text{ such that } u(\mathbf{x}) = w_{\mathbf{x}} \qquad \forall \mathbf{x} \in \mathcal{I}$$

has a unique solution for all mesh functions $\mathbf{w} = (w_{\mathbf{x}})_{\mathbf{x}\in\mathcal{I}} \in \mathbb{C}^{\mathcal{I}}$.

The collocation solution $u_S^{coll} \in S$ for the boundary integral equation

$$Au = f$$

is defined by

$$\left(Au_S^{coll}\right)(\mathbf{x}) = f(\mathbf{x}) \qquad \forall \mathbf{x} \in \mathcal{I}. \tag{7.90}$$

Let A be a boundary integral operator of the form

$$A = \lambda I + K \quad \text{with} \quad (Ku)(\mathbf{x}) = \int_\Gamma k(\mathbf{x}, \mathbf{y})\, u(\mathbf{y})\, ds_{\mathbf{y}} \qquad \forall \mathbf{x} \in \Gamma.$$

In order to ensure that the collocation method is well defined we assume that $AS \subset C^0(\Gamma)$.

The cluster method for Galerkin discretizations can be transferred to collocation methods with small modifications.

For this we define the space V as the span of delta distributions in the collocation points

$$V := \operatorname{span}\{\delta_{\mathbf{x}} : \mathbf{x} \in \mathcal{I}\}$$

and use $(\delta_{\mathbf{x}})_{\mathbf{x}\in\mathcal{I}}$ as a basis for V. Then the collocation method can be written in the "variational form": Find u_S^{coll} such that

$$a(u_S, v) := v\left(Au_S^{coll}\right) = v(f) \qquad \forall v \in V.$$

The cluster algorithm can be easily generalized to handle the bilinear form $a : S \times V \to \mathbb{C}$. It turns out that for the collocation method only the definition of the left far-field coefficients $L_{\{i\}}^{\nu}$ in (7.34) has to be replaced by

$$L_{\{i\}}^{(\nu)} := \Phi_{\{i\}}^{(\nu)}(\mathbf{x}_i) \qquad \forall \nu \in \mathcal{L}_m$$

and the definition of the near-field matrix has to be replaced by

$$(\mathbf{K}_{near})_{i,j} := \begin{cases} \int_\Gamma k(\mathbf{x}_i, \mathbf{y})\, b_j(\mathbf{y})\, ds_{\mathbf{y}} & \text{if } (\{i\}, \{j\}) \in P^{near}, \\ 0 & \text{otherwise.} \end{cases}$$

7.6 Remarks and Additional Results

In the present chapter, we introduced the panel-clustering methods for the fast numerical evaluation of discretized integral operators applied to vectors; this is the key step in the iterative numerical solution of the large, densely populated matrix equations which result from the boundary element discretization of BIEs. Formally, these methods allow the reduction of the matrix-vector multiplication from $O(N^2)$ to log-linear in N complexity, and are the key reason why boundary integral equation based methods are today highly competitive solvers for linear elliptic problems. In [104], a version of the panel-clustering method for optimally refined, anisotropic meshes has been developed.

Naturally, as in previous chapters, problems with the panel clustering and the related, *Fast-Multipole-Method* (FMM) type approach arise with high frequency scattering, since the truncation error introduced in the complexity reduction is not uniformly small with respect to the wave number. Special versions of the FMM are available, however, which do exhibit log-linear scaling complexity independent of the wave number. We refer to [77, 78, 194], for example. All these references have in common a very carefully tailored multipole expansion in different regimes of the frequency.

An alternative approach which has emerged in recent years and which is, at least in key components, related to FMM and panel clustering, is the concept of so-called hierarchical matrices ($\mathcal{H}$- and $\mathcal{H}^2$-Matrices). We refer to [19, 24, 118–121]. Here, no explicit, wave number dependent FMM expansion is required, as optimal, separated approximations of the kernel of the BIE are generated by means of recursive (over all clusters) adaptive low-rank matrix approximations. Another automatic and purely algebraic approach for the sparse representation of non-local operators is the Adaptive Cross Approximation (ACA). We refer to [19, 20, 25, 103] for details.

Chapter 8
p-Parametric Surface Approximation

In practice, the description of the "true" physical surface might be very complicated or even not available as an exact analytic function and has to be approximated by using, e.g., pointwise measurements of the surface or some geometric modelling software. In this chapter, we will address the question how to approximate quite general surfaces in a flexible way by p-parametric boundary elements. Surface approximations for integral equations and their influence on the discretization error have first been studied systematically in [167]. Further papers on this topic are [168], [80, Chap. XIII, Sect. 2], [84], [21], [63, Sect. 1.4].

For the error estimates, we will need some tools from elementary differential geometry. In order to keep this book self contained we have included Sect. 8.4, where these tools are developed.

Readers who are interested in the practical application of p-parametric surface approximations and not so much in the convergence analysis will find in Sect. 8.3 an overview of the required polynomial orders for the p-parametric approximations.

8.1 Discretization of Boundary Integral Equations with Surface Approximations

In this section, we assume that Γ is the surface of a bounded domain $\Omega \subset \mathbb{R}^3$. We emphasize that the concept of p-parametric surface approximation can be applied verbatim to one-dimensional boundaries of two-dimensional domains $\Omega \subset \mathbb{R}^2$ and in higher dimensions as well.

8.1.1 p-Parametric Surface Meshes for Globally Smooth Surfaces

The approximation of Γ starts with the construction of a mesh $\mathcal{G}^{\text{affine}}$ which consists of plane, open triangles with straight edges.

S.A. Sauter and C. Schwab, *Boundary Element Methods*, Springer Series in Computational Mathematics 39, DOI 10.1007/978-3-540-68093-2_8,

Definition 8.1.1 (Affine Surface Approximation). Let Γ be the boundary of a bounded domain Ω. A set $\mathcal{G}^{\text{affine}} := \{\tau_1, \tau_2, \ldots, \tau_N\}$ consisting of plane open triangles with straight edges in $\mathbb{R}^3$ is an affine surface mesh for Γ if Conditions 1–3 are satisfied:

1. $\mathcal{G}^{\text{affine}}$ is a regular boundary element mesh for
$$\Gamma^{\text{affine}} := \bigcup_{\tau \in \mathcal{G}^{\text{affine}}} \overline{\tau}.$$
2. Γ^{affine} interpolates the surface Γ at the vertices of the triangles.
3. The reference mappings $\chi_\tau^{\text{affine}} : \hat{\tau} \to \tau$ are affine.
4. There exists a bi-Lipschitz continuous lifting $\theta^{\text{affine}} : \Gamma \to \Gamma^{\text{affine}}$, i.e., there exist constants $c^{\text{affine}}, C^{\text{affine}} > 0$ such that
$$c^{\text{affine}} \left\| \theta^{\text{affine}}(\mathbf{x}) - \theta^{\text{affine}}(\mathbf{y}) \right\| \leq \|\mathbf{x} - \mathbf{y}\| \leq C^{\text{affine}} \left\| \theta^{\text{affine}}(\mathbf{x}) - \theta^{\text{affine}}\mathbf{y} \right\| \quad \forall \mathbf{x}, \mathbf{y} \in \Gamma. \tag{8.1}$$

We refer to Γ^{affine} as the **affine surface approximation**.

Remark 8.1.2. *We restrict ourselves to simplicial surface meshes because then $\mathcal{G}^{\text{affine}}$ consists of plane triangles while the bilinear images of the unit square would be, in general, curved quadrilaterals.*

The definition of the abstract surface lifting $\theta^{\text{affine}} : \Gamma \to \Gamma^{\text{affine}}$ typically depends on the concrete application. We will consider a construction which is based on the following assumption.

Assumption 8.1.3. *There is a neighborhood U_Γ of Γ and a mapping $P : U_\Gamma \to \Gamma$ such that:*

(a) $\Gamma^{\text{affine}} \subset U_\Gamma$.
(b) $P|_{\Gamma^{\text{affine}}} : \Gamma^{\text{affine}} \to \Gamma$ *is bi-Lipschitz continuous, i.e., there exists a constant* $c_P > 0$ *such that*
$$c_P \|\mathbf{x} - \mathbf{y}\| \leq \|P(\mathbf{x}) - P(\mathbf{y})\| \leq c_P^{-1} \|\mathbf{x} - \mathbf{y}\| \qquad \forall \mathbf{x}, \mathbf{y} \in \Gamma^{\text{affine}}.$$
(c) For any $\tau \in \mathcal{G}^{\text{affine}}$ *and* $\check{\tau} := P(\tau)$, *the restriction* $P|_\tau : \tau \to \check{\tau}$ *is a* C^{k-1} *diffeomorphism for some* $k \geq 2$.

If Assumption 8.1.3 is satisfied, we may simply set
$$\theta^{\text{affine}} := \left(P|_{\Gamma^{\text{affine}}}\right)^{-1} : \Gamma \to \Gamma^{\text{affine}}.$$

The p-th order parametric surface approximation will be defined by employing the mapping P as in Assumption 8.1.3. Recall that $\left\{\widehat{P}_{(i,j)}^{(p)} : (i,j) \in \iota_{\hat{\tau}}^{(p)}\right\}$ denotes the set of nodal points on the reference element and $\widehat{N}_{(i,j)}^{(p)}$ is the Lagrange basis on $\hat{\tau}$ corresponding to the nodal point $\widehat{P}_{(i,j)}$.

Definition 8.1.4. For given panel $\tau \in \mathcal{G}^{\text{affine}}$ and degree $p \geq 1$, the **p–parametric reference mapping** $\chi_{\tau,p}$ of degree p is (componentwise) the p-th order nodal interpolation of $P \circ \chi_\tau^{\text{affine}}$, i.e., $\chi_{\tau,p} = \widehat{I}^p \left(P \circ \chi_\tau^{\text{affine}}\right)$. It has the explicit representation

$$\chi_{\tau,p}(\hat{x}) := \sum_{(i,j)\in\iota_p^{\hat{\tau}}} \mathbf{P}_{(i,j)}^{(p)} \widehat{N}_{(i,j)}^{(p)}(\hat{x}) \qquad \forall \hat{x} \in \hat{\tau}, \text{ where } \mathbf{P}_{(i,j)}^{(p)} := P \circ \chi_\tau^{\text{affine}}\left(\widehat{P}_{(i,j)}^{(p)}\right).$$

For $\tau \in \mathcal{G}^{\text{affine}}$, the image $\chi_{\tau,p}(\hat{\tau})$ is the **p–parametric panel** of degree p. The mesh $\mathcal{G}^p := \left\{\chi_{\tau,p}(\hat{\tau}) : \tau \in \mathcal{G}\right\}$ is the **p–parametric surface mesh** of degree p. The **p–parametric surface approximation** of degree p corresponding to $\mathcal{G}^p$ is

$$\Gamma^p := \overline{\bigcup_{\tilde{\tau}\in\mathcal{G}^p} \tilde{\tau}}.$$

Remark 8.1.5. *The p-parametric reference mapping $\chi_{\tau,p} : \hat{\tau} \to \tilde{\tau}$ is componentwise a polynomial of maximal degree p. The definition of $\chi_{\tau,p}$ does* **not** *require the analytic knowledge of the pullbacks $P \circ \chi_\tau^{\text{affine}}$ for all $\hat{\mathbf{x}} \in \hat{\tau}$ but only the discrete images of the nodal points $P \circ \chi_\tau^{\text{affine}}\left(\widehat{P}_{(i,j)}^{(p)}\right)$ for all reference indices $(i,j) \in \iota_p^{\hat{\tau}}$. In practice, such discrete images can be obtained either by measurements or by using geometric modelling provided by CAD-programs.*

For $p = 1$, the mesh $\mathcal{G}^1 = \mathcal{G}^{\text{affine}}$ is independent of the choice of the surface projection P since all nodal points of $\mathcal{G}^{\text{affine}}$ lie on the true surface.

The analogue to Assumption 8.1.3 for p-parametric surface approximations of higher order reads as follows.

Assumption 8.1.6. *The mapping $P : U_\Gamma \to \Gamma$ as in Assumption 8.1.3 satisfies:*

(a) $\Gamma^p \subset U_\Gamma$.
(b) $P|_{\Gamma^p} : \Gamma^p \to \Gamma$ *is bi-Lipschitz continuous, i.e., there exists a constant* $c_P > 0$ *such that*

$$c_P \|\mathbf{x} - \mathbf{y}\| \leq \|P(\mathbf{x}) - P(\mathbf{y})\| \leq c_P^{-1} \|\mathbf{x} - \mathbf{y}\| \qquad \forall \mathbf{x}, \mathbf{y} \in \Gamma^p.$$

(c) For any $\tilde{\tau} \in \mathcal{G}^p$ and $\check{\tau} := P(\tilde{\tau})$, the restriction $P|_{\tilde{\tau}} : \tilde{\tau} \to \check{\tau}$ is a C^{k-1} diffeomorphism for some $k \geq 2$.

If Assumption 8.1.6 is satisfied we set

$$\theta^p = \left(P|_{\Gamma^p}\right)^{-1}.$$

Notation 8.1.7. *For the panels in $\mathcal{G}^{\text{affine}}$ we write τ and use them as counting indices. For the lifted panel on $\mathcal{G}^p$, we use a "$\tilde{\cdot}$" notation and write $\tilde{\tau} = \operatorname{lift}_{\tau,p}(\tau)$, where $\operatorname{lift}_{\tau,p} := \chi_{\tau,p} \circ \left(\chi_\tau^{\text{affine}}\right)^{-1} : \tau \to \tilde{\tau}$. For the corresponding panel on the* *original surface we use a superscript "$\check{\cdot}$" and write* $\check{\tau} = P(\tilde{\tau}) = \text{lift}_\tau(\tau)$, *where* $\text{lift}_\tau := P \circ \text{lift}_{\tau,p}$.

In the case of globally smooth surfaces, the orthogonal surface projection is well defined in a neighborhood of Γ. This choice will lead to estimates in the analysis of the surface approximation errors which are improved by one order of h compared to more general mappings P. The proof of the following example can be found in, e.g., [100, Lemma 14.16].

Example 8.1.8 (Orthogonal surface projection). *Let Γ be the surface of a bounded domain $\Omega \subset \mathbb{R}^3$ of class C^k for some $k \geq 2$. This implies that there exist $\delta > 0$ and a tubular neighborhood V of Γ such that the mapping*

$$\psi : \Gamma \times]-\delta, \delta[\to V \qquad \psi(\mathbf{x}, s) = \mathbf{x} + s\mathbf{n}(\mathbf{x}) \tag{8.2}$$

is a C^{k-1} diffeomorphism. Thus the orthogonal projection $P : V \to \Gamma$ resp. the oriented distance function $\text{dist} : V \to]-\delta, \delta[$ *are well defined by*

$$P(\mathbf{x} + s\mathbf{n}(\mathbf{x})) = \mathbf{x} \quad \text{and} \quad \text{dist}(\mathbf{x} + s\mathbf{n}(\mathbf{x})) = s$$

and of class C^{k-1} resp. C^k.

Remark 8.1.9. *The case $d = 2$ is special because, for sufficiently small mesh width h, the projection P can always be chosen, locally, as the orthogonal surface projection, i.e.,*

$$\forall \mathbf{x} \in \tau \in \mathcal{G}^{\text{affine}} \qquad \check{\mathbf{x}} - \mathbf{x} \perp T_{\check{\mathbf{x}}},$$

where $\check{\mathbf{x}} = P(\mathbf{x})$ and $T_{\check{\mathbf{x}}}$ denotes the tangential plane at $\check{\tau} = P(\tau)$ in $\check{\mathbf{x}}$. This is due to the fact that for sufficiently small h_τ, δ, and any $\check{\mathbf{x}} \in \check{\tau}$, the line $\{\check{\mathbf{x}} + s\mathbf{n}(\check{\mathbf{x}}) : s \in [-\delta, \delta]\}$ and τ have a unique intersection point.

In the case of anisotropic boundary value problems, where the principal part of L is given by $-\text{div}(\mathbf{A} \,\text{grad}\, u)$ with some positive definite $\mathbf{A} \neq \mathbf{I}$, the rôle of the orthogonal surface projection is played by the conormal surface projection.

Example 8.1.10 (Conormal surface projection). *Let Γ be the boundary of a bounded domain $\Omega \subset \mathbb{R}^3$ of class C^k for some $k \geq 2$. Let $\mathbf{A} \in \mathbb{R}^{3\times 3}$ be positive definite and define the exterior conormal vector at $\mathbf{z} \in \Gamma$ by $\mathbf{v}(\mathbf{z}) = \mathbf{A}\mathbf{n}(\mathbf{z})$. This implies that there exist $\delta > 0$ and a tubular neighborhood V of Γ such that the mapping*

$$\psi : \Gamma \times]-\delta, \delta[\to \widetilde{V} \qquad \psi(\mathbf{x}, s) = \mathbf{x} + s\mathbf{v}(\mathbf{x})$$

is a C^{k-1} diffeomorphism. Thus the conormal projection $P : V \to \Gamma$ resp. the function $\text{dist}_\mathbf{A} : V \to]-\delta, \delta[$ *are well defined by*

$$P(\mathbf{x} + s\mathbf{v}(\mathbf{x})) = \mathbf{x} \quad \text{and} \quad \text{dist}_\mathbf{A}(\mathbf{x} + s\mathbf{v}(\mathbf{x})) = s \tag{8.3}$$

and of class C^{k-1} resp. C^k.

Proof. The proof of Example 8.1.8 is based on the introduction of a local coordinate system $(\mathbf{t}_1(\mathbf{x}), \mathbf{t}_2(\mathbf{x}), \mathbf{n}(\mathbf{x}))$ at some points $\mathbf{x} \in \Gamma$ and the application of the inverse mapping theorem along with the compactness of Γ. The positive definiteness of the diffusion matrix $\mathbf{A}$ implies $\langle \mathbf{v}(\mathbf{x}), \mathbf{n}(\mathbf{x}) \rangle = \langle \mathbf{A}\mathbf{n}(\mathbf{x}), \mathbf{n}(\mathbf{x}) \rangle \geq \lambda_{\min}$, where $\lambda_{\min}$ is the smallest eigenvalue of $\mathbf{A}$. As a consequence, the transformation $(\mathbf{t}_1(\mathbf{x}), \mathbf{t}_2(\mathbf{x}), \mathbf{n}(\mathbf{x})) \to (\mathbf{t}_1(\mathbf{x}), \mathbf{t}_2(\mathbf{x}), \mathbf{v}(\mathbf{x}))$ is regular and hence the asserted properties of the conormal projection (8.3) are inherited from the analogous properties of the orthogonal projection. □

Remark 8.1.11. *In some applications, the function $P : U_\Gamma \to \Gamma$, which is used in Definitions 8.1.1 and 8.1.4 for the construction of the p-parametric surface approximations, is not explicitly given. Instead the surface mesh is given in parametrized form by a set of pullbacks $\chi_k : \hat{\tau} \to \mathbb{R}^3$, $k \in I$, where I is some finite index set. We assume here that*

$$\check{\mathcal{G}} := \{\chi_k(\hat{\tau}) : k \in I\}$$

is a surface mesh for Γ in the sense of Definition 4.1.2. If the evaluation of the mappings χ_k and their derivatives are costly it is recommended that χ_k be replaced by a p-parametric approximation

$$\chi_{k,p} := \widehat{I}^p(\chi_k) \quad \text{and} \quad \mathcal{G}^p := \left\{\theta_k^p(\check{\tau}) : \check{\tau} \in \check{\mathcal{G}}\right\},$$

where the lifting is defined by $\theta_k^p := \chi_{k,p} \circ \chi_k^{-1}$. Note that, in general, θ_τ^p is not the orthogonal surface projection.

8.1.2 (k,p)-Boundary Element Spaces with p-Parametric Surface Approximation

The definition of boundary element spaces as introduced in Chap. 4 has to be modified slightly if the original surface is replaced by its p-parametric approximation.

The p-parametric surface mesh $\mathcal{G}^p$ is characterized by the set of parametrizations

$$\vec{\chi}_p := \left\{\chi_{\tau,p} : \tau \in \mathcal{G}^{\text{affine}}\right\}$$

which map the reference element $\hat{\tau}$ to the panel $\chi_{\tau,p}(\hat{\tau}) \in \mathcal{G}^p$. (Recall that we always use the elements of the affine mesh $\mathcal{G}^{\text{affine}}$ as the counting index for mesh-related quantities such as the pullbacks $\left\{\chi_{\tau,p} : \tau \in \mathcal{G}^{\text{affine}}\right\}$.)

Definition 8.1.12. For given $p \geq 1$, let $\mathcal{G}^p$ denote a p-parametric surface approximation of degree p. The space of (k,p)-discontinuous boundary elements of **algebraic degree** $k \in \mathbb{N}_0$ and **geometric degree** p is given by

$$S_{\vec{\chi}_p}^{k,-1} := \left\{\psi : \Gamma^p \to \mathbb{K} \mid \forall \tau \in \mathcal{G}^{\text{affine}} : \psi \circ \chi_{\tau,p} \in \mathbb{P}_k^\Delta\right\}.$$

The space of (k, p)-continuous boundary elements of algebraic degree $k \in \mathbb{N}$ and geometric degree p is given by

$$S^{k,0}_{\vec{\chi}_p} := \left\{\psi \in C^0\left(\Gamma^p\right) \mid \forall \tau \in \mathcal{G}^{\text{affine}} : \psi \circ \chi_{\tau,p} \in \mathbb{P}^{\tau}_{k}\right\}.$$

Remark 8.1.13. *If parts of the sub- and superscripts in the notation* $S^{k,r}_{\vec{\chi}_p}$ *are clear from the context, we will write for short* $S^{k,0,p}$, $S^{k,1,p}$, $S^{k,p}$, *or simply* S.

Note that the functions $u \in S^{k,r,p}$, in general, are not defined on the true surface. The approximations to the exact solution are obtained by using the surface projection by means of $u \circ \theta^p$. For a function $v : \Gamma^p \to \mathbb{C}$, we denote by $\check{v}$ the lifted function

$$\check{v} := v \circ \theta^p : \Gamma \to \mathbb{C} \tag{8.4}$$

8.1.3 Discretization of Boundary Integral Equations with p-Parametric Surface Approximation

We will consider the abstract problem:

$$\text{Find } u \in H : \qquad b\left(u, v\right) = F\left(v\right) \qquad \forall v \in H. \tag{8.5}$$

The Hilbert space H, typically, is a Sobolev space $H^s\left(\Gamma\right)$, $s = -1/2, 0, 1/2$ resp. a suitable closed subspace. The functional $F \in H'$ denotes a given right-hand side which, possibly as, e.g., in the case of the direct method (cf. Sect. 3.4.2), might be defined via integral operators. For the boundary integral operators V, K, and K' (cf. Chap. 3), the sesquilinear form $b\left(\cdot,\cdot\right)$ has the abstract form

$$b\left(u, v\right) = \left(Bu, v\right)_{L^2(\Gamma)} \tag{8.6}$$

with the boundary integral operator

$$\left(Bu\right)\left(\mathbf{x}\right) = \lambda_1\left(\mathbf{x}\right) u\left(\mathbf{x}\right) + \lambda_2\left(\mathbf{x}\right) \int_\Gamma k\left(\mathbf{x},\mathbf{y}\right) u\left(\mathbf{y}\right) ds_{\mathbf{y}} \qquad \mathbf{x} \in \Gamma \text{ a.e.} \tag{8.7}$$

For the integral operator W, the kernel function is hypersingular and we choose the regularization via partial integration (cf. Sect. 3.3.4). The sesquilinear form is given by [cf. (3.3.22)]

$$b\left(u, v\right) := \int_{\Gamma\times\Gamma} \left\{k_1\left(\mathbf{x},\mathbf{y}\right) \left\langle \operatorname{curl}_{\Gamma,\mathbf{A},\mathbf{0}} u\left(\mathbf{y}\right), \operatorname{curl}_{\Gamma,\mathbf{A},\mathbf{2b}} \overline{v}\left(\mathbf{x}\right)\right\rangle \right.$$
$$\left. + c k_2\left(\mathbf{x},\mathbf{y}\right) u\left(\mathbf{y}\right) \overline{v}\left(\mathbf{x}\right)\right\} ds_{\mathbf{y}} ds_{\mathbf{x}} \tag{8.8}$$

with

$$k_1(\mathbf{x},\mathbf{y}) := G(\mathbf{x}-\mathbf{y}) \quad \text{and} \quad k_2(\mathbf{x},\mathbf{y}) = G(\mathbf{x}-\mathbf{y}) \left\langle \mathbf{A}^{1/2}\mathbf{n}(\mathbf{x}), \mathbf{A}^{1/2}\mathbf{n}(\mathbf{y}) \right\rangle.$$

Assumption 8.1.14. *The kernel function $k : \Gamma \times \Gamma \to \mathbb{C}$ is either a fundamental solution $G : \mathbb{R}^3 \to \mathbb{C}$ [cf. (3.3)]*

$$G(\mathbf{z}) = \frac{1}{4\pi\sqrt{\det \mathbf{A}}} \frac{e^{\langle \mathbf{b},\mathbf{z}\rangle_{\mathbf{A}} - \lambda \|\mathbf{z}\|_{\mathbf{A}}}}{\|\mathbf{z}\|_{\mathbf{A}}} \tag{8.9}$$

or a suitable Gâteaux derivative:

1. $k(\mathbf{x},\mathbf{y}) = G(\mathbf{x}-\mathbf{y})$,
2. $k(\mathbf{x},\mathbf{y}) = \partial G(\mathbf{x}-\mathbf{y}) / \partial \mathbf{v}_{\mathbf{x}}$,
3. $k(\mathbf{x},\mathbf{y}) = \frac{\partial}{\partial \mathbf{v}_{\mathbf{y}}} G(\mathbf{x}-\mathbf{y}) + 2\langle \mathbf{b}, \mathbf{v}_{\mathbf{y}}\rangle G(\mathbf{x}-\mathbf{y})$,
4. $k(\mathbf{x},\mathbf{y}) = G(\mathbf{x}-\mathbf{y}) \langle \mathbf{A}^{1/2}\mathbf{n}(\mathbf{x}), \mathbf{A}^{1/2}\mathbf{n}(\mathbf{y})\rangle$,

where $\mathbf{v} = \mathbf{A}\mathbf{n}$ denotes the conormal vector [cf. (2.103)].

Notation 8.1.15. *Existence and uniqueness of the Galerkin solution on the surface Γ^p will be proved for sufficiently fine mesh width. In this light, we consider a sequence $(\mathcal{G}_\ell^p)_{\ell\in\mathbb{N}}$ of p-parametric surface meshes with corresponding approximate surfaces $(\Gamma_\ell^p)_{\ell\in\mathbb{N}}$. We write $\left(S_\ell^{k,r,p}\right)_{\ell\in\mathbb{N}}$ for the corresponding sequence of boundary element spaces.*

If the polynomial orders (k,p) and the regularity index $r \in \{-1,0\}$ are clear from the context, we simply write $\mathcal{G}_\ell$, Γ_ℓ, S_ℓ, θ_ℓ. For the mesh sequence $(\mathcal{G}_\ell^{\text{affine}})_{\ell\in\mathbb{N}}$, the constants in Definition 8.1.1 and Assumption 8.1.3, resp. Assumption 8.1.6, in general, depend on ℓ and we write c_ℓ^{affine}, C_ℓ^{affine}, $c_{P,\ell}$.

The following Assumption expresses the requirement that these constants are uniformly bounded.

Assumption 8.1.16. *There exists constants $c^{\text{affine}}, c_P > 0$, $C^{\text{affine}} < \infty$ such that the constants c_ℓ^{affine}, C_ℓ^{affine}, $c_{P,\ell}$ in Definition 8.1.1 and Assumption 8.1.3, resp. Assumption 8.1.6 satisfy*

$$\forall \ell \in \mathbb{N} \qquad c^{\text{affine}} < c_\ell^{\text{affine}} \le C_\ell^{\text{affine}} \le C^{\text{affine}} \quad \text{and} \quad c_P \le c_{P,\ell}.$$

Example 8.1.17. *Let $\Gamma = \mathbb{S}_2$ and let Γ^{affine} be the double pyramid with vertices $(1,0,0)^\intercal$, $(0,1,0)^\intercal$, $(-1,0,0)^\intercal$, $(0,-1,0)^\intercal$, $(0,0,1)^\intercal$, $(0,0,-1)^\intercal$. The mesh $\mathcal{G}_0^{\text{affine}}$ is the set of the eight triangular faces of Γ^{affine}. The orthogonal projection $P : \Gamma^{\text{affine}} \to \Gamma$ is given by $P(\mathbf{x}) = \mathbf{x}/\|\mathbf{x}\|$ and can be extended to an appropriate neighborhood U_Γ as required in Assumption 8.1.3.*

Recursively, we assume $\mathcal{G}_{\ell-1}^{\text{affine}}$ has been generated for some $\ell \ge 1$. A finer mesh $\mathcal{G}_\ell^{\text{affine}}$ is constructed by (a) connecting the midpoints of edges of the panels in $\mathcal{G}_{\ell-1}$ and (b) projecting these midpoints to the surface by means of the mapping P.

Because the projection P is given explicitly, the construction of the p-parametric surface approximations Γ_ℓ^p *is without any difficulty. This leads to meshes* $\mathcal{G}_\ell^p$ *which satisfy Assumption 8.1.16. The proof of this statement is straightforward but somewhat lengthy and will be skipped.*

The Galerkin boundary element method of algebraic degree k and geometric degree p for the discretization of (8.7) consists of two steps (cf. Notation 8.1.15):

- *Replacing the true surface by the* **p***-parametric approximation:*

$$\text{Find } u_\ell \in S_\ell : \qquad b_\ell (u_\ell, v) = F_\ell (v) \qquad \forall v \in S_\ell. \tag{8.10}$$

The perturbed sesquilinear form is given for the boundary integral operators V, K, and K' by

$$b_\ell (u, v) = (B_\ell u, v)_{L^2(\Gamma_\ell)},$$

where

$$(B_\ell u)(\mathbf{x}) = \lambda_{1,\ell}(\mathbf{x})\, u(\mathbf{x}) + \lambda_{2,\ell}(\mathbf{x}) \int_{\Gamma_\ell} k_\ell (\mathbf{x}, \mathbf{y})\, u(\mathbf{y})\, ds_{\mathbf{y}} \qquad \mathbf{x} \in \Gamma_\ell \text{ a.e.}$$

and, for the hypersingular operator W, by

$$b_\ell (u, v) := \int_{\Gamma_\ell \times \Gamma_\ell} \left\{ k_{1,\ell} (\mathbf{x}, \mathbf{y}) \left\langle \operatorname{curl}_{\Gamma_\ell, \mathbf{A}, \mathbf{0}}\, u(\mathbf{y}), \operatorname{curl}_{\Gamma_\ell, \mathbf{A}, \mathbf{2b}}\, \overline{v}(\mathbf{x}) \right\rangle \right. \\ \left. + c k_{2,\ell} (\mathbf{x}, \mathbf{y})\, u(\mathbf{y})\, \overline{v}(\mathbf{x}) \right\} ds_{\mathbf{y}} ds_{\mathbf{x}}.$$

The kernels $k_\ell : \Gamma_\ell \times \Gamma_\ell \to \mathbb{C}$ which correspond to the four cases in Assumption 8.1.14 are given by:

1. $k_\ell (\mathbf{x}, \mathbf{y}) = k_{1,\ell} (\mathbf{x}, \mathbf{y}) = G(\mathbf{x} - \mathbf{y})$.
2. $k_\ell (\mathbf{x}, \mathbf{y}) = \partial G (\mathbf{x} - \mathbf{y}) / \partial \mathbf{v}_{\ell, \mathbf{x}}$, where $\mathbf{v}_\ell := \mathbf{A} \mathbf{n}_\ell$. The matrix $\mathbf{A}$ is as in (2.98) and $\mathbf{n}_\ell$ denotes the normal field at the p-parametric surface Γ_ℓ pointing towards the unbounded exterior domain. The notation $\mathbf{v}_{\ell,\mathbf{x}}$ indicates that the Gâteaux derivative is applied with respect to the variable $\mathbf{x}$.
3. $k_\ell (\mathbf{x}, \mathbf{y}) = \partial G (\mathbf{x} - \mathbf{y}) / \partial \mathbf{v}_{\ell, \mathbf{y}} + 2 \langle \mathbf{b}, \mathbf{v}_{\ell,\mathbf{y}} \rangle G (\mathbf{x} - \mathbf{y})$.
4. $k_{2,\ell} (\mathbf{x}, \mathbf{y}) = G (\mathbf{x} - \mathbf{y}) \langle \mathbf{A}^{1/2} \mathbf{n}_\ell (\mathbf{x}), \mathbf{A}^{1/2} \mathbf{n}_\ell (\mathbf{y}) \rangle$.

Furthermore, the (real-valued) coefficients $\lambda_{1,\ell}$ and $\lambda_{2,\ell}$ are defined by $\lambda_{1,\ell} := \lambda_1 \circ \theta_\ell^{-1}$, $\lambda_{2,\ell} := \lambda_2 \circ \theta_\ell^{-1}$. Note that in most cases λ_1, λ_2 are piecewise constant implying that $\lambda_{1,\ell} = \lambda_1$ and $\lambda_{2,\ell} = \lambda_2$.

Remark 8.1.18. *We do not discuss the approximation of the right-hand side* $F(v)$ *in detail, since its accurate evaluation strongly depends on the specific problem. As a rule of thumb, one replaces integrals over* Γ *by integrals over* Γ_ℓ *and (expressions containing) the conormal vector field* $\mathbf{v}$ *by* $\mathbf{A}\mathbf{n}_\ell$*. The arising approximation of the functional* F *in (8.10) is denoted by* F_ℓ*.*

- *Lifting the approximation u_ℓ to the true surface:*
  The p-parametric Galerkin approximation to the exact solution of (8.5) is given by
$$\check{u}_\ell := u_\ell \circ \theta_\ell. \tag{8.11}$$

8.2 Convergence Analysis

In this section, we will address the question how the degree p in the p-parametric surface approximation has to be chosen such that the convergence rates of the unperturbed Galerkin discretization (on the exact surface) are preserved. The discretization parameter for the mesh $\mathcal{G}_\ell$ is the mesh width h_ℓ. In this section, we will investigate the convergence of the corresponding sequence $(u_\ell)_\ell$ of Galerkin solutions.

For the continuous problem (8.5) we assume existence and uniqueness via the following conditions.

Assumption 8.2.1. *1. Continuity: There exists $C > 0$ such that*

$$\forall u, v \in H: \qquad |b(u,v)| \leq C \|u\|_H \|v\|_H.$$

2. Gårdings's inequality: There exist a constant $c > 0$ and a compact operator $T : H \to H'$ such that

$$\forall u \in H: \qquad |b(u,u) - \langle Tu, u\rangle_{H' \times H}| \geq c \|u\|_H^2.$$

3. Injectivity: For all $v \in H \backslash \{0\}$,

$$b(u,v) = 0 \implies u = 0.$$

In order to compare the sesquilinear forms $b(\cdot,\cdot)$ and $b_\ell(\cdot,\cdot)$, we will lift the sesquilinear form $b_\ell(\cdot,\cdot)$ to the true surface Γ. For $\tau \in \mathcal{G}^{\text{affine}}$ and $\tilde{\tau} = \text{lift}_{\tau,p}(\tau)$ (cf. Notation 8.1.7), the pullback of the corresponding panel $\check{\tau} = P(\tilde{\tau}) \subset \Gamma$ to the reference element is the composition

$$\chi_\tau = P \circ \chi_{\tau,p} : \hat{\tau} \to \check{\tau}. \tag{8.12}$$

The function $\rho_\ell : \Gamma \to \mathbb{R}_{>0}$ is the quotient of the square roots of Gram's determinants for Γ and Γ_ℓ. We define $\rho_\ell : \Gamma \to \mathbb{R}$ piecewise, for $\mathbf{x} \in \tau \subset \mathcal{G}_\ell$, by

$$\rho_\ell(\mathbf{x}) := \frac{\left(g_{\tau,p} \circ \chi_\tau^{-1}\right)(\mathbf{x})}{g_\tau \circ \chi_\tau^{-1}(\mathbf{x})}.$$

Here, $g_\tau : \hat{\tau} \to \mathbb{R}_{>0}$ denotes the surface element of $\tau \in \mathcal{G}_\ell$ and $g_{\tau,p}$ is the surface element of the corresponding p-parametric panel. Explicitly, we have

$$g_\tau(\hat{\mathbf{x}}) = \sqrt{\det \mathbf{G}_\tau(\hat{\mathbf{x}})} \quad \text{with} \quad \mathbf{G}_\tau(\hat{\mathbf{x}}) := \mathbf{J}_\tau^\intercal(\hat{\mathbf{x}})\,\mathbf{J}_\tau(\hat{\mathbf{x}}), \tag{8.13a}$$

$$g_{\tau,p}(\hat{\mathbf{x}}) = \sqrt{\det \mathbf{G}_{\tau,p}(\hat{\mathbf{x}})} \quad \text{with} \quad \mathbf{G}_{\tau,p}(\hat{\mathbf{x}}) := \mathbf{J}_{\tau,p}^\intercal(\hat{\mathbf{x}})\,\mathbf{J}_{\tau,p}(\hat{\mathbf{x}}) \tag{8.13b}$$

where $\mathbf{J}_\tau$ (resp. $\mathbf{J}_{\tau,p}$) is the Jacobi matrix of χ_τ (resp. $\chi_{\tau,p}$).

The following quantities will enter the error estimates

$$\bar{\rho}_\ell := \|\rho_\ell\|_{L^\infty(\Gamma)} \quad \text{and} \quad \bar{\rho}_\ell^\Delta := \|\rho_\ell - 1\|_{L^\infty(\Gamma)}\,. \tag{8.14}$$

By employing the function ρ_ℓ, we may define the *lifted* p-parametric boundary element space by

$$\check{S}_\ell := \{u \circ \theta_\ell : u \in S_\ell\}\,,$$

and the *lifted* sesquilinear form $\check{b}_\ell : \check{S}_\ell \times \check{S}_\ell \to \mathbb{C}$ as follows:

- For the boundary integral operators V, K, and K',

$$\check{b}_\ell(\check{u},\check{v}) := \int_\Gamma \lambda_1(\mathbf{x})\,\overline{\check{v}(\mathbf{x})}\check{u}(\mathbf{x})\,\rho_\ell(\mathbf{x})\,ds_\mathbf{x}$$
$$+ \int_\Gamma \lambda_2(\mathbf{x})\,\overline{\check{v}(\mathbf{x})}\rho_\ell(\mathbf{x}) \left(\int_\Gamma \check{k}_\ell(\mathbf{x},\mathbf{y})\,\check{u}(\mathbf{y})\,\rho_\ell(\mathbf{y})\,ds_\mathbf{y} \right) ds_\mathbf{x},$$

  where $\check{k}_\ell(\mathbf{x},\mathbf{y}) := k_\ell(\theta_\ell(\mathbf{x}), \theta_\ell(\mathbf{y}))$.
- For the hypersingular integral operator W,

$$\check{b}_\ell(\check{u},\check{v}) := \int_{\Gamma\times\Gamma} \rho_\ell(\mathbf{x})\,\rho_\ell(\mathbf{y}) \left\{ \check{k}_{1,\ell}(\mathbf{x},\mathbf{y}) \left\langle \left(\operatorname{curl}_{\Gamma_\ell,\mathbf{A},\mathbf{0}}\tilde{u}\right)(\tilde{\mathbf{y}}), \left(\operatorname{curl}_{\Gamma_\ell,\mathbf{A},\mathbf{2b}}\bar{\tilde{v}}\right)(\tilde{\mathbf{x}}) \right\rangle \right.$$
$$\left. + c\check{k}_{2,\ell}(\mathbf{x},\mathbf{y})\,\check{u}(\mathbf{y})\,\bar{\check{v}}(\mathbf{x}) \right\} ds_\mathbf{y} ds_\mathbf{x}, \tag{8.15}$$

  where $\check{k}_{i,\ell}(\mathbf{x},\mathbf{y}) := k_{i,\ell}(\theta_\ell(\mathbf{x}), \theta_\ell(\mathbf{y}))$, $i = 1,2$, and $\tilde{v} := \check{v} \circ \theta_\ell^{-1}$, $\bar{\tilde{v}} := \bar{\check{v}} \circ \theta_\ell^{-1}$, and $\tilde{\mathbf{x}} := \theta_\ell(\mathbf{x})$, $\tilde{\mathbf{y}} := \theta_\ell(\mathbf{y})$.
- The *lifted* right-hand side $\check{F}_\ell : \check{S}_\ell \to \mathbb{C}$ is given by

$$\check{F}_\ell(\check{u}) := F_\ell\left(\check{u} \circ \theta_\ell^{-1}\right).$$

Remark 8.2.2. *(a) For all $u, v \in S_\ell$ and corresponding lifted $\check{u}, \check{v}$ [cf. (8.4)],*

$$b_\ell(u,v) = \check{b}_\ell(\check{u},\check{v}) \quad \text{and} \quad \check{F}_\ell(\check{u}) = F_\ell(u)\,.$$

(b) The Galerkin solution $\check{u}_\ell$ in (8.11) can be characterized equivalently: Find $\check{u}_\ell \in \check{S}_\ell$, such that

$$\check{b}_\ell(\check{u}_\ell, \check{v}_\ell) = \check{F}_\ell(\check{v}_\ell) \qquad \forall \check{v}_\ell \in \check{S}_\ell.$$

Since the sesquilinear form $\check{b}_\ell$ and the right-hand side $\check{F}_\ell$ are both defined for functions on the true surface Γ, we may compare them with the original sesquilinear form b and right-hand side F. In order to ensure well-posedness of this problem and quasi-optimal error estimates, we employ the concepts of *uniform continuity*, *consistency*, and *stability* as introduced in Sect. 4.2.4.

1. The family of sesquilinear forms $b_\ell(\cdot,\cdot) : S_\ell \times S_\ell \to \mathbb{C}$ is uniformly continuous if there is a constant C such that
$$\forall u, v \in S_\ell : \quad |b_\ell(u,v)| \leq C \|u\|_H \|v\|_H . \tag{8.16a}$$

2. The family of lifted sesquilinear forms $\check{b}_\ell(\cdot,\cdot) : \check{S}_\ell \times \check{S}_\ell \to \mathbb{C}$ is uniformly stable if there exists a null sequence $(c_\ell)_\ell$ such that
$$\forall \check{u}, \check{v} \in \check{S}_\ell : \quad \left| b(\check{u},\check{v}) - \check{b}_\ell(\check{u},\check{v}) \right| \leq c_\ell \|\check{u}\|_H \|\check{v}\|_H . \tag{8.16b}$$

3. Let U be a subspace of H with continuous embedding and $\check{S}_\ell \subset U$. (Note that the trivial choice $U = H$ is always possible.) Let the exact solution be U-regular, i.e, the solution u of (8.5) satisfies $u \in U$. The family of lifted sesquilinear forms $\check{b}_\ell(\cdot,\cdot) : \check{S}_\ell \times \check{S}_\ell \to \mathbb{C}$ is uniformly consistent with respect to $\|\cdot\|_U$ if there exists a null sequence $\left(\delta_\ell^{\mathrm{I}}\right)_\ell$ such that
$$\forall \check{u}, \check{v} \in \check{S}_\ell : \quad \left| b(\check{u},\check{v}) - \check{b}_\ell(\check{u},\check{v}) \right| \leq \delta_\ell^{\mathrm{I}} \|\check{u}\|_U \|\check{v}\|_H . \tag{8.16c}$$

4. The family of right-hand sides $\check{F}_\ell(\cdot) : \check{S}_\ell \to \mathbb{C}$ is uniformly consistent if there exist a constant C_F which depends only on F and a null sequence $\left(\delta_\ell^{\mathrm{II}}\right)_\ell$ such that
$$\forall \check{u} \in \check{S}_\ell : \quad \left| F(\check{u}) - \check{F}_\ell(\check{u}) \right| \leq C_F \delta_\ell^{\mathrm{II}} \|\check{u}\|_H . \tag{8.16d}$$

Remark 8.2.3. *Let the exact solution satisfy $u \in U$. Conditions (8.16) imply, by means of Strang's lemma (cf. Sect. 4.2.4), that the error of the exact solution u and the (k,p)-boundary element approximation $\check{u}_\ell$ can be estimated by*
$$\|u - \check{u}_\ell\|_H \leq C \left(\|u - \Pi_\ell u\|_H + \delta_\ell^{\mathrm{I}} \|u\|_U + C_F \delta_\ell^{\mathrm{II}} \right), \tag{8.17}$$
where $\Pi_\ell : H \to S_\ell$ denotes the orthogonal projection with respect to the scalar product $(\cdot,\cdot)_U$. The first term on the right-hand side of (8.17) can be estimated as in (4.158). If $H = H^s(\Gamma)$ for some $s \in \{-1/2, 0, 1/2\}$, if the polynomial order for the boundary element functions equals k, and if problem (8.5) has full H^{k+1}-regularity, the error estimate
$$\|u - \check{u}_\ell\|_{H^s(\Gamma)} \leq C \left(h_\ell^{k+1-s} \|u\|_{H^{k+1}(\Gamma)} + \delta_\ell^{\mathrm{I}} \|u\|_U + C_F \delta_\ell^{\mathrm{II}} \right)$$

follows. Hence the optimal convergence order is preserved if the family of sesquilinear forms is uniformly stable, and both δ_ℓ^{I} *and* $\delta_\ell^{\mathrm{II}}$ *are bounded from above by* Ch^{k+1-s}.

In view of the Strang's lemma it suffices to estimate the constants in the estimates (8.16). We write the sesquilinear form $b(\cdot,\cdot)$ as a sum

$$b(u,v) = b^{\mathrm{I}}(1,u,1,v) + b^{\mathrm{II}}(1,u,1,v),$$

where

$$b^{\mathrm{I}}(w_1,u,w_2,v) := \int_\Gamma \lambda_1 w_2(\mathbf{x})\,\overline{v}(\mathbf{x})\,w_1(\mathbf{x})\,u(\mathbf{x})\,ds_{\mathbf{x}}$$

and

- For the boundary integral operators V, K, and K',

$$b^{\mathrm{II}}(w_1,u,w_2,v) := \int_\Gamma \lambda_2(\mathbf{x})\,w_2(\mathbf{x})\,\overline{v}(\mathbf{x})\left(\int_\Gamma k(\mathbf{x},\mathbf{y})\,u(\mathbf{y})\,w_1(\mathbf{y})\,ds_{\mathbf{y}}\right)ds_{\mathbf{x}}.$$

- For the boundary integral operator W, we set $b^{\mathrm{I}} = 0$ and

$$\begin{aligned} b^{\mathrm{II}}(w_1,\check{u},w_2,\check{v}) := \int_{\Gamma\times\Gamma} & w_2(\mathbf{x})\,w_1(\mathbf{y})\left\{\check{k}_1(\mathbf{x},\mathbf{y})\left\langle\left(\mathrm{curl}_{\Gamma,\mathbf{A},\mathbf{0}}\,\check{u}\right)(\tilde{\mathbf{y}}),\left(\mathrm{curl}_{\Gamma,\mathbf{A},\mathbf{2b}}\,\overline{\check{v}}\right)(\tilde{\mathbf{x}})\right\rangle\right.\\ &\left.+c\check{k}_2(\mathbf{x},\mathbf{y})\,\check{u}(\mathbf{y})\,\overline{\check{v}}(\mathbf{x})\right\}ds_{\mathbf{y}}ds_{\mathbf{x}}.\end{aligned}$$

Note that the use of the real-valued weights w_1, w_2 in the sesquilinear forms b^{I} and b^{II} allows us to express some error splittings in a more compact way.

An analogous splitting is employed for the sesquilinear form $\check{b}_\ell(\cdot,\cdot)$

$$\check{b}_\ell(u,v) = \check{b}_\ell^{\mathrm{I}}(1,u,\rho_\ell,v) + \check{b}_\ell^{\mathrm{II}}(\rho_\ell,u,\rho_\ell,v).$$

The difference of the sesquilinear forms $b(\cdot,\cdot)$ and $b_\ell(\cdot,\cdot)$ can be written in the form

$$b(\check{u},\check{v}) - \check{b}_\ell(\check{u},\check{v}) = \sum_{i=1}^{3} e_i(\check{u},\check{v}),$$

where

$$\begin{aligned} e_1(\check{u},\check{v}) &= b^{\mathrm{I}}(1,\check{u},(1-\rho_\ell),\check{v}) + b^{\mathrm{II}}(1,\check{u},(1-\rho_\ell),\check{v})\\ e_2(\check{u},\check{v}) &= b^{\mathrm{II}}(\rho_\ell,\check{u},\rho_\ell,\check{v}) - b_\ell^{\mathrm{II}}(\rho_\ell,\check{u},\rho_\ell,\check{v})\\ e_3(\check{u},\check{v}) &= b^{\mathrm{II}}((1-\rho_\ell),\check{u},\rho_\ell,\check{v}).\end{aligned} \tag{8.18}$$

Lemma 8.2.4. *For all $u \in L^2(\Gamma)$ and $w \in L^\infty(\Gamma)$,*

$$\|wu\|_{L^2(\Gamma)} \leq C \|w\|_{L^\infty(\Gamma)} \|u\|_{L^2(\Gamma)} .$$

The estimate of the sesquilinear form e_2 is based on an estimate of the difference of the kernel functions k and $\check{k}_\ell$. Let k be defined as in Assumption 8.1.14. Since norms are equivalent in finite-dimensional spaces there exist constants $c_{\mathbf{A}}, C_{\mathbf{A}} > 0$ such that

$$c_{\mathbf{A}} \|\mathbf{u}\| \leq \|\mathbf{u}\|_{\mathbf{A}} \leq C_{\mathbf{A}} \|\mathbf{u}\| \qquad \forall \mathbf{u} \in \mathbb{R}^3. \tag{8.19}$$

In the following convergence analysis, certain geometric quantities $d_\ell, \tilde{d}_\ell, \varepsilon_\ell$ will determine the convergence rates of the (k, p)-boundary element method. Let

$$d_\ell := \max_{\alpha \in \{-1,1,2\}} \left\{ \sup_{\substack{\mathbf{x},\mathbf{y}\in\Gamma \\ \mathbf{x}\neq\mathbf{y}}} \frac{\left| \|\mathbf{x}-\mathbf{y}\|_{\mathbf{A}}^\alpha - \|\theta_\ell(\mathbf{x}) - \theta_\ell(\mathbf{y})\|_{\mathbf{A}}^\alpha \right|}{\|\mathbf{x}-\mathbf{y}\|^\alpha} \right\} \tag{8.20a}$$

and

$$\tilde{d}_\ell := \sup_{\substack{\mathbf{x},\mathbf{y}\in\Gamma \\ \mathbf{x}\neq\mathbf{y}}} \|\mathbf{x} - \mathbf{y} - (\theta_\ell(\mathbf{x}) - \theta_\ell(\mathbf{y}))\| \quad \text{and} \quad \varepsilon_\ell := C_{\mathbf{A}} \|\mathbf{b}\|_{\mathbf{A}} \tilde{d}_\ell + (\operatorname{diam}\Gamma) |\lambda| d_\ell, \tag{8.20b}$$

where $\mathbf{A}, \mathbf{b}, \lambda$ are as in (3.3), (8.9).

First, we prove convergence estimates in terms of $d_\ell, \tilde{d}_\ell, \varepsilon_\ell$, while the convergence rates of these three geometric quantities will be estimated in Sect. 8.4.

Lemma 8.2.5. *Let the kernel function be defined as in Assumption 8.1.14(1). Then there exists a constant $C > 0$ which depends only on the coefficients of the differential operator L and the diameter of Γ such that*

$$\forall \mathbf{x}, \mathbf{y} \in \Gamma, \mathbf{x} \neq \mathbf{y}, \qquad \left| k(\mathbf{x}, \mathbf{y}) - \check{k}_\ell(\mathbf{x}, \mathbf{y}) \right| \leq C(\varepsilon_\ell + d_\ell) \frac{1}{4\pi \|\mathbf{x} - \mathbf{y}\|}. \tag{8.21}$$

Proof. For $\mathbf{x}, \mathbf{y} \in \Gamma$, let $\tilde{\mathbf{x}} := \theta_\ell(\mathbf{x})$ and $\tilde{\mathbf{y}} \in \theta_\ell(\mathbf{y})$. The differences are denoted by $\mathbf{z} = \mathbf{y} - \mathbf{x}$ and $\tilde{\mathbf{z}} := \tilde{\mathbf{y}} - \tilde{\mathbf{x}}$. Then

$$k(\mathbf{x}, \mathbf{y}) - \check{k}_\ell(\mathbf{x}, \mathbf{y}) = \frac{e^{\langle \mathbf{b}, \mathbf{z} \rangle_{\mathbf{A}} - \lambda \|\mathbf{z}\|_{\mathbf{A}}}}{4\pi \sqrt{\det \mathbf{A}} \|\mathbf{z}\|_{\mathbf{A}}} \zeta_2(\mathbf{z}, \tilde{\mathbf{z}}) + \frac{e^{\langle \mathbf{b}, \tilde{\mathbf{z}} \rangle_{\mathbf{A}} - \lambda \|\tilde{\mathbf{z}}\|_{\mathbf{A}}}}{4\pi \sqrt{\det \mathbf{A}}} \zeta_1(\mathbf{z}, \tilde{\mathbf{z}}).$$

where

$$\zeta_1(\mathbf{z}, \tilde{\mathbf{z}}) := \|\mathbf{z}\|_{\mathbf{A}}^{-1} - \|\tilde{\mathbf{z}}\|_{\mathbf{A}}^{-1} \quad \text{and} \quad \zeta_2(\mathbf{z}, \tilde{\mathbf{z}}) := 1 - e^{\langle \mathbf{b}, \tilde{\mathbf{z}} - \mathbf{z} \rangle_{\mathbf{A}} - \lambda(\|\tilde{\mathbf{z}}\|_{\mathbf{A}} - \|\mathbf{z}\|_{\mathbf{A}})}.$$

The definition of d_ℓ immediately implies that

$$|\zeta_1(\mathbf{z},\tilde{\mathbf{z}})| \leq d_\ell / \|\tilde{\mathbf{z}}\|_{\mathbf{A}} \leq d_\ell / \|\tilde{\mathbf{z}}\|_{\mathbf{A}} \overset{\text{Assumption 8.1.6b}}{\leq} \frac{d_\ell}{c_P c_{\mathbf{A}} \|\mathbf{z}\|_{\mathbf{A}}} \overset{(8.19)}{\leq} \frac{d_\ell}{c_P c_{\mathbf{A}}^2 \|\mathbf{z}\|}, \tag{8.22}$$

while the exponent in the definition of ζ_2 can be estimated by

$$|\langle \mathbf{b}, \tilde{\mathbf{z}} - \mathbf{z}\rangle_{\mathbf{A}} - \lambda (\|\tilde{\mathbf{z}}\|_{\mathbf{A}} - \|\mathbf{z}\|_{\mathbf{A}})| \leq \|\mathbf{b}\|_{\mathbf{A}} \|\tilde{\mathbf{z}} - \mathbf{z}\|_{\mathbf{A}} + |\lambda| d_\ell \|\mathbf{z}\| \leq \varepsilon_\ell \leq \tilde{c}.$$

This leads to

$$|\zeta_2(\mathbf{z},\tilde{\mathbf{z}})| \leq e^{\tilde{c}} \varepsilon_\ell.$$

Since the surface is bounded there is a constant $\widetilde{C} > 0$ such that

$$\sup_{\substack{\mathbf{x},\mathbf{y}\in\Gamma \\ \mathbf{z}=\mathbf{x}-\mathbf{y}}} \frac{e^{\langle \mathbf{b},\mathbf{z}\rangle_{\mathbf{A}} - \lambda\|\mathbf{z}\|_{\mathbf{A}}}}{\sqrt{\det \mathbf{A}}} \leq \widetilde{C} \quad \text{and} \quad \sup_{\substack{\mathbf{x},\mathbf{y}\in\Gamma \\ \tilde{\mathbf{z}}=\theta_\ell(\mathbf{x})-\theta_\ell(\mathbf{y})}} \frac{e^{\langle \mathbf{b},\tilde{\mathbf{z}}\rangle_{\mathbf{A}} - \lambda\|\tilde{\mathbf{z}}\|_{\mathbf{A}}}}{\sqrt{\det \mathbf{A}}} \leq \widetilde{C}. \tag{8.23}$$

Thus we have proved that

$$\left|k(\mathbf{x},\mathbf{y}) - \check{k}_\ell(\mathbf{x},\mathbf{y})\right| \leq \frac{C}{4\pi\|\mathbf{z}\|}(\varepsilon_\ell + d_\ell) \qquad \forall x, y \in \Gamma, x \neq y,$$

where C depends on the coefficients of the differential operator. □

For the proof of the following lemma, we will employ an inverse inequality in the form

$$\forall u \in \check{S} \qquad \|u\|_{L^2(\Gamma)} \leq C_{\text{inv}} h_\ell^{-1/2} \|u\|_{H^{-1/2}(\Gamma)}. \tag{8.24}$$

The constant C_{inv} is moderately bounded for shape-regular and quasi-uniform meshes. In addition, it depends on the polynomial degree of the shape functions (cf. Theorem 4.4.3).

Corollary 8.2.6. *Let the kernel function be defined by Assumption 8.1.14(1). Then there exist a constant $C > 0$ which depends only on the coefficients of the differential operator L, the diameter of Γ and C_{inv} in (8.24) such that for $s = 0, 1/2$*

$$\forall \check{u}, \check{v} \in \check{S}_\ell \quad \left|b(\check{u},\check{v}) - \check{b}_\ell(\check{u},\check{v})\right| \leq C h_\ell^{-1/2-s} \left\{(\varepsilon_\ell + d_\ell)\bar{\rho}_\ell^2 + (1+\bar{\rho}_\ell)\bar{\rho}_\ell^{\Delta}\right\} \|\check{u}\|_{H^{-s}(\Gamma)} \|\check{v}\|_{H^{-1/2}(\Gamma)}.$$

Proof. We start with the estimate of the sesquilinear form e_2 as in (8.18). The monotonicity of the Riemann integral allows us to use (8.21) to get

$$|e_2(\check{u},\check{v})| \leq C(\varepsilon_\ell + d_\ell) \int_\Gamma \int_\Gamma \frac{|\rho_\ell(\mathbf{y})\check{u}(\mathbf{y})| |\rho_\ell(\mathbf{x})\check{v}(\mathbf{x})|}{4\pi\|\mathbf{x}-\mathbf{y}\|} ds_{\mathbf{y}} ds_{\mathbf{x}}.$$

The continuity of the single layer potential for the Laplace operator [from $L^2(\Gamma)$ to $H^1(\Gamma)$ (cf. Theorem 3.1.16)] implies the continuity from $L^2(\Gamma)$ to $L^2(\Gamma)$. Thus

$$|e_2(\check{u},\check{v})| \leq C(\varepsilon_\ell + d_\ell)\int_\Gamma\int_\Gamma \frac{|\rho_\ell(\mathbf{y})\check{u}(\mathbf{y})|\,|\rho_\ell(\mathbf{x})\check{v}(\mathbf{x})|}{4\pi\|\mathbf{x}-\mathbf{y}\|} ds_{\mathbf{y}} ds_{\mathbf{x}} \leq C(\varepsilon_\ell + d_\ell)$$
$$\|\rho_\ell\check{u}\|_{L^2(\Gamma)}\|\rho_\ell\check{v}\|_{L^2(\Gamma)}$$
$$\overset{\text{Lemma 8.2.4}}{\leq} C(\varepsilon_\ell + d_\ell)\bar{\rho}_\ell^2\|\check{u}\|_{L^2(\Gamma)}\|\check{v}\|_{L^2(\Gamma)}, \tag{8.25}$$

where $\bar{\rho}_\ell$ is as in (8.14). The inverse inequality (cf. Remark 4.4.4) finishes the estimate of the sesquilinear form e_2.

The estimates of the sesquilinear forms e_1, e_3 are simpler because no perturbation of the kernel function appears therein and we may directly apply the continuity of the original boundary integral operator to obtain

$$|e_1(\check{u},\check{v})| = |b(\check{u},(1-\rho_\ell)\check{v})| \leq C\|\check{u}\|_{L^2(\Gamma)}\|(1-\rho_\ell)\check{v}\|_{L^2(\Gamma)}$$
$$\overset{\text{Lemma 8.2.4}}{\leq} C\bar{\rho}_\ell^\Delta\|\check{u}\|_{L^2(\Gamma)}\|\check{v}\|_{L^2(\Gamma)}. \tag{8.26}$$

In a similar fashion, the estimate

$$|e_3(\check{u},\check{v})| \leq C\bar{\rho}_\ell\bar{\rho}_\ell^\Delta\|\check{u}\|_{L^2(\Gamma)}\|\check{v}\|_{L^2(\Gamma)} \tag{8.27}$$

is derived and an inverse inequality finishes the proof. □

In the following we will briefly explain why our proof does not give a better estimate if we employ the $H^{-1/2}(\Gamma)$-continuity of the single layer potential directly instead of the $L^2(\Gamma)$-continuity along with an inverse inequality.

Note that the inclusion $\check{S} \subset L^2(\Gamma)$ implies that $|\rho_\ell\check{u}|, |\rho_\ell\check{v}| \in L^2(\Gamma) \subset H^{-1/2}(\Gamma)$ holds for all $\check{u} \in \check{S}$. Hence in (8.25) we could employ the $H^{-1/2}(\Gamma)$-continuity of the single layer potential to obtain

$$|e_2(\check{u},\check{v})| \leq C(\varepsilon_\ell + d_\ell)\,\||\rho_\ell\check{u}|\|_{H^{-1/2}(\Gamma)}\,\||\rho_\ell\check{v}|\|_{H^{-1/2}(\Gamma)}. \tag{8.28}$$

By the continuity of the embedding $L^2(\Gamma) \hookrightarrow H^{-1/2}(\Gamma)$ and an inverse inequality the estimate

$$\||\rho_\ell\check{u}|\|_{H^{-1/2}(\Gamma)} \leq C\,\||\rho_\ell\check{u}|\|_{L^2(\Gamma)} \leq C\bar{\rho}_\ell\,\||\check{u}|\|_{L^2(\Gamma)}$$
$$= C\bar{\rho}_\ell\|\check{u}\|_{L^2(\Gamma)} \leq \widetilde{C}h_\ell^{-1/2}\bar{\rho}_\ell\|\check{u}\|_{H^{-1/2}(\Gamma)}$$

follows. The combination with (8.28) yields

$$|e_2(\check{u},\check{v})| \leq Ch_\ell^{-1}(\varepsilon_\ell + d_\ell)\|\check{u}\|_{H^{-1/2}(\Gamma)}\|\check{v}\|_{H^{-1/2}(\Gamma)}. \tag{8.29}$$

Example 8.2.7 explains why the estimate

$$\||\check{u}|\|_{H^{-1/2}(\Gamma)} \leq C h_\ell^{-1/2} \|\check{u}\|_{H^{-1/2}(\Gamma)} \qquad \forall \check{u} \in \check{S},$$

in general, is sharp so that (8.29) cannot be improved along the lines of our proof. In order to avoid technicalities we consider the $H^{-1}(\Gamma)$-norm instead of the $H^{-1/2}(\Gamma)$-norm in the following example.

Example 8.2.7. *Let $\Omega = (0,1)$ and $N \in \mathbb{N}$. For $h = N^{-1}$, we set $x_i := ih$, $i = 0,1,\ldots,N$, and define the intervals $\tau_i = (x_{i-1}, x_i)$, $1 \leq i \leq N$, which define the mesh $\mathcal{G} := \{\tau_i : 1 \leq i \leq N\}$. For a function $f : \Omega \to \mathbb{R}$, we define its scaled and periodic version $f^{\text{per}} : \Omega \to \mathbb{R}$ elementwise by*

$$f^{\text{per}}|_{\tau_i} = f \circ \chi_i^{-1} \quad \text{with} \quad \chi_i^{-1}(x) = \frac{x - x_{i-1}}{h}, \qquad 1 \leq i \leq N.$$

We consider the function

$$u(x) := \begin{cases} -1 & x \in (0, 1/2), \\ 1 & x \in (1/2, 1). \end{cases}$$

Note that

$$u^{(-1)}(x) := \int_0^x u(s)\, ds = \left|\frac{1}{2} - x\right| - \frac{1}{2},$$

$$u^{(-2)}(x) = \int_0^x u^{(-1)}(s)\, ds = \frac{1}{2}\left(\frac{1}{4} - x + \left|x - \frac{1}{2}\right|\left(x - \frac{1}{2}\right)\right).$$

Note that $u^{\text{per}} \in S_{\mathcal{G}}^{0,-1}$ is piecewise constant. Our goal is to compute $\|u^{\text{per}}\|_{H^{-1}(\Omega)}$, where $H^{-1}(\Omega) = \left(H_0^1(\Omega)\right)'$. By definition, we have[1]

$$\|u^{\text{per}}\|_{H^{-1}(\Omega)} = \sup_{v \in H_0^1(\Omega)\backslash\{0\}} \frac{(u^{\text{per}}, v)_{L^2(\Omega)}}{|v|_{H^1(\Omega)}}. \tag{8.30}$$

For the computation of the maximizer v_0 we define the Lagrange function

$$J(v) = (u, v)_{L^2(\Omega)} - \lambda\left(|v|_{H^1(\Omega)}^2 - 1\right).$$

The equations for a stationary point (v_0, λ_0) are given by

$$\left(v_0', w'\right)_{L^2(\Omega)} = \frac{1}{2\lambda_0}(u^{\text{per}}, w)_{L^2(\Omega)} \qquad \forall w \in H_0^1(\Omega) \quad \wedge \quad |v_0|_{H^1(\Omega)} = 1.$$

[1] For simplicity, we employ the H^1-seminorm as the norm in $H_0^1(\Omega)$ which is equivalent to the standard norm.

The strong formulation is

$$-\left(v_0^{\text{per}}\right)'' = \frac{1}{2\lambda_0} u^{\text{per}} \quad \wedge \quad |v_0|_{H^1(\Omega)} = 1.$$

The solution is the scaled and period version v_0^{per} *of*

$$v_0(x) = -\frac{h^2}{2\lambda_0}\left(-x u^{(-2)}(1) + u^{(-2)}(x)\right),$$

where λ_0 *is chosen such that* $\left\| \left(v_0^{\text{per}}\right)' \right\|_{L^2(\Omega)} = 1$. *Since the scaling in (8.30) cancels we set* $\lambda_0 = 1$. *We get*

$$\left\| \left(v_0^{\text{per}}\right)' \right\|_{L^2(\Omega)}^2 = \sum_{i=1}^{N} \int_{\tau_i} \left| \left(v_0^{\text{per}}\right)' \right|^2 = h^{-2} \int_0^1 |v_0'|^2 = \frac{h^2}{192}$$

and

$$\left(u^{\text{per}}, v_0^{\text{per}}\right)_{L^2(\Omega)} = \sum_{i=1}^{N} \int_{\tau_i} u^{\text{per}} v_0^{\text{per}} = \int_0^1 u v_0 = \frac{h^2}{96}.$$

Hence

$$\|u^{\text{per}}\|_{H^{-1}(\Omega)} = \frac{\left(u^{\text{per}}, v_0^{\text{per}}\right)_{L^2(\Omega)}}{\left|v_0^{\text{per}}\right|_{H^1(\Omega)}} = \frac{\sqrt{3}}{12} h.$$

Since $|u^{\text{per}}| \equiv 1$ *we get*

$$|||u^{\text{per}}|||_{H^{-1}(\Omega)} = \|1\|_{H^{-1}(\Omega)} = C = O(1)$$

so that

$$|||u^{\text{per}}|||_{H^{-1}(\Omega)} = C \frac{12}{\sqrt{3}} h^{-1} \|u^{\text{per}}\|_{H^{-1}(\Omega)}.$$

Hence the estimate

$$|||u|||_{H^{-1}(\Omega)} \le \|u\|_{L^2(\Omega)} \le C h^{-1} \|u\|_{H^{-1}(\Omega)}$$

cannot be improved in general for piecewise constant functions.

We turn to the estimate of the derivatives of the fundamental solution. For $\mathbf{x}, \mathbf{y} \in \Gamma$, let $\tilde{\mathbf{x}} = \theta(\mathbf{x})$ and $\tilde{\mathbf{y}} \in \theta(\mathbf{y})$. The error estimates depend on the following quantities

$$n_{1,\ell}^{\mathrm{I}} := \max_{\substack{\check{\tau}_1,\check{\tau}_2 \in \check{\mathcal{G}}_\ell \\ \check{\tau}_1 \neq \check{\tau}_2}} \sup_{\substack{\mathbf{x}\in\check{\tau}_1 \\ \mathbf{y}\in\check{\tau}_2}} \frac{\left|\langle \mathbf{n}(\mathbf{x}), \mathbf{y}-\mathbf{x}\rangle - \langle \mathbf{n}_\ell(\tilde{\mathbf{x}}), \tilde{\mathbf{y}}-\tilde{\mathbf{x}}\rangle\right|}{\|\mathbf{y}-\mathbf{x}\|}, \quad n_\ell^{\mathrm{II}} := \max_{\check{\tau}\in\check{\mathcal{G}}_\ell} \sup_{\mathbf{x}\in\check{\tau}} \|\mathbf{n}(\mathbf{x}) - \mathbf{n}_\ell(\tilde{\mathbf{x}})\|,$$

$$n_{2,\ell}^{\mathrm{I}} := \max_{\check{\tau}\in\check{\mathcal{G}}_\ell} \sup_{\mathbf{x},\mathbf{y}\in\check{\tau}} \frac{\left|\langle \mathbf{n}(\mathbf{x}), \mathbf{y}-\mathbf{x}\rangle - \langle \mathbf{n}_\ell(\tilde{\mathbf{x}}), \tilde{\mathbf{y}}-\tilde{\mathbf{x}}\rangle\right|}{\|\mathbf{y}-\mathbf{x}\|^2}. \tag{8.31}$$

The proof of the following lemma employs two estimates which will be proved in Sect. 8.4:

$$\max_{|\beta|=2} \left\|\partial^\beta \chi_{\tau,p}\right\|_{C^0(\hat{\tau})} \overset{\text{Lemma 8.4.2}}{\leq} C_{\mathrm{stab}} h_\tau^2 \quad \text{and} \quad \sup_{\substack{\hat{\mathbf{x}},\hat{\mathbf{y}}\in\hat{\tau},\ \hat{\mathbf{x}}\neq\hat{\mathbf{y}} \\ \mathbf{x}=\chi_\tau(\hat{\mathbf{x}}),\ \mathbf{y}=\chi_\tau(\hat{\mathbf{y}})}} \frac{\|\mathbf{y}-\mathbf{x}\|}{\|\hat{\mathbf{y}}-\hat{\mathbf{x}}\|} \overset{(8.85)}{\geq} c h_\tau. \tag{8.32}$$

Lemma 8.2.8. *Let the kernel function be defined by Assumption 8.1.14(2). Then there exists a constant $C > 0$ which depends only on the coefficients of the differential operator L, the diameter of Γ, and the constant in Lemma 2.2.14 such that, for sufficiently small $0 < h \leq \underline{h}$,*

1. $\forall \mathbf{x},\mathbf{y} \in \check{\tau} \in \check{\mathcal{G}}$:

$$\left|k(\mathbf{x},\mathbf{y}) - \check{k}_\ell(\mathbf{x},\mathbf{y})\right| \leq \frac{C}{4\pi\|\mathbf{x}-\mathbf{y}\|}\left(n_{2,\ell}^{\mathrm{I}} + d_\ell + \varepsilon_\ell + \|\mathbf{b}\|\, n_\ell^{\mathrm{II}}\right). \tag{8.33a}$$

2. $\forall \check{\tau}_1, \check{\tau}_2 \in \check{\mathcal{G}}$, $\check{\tau}_1 \neq \check{\tau}_2$, *and* $\mathbf{x} \in \check{\tau}_1$ *and* $\mathbf{y} \in \check{\tau}_2$,

$$\left|k(\mathbf{x},\mathbf{y}) - \check{k}_\ell(\mathbf{x},\mathbf{y})\right| \leq \frac{C}{4\pi\|\mathbf{x}-\mathbf{y}\|}\left(\frac{n_{1,\ell}^{\mathrm{I}} + d_\ell + \varepsilon_\ell}{\|\mathbf{x}-\mathbf{y}\|} + \|\mathbf{b}\|\, n_\ell^{\mathrm{II}}\right). \tag{8.33b}$$

Proof. For $\mathbf{x},\mathbf{y} \in \Gamma$, let $\tilde{\mathbf{x}} := \theta_\ell(\mathbf{x})$ and $\tilde{\mathbf{y}} \in \theta_\ell(\mathbf{y})$. The differences are denoted by $\mathbf{z} = \mathbf{x}-\mathbf{y}$ and $\tilde{\mathbf{z}} := \tilde{\mathbf{x}} - \tilde{\mathbf{y}}$. We start with the splitting

$$\begin{aligned} k(\mathbf{x},\mathbf{y}) - \check{k}_\ell(\mathbf{x},\mathbf{y}) &= G(\mathbf{z})\left(-\lambda\eta_{1,\ell}^{\Delta}(\mathbf{x},\mathbf{y}) - \eta_{2,\ell}^{\Delta}(\mathbf{x},\mathbf{y}) + \varphi_\ell^{\Delta}(\mathbf{x})\right) \\ &\quad + \left(G(\mathbf{z}) - G(\tilde{\mathbf{z}})\right)\left(-\eta_{2,\ell}(\tilde{\mathbf{x}},\tilde{\mathbf{y}}) + \varphi_\ell(\tilde{\mathbf{x}}) - \lambda\eta_{1,\ell}(\tilde{\mathbf{x}},\tilde{\mathbf{y}})\right), \end{aligned} \tag{8.34}$$

where, for $\alpha \in \{1,2\}$,

$$\eta_\alpha(\mathbf{x},\mathbf{y}) := \frac{\langle \mathbf{x}-\mathbf{y}, \mathbf{n}(\mathbf{x})\rangle}{\|\mathbf{x}-\mathbf{y}\|_{\mathbf{A}}^{\alpha}}, \qquad \varphi(\mathbf{x}) := \langle \mathbf{b}, \mathbf{n}(\mathbf{x})\rangle,$$

$$\eta_{\alpha,\ell}(\mathbf{x},\mathbf{y}) := \frac{\langle \mathbf{x}-\mathbf{y}, \mathbf{n}_\ell(\mathbf{x})\rangle}{\|\mathbf{x}-\mathbf{y}\|_{\mathbf{A}}^{\alpha}}, \qquad \varphi_\ell(\mathbf{x}) := \langle \mathbf{b}, \mathbf{n}_\ell(\mathbf{x})\rangle,$$

$$\eta_{\alpha,\ell}^{\Delta}(\mathbf{x},\mathbf{y}) := \eta_\alpha(\mathbf{x},\mathbf{y}) - \eta_{\alpha,\ell}(\tilde{\mathbf{x}},\tilde{\mathbf{y}}), \quad \varphi_\ell^{\Delta}(\mathbf{x}) := \varphi(\mathbf{x}) - \varphi_\ell(\tilde{\mathbf{x}}).$$

Next, we will estimate the different terms in this splitting separately.

Estimate of $\eta^{\Delta}_{\alpha,\ell}$:
We distinguish two cases:

1. For $\mathbf{x}, \mathbf{y} \in \check{\tau} \in \check{\mathcal{G}}$, let $\tilde{\mathbf{x}}, \tilde{\mathbf{y}}$ be defined as before. Note that $\tilde{\mathbf{x}} = \chi_{\tau,p}(\hat{\mathbf{x}})$, $\tilde{\mathbf{y}} = \chi_{\tau,p}(\hat{\mathbf{y}})$ holds for $\hat{\mathbf{x}} := \chi_{\tau}^{-1}(\mathbf{x})$ and $\hat{\mathbf{y}} := \chi_{\tau}^{-1}(\mathbf{y})$. By using the quantities $n^{\mathrm{I}}_{2,\ell}$, d_ℓ we obtain

$$\left|\eta^{\Delta}_{\alpha,\ell}(\mathbf{x},\mathbf{y})\right| = \left|\frac{\langle \mathbf{x}-\mathbf{y}, \mathbf{n}(\mathbf{x})\rangle}{\|\mathbf{x}-\mathbf{y}\|^{\alpha}_{\mathbf{A}}} - \frac{\langle \tilde{\mathbf{x}}-\tilde{\mathbf{y}}, \mathbf{n}_\ell(\tilde{\mathbf{x}})\rangle}{\|\tilde{\mathbf{x}}-\tilde{\mathbf{y}}\|^{\alpha}_{\mathbf{A}}}\right|$$
$$\leq c_{\mathbf{A}}^{-\alpha}\|\mathbf{x}-\mathbf{y}\|^{2-\alpha} n^{\mathrm{I}}_{2,\ell} + c_{\mathbf{A}}^{-2\alpha} c_P^{-\alpha} \frac{|\langle \tilde{\mathbf{x}}-\tilde{\mathbf{y}}, \mathbf{n}_\ell(\tilde{\mathbf{x}})\rangle|}{\|\tilde{\mathbf{x}}-\tilde{\mathbf{y}}\|^{\alpha}} d_\ell$$
$$\leq C\left(h_\tau^{2-\alpha} n^{\mathrm{I}}_{2,\ell} + \frac{\left|\left\langle \chi_{\tau,p}(\hat{\mathbf{x}}) - \chi_{\tau,p}(\hat{\mathbf{y}}), \mathbf{n}_\ell \circ \chi_{\tau,p}(\hat{\mathbf{x}})\right\rangle\right|}{\|\tilde{\mathbf{x}}-\tilde{\mathbf{y}}\|^{\alpha}} d_\ell\right), \tag{8.35}$$

where C depends only on $c_{\mathbf{A}}$, α, and c_P.
The first factor in the second term in the right-hand side of (8.35) can be estimated by repeating the steps in the proof of Lemma 2.2.14. We obtain

$$\frac{\left|\left\langle \chi_{\tau,p}(\hat{\mathbf{x}}) - \chi_{\tau,p}(\hat{\mathbf{y}}), \mathbf{n}_\ell \circ \chi_{\tau,p}(\hat{\mathbf{x}})\right\rangle\right|}{\|\tilde{\mathbf{x}}-\tilde{\mathbf{y}}\|^{\alpha}} \leq C \max_{|\beta|=2}\left\|\partial^\beta \chi_{\tau,p}\right\|_{C^0(\hat{\tau})} c_P^{\alpha} \frac{\|\hat{\mathbf{x}}-\hat{\mathbf{y}}\|^2}{\|\mathbf{x}-\mathbf{y}\|^{\alpha}}, \tag{8.36}$$

where c_P is as in Assumption 8.1.6. From (8.32) we conclude that

$$\left|\eta^{\Delta}_{\alpha,\ell}(\mathbf{x},\mathbf{y})\right| \leq \widetilde{C} h_\tau^{2-\alpha}\left(n^{\mathrm{I}}_{2,\ell} + d_\ell\right), \tag{8.37a}$$

where $\widetilde{C}$ depends only on C in (8.36), C_{stab}, c_P, c in (8.32), and $c_{\mathbf{A}}$.
2. Let $\check{\tau}_1, \check{\tau}_2 \in \check{\mathcal{G}}$ with $\check{\tau}_1 \neq \check{\tau}_2$ and $\mathbf{x} \in \check{\tau}_1$, $\mathbf{y} \in \check{\tau}_2$. In the same way as before we derive

$$\left|\eta^{\Delta}_{\alpha,\ell}(\mathbf{x},\mathbf{y})\right| = \left|\frac{\langle \mathbf{x}-\mathbf{y}, \mathbf{n}(\mathbf{x})\rangle}{\|\mathbf{x}-\mathbf{y}\|^{\alpha}_{\mathbf{A}}} - \frac{\langle \tilde{\mathbf{x}}-\tilde{\mathbf{y}}, \mathbf{n}_\ell(\tilde{\mathbf{x}})\rangle}{\|\tilde{\mathbf{x}}-\tilde{\mathbf{y}}\|^{\alpha}_{\mathbf{A}}}\right| \leq C\|\mathbf{x}-\mathbf{y}\|^{1-\alpha}\left(n^{\mathrm{I}}_{1,\ell} + d_\ell\right). \tag{8.37b}$$

Estimate of $\eta_{\alpha,\ell}$:
As before, we obtain improved estimates if $\mathbf{x}, \mathbf{y}$ belong to the same panel

$$\eta_{\alpha,\ell}(\mathbf{x},\mathbf{y}) = \frac{|\langle \tilde{\mathbf{x}}-\tilde{\mathbf{y}}, \mathbf{n}_\ell(\tilde{\mathbf{x}})\rangle|}{\|\tilde{\mathbf{x}}-\tilde{\mathbf{y}}\|^{\alpha}_{\mathbf{A}}} \leq \begin{cases} C h_\tau^{2-\alpha} & \text{if } \mathbf{x},\mathbf{y} \in \check{\tau} \in \check{\mathcal{G}}, \\ C\|\mathbf{x}-\mathbf{y}\|^{1-\alpha} & \text{if } \mathbf{x} \in \check{\tau}_1, \mathbf{y} \in \check{\tau}_2. \end{cases} \tag{8.37c}$$

Estimate of φ_ℓ^Δ:
By employing the quantity n_ℓ^{II} we obtain

$$\left|\varphi_\ell^\Delta(\mathbf{x})\right| = \left|\langle \mathbf{b}, \mathbf{n}(\mathbf{x}) - \mathbf{n}_\ell(\tilde{\mathbf{x}})\rangle\right| \leq \|\mathbf{b}\| \, \|\mathbf{n}(\mathbf{x}) - \mathbf{n}_\ell(\tilde{\mathbf{x}})\| \leq \|\mathbf{b}\| \, n_\ell^{\mathrm{II}}. \tag{8.37d}$$

Estimate of φ_ℓ:
Since $\mathbf{n}_\ell$ has length 1,

$$|\varphi_\ell(\tilde{\mathbf{x}})| \leq \|\mathbf{b}\|. \tag{8.37e}$$

We have all ingredients to estimate the perturbation error (8.34) of the kernel function due to boundary approximation. Again, we will distinguish two cases:

(a) Let $\mathbf{x}, \mathbf{y} \in \check{\tau} \in \check{\mathcal{G}}$.
The combination of (8.21), (8.23), and (8.37) yields

$$\left|k(\mathbf{x},\mathbf{y}) - \check{k}_\ell(\mathbf{x},\mathbf{y})\right| \leq \frac{C}{4\pi \|\mathbf{x}-\mathbf{y}\|} \left(n_{2,\ell}^{\mathrm{I}} + d_\ell + \varepsilon_\ell + \|\mathbf{b}\| \, n_\ell^{\mathrm{II}}\right).$$

(b) Let $\mathbf{x} \in \check{\tau}_1$ and $\mathbf{y} \in \check{\tau}_2$ for some non-identical panels $\check{\tau}_1, \check{\tau}_2 \in \check{\mathcal{G}}$.
In this case, we obtain

$$\left|k(\mathbf{x},\mathbf{y}) - \check{k}_\ell(\mathbf{x},\mathbf{y})\right| \leq \frac{C}{4\pi \|\mathbf{x}-\mathbf{y}\|} \left(\frac{n_{1,\ell}^{\mathrm{I}} + d_\ell + \varepsilon_\ell}{\|\mathbf{x}-\mathbf{y}\|} + \|\mathbf{b}\| \, n_\ell^{\mathrm{II}}\right),$$

where C depends on the coefficients $\mathbf{A}$, $\mathbf{b}$, c of the differential operator L. □

Corollary 8.2.9. *Let the kernel function be defined by Assumption 8.1.14(2). Then there exists a constant $C > 0$ which depends only on the coefficients of the differential operator L, the diameter of Γ, the shape-regularity of the mesh, and the constant in Lemma 2.2.14 such that*

$$\forall \check{u}, \check{v} \in \check{S}_\ell: \qquad \left|b(\check{u},\check{v}) - \check{b}_\ell(\check{u},\check{v})\right| \leq C\delta_\ell \|\check{u}\|_{L^2(\Gamma)} \|\check{v}\|_{L^2(\Gamma)},$$

where

$$\delta_\ell := \bar{\rho}_\ell^\Delta + \bar{\rho}_\ell^{-2} \left\{n_{2,\ell}^{\mathrm{I}} h_\ell + \left(n_{1,\ell}^{\mathrm{I}} + d_\ell + \varepsilon_\ell\right)(1 + |\log h_\ell|)\right\}.$$

Proof. We consider first the term e_2 in (8.18). For any $\check{\tau} \in \check{\mathcal{G}}$, we introduce a partition of Γ and of $\check{\mathcal{G}}$ by

$$\begin{aligned}
\check{\mathcal{G}}_{\check{\tau}}^{\mathrm{I}} &:= \{\check{\tau}\}, & \Gamma_{\check{\tau}}^{\mathrm{I}} &:= \check{\tau}, \\
\check{\mathcal{G}}_{\check{\tau}}^{\mathrm{II}} &:= \left\{\check{\tau}_1 \in \check{\mathcal{G}} \backslash \{\check{\tau}\} : \overline{\check{\tau}_1} \cap \overline{\check{\tau}} \neq \emptyset\right\}, & \Gamma_{\check{\tau}}^{\mathrm{II}} &:= \bigcup_{\check{\tau}_1 \in \check{\mathcal{G}}_{\check{\tau}}^{\mathrm{II}}} \overline{\check{\tau}_1}, \\
\check{\mathcal{G}}_{\check{\tau}}^{\mathrm{III}} &:= \left\{\check{\tau}_1 \in \check{\mathcal{G}} : \overline{\check{\tau}_1} \cap \overline{\check{\tau}} = \emptyset\right\}, & \Gamma_{\check{\tau}}^{\mathrm{III}} &:= \bigcup_{\check{\tau}_1 \in \check{\mathcal{G}}_{\check{\tau}}^{\mathrm{III}}} \overline{\check{\tau}_1}.
\end{aligned}$$

With this partition at hand, we employ (8.33) to obtain

$$\begin{aligned}
|e_2(\check{u},\check{v})| &\le \sum_{\check{\tau}\in\check{\mathcal{G}}} \int_{\check{\tau}} \sum_{R\in\{\mathrm{I},\mathrm{II},\mathrm{III}\}} \int_{\Gamma_{\check{\tau}}^R} |\rho_\ell(\mathbf{x})\, \check{v}_\ell(\mathbf{x})| \left| k(\mathbf{x},\mathbf{y}) - \check{k}_\ell(\mathbf{x},\mathbf{y}) \right| \\
&\quad \times |\rho_\ell(\mathbf{y})\, \check{u}(\mathbf{y})|\, ds_{\mathbf{y}} ds_{\mathbf{x}} \\
&\le C\left(n_{2,\ell}^{\mathrm{I}} + d_\ell + \varepsilon_\ell\right) \bar{\rho}_\ell^2 \sum_{\check{\tau}\in\check{\mathcal{G}}} \|\check{v}_\ell\|_{L^\infty(\check{\tau})} \|\check{u}_\ell\|_{L^\infty(\check{\tau})} \\
&\quad \times \int_{\check{\tau}\times\check{\tau}} \frac{1}{4\pi\|\mathbf{x}-\mathbf{y}\|} ds_{\mathbf{y}} ds_{\mathbf{x}} \\
&\quad + C\bar{\rho}_\ell^2 \left(n_{1,\ell}^{\mathrm{I}} + d_\ell + \varepsilon_\ell\right) \sum_{\check{\tau}\in\check{\mathcal{G}}} \|\check{v}_\ell\|_{L^\infty(\check{\tau})} \sum_{R\in\{\mathrm{II},\mathrm{III}\}} \|\check{u}_\ell\|_{L^\infty(\Gamma_{\check{\tau}}^R)} \\
&\quad \times \int_{\check{\tau}} \int_{\Gamma_{\check{\tau}}^R} \frac{1}{4\pi\|\mathbf{x}-\mathbf{y}\|^2} ds_{\mathbf{y}} ds_{\mathbf{x}} \\
&\quad + C\bar{\rho}_\ell^2 \|\mathbf{b}\|\, n_\ell^{\mathrm{II}} \int_\Gamma \int_\Gamma \frac{|\check{u}_\ell(\mathbf{y})|\, |\check{v}_\ell(\mathbf{x})|}{4\pi\|\mathbf{x}-\mathbf{y}\|} ds_{\mathbf{y}} ds_{\mathbf{x}} \\
&= S_1 + S_2 + S_3.
\end{aligned}$$

Estimate of S_1:
By introducing local polar coordinates as in (5.53) with $s = 1$ one proves for any $\check{\tau} \in \check{\mathcal{G}}$

$$\int_{\check{\tau}\times\check{\tau}} \frac{1}{\|\mathbf{x}-\mathbf{y}\|} ds_{\mathbf{y}} ds_{\mathbf{x}} \le C h_\tau^3.$$

The combination of this estimate and Corollary 4.4.6 yields

$$\begin{aligned}
S_1 &\le C\left(n_{2,\ell}^{\mathrm{I}} + d_\ell + \varepsilon_\ell\right) \bar{\rho}_\ell^2 h_\ell \sum_{\check{\tau}\in\check{\mathcal{G}}} \|\check{v}_\ell\|_{L^2(\check{\tau})} \|\check{u}_\ell\|_{L^2(\check{\tau})} \\
&\le C\left(n_{2,\ell}^{\mathrm{I}} + d_\ell + \varepsilon_\ell\right) \bar{\rho}_\ell^2 h_\ell \|\check{v}_\ell\|_{L^2(\Gamma)} \|\check{u}_\ell\|_{L^2(\Gamma)}.
\end{aligned}$$

Estimate of S_3:
The continuity of the single layer potential in $L^2(\Gamma)$ results in

$$S_3 \le C\bar{\rho}_\ell^2 \|\mathbf{b}\|\, n_\ell^{\mathrm{II}} \|\check{u}_\ell\|_{L^2(\Gamma)} \|\check{v}_\ell\|_{L^2(\Gamma)}.$$

Estimate of S_2:
For non-identical panels $\check{\tau}_1, \check{\tau}_2 \in \check{\mathcal{G}}$ which have, at least, one common point, i.e., $\overline{\check{\tau}_1} \cap \overline{\check{\tau}_2} \neq \emptyset$, we may use (5.59), (5.60), resp. (5.62) to get

$$\int_{\check{\tau}_1\times\check{\tau}_2} \frac{1}{\|\mathbf{x}-\mathbf{y}\|^2} ds_{\mathbf{y}} ds_{\mathbf{x}} \le C h_{\tau_1}^2,$$

where the shape-regularity of the panels implies $h_{\tau_1} \sim h_{\tau_2}$.

The shape-regularity also implies that $\sharp\check{\mathcal{G}}^{\text{II}}_{\check{\tau}} = O(1)$ and hence the summation over all panels in $\check{\mathcal{G}}^{\text{II}}_{\check{\tau}}$ yields

$$\int_{\check{\tau}}\int_{\Gamma^{\text{II}}_{\check{\tau}}} \frac{1}{4\pi\,\|\mathbf{x}-\mathbf{y}\|^2} ds_{\mathbf{y}} ds_{\mathbf{x}} \leq C h^2_{\check{\tau}}.$$

For the integral over $\check{\tau} \times \Gamma^{\text{III}}_{\check{\tau}}$ we argue as in the proof (Part b) of Lemma 7.3.19 to obtain

$$\int_{\check{\tau}}\int_{\Gamma^{\text{III}}_{\check{\tau}}} \frac{1}{4\pi\,\|\mathbf{x}-\mathbf{y}\|^2} ds_{\mathbf{y}} ds_{\mathbf{x}} \leq C h^2_{\tau}\left(1+\left|\log h_{\tau}\right|\right).$$

This leads to the estimate of S_2

$$S_2 \leq C \bar{\rho}_\ell^{2}\left(n^{\text{I}}_{1,\ell} + d_\ell + \varepsilon_\ell\right) C h^2_\ell \left(1+\left|\log h_\ell\right|\right) \sum_{\check{\tau}\in\check{\mathcal{G}}} \|\check{v}_\ell\|_{L^\infty(\check{\tau})} \|\check{u}_\ell\|_{L^\infty(\Gamma)}.$$

Let $\check{\tau}_1 \in \check{\mathcal{G}}$ denote the triangle with $\|\check{u}_\ell\|_{L^\infty(\Gamma)} = \|\check{u}_\ell\|_{L^\infty(\check{\tau}_1)}$. By using the inverse inequality (cf. Corollary 4.4.6) we obtain

$$\|\check{u}_\ell\|_{L^\infty(\Gamma)} = \|\check{u}_\ell\|_{L^\infty(\check{\tau}_1)} \leq h^{-1}_{\tau_1} \|\check{u}_\ell\|_{L^2(\check{\tau}_1)} \leq h^{-1}_\ell \|\check{u}_\ell\|_{L^2(\Gamma)}$$

from which we conclude with a Cauchy–Schwarz inequality that

$$S_2 \leq C \bar{\rho}_\ell^{2}\left(n^{\text{I}}_{1,\ell} + d_\ell + \varepsilon_\ell\right)\left(1+\left|\log h_\ell\right|\right) \|\check{u}_\ell\|_{L^2(\Gamma)} \|\check{v}_\ell\|_{L^2(\Gamma)}.$$

This finishes the estimate of e_2

$$\left|e_2(\check{u},\check{v})\right| \leq C \bar{\rho}_\ell^{2}\left(n^{\text{I}}_{2,\ell} h_\ell + \left(n^{\text{I}}_{1,\ell} + d_\ell + \varepsilon_\ell\right)\left(1+\left|\log h_\ell\right|\right)\right) \|\check{v}_\ell\|_{L^2(\Gamma)} \|\check{u}_\ell\|_{L^2(\Gamma)}.$$

The estimates of e_1 and e_3 are based on the continuity of the integral operator and derived as estimates (8.26), (8.27). □

Corollary 8.2.10. *Let the kernel function be defined by Assumption 8.1.14(3). Then there exists a constant $C > 0$ which depends only on the coefficients of the differential operator L, the diameter of Γ, the shape-regularity of the mesh, and the constant in Lemma 2.2.14 such that*

$$\forall \check{u},\check{v} \in \check{S}_\ell : \qquad \left|b(\check{u},\check{v}) - \check{b}_\ell(\check{u},\check{v})\right| \leq C \tilde{\delta}_\ell \|\check{u}\|_{L^2(\Gamma)} \|\check{v}\|_{L^2(\Gamma)},$$

where (recall δ_ℓ as in Corollary 8.2.9)

$$\tilde{\delta}_\ell := C \bar{\rho}_\ell^{2}\left(n^{\text{II}}_\ell + d_\ell + \varepsilon_\ell\right) + \delta_\ell.$$

Proof. The kernel function for the double layer potential is (recall: $\mathbf{v_y} = \mathbf{An_y}$)

$$k(\mathbf{x}, \mathbf{y}) = \frac{\partial}{\partial \mathbf{v_y}} G(\mathbf{x}-\mathbf{y}) + 2\langle \mathbf{b}, \mathbf{v_y}\rangle G(\mathbf{x}-\mathbf{y}).$$

The perturbation due to the first part of the kernel function can be estimated as in Corollary 8.2.9 and we consider only the second summand. Let $\mathbf{z} := \mathbf{x} - \mathbf{y}$ and $\tilde{\mathbf{z}} := \tilde{\mathbf{x}} - \tilde{\mathbf{y}} := \theta_\ell(\mathbf{x}) - \theta_\ell(\mathbf{y})$. The splitting

$$\begin{aligned} k_\Delta^{\mathrm{I}}(\mathbf{x}, \mathbf{y}) &:= 2\langle \mathbf{b}, \mathbf{v_y}\rangle G(\mathbf{z}) - 2\langle \mathbf{b}, \mathbf{v}_{\ell,\tilde{\mathbf{y}}}\rangle G(\tilde{\mathbf{z}}) \\ &= 2\langle \mathbf{Ab}, \mathbf{n_y} - \mathbf{n}_{\ell,\tilde{\mathbf{y}}}\rangle G(\mathbf{z}) + 2\langle \mathbf{Ab}, \mathbf{n}_{\ell,\tilde{\mathbf{y}}}\rangle (G(\mathbf{z}) - G(\tilde{\mathbf{z}})) \end{aligned}$$

leads to

$$\begin{aligned} \left|k_\Delta^{\mathrm{I}}(\mathbf{x}, \mathbf{y})\right| &\overset{(8.23)}{\leq} C\left(\frac{\left\|\mathbf{n_y} - \mathbf{n}_{\ell,\tilde{\mathbf{y}}}\right\|}{4\pi\|\mathbf{z}\|} + |G(\mathbf{x}-\mathbf{y}) - G(\tilde{\mathbf{x}}-\tilde{\mathbf{y}})|\right) \\ &\overset{(8.37\mathrm{d}),\,(8.22),\,(8.23)}{\leq} C\frac{n_\ell^{\mathrm{II}} + d_\ell + \varepsilon_\ell}{4\pi\|\mathbf{z}\|}. \end{aligned}$$

For the estimate

$$\begin{aligned} &\left|\int_{\Gamma\times\Gamma} k_\Delta^{\mathrm{I}}(\mathbf{x}, \mathbf{y})\, \check{u}(\mathbf{y})\, \rho_\ell(\mathbf{y})\, \overline{\check{v}}(\mathbf{x})\, \rho_\ell(\mathbf{x})\, ds_{\mathbf{y}} ds_{\mathbf{x}}\right| \\ &\quad\leq C\bar{\rho}_\ell^2\left(n_\ell^{\mathrm{II}} + d_\ell + \varepsilon_\ell\right)\|\check{u}\|_{L^2(\Gamma)}\|\check{v}\|_{L^2(\Gamma)} \end{aligned}$$

we may argue as in (8.25) and this finishes the proof. □

It remains to estimate the hypersingular boundary integral equation. Let $\tau \in \mathcal{G}^{\text{affine}}$. The pullback of $\check{\tau} = \text{lift}_\tau(\tau) \in \check{\mathcal{G}}$ is χ_τ and its Jacobi matrix is denoted by $\mathbf{J}_\tau$. Similarly the Jacobi matrix for $\chi_{\tau,p}$ is $\mathbf{J}_{\tau,p}$.

Let

$$J_\ell := \max_{\tau\in\mathcal{G}^{\text{affine}}} \sup_{\hat{\mathbf{x}}\in\hat{\tau}} \frac{\left\|\mathbf{J}_\tau(\hat{\mathbf{x}}) - \mathbf{J}_{\tau,p}(\hat{\mathbf{x}})\right\|}{h_\tau}. \tag{8.38}$$

$\|\cdot\|$ denotes the matrix norm which is induced by the Euclidean vector norm. The proof of the following lemma requires an estimate of the number of mesh cells. Under a weak assumption on the mesh, there exists a constant $C_\sharp$ such that

$$\sharp\mathcal{G}^{\text{affine}} \leq C_\sharp h_\ell^{-2}. \tag{8.39}$$

The proof of the following lemma uses two results from Sect. 8.4 which reads as follows. Corollary 8.4.3 states that

$$\forall\hat{\mathbf{x}} \in \hat{\tau} \qquad \|\mathbf{G}_\tau(\hat{\mathbf{x}})\| \leq Ch_\tau^2 \quad \text{and} \quad \left\|\mathbf{G}_{\tau,p}(\hat{\mathbf{x}})\right\| \leq Ch_\tau^2 \tag{8.40a}$$

while Lemma 8.4.4 implies that

$$\left|g_{\tau,p}(\hat{\mathbf{x}})\right| \leq Ch_\tau^2 \quad \text{and} \quad |g_\tau(\hat{\mathbf{x}})| \leq Ch_\tau^2. \tag{8.40b}$$

Lemma 8.2.11. *Assume that $\check{S}_\ell \subset H^1(\Gamma)$. Let the sesquilinear forms b and $\check{b}_\ell$ be defined by (8.8) and (8.15). Let Assumption 8.1.6 and the estimates (8.40) be satisfied. Then there exists a constant $C > 0$ which depends only on the coefficients of the differential operator L, the diameter of Γ, the shape-regularity of the mesh, the constants in (8.40) and in Lemma 2.2.14, and the constants in (8.39) and the inverse inequality, such that*

$$\left|b(\check{u},\check{v}) - \check{b}_\ell(\check{u},\check{v})\right| \leq C\left(d_\ell + \varepsilon_\ell + n_\ell^{\mathrm{II}} + J_\ell + \bar{\rho}_\ell^{\Delta}\right) h_\ell^{-\frac{1}{2}-s} \|\check{u}\|_{H^{1-s}(\Gamma)} \|\check{v}\|_{H^{1/2}(\Gamma)}$$

for $s = 0, 1/2$ and any $\check{u}, \check{v} \in \check{S}_\ell$.

Proof. We will estimate the sesquilinear forms e_1, e_2, e_3 as in (8.18) and start with e_2.

The kernel function (cf. (8.8) consists of two parts. The estimate of the second part is simple by using the previous results. Again, let $\mathbf{z} := \mathbf{x} - \mathbf{y}$ and $\tilde{\mathbf{z}} := \tilde{\mathbf{x}} - \tilde{\mathbf{y}} := \theta_\ell(\mathbf{x}) - \theta_\ell(\mathbf{y})$. We have

$$\begin{aligned} k_\Delta^{\mathrm{II}}(\mathbf{x},\mathbf{y}) &:= G(\mathbf{z})\left\langle \mathbf{A}^{1/2}\mathbf{n}(\mathbf{x}), \mathbf{A}^{1/2}\mathbf{n}(\mathbf{y})\right\rangle - G(\tilde{\mathbf{z}})\left\langle \mathbf{A}^{1/2}\mathbf{n}_\ell(\tilde{\mathbf{x}}), \mathbf{A}^{1/2}\mathbf{n}_\ell(\tilde{\mathbf{y}})\right\rangle \\ &= G(\mathbf{z})\left(\langle \mathbf{A}\mathbf{n}(\mathbf{x}), \mathbf{n}(\mathbf{y})\rangle - \langle \mathbf{A}\mathbf{n}_\ell(\tilde{\mathbf{x}}), \mathbf{n}_\ell(\tilde{\mathbf{y}})\rangle\right) \\ &\quad + \left(G(\mathbf{z}) - G(\tilde{\mathbf{z}})\right)\langle \mathbf{A}\mathbf{n}_\ell(\tilde{\mathbf{x}}), \mathbf{n}_\ell(\tilde{\mathbf{y}})\rangle. \end{aligned}$$

The estimates

$$|G(\mathbf{z})| \leq \frac{C}{4\pi\|\mathbf{z}\|}, \qquad |G(\mathbf{z}) - G(\tilde{\mathbf{z}})| \leq \frac{C}{4\pi\|\mathbf{z}\|}(d_\ell + \varepsilon_\ell)$$

follow from Lemma 8.2.5, while

$$|\langle \mathbf{A}\mathbf{n}_\ell(\tilde{\mathbf{x}}), \mathbf{n}_\ell(\tilde{\mathbf{y}})\rangle| \leq C$$

and

$$|\langle \mathbf{A}\mathbf{n}(\mathbf{x}), \mathbf{n}(\mathbf{y})\rangle - \langle \mathbf{A}\mathbf{n}_\ell(\tilde{\mathbf{x}}), \mathbf{n}_\ell(\tilde{\mathbf{y}})\rangle| \leq C n_\ell^{\mathrm{II}}$$

are obvious. Hence we may argue as in the proof of Lemma 8.2.5 to obtain the estimate of the perturbation in the second part of the sesquilinear form associated with the operator W

$$\begin{aligned} &\left|\int_{\Gamma\times\Gamma} k_\Delta^{\mathrm{II}}(\mathbf{x},\mathbf{y})\,\rho_\ell(\mathbf{y})\,\check{u}_\ell(\mathbf{y})\,\rho_\ell(\mathbf{x})\,\overline{\check{v}_\ell}(\mathbf{x})\,ds_{\mathbf{y}}ds_{\mathbf{x}}\right| \\ &\quad \leq C\bar{\rho}_\ell^{-2}\left(d_\ell + \varepsilon_\ell + n_\ell^{\mathrm{II}}\right)\|\check{u}_\ell\|_{L^2(\Gamma)}\|\check{v}_\ell\|_{L^2(\Gamma)} \\ &\quad \leq C\bar{\rho}_\ell^{-2}\left(d_\ell + \varepsilon_\ell + n_\ell^{\mathrm{II}}\right)\|\check{u}_\ell\|_{H^{1/2}(\Gamma)}\|\check{v}_\ell\|_{H^{1/2}(\Gamma)}. \end{aligned} \tag{8.41}$$

We turn to the first part of the sesquilinear form (8.8). For $i = 1, 2$, fix some $\tau_i \in \mathcal{G}^{\text{affine}}$ and let $\check{\tau}_i := \operatorname{lift}_{\tau_i}(\tau_i)$. Define the local sesquilinear form b_{τ_1,τ_2} by

$$b_{\tau_1,\tau_2}(\check{u}, \check{v}) = \int_{\check{\tau}_1 \times \check{\tau}_2} G(\mathbf{x}-\mathbf{y}) \left\langle \operatorname{curl}_{\check{\tau}_2,\mathbf{A},\mathbf{0}} \check{u}(\mathbf{y}), \operatorname{curl}_{\check{\tau}_1,\mathbf{A},2\mathbf{b}} \overline{\check{v}}(\mathbf{x}) \right\rangle \rho_\ell(\mathbf{y}) \rho_\ell(\mathbf{x})\, ds_{\mathbf{y}} ds_{\mathbf{x}}.$$

We will transform the integrals to the reference element. For $i = 1, 2$, the local pullbacks of the normal vectors are

$$\hat{\mathbf{n}}_i = \mathbf{n} \circ \chi_{\tau_i}.$$

The column vectors of $\mathbf{A}^{1/2}$ are denoted by $\mathbf{a}_j \in \mathbb{R}^3$, $1 \leq j \leq 3$. Let $\mathbf{A}^{\times} := \mathbf{A}^{1/2} \times \mathbf{A}^{1/2} \in \mathbb{R}^{3\times 3}$ be the matrix formed by the column vectors $\mathbf{a}_2 \times \mathbf{a}_3$, $\mathbf{a}_3 \times \mathbf{a}_1$, $\mathbf{a}_1 \times \mathbf{a}_2$. Then, for any $\tau \in \mathcal{G}^{\text{affine}}$ and $\check{\tau} = \operatorname{lift}_\tau(\tau) \in \check{\mathcal{G}}$,

$$\operatorname{curl}_{\check{\tau},\mathbf{A},\mathbf{c}} \check{v} = \mathbf{A}^{\times} \operatorname{curl}_{\check{\tau},\mathbf{I},\mathbf{0}} \check{v} + \check{v} \left(\mathbf{A}^{-1/2}\mathbf{c} \times \mathbf{A}^{1/2} \mathbf{n}_{\check{\tau}} \right) \qquad \forall \check{v} \in H^1(\check{\tau}),$$

where $\mathbf{I}$ is the 3×3 identity matrix. The transformation formula (3.79) leads to

$$\left(\operatorname{curl}_{\check{\tau},\mathbf{I},\mathbf{0}} \check{v}\right) \circ \chi_\tau = \mathbf{J}_\tau \mathbf{G}_\tau^{-1} \widehat{\nabla} \hat{u} \times \mathbf{n}_{\check{\tau}},$$

where $\mathbf{J}_\tau$ is the Jacobi matrix of χ_τ and $\mathbf{G}_\tau := \mathbf{J}_\tau^{\mathsf{T}} \mathbf{J}_\tau$ denotes Gram's matrix. Some tedious but elementary tensor analysis leads to

$$\left(\operatorname{curl}_{\check{\tau},\mathbf{I},\mathbf{0}} \check{v}\right) \circ \chi_\tau = g_\tau^{-1} \mathbf{J}_\tau \hat{\nabla}^{\perp} \hat{v},$$

where $\widehat{\nabla}^{\perp} \hat{v} := (\partial_2 \hat{v}, -\partial_1 \hat{v})^{\mathsf{T}}$ and $g_\tau := \sqrt{\det \mathbf{G}_\tau}$ is the surface element. We introduce the matrix $\mathbf{A}^{\times\times}$ and the vector $\mathbf{w}_1$ by

$$\mathbf{A}^{\times\times} := \left(\mathbf{A}^{\times}\right)^{\mathsf{T}} \mathbf{A}^{\times} \quad \text{and} \quad \mathbf{w}_1 := \left(\mathbf{A}^{\times}\right)^{\mathsf{T}} \left(2\mathbf{A}^{-1/2}\mathbf{b} \times \mathbf{A}^{1/2} \hat{\mathbf{n}}_1 \right). \tag{8.42}$$

(If $\mathbf{A}$ is the identity matrix we have $\mathbf{A}^{1/2} = \mathbf{A}^{\times} = \mathbf{A}^{\times\times} = \mathbf{I}$.) Thus the pullback of $b_{\tau_1,\tau_2}(\check{u}, \check{v})$ can be written as

$$\begin{aligned} b_{\tau_1,\tau_2}(\check{u}, \check{v}) = \int_{\hat{\tau}\times\hat{\tau}} \hat{k}(\hat{\mathbf{x}}, \hat{\mathbf{y}}) \Big\{ & \left\langle \mathbf{A}^{\times\times} \widehat{\operatorname{curl}}_2 \hat{u}(\hat{\mathbf{y}}), \widehat{\operatorname{curl}}_1 \overline{\hat{v}}(\hat{\mathbf{x}}) \right\rangle \\ & + \left\langle \widehat{\operatorname{curl}}_2 \hat{u}(\hat{\mathbf{y}}), \mathbf{w}_1(\hat{\mathbf{x}}) \right\rangle g_1(\hat{\mathbf{x}}) \overline{\hat{v}}(\hat{\mathbf{x}}) \Big\}\, ds_{\mathbf{y}} ds_{\mathbf{x}}, \end{aligned} \tag{8.43a}$$

where

$$\hat{k}(\hat{\mathbf{x}}, \hat{\mathbf{y}}) = \hat{\rho}_\ell(\hat{\mathbf{x}}) \hat{\rho}_\ell(\hat{\mathbf{y}}) G\left(\chi_{\tau_1}(\hat{\mathbf{x}}) - \chi_{\tau_2}(\hat{\mathbf{y}})\right) \quad \text{and} \quad \widehat{\operatorname{curl}}_i \hat{w}(\hat{\mathbf{y}}) := \mathbf{J}_{\tau_i}(\hat{\mathbf{y}}) \hat{\nabla}^{\perp} \hat{w}(\hat{\mathbf{y}}).$$

The pullbacks of the finite element functions are

$$\hat{v} = \check{v} \circ \chi_{\tau_1} \quad \text{and} \quad \hat{u} = \check{u} \circ \chi_{\tau_2}.$$

To estimate the effect of the perturbation we have to compare b_{τ_1,τ_2} with

$$\tilde{b}_{\tau_1,\tau_2}(\check{u},\check{v}) = \int_{\check{\tau}_1 \times \check{\tau}_2} \hat{k}_\ell(\hat{\mathbf{x}},\hat{\mathbf{y}}) \left\{ \left\langle \mathbf{A}^{\times\times} \widehat{\operatorname{curl}_{2,\ell} \hat{u}}(\hat{\mathbf{y}}), \widehat{\operatorname{curl}_{1,\ell} \bar{\hat{v}}}(\hat{\mathbf{x}}) \right\rangle \right.$$
$$\left. + \left\langle \widehat{\operatorname{curl}_{2,\ell} \hat{u}}(\hat{\mathbf{y}}), \mathbf{w}_{1,\ell}(\hat{\mathbf{x}}) \right\rangle g_{1,\ell}(\hat{\mathbf{x}}) \bar{\hat{v}}(\hat{\mathbf{x}}) \right\} ds_{\mathbf{y}} ds_{\mathbf{x}}, \qquad (8.43b)$$

where

$$\hat{k}_\ell(\hat{\mathbf{x}},\hat{\mathbf{y}}) := G\left(\chi_{\tau_1,p}(\hat{\mathbf{x}}) - \chi_{\tau_2,p}(\hat{\mathbf{y}})\right) \hat{\rho}_\ell(\hat{\mathbf{x}}) \hat{\rho}_\ell(\hat{\mathbf{y}})$$

and the remaining quantities are defined by replacing χ_{τ_i} by $\chi_{\tau_i,p}$ (recall the relation (8.12) between $\chi_{\tau_i,p}$ and χ_{τ_i}).

We have to estimate corresponding quantities in (8.43) and their differences:

1. *Estimate of (the difference of) the kernel function*
   Lemma 8.2.5 implies that

$$|\rho_\ell(\mathbf{x}) \rho_\ell(\mathbf{y}) G(\tilde{\mathbf{x}} - \tilde{\mathbf{y}}) - \rho_\ell(\mathbf{x}) \rho_\ell(\mathbf{y}) G(\mathbf{x} - \mathbf{y})| \leq C(\varepsilon_\ell + d_\ell) \bar{\rho}_\ell^2 \frac{1}{4\pi \|\mathbf{x} - \mathbf{y}\|}$$

   with $\bar{\rho}_\ell$ as in (8.14).
   For the kernel function itself, we obtain (cf. Proof of Lemma 8.2.5)

$$|\rho_\ell(\mathbf{x}) \rho_\ell(\mathbf{y}) G(\mathbf{x} - \mathbf{y})| \leq C \bar{\rho}_\ell^2 \frac{1}{4\pi \|\mathbf{x} - \mathbf{y}\|}.$$

2. *Estimate of (the difference of) the first part of the sesquilinear form (8.43)*
   We obtain

$$\left\langle \mathbf{A}^{\times\times} \widehat{\operatorname{curl}_2 \hat{u}}(\hat{\mathbf{y}}), \widehat{\operatorname{curl}_1 \bar{\hat{v}}}(\hat{\mathbf{x}}) \right\rangle - \left\langle \mathbf{A}^{\times\times} \widehat{\operatorname{curl}_{2,\ell} \hat{u}}(\hat{\mathbf{y}}), \widehat{\operatorname{curl}_{1,\ell} \bar{\hat{v}}}(\hat{\mathbf{x}}) \right\rangle$$
$$= \left\langle \mathbf{A}^{\times\times} \left(\widehat{\operatorname{curl}_2 \hat{u}}(\hat{\mathbf{y}}) - \widehat{\operatorname{curl}_{2,\ell} \hat{u}}(\hat{\mathbf{y}}) \right), \widehat{\operatorname{curl}_1 \bar{\hat{v}}}(\hat{\mathbf{x}}) \right\rangle$$
$$+ \left\langle \mathbf{A}^{\times\times} \widehat{\operatorname{curl}_{2,\ell} \hat{u}}(\hat{\mathbf{y}}), \widehat{\operatorname{curl}_1 \bar{\hat{v}}}(\hat{\mathbf{x}}) - \widehat{\operatorname{curl}_{1,\ell} \bar{\hat{v}}}(\hat{\mathbf{x}}) \right\rangle. \qquad (8.44)$$

   The definition of J_ℓ [cf. (8.38)] implies

$$\left\| \widehat{\operatorname{curl}_1 \hat{v}}(\hat{\mathbf{x}}) - \widehat{\operatorname{curl}_{1,\ell} \hat{v}}(\hat{\mathbf{x}}) \right\| = \left\| \left(\mathbf{J}_{\tau_1}(\hat{\mathbf{x}}) - \mathbf{J}_{\tau_1,p}(\hat{\mathbf{x}}) \right) \widehat{\nabla}^\perp \hat{v}(\hat{\mathbf{x}}) \right\| \leq J_\ell h_{\tau_1} \left\| \widehat{\nabla} \hat{v}(\hat{\mathbf{x}}) \right\|.$$

   Recall that $\hat{v} = \check{v} \circ P \circ \chi_{\tau_2,p}$ so that the chain rule gives us

$$\widehat{\nabla} \hat{v}(\hat{\mathbf{x}}) = \mathbf{J}_{\tau_1,p}^{\mathsf{T}}(\hat{\mathbf{x}}) \mathbf{J}_P^{\mathsf{T}}(\tilde{\mathbf{x}}) \nabla_\Gamma \check{v}(\mathbf{x}),$$

where $\tilde{\mathbf{x}} := \chi_{\tau_1,p}(\hat{\mathbf{x}})$, $\mathbf{x} := P(\tilde{\mathbf{x}})$ and $\mathbf{J}_P$ denotes the Jacobi matrix of P. The surface gradient is defined as the composition $\nabla_\Gamma \check{v} = \gamma_0 \nabla Z_- \check{v}$, where Z_- denotes the lifting of $\check{v}$ to Ω^- (cf. Remark 2.6.12).
The Euclidean norm of $\mathbf{J}_P$ is bounded

$$\left\| \mathbf{J}_P^\intercal(\tilde{\mathbf{x}}) \right\| \leq C,$$

where C depends only on c_P in Assumption 8.1.6. Hence

$$\left\| \widehat{\nabla} \hat{v}(\hat{\mathbf{x}}) \right\| \leq C \left\| \mathbf{J}_{\tau_1,p}(\hat{\mathbf{x}}) \mathbf{J}_{\tau_1,p}^\intercal(\hat{\mathbf{x}}) \right\|^{1/2} \left\| \nabla_\Gamma \check{v}(\mathbf{x}) \right\|.$$

From elementary linear algebra follows the first equality in

$$\left\| \mathbf{J}_{\tau_1,p}(\hat{\mathbf{x}}) \mathbf{J}_{\tau_1,p}^\intercal(\hat{\mathbf{x}}) \right\|^{1/2} = \left\| \mathbf{G}_{\tau_1,p}(\mathbf{x}) \right\|^{1/2} \overset{(8.40a)}{\leq} C h_{\tau_1}.$$

Thus we have shown that

$$\left\| \widehat{\operatorname{curl}_1} \hat{v}(\hat{\mathbf{x}}) - \widehat{\operatorname{curl}_{1,\ell}} \hat{v}(\hat{\mathbf{x}}) \right\| \leq J_\ell h_{\tau_1}^2 \left\| \widehat{\nabla}_\Gamma \check{v}(\mathbf{x}) \right\|. \tag{8.45}$$

Similarly, we obtain

$$\left\| \widehat{\operatorname{curl}_{2,\ell}} \hat{u}(\hat{\mathbf{y}}) \right\| \leq \left\| \mathbf{G}_{\tau_2,p}(\hat{\mathbf{y}}) \right\|^{1/2} \left\| \widehat{\nabla} \hat{u}(\hat{\mathbf{y}}) \right\| \leq C h_{\tau_2}^2 \left\| \nabla_\Gamma \check{u}(\mathbf{y}) \right\|. \tag{8.46a}$$

The estimate

$$\left\| \widehat{\operatorname{curl}_2} \hat{u}(\hat{\mathbf{y}}) \right\| \leq C h_{\tau_2}^2 \left\| \nabla_\Gamma \check{u}(\hat{\mathbf{y}}) \right\| \tag{8.46b}$$

is derived analogously to (8.46a). From (8.46), one derives

$$\left| \left\langle \mathbf{A}^{\times\times} \widehat{\operatorname{curl}_{2,\ell}} \hat{u}(\hat{\mathbf{y}}), \widehat{\operatorname{curl}_1} \bar{\hat{v}}(\hat{\mathbf{x}}) - \widehat{\operatorname{curl}_{1,\ell}} \bar{\hat{v}}(\hat{\mathbf{x}}) \right\rangle \right| \leq C J_\ell h_{\tau_1}^2 h_{\tau_2}^2 \left\| \nabla_\Gamma \check{v}(\mathbf{x}) \right\| \left\| \nabla_\Gamma \check{u}(\mathbf{y}) \right\|.$$

The same arguments, applied to the first term on the right-hand side in (8.44), altogether leads to the estimate

$$\begin{aligned} &\left| \left\langle \mathbf{A}^{\times\times} \widehat{\operatorname{curl}_2} \hat{u}(\hat{\mathbf{y}}), \widehat{\operatorname{curl}_1} \bar{\hat{v}}(\hat{\mathbf{x}}) \right\rangle - \left\langle \mathbf{A}^{\times\times} \widehat{\operatorname{curl}_{2,\ell}} \hat{u}(\hat{\mathbf{y}}), \widehat{\operatorname{curl}_{1,\ell}} \bar{\hat{v}}(\hat{\mathbf{x}}) \right\rangle \right| \\ &\quad \leq C J_\ell h_{\tau_1}^2 h_{\tau_2}^2 \left\| \nabla_\Gamma \check{v}(\mathbf{x}) \right\| \left\| \nabla_\Gamma \check{u}(\mathbf{y}) \right\|. \end{aligned}$$

For the term $\left\langle \mathbf{A}^{\times\times} \widehat{\operatorname{curl}_2} \hat{u}(\hat{\mathbf{y}}), \widehat{\operatorname{curl}_1} \bar{\hat{v}}(\hat{\mathbf{x}}) \right\rangle$, we apply (8.46) twice to obtain

$$\left| \left\langle \mathbf{A}^{\times\times} \widehat{\operatorname{curl}_2} \hat{u}(\hat{\mathbf{y}}), \widehat{\operatorname{curl}_1} \bar{\hat{v}}(\hat{\mathbf{x}}) \right\rangle \right| \leq C h_{\tau_1}^2 h_{\tau_2}^2 \left\| \nabla_\Gamma \check{v}(\mathbf{x}) \right\| \left\| \nabla_\Gamma \check{u}(\mathbf{y}) \right\|.$$

3. *Estimate of (the difference in) the second part of the sesquilinear forms (8.43)* The estimate of the difference

$$\Delta := \left\langle \widehat{\operatorname{curl}_2 \hat{u}}(\hat{\mathbf{y}}), \mathbf{w}_1(\hat{\mathbf{x}}) \right\rangle g_1(\hat{\mathbf{x}}) \,\bar{\hat{v}}(\hat{\mathbf{x}}) - \left\langle \widehat{\operatorname{curl}_{2,\ell} \hat{u}}(\hat{\mathbf{y}}), \mathbf{w}_{1,\ell}(\hat{\mathbf{x}}) \right\rangle g_{1,\ell}(\hat{\mathbf{x}}) \,\bar{\hat{v}}(\hat{\mathbf{x}})$$

requires us to investigate corresponding quantities in both terms and their differences:

(a) As in (8.45) and (8.46) we get

$$\left\| \widehat{\operatorname{curl}_2 \hat{u}}(\hat{\mathbf{y}}) - \widehat{\operatorname{curl}_{2,\ell} \hat{u}}(\hat{\mathbf{y}}) \right\| \leq C J_\ell h_{\tau_2}^2 \left\| \nabla_\Gamma \check{u}(\mathbf{y}) \right\|,$$
$$\left\| \widehat{\operatorname{curl}_{2,\ell} \hat{u}}(\hat{\mathbf{y}}) \right\| \leq C h_{\tau_2}^2 \left\| \nabla_\Gamma \check{u}(\mathbf{y}) \right\|, \qquad \left\| \widehat{\operatorname{curl}_2 \hat{u}}(\hat{\mathbf{y}}) \right\| \leq C h_{\tau_2}^2 \left\| \nabla_\Gamma \check{u}(\mathbf{y}) \right\|.$$

(b) The definition of the vector $\mathbf{w}_1$ [cf. (8.42)], directly implies that

$$\left\| \mathbf{w}_1(\hat{\mathbf{x}}) \right\| \leq C, \quad \left\| \mathbf{w}_{1,\ell}(\hat{\mathbf{x}}) \right\| \leq C, \quad \left\| \mathbf{w}_1(\hat{\mathbf{x}}) - \mathbf{w}_{1,\ell}(\hat{\mathbf{x}}) \right\| \leq C n_\ell^{\mathrm{II}},$$

where n_ℓ^{II} is as in (8.31) and C depends only on the coefficients $\mathbf{A}$ and $\mathbf{b}$.

(c) It remains to consider the surface element g_1. From (8.40b) we conclude that

$$\left| g_1(\hat{\mathbf{x}}) \right| \leq C h_{\tau_1}^2.$$

Recall the definition of $\bar{\rho}_\ell^\Delta$ as in (8.14). The difference can be estimated by

$$\left| g_1(\hat{\mathbf{x}}) - g_{1,\ell}(\hat{\mathbf{x}}) \right| = \left| \rho_\ell(\hat{\mathbf{x}}) - 1 \right| \left| g_1(\hat{\mathbf{x}}) \right| \leq C \bar{\rho}_\ell^\Delta h_{\tau_1}^2.$$

By combining these inequalities we obtain

$$|\Delta| \leq C \left(J_\ell + n_\ell^{\mathrm{II}} + \bar{\rho}_\ell^\Delta \right) h_{\tau_1}^2 h_{\tau_2}^2 \left\| \nabla_\Gamma \check{u}(\mathbf{y}) \right\| \left| \check{v}(\mathbf{x}) \right|.$$

Since the function $\left\| \chi_{\tau_1}(\hat{\mathbf{x}}) - \chi_{\tau_2}(\hat{\mathbf{y}}) \right\|^{-1}$ is integrable there exists a constant such that

$$\int_{\hat{\tau} \times \hat{\tau}} \frac{1}{4\pi \left\| \chi_{\tau_1}(\hat{\mathbf{x}}) - \chi_{\tau_2}(\hat{\mathbf{y}}) \right\|} d\hat{\mathbf{x}} d\hat{\mathbf{y}} \leq C h_{\tau_2}^{-2} h_{\tau_1}^{-2} \int_{\check{\tau}_1 \times \check{\tau}_2} \frac{1}{4\pi \left\| \mathbf{x} - \mathbf{y} \right\|} ds_{\mathbf{y}} ds_{\mathbf{x}}.$$

In summary, we have estimated

$$\left| b_{\tau_1,\tau_2}(\check{u}, \check{v}) - \tilde{b}_{\tau_1,\tau_2}(\check{u}, \check{v}) \right| \leq C \bar{\rho}_\ell^2 \left(\varepsilon_\ell + d_\ell + J_\ell + n_\ell^{\mathrm{II}} + \bar{\rho}_\ell^\Delta \right)$$
$$\times \int_{\check{\tau}_1 \times \check{\tau}_2} \frac{\left\| \nabla_\Gamma \check{u}(\mathbf{y}) \right\| \left(\left\| \nabla_\Gamma \check{v}(\mathbf{x}) \right\| + \left| \check{v}(\mathbf{x}) \right| \right)}{4\pi \left\| \mathbf{x} - \mathbf{y} \right\|} ds_{\mathbf{y}} ds_{\mathbf{y}}.$$

Let $b(\cdot,\cdot) := \sum_{\check{\tau}_1,\check{\tau}_2\in\check{\mathcal{G}}} b_{\tau_1,\tau_2}(\cdot,\cdot)$ and define $\tilde{b}(\cdot,\cdot)$ analogously. A triangle inequality leads to

$$\left|b(\check{u},\check{v}) - \tilde{b}(\check{u},\check{v})\right| \leq C\bar{\rho}_\ell^2\left(\varepsilon_\ell + d_\ell + J_\ell + n_\ell^{\mathrm{II}} + \bar{\rho}_\ell^{\Delta}\right)$$
$$\int_{\Gamma\times\Gamma} \frac{\|\nabla_\Gamma \check{u}(\mathbf{y})\|\left(\|\nabla_\Gamma \check{v}(\mathbf{x})\| + |\check{v}(\mathbf{x})|\right)}{4\pi\|\mathbf{x}-\mathbf{y}\|} ds_{\mathbf{y}} ds_{\mathbf{y}}.$$

Our assumption $\check{S}_\ell \subset H^1(\Gamma)$ implies that the functions $\|\nabla_\Gamma \check{u}(\cdot)\|$ and $\|\nabla_\Gamma \check{v}(\cdot)\| + |\check{v}(\cdot)|$ are in $L^2(\Gamma)$. From the continuity of the single layer potential we obtain

$$\left|b(\check{u},\check{v}) - \tilde{b}(\check{u},\check{v})\right| \leq C\bar{\rho}_\ell^2\left(\varepsilon_\ell + d_\ell + J_\ell + n_\ell^{\mathrm{II}} + \bar{\rho}_\ell^{\Delta}\right)\|\check{u}\|_{H^1(\Gamma)}\|\check{v}\|_{H^1(\Gamma)}. \tag{8.47}$$

The combination of (8.41) and (8.47) yields the estimate of the perturbation e_2

$$|e_2(\check{u},\check{v})| \leq C\left(d_\ell + \varepsilon_\ell + n_\ell^{\mathrm{II}} + h_\ell^{-1/2-s}\left(\varepsilon_\ell + d_\ell + J_\ell + \bar{\rho}_\ell^{\Delta} + n_\ell^{\mathrm{II}}\right)\right)$$
$$\|\check{u}\|_{H^{1-s}(\Gamma)}\|\check{v}\|_{H^{1/2}(\Gamma)} \tag{8.48a}$$

for $s = 0, 1/2$.

The estimates of the perturbations e_1 and e_3 are based on the L^2-continuity of the single layer potential. Lemma 8.2.4 and an inverse inequality imply for $s = 0, 1/2$ that

$$|e_1(\check{u},\check{v})| \leq C\bar{\rho}_\ell^{\Delta}\|\check{u}\|_{H^1(\Gamma)}\|\check{v}\|_{H^1(\Gamma)} \leq C\bar{\rho}_\ell^{\Delta}h_\ell^{-1/2-s}\|\check{u}\|_{H^{1-s}(\Gamma)}\|\check{v}\|_{H^{1/2}(\Gamma)}. \tag{8.48b}$$

In a similar fashion one proves that

$$|e_3(\check{u},\check{v})| \leq C\bar{\rho}_\ell\bar{\rho}_\ell^{\Delta}h_\ell^{-1/2-s}\|\check{u}\|_{H^{1-s}(\Gamma)}\|\check{v}\|_{H^{1/2}(\Gamma)}. \tag{8.48c}$$

The assertion follows from (8.48). □

8.3 Overview of the Orders of the p-Parametric Surface Approximations

In the previous section, we have analyzed the perturbations which are introduced by the p-parametric surface approximation. The errors are expressed in terms of geometric quantities. In Sect. 8.4, we will estimate these geometric perturbations by using elementary tools of differential geometry. Here, we will summarize these results and combine them with the perturbation analysis of the previous section to derive conditions for the orders of the p-parametric surface approximation such that the optimal convergence rates of the original Galerkin method are preserved.

First, we introduce the appropriate notion of smoothness of the surface Γ.

Definition 8.3.1. The mapping P in (8.12) belongs to the class $\mathcal{A}_{\mathrm{pw}}^{m}$ if there is a neighborhood U_Γ of Γ such that $P : U_\Gamma \to \Gamma$ is Lipschitz continuous and:

1. $\Gamma^p \subset U_\Gamma$.
2. $P|_{\Gamma^p} : \Gamma^p \to \Gamma$ is bi-Lipschitz continuous:

$$c_P \|\mathbf{x} - \mathbf{y}\| \leq \|P(\mathbf{x}) - P(\mathbf{y})\| \leq c_P^{-1} \|\mathbf{x} - \mathbf{y}\| \qquad \forall \mathbf{x}, \mathbf{y} \in \Gamma^p.$$

3. For any $\tau \in \mathcal{G}^{\mathrm{affine}}$ and $\tilde{\tau} = P(\tau)$, the restriction $P|_\tau : \tau \to \tilde{\tau}$ is a C^m diffeomorphism for some $m \geq 1$.

A mapping $P \in \mathcal{A}_{\mathrm{pw}}^{m}$ belongs to $\mathcal{A}^m$ if $\Gamma \in C^2$ and if Condition 2 can be replaced by

$$P|_{\Gamma^p} : \Gamma^p \to \Gamma \text{ is a } C^1\text{-diffeomorphism.}$$

For the error analysis, we assume that χ_τ is as in (8.12) and that the mapping P satisfies Assumptions 8.1.3, resp. 8.1.6. If $P : U_\Gamma \to \Gamma$ is m times differentiable the quantity

$$C_{\Gamma,m} := \max_{\substack{\alpha \in \mathbb{N}_0^3 \\ |\alpha| \leq m}} \sup_{\mathbf{x} \in U_\Gamma} \|\partial^\alpha P(\mathbf{x})\| \tag{8.49}$$

is well defined. We distinguish between the following two cases

$$P \text{ belongs to } \mathcal{A}_{\mathrm{pw}}^{m} \text{ for } m \geq k+2, \tag{8.50a}$$

$$P \text{ belongs to } \mathcal{A}^{m} \text{ for } m \geq k+2 \text{ and is chosen as the conormal projection,} \tag{8.50b}$$

where k denotes the algebraic polynomial degree of the finite element space (cf. Definition 8.1.12).

In Sect. 8.4, we will prove that, for the p-parametric surface approximation of degree p, the quantities $d_\ell, \tilde{d}_\ell, \ldots$ can be estimated from above by Ch_ℓ^q, where the corresponding values of q are listed in Table 8.1. Hence the perturbations in the sesquilinear forms can be estimated by using Corollaries 8.2.6, 8.2.9, 8.2.10 and Lemma 8.2.11.

Exemplarily, we discuss the integral equation for the single layer potential [cf. (4.9)]

$$\text{Find } \varphi \in H^{-1/2}(\Gamma) : (V\varphi, \eta)_{L^2(\Gamma)} = (g_D, \eta)_{L^2(\Gamma)} \qquad \forall \eta \in H^{-1/2}(\Gamma). \tag{8.51}$$

We assume that the coefficients $\mathbf{A}$, $\mathbf{b}$, c are such that Assumption 8.2.1 is satisfied. We consider the discretization of (8.51) by (k, p)-boundary elements. Corollary 8.2.6 implies that

$$\left| b(\check{u}, \check{v}) - \breve{b}_\ell(\check{u}, \check{v}) \right| \leq Ch_\ell^{-1/2-s} \left\{ (\varepsilon_\ell + d_\ell)\, \bar{\rho}_\ell^2 + \bar{\rho}_\ell^\Delta \right\} \|\check{u}\|_{H^{-s}(\Gamma)} \|\check{v}\|_{H^{-1/2}(\Gamma)}$$

Table 8.1 Convergence orders of geometric quantities which are related to the surface approximation

	Case (8.50a)		Case (8.50b)	
	$\mathbf{b} = \mathbf{0}$	$\mathbf{b} \neq \mathbf{0}$	$\mathbf{b} = \mathbf{0}$	$\mathbf{b} \neq \mathbf{0}$
d_ℓ (8.20a)	p Lem. 8.4.12 Cor. 8.4.13	p Lem. 8.4.12 Cor. 8.4.13	$p+1$ Lem. 8.4.12 Cor. 8.4.13	$p+1$ Lem. 8.4.12 Cor. 8.4.13
$\tilde{d}_\ell$ (8.20b)	p Lem. 8.4.11	p Lem. 8.4.11	p Lem. 8.4.11	p Lem. 8.4.11
ε_ℓ (8.20b)	p	p	$p+1$	p
$\bar{\rho}_\ell$ (8.14)	0 Lem. 8.4.4	0 Lem. 8.4.4	0 Lem. 8.4.4	0 Lem. 8.4.4
$\bar{\rho}_\ell^\Delta$ (8.14)	p Cor. 8.4.7	p Cor. 8.4.7	$p+1$ Lem. 8.4.10	$p+1$ Lem. 8.4.10
n_ℓ^{II} (8.31)	p (8.95b)	p (8.95b)	p (8.95b)	p (8.95b)
$n_{1,\ell}^{\mathrm{I}}$ (8.31)	p Lem. 8.4.14	p Lem. 8.4.14	p Lem. 8.4.14	p Lem. 8.4.14
$n_{2,\ell}^{\mathrm{I}}$ (8.31)	$p-1$ Lem. 8.4.14	$p-1$ Lem. 8.4.14	$p-1$ Lem. 8.4.14	$p-1$ Lem. 8.4.14
J_ℓ (8.38)	p (8.71)	p (8.71)	p (8.71)	p (8.71)

for $s \in \{0, 1/2\}$ and all $\check{u}, \check{v} \in \check{S}_\ell$. For the piecewise smooth case, i.e., Case (8.50a), we obtain from Table 8.1

$$\left| b(\check{u}, \check{v}) - \check{b}_\ell(\check{u}, \check{v}) \right| \leq C h_\ell^{p-1/2-s} \|\check{u}\|_{H^{-s}(\Gamma)} \|\check{v}\|_{H^{-1/2}(\Gamma)} .$$

Hence the sequences in (8.16) are given by $c_\ell = h_\ell^{p-1}$, $\delta_\ell^{\mathrm{I}} = h_\ell^{p-1/2}$, and the uniform stability follows for sufficiently large $\ell \geq \ell_0$ from

$$\left| \check{b}_\ell(\check{u}, \check{v}) \right| \leq |b(\check{u}, \check{v})| + \left| b(\check{u}, \check{v}) - \check{b}_\ell(\check{u}, \check{v}) \right| \leq C \left(1 + h_\ell^{p-1} \right) \|\check{u}\|_{H^{-1/2}(\Gamma)} \|\check{v}\|_{H^{-1/2}(\Gamma)}$$

by taking into account Assumption 8.2.1. If the boundary element space consists of piecewise polynomials of degree k on a piecewise smooth surface, the optimal choice of p is given by

$$p = k + 2.$$

All other cases can be treated in an analogous way. Table 8.2 lists the values of c_ℓ and δ_ℓ^{I}, where $s = 1/2$ corresponds to c_ℓ and $s = 0$ to δ_ℓ^{I}. Finally, Table 8.3 lists the orders of the p-parametric surface approximation so that the convergence order is as in the unperturbed case (up to, possibly, a logarithmic factor $(1 + \log |h|)$.

To preserve the convergence rates of the perturbed Galerkin method with respect to weaker norms or, e.g., field point evaluations, the order of the p-parametric surface approximation has to chosen higher as for the energy norm. The error due to the geometric perturbation has to be in balance with the convergence order of the unperturbed Galerkin method. For field point evaluations, the convergence rates are twice as high as the rates with respect to the energy norm provided the problem is sufficiently regular (cf. Sect. 4.2.5.1), $|u(\mathbf{x}) - u_\ell(\mathbf{x})| \leq Ch^{2k+2-2\mu}$ for any $\mathbf{x} \in \Omega$. The constant C is independent of the mesh width but, possibly, depends on $\mathbf{x}$. Table 8.4 lists the required orders of the p-parametric approximation to preserve these higher convergence rates.

Table 8.2 Values of c_ℓ and δ_ℓ^I in the Strang Lemma (cf. Remark 8.2.3) where $s = 1/2$ corresponds to c_ℓ and $s = 0$ to δ_ℓ^I

Operator	Case (8.50a)		Case (8.50b)	
	$\mathbf{b} = 0$	$\mathbf{b} \neq 0$	$\mathbf{b} = 0$	$\mathbf{b} \neq 0$
V	$Ch_\ell^{p-1/2-s}$	$Ch_\ell^{p-1/2-s}$	$Ch_\ell^{p+1/2-s}$	$Ch_\ell^{p-1/2-s}$
K	$Ch_\ell^p (1 + \|\log h_\ell\|)$	$Ch_\ell^p (1 + \|\log h_\ell\|)$	$Ch_\ell^p (1 + \|\log h_\ell\|)$	$Ch_\ell^p (1 + \|\log h_\ell\|)$
K'	$Ch_\ell^p (1 + \|\log h_\ell\|)$	$Ch_\ell^p (1 + \|\log h_\ell\|)$	$Ch_\ell^p (1 + \|\log h_\ell\|)$	$Ch_\ell^p (1 + \|\log h_\ell\|)$
W	$Ch_\ell^{p-1/2-s}$	$Ch_\ell^{p-1/2-s}$	Ch_ℓ^p	Ch_ℓ^p

Table 8.3 Required orders p of the p-parametric surface approximation so that the overall convergence rates are preserved (up to, possibly, a logarithmic factor $(1 + \log |h|)$)

Operator	Case (8.50a)		Case (8.50b)	
	$\mathbf{b} = 0$	$\mathbf{b} \neq 0$	$\mathbf{b} = 0$	$\mathbf{b} \neq 0$
V	$p = k + 2$	$p = k + 2$	$p = k + 1$	$p = k + 2$
K	$p = k + 1$	$p = k + 1$	$p = k + 1$	$p = k + 1$
K'	$p = k + 1$	$p = k + 1$	$p = k + 1$	$p = k + 1$
W	$p = k + 1$	$p = k + 1$	$p = k + 1$	$p = k + 1$

Table 8.4 Required orders p of the p-parametric surface approximation so that the overall convergence rates with respect to field point evaluations are preserved

Operator	Case (8.50a)		Case (8.50b)	
	$\mathbf{b} = 0$	$\mathbf{b} \neq 0$	$\mathbf{b} = 0$	$\mathbf{b} \neq 0$
V	$p = 2k + 4$	$p = 2k + 4$	$p = 2k + 3$	$p = 2k + 4$
K	$p = 2k + 2$	$p = 2k + 2$	$p = 2k + 2$	$p = 2k + 2$
K'	$p = 2k + 2$	$p = 2k + 2$	$p = 2k + 2$	$p = 2k + 2$
W	$p = 2k + 2$	$p = 2k + 2$	$p = 2k + 1$	$p = 2k + 1$

8.4 Elementary Differential Geometry

The convergence analysis of boundary element methods with p-parametric surface approximation requires some estimates for d_ℓ, $\tilde{d}_\ell$, ε_ℓ in (8.20) which, in turn, require the approximation of the normal vector, the surface element, and the kernel function. In this section, we will develop these results from basic calculus. We avoid the use of the intrinsic calculus of differential geometry but employ the parametrizations which are also used for the numerical realization of boundary element methods.

Some of the results in this section go back to [167]. Further papers on this topic are [168], [80, Chap. XIII, Sect. 2], [84], [21], [63, Sect. 1.4].

Recall that $\breve{\mathcal{G}}$ denotes the boundary element mesh on the original surface Γ and $\mathcal{G}^{\text{affine}}$ is the corresponding affine surface mesh. The p-parametric surface mesh is $\mathcal{G}^p$. For $\tau \in \mathcal{G}^{\text{affine}}$, the lifted panel in $\mathcal{G}^p$ is denoted by $\tilde{\tau} = \text{lift}_{\tau,p}(\tau)$ (cf. Notation 8.1.7) and the corresponding panel in $\breve{\mathcal{G}}$ by $\breve{\tau} = P(\tilde{\tau}) = \chi_\tau(\hat{\tau})$, where

$$\chi_\tau = P \circ \chi_{\tau,p} : \hat{\tau} \to \breve{\tau}. \tag{8.52}$$

For the convergence analysis, we consider a sequence of meshes $\mathcal{G}_\ell^{\text{affine}}, \mathcal{G}_\ell^p, \ldots$ (cf. Sect. 4 and Notation 8.1.15), while we skip the index ℓ here for ease of notation. Throughout this section we assume that Assumption 8.1.16 is satisfied.

Let $\tau \in \mathcal{G}^{\text{affine}}$ and $\check{\tau} = \text{lift}_\tau(\tau) \in \check{\mathcal{G}}$. Recall that $\chi_{\tau,p} = \widehat{I}^p\left(P \circ \chi_\tau^{\text{affine}}\right)$, while the pullback of $\check{\tau}$ is given by $\chi_\tau = P \circ \chi_{\tau,p}$. We will apply the interpolation error estimates as derived in Sect. 4.3.

Lemma 8.4.1. *Let P in (8.52) belong to the class $\mathcal{A}_{\text{pw}}^{p+1}$ (cf. Definition 8.3.1) for some $p \geq 1$. Then, for $1 \leq i \leq 3$ and $\alpha \in \mathbb{N}_0^2$ with $|\alpha| \leq p$, we have*

$$\left\| \left(P \circ \chi_{\tau,p} - \chi_{\tau,p}\right)_i \right\|_{C^0(\widehat{\tau})} \leq C_0 C_{\Gamma,p+1} h_\tau^{p+1}, \tag{8.53a}$$

$$\left\| \partial^\alpha \left(P \circ \chi_{\tau,p} - \chi_{\tau,p}\right)_i \right\|_{C^0(\widehat{\tau})} \leq C_1 C_{\Gamma,p+1} h_\tau^{p+1}, \tag{8.53b}$$

where C_0, C_1 depend only on p and $C_{\Gamma,p+1}$ is as in (8.49).

Proof. We have $\chi_{\tau,p} = \widehat{I}^p\left(P \circ \chi_\tau^{\text{affine}}\right)$. For any nodal point $\widehat{P}_{(i,j)}^{(p)}$, we have $\chi_{\tau,p}\left(\widehat{P}_{(i,j)}^{(p)}\right) \in \Gamma$ and thus $\chi_{\tau,p} = \widehat{I}^p\left(P \circ \chi_{\tau,p}\right)$. This leads to

$$\left(P \circ \chi_{\tau,p} - \chi_{\tau,p}\right)_i(\hat{\mathbf{x}}) = \left(P \circ \chi_{\tau,p} - \widehat{I}^p\left(P \circ \chi_{\tau,p}\right)\right)_i(\hat{\mathbf{x}})$$

and we may apply standard interpolation estimates to bound the right-hand side and its derivatives.

Fix $1 \leq i \leq 3$ and write $\check{\eta}_p := \left(P \circ \chi_{\tau,p}\right)_i$ resp. $\eta_p := \left(\chi_{\tau,p}\right)_i$. Note that $\widehat{I}^p \check{\eta}_p = \eta_p$. Let $T_p\left(\check{\eta}_p\right)$ denote the p-th order Taylor expansion of $\check{\eta}_p$ about the barycenter $\widehat{\mathbf{M}}$ of $\hat{\tau}$. Then

$$\check{\eta}_p - \eta_p = \left(\check{\eta}_p - T_p\left(\check{\eta}_p\right)\right) + \left(T_p\left(\check{\eta}_p\right) - \widehat{I}^p \check{\eta}_p\right) = \left(\check{\eta}_p - T_p\left(\check{\eta}_p\right)\right) + \widehat{I}^p\left(T_p\left(\check{\eta}_p\right) - \check{\eta}_p\right).$$

This leads to

$$\left\|\check{\eta}_p - \eta_p\right\|_{C^0(\widehat{\tau})} \leq \left(1 + \left\|\widehat{I}^p\right\|_{C^0(\widehat{\tau}) \leftarrow C^0(\widehat{\tau})}\right) \left\|T_p\left(\check{\eta}_p\right) - \check{\eta}_p\right\|_{C^0(\widehat{\tau})}. \tag{8.54}$$

Standard error estimates for two-dimensional Taylor expansions lead to

$$\left\|T_p\left(\check{\eta}_p\right) - \check{\eta}_p\right\|_{C^0(\widehat{\tau})} \leq \frac{1}{(p+1)!} \max_{0 \leq j \leq p+1} \left\|\partial_1^j \partial_2^{p+1-j} \check{\eta}_p\right\|_{C^0(\widehat{\tau})}.$$

In Lemma 8.4.2, we will prove that

$$\max_{0 < j \leq p+1} \left\|\partial_1^j \partial_2^{p+1-j} \check{\eta}_p\right\|_{C^0(\widehat{\tau})} \leq C h_\tau^{p+1},$$

where C depends only on p and $C_{\Gamma,p+1}$.

It remains to estimate the first factor in (8.54). For any $w \in C^0\left(\overline{\hat{\tau}}\right)$, let $w_p := \widehat{I}^p(w)$. The coefficient vector of w_p with respect to the nodal basis is denoted by $\mathbf{w}_p$. We argue as in the proof of Corollary 4.4.6 to obtain

$$\left\|w_p\right\|_{C^0\left(\overline{\hat{\tau}}\right)} \overset{\text{Theorem 2.5.4}}{\leq} C\left\|w_p\right\|_{H^{p+1}(\hat{\tau})} \overset{\text{Lemma 4.3.1}}{\leq} C'\left[w_p\right]_{p+1}$$
$$\overset{(4.262)}{=} C'\left\|\mathbf{w}_p\right\|_{\ell^1} \leq C''\left\|\mathbf{w}_p\right\|_{\ell^\infty} \leq C''\|w\|_{C^0\left(\overline{\hat{\tau}}\right)}, \tag{8.55}$$

where C'' depends only on p.

For the estimate of the derivatives (8.53b), we employ

$$\begin{aligned}\partial^\alpha\left(\check{\eta}_p - \eta_p\right) &= \left\{\partial^\alpha\left(\check{\eta}_p - T_p\left(\check{\eta}_p\right)\right) + \partial^\alpha\left(T_p\left(\check{\eta}_p\right) - \widehat{I}^p\check{\eta}_p\right)\right\} \\ &= \left(\partial^\alpha\check{\eta}_p - \partial^\alpha T_p\left(\check{\eta}_p\right)\right) + \partial^\alpha\widehat{I}^p\left(T_p\left(\check{\eta}_p\right) - \check{\eta}_p\right).\end{aligned}$$

The inverse inequality (cf. Theorem 4.4.2), applied to the reference element $\hat{\tau}$, results in

$$\left\|\partial_j w_p\right\|_{C^0\left(\overline{\hat{\tau}}\right)} \leq \widetilde{C}_{\text{inv}}\left\|w_p\right\|_{C^0\left(\overline{\hat{\tau}}\right)} \overset{(8.55)}{\leq} \widetilde{C}_{\text{inv}}C''\|w\|_{C^0\left(\overline{\hat{\tau}}\right)} \tag{8.56}$$

and a recursive application of the first estimate in (8.56) yields

$$\left\|\partial^\alpha w_p\right\|_{C^0\left(\overline{\hat{\tau}}\right)} \leq \widetilde{C}_{\text{inv}}^{|\alpha|}C''\|w\|_{C^0\left(\overline{\hat{\tau}}\right)}.$$

The Taylor expansion commutes with the derivatives, more precisely, $\partial^\alpha T_p\left(\check{\eta}_p\right) = T_{p-|\alpha|}\left(\partial^\alpha\check{\eta}_p\right)$, where formally, we set $T_k(f) = 0$ if $k < 0$. Thus we obtain

$$\begin{aligned}\left|\partial^\alpha\left(\check{\eta}_p - \eta_p\right)(\hat{\mathbf{x}})\right| &\leq \left|\left(\partial^\alpha\check{\eta}_p - T_{p-|\alpha|}\left(\partial^\alpha\check{\eta}_p\right)\right)(\hat{\mathbf{x}})\right| + \left(\widetilde{C}_{\text{inv}}C''\right)^{|\alpha|}\frac{C_{\Gamma,p+1}}{(p+1)!}\left(3h_\tau\right)^{p+1} \\ &\leq C_1 C_{\Gamma,p+1}h_\tau^{p+1},\end{aligned}$$

where C_1 depends only on p. □

Lemma 8.4.2. *Let $p \geq 1$ and let P in (8.52) belong to the class $\mathcal{A}_{\text{pw}}^k$ for some k. For any $\mu \in \mathbb{N}_0^2$ with $|\mu| = k$,*[2]

$$\left\|\partial^\mu \chi_{\tau,p}\right\|_{C^0\left(\overline{\hat{\tau}}\right)} \leq C_{\text{stab}}h_\tau^k, \tag{8.57a}$$

[2] Note that, for $k > p$ and $|\mu| = k$, the derivatives $\partial^\mu \chi_{\tau,p}$ vanish and the right-hand side in (8.57a) may even be replaced by 0.

where C_{stab} depends only on p, k, and $C_{\Gamma,k}$. The derivatives of the pullback $\chi_\tau = P \circ \chi_{\tau,p}$ satisfy

$$\|\partial^\mu \chi_\tau\|_{C^0(\widehat{\tau})} \leq \widetilde{C}_{\mathrm{stab}} h_\tau^k, \tag{8.57b}$$

where $\widetilde{C}_{\mathrm{stab}}$ depends only on p, k, and $C_{\Gamma,k}$.

Proof. First, we will prove (8.57b) for the case $p = 1$, i.e., $\chi_{\tau,p} = \chi_\tau^{\mathrm{affine}}$. The mapping $\chi_\tau^{\mathrm{affine}}$ has the form

$$\chi_\tau^{\mathrm{affine}}(\hat{\mathbf{x}}) = \mathbf{B}_\tau \hat{\mathbf{x}} + \mathbf{b}_\tau$$

with some $\mathbf{B}_\tau \in \mathbb{R}^{3\times 2}$ and $\mathbf{b}_\tau \in \mathbb{R}^3$. The columns of $\mathbf{B}_\tau$ are denoted by $\mathbf{a}_1, \mathbf{a}_2 \in \mathbb{R}^3$. As in the proof of Lemma 4.3.6, we derive

$$\partial^\mu \left(P_i \circ \chi_\tau^{\mathrm{affine}}\right) = \sum_{\substack{\beta\in\mathbb{N}_0^3\\ |\beta|=\mu_1}} \sum_{\substack{\nu\in\mathbb{N}_0^3\\ |\nu|=\mu_2}} \frac{\mu!}{\beta!\nu!} \mathbf{a}_1^\beta \mathbf{a}_2^\nu \left(\partial^{\beta+\nu} P_i\right) \circ \chi_\tau^{\mathrm{affine}}.$$

Next, we employ $\left|(\mathbf{B}_\tau)_{i,j}\right| \leq h_\tau$ and obtain

$$\sup_{\hat{\mathbf{x}}\in\hat{\tau}} \left|\partial^\mu \left(P_i \circ \chi_\tau^{\mathrm{affine}}\right)(\hat{\mathbf{x}})\right| \leq C_{\Gamma,k} (3h_\tau)^k, \tag{8.58}$$

where $C_{\Gamma,k}$ is as in (8.49).

Next, we will prove (8.57a) for general $p \in \mathbb{N}$. The result is trivial for $k > p$ and we restrict to the case $0 \leq k \leq p$. We use the same notation as in the previous proof. (Recall that $\breve{\eta}_1 = P_i \circ \chi_\tau^{\mathrm{affine}}$ and $\eta_p = \widehat{I}^p \breve{\eta}_1$.) We obtain

$$\partial^\mu \eta_p = \partial^\mu \left(\widehat{I}^p \breve{\eta}_1\right) = \partial^\mu \left(I - \widehat{I}^p\right)(T_{k-1}\breve{\eta}_1 - \breve{\eta}_1) + \partial^\mu \breve{\eta}_1, \tag{8.59}$$

where I is the identity and, formally, we put $T_{-1}\breve{\eta}_1 = 0$. Observe that $\partial^\mu T_{k-1}\breve{\eta}_1 = 0$. By taking norms on both sides in (8.59) and by using (8.58), we obtain

$$\begin{aligned}\left\|\partial^\mu \eta_p\right\|_{C^0(\widehat{\tau})} &\leq \|\partial^\mu \breve{\eta}_1\|_{C^0(\widehat{\tau})} + \gamma_\mu \|T_{k-1}\breve{\eta}_1 - \breve{\eta}_1\|_{C^0(\widehat{\tau})} + C_{\Gamma,k}(3h_\tau)^k \\ &\leq \gamma_\mu \|T_{k-1}\breve{\eta}_1 - \breve{\eta}_1\|_{C^0(\widehat{\tau})} + 2C_{\Gamma,k}(3h_\tau)^k\end{aligned}$$

where [cf. (8.56)]

$$\gamma_\mu := \sup_{w\in C^0(\widehat{\tau})\setminus\{0\}} \frac{\left\|\partial^\mu \widehat{I}^p w\right\|_{C^0(\widehat{\tau})}}{\|w\|_{C^0(\widehat{\tau})}} \leq \left(\widetilde{C}_{\mathrm{inv}} C''\right)^k.$$

For the estimate of the error in the Taylor expansion, we employ (8.58) and get

$$\|T_{k-1}\breve{\eta}_1 - \breve{\eta}_1\|_{C^0(\hat{\tau})} \leq \frac{C_{\Gamma,k}}{k!}(3h_\tau)^k. \tag{8.60}$$

The combination of these estimates proves assertion (8.57a).

Next, we will estimate the derivatives of the pullback $\chi_\tau = P \circ \chi_{\tau,p}$ of a surface panel $\breve{\tau} = \operatorname{lift}_\tau(\tau) \in \breve{\mathcal{G}}$ for general $p \geq 1$ and employ Lemma 4.3.11 (with the substitutions $u \leftarrow P_i$ and $\eta \leftarrow \chi_{\tau,p}$). We obtain

$$\left(\partial^\mu \breve{\eta}_p\right) = \sum_{|\beta|=1}^{|\mu|} c_\beta \left(\partial^\beta P_i\right) \circ \chi_{\tau,p}$$

where the coefficients c_β are real linear combinations of the terms

$$\prod_{r=1}^{|\beta|} \partial^{\lambda_r} \left(\chi_{\tau,p}\right)_{n_r},$$

where, for $1 \leq r \leq |\beta|$, the indices obey the relations $1 \leq n_r \leq 3$, $\lambda_r \in \mathbb{N}_0^3$, and $\sum_{r=1}^{|\beta|} |\lambda_r| = |\mu|$. By using (8.57a) we get

$$\left|\prod_{r=1}^{|\beta|} \partial^{\lambda_r} \left(\chi_{\tau,p}\right)_{n_r}(\hat{\mathbf{x}})\right| \leq \prod_{r=1}^{|\beta|} \left(C_{\text{stab}} h_\tau^{|\lambda_r|}\right) \leq \left(C_{\text{stab}} h_\tau\right)^{|\mu|}.$$

Thus the coefficients c_β satisfy the same estimate (with a different constant C) and we obtain

$$\left|\left(\partial^\mu \breve{\eta}_p\right)(\hat{\mathbf{x}})\right| = C h_\tau^{|\mu|},$$

where C depends only on k, p, and $C_{\Gamma,k}$.

□

Corollary 8.4.3. *The Gram matrices for χ_τ and $\chi_{\tau,p}$ satisfy*

$$\forall \hat{\mathbf{x}} \in \hat{\tau} \qquad \|\mathbf{G}_\tau(\hat{\mathbf{x}})\| \leq C C_{\Gamma,1}^2 h_\tau^2 \quad \textit{and} \quad \left\|\mathbf{G}_{\tau,p}(\hat{\mathbf{x}})\right\| \leq C C_{\Gamma,1}^2 h_\tau^2. \tag{8.61}$$

Proof. Let $\hat{\mathbf{t}}_1(\hat{\mathbf{x}})$, $\hat{\mathbf{t}}_2(\hat{\mathbf{x}})$ denote the column vectors of $\mathbf{J}_\tau(\hat{\mathbf{x}})$. The Gram matrix can be written in the form $\mathbf{G}_\tau(\hat{\mathbf{x}}) = \left(\left\langle \hat{\mathbf{t}}_i(\hat{\mathbf{x}}), \hat{\mathbf{t}}_j(\hat{\mathbf{x}})\right\rangle_{i,j=1,2}\right)$. Lemma 8.4.2 implies that

$$\left|\left\langle \hat{\mathbf{t}}_i(\hat{\mathbf{x}}), \hat{\mathbf{t}}_j(\hat{\mathbf{x}})\right\rangle\right| \leq C h_\tau^2.$$

Hence the matrix norm

$$\|\mathbf{G}_\tau(\hat{\mathbf{x}})\| := \max_{1\le i,j\le 2} \left|(\mathbf{G}_\tau(\hat{\mathbf{x}}))_{i,j}\right|$$

is bounded by Ch_τ^2. Since $\mathbf{G}_\tau(\hat{\mathbf{x}})$ is a 2×2 matrix and norms in finite-dimensional spaces are equivalent the first assertion follows.

The proof of the estimate for $\mathbf{G}_{\tau,p}$ is analogous. □

For $\chi_{\tau,p}$ and $\chi_\tau = P\circ\chi_{\tau,p}$, the surface elements g_τ and $g_{\tau,p}$ can be written in the form

$$g_\tau(\hat{\mathbf{x}}) = \|\partial_1\chi_\tau(\hat{\mathbf{x}})\times\partial_2\chi_\tau(\hat{\mathbf{x}})\| \quad \text{and} \quad g_{\tau,p}(\hat{\mathbf{x}}) = \left\|\partial_1\chi_{\tau,p}(\hat{\mathbf{x}})\times\partial_2\chi_{\tau,p}(\hat{\mathbf{x}})\right\|.$$

Lemma 8.4.4. *Let P in (8.52) belong to the class $\mathcal{A}_{\text{pw}}^{p+1}$ for some $p\ge 1$. Then, for all $\hat{\mathbf{x}}\in\hat{\tau}$,*

$$\left|g_{\tau,p}(\hat{\mathbf{x}}) - g_\tau(\hat{\mathbf{x}})\right| \le C_2 h_\tau^{p+2} \quad \text{and} \quad \left|g_{\tau,p}(\hat{\mathbf{x}})\right| \le C_3 h_\tau^2. \tag{8.62a}$$

For sufficiently small $h_\tau \le \underline{h}$, we have the lower estimates

$$c_3 h_\tau^2 \le \left|g_{\tau,p}(\hat{\mathbf{x}})\right| \quad \text{and} \quad \tilde{c}_3 h_\tau^2 \le \left|g_\tau(\hat{\mathbf{x}})\right|. \tag{8.62b}$$

The constants $C_2, c_3, \tilde{c}_3, C_3$ depend only on $C_{\text{stab}}, C_{\Gamma,p+1}$, and p while $\underline{h}$ may in addition depend on the shape-regularity of the mesh [cf. (4.17)].

Proof. The reverse triangle inequality leads to

$$\begin{aligned}\left|g_{\tau,p}(\hat{\mathbf{x}}) - g_\tau(\hat{\mathbf{x}})\right| &\le \left\|\partial_1\chi_\tau(\hat{\mathbf{x}})\times\partial_2\chi_\tau(\hat{\mathbf{x}}) - \partial_1\chi_{\tau,p}(\hat{\mathbf{x}})\times\partial_2\chi_{\tau,p}(\hat{\mathbf{x}})\right\| \\ &\le \left\|\partial_1\left(\chi_\tau(\hat{\mathbf{x}}) - \chi_{\tau,p}(\hat{\mathbf{x}})\right)\times\partial_2\chi_\tau(\hat{\mathbf{x}})\right\| \\ &\quad + \left\|\partial_1\chi_{\tau,p}(\hat{\mathbf{x}})\times\partial_2\left(\chi_\tau(\hat{\mathbf{x}}) - \chi_{\tau,p}(\hat{\mathbf{x}})\right)\right\| \\ &=: S_1 + S_2.\end{aligned} \tag{8.63}$$

We start with the estimate of S_2 and employ (8.53) to obtain

$$S_2 \le \left\|\partial_1\chi_{\tau,p}(\hat{\mathbf{x}})\right\| \left\|\partial_2\left(\chi_\tau(\hat{\mathbf{x}}) - \chi_{\tau,p}(\hat{\mathbf{x}})\right)\right\| \le C_1 C_{\Gamma,p+1} h_\tau^{p+1} \left\|\partial_1\chi_{\tau,p}\right\|.$$

The quantity $\partial_1\chi_{\tau,p}$ can be estimated by means of Lemma 8.4.2

$$\left\|\partial_1\chi_{\tau,p}\right\| \le C_{\text{stab}} h_\tau. \tag{8.64}$$

The combination of these estimates leads to

$$S_2 \le C h_\tau^{p+2}, \tag{8.65}$$

where C depend only on C_{stab}, $C_{\Gamma,p+1}$ and p. The estimate of S_1 is just a repetition of the previous arguments and is skipped here.

The estimate of $g_{\tau,p}$ from above follows from (8.64)

$$\left|g_{\tau,p}(\hat{\mathbf{x}})\right| \le \left\|\partial_1 \chi_{\tau,p}(\hat{\mathbf{x}})\right\| \left\|\partial_2 \chi_{\tau,p}(\hat{\mathbf{x}})\right\| \le C_{\mathrm{stab}}^2 h_\tau^2.$$

For the estimate from below, we start with

$$\begin{aligned}\left|g_{\tau,p}(\hat{\mathbf{x}})\right| &\ge \left|g_\tau^{\mathrm{affine}}(\hat{\mathbf{x}})\right| - \left|\left(g_\tau^{\mathrm{affine}} - g_{\tau,p}\right)(\hat{\mathbf{x}})\right| \\ &\ge \left|g_\tau^{\mathrm{affine}}(\hat{\mathbf{x}})\right| - \left|\left(g_\tau^{\mathrm{affine}} - g_\tau\right)(\hat{\mathbf{x}})\right| - \left|\left(g_\tau - g_{\tau,p}\right)(\hat{\mathbf{x}})\right| \\ &\ge 2\left|\tau^{\mathrm{affine}}\right| - C_2\left(h_\tau^3 + h_\tau^{p+2}\right).\end{aligned}$$

The shape-regularity of $\mathcal{G}^{\mathrm{affine}}$ implies that

$$\left|\tau^{\mathrm{affine}}\right| \ge c h_\tau^2,$$

where c depends only on the shape-regularity of the mesh [cf. (4.17)]. Thus, for sufficiently small $h_\tau \le \underline{h}$, the first estimate in (8.62b) is proved.

The second one simply follows by combining the first estimates in (8.62a) and (8.62b)

$$\left|g_\tau(\hat{\mathbf{x}})\right| \ge \left|g_{\tau,p}(\hat{\mathbf{x}})\right| - \left|g_\tau(\hat{\mathbf{x}}) - g_{\tau,p}(\hat{\mathbf{x}})\right| \ge c_3 h_\tau^2 - C_2 h_\tau^{p+2} \ge \tilde{c}_3 h_\tau^2$$

for sufficiently small h_τ. □

Next, we will transport estimates (8.53) to the surface. For this we have to introduce some notation. Recall that, for $\tau \in \mathcal{G}^{\mathrm{affine}}$, the Jacobi matrices of $\chi_{\tau,p}$ resp. $\chi_\tau = P \circ \chi_{\tau,p}$ are denoted by $\mathbf{J}_{\tau,p}$ resp. $\mathbf{J}_\tau$ and the corresponding Gram matrices by $\mathbf{G}_{\tau,p}$ resp. $\mathbf{G}_\tau$ [cf. (8.13)].

Notation 8.4.5. *Let $\tau \in \mathcal{G}^{\mathrm{affine}}$ and $\check{\tau} = \mathrm{lift}_\tau(\tau) \in \check{\mathcal{G}}$. For $\mathbf{x} \in \check{\tau}$, let $T_{\mathbf{x}}$ be the tangential plane to Γ at $\mathbf{x}$ which is spanned by[3] $\hat{\mathbf{t}}_i(\hat{\mathbf{x}}) = \partial_i \chi_\tau(\hat{\mathbf{x}})$, $i = 1, 2$. For a tangential vector $\xi \in T_{\mathbf{x}}$, the Gâteaux derivative of some sufficiently smooth function $\eta : \check{\tau} \to \mathbb{R}^3$ is defined by*

$$D\eta(\mathbf{x})\,\xi := (D\hat{\eta})(\hat{\mathbf{x}})\,\mathbf{G}_\tau^{-1}(\hat{\mathbf{x}})\,\mathbf{J}_\tau^{\mathsf{T}}(\hat{\mathbf{x}})\,\xi, \tag{8.66}$$

where $\hat{\eta} = \eta \circ \chi_\tau$.

Corollary 8.4.6. *Let P in (8.52) belong to the class $\mathcal{A}_{\mathrm{pw}}^{p+1}$ for some $p \ge 1$. For $\tau \in \mathcal{G}^{\mathrm{affine}}$, let $\check{\tau} = \mathrm{lift}_\tau(\tau) \in \check{\mathcal{G}}$. If h_τ is sufficiently small $h_\tau \le h_0$ then, for any $\mathbf{x} \in \check{\tau}$ and $\xi \in T_{\mathbf{x}}$,*

$$\left\|\theta_\tau(\mathbf{x}) - \mathbf{x}\right\| \le \widetilde{C}_0 h_\tau^{p+1}, \tag{8.67a}$$

$$\left\|D\theta_\tau(\mathbf{x})\,\xi - \xi\right\| \le C_4 h_\tau^{p}\left\|\xi\right\|, \tag{8.67b}$$

---

[3] *Convention:* If $\mathbf{x} \in \check{\tau} \in \check{\mathcal{G}}$ and $\hat{\mathbf{x}} \in \hat{\tau}$ appear in the same context, they are related by $\mathbf{x} = \chi_\tau(\hat{\mathbf{x}})$.

where θ_τ is as in Remark 8.1.11. The constant $\widetilde{C}_0$ depends only on C_{stab}, $C_{\Gamma,p+1}$, and p, while C_4 and $h_0 > 0$ may in addition depend on the shape-regularity of the mesh [cf. (4.17)].

Proof. For $\tau \in \mathcal{G}^{\text{affine}}$, let $\check{\tau} = \text{lift}_\tau(\tau) \in \check{\mathcal{G}}$. The first estimate follows from (8.53a) by

$$\sup_{\mathbf{x}\in\check{\tau}} \|\theta_\tau(\mathbf{x}) - \mathbf{x}\| = \sup_{\hat{\mathbf{x}}\in\hat{\tau}} \left\| \chi_{\tau,p}(\hat{\mathbf{x}}) - \chi_\tau(\hat{\mathbf{x}}) \right\| \leq \sqrt{3} C_0 C_{\Gamma,p+1} h_\tau^{p+1}.$$

We turn to the second estimate. Let $\mathbf{I} : \mathbb{R}^3 \to \mathbb{R}^3$ denote the identity. Then (8.66) simplifies to[4]

$$(D\mathbf{I})(\mathbf{x})\,\xi = \left(\mathbf{J}_\tau(\hat{\mathbf{x}})\,\mathbf{G}_\tau^{-1}(\hat{\mathbf{x}})\,\mathbf{J}_\tau^{\intercal}(\hat{\mathbf{x}})\right)\xi = \xi \qquad \forall \xi \in T_{\mathbf{x}} \quad \text{and } \mathbf{x} = \chi_\tau(\hat{\mathbf{x}}). \tag{8.68}$$

Note that $\theta_\tau = \chi_{\tau,p} \circ \chi_\tau^{-1}$ so that $\hat{\theta}_\tau = \chi_{\tau,p}$ and $\left(D\hat{\theta}_\tau\right)(\hat{\mathbf{x}}) = \mathbf{J}_{\tau,p}(\hat{\mathbf{x}})$. Thus and in view of (8.66) we derive

$$\|D\theta_\tau(\mathbf{x})\,\xi - \xi\| = \left\| \left(\mathbf{J}_\tau - \mathbf{J}_{\tau,p}\right)(\hat{\mathbf{x}})\,\mathbf{G}_\tau^{-1}(\hat{\mathbf{x}})\,\mathbf{J}_\tau^{\intercal}\xi \right\|. \tag{8.69}$$

For a matrix-valued function $\mathbf{A} : \hat{\tau} \to \mathbb{R}^{m\times n}$, $\mathbf{A}(\hat{\mathbf{x}}) = \left(a_{i,j}(\hat{\mathbf{x}})\right)_{\substack{1\leq i\leq m,\\ 1\leq j\leq n}}$ we define the norm

$$\|\mathbf{A}\|_{\hat{\tau}} := \sup_{\hat{\mathbf{x}}\in\hat{\tau}} \max_{\substack{1\leq i\leq m\\ 1\leq j\leq n}} \left|a_{i,j}(\hat{\mathbf{x}})\right|. \tag{8.70}$$

Since all norms are equivalent in finite-dimensional spaces (the matrix dimensions occurring in (8.69) are 3×2, 2×2, and 2×3), there is a constant $C > 0$ such that

$$\begin{aligned} \|D\theta_\tau(\mathbf{x})\,\xi - \xi\| &\leq C \left\| \left(\mathbf{J}_\tau - \mathbf{J}_{\tau,p}\right) \right\|_{\hat{\tau}} \left\|\mathbf{G}_\tau^{-1}\right\|_{\hat{\tau}} \left\|\mathbf{J}_\tau^{\intercal}\right\|_{\hat{\tau}} \|\xi\| \\ &\overset{(8.53b)}{\leq} C C_1 C_{\Gamma,p+1} h_\tau^{p+1} \left\|\mathbf{G}_\tau^{-1}\right\|_{\hat{\tau}} \left\|\mathbf{J}_\tau^{\intercal}\right\|_{\hat{\tau}} \|\xi\|. \end{aligned} \tag{8.71}$$

The chain rule implies that

$$\mathbf{J}_\tau(\hat{\mathbf{x}}) = \left(\mathbf{J}_P \circ \chi_{\tau,p}(\hat{\mathbf{x}})\right) \mathbf{J}_{\tau,p},$$

where $\mathbf{J}_P$ is the 3×3 Jacobi matrix of P. By using Lemma 8.4.2 we obtain

$$\left\|\mathbf{J}_\tau^{\intercal}\right\|_{\hat{\tau}} \leq C C_{\Gamma,1} h_\tau. \tag{8.72}$$

It remains to estimate the inverse of the Gram matrix. We have

$$\mathbf{G}_\tau^{-1} = \frac{1}{g_\tau^2} \begin{bmatrix} \left\|\hat{\mathbf{t}}_2\right\|^2 & -\left\langle\hat{\mathbf{t}}_1, \hat{\mathbf{t}}_2\right\rangle \\ \left\langle\hat{\mathbf{t}}_1, \hat{\mathbf{t}}_2\right\rangle & \left\|\hat{\mathbf{t}}_1\right\|^2 \end{bmatrix}.$$

---

[4] This can easily be seen by writing $\xi = \mathbf{J}_\tau\hat{\mathbf{z}}$ and inserting this into (8.68).

Lemma 8.4.4 implies that

$$g_\tau(\hat{\mathbf{x}}) \geq \tilde{c}_3 h_\tau^2$$

for sufficiently small h_τ. The estimate [cf. (8.72)]

$$\left\langle \hat{\mathbf{t}}_i(\hat{\mathbf{x}}), \hat{\mathbf{t}}_j(\hat{\mathbf{x}})\right\rangle \leq C C_{\Gamma,1}^2 h_\tau^2 \tag{8.73}$$

leads to

$$\left\| \mathbf{G}_\tau^{-1} \right\|_{\hat{\tau}} \leq C h_\tau^{-2}. \tag{8.74}$$

The combination of (8.71), (8.72), and (8.74) leads to the assertion. □

The following statement is a Corollary of Lemma 8.4.4 and applies for any sufficiently smooth projection $P : \Gamma_\ell \to \Gamma$. In the case that P is the *orthogonal surface projection*, estimate (8.75) can be improved by one order (cf. Lemma 8.4.10).

Corollary 8.4.7. *Let the Assumptions of Lemma 8.4.4 be satisfied. For sufficiently small $h_\tau \leq \underline{h}$, we have*

$$\left| \frac{\left(g_{\tau,p} \circ \chi_\tau^{-1}\right)(\mathbf{x})}{g_\tau \circ \chi_\tau^{-1}(\mathbf{x})} - 1 \right| \leq \frac{C_2}{\tilde{c}_3} h_\tau^p, \tag{8.75}$$

where $\underline{h}$, C_2, $\tilde{c}_3$ are as in Lemma 8.4.4.

Proof. By using Lemma 8.4.4 we obtain

$$\left| \frac{\left(g_{\tau,p} \circ \chi_\tau^{-1}\right)(\mathbf{x})}{g_\tau \circ \chi_\tau^{-1}(\mathbf{x})} - 1 \right| = \left| \frac{\left(g_{\tau,p} \circ \chi_\tau^{-1}\right)(\mathbf{x}) - g_\tau \circ \chi_\tau^{-1}(\mathbf{x})}{g_\tau \circ \chi_\tau^{-1}(\mathbf{x})} \right| \leq \sup_{\hat{\mathbf{x}} \in \hat{\tau}} \left| \frac{g_{\tau,p}(\hat{\mathbf{x}}) - g_\tau(\hat{\mathbf{x}})}{g_\tau(\hat{\mathbf{x}})} \right|$$
$$\leq \frac{C_2}{\tilde{c}_3} h_\tau^p$$

□

We will derive an estimate for the error in the approximation of tangential vectors and surface elements under the assumption that P is the orthogonal resp. conormal surface projection (cf. Example 8.1.8). We start by writing the ratio in (8.75) in an alternative way. Note that the Gram matrix $\mathbf{G}_\tau(\hat{\mathbf{x}}) = \mathbf{J}_\tau^\intercal(\hat{\mathbf{x}}) \mathbf{J}_\tau(\hat{\mathbf{x}}) \in \mathbb{R}^{2\times 2}$ is symmetric and positive definite so that there exists a positive definite square root $\mathbf{G}_\tau^{1/2}(\hat{\mathbf{x}}) \in \mathbb{R}^{2\times 2}$ such that $\mathbf{G}_\tau(\hat{\mathbf{x}}) = \mathbf{G}_\tau^{1/2}(\hat{\mathbf{x}}) \mathbf{G}_\tau^{1/2}(\hat{\mathbf{x}})$. The inverse of $\mathbf{G}_\tau^{1/2}(\hat{\mathbf{x}})$ is denoted by $\mathbf{G}_\tau^{-1/2}(\hat{\mathbf{x}})$.

Lemma 8.4.8. *For any $\mathbf{x} \in \tau \in \mathcal{G}$, the ratio of the original surface element and its p-parametric approximation satisfies*

$$\frac{\left(g_{\tau,p} \circ \chi_\tau^{-1}\right)(\mathbf{x})}{g_\tau \circ \chi_\tau^{-1}(\mathbf{x})} = \sqrt{\det\left(\mathbf{G}_\tau^{-1/2}(\hat{\mathbf{x}})\, \mathbf{G}_{\tau,p}(\hat{\mathbf{x}})\, \mathbf{G}_\tau^{-1/2}(\hat{\mathbf{x}})\right)},$$

where $\hat{\mathbf{x}} := \chi_\tau^{-1}(\mathbf{x})$.

Proof. The assertion follows from Exercise 8.4.9 and the multiplication theorem for determinants of matrix products. □

Exercise 8.4.9. *For* $\mathbf{a}_1, \mathbf{a}_2 \in \mathbb{R}^3$ *and* $\mathbf{A} = [\mathbf{a}_1, \mathbf{a}_2] \in \mathbb{R}^{3\times 2}$,

$$\det\left(\mathbf{A}^{\intercal}\mathbf{A}\right) = \|\mathbf{a}_1 \times \mathbf{a}_2\|^2 .$$

Let $\tau \in \mathcal{G}^{\mathrm{affine}}$. For any $\mathbf{x} \in \check{\tau} = \operatorname{lift}_\tau(\tau) \in \check{\mathcal{G}}$ and $\xi \in T_{\mathbf{x}}$, we introduce the deformed tangential vector by $\tilde{\xi} := D\theta_\tau(\mathbf{x})\,\xi$.

Lemma 8.4.10. *Let P in (8.52) belong to the class $\mathcal{A}^{p+1}$ for some $p \geq 1$ and be chosen as the orthogonal surface projection (cf. Example 8.1.8). Let $\check{\tau} = \operatorname{lift}_\tau(\tau) \in \check{\mathcal{G}}$ for some $\tau \in \mathcal{G}^{\mathrm{affine}}$ and assume that $h_\tau \leq h_0$ with h_0 as in Corollary 8.4.6. Then, for all $\mathbf{x} \in \tau$,*

$$\left|\left\langle\tilde{\xi}_1, \tilde{\xi}_2\right\rangle - \langle\xi_1, \xi_2\rangle\right| \leq C_5 h_\tau^{p+1}\|\xi_1\|\,\|\xi_2\| \qquad \forall \xi_1, \xi_2 \in T_{\mathbf{x}}, \tag{8.76a}$$

$$\left|\frac{\left(g_{\tau,p}\circ\chi_\tau^{-1}\right)(\mathbf{x})}{g_\tau\circ\chi_\tau^{-1}(\mathbf{x})} - 1\right| \leq C_6 h_\tau^{p+1}, \tag{8.76b}$$

where C_5, C_6 depend on $\widetilde{C}_0$, C_4 as in Corollary 8.4.6.

Proof. For $\tau \in \mathcal{G}^{\mathrm{affine}}$, let $\mathbf{x} \in \check{\tau} = \operatorname{lift}_\tau(\tau) \in \check{\mathcal{G}}$. The difference in (8.76a) is split into

$$\left\langle\tilde{\xi}_1, \tilde{\xi}_2\right\rangle - \langle\xi_1, \xi_2\rangle = \left\langle\xi_1, \tilde{\xi}_2 - \xi_2\right\rangle + \left\langle\tilde{\xi}_1 - \xi_1, \xi_2\right\rangle + \left\langle\tilde{\xi}_1 - \xi_1, \tilde{\xi}_2 - \xi_2\right\rangle. \tag{8.77}$$

The estimate for the last term follows from Corollary 8.4.6

$$\left\|\tilde{\xi}_i - \xi_i\right\| = \|D\theta_\tau(\mathbf{x})\,\xi_i - \xi_i\| \leq C_4 h_\tau^p$$

since $2p \geq p+1$.

Next, we consider the first term on the right-hand side of (8.77). The vector $\theta_\tau(\mathbf{x}) - \mathbf{x} = \chi_{\tau,p}(\hat{\mathbf{x}}) - \chi_\tau(\hat{\mathbf{x}})$ is collinear to the normal vector $\mathbf{n}(\mathbf{x})$. Since, for any $\hat{\mathbf{y}} \in \mathbb{R}^2$, the vector $\mathbf{J}_\tau(\hat{\mathbf{x}})\,\hat{\mathbf{y}} \in T_{\mathbf{x}}$, we have

$$\left\langle\chi_{\tau,p}(\hat{\mathbf{x}}) - \chi_\tau(\hat{\mathbf{x}}), \mathbf{J}_\tau(\hat{\mathbf{x}})\,\hat{\mathbf{y}}\right\rangle = 0.$$

Applying the Gâteaux derivative in the direction of some $\hat{\mathbf{z}} \in \mathbb{R}^2$ results in

$$\left\langle\left(\mathbf{J}_{\tau,p}(\hat{\mathbf{x}}) - \mathbf{J}_\tau(\hat{\mathbf{x}})\right)\hat{\mathbf{z}}, \mathbf{J}_\tau(\hat{\mathbf{x}})\,\hat{\mathbf{y}}\right\rangle + \left\langle\chi_{\tau,p}(\hat{\mathbf{x}}) - \chi_\tau(\hat{\mathbf{x}}), \mathbf{DJ}_\tau(\hat{\mathbf{x}})(\hat{\mathbf{z}})(\hat{\mathbf{y}})\right\rangle = 0, \tag{8.78}$$

where we employed the notation

$$\mathbf{DJ}_\tau(\hat{\mathbf{x}})(\hat{\mathbf{z}})(\hat{\mathbf{y}}) := \sum_{k=1}^{2}\sum_{j=1}^{2}\hat{\mathbf{z}}_k\hat{\mathbf{y}}_j\partial_k\partial_j\chi_\tau(\hat{\mathbf{x}}) \in \mathbb{R}^3.$$

Any $\xi_1 \in T_{\mathbf{x}}$ can be written in the form $\xi_1 = \mathbf{J}_\tau(\hat{\mathbf{x}})\,\hat{\mathbf{y}}$ for some $\hat{\mathbf{y}} \in \mathbb{R}^2$. Hence (8.78) is equivalent to

$$\left\langle \left(\mathbf{J}_{\tau,p}(\hat{\mathbf{x}}) - \mathbf{J}_\tau(\hat{\mathbf{x}})\right)\hat{\mathbf{z}}, \xi_1\right\rangle + \left\langle \chi_{\tau,p}(\hat{\mathbf{x}}) - \chi_\tau(\hat{\mathbf{x}}), \mathbf{DJ}_\tau(\hat{\mathbf{x}})(\hat{\mathbf{z}})(\hat{\mathbf{y}})\right\rangle = 0. \tag{8.79}$$

By substituting ξ by $\mathbf{J}_\tau(\hat{\mathbf{x}})\,\hat{\mathbf{z}}$ and η by θ_τ in (8.66), we derive

$$\mathbf{J}_{\tau,p}(\hat{\mathbf{x}})\,\hat{\mathbf{z}} = D\theta_\tau(\mathbf{x})\left(\mathbf{J}_\tau(\hat{\mathbf{x}})\,\hat{\mathbf{z}}\right).$$

Let $\xi_2 = \mathbf{J}_\tau(\hat{\mathbf{x}})\,\hat{\mathbf{z}}$. Then (8.79) can be written equivalently as

$$\left\langle \tilde{\xi}_2 - \xi_2, \xi_1 \right\rangle = -\left\langle \theta_\tau(\mathbf{x}) - \mathbf{x}, \mathbf{DJ}_\tau(\hat{\mathbf{x}})(\hat{\mathbf{z}})(\hat{\mathbf{y}})\right\rangle.$$

The chain rule yields (with $\tilde{\mathbf{x}} = \chi_{\tau,p}(\hat{\mathbf{x}})$)

$$\begin{aligned}\partial_k \partial_j \left(\chi_\tau(\hat{\mathbf{x}})\right)_i &= \sum_{m,n=1}^{3} \left(\partial_m \partial_n P_i\right)(\tilde{\mathbf{x}}) \left(\partial_j \left(\chi_{\tau,p}(\hat{\mathbf{x}})\right)_m\right)\left(\partial_k \left(\chi_{\tau,p}(\hat{\mathbf{x}})\right)_n\right) \\ &\quad + \sum_{m=1}^{3} \left(\partial_m P_i\right)(\tilde{\mathbf{x}})\, \partial_k \partial_j \left(\chi_{\tau,p}(\hat{\mathbf{x}})\right)_m.\end{aligned}$$

Lemma 8.4.2 implies that

$$\left\| \partial_k \partial_j \chi_{\tau,p}(\hat{\mathbf{x}}) \right\| \leq C h_\tau^2,$$

where C depends on C_{stab} and $C_{\Gamma,2}$. Since the dimension of the tensor $\mathbf{DJ}_\tau(\mathbf{x})$ is finite and equals $3 \times 2 \times 2$, the equivalence of norms in finite-dimensional spaces implies that there is a constant $C > 0$ such that

$$\left|\left\langle \tilde{\xi}_2 - \xi_2, \xi_1 \right\rangle\right| \leq C h_\tau^2 \left\|\theta_\tau(\mathbf{x}) - \mathbf{x}\right\| \|\hat{\mathbf{z}}\| \|\hat{\mathbf{y}}\| \overset{(8.67a)}{\leq} \widetilde{C} h_\tau^{p+3} \|\hat{\mathbf{z}}\| \|\hat{\mathbf{y}}\|.$$

Finally, we employ

$$\|\xi_1\| = \left\|\mathbf{J}_\tau(\hat{\mathbf{x}})\,\hat{\mathbf{y}}\right\| = \left\langle \mathbf{G}_\tau(\hat{\mathbf{x}})\,\hat{\mathbf{y}}, \hat{\mathbf{y}}\right\rangle^{1/2} \geq C \frac{1}{\|\mathbf{G}_\tau^{-1}\|_{\hat{\tau}}^{1/2}} \|\hat{\mathbf{y}}\| \overset{(8.74)}{\geq} C h_\tau \|\hat{\mathbf{y}}\|.$$

The combination with the analogous estimate for $\xi_2 = \mathbf{J}_\tau(\hat{\mathbf{x}})\,\hat{\mathbf{z}}$ leads to

$$\left|\left\langle \tilde{\xi}_2 - \xi_2, \xi_1 \right\rangle\right| \leq \widehat{C} h_\tau^{p+1} \|\xi_1\| \|\xi_2\|.$$

Repeating all arguments for the second term on the right-hand side in (8.77) completes the proof of (8.76a).

It remains to prove (8.76b). In view of Lemma 8.4.8, we consider

$$\mathbf{G}_\tau^{-1/2}(\hat{\mathbf{x}})\,\mathbf{G}_{\tau,p}(\hat{\mathbf{x}})\,\mathbf{G}_\tau^{-1/2}(\hat{\mathbf{x}}) = \mathbf{I} + \mathbf{E}(\mathbf{x}),$$

where $\mathbf{I}$ is the 2×2 identity matrix and $\mathbf{E}(\mathbf{x}) := \mathbf{G}_\tau^{-1/2}(\hat{\mathbf{x}})\,\mathbf{G}_{\tau,p}(\hat{\mathbf{x}})\,\mathbf{G}_\tau^{-1/2}(\hat{\mathbf{x}}) - \mathbf{I}$. From the continuity of the determinant we conclude that

$$\det(\mathbf{I}+\mathbf{E}) = 1 + O(\|\mathbf{E}\|), \tag{8.80}$$

where $\|\cdot\|$ denotes the spectral norm of $\mathbf{E}$. Since $\mathbf{E}$ is symmetric we have

$$\begin{aligned}\|\mathbf{E}\| &= \sup_{\mathbf{v}\in\mathbb{R}^2\backslash\{0\}} \frac{|\langle \mathbf{v}, \mathbf{E}\mathbf{v}\rangle|}{\|\mathbf{v}\|^2} = \sup_{\mathbf{w}\in\mathbb{R}^2\backslash\{0\}} \frac{\left|\langle \mathbf{w}, \mathbf{G}_{\tau,p}(\hat{\mathbf{x}})\,\mathbf{w}\rangle - \langle \mathbf{w}, \mathbf{G}_\tau(\hat{\mathbf{x}})\,\mathbf{w}\rangle\right|}{\langle \mathbf{w}, \mathbf{G}_\tau(\hat{\mathbf{x}})\,\mathbf{w}\rangle} \\ &\le \sup_{\substack{\xi\in T_{\mathbf{x}}\\ \tilde{\xi}=D\theta_\tau(\mathbf{x})\xi}} \frac{\left|\|\tilde{\xi}\|^2 - \|\xi\|^2\right|}{\|\xi\|^2} \le C h_\tau^{p+1}.\end{aligned}$$

Thus (8.76b) is proved. □

Next, we will investigate the difference of the distance $\|\mathbf{x}-\mathbf{y}\|$ between two points on the original surface Γ and the distance $\|\tilde{\mathbf{x}}-\tilde{\mathbf{y}}\|$ between the corresponding points on the approximate surface. Recall the definition of g_Γ which measures the ratio between the geodetic distance and the Euclidean distance of surface points [cf. (4.80)].

Lemma 8.4.11. *Let P in (8.52) belong to the class $\mathcal{A}_{\mathrm{pw}}^{p+1}$ for some $p \ge 1$. Let the mesh width h be sufficiently small $h \le h_0$ with h_0 as in Corollary 8.4.6. For any $\mathbf{x}, \mathbf{y} \in \Gamma$, we have*

$$\|\theta(\mathbf{x}) - \theta(\mathbf{y}) - (\mathbf{x}-\mathbf{y})\| \le C_7 h^p \|\mathbf{x}-\mathbf{y}\|.$$

The constant C_7 depends only on C_{stab} [cf. (8.57)], C_4 as in Corollary 8.4.6, c_P as in Assumption 8.1.16, g_Γ [cf. (4.80)] and the shape-regularity of the mesh.

Proof. First, we consider two points $\mathbf{x}, \mathbf{y} \in \check{\tau} = \mathrm{lift}_\tau(\tau) \in \check{\mathcal{G}}$ for some $\tau \in \mathcal{G}^{\mathrm{affine}}$. Let $\hat{\mathbf{x}} = \chi_\tau^{-1}(\mathbf{x})$, $\hat{\mathbf{y}} = \chi_\tau^{-1}(\mathbf{y})$. For $t \in [0,1]$ we write $\hat{\mathbf{z}}(t) := \hat{\mathbf{y}} + t(\hat{\mathbf{x}} - \hat{\mathbf{y}})$ and $\mathbf{z}(t) := \chi_\tau(\hat{\mathbf{z}}(t))$. Then

$$\begin{aligned}\theta_\tau(\mathbf{x}) - \theta_\tau(\mathbf{y}) - (\mathbf{x}-\mathbf{y}) &= \chi_{\tau,p}(\hat{\mathbf{x}}) - \chi_{\tau,p}(\hat{\mathbf{y}}) - (\chi_\tau(\hat{\mathbf{x}}) - \chi_\tau(\hat{\mathbf{y}})) \\ &= \int_0^1 \frac{d\left((\chi_{\tau,p} - \chi_\tau)(\hat{\mathbf{z}}(t))\right)}{dt}dt = \int_0^1 \left(\mathbf{J}_{\tau,p} - \mathbf{J}_\tau\right)(\hat{\mathbf{z}}(t))\,(\hat{\mathbf{x}} - \hat{\mathbf{y}})\,dt.\end{aligned}$$

Note that $\xi := \mathbf{J}_\tau(\hat{\mathbf{z}}(t))(\hat{\mathbf{x}} - \hat{\mathbf{y}}) \in T_{\mathbf{z}(t)}$ and, thus, we obtain

$$\theta_\tau(\mathbf{x}) - \theta_\tau(\mathbf{y}) - (\mathbf{x}-\mathbf{y}) = \int_0^1 \left(D\theta_\tau(\mathbf{z}(t))\,\xi - \xi\right)dt. \tag{8.81}$$

Estimate (8.67b) implies that

$$\|\theta_\tau(\mathbf{x}) - \theta_\tau(\mathbf{y}) - (\mathbf{x}-\mathbf{y})\| \leq C_4 h_\tau^p \|\xi\|. \tag{8.82}$$

The norm of ξ can be estimated by [cf. (8.73)]

$$\|\xi\|^2 = \langle(\hat{\mathbf{x}}-\hat{\mathbf{y}}), \mathbf{G}_\tau(\hat{\mathbf{z}}(t))(\hat{\mathbf{x}}-\hat{\mathbf{y}})\rangle \leq C C_{\Gamma,1}^2 h_\tau^2 \|\hat{\mathbf{x}}-\hat{\mathbf{y}}\|^2. \tag{8.83}$$

Let $\mathbf{x} = \chi_\tau(\hat{\mathbf{x}})$ and $\mathbf{y} = \chi_\tau(\hat{\mathbf{y}})$. By using the bi-Lipschitz continuity of P [cf. Definition 8.3.1(2)] we get

$$\|\mathbf{x}-\mathbf{y}\| = \left\|P\left(\chi_{\tau,p}(\hat{\mathbf{x}})\right) - P\left(\chi_{\tau,p}(\hat{\mathbf{y}})\right)\right\| \geq c_P \left\|\chi_{\tau,p}(\hat{\mathbf{x}}) - \chi_{\tau,p}(\hat{\mathbf{y}})\right\|.$$

Taylor expansion of $\chi_{\tau,p}$ about $\hat{\mathbf{y}}$ yields

$$\chi_{\tau,p}(\hat{\mathbf{x}}) - \chi_{\tau,p}(\hat{\mathbf{y}}) = \mathbf{J}_{\tau,p}(\hat{\mathbf{y}})(\hat{\mathbf{x}}-\hat{\mathbf{y}}) + R(\hat{\mathbf{x}},\hat{\mathbf{y}})$$

where

$$\|R(\hat{\mathbf{x}},\hat{\mathbf{y}})\| \leq \frac{\|\hat{\mathbf{y}}-\hat{\mathbf{x}}\|^2}{2} \max_{|\alpha|=2} \left\|\partial^\alpha \chi_{\tau,p}\right\|_{C^0(\hat{\tau})} \overset{\text{Lemma 8.4.2}}{\leq} C_{\text{stab}} \frac{\|\hat{\mathbf{y}}-\hat{\mathbf{x}}\|^2}{2} h_\tau^2$$

This leads to

$$\|\mathbf{x}-\mathbf{y}\| \geq c_P \left(\left\|\mathbf{J}_{\tau,p}(\hat{\mathbf{y}})(\hat{\mathbf{x}}-\hat{\mathbf{y}})\right\| - C_{\text{stab}} \frac{\|\hat{\mathbf{y}}-\hat{\mathbf{x}}\|^2}{2} h_\tau^2\right). \tag{8.84}$$

For the first term on the right-hand side, we proceed with

$$\left\|\mathbf{J}_{\tau,p}(\hat{\mathbf{y}})(\hat{\mathbf{x}}-\hat{\mathbf{y}})\right\|^2 = \left\langle\hat{\mathbf{x}}-\hat{\mathbf{y}}, \mathbf{G}_{\tau,p}(\hat{\mathbf{y}})(\hat{\mathbf{x}}-\hat{\mathbf{y}})\right\rangle.$$

The inverse of the Gram matrix has the form

$$\mathbf{G}_{\tau,p}^{-1} = \frac{1}{g_{\tau,p}^2} \begin{bmatrix} \left\|\partial_2 \chi_{\tau,p}\right\|^2 & -\left\langle\partial_1 \chi_{\tau,p}, \partial_2 \chi_{\tau,p}\right\rangle \\ -\left\langle\partial_1 \chi_{\tau,p}, \partial_2 \chi_{\tau,p}\right\rangle & \left\|\partial_1 \chi_{\tau,p}\right\|^2 \end{bmatrix}.$$

The combination of (8.62b) and (8.57a) leads to

$$\left\|\mathbf{G}_{\tau,p}^{-1}\right\|_{\hat{\tau}} \leq C \frac{C_{\text{stab}}}{c_3^2 h_\tau^2} \quad \text{and} \quad \left\|\mathbf{J}_{\tau,p}(\hat{\mathbf{y}})(\hat{\mathbf{x}}-\hat{\mathbf{y}})\right\|$$
$$\geq \frac{1}{\sqrt{\left\|\mathbf{G}_{\tau,p}^{-1}\right\|_{\hat{\tau}}}} \|\hat{\mathbf{x}}-\hat{\mathbf{y}}\| \geq \frac{c_3 h_\tau}{\sqrt{C C_{\text{stab}}}} \|\hat{\mathbf{x}}-\hat{\mathbf{y}}\|.$$

Inserting this into (8.84) results in

$$\|\mathbf{x}-\mathbf{y}\| \geq c_P \|\hat{\mathbf{x}}-\hat{\mathbf{y}}\| \left(\frac{c_3 h_\tau}{\sqrt{C C_{\text{stab}}}} - \frac{C_{\text{stab}}}{2} h_\tau^2 \right).$$

For sufficiently small h_τ we have proved that

$$\|\mathbf{y}-\mathbf{x}\| \geq c_8 h_\tau \|\hat{\mathbf{y}}-\hat{\mathbf{x}}\|, \tag{8.85}$$

where $c_8 > 0$ depends on C_{stab}, c_3 and the shape-regularity of the mesh.

The combination of (8.1) and (8.82)–(8.85) leads to

$$\|\theta_\tau(\mathbf{x}) - \theta_\tau(\mathbf{y}) - (\mathbf{x}-\mathbf{y})\| \leq C C_4 C_{\Gamma,1} h_\tau^{p+1} \|\hat{\mathbf{x}}-\hat{\mathbf{y}}\| \leq C_9 h_\tau^p \|\mathbf{x}-\mathbf{y}\| \tag{8.86}$$

with $C_9 := \left(C C_4 C_{\Gamma,1} C^{\text{affine}}\right)/c_8$.

Next, consider some $\mathbf{x},\mathbf{y} \in \Gamma$, $\mathbf{x} \neq \mathbf{y}$, and choose a path $\mathbf{s} := \{s(t) : t \in [0,1]\} \in C^{0,1}([0,1], \Gamma)$ of minimal geodesic length, which connects $\mathbf{y}$ and $\mathbf{x}$ and satisfies $\mathbf{y} = s(0)$ and $\mathbf{x} = s(1)$. Let $(\check{\tau}_k)_{k=1}^q$ denote a minimal sequence of triangles in $\check{\mathcal{G}}$ such that

$$\mathbf{s} = \{s(t) : t \in [0,1]\} \subset \bigcup_{k=1}^{q} \overline{\check{\tau}_k}.$$

Choose a minimal number of points

$$0 = t_0 < t_1 < \ldots < t_m = 1$$

such that $\mathbf{s}_1 := \{s(t) : 0 \leq t \leq t_1\}$ and

$$\forall 2 \leq j \leq m \qquad \mathbf{s}_j := \left\{s(t) : t_{j-1} < t \leq t_j\right\}$$

define a disjoint partitioning of $\mathbf{s}$ which satisfies

$$\forall 1 \leq j \leq q \quad \exists 1 \leq k(j) \leq q \qquad \mathbf{s}_j \subset \overline{\check{\tau}_{k(j)}}.$$

Let $|\mathbf{s}|$ denote the length of $\mathbf{s}$ while the length of $\mathbf{s}_j$ is called $\left|\mathbf{s}_j\right|$. Then

$$\begin{aligned}
\|\theta(\mathbf{x}) - \theta(\mathbf{y}) - (\mathbf{x}-\mathbf{y})\| &\leq \sum_{j=1}^{m} \left\|\theta_{\tau_{k(j)}}\left(\mathbf{z}_j\right) - \theta_{\tau_{k(j)}}\left(\mathbf{z}_{j-1}\right) - \left(\mathbf{z}_j - \mathbf{z}_{j-1}\right)\right\| \\
&\leq C_9 \sum_{j=1}^{m} h_\tau^p \left\|\mathbf{z}_j - \mathbf{z}_{j-1}\right\| \leq C_9 h^p \sum_{j=1}^{m} \left|\mathbf{s}_j\right| \leq C\,|\mathbf{s}|\, h^p \\
&\overset{(4.80)}{\leq} C_9 g_\Gamma h^p \|\mathbf{x} \quad \mathbf{y}\|.
\end{aligned}$$

□

Lemma 8.4.12. *Let P in (8.52) belong to the class $\mathcal{A}_{\mathrm{pw}}^{p+1}$ for some $p \geq 1$. Let the mesh width h of $\mathcal{G}$ be sufficiently small $h_\tau \leq h_1$, where $h_1 > 0$ depends on h_0 as in Corollary 8.4.6 and C_7 as in Lemma 8.4.11. For any $\alpha \in \mathbb{R}$, there exist $C_{10}, C_{11} > 0$ such that, for any $\mathbf{x}, \mathbf{y} \in \Gamma$, we have*

$$\left| \|\theta(\mathbf{x}) - \theta(\mathbf{y})\|^\alpha - \|\mathbf{x} - \mathbf{y}\|^\alpha \right| \leq C_{10} h^p \|\mathbf{x} - \mathbf{y}\|^\alpha . \tag{8.87a}$$

If $P \in \mathcal{A}^{p+1}$ is chosen as the orthonormal projection (cf. Exercise 8.1.8) the estimate can be improved by one order of h

$$\left| \|\theta(\mathbf{x}) - \theta(\mathbf{y})\|^\alpha - \|\mathbf{x} - \mathbf{y}\|^\alpha \right| \leq C_{11} h^{p+1} \|\mathbf{x} - \mathbf{y}\|^\alpha . \tag{8.87b}$$

The constants C_{10} depends only on C_7 while C_{11} depends on $\widetilde{C}_0, C_7, h_1$, and the constant C as in Lemma 2.2.14.

Proof. The function $p_\alpha : \mathbb{R}^3 \backslash \{0\} \to \mathbb{R}$, $p_\alpha(\mathbf{x}) := \|\mathbf{x}\|^\alpha$ is infinitely differentiable. Let further

$$\mathbf{q}(t) := \mathbf{x} + t(\theta(\mathbf{x}) - \mathbf{x}) - (\mathbf{y} + t(\theta(\mathbf{y}) - \mathbf{y})) .$$

We obtain

$$\begin{aligned} \|\theta(\mathbf{x}) - \theta(\mathbf{y})\|^\alpha - \|\mathbf{x} - \mathbf{y}\|^\alpha &= \int_0^1 \frac{d}{dt} p_\alpha(\mathbf{q}(t))\, dt \\ &= \int_0^1 \alpha p_{\alpha-2}(\mathbf{q}(t)) \langle \mathbf{q}(t), \theta(\mathbf{x}) - \mathbf{x} - (\theta(\mathbf{y}) - \mathbf{y}) \rangle\, dt . \end{aligned}$$

The norm of $\mathbf{q}$ can be estimated by using Lemma 8.4.11.

$$\begin{aligned} \|\mathbf{q}(t)\| &= \|\mathbf{x} - \mathbf{y} + t(\theta(\mathbf{x}) - \theta(\mathbf{y}) - (\mathbf{x} - \mathbf{y}))\| \geq \|\mathbf{x} - \mathbf{y}\| - tC_7 h^p \|\mathbf{x} - \mathbf{y}\| \\ &\geq (1 - C_7 h^p) \|\mathbf{x} - \mathbf{y}\| . \end{aligned} \tag{8.88}$$

Analogously, the estimate $\|\mathbf{q}(t)\| \leq (1 + C_7 h^p) \|\mathbf{x} - \mathbf{y}\|$ is derived. Hence, for sufficiently small $h \leq h_1$, we have

$$\frac{1}{2} \|\mathbf{x} - \mathbf{y}\| \leq \|\mathbf{q}(t)\| \leq \frac{3}{2} \|\mathbf{x} - \mathbf{y}\| \quad \text{and} \quad |\alpha p_{\alpha-2}(\mathbf{q}(t))| \leq C_{12} \|\mathbf{x} - \mathbf{y}\|^{\alpha-2} , \tag{8.89}$$

where C_{12} depends only on $\alpha \in \mathbb{R}$.

The combination of the left estimate in (8.89) and Lemma 8.4.11 leads to

$$\begin{aligned} |\langle \mathbf{q}(t), \theta(\mathbf{x}) - \mathbf{x} - (\theta(\mathbf{y}) - \mathbf{y}) \rangle| &\leq \|\mathbf{q}(t)\| \|\theta(\mathbf{x}) - \theta(\mathbf{y}) - (\mathbf{x} - \mathbf{y})\| \\ &\leq \frac{3}{2} C_7 h^p \|\mathbf{x} - \mathbf{y}\|^2 . \end{aligned} \tag{8.90}$$

Hence (8.87a) follows from (8.89) and (8.90).

If $P \in \mathcal{A}^{p+1}$ is the orthonormal surface projection the estimate can be improved. We write

$$\langle \mathbf{q}(t), \theta(\mathbf{x}) - \mathbf{x} - (\theta(\mathbf{y}) - \mathbf{y})\rangle = \langle \mathbf{x} - \mathbf{y}, \theta(\mathbf{x}) - \mathbf{x} - (\theta(\mathbf{y}) - \mathbf{y})\rangle + t\,\|\theta(\mathbf{x}) - \mathbf{x} - (\theta(\mathbf{y}) - \mathbf{y})\|^2. \quad (8.91)$$

The second term on the right-hand side can be estimated by using Lemma 8.4.11

$$\|\theta(\mathbf{x}) - \mathbf{x} - (\theta(\mathbf{y}) - \mathbf{y})\|^2 \le C_7^2 h^{2p}\,\|\mathbf{x} - \mathbf{y}\|^2 \le C_{13} h^{p+1}\,\|\mathbf{x} - \mathbf{y}\|^2, \quad (8.92)$$

where C_{13} depends only on h_1 and C_7. For the first term in (8.91), we employ $\theta(\mathbf{z}) - \mathbf{z} = \pm\,\|\mathbf{z} - \theta(\mathbf{z})\|\,\mathbf{n}(\mathbf{z})$ for $\mathbf{z} \in \{\mathbf{x}, \mathbf{y}\}$. Thus

$$|\langle \mathbf{x} - \mathbf{y}, \theta(\mathbf{x}) - \mathbf{x} - (\theta(\mathbf{y}) - \mathbf{y})\rangle| \le \|\theta(\mathbf{x}) - \mathbf{x}\|\,|\langle \mathbf{x} - \mathbf{y}, \mathbf{n}(\mathbf{x})\rangle| + \|\theta(\mathbf{y}) - \mathbf{y}\|\,|\langle \mathbf{x} - \mathbf{y}, \mathbf{n}(\mathbf{y})\rangle|.$$

The combination of Lemma 2.2.14 and (8.67a) leads to

$$|\langle \mathbf{x} - \mathbf{y}, \theta(\mathbf{x}) - \mathbf{x} - (\theta(\mathbf{y}) - \mathbf{y})\rangle| \le 2C\widetilde{C}_0 h_\tau^{p+1}\,\|\mathbf{x} - \mathbf{y}\|^2,$$

where C is the constant as in Lemma 2.2.14. □

For anisotropic problems, where the principal part of L is given by $-\operatorname{div}(\mathbf{A}\operatorname{grad} u)$ with some positive definite $\mathbf{A} \neq \mathbf{I}$, the fundamental solution G [cf. (3.3)] contains the anisotropic distance $\|\cdot\|_{\mathbf{A}} = \langle\cdot,\cdot\rangle_{\mathbf{A}}^{1/2}$, where

$$\langle \mathbf{u}, \mathbf{v}\rangle_{\mathbf{A}} := \left\langle \mathbf{A}^{-1}\mathbf{u}, \mathbf{v}\right\rangle.$$

In this case, the following corollary of Lemma 8.4.12 will be employed for the error estimates.

Corollary 8.4.13. *Let the assumptions of Lemma 8.4.12 be satisfied. For any $\alpha \in \mathbb{R}$ and $\mathbf{x}, \mathbf{y} \in \Gamma$, we have*

$$\left|\|\theta(\mathbf{x}) - \theta(\mathbf{y})\|_{\mathbf{A}}^\alpha - \|\mathbf{x} - \mathbf{y}\|_{\mathbf{A}}^\alpha\right| \le \widetilde{C}_{10} h^p\,\|\mathbf{x} - \mathbf{y}\|^\alpha. \quad (8.93a)$$

If $P \in \mathcal{A}^{p+1}$ is chosen as the conormal projection (cf. Exercise 8.1.8) the estimate can be improved by one order of h

$$\left|\|\theta(\mathbf{x}) - \theta(\mathbf{y})\|_{\mathbf{A}}^\alpha - \|\mathbf{x} - \mathbf{y}\|_{\mathbf{A}}^\alpha\right| \le \widetilde{C}_{11} h^{p+1}\,\|\mathbf{x} - \mathbf{y}\|^\alpha. \quad (8.93b)$$

The constants $\widetilde{C}_{10}$ (resp. $\widetilde{C}_{11}$) depend only on $c_{\mathbf{A}}$, $C_{\mathbf{A}}$ [cf. (8.19)], and C_{10} (resp. C_{11}).

Proof. Instead of the function p_α as in the proof of Lemma 8.4.12 we employ $\tilde{p}_\alpha : \mathbb{R}^3 \backslash \{0\} \to \mathbb{R}$, $\tilde{p}_\alpha(\mathbf{x}) := \|\mathbf{x}\|_{\mathbf{A}}^\alpha$. By repeating all steps of the previous proof we derive

$$\|\theta(\mathbf{x}) - \theta(\mathbf{y})\|_{\mathbf{A}}^{\alpha} - \|\mathbf{x} - \mathbf{y}\|_{\mathbf{A}}^{\alpha} = \int_0^1 \alpha \tilde{p}_{\alpha-2}(\mathbf{q}(t)) \langle \mathbf{q}(t), \theta(\mathbf{x}) - \mathbf{x} - (\theta(\mathbf{y}) - \mathbf{y})\rangle_{\mathbf{A}} \, dt.$$

The equivalence of norms [cf. (8.19)] implies that

$$\tilde{p}_{\alpha-2}(\mathbf{z}) \leq C_{14} p_{\alpha-2}(\mathbf{z}),$$
$$|\langle \mathbf{q}(t), \theta(\mathbf{x}) - \mathbf{x} - (\theta(\mathbf{y}) - \mathbf{y})\rangle_{\mathbf{A}}| \leq C_{\mathbf{A}}^2 \|\mathbf{q}(t)\| \, \|\theta(\mathbf{x}) - \theta(\mathbf{y}) - (\mathbf{x} - \mathbf{y})\|.$$

where C_{14} depends only on $c_{\mathbf{A}}$, $C_{\mathbf{A}}$, and α. We use (8.88)–(8.90) to obtain

$$\left| \|\theta(\mathbf{x}) - \theta(\mathbf{y})\|_{\mathbf{A}}^{\alpha} - \|\mathbf{x} - \mathbf{y}\|_{\mathbf{A}}^{\alpha} \right| \leq \frac{3C_{\mathbf{A}}^2 C_7 C_{12} C_{14}}{2} h^p \|\mathbf{x} - \mathbf{y}\|^{\alpha}.$$

If $P \in \mathcal{A}^{p+1}$ is the conormal surface projection the estimate can be improved. We write

$$\begin{aligned} \langle \mathbf{q}(t), \theta(\mathbf{x}) - \mathbf{x} - (\theta(\mathbf{y}) - \mathbf{y})\rangle_{\mathbf{A}} &= \langle \mathbf{x} - \mathbf{y}, \theta(\mathbf{x}) - \mathbf{x} - (\theta(\mathbf{y}) - \mathbf{y})\rangle_{\mathbf{A}} \\ &\quad + t \, \|\theta(\mathbf{x}) - \mathbf{x} - (\theta(\mathbf{y}) - \mathbf{y})\|_{\mathbf{A}}^2. \end{aligned} \tag{8.94}$$

The second term on the right-hand side can be estimated by using (8.92)

$$\|\theta(\mathbf{x}) - \mathbf{x} - (\theta(\mathbf{y}) - \mathbf{y})\|_{\mathbf{A}}^2 \leq C_{\mathbf{A}} C_{13} h^{p+1} \|\mathbf{x} - \mathbf{y}\|^2.$$

For the first term in (8.94), we employ $\theta(\mathbf{z}) - \mathbf{z} = \pm \|\mathbf{z} - \theta(\mathbf{z})\| \, \mathbf{v}(\mathbf{z})$ for $\mathbf{z} \in \{\mathbf{x}, \mathbf{y}\}$ with the conormal vector $\mathbf{v}(\mathbf{z}) = \mathbf{A}\mathbf{n}(\mathbf{z})$. Taking into account the definition of the $\langle \cdot, \cdot \rangle_{\mathbf{A}}$ we derive

$$\begin{aligned} |\langle \mathbf{x} - \mathbf{y}, \theta(\mathbf{x}) - \mathbf{x} - (\theta(\mathbf{y}) - \mathbf{y})\rangle_{\mathbf{A}}| &= |(\pm) \|\mathbf{x} - \theta(\mathbf{x})\| \langle \mathbf{x} - \mathbf{y}, \mathbf{n}(\mathbf{x})\rangle \\ &\quad - (\pm) \|\mathbf{y} - \theta(\mathbf{y})\| \langle \mathbf{x} - \mathbf{y}, \mathbf{n}(\mathbf{y})\rangle| \\ &\leq \|\theta(\mathbf{x}) - \mathbf{x}\| \, |\langle \mathbf{x} - \mathbf{y}, \mathbf{n}(\mathbf{x})\rangle| \\ &\quad + \|\theta(\mathbf{y}) - \mathbf{y}\| \, |\langle \mathbf{x} - \mathbf{y}, \mathbf{n}(\mathbf{y})\rangle| \\ &\leq 2C\widetilde{C}_0 h_\tau^{p+1} \|\mathbf{x} - \mathbf{y}\|^2. \end{aligned}$$

□

In order to analyze the perturbations for the kernel of the double layer potential and its adjoint, we will use the following lemma.

Lemma 8.4.14. *Let the assumptions of Lemma 8.4.12 be satisfied. Then*

$$\forall \mathbf{x}, \mathbf{y} \in \check{\tau} = \operatorname{lift}_\tau(\tau) \in \check{\mathcal{G}}: \quad |\langle \mathbf{n}(\mathbf{x}), \mathbf{y} - \mathbf{x}\rangle - \langle \mathbf{n}_\ell(\tilde{\mathbf{x}}), \tilde{\mathbf{y}} - \tilde{\mathbf{x}}\rangle| \leq C_{14} h_\tau^{p-1} \|\mathbf{y} - \mathbf{x}\|^2.$$

For any pair of non-identical panels $\tau_1, \tau_2 \in \mathcal{G}^{\text{affine}}$ *and* $\mathbf{x} \in \check{\tau}_1 = \operatorname{lift}_{\tau_1}(\tau_1)$, $\mathbf{y} \in \check{\tau}_2 = \operatorname{lift}(\tau_2)$ *we have*

$$\left|\langle \mathbf{n}(\mathbf{x}), \mathbf{y}-\mathbf{x}\rangle - \langle \mathbf{n}_\ell(\tilde{\mathbf{x}}), \tilde{\mathbf{y}}-\tilde{\mathbf{x}}\rangle\right| \le C_{15} h_\tau^p \|\mathbf{y}-\mathbf{x}\|.$$

where C_{14} *and* C_{15} *depend on* c_8, C_{stab}, $\widetilde{C}_{\text{stab}}$, $C_{\Gamma,p+1}$, C_2, c_3, p, *and the shape-regularity of the mesh.*

Proof. First, we will consider the case $\mathbf{x}, \mathbf{y} \in \check{\tau} = \operatorname{lift}_\tau(\tau) \in \check{\mathcal{G}}$ for some $\tau \in \mathcal{G}^{\text{affine}}$. Let $\tilde{\mathbf{x}} = \theta(\mathbf{x})$ and $\tilde{\mathbf{y}} = \theta(\mathbf{y})$. Taylor expansion on the reference triangle about $\hat{\mathbf{x}}$ leads to (cf. (5.3) and Definition 5.1.2)

$$\begin{aligned}
\langle \mathbf{n}(\mathbf{x}), \mathbf{y}-\mathbf{x}\rangle - \langle \mathbf{n}_\ell(\tilde{\mathbf{x}}), \tilde{\mathbf{y}}-\tilde{\mathbf{x}}\rangle &= \langle \mathbf{n}(\mathbf{x}), \chi_\tau(\hat{\mathbf{y}}) - \chi_\tau(\hat{\mathbf{x}})\rangle - \left\langle \mathbf{n}_\ell(\tilde{\mathbf{x}}), \chi_{\tau,p}(\hat{\mathbf{y}}) - \chi_{\tau,p}(\hat{\mathbf{x}})\right\rangle \\
&= \left\langle \mathbf{n}(\mathbf{x}), \sum_{k=1}^{p} \frac{\left\langle \hat{\mathbf{y}}-\hat{\mathbf{x}}, \widehat{\nabla}\right\rangle^k \chi_\tau(\hat{\mathbf{x}})}{k!} + R_1(\hat{\mathbf{x}}, \hat{\mathbf{y}})\right\rangle \\
&\quad - \left\langle \mathbf{n}_\ell(\tilde{\mathbf{x}}), \sum_{k=1}^{p} \frac{\left\langle \hat{\mathbf{y}}-\hat{\mathbf{x}}, \widehat{\nabla}\right\rangle^k \chi_{\tau,p}(\hat{\mathbf{x}})}{k!} + R_2(\hat{\mathbf{x}}, \hat{\mathbf{y}})\right\rangle.
\end{aligned}$$

Because $\left\langle \hat{\mathbf{y}}-\hat{\mathbf{x}}, \hat{\nabla}\right\rangle \chi_\tau(\hat{\mathbf{x}}) = \mathbf{J}_\tau(\hat{\mathbf{x}})(\hat{\mathbf{y}}-\hat{\mathbf{x}}) \in T_{\mathbf{x}}$, the term $k = 1$ in the first sum vanishes and the same holds for the second sum and the summand with index $k = 1$. Thus

$$\begin{aligned}
\langle \mathbf{n}(\mathbf{x}), \mathbf{y}-\mathbf{x}\rangle - \langle \mathbf{n}_\ell(\tilde{\mathbf{x}}), \tilde{\mathbf{y}}-\tilde{\mathbf{x}}\rangle &= \left\langle \mathbf{n}(\mathbf{x}) - \mathbf{n}_\ell(\tilde{\mathbf{x}}), \sum_{k=2}^{p} \frac{\left\langle \hat{\mathbf{y}}-\hat{\mathbf{x}}, \widehat{\nabla}\right\rangle^k \chi_\tau(\hat{\mathbf{x}})}{k!}\right\rangle \\
&\quad + \langle \mathbf{n}(\mathbf{x}), R_1(\hat{\mathbf{x}}, \hat{\mathbf{y}})\rangle - \langle \mathbf{n}_\ell(\tilde{\mathbf{x}}), R_2(\hat{\mathbf{x}}, \hat{\mathbf{y}})\rangle \\
&\quad + \left\langle \mathbf{n}_\ell(\tilde{\mathbf{x}}), \sum_{k=2}^{p} \frac{\left\langle \hat{\mathbf{y}}-\hat{\mathbf{x}}, \widehat{\nabla}\right\rangle^k \left(\chi_\tau(\hat{\mathbf{x}}) - \chi_{\tau,p}(\hat{\mathbf{x}})\right)}{k!}\right\rangle.
\end{aligned}$$

We will estimate the various quantities in the above expression in the following.

Estimate of R_1, R_2:
Since R_1, R_2 are the remainders of the p-th order Taylor expansion, we obtain as in (8.60)

$$\begin{aligned}
\|R_1(\mathbf{x}, \mathbf{y})\| + \|R_2(\mathbf{x}, \mathbf{y})\| &\le C \frac{(h_\tau \|\hat{\mathbf{y}}-\hat{\mathbf{x}}\|)^{p+1}}{(p+1)!} \\
&\overset{(8.85)}{\le} \frac{C}{c_8^{p+1}} \frac{\|\mathbf{y}-\mathbf{x}\|^{p+1}}{(p+1)!} \le \widetilde{C} \|\mathbf{y}-\mathbf{x}\|^2 h_\tau^{p-1}, \qquad (8.95a)
\end{aligned}$$

where $\widetilde{C}$ depends only on p, c_8, C_{stab}, $\widetilde{C}_{\text{stab}}$, and $C_{\Gamma,p+1}$.

Estimate of $\mathbf{n}-\mathbf{n}_\ell$:

Next, we consider the difference of the normal vectors and derive, from Lemma 8.4.4 and its proof, the estimate

$$\begin{aligned}\|\mathbf{n}(\mathbf{x})-\mathbf{n}_\ell(\tilde{\mathbf{x}})\| &= \left\|\frac{\hat{\mathbf{t}}_1(\hat{\mathbf{x}})\times\hat{\mathbf{t}}_2(\hat{\mathbf{x}})}{g_\tau(\hat{\mathbf{x}})}-\frac{\hat{\mathbf{t}}_{p,1}(\hat{\mathbf{x}})\times\hat{\mathbf{t}}_{p,2}(\hat{\mathbf{x}})}{g_{\tau,p}(\hat{\mathbf{x}})}\right\| \\ &\le \left\|\frac{g_{\tau,p}(\hat{\mathbf{x}})-g_\tau(\hat{\mathbf{x}})}{g_{\tau,p}(\hat{\mathbf{x}})}\mathbf{n}(\mathbf{x})\right\| \\ &\quad+\frac{\left\|\hat{\mathbf{t}}_1(\hat{\mathbf{x}})\times\hat{\mathbf{t}}_2(\hat{\mathbf{x}})-\hat{\mathbf{t}}_{p,1}(\hat{\mathbf{x}})\times\hat{\mathbf{t}}_{p,2}(\hat{\mathbf{x}})\right\|}{g_{\tau,p}(\hat{\mathbf{x}})}\le Ch_\tau^p, \end{aligned}\tag{8.95b}$$

where C depends only on C_2 and c_3.

Estimate of derivatives of the pullback χ_τ:

We employ estimate (8.57b) and (8.85) to derive, for $k\ge 2$,

$$\left\|\frac{\left\langle \hat{\mathbf{y}}-\hat{\mathbf{x}},\hat{\nabla}\right\rangle^k\chi_\tau(\hat{\mathbf{x}})}{k!}\right\|=\left\|\sum_{\substack{\alpha\in\mathbb{N}_0^2\\|\alpha|=k}}\frac{1}{\alpha!}(\hat{\mathbf{y}}-\hat{\mathbf{x}})^\alpha\partial^\alpha\chi_\tau(\hat{\mathbf{x}})\right\|\le C(h_\tau\|\hat{\mathbf{y}}-\hat{\mathbf{x}}\|)^k\le C\|\mathbf{y}-\mathbf{x}\|^2, \tag{8.95c}$$

where C depends on $\widetilde{C}_{\text{stab}}$, c_8, and k.

Estimate of derivatives of the p-parametric approximation error:

By using (8.53b), we obtain

$$\left\|\frac{\left\langle \hat{\mathbf{y}}-\hat{\mathbf{x}},\hat{\nabla}\right\rangle^k\left(\chi_\tau(\hat{\mathbf{x}})-\chi_{\tau,p}(\hat{\mathbf{x}})\right)}{k!}\right\|\le C_1C_{\Gamma,p+1}h_\tau^{p+1}\|\hat{\mathbf{y}}-\hat{\mathbf{x}}\|^k\le Ch_\tau^{p-1}\|\mathbf{y}-\mathbf{x}\|^2, \tag{8.95d}$$

where C depends only on p and $C_{\Gamma,p+1}$.

The combination of estimates (8.95) with some Cauchy–Schwarz inequalities leads to

$$|\langle\mathbf{n}(\mathbf{x}),\mathbf{y}-\mathbf{x}\rangle-\langle\mathbf{n}_\ell(\tilde{\mathbf{x}}),\tilde{\mathbf{y}}-\tilde{\mathbf{x}}\rangle|=Ch_\tau^{p-1}\|\mathbf{y}-\mathbf{x}\|^2.$$

We turn now to the case that $\mathbf{x}\in\check{\tau}_1=\text{lift}_{\tau_1}(\tau_1)$ and $\mathbf{y}\in\check{\tau}_2=\text{lift}_{\tau_2}(\tau_2)$ for some non-identical $\tau_1,\tau_2\in\mathcal{G}^{\text{affine}}$. Lemma 8.4.11 and (8.95b) imply that

$$\begin{aligned}|\langle\mathbf{n}(\mathbf{x}),\mathbf{y}-\mathbf{x}\rangle-\langle\mathbf{n}_\ell(\tilde{\mathbf{x}}),\tilde{\mathbf{y}}-\tilde{\mathbf{x}}\rangle| &= |\langle\mathbf{n}(\mathbf{x})-\mathbf{n}_\ell(\tilde{\mathbf{x}}),\mathbf{y}-\mathbf{x}\rangle+\langle\mathbf{n}_\ell(\tilde{\mathbf{x}}),\mathbf{y}-\mathbf{x}-(\tilde{\mathbf{y}}-\tilde{\mathbf{x}})\rangle| \\ &\le C_{15}h_\tau^p\|\mathbf{y}-\mathbf{x}\|,\end{aligned}$$

where C depends on the constant in (8.95b) and C_7.

Chapter 9
A Posteriori Error Estimation

The error analysis for the Galerkin discretization exhibits the asymptotic convergence rates for the boundary element method which depend on the regularity of the underlying integral equation. These estimates are called a priori estimates because they hold for large classes of problems which are characterized by their regularity. They are important because they show the *asymptotic quality* of the Galerkin boundary method. However, for a concrete problem these estimates could be by far too pessimistic and do not allow answers to the following questions:

- Is the size of the error $u_\ell - u$ with respect to some norm or to some other measure below some given error tolerance ε?
- If the numerical solution $u_\ell \in S_\ell$ is not accurate enough, what is a good strategy to enrich the space S_ℓ in a problem-oriented way? Is the uniform refinement as described in Remark 4.1.8 a good strategy?

The a posteriori error estimation uses the computed numerical solution u_ℓ and the given data (such as the right-hand side or the integral operator) and computes non-negative *indicators* $(\eta_i)_{i=1}^n$ which have the property that the (weighted) sum is an upper bound for the true error. The quantities will be local in the sense that their computation involves integrals over small patches $\omega_i \subset \Gamma$ and their number n depends linearly on the number of panels.

Furthermore, the size of these local quantities can be used directly to detect subregions on the surface Γ where the error is large and which should then be *locally* refined.

The development of a posteriori error estimation for finite element discretizations of partial differential equations started with the pioneering papers [10, 11]. Since then the number of publications in this field has grown enormously and we refer to the monographs [2, 12, 14, 172, 232] for a thorough treatment of this topic and further references.

However, for boundary element methods, the nonlocal character of the integral operator and the nonlocal fractional Sobolev norms cause difficulties in the mathematical derivation of local error indicators and much fewer authors have investigated local a posteriori error estimates for integral equations [47, 48, 50–53, 89, 90, 92, 93, 189, 196, 209, 210, 240, 244].

S.A. Sauter and C. Schwab, *Boundary Element Methods*, Springer Series in Computational Mathematics 39, DOI 10.1007/978-3-540-68093-2_9,

In this chapter, we will develop and analyze a posteriori error estimators for boundary integral operators. We will follow the approach and the analysis as introduced in [89, 90, 92].

9.1 Preliminaries

In Chap. 4 we introduced the Galerkin boundary element method for the abstract variational problem: For given $F \in H'$, find $u \in H$ such that

$$a(u, v) = F(v) \qquad \forall v \in H. \tag{9.1}$$

Let $A : H \to H'$ denote the operator associated with the sesquilinear form $a(\cdot,\cdot)$ (cf. Lemma 2.1.38). Throughout this chapter we will assume that the operator A is either of negative order and maps into $H^s(\Gamma)$ for some positive s or is of non-negative order. Note that we always require throughout this chapter that the range for the differentiability indices s in $H^s(\Gamma)$ obeys condition (2.84) depending on the smoothness of Γ. A first assumption on the operator A is stated next.

Assumption 9.1.1. *$A : H^s(\Gamma) \to H^{-s}(\Gamma)$ is an isomorphism for some order $2s \in \mathbb{R}$, i.e., there exist constants $C_1, C_2 > 0$ such that*

$$\|A\|_{H^{-s}(\Gamma)\leftarrow H^s(\Gamma)} \leq C_1 \quad \text{and} \quad \left\|A^{-1}\right\|_{H^s(\Gamma)\leftarrow H^{-s}(\Gamma)} \leq C_2.$$

The boundary element space S is composed by local polynomials which are lifted to the surface Γ via local charts and put together either in a continuous or discontinuous way. The Galerkin discretization is given by seeking $u_S \in S$ such that

$$a(u_S, v) = F(v) \qquad \forall v \in S. \tag{9.2}$$

The boundary element mesh is denoted by $\mathcal{G}$ consisting of surface panels τ (cf. Chap. 4).

Typically, the error $u - u_S$ will not be distributed uniformly over the surface Γ, and adaptive refinement aims at refining the mesh in regions where the error is larger than some threshold. In this chapter, we will introduce *local a posteriori refinement indicators* for the detection of such regions (and for the estimation of the total error). In this light, the goal of this chapter is to define computable quantities η_i which will depend on the discrete solution u_S such that the estimates

$$C_{\text{eff}} \sum_{i=1}^{n} \eta_i^2 \leq \|u_S - u\|_{H^s(\Gamma)}^2 \leq C_{\text{rel}} \sum_{i=1}^{n} \eta_i^2 \tag{9.3}$$

hold. The upper estimate is called "reliability" because it guarantees a prescribed given accuracy while the lower estimate is called "efficiency" because it implies that the qualitative behavior of the error is reflected by the error indicators and not overestimated in an unspecified way. The quantities η_i will be local in the sense that their computation involves integrals over small patches $\omega_i \subset \Gamma$. The number n in (9.3) will depend linearly on the number of panels in $\mathcal{G}$.

Remark 9.1.2. *The size of the constant C_{rel} is important for the practical use of a posteriori error estimation. In order to guarantee a prescribed accuracy for the energy error $\|u_S - u\|_{H^s(\Gamma)}$ one has to refine the discretization as long as the upper bound $\left(C_{\text{rel}} \sum_{i=1}^n \eta_i^2\right)^{1/2}$ is below the given threshold. In this light, we will track the dependence of C_{rel} on more elementary constants as clearly as possible.*

The adaptive solution of the problem (9.1) is structured in the following abstract way. Let $\varepsilon > 0$ be a given tolerance for the *energy error* $\|u_S - u\|_{H^s(\Gamma)}$ and let $\delta \in (0, 1)$ be a control parameter for the refinement. Let $\mathcal{G}_0$ be a coarse surface mesh and let S_0 denote the corresponding boundary element space.

Algorithm 9.1.3 (Adaptive Refinement).

begin

if $\ell = 0$ ***then*** *compute the solution u_0 of problem (9.2) for $S = S_0$;*

1: *compute the error indicators η_i, $1 \le i \le n_\ell$;*

if $\sum_{i=1}^n \eta_i^2 \le \varepsilon$ ***then STOP:*** *Solution is u_ℓ;*

else begin

$\ell := \ell + 1;$

refine all panels $\tau \in \mathcal{G}_{\ell-1}$ with $\tau \subset \omega_i$ and $\eta_i > \delta \max_{1 \le j \le n} \eta_j$; (9.4)

subdivide (if necessary) further panels $\tau \in \mathcal{G}_{\ell-1}$ such that the mesh $\mathcal{G}_\ell$ (9.5a)

becomes regular (cf. Definition 4.1.4); (9.5b)

solve problem (9.2) for $S = S_\ell$,

where S_ℓ corresponds to the new mesh $\mathcal{G}_\ell$;

goto 1;

end;

end;

The realization of the algorithm requires some mesh refinement techniques and we will present some basic principles in the sequel.

We assume that a coarse mesh $\mathcal{G}_0$ is given. This initial mesh is used to generate finer ones $(\mathcal{G}_\ell)_{\ell=1}^{\ell_{\max}}$ in a recursive way by applying different refinement patterns. For triangles, there exist various refinement patterns, some of them are depicted in Fig. 9.1.

If discontinuous boundary elements are employed the mesh $\mathcal{G}$ is not required to be regular (cf. Definition 4.1.4). In this case, step (9.4) in the adaptive algorithm is realized by refining all panels which satisfy $\eta_\tau \le \delta \max_{t \in \mathcal{G}} \eta_t$ by the red refinement pattern and skip step (9.5).

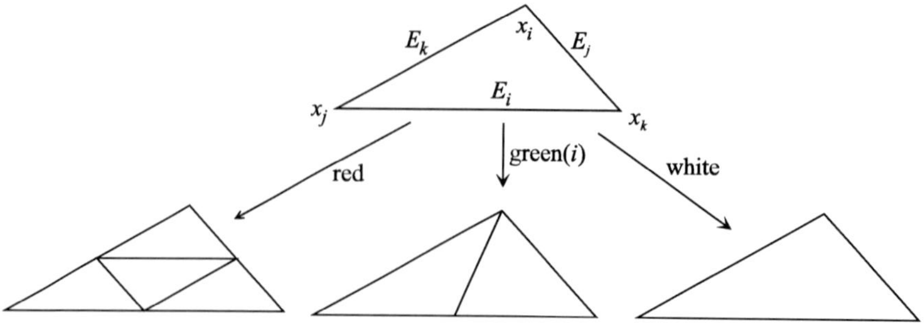

Fig. 9.1 Different refinement patterns for triangles. *Red*: Subdivision into four concruent triangles, *Green* (i): Connecting the i-th vertex with the midpoint of the i-th edge, *White*: No refinement

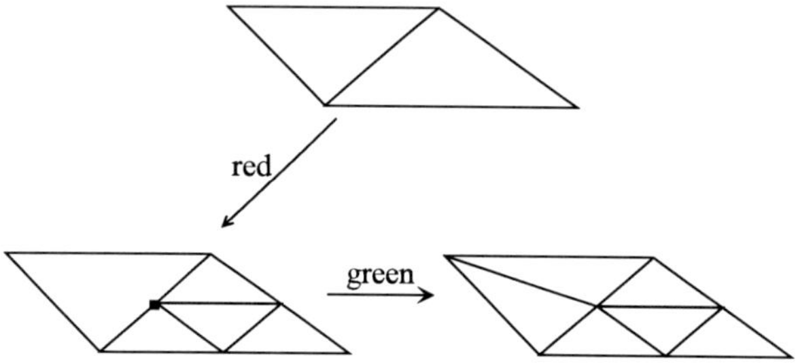

Fig. 9.2 Red refinement of a panel, typically, results in a hanging node (here marked by *filled square*). Connecting this node with the opposite vertex of the neighboring panel results in a regular mesh

If continuous boundary elements are employed the mesh $\mathcal{G}$ must be regular. In particular, *hanging nodes* have to be avoided which typically arise if some panels are refined by the red pattern. A panel vertex is called a hanging node if it lies in the interior of an edge of some other panel. A typical situation is depicted in Fig. 9.2.

The *green closure* algorithm refines some further panels so that the resulting mesh becomes regular. It uses a function "mark" which contains the refinement patterns for the panels in $\mathcal{G}_{\ell-1}$ and is initialized by mark (τ) = "white" for all $\tau \in \mathcal{G}_{\ell-1}$. Hence it is straightforward to check from the function mark $(\cdot)$ whether a mesh, which would result by refining $\mathcal{G}_{\ell-1}$ according to mark $(\cdot)$, is regular or not. We set the function regular $(\mathcal{G}_{\ell-1}, \text{mark})$ = "true" in the first case and "false" for the other case.

The green closure algorithm is realized by replacing steps (9.4) and (9.5) by the following piece of code.

for all panels $\tau \in \mathcal{G}_{\ell-1}$ with: $\tau \subset \omega_i$ and $\eta_i > \delta \max_{1 \leq j \leq n_\ell} \eta_j$ for some $1 \leq i \leq n$ **do**

mark $(\tau) :=$ red;

while regular $(\mathcal{G}_{\ell-1}, \text{mark}) =$ false **do**

for all $\tau \in \mathcal{G}_{\ell-1}$ **do begin**

if mark $(\tau) =$ green $\wedge \tau$ contains one additional hanging node **then** mark $(\tau) :=$ red;

elseif mark $(\tau) =$ white **then begin**

if τ contains more than one hanging node **then** mark $(\tau) :=$ red;

elseif τ contains one hanging node **then** mark $(\tau) :=$ green;

end;

end;

Remark 9.1.4. *In the literature, there exist many advanced versions of the green closure algorithm (see, e.g., [17] and the update [18]), which, e.g., use a larger class of refinement patterns or guarantee that the shape-regularity of the panels is preserved through the refinement process.*

The efficient algorithmic realization of the refinement algorithm requires appropriate data structures for the mesh handling. We refer, e.g., to [17] and the update [18] for the details.

9.2 Local Error Indicators and A Posteriori Error Estimators

The definition of local error indicators and a posteriori error estimators for operators of negative order will differ from those of non-negative order.

9.2.1 *Operators of Negative Order*

We start with the case that the operator $A : H^s(\Gamma) \to H^{-s}(\Gamma)$ which is associated with the sesquilinear form $a(\cdot,\cdot)$ in (9.1) is of negative order $2s \in [-4, 0]$ and, in addition, s has to satisfy condition (2.84) depending on the smoothness of Γ. Note that the boundary integral operator V for the single layer potential for the Laplacian satisfies this condition for $2s = -1$ (cf. Theorem 3.1.16).

Let u denote the exact solution to problem (9.1) while the Galerkin solution u_S is the solution of (9.2). Our goal is to estimate the Galerkin error

$$e = u_S - u$$

by computable local error indicators. The image of the error under A is denoted as the *residual*

$$r := Ae = Au_S - F \in H^{-s}(\Gamma).$$

The continuity (cf. Assumption 9.1.1) of A and A^{-1} leads to the estimates

$$C_1^{-2} \|Av\|_{H^{-s}(\Gamma)}^2 \leq \|v\|_{H^s(\Gamma)}^2 \leq C_2^2 \|Av\|_{H^{-s}(\Gamma)}^2 \qquad \forall v \in H^s(\Gamma).$$

The choice $v = e$ shows that the Galerkin error is bounded from below and from above in terms of the residual

$$C_1^{-2} \|r\|_{H^{-s}(\Gamma)}^2 \leq \|e\|_{H^s(\Gamma)}^2 \leq C_2^2 \|r\|_{H^{-s}(\Gamma)}^2 \qquad \forall v \in H^s(\Gamma).$$

We will show in Sect. 9.3 that the norm of the residuals can be localized under very mild assumptions on the mesh.

Let $\mathcal{I}$ denote the set of counting indices for the basis functions b_i, $i \in \mathcal{I}$. The nodal point associated with b_i is denoted by $\mathbf{z}_i \in \Gamma$. For any $i \in \mathcal{I}$, we introduce panel neighborhoods ω_i about $\mathbf{z}_i$ and, for any $\tau \in \mathcal{G}$, neighborhoods ω_τ about τ by

$$\omega_i := \operatorname{supp} b_i \quad \text{and} \quad \omega_\tau := \overline{\bigcup_{\substack{t \in \mathcal{G} \\ \bar{t} \cap \bar{\tau} \neq \emptyset}} t}. \tag{9.6}$$

The distance of a panel τ from $\Gamma \backslash \omega_\tau$ is denoted by

$$d_\tau := \operatorname{dist}(\tau, \Gamma \backslash \omega_\tau). \tag{9.7}$$

The definition of the local error indicators will be based on the explicit representation of the seminorm $|\cdot|_{H^\sigma(\omega_i)}$ for non-integers $\sigma > 0$:

$$|\varphi|_{H^\sigma(\omega_i)}^2 = \sum_{\alpha = \lfloor \sigma \rfloor} \int_{\omega_i} \int_{\omega_i} \frac{|\partial^\alpha \varphi(\mathbf{x}) - \partial^\alpha \varphi(\mathbf{y})|^2}{\|\mathbf{x} - \mathbf{y}\|^{2+2\lambda}} ds_{\mathbf{y}} ds_{\mathbf{x}},$$

where $\sigma = \lfloor \sigma \rfloor + \lambda$. The surface derivatives $\partial^\alpha \varphi$ are defined via local pullbacks to two-dimensional parameter domains as in (2.85). Recall that the range of σ is restricted depending on the smoothness of the surface [cf. (2.84)].

Definition 9.2.1. Let u_S denote the Galerkin solution to problem (9.2) and let the residual be given by

$$r = Au_S - F.$$

The local error indicators are given by

$$\eta_i := |r|_{H^{-s}(\omega_i)} \qquad i \in \mathcal{I}$$

and the global error estimator is

$$\eta := \sqrt{\sum_{i \in \mathcal{I}} \eta_i^2}.$$

9.2.2 Operators of Non-negative Order

If the operator $A : H^s(\Gamma) \to H^{-s}(\Gamma)$ which is associated with the sesquilinear form in (9.1) is an isomorphism of non-negative order $2s \geq 0$, the residual

$$r = Au_S - F \in H^{-s}(\Gamma)$$

is a functional. In this light, the operator A acts as a differential operator in a broader sense. For this case, we will derive error indicators which are of *residual type*. For partial differential equations this type of error estimators goes back to the pioneering papers [10, 11]. For integral equations they have been developed in [89].

Definition 9.2.2. Let $(b_i)_{i\in\mathcal{I}}$ denote the basis of the boundary element space. The Galerkin solution is denoted by u_S, the exact solution by u, and the corresponding error by $e = u_S - u$. For $i \in \mathcal{I}$, the local error indicators are given by

$$\eta_i := \sup_{\substack{v\in H^s(\Gamma)\\ b_i v\neq 0}} \frac{|a(e, b_i v)|}{\|b_i v\|_{H^s(\Gamma)}}. \tag{9.8a}$$

Note that

$$\sup_{\substack{v\in H^s(\Gamma)\\ b_i v\neq 0}} \frac{|a(e, b_i v)|}{\|b_i v\|_{H^s(\Gamma)}} = \sup_{\substack{v\in H^s(\Gamma)\\ b_i v\neq 0}} \frac{\left|\langle r, b_i v\rangle_{H^{-s}(\Gamma)\times H^s(\Gamma)}\right|}{\|b_i v\|_{H^s(\Gamma)}}. \tag{9.8b}$$

Remark 9.2.3. *Note that the quantities η_i are not computable because the supremum (9.8) is taken over an infinite-dimensional space. Under the assumption that the residuum is in $L^2(\Gamma)$, computable lower and upper bounds which can be approximated by quadrature formulae are given by*

$$\frac{\left|\left(r, b_i^2\right)_{H^{-s}(\Gamma)\times H^s(\Gamma)}\right|}{\left\|b_i^2\right\|_{H^s(\Gamma)}} \leq \eta_i \leq C \sqrt{\sum_{\substack{\tau\in\mathcal{G}\\ \tau\subset\omega_i^\star}} h_\tau^{2s} \|r\|_{L^2(\tau)}^2}$$

for all $i \in \mathcal{I}$, where

$$\omega_i^\star := \{\tau \in \mathcal{G} : \overline{\tau} \cap \overline{\omega_i} \neq \emptyset\}. \tag{9.9}$$

The constant C depends only on the shape-regularity of $\mathcal{G}$.

9.3 Proof of Efficiency and Reliability

This section is devoted to the proof of efficiency and reliability of the error estimators which have been presented in the previous sections. Again we will distinguish between operators of negative order and operators of non-negative order. Throughout this section, we will assume that the surface Γ is of class $\mathcal{A}^2_{\mathrm{pw}}$ (cf. Definition 8.3.1) and we will restrict to the three-dimensional case.

9.3.1 Analysis of Operators of Negative Order

The analysis of the a posteriori error indicator as in Definition 9.2.1 will be based on the localization of the globally defined norms of Sobolev spaces of positive fractional order [cf. (2.85)]. Although this can be worked out for any Sobolev index $s \in \mathbb{R}_{>0} \backslash \mathbb{N}$ we restrict the presentation to $s \in]0,1[$ which includes the most important case $s = 1/2$ for integral operators of negative order corresponding to elliptic boundary value problems of second order. For the general case we refer to [92].

We start with two preparatory lemmata.

Lemma 9.3.1. *For $\lambda > 0$ and for any $\mathbf{y} \in \mathbb{R}^2$ and $\varepsilon > 0$, we have*

$$\int_{\mathbb{R}^2 \backslash B_\varepsilon(\mathbf{y})} \frac{1}{\|\mathbf{y}-\mathbf{x}\|^{2+\lambda}} d\mathbf{x} = \frac{2\pi}{\lambda} \varepsilon^{-\lambda}.$$

Proof. We introduce polar coordinates centered at $\mathbf{y}$ by $\mathbf{x} = \mathbf{y} + r(\cos\varphi, \sin\varphi)^\intercal$ and obtain

$$\int_{\mathbb{R}^2 \backslash B_\varepsilon(\mathbf{y})} \frac{1}{\|\mathbf{y}-\mathbf{x}\|^{2+\lambda}} d\mathbf{x} = 2\pi \int_\varepsilon^\infty \frac{1}{r^{2+\lambda}} r dr = \frac{2\pi}{\lambda} \varepsilon^{-\lambda}.$$

□

Lemma 9.3.2. *For $\lambda > 0$ there exists a constant C_λ depending only on λ and the geometry of Γ such that*

$$\int_{\Gamma \backslash B_\varepsilon(\mathbf{z})} \frac{1}{\|\mathbf{y}-\mathbf{x}\|^{2+\lambda}} ds_{\mathbf{x}} \leq C_\lambda \varepsilon^{-\lambda}$$

for all $\mathbf{z} \in \mathbb{R}^3$ and $\varepsilon > 0$.

The proof is completely analogous to the estimate of the quantity $S_2(\mathbf{x})$ in the proof of Theorem 3.3.5 and we leave the details as an exercise.

We start with the derivation of estimates of the $H^s(\Gamma)$-norm by a sum of local integrals for the case $s \in]0,1[$. Recall the definition

$$\|v\|^2_{H^s(\omega)} = \|v\|^2_{L^2(\omega)} + |v|^2_{H^s(\omega)}$$

for any measurable $\omega \subset \Gamma$ and note that, for $s \in]0,1[$, we have

$$|v|^2_{H^s(\omega)} = \int_\omega \int_\omega \frac{|v(\mathbf{x}) - v(\mathbf{y})|^2}{\|\mathbf{x}-\mathbf{y}\|^{2+2s}} ds_{\mathbf{y}} ds_{\mathbf{x}}.$$

Lemma 9.3.3. *Let $s \in]0,1[$. For any function $v \in H^s(\Gamma)$ and any boundary element mesh $\mathcal{G}$,*

$$|v|^2_{H^s(\Gamma)} \leq \sum_{\tau\in\mathcal{G}} \left\{ \int_\tau \int_{\omega_\tau} \frac{|v(\mathbf{x}) - v(\mathbf{y})|^2}{\|\mathbf{x}-\mathbf{y}\|^{2+2s}} ds_{\mathbf{x}} ds_{\mathbf{y}} + 4C_{2s} d_\tau^{-2s} \|v\|^2_{L^2(\tau)} \right\} \quad (9.10a)$$

$$\leq \sum_{i\in\mathcal{I}} |v|^2_{H^s(\omega_i)} + 4C_{2s} \sum_{\tau\in\mathcal{G}} d_\tau^{-2s} \|v\|^2_{L^2(\tau)}, \quad (9.10b)$$

where ω_τ, ω_i are as in (9.6) and d_τ is as in (9.7). The constant C_{2s} depends only on s and on the geometry of Γ.

Proof. The additivity of the integral leads to the splitting

$$|v|^2_{H^s(\Gamma)} = \sum_{\tau\in\mathcal{G}} \int_\tau \int_\Gamma \frac{|v(\mathbf{x}) - v(\mathbf{y})|^2}{\|\mathbf{x}-\mathbf{y}\|^{2+2s}} ds_{\mathbf{x}} ds_{\mathbf{y}} = \sum_{\tau\in\mathcal{G}} \left\{ \int_\tau \int_{\omega_\tau} \cdots + \int_\tau \int_{\Gamma\backslash\omega_\tau} \cdots \right\}. \quad (9.11)$$

The second term can be estimated by

$$\int_\tau \int_{\Gamma\backslash\omega_\tau} \frac{|v(\mathbf{x}) - v(\mathbf{y})|^2}{\|\mathbf{x}-\mathbf{y}\|^{2+2s}} ds_{\mathbf{x}} ds_{\mathbf{y}} \leq 2J_\tau^{\mathrm{I}} + 2J_\tau^{\mathrm{II}}, \quad (9.12)$$

where

$$J_\tau^{\mathrm{I}} := \int_\tau \int_{\Gamma\backslash\omega_\tau} \frac{|v(\mathbf{y})|^2}{\|\mathbf{x}-\mathbf{y}\|^{2+2s}} ds_{\mathbf{x}} ds_{\mathbf{y}} \quad \text{and} \quad J_\tau^{\mathrm{II}} := \int_{\Gamma\backslash\omega_\tau} \int_\tau \frac{|v(\mathbf{x})|^2}{\|\mathbf{x}-\mathbf{y}\|^{2+2s}} ds_{\mathbf{y}} ds_{\mathbf{x}}.$$

Note that for any $\tau \in \mathcal{G}$ we have $\{t \in \mathcal{G} : \tau \subset \Gamma\backslash\omega_t\} = \{t \in \mathcal{G} : t \subset \Gamma\backslash\omega_\tau\}$ and, thus, the summations of J_τ^{I} and J_τ^{II} over all panels coincide as can be seen from

$$\sum_{\tau\in\mathcal{G}} J_\tau^{\mathrm{II}} = \sum_{\tau\in\mathcal{G}} \sum_{\substack{t\in\mathcal{G} \\ t\subset\Gamma\backslash\omega_\tau}} \int_t \int_\tau \frac{|v(\mathbf{x})|^2}{\|\mathbf{x}-\mathbf{y}\|^{2+2s}} ds_{\mathbf{y}} ds_{\mathbf{x}}$$

$$= \sum_{\tau\in\mathcal{G}} \sum_{\substack{t\in\mathcal{G} \\ \tau\subset\Gamma\backslash\omega_t}} \int_\tau \int_t \frac{|v(\mathbf{y})|^2}{\|\mathbf{y}-\mathbf{x}\|^{2+2s}} ds_{\mathbf{x}} ds_{\mathbf{y}}$$

$$= \sum_{\tau\in\mathcal{G}} \sum_{\substack{t\in\mathcal{G} \\ t\subset\Gamma\backslash\omega_\tau}} \int_\tau \int_t \frac{|v(\mathbf{y})|^2}{\|\mathbf{y}-\mathbf{x}\|^{2+2s}} ds_{\mathbf{x}} ds_{\mathbf{y}} = \sum_{\tau\in\mathcal{G}} J_\tau^{\mathrm{I}}.$$

The combination of this equality with (9.11) and (9.12) implies that

$$|v|^2_{H^s(\Gamma)} = \sum_{\tau\in\mathcal{G}} \left\{ \int_\tau \int_{\omega_\tau} \frac{|v(\mathbf{x}) - v(\mathbf{y})|^2}{\|\mathbf{x}-\mathbf{y}\|^{2+2s}} ds_{\mathbf{x}} ds_{\mathbf{y}} + 4J^{\mathrm{I}}_\tau \right\}. \tag{9.13}$$

Observe that $\Gamma\backslash\omega_\tau \subset \Gamma\backslash B_{d_\tau}(\mathbf{y})$ for any $\mathbf{y}\in\tau$ and Lemma 9.3.2 implies

$$J^{\mathrm{I}}_\tau = \int_\tau |v(\mathbf{y})|^2 \left(\int_{\Gamma\backslash\omega_\tau} \frac{1}{\|\mathbf{x}-\mathbf{y}\|^{2+2s}} ds_{\mathbf{x}} \right) ds_{\mathbf{y}} \leq C_{2s} d_\tau^{-2s} \|v\|^2_{L^2(\tau)}.$$

Thus the first estimate (9.10a) is proved.

For the second estimate (9.10b), it remains to consider the first term in (9.13). Note that for any pair of panels $\tau, t \in \mathcal{G}$ with $\overline{\tau}\cap\overline{t} \neq \emptyset$ there exists at least one $i \in \mathcal{I}$ such that $\tau \cup \tau' \subset \omega_i$. Hence we are led to the final estimate

$$\sum_{\tau\in\mathcal{G}} \int_\tau \int_{\omega_\tau} \frac{|v(\mathbf{x}) - v(\mathbf{y})|^2}{\|\mathbf{x}-\mathbf{y}\|^{2+2s}} ds_{\mathbf{x}} ds_{\mathbf{y}} = \sum_{\substack{\tau,t\in\mathcal{G}\\ \overline{\tau}\cap\overline{t}\neq\emptyset}} \int_\tau \int_t \cdots \leq \sum_{i\in\mathcal{I}} \sum_{\tau,\tau'\subset\omega_i} \int_\tau \int_t \cdots$$
$$= \sum_{i\in\mathcal{I}} \int_{\omega_i} \int_{\omega_i} \frac{|v(\mathbf{x}) - v(\mathbf{y})|^2}{\|\mathbf{x}-\mathbf{y}\|^{2+2s}} ds_{\mathbf{x}} ds_{\mathbf{y}}.$$

□

The second term on the right-hand sides of (9.10) is a weighted sum of local L^2-norms of the function v and our goal is to estimate this terms by a sum of local $H^s(\omega_i)$-norms. Note that an estimate of the form

$$d_\tau^{-2s} \|v\|^2_{L^2(\tau)} \leq C \|v\|^2_{H^s(\mathcal{U}_\tau)} \tag{9.14}$$

for some small Γ-neighborhood $\mathcal{U}_\tau$ of τ cannot hold for arbitrary functions $v \in H^s(\Gamma)$ as can be seen from the counter example $v \equiv 1$.

$$\frac{d_\tau^{-2s}\|v\|^2_{L^2(\tau)}}{\|v\|^2_{H^s(\mathcal{U}_\tau)}} = \frac{d_\tau^{-2s}|\tau|}{|\mathcal{U}_\tau|} \overset{h_{\mathcal{G}}\to 0}{\longrightarrow} \infty \quad \text{if the ratio } |\tau| / |\mathcal{U}_\tau| \text{ is bounded away from zero.}$$

However, we will prove an estimate similar to (9.14) for functions being orthogonal to some boundary element functions. First, the result will be proved in some two-dimensional parameter planes and then lifted to the surface. Let ω_i be as in (9.6). We introduce a lifting $\chi_i : \widetilde{\omega_i} \to \omega_i$ from a two-dimensional convex polygonal domain to ω_i as follows:

- If $\omega_i = \tau$ for some panel $\tau \in \mathcal{G}$ the mapping χ_i is chosen as the reference mapping χ_τ of τ (cf. Definition 4.1.2).
- For continuous boundary elements, ω_i is the union of a few panels, i.e., there exist subsets $\mathcal{G}_i \subset \mathcal{G}$ – the cardinality $|\mathcal{G}_i|$ of which is bounded by a constant independent of the refinement level – such that

$$\overline{\omega_i} = \bigcup_{\tau \in \mathcal{G}_i} \overline{\tau}.$$

In this case, the lifting $\chi_i : \tilde{\omega_i} \to \omega_i$ is as introduced in Sect. 4.3.5 (cf. Assumption 4.3.25). Recall that $\tilde{\omega_i}$ is a two-dimensional convex and polygonal parameter domain. Note that, for any $\tau \in \mathcal{G}_i$, the pullback $\tilde{\tau} := \chi_i^{-1}(\tau)$ is a plane panel with straight edges.

The following Lemma is a refined version of the second Poincaré inequality (cf. Theorem 2.5.9), where the dependence of the constants on the geometry is traced more explicitly.

Lemma 9.3.4. *Let $D \subset \mathbb{R}^2$ be a polygonal domain. Then we obtain for $s \in]0, 1[$ and any functions $w \in H^s(D)$ the estimate*

$$\|w\|_{L^2(D)}^2 \leq \frac{1}{2} \frac{(\operatorname{diam} D)^{2+2s}}{|D|} |w|_{H^s(D)}^2 + \frac{1}{|D|} \left| \int_D w(\mathbf{x})\, d\mathbf{x} \right|^2 . \tag{9.15}$$

Proof. We proceed with

$$\begin{aligned}
&\int_D \int_D |w(\mathbf{x}) - w(\mathbf{y})|^2 \, d\mathbf{x} d\mathbf{y} \\
&\quad = \int_D \int_D |w(\mathbf{x})|^2 \, d\mathbf{x} d\mathbf{y} + \int_D \int_D |w(\mathbf{y})|^2 \, d\mathbf{x} d\mathbf{y} \\
&\quad - \int_D \int_D \left(w(\mathbf{x}) \overline{w(\mathbf{y})} + \overline{w(\mathbf{x})} w(\mathbf{y}) \right) d\mathbf{x} d\mathbf{y} \\
&\quad = 2|D| \, \|w\|_{L^2(D)}^2 - 2|J|^2
\end{aligned} \tag{9.16}$$

where $J := \int_D w$. Hence (9.15) follows from

$$\begin{aligned}
2|D| \, \|w\|_{L^2(D)}^2 - 2|J|^2 &= \int_D \int_D \frac{|w(\mathbf{x}) - w(\mathbf{y})|^2}{\|\mathbf{x} - \mathbf{y}\|^{2+2s}} \underbrace{\|\mathbf{x} - \mathbf{y}\|^{2+2s}}_{\leq (\operatorname{diam} D)^{2+2s}} d\mathbf{x} d\mathbf{y} \\
&\leq (\operatorname{diam} D)^{2+2s} \, |w|_{H^s(D)}^2 .
\end{aligned}$$

□

In the following we will estimate the term $|J|^2$ in (9.16) in more detail. Note that, for general functions $w \in L^2(D)$, the Cauchy–Schwarz inequality

$$|J| = \left| \int_D w \right| \leq |D|^{1/2} \, \|w\|_{L^2(D)} \tag{9.17}$$

is sharp. In the next lemma, we will prove that, for functions $w \in L^2$ which are orthogonal to finite element basis functions, the estimate (9.17) can be strengthened.

Recall that, for the boundary element space S, the local nodal basis is denoted by $(b_i)_{i\in\mathcal{I}}$. First, we will prove the result for the pullbacks $\tilde{\omega}_i := \chi_i^{-1}(\omega_i)$ and then lift it to the surface. For $s > 0$, we introduce the mesh-dependent Sobolev space

$$H_\perp^s(\tilde{\omega}_i) := \left\{ v \in H^s(\tilde{\omega}_i) : (v, b_i \circ \chi_i)_{L^2(\tilde{\omega}_i)} = 0 \right\}. \tag{9.18}$$

Definition 9.3.5 (Local Residual Property). A boundary element space along with a basis $(b_i)_{i\in\mathcal{I}}$ satisfies the "local residual property" if there is some constant $0 \le \gamma < 1$ such that

$$\left| \int_{\tilde{\omega}_i} w \right| \le \gamma \, |\tilde{\omega}_i|^{1/2} \, \|w\|_{L^2(\tilde{\omega}_i)} \qquad \forall w \in H_\perp^s(\tilde{\omega}_i) \quad \forall i \in \mathcal{I}. \tag{9.19}$$

Later, we will need the local residual property only for the two basic boundary element spaces $S_{\mathcal{G}}^{0,-1}$, which consists of discontinuous piecewise constant functions, and $S_{\mathcal{G}}^{1,0}$ which consists of continuous, piecewise affine boundary elements.

Lemma 9.3.6. *For $S = S_{\mathcal{G}}^{0,-1}$ the local residual property is satisfied with $\gamma = 0$ while, for $S = S_{\mathcal{G}}^{1,0}$, the estimate (9.19) holds for $\gamma = 1/\sqrt{2}$.*

Proof. Let $S = S_{\mathcal{G}}^{0,-1}$. Then it is sufficient to prove (9.19) for the unit panel $\hat{\tau}$, i.e., either the unit triangle or the unit square [cf. (4.13)]. In this case, we have $H_\perp^s(\hat{\tau}) = \left\{ w \in H^s(\hat{\tau}) : \int_{\hat{\tau}} w = 0 \right\}$ and (9.19) holds with $\gamma = 0$.

Let $S = S_{\mathcal{G}}^{1,0}$. For $i \in \mathcal{I}$, let $\tilde{\omega}_i$ again denote the pullback to the two-dimensional parameter domain and let $\widetilde{b_i} := b_i \circ \chi_i$. For any $w \in H_\perp^s(\tilde{\omega}_i)$, we have

$$\left| \int_{\tilde{\omega}_i} w \right| = \left| \int_{\tilde{\omega}_i} \left(1 - \widetilde{b_i} \right) w \right| \le \left\| 1 - \widetilde{b_i} \right\|_{L^2(\tilde{\omega}_i)} \|w\|_{L^2(\widetilde{\omega_i})}. \tag{9.20}$$

Pick some $\tau \subset \omega_i$ and let $\tilde{\tau} = \chi_i^{-1}(\tau) \subset \tilde{\omega}_i$. Then $\hat{b}_\tau := \left. \widetilde{b_i} \right|_{\tilde{\tau}}$ is the affine function which equals one at some vertex of $\tilde{\tau}$ and zero at the others. Note that $\left(1 - \hat{b}_\tau\right)^2$ is a quadratic polynomial and the following quadrature rule is exact. Let m_i, $1 \le i \le 3$, denote the midpoints of the edges of $\tilde{\tau}$. Then

$$\int_{\tilde{\tau}} \left(1 - \hat{b}_\tau\right)^2 = \frac{|\tilde{\tau}|}{3} \sum_{i=1}^{3} \left(1 - \hat{b}_\tau\right)^2 (m_i) = \frac{|\tilde{\tau}|}{2}$$

and for the integral over $\widetilde{\omega_i}$ we get

$$\left\| 1 - \widetilde{B}_i \right\|_{L^2(\widetilde{\omega_i})}^2 \le \frac{|\widetilde{\omega_i}|}{2}. \tag{9.21}$$

The combination of (9.21) with (9.20) yields

$$\left|\int_{\widetilde{\omega}_i} w\right| \leq \frac{1}{\sqrt{2}} \left|\widetilde{\omega}_i\right|^{1/2} \|w\|_{L^2(\widetilde{\omega}_i)}$$

and γ equals $1/\sqrt{2}$. □

Lemma 9.3.7. *Let $s \in]0,1[$ and assume that the boundary element space has the local residual property. Then, for any $i \in \mathcal{I}$ and $w \in H^s_\perp(\widetilde{\omega}_i)$, we have*

$$\|w\|^2_{L^2(\widetilde{\omega}_i)} \leq \frac{(\operatorname{diam}\widetilde{\omega}_i)^{2+2s}}{2\left(1-\gamma^2\right)\left|\widetilde{\omega}_i\right|} |w|^2_{H^s(\widetilde{\omega}_i)}.$$

Proof. Let $J_i := \int_{\widetilde{\omega}_i} w$. By (9.15) and the local residual property of S we obtain, for any $w \in H^s_\perp(\widetilde{\omega}_i)$, the estimate

$$\begin{aligned}
\|w\|^2_{L^2(\widetilde{\omega}_i)} &\leq \frac{1}{2}\frac{(\operatorname{diam}\widetilde{\omega}_i)^{2+2s}}{\left|\widetilde{\omega}_i\right|} |w|^2_{H^s(\widetilde{\omega}_i)} + \frac{1}{\left|\widetilde{\omega}_i\right|}\left|J_i\right|^2 \\
&\leq \frac{1}{2}\frac{(\operatorname{diam}\widetilde{\omega}_i)^{2+2s}}{\left|\widetilde{\omega}_i\right|} |w|^2_{H^s(\widetilde{\omega}_i)} + \gamma^2 \|w\|^2_{L^2(\widetilde{\omega}_i)}.
\end{aligned}$$

From $\gamma \in]0,1[$, the assertion follows. □

Now, we are in the position to prove a Poincaré-type inequality similar to (9.14). The global version of $H^s_\perp(\widetilde{\omega}_i)$, for a boundary element space S, is given by

$$H^s_\perp(S,\Gamma) := \left\{w \in H^s(\Gamma) \mid \forall v \in S : (v,w)_{L^2(\Gamma)} = 0\right\}, \qquad (9.22)$$

i.e., $H^s_\perp(S,\Gamma) = S^\perp \cap H^s(\Gamma)$. The following remark states that piecewise smooth surfaces can be covered by a selection of smooth patches ω_i provided that the mesh width of the surface mesh is small enough. This property will be needed in the proof of Lemma 9.3.9.

Remark 9.3.8. *For a given surface Γ of class $\mathcal{A}^2_{\mathrm{pw}}$ (cf. Definition 8.3.1) and any regular surface mesh $\mathcal{G}$ with sufficiently small mesh width $h_{\mathcal{G}} \leq h_\Gamma$ the following property holds. For all $\tau \in \mathcal{G}$ there exists $i \in \mathcal{I}$ such that $\tau \subset \omega_i$ and the mapping χ_i can be chosen such that $\chi_i \in C^2(\widetilde{\omega}_i)$.*

We always choose $\chi_i \in C^2(\widetilde{\omega}_i)$ if possible. The subset $\mathcal{I}^{\mathrm{smooth}} \subset \mathcal{I}$ contains all indices $\mathcal{I}$ such that $\chi_i \in C^2(\widetilde{\omega}_i)$. Note that $\Gamma = \bigcup_{i\in\mathcal{I}^{\mathrm{smooth}}} \omega_i$.

Lemma 9.3.9. *Let the surface Γ be of class $\mathcal{A}^2_{\mathrm{pw}}$ and $h_{\mathcal{G}} \leq h_\Gamma$ (cf. Remark 9.3.8). Let Assumption 4.3.29 hold and let Assumption 4.3.17 or Assumption 4.3.18 hold. Further, let $s \in]0,1[$. Then there exists a constant $C_\perp$ depending on the shape-regularity constant and on Γ such that*

$$\sum_{\tau\in\mathcal{G}} d_\tau^{-2s} \|v\|^2_{L^2(\tau)} \leq C_\perp \sum_{i\in\mathcal{I}} |v|^2_{H^s(\omega_i)}$$

for all $v \in H^s_\perp(S,\Gamma)$.

Proof. Let $\tau \in \mathcal{G}$ and choose $i \in \mathcal{I}_{\text{smooth}}$ such that $\tau \subset \omega_i$. For $v \in H_{\perp}^{s}(S,\Gamma)$ we have

$$0 = (v, b_i)_{L^2(\Gamma)} = \int_{\widetilde{\omega}_i} \widetilde{B}_i \tilde{v} g_i \, d\mathbf{x},$$

where $\widetilde{B}_i := b_i \circ \chi_i$, $\tilde{v} := v \circ \chi_i$, and g_i is the surface element. We conclude that $\tilde{v} g_i \in H_{\perp}^{s}(\widetilde{\omega}_i)$ and apply Lemma 9.3.7 to obtain

$$\sum_{\substack{\tau \in \mathcal{G} \\ \tau \subset \omega_i}} d_\tau^{-2s} \|v\|_{L^2(\tau)}^2 = \sum_{\substack{\tau \in \mathcal{G} \\ \tau \subset \omega_i}} d_\tau^{-2s} \int_{\tilde{\tau}} |\tilde{v}|^2 g_i = \sum_{\substack{\tau \in \mathcal{G} \\ \tau \subset \omega_i}} d_\tau^{-2s} \int_{\tilde{\tau}} |\tilde{v} g_i|^2 g_i^{-1}, \tag{9.23}$$

where $\tilde{\tau} = \chi_i^{-1}(\tau)$. By using $\left\| g_i^{-1} \right\|_{L^\infty(\widetilde{\omega}_i)} = \theta \, |\widetilde{\omega}_i| / |\omega_i|$ (cf. (4.227), Lemma 4.3.27) we get

$$\sum_{\substack{\tau \in \mathcal{G} \\ \tau \subset \omega_i}} d_\tau^{-2s} \|v\|_{L^2(\tau)}^2 \leq \theta \frac{|\widetilde{\omega}_i|}{|\omega_i|} \sum_{\substack{\tau \in \mathcal{G} \\ \tau \subset \omega_i}} d_\tau^{-2s} \int_{\tilde{\tau}} |\tilde{v} g_i|^2 \leq \theta \frac{|\widetilde{\omega}_i|}{|\omega_i|} \left(\max_{\substack{\tau \in \mathcal{G} \\ \tau \subset \omega_i}} d_\tau^{-2s} \right) \int_{\widetilde{\omega}_i} |\tilde{v} g_i|^2 .$$

Note that $d_\tau \geq c_4 (\operatorname{diam} \omega_i)$ where c_4 depends only on the constant of shape-regularity and on the global chart χ (cf. Assumption 4.3.10). This leads to

$$\begin{aligned}
\sum_{\substack{\tau \in \mathcal{G} \\ \tau \subset \omega_i}} d_\tau^{-2s} \|v\|_{L^2(\tau)}^2 &\leq \frac{\theta \, |\widetilde{\omega}_i|}{|\omega_i| \, (c_4 \operatorname{diam} \omega_i)^{2s}} \|\tilde{v} g_i\|_{L^2(\widetilde{\omega}_i)}^2 \\
&\overset{\text{Lemma 9.3.7}}{\leq} \frac{\theta \, |\widetilde{\omega}_i|}{|\omega_i| \, (c_4 \operatorname{diam} \omega_i)^{2s}} \frac{(\operatorname{diam} \widetilde{\omega}_i)^{2+2s}}{2(1-\gamma^2) \, |\widetilde{\omega}_i|} |\tilde{v} g_i|_{H^s(\widetilde{\omega}_i)}^2 .
\end{aligned}$$

There is a constant C_5 depending only on the shape-regularity of the mesh and the global chart χ such that

$$(\operatorname{diam} \widetilde{\omega}_i)^2 / |\widetilde{\omega}_i| \leq C_5.$$

This leads to

$$\sum_{\substack{\tau \in \mathcal{G} \\ \tau \subset \omega_i}} d_\tau^{-2s} \|v\|_{L^2(\tau)}^2 \leq \frac{C_5 \theta \, |\widetilde{\omega}_i|}{2(1-\gamma^2) \, |\omega_i|} \left(\frac{\operatorname{diam} \widetilde{\omega}_i}{c_4 \operatorname{diam} \omega_i} \right)^{2s} |\tilde{v} g_i|_{H^s(\widetilde{\omega}_i)}^2 . \tag{9.24}$$

Next, we will transform the $H^s(\widetilde{\omega}_i)$-seminorm back to the surface. Corollary 4.3.30 implies that

$$\|\tilde{\mathbf{x}} - \tilde{\mathbf{y}}\| \geq c \frac{\operatorname{diam} \widetilde{\omega}_i}{h_i} \|\chi_i(\tilde{\mathbf{x}}) - \chi_i(\tilde{\mathbf{y}})\|$$

for all $\tilde{\mathbf{x}}, \tilde{\mathbf{y}} \in \widetilde{\omega}_i$. Hence we obtain (with $\breve{g}_i = g_i \circ \chi_i^{-1}$)

$$
\begin{aligned}
|\tilde{v}g_i|^2_{H^s(\widetilde{\omega_i})} &= \int_{\tilde{\omega}_i}\int_{\tilde{\omega}_i} \frac{|\tilde{v}g_i(\tilde{\mathbf{x}}) - \tilde{v}g_i(\tilde{\mathbf{y}})|^2}{\|\tilde{\mathbf{x}} - \tilde{\mathbf{y}}\|^{2+2s}} d\tilde{\mathbf{x}} d\tilde{\mathbf{y}} \\
&\le 2\left(\frac{h_i}{c \operatorname{diam} \widetilde{\omega_i}}\right)^{2+2s} \int_{\omega_i} |v(\mathbf{x})|^2 \int_{\omega_i} \frac{|(\breve{g}_i(\mathbf{x}) - \breve{g}_i(\mathbf{y}))|^2}{\|\mathbf{x} - \mathbf{y}\|^{2+2s} \breve{g}_i(\mathbf{x}) \breve{g}_i(\mathbf{y})} ds_{\mathbf{y}} ds_{\mathbf{x}} \\
&\quad + 2\left(\frac{h_i}{c \operatorname{diam} \widetilde{\omega_i}}\right)^{2+2s} \int_{\omega_i}\int_{\omega_i} \frac{|v(\mathbf{x}) - v(\mathbf{y})|^2}{\|\mathbf{x} - \mathbf{y}\|^{2+2s}} \frac{\breve{g}_i(\mathbf{y})}{\breve{g}_i(\mathbf{x})} ds_{\mathbf{x}} ds_{\mathbf{y}}.
\end{aligned}
\tag{9.25}
$$

Recall (cf. (4.227) and Lemma 4.3.27) that

$$
\left\| g_i^{-1} \right\|_{L^\infty(\widetilde{\omega_i})} = \theta \frac{|\widetilde{\omega_i}|}{|\omega_i|} \quad \text{and} \quad \|g_i\|_{L^\infty(\tilde{\omega}_i)} = \Theta \frac{|\omega_i|}{|\widetilde{\omega_i}|}. \tag{9.26}
$$

The difference $\breve{g}_i(\mathbf{x}) - \breve{g}(\mathbf{y})$ can be estimated by

$$
\begin{aligned}
|\breve{g}_i(\mathbf{x}) - \breve{g}_i(\mathbf{y})| &\overset{\mathbf{x}=\chi_i(\tilde{\mathbf{x}}),\, \mathbf{y}=\chi_i(\tilde{\mathbf{y}})}{=} |g_i(\tilde{\mathbf{x}}) - g_i(\tilde{\mathbf{y}})| \le \|\nabla g_i\|_{L^\infty(\tilde{\omega}_i)} \|\tilde{\mathbf{x}} - \tilde{\mathbf{y}}\| \\
&\overset{\text{Corollary 4.3.30}}{\le} \frac{\operatorname{diam} \widetilde{\omega_i}}{c h_i} \|\nabla g_i\|_{L^\infty(\tilde{\omega}_i)} \|\mathbf{y} - \mathbf{x}\|,
\end{aligned}
$$

where c is as in Corollary 4.3.30. Recall that, for $\tilde{\tau} \subset \widetilde{\omega_i}$, the gradient of $g_{\tilde{\tau}} := g_i|_{\tilde{\tau}}$ can be written as

$$
\nabla g_{\tilde{\tau}} = \nabla \|\tilde{\mathbf{n}}_{\tilde{\tau}}\| = g_{\tilde{\tau}}^{-1} \mathbf{H}_{\tilde{\tau}} \tilde{\mathbf{n}}_{\tilde{\tau}}, \tag{9.27}
$$

where $\tilde{\mathbf{n}}_{\tilde{\tau}} := \mathbf{t}^1 \times \mathbf{t}^2$ with $\mathbf{t}^k := \partial_k(\chi_i|_\tau)$, $k = 1, 2$, and

$$
(\mathbf{H}_{\tilde{\tau}})_{i,j} := \partial_i (\tilde{\mathbf{n}}_{\tilde{\tau}})_j \qquad 1 \le j \le 3, \qquad 1 \le i \le 2.
$$

Applying norms to both sides in (9.27) results in

$$
\|\nabla g_{\tilde{\tau}}\|_{L^\infty(\tilde{\tau})} \le \left\| g_{\tilde{\tau}}^{-1} \right\|_{L^\infty(\tilde{\tau})} \|g_{\tilde{\tau}}\|_{L^\infty(\tilde{\tau})} \|\mathbf{H}_{\tilde{\tau}}\|_{\tilde{\tau}},
$$

where $\|\cdot\|_{\tilde{\tau}}$ is as in (8.70). The combination of (9.27) and Lemma 4.3.31 leads to

$$
\|\nabla g_{\tilde{\tau}}\|_{L^\infty(\tilde{\tau})} \le C_9 \left(\frac{h_i}{\operatorname{diam} \widetilde{\omega_i}}\right)^3,
$$

where C_9 depends only on $\theta, \Theta, C_{\Gamma,2}$ [cf. (8.49)], and the shape-regularity of the surface mesh. Because $i \in \mathcal{I}_{\text{smooth}}$ we have $\nabla g_i \in C^0(\widetilde{\omega_i})$ and

$$
\|\nabla g_i\|_{L^\infty(\widetilde{\omega_i})} \le C_9 \left(\frac{h_i}{\operatorname{diam} \widetilde{\omega_i}}\right)^3.
$$

Thus we have proved that

$$
|\breve{g}_i(\mathbf{x}) - \breve{g}_i(\mathbf{y})| \le \frac{C_9}{c} \left(\frac{h_i}{\operatorname{diam} \widetilde{\omega_i}}\right)^2 \|\mathbf{y} - \mathbf{x}\|.
$$

Substituting this and (9.26) into (9.25) yields

$$|\tilde{v}g_i|^2_{H^s(\widetilde{\omega_i})} \leq C_{10} \left(\frac{h_i}{\operatorname{diam} \widetilde{\omega_i}} \right)^{2+2s} \left\{ \int_{\omega_i} |v(\mathbf{x})|^2 \int_{\omega_i} \frac{1}{\|\mathbf{x}-\mathbf{y}\|^{2s}} ds_{\mathbf{y}} ds_{\mathbf{x}} + \int_{\omega_i} \int_{\omega_i} \frac{|v(\mathbf{x}) - v(\mathbf{y})|^2}{\|\mathbf{x}-\mathbf{y}\|^{2+2s}} ds_{\mathbf{x}} ds_{\mathbf{y}} \right\},$$

where C_{10} depends only on c, θ, Θ, C_9, the shape-regularity of the surface mesh, and the global chart χ. Since $\|\mathbf{x} - \cdot\|^{-2s}$ is weakly singular, there exists a constant C_{11} depending only on $s \in]0, 1[$ such that

$$\sup_{\mathbf{x}\in\omega_i} \int_{\omega_i} \|\mathbf{x}-\mathbf{y}\|^{-2s} \, ds_{\mathbf{y}} \leq C_{11}. \tag{9.28}$$

Note that the improved estimate $\leq C_{11} h_i^{2-2s}$ is possible by introducing local polar coordinates in the parameter plane, while it turns out that (9.28) is sufficiently sharp for our purpose.

We have proved that

$$|\tilde{v}g_i|^2_{H^s(\widetilde{\omega_i})} \leq C_{10} \left(\frac{h_i}{\operatorname{diam} \widetilde{\omega_i}} \right)^{2+2s} \left(C_{11} \|v\|^2_{L^2(\omega_i)} + |v|^2_{H^s(\omega_i)} \right)$$

and the combination with (9.24) leads to

$$\begin{aligned} \sum_{\substack{\tau\in\mathcal{G} \\ \tau\subset\omega_i}} d_\tau^{-2s} \|v\|^2_{L^2(\tau)} &\leq C_{12} \|v\|^2_{L^2(\omega_i)} + C_{13} |v|^2_{H^s(\omega_i)} \\ &\leq C_{12} h_{\mathcal{G}}^{2s} \sum_{\substack{\tau\in\mathcal{G} \\ \tau\subset\omega_i}} d_\tau^{-2s} \|v\|^2_{L^2(\tau)} + C_{13} |v|^2_{H^s(\omega_i)}, \end{aligned}$$

where C_{12}, C_{13} depends only on Γ, on γ [cf. (9.19)], and on the shape-regularity of the mesh. Thus, for sufficiently small mesh width $h_{\mathcal{G}} \leq (2C_{12})^{-\frac{1}{2s}}$, we have

$$\sum_{\substack{\tau\in\mathcal{G} \\ \tau\subset\omega_i}} d_\tau^{-2s} \|v\|^2_{L^2(\tau)} \leq 2C_{13} |v|^2_{H^s(\omega_i)}.$$

Finally, we obtain

$$\sum_{\tau\in\mathcal{G}} d_\tau^{-2s} \|v\|^2_{L^2(\tau)} \leq \sum_{i\in\mathcal{I}^{\text{smooth}}} \sum_{\substack{\tau\in\mathcal{G} \\ \tau\subset\omega_i}} d_\tau^{-2s} \|v\|^2_{L^2(\tau)} \leq 2C_{13} \sum_{i\in\mathcal{I}} |v|^2_{H^s(\omega_i)}.$$

□

The combination of Lemmas 9.3.3 and 9.3.9 leads to the following localization of the $H^s(\Gamma)$-norm for fractional Sobolev indices $s \in]0,1[$.

Theorem 9.3.10. *Let Γ be in the class $\mathcal{A}^2_{\mathrm{pw}}$. Let the boundary element mesh be shape-regular and satisfy the local residual property. Then*

$$\|v\|_{H^s(\Gamma)} \leq (1 + 4C_{2s}C_\perp) \left(\sum_{i \in \mathcal{I}} |v|_{H^s(\omega_i)} \right) \qquad \forall v \in H^s_\perp(S, \Gamma),$$

where C_{2s} is as in (9.10a) and $C_\perp$ is as in Lemma 9.3.9.

Proof. Combine (9.10a) and Lemma 9.3.9. □

We come to the main result of this section stating that the local error indicators as in Definition 9.2.1 are reliable and efficient for shape-regular meshes. Recall that all boundary element spaces S which have been introduced in Chap. 4 contain either the space of discontinuous, piecewise constant boundary elements or the space of continuous, piecewise affine boundary elements, in short, the following condition is satisfied

$$\exists S_{\min} \in \left\{ S^{0,-1}_{\mathcal{G}}, S^{1,0}_{\mathcal{G}} \right\} \quad \text{such that} \quad S_{\min} \subset S. \tag{9.29}$$

Furthermore, the overlap constant

$$C_{\mathrm{ol}} := \max_{\tau \in \mathcal{G}} \{ i \in \mathcal{I} : \tau \subset \omega_i \} \tag{9.30}$$

depends only on the *kind* of boundary element, i.e., the polynomial degree, the basis functions, the type of panels (triangular/quadrilateral), and on the shape-regularity but not on the mesh width.

Theorem 9.3.11. *Let $s \in]-1,0[$ and assume that A satisfies Assumption 9.1.1. Let S be a boundary element space satisfying (9.29). Then the a posteriori estimate for the error $\|u - u_S\|_{H^s(\Gamma)}$ of the Galerkin solution holds:*

$$C_{\mathrm{eff}} \sum_{i \in \mathcal{I}} \eta_i^2 \leq \|u - u_S\|^2_{H^s(\Gamma)} \leq C_{\mathrm{rel}} \sum_{i \in \mathcal{I}} \eta_i^2 \tag{9.31}$$

with η_i as in Definition 9.2.1.

The efficiency estimate holds for any boundary element mesh $\mathcal{G}$ and the estimate of reliability holds for shape-regular meshes.

The "efficiency" constant is given by $C_{\mathrm{eff}} = C_1^2 / C_{\mathrm{ol}}$ with C_1 as in Assumption 9.1.1 and the "reliability" constant C_{rel} depends only on Γ and the shape-regularity of the mesh.

Proof. Let $e = u_S - u$ and recall the definition of the residual $r = Ae$. The lower estimate follows from

$$\sum_{i\in\mathcal{I}} |r|^2_{H^s(\omega_i)} \leq \sum_{i\in\mathcal{I}} \sum_{\substack{\tau\in\mathcal{G}\\ \tau\subset\omega_i}} \int_\tau \int_{\omega_i} \frac{|r(\mathbf{y}) - r(\mathbf{x})|^2}{\|\mathbf{x}-\mathbf{y}\|^{2+2s}} ds_{\mathbf{y}} ds_{\mathbf{x}}$$
$$\leq C_{\mathrm{ol}} \sum_{\tau\in\mathcal{G}} \int_\tau \int_\Gamma \frac{|r(\mathbf{y}) - r(\mathbf{x})|^2}{\|\mathbf{x}-\mathbf{y}\|^{2+2s}} ds_{\mathbf{y}} ds_{\mathbf{x}}$$
$$\leq C_{\mathrm{ol}} \|r\|^2_{H^{-s}(\Gamma)} \leq \frac{C_{\mathrm{ol}}}{C_1^2} \|e\|^2_{H^s(\Gamma)}.$$

For the proof of the upper estimate we assume that the mesh is shape-regular. Note that the Galerkin orthogonality (4.120) implies that

$$\langle r, v\rangle_{H^{-s}(\Gamma)\times H^s(\Gamma)} = \langle Ae, v\rangle_{H^{-s}(\Gamma)\times H^s(\Gamma)} = a(e, v) = 0 \qquad \forall v \in S. \quad (9.32)$$

In particular, (9.32) holds for the space $S_{\min} \subset S$. Since $S_{\min}$ has the local residual property, we conclude that $r \in H_\perp^{-s}(S_{\min}, \Gamma)$. Hence we may apply Theorem 9.3.10 to obtain

$$\|e\|_s^2 \overset{\text{Assumption 9.1.1}}{\leq} C_2^2 \|r\|^2_{H^{-s}(\Gamma)} \leq C_{14} C_2^2 \sum_{i\in\mathcal{I}} |r|^2_{H^{-s}(\omega_i)}.$$

This is the upper estimate in (9.31) with $C_{\mathrm{rel}} = C_{14} C_2^2$. □

Remark 9.3.12. *(a) The above theorem expresses the efficiency and reliability of the local a posteriori error indicators. The upper estimate requires shape-regular meshes. However, it is possible to modify the error indicators so that the upper estimate becomes independent of the shape-regularity of the mesh while then the efficiency estimate depends on that property. For the details, we refer to [92, Theorem 5.2].*

(b) In this section, we considered isomorphisms $A : H^s(\Gamma) \to H^{-s}(\Gamma)$ for some negative s. In [92, Theorem 5.2], the theory has been developed for more general isomorphisms $A : H^{\alpha+2s}(\Gamma) \to H^\alpha(\Gamma)$ for some $\alpha \geq 0$ and arbitrary $s \in \mathbb{R}$.

9.3.2 Analysis of Operators of Non-negative Order

We start by introducing the assumptions for the main theorem of this section.

Assumption 9.3.13. *The sesquilinear form $a : H^s(\Gamma) \times H^s(\Gamma) \to \mathbb{C}$ satisfies a Gårding inequality of the form: There exist a constant $C_G \geq 0$ and an index $\sigma < s$ such that, for all $u \in H^s(\Gamma)$,*

$$\left| a(u, u) + C_G \|u\|^2_{H^\sigma(\Gamma)} \right| \geq \gamma \|u\|^2_{H^s(\Gamma)}. \quad (9.33)$$

Furthermore, $A : H^s(\Gamma) \to H^{-s}(\Gamma)$ *is an isomorphism for some* $s \in [0, 1]$.

The next assumption is concerned with the boundary element space S.

Assumption 9.3.14. *If* $s \geq 1/2$*, the continuous, piecewise affine functions* $S_{\mathcal{G}}^{1,0}$ *are contained in the boundary element space* S *and*

$$S_{\mathcal{G}}^{1,0} \subset S \subset C^0(\Gamma). \tag{9.34}$$

For $S_{\mathcal{G}}^{1,0} \not\subset S$*, there holds* $0 \leq s < 1/2$ *and* $S = S_{\mathcal{G}}^{p,-1}$ *for some* $p \in \mathbb{N}$.
In both cases, the basis $(b_i)_{i \in \mathcal{I}}$ *satisfies* $\sum_{i \in \mathcal{I}} b_i = 1$.

Note that the standard case of an integral operator of positive order is the hypersingular operator which maps $H^{1/2}(\Gamma) \to H^{-1/2}(\Gamma)$. The conformity condition $S \subset H^{1/2}(\Gamma)$ implies that the functions in S are continuous and the inclusion (9.34) holds for all practical cases. The proof of the second part of the following lemma is based on the Aubin–Nitsche duality technique (cf. Sect. 4.2.5). In this light, we assume some regularity for the adjoint problem:

$$\text{For } v \in H^{-s}(\Gamma) \text{ find } w_v \in H^s(\Gamma) \text{ such that } a(w, w_v) = (v, w)_{L^2(\Gamma)} \quad \forall w \in H^s(\Gamma). \tag{9.35}$$

Assumption 9.3.15. *There exist some* $t > 0$ *and some constant* $C_{\text{adj}} > 0$ *such that, for any* $v \in H^{-s+t}(\Gamma)$*, the solution of (9.35) satisfies* $w_v \in H^{s+t}(\Gamma)$ *and the estimate*

$$\|w_v\|_{H^{s+t}(\Gamma)} \leq C_{\text{adj}} \|v\|_{H^{-s+t}(\Gamma)}$$

holds.

Example 9.3.16. Let Ω be a Lipschitz domain and consider the operator $-\Delta u + cu$ for some $c > 0$. Theorem 3.2.3 implies that the sesquilinear form for the corresponding hypersingular integral operator $W : H^{1/2}(\Gamma) \to H^{1/2}(\Gamma)$ satisfies Assumption 9.3.15 for any $0 < t < 1/2$.

Lemma 9.3.17. *Let* $\mathcal{G}$ *be a boundary element mesh and let* S *denote a boundary element space which satisfies Assumption 9.3.14:*

(a) There exists a constant C_s^{stab} *such that*

$$\forall v \in H^s(\Gamma) \quad \exists \varphi \in S : \quad \sum_{i \in \mathcal{I}} \|b_i (v - \varphi)\|^2_{H^s(\Gamma)} \leq C_s^{\text{stab}} \|v\|^2_{H^s(\Gamma)}. \tag{9.36}$$

(b) Let $u_S \in S$ *denote the solution of (9.2). For* $s = 1$*, we choose* $0 < t \leq 1$ *such that Assumption 9.3.15 holds and, for* $0 \leq s < 1$*, we choose* t *such* $s + t \leq 1$ *and Assumption 9.3.15 holds. Let Assumption 9.3.13 hold.*
Then there exists a constant C_{dual} *such that*

$$\|u - u_S\|_{H^{s-t}(\Gamma)} \leq C_{\text{dual}} h_{\mathcal{G}}^t \|u - u_S\|_{H^s(\Gamma)}. \tag{9.37}$$

Proof. Part a:
Case 1: $s = 0$.
The choice $\varphi = 0$ leads to

$$\sum_{i\in\mathcal{I}} \|b_i (v-\varphi)\|^2_{L^2(\Gamma)} = \sum_{i\in\mathcal{I}} \|b_i v\|^2_{L^2(\omega_i)} \leq \max_{i\in\mathcal{I}} \|b_i\|^2_{L^\infty(\omega_i)} \sum_{i\in\mathcal{I}} \|v\|^2_{L^2(\omega_i)} \tag{9.38}$$
$$\leq C_\sharp \breve{C}_1 \|v\|^2_{L^2(\Gamma)},$$

where

$$C_\sharp := \max_{\tau\in\mathcal{G}} \operatorname{card}\{i \in \mathcal{I} : \tau \subset \omega_i\}$$

depends only on the shape-regularity of the mesh and the polynomial degree of the space S and $\breve{C}_1$ is as in Corollary 4.4.8.

Case 2: $s = 1$:
Let $Q_\mathcal{G} : L^1(\Gamma) \to S^{1,0}_\mathcal{G} \subset S$ denote the Clément interpolation operator as in Sect. 4.3.5. For $v \in H^1(\Gamma)$, let $\varphi := Q_\mathcal{G}(v)$ and $e := v - \varphi$. Then

$$\begin{aligned}
|b_i e|^2_{H^1(\omega_i)} &= \|e\nabla b_i + b_i \nabla e\|^2_{\mathbf{L}^2(\omega_i)} \leq 2\|\nabla b_i\|^2_{\mathbf{L}^\infty(\omega_i)} \|e\|^2_{L^2(\omega_i)} \\
&\quad + 2\|b_i\|^2_{L^\infty(\omega_i)} \|\nabla e\|^2_{\mathbf{L}^2(\omega_i)} \\
&\overset{\text{Corollary 4.4.8}}{\leq} 2\breve{C}_2^2 h_i^{-2} \|e\|^2_{L^2(\omega_i)} + 2\breve{C}_1^2 \|\nabla e\|^2_{\mathbf{L}^2(\omega_i)} \\
&\overset{\text{Theorem 4.3.28}}{\leq} \underbrace{2\left(C\breve{C}_2^2 c_1^2 + \breve{C}_1^2 \tilde{c}_1^2\right)}_{=:\overline{C}} \|v\|^2_{H^1(\omega_i^\star)},
\end{aligned}$$

where the constants are as in the quoted Corollary and Theorem and $\omega_i^\star$ is as in (9.9). A summation over all $i \in \mathcal{I}$ and using (9.38) gives

$$\begin{aligned}
\sum_{i\in\mathcal{I}} \|b_i (v-\varphi)\|^2_{H^1(\Gamma)} &\leq \sum_{i\in\mathcal{I}} \left(\|b_i (v-\varphi)\|^2_{L^2(\omega_i)} + |b_i e|^2_{H^1(\omega_i)} \right) \\
&\leq C_\sharp \breve{C}_1 \|v\|^2_{L^2(\Gamma)} + \overline{C} \sum_{i\in\mathcal{I}} \|v\|^2_{H^1(\omega_i^\star)} \\
&\leq \left(C_\sharp \breve{C}_1 + C\overline{C} \right) \|v\|^2_{H^1(\Gamma)},
\end{aligned}$$

where $C := \max_{\tau\in\mathcal{G}} \operatorname{card}\{i \in \mathcal{I} : \tau \subset \omega_i^\star\}$ again depends only on the shape-regularity of the mesh.

Case 3: $s \in]0,1[$ *and* $S^{1,0}_\mathcal{G} \subset S$.
For $v \in H^s(\Gamma)$, let $\varphi := Q_\mathcal{G}(v)$ and $e := v - \varphi$. We have

$$
\begin{aligned}
\|b_i e\|_{H^s(\Gamma)}^2 &\overset{(9.10a)}{\leq} \sum_{\tau \in \mathcal{G}} \left\{ 4C_{2s} d_\tau^{-2s} \|b_i e\|_{L^2(\tau)}^2 \right. \\
&\quad \left. + \int_\tau \int_{\omega_\tau} \frac{|(b_i e)(\mathbf{x}) - (b_i e)(\mathbf{y})|^2}{\|\mathbf{x}-\mathbf{y}\|^{2+2s}} ds_{\mathbf{x}} ds_{\mathbf{y}} \right\} \\
&\leq 4C_{2s} \sum_{\substack{\tau \in \mathcal{G} \\ \tau \subset \omega_i}} d_\tau^{-2s} \|b_i e\|_{L^2(\tau)}^2 \\
&\quad + 2 \sum_{\substack{\tau \in \mathcal{G} \\ \tau \subset \omega_i}} \int_\tau \int_{\omega_\tau} |b_i(\mathbf{y})|^2 \frac{|e(\mathbf{x}) - e(\mathbf{y})|^2}{\|\mathbf{x}-\mathbf{y}\|^{2+2s}} ds_{\mathbf{x}} ds_{\mathbf{y}} \\
&\quad + 2 \sum_{\tau \in \mathcal{G}} \int_{\omega_\tau} |e(\mathbf{x})|^2 \underbrace{\int_\tau \frac{|b_i(\mathbf{x}) - b_i(\mathbf{y})|^2}{\|\mathbf{x}-\mathbf{y}\|^{2+2s}} ds_{\mathbf{y}}}_{=:Q(\mathbf{x})} ds_{\mathbf{x}}.
\end{aligned} \tag{9.39}
$$

Note that the term $Q(\mathbf{x})$ vanishes if $|\omega_i \cap \omega_\tau| = 0$. Otherwise, we employ the Lipschitz continuity of the basis functions to obtain

$$
|b_i(\mathbf{x}) - b_i(\mathbf{y})| \leq \|b_i\|_{W^{1,\infty}(\omega_\tau)} \|\mathbf{x}-\mathbf{y}\| \overset{\text{Corollary 4.4.8}}{\leq} C h_\tau^{-1} \|\mathbf{x}-\mathbf{y}\| \quad \forall \mathbf{x}, \mathbf{y} \in \omega_\tau.
$$

Thus

$$
|Q(\mathbf{x})| \leq C h_\tau^{-2} \int_\tau \frac{1}{\|\mathbf{x}-\mathbf{y}\|^{2s}} ds_{\mathbf{y}} \leq C h_\tau^{-2s}, \tag{9.40}
$$

where the last inequality is proved in the same way as Theorem 3.3.5 by transforming to a two-dimensional parameter plane, introducing local polar coordinates as in (5.20a)–(5.20c) and then integrating with respect to the radial coordinate.

This leads to

$$
\begin{aligned}
\|b_i e\|_{H^s(\Gamma)}^2 &\leq C \left(\sum_{\substack{\tau \in \mathcal{G} \\ \tau \subset \omega_i}} h_\tau^{-2s} \|e\|_{L^2(\tau)}^2 + \int_{\omega_i} \int_{\omega_i^\star} \frac{|e(\mathbf{x}) - e(\mathbf{y})|^2}{\|\mathbf{x}-\mathbf{y}\|^{2+2s}} ds_{\mathbf{x}} ds_{\mathbf{y}} \right. \\
&\quad \left. + \sum_{\substack{\tau \in \mathcal{G} \\ |\omega_i \cap \omega_\tau| > 0}} h_\tau^{-2s} \|e\|_{L^2(\omega_\tau)}^2 \right) \\
&\leq C \left(\sum_{\substack{\tau \in \mathcal{G} \\ \tau \subset \omega_i}} h_\tau^{-2s} \|e\|_{L^2(\tau)}^2 + |e|_{H^s(\omega_i^\star)} + \sum_{\substack{\tau \in \mathcal{G} \\ \tau \subset \omega_i^{\star\star}}} h_\tau^{-2s} \|e\|_{L^2(\tau)}^2 \right),
\end{aligned}
$$

where $\omega_i^\star := \bigcup_{\substack{\tau \in \mathcal{G} \\ \overline{\tau} \cap \overline{\omega_i} \neq \emptyset}} \overline{\tau}$ and $\omega_i^{\star\star} := \bigcup_{\substack{\tau \in \mathcal{G} \\ \overline{\tau} \cap \omega_i^\star \neq \emptyset}} \overline{\tau}$. A summation over all $i \in \mathcal{I}$ yields

$$\|b_i e\|_{H^s(\Gamma)}^2 \le C' \left(|e|_{H^s(\Gamma)}^2 + \sum_{\tau\in\mathcal{G}} h_\tau^{-2s} \|e\|_{L^2(\tau)}^2 \right) \overset{(4.237b)}{\le} C'' \|v\|_{H^s(\Gamma)}^2 .$$

Case 4: $s \in]0,1[$ and $S_{\mathcal{G}}^{1,0} \not\subset S$.
From Assumption 9.3.14 it follows that $0 \le s < 1/2$ and $S = S_{\mathcal{G}}^{p,-1}$. We choose $\varphi = I_{\mathcal{G}}^{p,-1}(v)$, where $I_{\mathcal{G}}^{p,-1}$ is as in Theorem 4.3.19 and obtain, analogously as (9.39),

$$\begin{aligned}\|b_i e\|_{H^s(\Gamma)}^2 \le 4C_{2s} &\sum_{\substack{\tau\in\mathcal{G}\\ \tau\subset\omega_i}} d_\tau^{-2s} \|b_i e\|_{L^2(\tau)}^2 \\ &+ 2 \sum_{\substack{\tau\in\mathcal{G}\\ \tau\subset\omega_i}} \int_\tau \int_{\omega_\tau} |b_i(\mathbf{y})|^2 \frac{|e(\mathbf{x}) - e(\mathbf{y})|^2}{\|\mathbf{x}-\mathbf{y}\|^{2+2s}} ds_{\mathbf{x}} ds_{\mathbf{y}} \\ &+ 2 \sum_{\tau\in\mathcal{G}} \int_{\omega_\tau} |e(\mathbf{x})|^2 Q(\mathbf{x})\, ds_{\mathbf{x}}.\end{aligned}$$

Since the basis functions are discontinuous we obtain (cf. (9.40)

$$|Q(\mathbf{x})| \le \int_\tau \frac{1}{\|\mathbf{x}-\mathbf{y}\|^{2+2s}} ds_{\mathbf{y}} \le C h^{-2s}.$$

The rest of the proof is a repetition of the arguments as for the previous case.

Part b:
Statement b is proved by the Aubin–Nitsche duality technique (cf. Sect. 4.2.5). We apply (4.177) to obtain

$$\|u - u_S\|_{H^{s-t}(\Gamma)} \le C \|u - u_S\|_{H^s(\Gamma)} \sup_{v\in H^{-s+t}(\Gamma)\setminus\{0\}} \inf_{w\in S} \frac{\|w_v - w\|_{H^s(\Gamma)}}{\|v\|_{H^{-s+t}(\Gamma)}}$$

with w_v as in (9.35). For $s = 1$, we have $s + t > 1$, where $0 < t \le 1$ is chosen such that Assumption 9.3.15 is satisfied. Hence $H^{s+t}(\Gamma) \subset C^0(\Gamma)$ and the nodal interpolant $I_{\mathcal{G}}^{1,0} : H^{s+t}(\Gamma) \to S$ is well defined. We employ Theorem 4.3.22 to obtain

$$\begin{aligned}\inf_{w\in S} \|w_v - w\|_{H^s(\Gamma)} \le \left\| w_v - I_{\mathcal{G}}^{1,0} w_v \right\|_{H^s(\Gamma)} &\le C h^t \|w_v\|_{H^{s+t}(\Gamma)} \\ &\le C C_{\mathrm{adj}} h^t \|v\|_{H^{-s+t}(\Gamma)},\end{aligned}$$

where C is as in Theorem 4.3.22.

For $0 \le s < 1$, we choose $t > 0$ such that Assumption 9.3.15 and $s + t < 1$ hold. Let $Q_{\mathcal{G}} : H^{s+t}(\Gamma) \to S$ denote the Clément interpolation operator as in

Definition 4.3.24. Then (4.237b) and Assumption 9.3.15 imply that

$$\inf_{w\in S} \|w_v - w\|_{H^s(\Gamma)} \leq \|w_v - Q_{\mathcal{G}} w_v\|_{H^s(\Gamma)} \leq C h^t \|w_v\|_{H^{s+t}(\Gamma)}$$
$$\leq C C_{\mathrm{adj}} h^t \|v\|_{H^{-s+t}(\Gamma)}.$$

□

For the proof of the next Theorem 9.3.18 we have to introduce further notation. We consider decompositions $\mathcal{D} = \{\mathcal{I}_m\}_{m=1}^{q_\mathcal{D}}$ of $\mathcal{I}$ into non-empty, pairwise disjoint subsets $\mathcal{I}_m \subset \mathcal{I}$, $1 \leq m \leq q_\mathcal{D}$, such that $\mathcal{I} = \bigcup_{m=1}^{q_\mathcal{D}} \mathcal{I}_m$ and

$$\forall 1 \leq m \leq q_\mathcal{D} \quad \forall i, j \in \mathcal{I}_m : \omega_i \cap \omega_j = \emptyset.$$

A decomposition $\mathcal{D}$ which satisfies these conditions and consists of a minimal number of subsets $\{\mathcal{I}_m\}_{m=1}^{q_\mathcal{D}}$ is called *minimal disjoint decomposition* of $\mathcal{I}$. For such a decomposition, we define a second overlap constant [cf. (9.30)] by

$$C_{\mathrm{ol}}^\star := q_\mathcal{D}.$$

Now we have all ingredients to prove the main theorem of this section which states that the error indicators as in Definition 9.2.2 for operators of positive order are reliable and efficient.

Theorem 9.3.18. *Let Assumptions 9.1.1, 9.3.13, 9.3.14 and 9.3.15 be satisfied for some $s \in [0,1]$. Let $e = u - u_S$ be the error for the Galerkin solution. Then the estimate*

$$C_{\mathrm{eff}} \sum_{i\in\mathcal{I}} \eta_i^2 \leq \|e\|_{H^s(\Gamma)}^2 \leq C_{\mathrm{rel}} \sum_{i\in\mathcal{I}} \eta_i^2 \tag{9.41}$$

holds, where the error indicators η_i are as in Definition 9.2.2 and

$$C_{\mathrm{eff}} := \frac{2}{5} \left(C_{\mathrm{ol}}^\star C_1^2\right)^{-1}$$

with C_1 is as Assumption 9.1.1. The constant C_{rel} is given by

$$C_{\mathrm{rel}} := \begin{cases} C_s^{\mathrm{stab}} \gamma^{-2} & \text{if } C_{\mathrm{G}} = 0 \text{ in (9.33)}, \\ 4 C_s^{\mathrm{stab}} \gamma^{-2} & \text{if } C_{\mathrm{G}} > 0, \end{cases} \tag{9.42}$$

where γ is as in (9.33) and C_s^{stab} as in (9.36). In the case $C_{\mathrm{G}} > 0$, we have to assume, in addition, that the mesh size is sufficiently small, i.e., $h_\mathcal{G} \leq h_0$, for some $h_0 > 0$ (cf. Theorem 4.2.7).

Proof. Part a: Proof of efficiency.
Let $\mathcal{D} = \{\mathcal{I}_m\}_{m=1}^{C_{\text{ol}}}$ denote a minimal disjoint decomposition of the index set $\mathcal{I}$ and set

$$D_m := \bigcup_{i\in\mathcal{I}_m} \operatorname{supp} b_i .$$

In the first step, we will prove the estimate

$$\begin{aligned}\sum_{i\in\mathcal{I}_m} \sup_{\substack{v\in H^s(\Gamma)\\ \|b_i v\|_{H^s(\Gamma)}=1}} & \left|\langle r, b_i v\rangle_{H^{-s}(\Gamma)\times H^s(\Gamma)}\right|^2 \\ \leq \sup_{\substack{w\in H^s(\Gamma)\\ |D_m\cap \operatorname{supp} w|>0}} & \frac{\left|\sum_{i\in\mathcal{I}_m}\langle r, b_i w\rangle_{H^{-s}(\Gamma)\times H^s(\Gamma)}\right|^2}{\sum_{i\in\mathcal{I}_m}\|b_i w\|^2_{H^s(\Gamma)}} =: M_m. \end{aligned} \tag{9.43}$$

Let $\varepsilon > 0$. For $i \in \mathcal{I}_m$, there exist $f_i \in H^s(\Gamma)$ with

$$\|b_i f_i\|_{H^s(\Gamma)} = 1 \tag{9.44}$$

and

$$\left|\langle r, b_i f_i\rangle_{H^{-s}(\Gamma)\times H^s(\Gamma)}\right|^2 \geq \sup_{\substack{v\in H^s(\Gamma)\\ \|b_i v\|_{H^s(\Gamma)}=1}} \left|\langle r, b_i v\rangle_{H^{-s}(\Gamma)\times H^s(\Gamma)}\right|^2 - \varepsilon_1 \quad\text{with}\quad \varepsilon_1 := \frac{\varepsilon}{\operatorname{card} I_m}. \tag{9.45}$$

Because the intersections $\operatorname{supp} b_i \cap \operatorname{supp} b_j$ have zero measure for all $i, j \in \mathcal{I}_m$ with $i \neq j$, there exists $w \in H^s(\Gamma)$ such that for all $i \in \mathcal{I}_m$ we have

$$w|_{\operatorname{supp} b_i} = c_i\ f_i|_{\operatorname{supp} b_i} \quad\text{with}\quad c_i := (r, b_i f_i)_{L^2(\Gamma)}. \tag{9.46}$$

This function satisfies

$$b_i w = c_i b_i f_i \quad \text{on } \Gamma \text{ for all } i \in \mathcal{I}. \tag{9.47}$$

The definition of M_m [cf. (9.43)] implies that

$$\begin{aligned} M_m &\geq \frac{\left|\sum_{i\in\mathcal{I}_m}\langle r, b_i w\rangle_{H^{-s}(\Gamma)\times H^s(\Gamma)}\right|^2}{\sum_{i\in\mathcal{I}_m}\|b_i w\|^2_{H^s(\Gamma)}} \overset{(9.47)}{=} \frac{\left|\sum_{i\in\mathcal{I}_m}\overline{c_i}\,\langle r, b_i f_i\rangle_{H^{-s}(\Gamma)\times H^s(\Gamma)}\right|^2}{\sum_{i\in\mathcal{I}_m}|c_i|^2\,\|b_i f_i\|^2_{H^s(\Gamma)}} \\ &\overset{(9.44),\,(9.46)}{=} \frac{\left|\sum_{i\in\mathcal{I}_m}|c_i|^2\right|^2}{\sum_{i\in\mathcal{I}_m}|c_i|^2} = \sum_{i\in\mathcal{I}_m}\left|\langle r, b_i f_i\rangle_{H^{-s}(\Gamma)\times H^s(\Gamma)}\right|^2 \\ &\overset{(9.45)}{\geq} \sum_{i\in\mathcal{I}_m} \sup_{\substack{v\in H^s(\Gamma)\\ \|b_i v\|_{H^s(\Gamma)}=1}} \left|\langle r, b_i v\rangle_{H^{-s}(\Gamma)\times H^s(\Gamma)}\right|^2 - \varepsilon. \end{aligned}$$

Since $\varepsilon > 0$ was arbitrary we have proved (9.43).

From (9.43) we derive

$$\sum_{i\in\mathcal{I}_m}\eta_i^2 \overset{(9.8)}{=} \sum_{i\in\mathcal{I}_m}\sup_{\substack{v\in H^s(\Gamma)\\ b_i v\neq 0}}\frac{\left|\langle r, b_i v\rangle_{H^{-s}(\Gamma)\times H^s(\Gamma)}\right|^2}{\|b_i v\|^2_{H^s(\Gamma)}}$$
$$= \sum_{i\in\mathcal{I}_m}\sup_{\substack{v\in H^s(\Gamma)\\ \|b_i v\|_{H^s(\Gamma)}=1}}\left|\langle r, b_i v\rangle_{H^{-s}(\Gamma)\times H^s(\Gamma)}\right|^2$$
$$\overset{(9.43)}{\leq} M_m.$$

This leads to

$$\sum_{i\in\mathcal{I}_m}\eta_i^2 \leq M_m = \sup_{\substack{w\in H^s(\Gamma)\\ |D_m\cap\operatorname{supp} w|>0}}\frac{\left|\left\langle Ae, \sum_{i\in\mathcal{I}_m} b_i w\right\rangle_{H^{-s}(\Gamma)\times H^s(\Gamma)}\right|^2}{\sum_{i\in\mathcal{I}_m}\|b_i w\|^2_{H^s(\Gamma)}}$$
$$\leq \|A\|^2_{H^{-s}(\Gamma)\leftarrow H^s(\Gamma)}\|e\|^2_{H^s(\Gamma)}\sup_{\substack{w\in H^s(\Gamma)\\ |D_m\cap\operatorname{supp} w|>0}}\frac{\left\|\sum_{i\in\mathcal{I}_m} b_i w\right\|^2_{H^s(\Gamma)}}{\sum_{i\in\mathcal{I}_m}\|b_i w\|^2_{H^s(\Gamma)}}$$
$$\overset{\text{Lemma 4.1.49b}}{\leq} \frac{5}{2}\|A\|^2_{H^{-s}(\Gamma)\leftarrow H^s(\Gamma)}\|e\|^2_{H^s(\Gamma)}$$

and we obtain the estimate for the error indicators by

$$\sum_{i\in\mathcal{I}}\eta_i^2 = \sum_{m=1}^{C_{\mathrm{ol}}}\sum_{i\in\mathcal{I}_m}\eta_i^2 \leq \frac{5C_{\mathrm{ol}}}{2}\|A\|^2_{H^{-s}(\Gamma)\leftarrow H^s(\Gamma)}\|e\|^2_{H^s(\Gamma)}.$$

Thus we have proved the left side in (9.41).

Part b: Proof of reliability.
The Galerkin orthogonality and the Definition 9.2.2 imply for any $v\in S$ that

$$|a(e,e)| = |a(e,e-v)| = \left|\sum_{i\in\mathcal{I}} a(e, b_i(e-\varphi))\right|$$
$$\leq \sum_{i\in\mathcal{I}}\eta_i\|b_i(e-\varphi)\|_{H^s(\Gamma)} \leq \sqrt{\sum_{i\in\mathcal{I}}\eta_i^2}\sqrt{\sum_{i\in\mathcal{I}}\|b_i(e-\varphi)\|^2_{H^s(\Gamma)}}. \quad (9.48)$$

From (9.36) we conclude that there exists $\varphi\in S$ such that

$$\sum_{i\in\mathcal{I}}\|b_i(e-\varphi)\|^2_{H^s(\Gamma)} \leq C_s^{\mathrm{stab}}\|e\|^2_{H^s(\Gamma)}.$$

Inserting this into (9.48) and using Gårding's inequality (9.33) yields

$$\gamma \|e\|^2_{H^s(\Gamma)} \le \left| a(e,e) + C_{\mathrm{G}} \|e\|^2_{H^\sigma(\Gamma)} \right| \le \sqrt{C_s^{\mathrm{stab}}} \|e\|_{H^s(\Gamma)} \sqrt{\sum_{i \in \mathcal{I}} \eta_i^2} + C_{\mathrm{G}} \|e\|^2_{H^\sigma(\Gamma)} \tag{9.49}$$

for some $\sigma < s$. If $C_{\mathrm{G}} = 0$ then the right estimate in (9.41) is proved with C_{rel} as in (9.42).

If $C_{\mathrm{G}} > 0$ we employ (9.37) to obtain

$$\|e\|_{H^{s-t}(\Gamma)} \le C_{\mathrm{dual}} h_{\mathcal{G}}^t \|e\|_{H^s(\Gamma)}.$$

We may choose $t > 0$ always sufficiently small such that $\sigma \le s - t$ and hence

$$\|e\|_{H^\sigma(\Gamma)} \le \|e\|_{H^{s-t}(\Gamma)} \le C_{\mathrm{dual}} h_{\mathcal{G}}^t \|e\|_{H^s(\Gamma)}.$$

Substituting this estimate into (9.49) yields

$$\begin{aligned} \gamma \|e\|^2_{H^s(\Gamma)} &\le \left| a(e,e) + C_{\mathrm{G}} \|u\|^2_{H^\sigma(\Gamma)} \right| \\ &\le \sqrt{C_s^{\mathrm{stab}}} \|e\|_{H^s(\Gamma)} \sqrt{\sum_{i \in \mathcal{I}} \eta_i^2} + C_{\mathrm{G}} C_{\mathrm{dual}}^2 h_{\mathcal{G}}^{2t} \|e\|^2_{H^s(\Gamma)}. \end{aligned}$$

If the minimal mesh width h_0 is chosen such that

$$C_{\mathrm{G}} C_{\mathrm{dual}}^2 h_0^{2t} \le \frac{1}{2}\gamma$$

we obtain

$$\|e\|_{H^s(\Gamma)} \le 4\gamma^{-2} C_s^{\mathrm{stab}} \sum_{i \in \mathcal{I}} \eta_i^2$$

and this is the assertion. □

In some cases, the boundary integral operator has a non-trivial, finite-dimensional null space as, e.g., the hypersingular operator for the Laplace operator on closed manifolds, where the null space is spanned by the constant functions. Then the operator $A : H^s(\Gamma) \to H^{-s}(\Gamma)$ is not an isomorphism. On the other hand, the restriction $A : \mathcal{H} \to \mathcal{H}'$ to some quotient space $\mathcal{H} := H^s(\Gamma)/\mathcal{N}$, where $\mathcal{N}$ is finite-dimensional, is an isomorphism with respect to the induced norms. Such operators are considered in the following remark.

Remark 9.3.19. *Assume that $A : H^s(\Gamma) \to H^{-s}(\Gamma)$ does* **not** *satisfy Assumption 9.1.1. Assume that, for a quotient space $\mathcal{H} := H^s(\Gamma)/\mathcal{N}$ with some finite-dimensional subspace $\mathcal{N} \subset H^s(\Gamma)$, Assumption 9.1.1 holds for the restricted operator $A : \mathcal{H} \to \mathcal{H}'$. Let S be a boundary element space for $H^s(\Gamma)$ and let the index set $\mathcal{I}$ correspond to the nodal points for S. Assume further that $\mathcal{N} \subset S \subset H^s(\Gamma)$ which, e.g., is satisfied for the hypersingular operator for the Laplacian on closed manifolds. Then $\mathcal{S} := S/\mathcal{N}$ is a finite-dimensional subspace of $\mathcal{H}$. For other operators such as, e.g., the Helmholtz operator, the null space of* *the corresponding hypersingular operator, in general, does not belong to* $\mathcal{S}$. *Then the following comments directly apply for this case* only *if the null space* $\mathcal{N}$*, resp., a basis thereof, is known.*

Assume that (9.33) holds for all $u \in \mathcal{H}$. *Let the error indicators be defined as before by (9.8a). For* $F \in \mathcal{H}'$, *let* $u_{\mathcal{S}}$ *denote the Galerkin solution of*

$$a(u_{\mathcal{S}}, v) = F(v) \qquad \forall v \in \mathcal{S}$$

and let $e := u - u_{\mathcal{S}}$ *denote the corresponding error. Then Theorem 9.3.18 and its proof remain valid without any changes. Note that the space* S *in Part b of the proof must not be replaced by* $\mathcal{S}$.

9.3.3 Bibliographical Remarks, Further Results and Open Problems

As we saw in Chaps. 3 and 4, boundary integral operators obtained via boundary reduction of elliptic boundary value problems via the direct method are boundedly invertible between fractional order Sobolev spaces on the boundary. The derivation of residual a posteriori error estimates in fractional Sobolev norms which are obtained as sums of $O(N)$ many scaled local residual error bounds is an easy consequence of the bounded invertibility of the boundary integral operators in Sobolev spaces which we established in Chaps. 3 and 4. The key to obtaining efficiently computable upper bounds for the residual in the relevant, fractional Sobolev norms is the localization of these norms, i.e., their representation as sums of $O(N)$ many scaled local residual error bounds. We showed, based on the work of B. Faermann [89, 90, 92] on the localization of fractional order Sobolev norms, how to obtain computable upper bounds for the intrinsically nonlocal fractional order Sobolev norms of the weak residual. While this is not possible for general functions, it is feasible for the weak residual by exploiting its Galerkin orthogonality as we explained in Sects. 9.2 and 9.3. This implies in particular that an analogous residual estimate for collocation BEM will require additional technical steps to achieve $O(N)$ complexity.

Due to the appearance of fractional order Sobolev norms, the derivation of upper bounds for the error of Galerkin discretizations of integral equations is substantially more involved than in the case of second order, elliptic partial differential equations, where it involves only elementwise integration by parts and, once more, Galerkin orthogonality.

Computable residual a posteriori error estimators are a convenient tool to decide when to terminate mesh refinement procedures in practical computations. If these estimators are obtained from sums of (squared) error contributions which are localized to (a patch of) elements, it is suggestive to use these contributions as *error indicators*, i.e., as a measure for the relative contributions to the global error bound from the element associated with the error indicator. This is usually successful in computational practice. Note, however, that this reasoning is completely heuristic:

there is a priori no reason at all why a localized quantity should be in one-to-one correspondence with the source of discretization error in the Finite Element Galerkin projection. In fact, counterexamples to this heuristic reasoning exist, even in the case of local operators, i.e., for finite element discretizations of elliptic partial differential equations. Nevertheless, in recent years, substantial progress in the analysis of adaptive Finite Element Methods for elliptic partial differential equations has been made. Proofs of *optimality* of adaptive finite element methods for partial differential equations are by now available.

While the mathematical understanding of adaptive Finite Element Methods (AFEM) has proceeded substantially in the past years and has, at least for second order elliptic PDEs and conforming FEM, reached a certain maturity, the corresponding situation for mathematical analysis of adaptive Galerkin BEM considered in this book is, at the time of writing, still considerably less developed. While reliability and efficiency of computable residual a posteriori error estimators is available in Sects. 9.2 and 9.3, neither a convergence result along the lines of [82] nor any kind of optimality result is known to us at time of writing *for Galerkin BEM based on the standard shape functions described in Chap. 4*.

There is, however, an alternative approach of *wavelet based Galerkin BEM* which does have a complete mathematical theory with optimality and convergence rates at linear computational complexity available. It is based on piecewise polynomial, *spline wavelets* as basis functions for the subspaces used in the Galerkin discretization. The construction of such spline wavelet basis functions on general polyhedra in $\mathbb{R}^3$ is involved, and their supports are considerably larger than the supports of basis functions described in Chap. 4 above. The effort in their construction and implementation is, however, worthwhile, since their use in Galerkin BEM achieves two purposes: (1) matrix compression and (2) optimal preconditioning in a unified fashion.

Multilevel Preconditioning in linear complexity is, for these basis functions, achieved by a simple diagonal scaling of the stiffness matrix due to the fact that the wavelets constitute *Riesz bases* of the energy spaces for the boundary integral operators.

Matrix compression implies that the Galerkin stiffness matrix in these wavelet bases is, while still being densely populated, *numerically sparse*. This means that all but $O(N)$ nonzero matrix entries out of the N^2 overall matrix entries need to be actually computed, and that *a mathematical analysis reveals the location and the accuracy of these $O(N)$ essential matrix entries*. We refer to [208] and to the recent paper [76] and the references therein for details and further results. Note that, in this setting, acceleration techniques of clustering or fast multipole type are not required any more. We remark, however, that fast multipole accelerations are naturally robust with respect to the complexity of the boundary surface, since they are based on the (coordinate free) approximation of the fundamental solution in ambient space, while the analysis of wavelet matrix compression methods reveals a substantial dependence of the matrix compression error on the surface parametrizations and their derivatives.

We note, in closing, that also in the wavelet Galerkin approach to the discretization of BIEs, the numerical quadrature of the diagonal entries of the Galerkin stiffness matrix requires the quadrature techniques presented in Chap. 5 above.

References

1. M. Abramowitz and I.A. Stegun. *Handbook of Mathematical Functions*. Applied Mathematics Series 55. National Bureau of Standards, U.S. Department of Commerce, 1972.
2. M. Ainsworth and J. T. Oden. *A Posteriori Error Estimation in Finite Element Analysis*. Wiley, New York, 2000.
3. H. W. Alt. *Lineare Funktionalanalysis*. Springer, Berlin, 1985.
4. F. P. Andriulli, K. Cools, H. Bağci, F. Olyslager, A. Buffa, S. H. Christiansen, and E. Michielssen. A multiplicative Calderon preconditioner for the electric field integral equation. *IEEE Trans. Antennas Propagation*, 56(8, part 1):2398–2412, 2008.
5. S. Arden, S. N. Chandler-Wilde, and S. Langdon. A collocation method for high-frequency scattering by convex polygons. *J. Comput. Appl. Math.*, 204(2):334–343, 2007.
6. D. N. Arnold and W. L. Wendland. On the asymptotic convergence of collocation methods. *Math. Comp.*, 41:197–242, 1983.
7. K. E. Atkinson. *A Survey of Numerical Methods for the Solution of Fredholm Integral Equations of the Second Kind*. SIAM, Philadelphia, 1976.
8. K. E. Atkinson. *The Numerical Solution of Integral Equations of the Second Kind*. Cambridge Univiversity Press, Cambridge, 1997.
9. I. Babuška and A. K. Aziz. The Mathematical Foundation of the Finite Element Method. In A. K. Aziz, editor, *The Mathematical Foundation of the Finite Element Method with Applications to Partial Differential Equations*, pages 5–359. Academic, New York, 1972.
10. I. Babuška and W. C. Rheinboldt. A-posteriori error estimates for the finite element method. *Int. J. Numer. Meth. Eng.*, 12:1597–1615, 1978.
11. I. Babuška and W. C. Rheinboldt. Error estimates for adaptive finite element computations. *SIAM J. Numer. Anal.*, 15:736–754, 1978.
12. I. Babuška and T. Strouboulis. *The Finite Element Method and Its Reliability*. Clarendon, Oxford, 2001.
13. P. K. Banerjee. *The Boundary Element Methods in Engineering*. McGraw-Hill, London, 1994.
14. W. Bangerth and R. Rannacher. *Adaptive Finite Element Methods for Differential Equations*. Birkhäuser, Basel, 2003.
15. L. Banjai and S. A. Sauter. A refined Galerkin error and stability analysis for highly indefinite variational problems. *SIAM J. Numer. Anal.*, 45(1):37–53, 2007.
16. R. Bank. The efficient implementation of local mesh refinement algorithms. In I. Babuška, J. Chandra, and J. E. Flaherty, editors, *Adaptive computational methods for partial differential equations, Proc. Workshop, College Park, MD, 1983*, pages 74–81. SIAM, Philadelphia, 1983.
17. R. E. Bank. *PLTMG User's Guide Version 6.0*. SIAM, Philadelphia, 1990.
18. R. E. Bank. PLTMG User's Guide Version 9.0. Technical report, University of San Diego, 2004.
19. M. Bebendorf. *Hierarchical Matrices*, volume 63 of *Lecture Notes in Computational Science and Engineering*. Springer, Berlin, 2008.
20. M. Bebendorf and S. Rjasanow. Adaptive low-rank approximation of collocation matrices. *Computing*, 70(1):1–24, 2003.

S.A. Sauter and C. Schwab, *Boundary Element Methods*, Springer Series in Computational Mathematics 39, DOI 10.1007/978-3-540-68093-2,

21. A. Bendali. Numerical analysis of the exterior boundary value problem for the time-harmonic Maxwell equations by a boundary finite element method. Part 2: The discrete problem. *Math. Comp.*, 43:47–68, 1984.
22. J. Bergh and J. Löfström. *Interpolation Spaces*. Springer, Berlin, 1976.
23. M. Bonnet. *Boundary Integral Equation Methods for Solids and Fluids*. Wiley, Chichester, 1999.
24. S. Börm. Data-sparse approximation of non-local operator by $\mathcal{H}^2$-matrices. *Linear Algebra Appl.*, 422(2–3):380–403, 2007.
25. S. Börm and L. Grasedyck. Hybrid cross approximation of integral operators. *Numer. Math.*, 101(2):221–249, 2005.
26. S. Börm and S. A. Sauter. BEM with linear complexity for the classical boundary integral operators. *Math. Comp.*, 74(251):1139–1177, 2005.
27. D. Braess. *Finite Elements*, 3rd edition. Cambridge University Press, Cambridge, 2007.
28. H. Brakhage and P. Werner. Über das Dirichletsche Außenraumproblem für die Helmholtzsche Schwingungsgleichung. *Arch. der Math.*, 16:325–329, 1965.
29. J. H. Bramble. *Multigrid Methods*. Pitman Research Notes in Mathematics. Longman Scientific & Technical, Essex, 1993.
30. J. H. Bramble, Z. Leyk, and J. E. Pasciak. The analysis of multigrid algorithms for pseudo-differential operators of order minus one. *Math. Comp.*, 63:461–478, 1994.
31. J. H. Bramble, J. E. Pasciak, and P. S. Vassilevksi. Computational scales of Sobolev norms with applications to preconditioning. *Math. Comp.*, 69:462–480, 1999.
32. C. A. Brebbia, J. C. F. Telles, and L. C. Wrobel. *Boundary Element Techniques: Theory and Applications in Engineering*. Springer, Berlin, 1984.
33. S. B. Brenner and L. R. Scott. *The Mathematical Theory of Finite Element Methods*. Springer, New York, 1994.
34. O. P. Bruno. Fast, high-order, high-frequency integral methods for computational acoustics and electromagnetics. In *Topics in Computational Wave Propagation*, volume 31 of *Lecture Notes in Computational Science and Engineering*, pages 43–82. Springer, Berlin, 2003.
35. O. P. Bruno. New high-order integral methods in computational electromagnetism. *CMES Comput. Model. Eng. Sci.*, 5(4):319–330, 2004.
36. O. P. Bruno and C. A. Geuzaine. An $O(1)$ integration scheme for three-dimensional surface scattering problems. *J. Comput. Appl. Math.*, 204(2):463–476, 2007.
37. A. Buffa and S. H. Christiansen. The electric field integral equation on Lipschitz screens: definitions and numerical approximation. *Numer. Math.*, 94(2):229–267, 2003.
38. A. Buffa and S. H. Christiansen. A dual finite element complex on the barycentric refinement. *Math. Comp.*, 76(260):1743–1769, 2007.
39. A. Buffa and P. Ciarlet, Jr. On traces for functional spaces related to Maxwell's equations. Part I: An integration by parts formula in Lipschitz polyhedra. *Math. Meth. Appl. Sci.*, 21: 9–30, 2001.
40. A. Buffa, M. Costabel, and C. Schwab. Boundary element methods for Maxwell's equations on non-smooth domains. *Numer. Math.*, 92(4):679–710, 2002.
41. A. Buffa and R. Hiptmair. Galerkin boundary element methods for electromagnetic scattering. In M. Ainsworth, P. Davis, D. Duncan, P. Martin, and B. Rynne, editors, *Topics in Computational Wave Propagation. Direct and inverse Problems*, volume 31 of *Lecture Notes in Computational Science and Engineering*, pages 83–124. Springer, Berlin, 2003.
42. A. Buffa and R. Hiptmair. A coercive combined field integral equation for electromagnetic scattering. *SIAM J. Numer. Anal.*, 42(2):621–640, 2004.
43. A. Buffa and R. Hiptmair. Regularized combined field integral equations. *Numer. Math.*, 100(1):1–19, 2005.
44. A. Buffa, R. Hiptmair, T. von Petersdorff, and C. Schwab. Boundary element methods for Maxwell transmission problems in Lipschitz domains. *Numer. Math.*, 95(3):459–485, 2003.
45. A. Buffa and S. A. Sauter. On the acoustic single layer potential: Stabilization and Fourier analysis. *SIAM J. Sci. Comput.*, 28(5):1974–1999, 2006.
46. A. P. Calderon and A. Zygmund. Singular integral operators and differential equations. *Am. J. Math.*, 79:901–921, 1957.

47. C. Carstensen. Efficiency of a-posteriori BEM-error estimates for first kind integral equations on uniform meshes. *Math. Comp.*, 65:69–84, 1996.
48. C. Carstensen. An a-posteriori error estimate for a first-kind integral equation. *Math. Comp.*, 66:139–155, 1997.
49. C. Carstensen and S. A. Funken. Coupling of mixed finite elements and boundary elements. *IMA J. Numer. Anal.*, 20(3):461–480, 2000.
50. C. Carstensen, S. A. Funken, and E. P. Stephan. An a-posteriori error estimates for *hp*-boundary element methods. *Appl. Anal.*, 61:233–253, 1996.
51. C. Carstensen, M. Maischak, D. Praetorius, and E. P. Stephan. Residual-based a posteriori error estimate for hypersingular equation on surfaces. *Numer. Math.*, 97:397–425, 2004.
52. C. Carstensen, M. Maischak, and E. P. Stephan. A posteriori error estimate and h-adaptive algorithm on surfaces for Symm's integral equation. *Numer. Math.*, 90(2):197–213, 2001.
53. C. Carstensen and E. P. Stephan. A-posteriori error estimates for boundary element methods. *Math. Comp.*, 64:483–500, 1995.
54. C. Carstensen. and E. P. Stephan. Coupling of FEM and BEM for a nonlinear interface problem: The h-p version. *Numer. Meth. Partial Diff. Eqns.*, 11(5):539–554, 1995.
55. S. Chandler-Wilde and I. Graham. Boundary integral methods in high frequency scattering. In B. Engquist, A. Fokas, E. Hairer, and A. Iserles, editors, *Highly Oscillatory Problems*, pages 154–193. Cambridge University Press, Cambridge, 2009.
56. S. Chandler-Wilde and P. Monk. Wave-number-eplicit bounds in time-harmonic scattering. *SIAM J. Math. Anal.*, 39:1428–1455, 2008.
57. S. N. Chandler-Wilde and S. Langdon. A Galerkin boundary element method for high frequency scattering by convex polygons. *SIAM J. Numer. Anal.*, 45(2):610–640 (electronic), 2007.
58. S. N. Chandler-Wilde and M. Lindner. Wave problems in unbounded domains: Fredholmness and the finite section method. In *Fluids and waves*, volume 440 of *Contemporary Mathematics*, pages 115–122. AMS, Providence, RI, 2007.
59. S. N. Chandler-Wilde and M. Lindner. Boundary integral equations on unbounded rough surfaces: Fredholmness and the finite section method. *J. Integral Equations Appl.*, 20(1):13–48, 2008.
60. G. Chen and J. Zhou. *Boundary Element Methods*. Academic, New York, 1992.
61. A. Chernov, M. Maischak, and E. P. Stephan. A priori error estimates for *hp* penalty BEM for contact problems in elasticity. *Comput. Meth. Appl. Mech. Eng.*, 196(37–40):3871–3880, 2007.
62. M. M. Chipot. *Elliptic Equations: An Introductory Course*. Birkhäuser, Basel, 2009.
63. S. H. Christiansen. *Résolution des équations intégrales pour la diffraction d'ondes acoustiques et électromagnetiques*. PhD thesis, l'Ecole Polytechnique, Paris, 2002.
64. S. H. Christiansen. Discrete Fredholm properties and convergence estimates for the electric field integral equation. *Math. Comp.*, 73(245):143–167, 2004.
65. S. H. Christiansen and J. C. Nédélec. Des préconditionneurs pour la résolution numérique des équations intégrales de frontière de l'acoustique. *C. R. Acad. Sci. Paris Sér. I Math.*, 330(7):617–622, 2000.
66. S. H. Christiansen and J. C. Nédélec. Des préconditionneurs pour la résolution numérique des équations intégrales de frontière de l'électromagnétisme. *C. R. Acad. Sci. Paris Sér. I Math.*, 331(9):733–738, 2000.
67. S. H. Christiansen and J. C. Nédélec. A preconditioner for the electric field integral equation based on Calderon formulas. *SIAM J. Numer. Anal.*, 40(3):1100–1135, 2002.
68. P. G. Ciarlet. *The Finite Element Method for Elliptic Problems*. North-Holland, Amsterdam, 1987.
69. P. Clément. Approximation by finite element functions using local regularization. *RAIRO, Sér. Rouge Anal. Numér.*, R-2:77–84, 1975.
70. D. L. Colton and R. Kress. *Integral Equation Methods in Scattering Theory*. Wiley, New York, 1983.

71. M. Costabel. Symmetric methods for the coupling of finite elements and boundary elements. In C. A. Brebbia et al., editor, *Boundary Elements IX*, volume 1, pages 411–420. Springer, Berlin, 1987.
72. M. Costabel. Boundary integral operators on Lipschitz domains: Elementary results. *SIAM J. Math. Anal.*, 19:613–626, 1988.
73. M. Costabel. A symmetric method for the coupling of finite elements and boundary elements. In J. R. Whiteman, editor, *Proc. 6th Conf. on the Mathematics of Finite Elements and Applications VI, Uxbridge 1987 (MAFELAP 1987)*, pages 281–288. Academic, New York, 1988.
74. M. Costabel and W. L. Wendland. Strong ellipticity of boundary integral operators. *J. Reine Angew. Math.*, 372:34–63, 1986.
75. W. Dahmen, B. Faermann, I. G. Graham, W. Hackbusch, and S. A. Sauter. Inverse inequalities on non-quasi-uniform meshes and application to the mortar element method. *Math. Comp.*, 73(247):1107–1138, 2004.
76. W. Dahmen, H. Harbrecht, and R. Schneider. Adaptive methods for boundary integral equations: Complexity and convergence estimates. *Math. Comp.*, 76(259):1243–1274, 2007.
77. E. Darve. The fast multipole method (I) : Error analysis and asymptotic complexity. *SIAM J. Numer. Anal.*, 38(1):98–128, 2000.
78. E. Darve. The fast multipole method: Numerical implementation. *J. Comput. Phys.*, 160(1): 195–240, 2000.
79. M. Dauge. *Elliptic Boundary Value Problems on Corner Domains*. Lecture Notes in Mathematics. Springer, New York, 1988.
80. R. Dautray and J. L. Lions. *Mathematical Analysis and Numerical Methods for Science and Technology*, volume 4. Springer, Berlin, 1990.
81. P. J. Davis. *Interpolation and Approximation*. Blaisdell, New York, 1963.
82. W. Dörfler. A convergent adaptive algorithm for Poisson's equation. *Numer. Math.*, 73:419–448, 1996.
83. M. G. Duffy. Quadrature over a pyramid or cube of integrands with a singularity at a vertex. *SIAM J. Numer. Anal.*, 19:1260–1262, 1982.
84. G. Dziuk. Finite elements for the Beltrami operator on arbitrary surfaces. In S. Hildebrandt and R. Leis, editors, *Partial Differential Equations and Calculus of Variations*, volume 1357 of *Lecture Notes in Mathematics*, pages 142–155. Springer, New York, 1988.
85. Y. V. Egorov and M. A. Shubin. *Partial Differential Equations I*, Encyclopaedia of Mathematical Sciences edition. Springer, Heidelberg, 1992.
86. J. Elschner. The double layer potential operator over polyhedral domains I: Solvability in weighted Sobolev spaces. *Appl. Anal.*, 45:117–134, 1992.
87. J. Elschner. The double layer potential operator over polyhedral domains II: Spline galerkin methods. *Math. Meth. Appl. Sci.*, 15:23–37, 1992.
88. S. Erichsen and S. A. Sauter. Efficient automatic quadrature in 3-D Galerkin BEM. *Comput. Meth. Appl. Mech. Eng.*, 157:215–224, 1998.
89. B. Faermann. Local a-posteriori error indicators for the Galerkin discretization of boundary integral equations. *Numer. Math.*, 79:43–76, 1998.
90. B. Faermann. Localization of the Aronszajn–Slobodeckij norm and application to adaptive boundary element methods. Part I. The two-dimensional case. *IMA J. Numer. Anal.*, 20: 203–234, 2000.
91. B. Faermann. *Lokalisierungstechniken für Sobolev–Normen und Anwendungen auf adaptive Randelementmethoden (in German)*. Habilitationsschrift, Universität Kiel, 2000.
92. B. Faermann. Localization of the Aronszajn–Slobodeckij norm and application to adaptive boundary element methods. Part II. The three-dimensional case. *Numer. Math.*, 92(3):467–499, 2002.
93. M. Feistauer, G. C. Hsiao, and R. E. Kleinman. Asymptotic and a posteriori error estimates for boundary element solutions of hypersingular integral equations. *SIAM J. Numer. Anal.*, 33:666–685, 1996.
94. R. Fletcher and C. H. Reeves. Function minimization by conjugate gradients. *Comput. J.*, 7:149–154, 1964.

95. O. Forster. *Analysis 2*. Vieweg, Brunswick, 1993. 5. Auflage.
96. I. Fredholm. Sur une classe d'equations fonctionelles. *Acta Matematica*, 27:365–390, 1903.
97. A. Friedman. *Partial Differential Equations*. Holt, Rinehart and Winston, New York, 1969.
98. A. Friedman. *Foundations of Modern Analysis*. Dover, New York, 1982. Reprint of the 1970 original.
99. W. Gautschi. *Numerical Analysis*. Birkhäuser, Basel, 1997.
100. D. Gilbarg and N. S. Trudinger. *Elliptic Partial Differential Equations of Second Order*. Springer, Berlin, 1983.
101. G. Giraud. Sur une classe generale d'equation a integrales principales. *C. R. Acad. Sci. Paris*, 202:2124–2126, 1936.
102. M. Gnewuch and S. A. Sauter. Boundary integral equations for second order elliptic boundary value problems. Technical Report 55, Max-Planck-Institut, Leipzig, Germany, 1999.
103. S. A. Goreinov, E. E. Tyrtyshnikov, and N. L. Zamarashkin. A theory of pseudoskeleton approximations. *Linear Algebra Appl.*, 261:1–21, 1997.
104. I. G. Graham, L. Grasedyck, W. Hackbusch, and S. A. Sauter. Optimal panel-clustering in the presence of anisotropic mesh refinement. *SIAM J. Numer. Anal.*, 46(1):517–543, 2008.
105. I. G. Graham, W. Hackbusch, and S. A. Sauter. Discrete boundary element methods on general meshes in 3D. *Numer. Math.*, 86:103–137, 2000.
106. I. G. Graham, W. Hackbusch, and S. A. Sauter. Hybrid Galerkin boundary elements: Theory and implementation. *Numer. Math.*, 86:139–172, 2000.
107. I. G. Graham, W. Hackbusch, and S. A. Sauter. Finite elements on degenerate meshes: Inverse-type inequalities and applications. *IMA J. Numer. Anal.*, 25(2):379–407, 2005.
108. I. G. Graham and W. McLean. Anisotropic mesh refinement: The conditioning of Galerkin boundary element matrices and simple preconditioners. *SIAM J. Numer. Anal.*, 44(4): 1487–1513, 2006.
109. L. Grasedyck and W. Hackbusch. Construction and arithmetics of $\mathcal{H}$-matrices. *Computing*, 70(4):295–334, 2003.
110. L. Greengard. *The Rapid Evaluation of Potential Fields in Particle Systems*. MIT, Cambridge, MA, 1988.
111. L. Greengard and V. Rokhlin. A new version of the fast multipole method for the Laplace equation in three dimensions. *Acta Numerica*, 6:229–269, 1997.
112. M. Griebel, P. Oswald, and T. Schiekofer. Sparse grids for boundary integral equations. *Numer. Math.*, 83(2):279–312, 1999.
113. P. G. Grisvard. *Singularities in Boundary Value Problems*, volume 22 of *Research Notes in Applied Mathematics*. Springer, Berlin, 1992.
114. W. Hackbusch. *Multi-Grid Methods and Applications*. Springer, Berlin, 1985.
115. W. Hackbusch. *Elliptic Differential Equations*. Springer, Berlin, 1992.
116. W. Hackbusch. *Iterative Solutions of Large Sparse Systems of Equations*. Springer, New York, 1994.
117. W. Hackbusch. *Integral Equations*. ISNM. Birkhäuser, Basel, 1995.
118. W. Hackbusch. A sparse matrix arithmetic based on $\mathcal{H}$-Matrices. Part I: Introduction to $\mathcal{H}$-Matrices. *Computing*, 62:89–108, 1999.
119. W. Hackbusch. *Hierarchische Matrizen*. Springer, Heidelberg, 2009.
120. W. Hackbusch, B. Khoromskij, and S. A. Sauter. On $\mathcal{H}^2$-matrices. In H.-J. Bungartz, R. H. W. Hoppe, and C. Zenger, editors, *Lectures on Applied Mathematics*, pages 9–30. Springer, Heidelberg, 2000.
121. W. Hackbusch and B. N. Khoromskij. A sparse $\mathcal{H}$-matrix arithmetic. Part II: Applications to multi-dimensional problems. *Computing*, 64(22):21–47, 2000.
122. W. Hackbusch, C. Lage, and S. A. Sauter. On the efficient realization of sparse matrix techniques for integral equations with focus on panel clustering, cubature and software design aspects. In W.L. Wendland, editor, *Boundary Element Topics*, pages 51–76. Springer, Berlin, 1997.
123. W. Hackbusch and Z. P. Nowak. On the Complexity of the Panel Method (in russian). In G.I. Marchuk, editor, *Proc. of the Conference: Modern Problems in Numerical Analysis, Nauka, Moskau, 1986*, pages 233–244, 1988.

124. W. Hackbusch and Z. P. Nowak. On the fast matrix multiplication in the boundary element method by panel-clustering. *Numer. Math.*, 54:463–491, 1989.
125. W. Hackbusch and S. A. Sauter. On the efficient use of the galerkin method to solve fredholm integral equations. *Appl. Math.*, 38(4–5):301–322, 1993.
126. J. Hadamard. *Lectures on Cauchy's Problem*. Yale University Press, New Haven, CT, 1923.
127. H. Han. The boundary integro-differential equations of three-dimensional neuman problem in linear elasticity. *Numer. Math.*, 68(2):269–281, 1994.
128. G. Hellwig. *Partielle Differentialgleichungen*. Teubner, Stuttgart, 1960.
129. M. R. Hestenes and E. Stiefel. Methods of conjugate gradients for solving linear systems. *J. Res. Nat. Bur. Stand.*, 49:409–436, 1952.
130. J. S. Hesthaven and T. Warburton. Nodal high-order methods on unstructured grids. *J. Comput. Phys.*, 181:186–221, 2002.
131. H. Heuser. *Funktionalanalysis*. Teubner, Stuttgart, 1986.
132. R. Hiptmair. Coercive combined field integral equations. *J. Numer. Math.*, 11(2):115–134, 2003.
133. L. Hörmander. *The Analysis of Linear Partial Differential Operators I: Distribution Theory and Fourier Analaysis*. Springer, Berlin, 1983.
134. G. C. Hsiao, P. Kopp, and W. L. Wendland. A Galerkin-collocation method for some integral equations of the first kind. *Computing*, 25:89–130, 1980.
135. G. C. Hsiao, O. Steinbach, and W. L. Wendland. Domain decomposition methods via boundary integral equations. *J. Comput. Appl. Math.*, 125:521–537, 2000.
136. G. C. Hsiao and W. L. Wendland. A finite element method for some integral equations of the first kind. *J. Math. Anal. Appl.*, 58:449–481, 1977.
137. G. C. Hsiao and W. L. Wendland. *Boundary Integral Equations*. Springer, Berlin, 2008.
138. J. D. Jackson. *Klassische Elektrodynamik*, 3rd edition. De Gruyter, Berlin, 2002.
139. F. John. *Plane Waves and Spherical Means*. Springer, New York, 1955.
140. C. G. L. Johnson and L. R. Scott. An analysis of quadrature errors in second kind boundary integral equations. *SIAM J. Numer. Anal.*, 26:1356–1382, 1989.
141. L. V. Kantorovich and G. P. Akilov. *Functional Analysis*, 2nd edition. Pergamon, Oxford, 1982.
142. O. D. Kellogg. *Foundations of Potential Theory*. Springer, Berlin, 1929.
143. C. Kenig. *Harmonic analysis techniques for second order elliptic boundary value problems*. In CBMS Regional Conference Series in Mathematics. AMS, Providence, RI, 1994.
144. R. Kieser. *Über einseitige Sprungrelationen und hypersinguläre Operatoren in der Methode der Randelemente*. PhD thesis, Mathematisches Institut A, Universität Stuttgart, Germany, 1991.
145. V. A. Kozlov, V. G. Mazya, and J. Rossmann. *Spectral Problems Associated with Corner Singularities of Solutions to Elliptic Equations*. AMS, Providence, RI, 2001.
146. R. Kress. Minimizing the condition number of boundary integral operators in acoustics and electromagnetic scattering. *Q. J. Mech. Appl. Math.*, 38:323–341, 1985.
147. R. Kress. *Linear Integral Equations*. Springer, Heidelberg, 1999.
148. N. Krzebek and S. A. Sauter. Fast Cluster Techniques for BEM. *Eng. Anal. Boundary Elements*, 27(5):455–467, 2003.
149. M. Kuhn and U. Langer. Adaptive domain decomposition methods in FEM and BEM. In J.R. Whiteman, editor, *Proc. 9th Conf. on the Mathematics of Finite Elements and Applications VI, Uxbridge 1996 (MAFELAP 1996)*, pages 103–122. Wiley, New York, 1997.
150. V. D. Kupradze, T. G. Gegelia, M. O. Basheleĭshvili, and T. V. Burchuladze. *Three-Dimensional Problems of the Mathematical Theory of Elasticity and Thermoelasticity*, volume 25 of *North-Holland Series in Applied Mathematics and Mechanics*. North-Holland, Amsterdam, Russian edition, 1979. Edited by V. D. Kupradze.
151. O. A. Ladyzhenskaya and N. N. Ural′tseva. *Linear and Quasilinear Elliptic Equations*. Academic, New York, 1968.
152. C. Lage. *Software development for boundary element mehtods: Analysis and design of efficient techniques (in German)*. PhD thesis, Lehrstuhl Prakt. Math., Universität Kiel, 1995.

153. S. Langdon and S. N. Chandler-Wilde. A wavennumber independent boundary element method for an acoustic scattering problem. *SIAM J. Numer. Anal.*, 43(6):2450–2477, 2006.
154. R. Leis. *Initial Boundary Value Problems in Mathematical Physics*. Teubner/Wiley, Stuttgart/Chichester, 1986.
155. J. L. Lions and E. Magenes. *Non-Homogeneous Boundary Value Problems and Applications*. Springer, Berlin, 1972.
156. M. Löhndorf and J. M. Melenk. Mapping properties of Helmholtz boundary integral operators and their application to the *hp*-BEM. Technical Report 34/2009, Institute for Analysis and Scientific Computing, TU Wien, 2009.
157. M. Maischak. *hp-Methoden für randintegralgleichungen bei 3D-problemen, theorie und implementierung*. PhD thesis, IFAM, Universität Hannover, Germany, 1995.
158. M. Maischak and E. P. Stephan. The *hp*-version of the boundary element method for the Lamé equation in 3D. In *Boundary Element Analysis*, volume 29 of *Lecture Notes in Applied and Computational Mechanics*, pages 97–112. Springer, Berlin, 2007.
159. H. W. Maue. Zur Formulierung eines allgemeinen beugungsproblems durch eine integralgleichung. *Z. Physik*, 126:601–618, 1949.
160. V. G. Mazya. Boundary integral equations. In V. G. Mazya and S. M. Nikolskiĭ, editors, *Encyclopaedia of Mathematical Sciences*, volume 27, pages 127–233. Springer, Heidelberg, 1991.
161. E. McKay Hyde and O. P. Bruno. A fast, higher-order solver for scattering by penetrable bodies in three dimensions. *J. Comput. Phys.*, 202(1):236–261, 2005.
162. W. McLean. *Strongly Elliptic Systems and Boundary Integral Equations*. Cambridge University Press, Cambridge, 2000.
163. M. Melenk and S. Langdon. An *hp*-boundary element method for high frequency scattering by convex polygons. In *Proc. 8th Int. Conf. on Mathematical and Numerical Aspects of Wave Propagation, Reading University*, pages 93–95. Academic, New York, 2007.
164. S. G. Mikhlin. *Integral Equations*. Pergamon, London, 1957.
165. N. I. Muskhelishvili. *Singular Integral Equations*. Noordhoff, Groningen, 1953.
166. J. Nečas. *Les Methodes Directes en Theorie des Equations Elliptiques*. Academia, Prague, 1967.
167. J. C. Nédélec. Curved finite element methods for the solution of singular integral equations on surfaces in R^3 *Comput. Meth. Appl. Mech. Eng.*, 8:61–80, 1976.
168. J. C. Nédélec. Cours de l'école d'été d'analyse numérique. Technical report, CEA-IRIA-EDF, 1977.
169. J. C. Nédélec. Integral equations with non integrable kernels. *Integral Equations Oper. Theory*, 5:562–572, 1982.
170. J. C. Nédélec. *Acoustic and Electromagnetic Equations*. Springer, New York, 2001.
171. J. C. Nédélec and J. Planchard. Une méthode variationelle d'élements finis pour la résolution numérique d'un problème extérieur dans R^3. *RAIRO*, 7(R3):105–129, 1973.
172. P. Neittaanmäki and S. Repin. *Reliable methods for computer simulation. Error control and a posteriori estimates*. Elsevier, New York, 2004.
173. C. Neumann. *Untersuchungen über das Logarithmische und Newtonsche Potential*. Teubner, Leipzig, 1877.
174. L. Nirenberg. Remarks on strongly elliptic partial differential equations. *Comm. Pure Appl. Math.*, 8:649–675, 1955.
175. F. Noether. Über eine Klasse singulärer Integralgleichungen. *Math. Ann.*, 82:42–63, 1921.
176. E. J. Nyström. Über die praktische auflösung von linearen integralgleichungen mit anwendungen auf randwertaufgaben der potentialtheorie. *Soc. Sci. Fenn. Comment. Phys.-Math.*, 4:1–52, 1928.
177. E. J. Nyström. Über die praktische auflösung von linearen integralgleichungen mit anwendungen auf randwertaufgaben. *Acta Math.*, 54:185–204, 1930.
178. P. Oswald. Multilevel norms for $H^{-1/2}$. *Computing*, 61, 1998.
179. O. I. Panich. On the question of the solvability of the exterior boundary-value problems for the wave equation and Maxwell's equations. *Usp. Mat. Nauk.*, 20A:221–226, 1965. in Russian.

180. L. E. Payne and H. F. Weinberger. An optimal Poincaré-inequality for convex domains. *Arch. Rational Mech. Anal.*, 5:286–292, 1960.
181. J. Plemelj. Ein ergänzungssatz zur cauchyschen integraldarstellung analytischer funktionen, randwerte betreffend. *Monatshefte f. Math. u. Phys.*, 19:205–210, 1908.
182. J. Plemelj. *Potentialtheoretische Untersuchungen*. Teubner, Leipzig, 1911. Preisschriften der Fürstlich Jablonowskischen Gesellschaft.
183. E. Polak. *Computational Methods in Optimization*. Academic, New York, 1971.
184. A. Pomp. *The Boundary-Domain Integral Method for Elliptic Systems*, volume 1683 of *Lecture Notes in Mathematics*. Springer, Berlin, 1998.
185. W. H. Press, S. A. Teukolsky, W. T. Vetterling, and B. P. Flannery. *Numerical Receipes in FORTRAN*, 2nd edition. Cambridge University Press, Cambridge, 1992.
186. M. D. Preston, P. G. Chamberlain, and S. N. Chandler-Wilde. An integral equation method for a boundary value problem arising in unsteady water wave problems. *J. Integral Equations Appl.*, 20(1):121–152, 2008.
187. S. Prößdorf and R. Schneider. A Spline Collocation Method for Multidimensional Strongly Elliptic Pseudodifferential Operators of Order Zero. *Integral Equations Oper. Theory*, 14:399–435, 1991.
188. J. Radon. Über die randwertaufgaben beim logarithmischen potential. *In: Sitzungsberichte der Akademie der Wissenschaften Wien. IIa*, 128:1123–1167, 1919.
189. E. Rank. Adaptive *h*-, *p*- and *hp*-versions for boundary integral method. *Int. J. Numer. Meth. Eng.*, 28:1335–1349, 1989.
190. A. Rathsfeld. A wavelet algorithm for the boundary element solution of a geodetic boundary value problem. *Comput. Meth. Appl. Mech. Eng.*, 157:267–287, 1998.
191. A. Rathsfeld. On a hierarchical three point basis of piecewise linear functions over smooth boundaries. In J. Elschner, I. Gohberg, and B. Silbermann, editors, *Operator Theory: Advances and Applications*, volume 121, pages 442–470. Birkhäuser, Basel, 2001.
192. T. J. Rivlin. *The Chebyshev Polynomials*. Wiley, New York, 1974.
193. V. Rokhlin. Rapid solutions of integral equations of classical potential theory. *J. Comput. Phys.*, 60(2):187–207, 1985.
194. V. Rokhlin. Diagonal forms of translation operators for the Helmholtz equation in three dimensions. *Appl. Comp. Harm. Anal.*, 1(1):82–93, 1993.
195. W. Rudin. *Functional Analysis*. McGraw-Hill, New York, 1991.
196. J. Saranen and W. L. Wendland. Local residual-type error estimates for adaptive boundary element methods on closed curves. *Appl. Anal.*, 48:37–50, 1993.
197. S. A. Sauter. *Über die effiziente verwendung des galerkinverfahrens zur lösung fredholmscher integralgleichungen*. PhD thesis, Inst. f. Prakt. Math., Universität Kiel, 1992.
198. S. A. Sauter. The panel clustering method in 3-D BEM. In G. Papanicolaou, editor, *Wave Propagation in Complex Media*, pages 199–224. Springer, Berlin, 1998. IMA-Volumes in Mathematics and Its Applications.
199. S. A. Sauter. Variable order panel clustering. *Computing*, 64:223–261, 2000.
200. S. A. Sauter and A. Krapp. On the effect of numerical integration in the Galerkin boundary element method. *Numer. Math.*, 74(3):337–360, 1996.
201. S. A. Sauter and C. Lage. Transformation of hypersingular integrals and black-box cubature. *Math. Comp.*, 70:223–250, 2001.
202. S. A. Sauter and C. Schwab. On the realization of *hp*-Galerkin BEM in 3-D. In W. Hackbusch and G. Wittum, editors, *BEM : Implementation and Analysis of Advanced Algorithms, Proceedings of the 12th GAMM-Seminar, Kiel*. Vieweg, Brunswick, 1996.
203. S. A. Sauter and C. Schwab. Quadrature for *hp*-Galerkin BEM in R^3. *Numer. Math.*, 78(2):211–258, 1997.
204. S. A. Sauter and C. Schwab. *Randelementmethoden*. Teubner, Leipzig, 2004.
205. G. Schmidlin, C. Lage, and C. Schwab. Rapid solution of first kind boundary integral equations in R^3. *Eng. Anal. Boundary Elements*, 27(5):469–490, 2003.
206. G. Schmidlin and C. Schwab. Wavelet agglomeration on unstructured meshes. In T. J. Barth, T. F. Chan, and R. Haimes, editors, *Lecture Notes in Computational Science and Engineering*, volume 20, pages 359–378. Springer, Heidelberg, 2002.

207. R. Schneider. Stability of a spline collocation method for strongly elliptic multidimensional singular integral equations. *Numer. Math.*, 58:855–873, 1991.
208. R. Schneider. *Multiskalen und wavelet-matrixkompression: Analysisbasierte methoden zur lösung großer vollbesetzter gleichungssysteme*. Teubner, Stuttgart, 1998.
209. H. Schulz and O. Steinbach. A new a-posteriori error estimator in adaptive direct boundary element methods. the dirichlet problem. *Calcolo*, 37:79–96, 2000.
210. H. Schulz and W. L. Wendland. Local residual-based a posteriori error estimates forcing adaptive boundary element methods. In W.L. Wendland, editor, *Boundary Element Topics*, pages 445–470. Springer, Berlin, 1997.
211. C. Schwab and W. L. Wendland. Kernel properties and representations of boundary integral operators. *Math. Nachr.*, 156:187–218, 1992.
212. C. Schwab and W. L. Wendland. On numerical cubatures of singular surface integrals in boundary element methods. *Numer. Math.*, 62:343–369, 1992.
213. C. Schwab and W. L. Wendland. On the extraction technique in boundary integral equations. *Math. Comp.*, 68(225):91–122, 1999.
214. R. T. Seeley. Singular integrals on compact manifolds. *Am. J. Math.*, 81:658–690, 1959.
215. I. H. Sloan. Error analysis of boundary integral methods. *Acta Numerica*, 92:287–339, 1992.
216. O. Steinbach. *Numerische näherungsverfahren für elliptische randwertprobleme*. Teubner, Stuttgart, 2003.
217. O. Steinbach and W. L. Wendland. The construction of some efficient preconditioners in the boundary element method. Numerical treatment of boundary integral equations. *Adv. Comput. Math.*, 9(1–2):191–216, 1998.
218. O. Steinbach and W. L. Wendland. Neumann's method for second-order elliptic systems in domains with non-smooth boundaries. *J. Math. Anal. Appl.*, 262(2):733–748, 2001.
219. E. P. Stephan. A boundary integral equation method for three-dimensional crack problems in elasticity. *Math. Meth. Appl. Sci.*, 8:609–623, 1986.
220. E. P. Stephan. Boundary integral equations for mixed boundary value problems in R^3. *Math. Nachr.*, 131:167–199, 1987.
221. E. P. Stephan. Improved Galerkin methods for integral equations on polygons and on polyheral surfaces. In *Proc. First Joint Japan/US Symposium on boundary element methods, Tokyo*, pages 73–80, 1988.
222. E. P. Stephan. The *h-p* boundary element method for solving 2- and 3-dimensional problems. *Comput. Meth. Appl. Mech. Eng.*, 133(3–4):183–208, 1996.
223. E. P. Stephan and W. L. Wendland. Remarks to Galerkin and least squares methods with finite elements for general elliptic problems. *Manuscripta Geodaetica*, 1:93–123, 1976.
224. J. Stoer and R. Bulirsch. *Numerische Mathematik II*, 3rd edition. Springer, Heidelberg, 1990.
225. J. Stoer and R. Bulirsch. *Introduction to Numerical Analysis*, 3rd edition. Springer, New York, 2002.
226. A. H. Stroud. *Approximate Calculations of Multiple Integrals*. Prentice Hall, Englewood Cliffs, 1973.
227. J. Tausch. Sparse BEM for potential theory and stokes flow using variable order wavelets. *Comput. Mech.*, 32(4–6):312–318, 2003.
228. J. Tausch and J. White. Multiscale bases for the sparse representation of boundary integral operators on complex geometry. *SIAM J. Sci Comp.*, 25(5):1610–1629, 2003.
229. F. G. Tricomi. *Integral Equations*. Interscience, New York, 1957.
230. H. Triebel. *Interpolation Theory, Function Spaces, Differential Operators*, 2nd edition. Johann Ambrosius Barth, Heidelberg, 1995.
231. G. Verchota. Layer potentials and regularity for the Dirichlet problem for the Laplace's equation in Lipschitz domains. *J. Funct. Anal.*, 59:572–611, 1984.
232. R. Verfürth. *A review of a posteriori error estimation and adaptive mesh refinement*. Wiley/Teubner, Chichester/Stuttgart, 1996.
233. T. von Petersdorff. Boundary integral equations for mixed Dirichlet , Neumann and transmission problems. *Math. Meth. Appl. Sci.*, 11:185–213, 1989.

234. T. von Petersdorff. *Elliptische randwertprobleme in polyedern: singularitäten und approximation mit randelementmethoden (in German)*. PhD thesis, Institut für Mathematik, Universität Darmstadt, Germany, 1989.
235. T. von Petersdorff and C. Schwab. Fully discrete multiscale Galerkin BEM. In W. Dahmen, P. Kurdila, and P. Oswald, editors, *Multiresolution Analysis and Partial Differential Equations*, pages 287–346. Academic, New York, 1997.
236. T. von Petersdorff, C. Schwab, and R. Schneider. Multiwavelets for second-kind integral equations. *SIAM J. Numer. Anal.*, 34(6):2212–2227, 1997.
237. T. von Petersdorff and Ch. Schwab. Wavelet discretization of first kind boundary integral equations of polygons. *Numer. Math.*, 74:479–519, 1996.
238. W. L. Wendland. Strongly elliptic boundary integral equations. In A. Iserles and M. Powell, editors, *The State of the Art in Numerical Analysis*, pages 511–561. Clarendon, Oxford, 1987.
239. W. L. Wendland, E. P. Stephan, and G. C. Hsiao. On the integral equation methods for plane mixed boundary value problems for the Laplacian. *Math. Meth. Appl. Sci.*, 1:265–321, 1979.
240. W. L. Wendland and D. Yu. Adaptive BEM for strongly elliptic integral equations. *Numer. Math.*, 53:539–558, 1988.
241. H. Whitney. *Geometric Integration Theory*. Princeton University Press, Princeton, NJ, 1957.
242. J. Wloka. *Partial Differential Equations*. Cambridge University Press, Cambridge, 1987. Translated from the German by C. B. Thomas and M. J. Thomas.
243. K. Yosida. *Functional Analysis*. Springer, Berlin, 1964.
244. D. Yu. A-posteriori estimates and adaptive approaches for some boundary element methods. In C. A. Brebbia, W. L. Wendland, and G. Kuhn, editors, *Boundary Elements 9*, pages 241–256. Springer, Berlin, 1987.
245. V. A. Zorich. *Mathematical Analysis. I*. Springer, Berlin, 2004. Translated from the 2002 fourth Russian edition by Roger Cooke.
246. V. A. Zorich. *Mathematical Analysis. II*. Springer, Berlin, 2004. Translated from the 2002 fourth Russian edition by Roger Cooke.

Index of Symbols

$\overline{V}^{\|\circ\|_W}$	Closure of a subset $V \subset W$ in a normed vector space $(W, \|\circ\|_W)$
T^*	Adjoint operator, 28
$\lvert\omega\rvert$	Volume measure of a measurable subset $\omega \subset \mathbb{R}^d$ or a surface measure of a measurable subset $\omega \subset \Gamma$ of a surface Γ
$\sharp\Theta$, $\lvert\Theta\rvert$, $\operatorname{card}\Theta$	Number of elements of a finite set Θ
$p.v.$	Cauchy principal value, 294
$\mathcal{A}^m_{\mathrm{pw}}$	Class of piecewise analytic surfaces, 496
$\mathcal{A}^m_{\mathrm{pw}}$	Class of smooth, piecewise analytic surfaces, 496
$B(\cdot,\cdot)$, $B_+(\cdot,\cdot)$, $B_-(\cdot,\cdot)$	Sesquilinear form that belongs to the elliptic differential equation. The indices $+$, $-$ indicate whether the exterior Ω^+ or the interior Ω^- is being considered
$B_r(\mathbf{x})$	Open ball with radius $r > 0$ around a point $\mathbf{x} \in X$ with respect to the norm in X, 50
B_r, B_r^+, B_r^-, B_r^0	Open ball with radius $r > 0$ around zero, upper half of B_r, lower half of B_r, middle plane of B_r, 50
$C^k(\Omega)$	Space of all k times continuously differentiable functions on Ω, 48
$C^k(\overline{\Omega})$	Space of all k times continuously differentiable functions on Ω with k times continuously differentiable extensions to $\overline{\Omega}$, 48
$C^{k,\lambda}(\overline{\Omega})$	Space of all k times Hölder continuously differentiable functions, 48
$C_0^\infty(\Omega)$	Space of all infinitely differentiable functions with compact support in Ω, 54
$C^\infty_{\mathrm{comp}}(\Omega)$	Restriction of $C_0^\infty(\mathbb{R}^d)$ to Ω, 54
$\mathbf{C}^k(\overline{\Omega_1}, \overline{\Omega_2})$	Space of all vector-valued, k times continuously differentiable functions, 49
$\mathbf{C}^{k,\lambda}(\overline{\Omega_1}, \overline{\Omega_2})$	Space of all vector-valued, k times Hölder continuously differentiable functions, 49

$H^{\ell}_{\mathrm{loc}}(\Omega)$	Fréchet space of all functions u with $\varphi u \in H^{\ell}(\Omega)$ for all $\varphi \in C^{\infty}_{\mathrm{comp}}(\Omega)$, 63
$H^{\ell}_{\mathrm{comp}}(\Omega)$	Fréchet space of all functions u with $\operatorname{supp} u \subset\subset \mathbb{R}^d$, 64
$H^{s}_{L}(\Omega)$	Sobolev space with $Lu \in L^2_{\mathrm{comp}}(\Omega)$, 69
$H^{1}_{L}(\mathbb{R}^d \backslash \Gamma)$	Sobolev space whose restrictions to Ω^-, Ω^+ are contained in $H^s_L(\Omega^-)$ $H^s_L(\Omega^+)$, 71
$H^1(L, \Omega)$	Sobolev spaces that have weight functions for the decay conditions that depend on the operator, 83
$H^1_T(L, \Omega)$	Associated test space, 83
$\cong$	Isomorphic
Σ_p	Set of nodal points on the reference element, 205, 205
$\mathbb{K} \in \{\mathbb{R}, \mathbb{C}\}$	Field
$L(X, Y)$	Set (vector space) of all bounded linear operators, 22
$L^{\infty}(\Omega)$	Lebesgue space of all measurable, almost everywhere bounded functions, 49
$L^2(\Omega)$	Lebesgue space of all measurable, square integrable functions, 54
$\mathbf{L}^{\infty}(\Omega)$	Lebesgue space of all d-valued functions with components in $L^{\infty}(\Omega)$, 67
$\mathrm{lift}_{\tau,p}$	Lifting operator from the affine surface approximation to the p-parametric one, 469
	$\mathbb{N} = \{1, 2, \ldots\}$
	$\mathbb{N}_0 = \{0, 1, 2, \ldots\}$
$\mathbf{n} : \Gamma \to \mathbb{S}_{d-1}$	Field of unit normals, 52
$\mathbf{n}_{\ell} : \Gamma^p_{\ell} \to \mathbb{S}_{d-1}$	Field of unit normals at p-parametric surface approximation, 474
	$\mathcal{J}_m := \{\mu \in \mathbb{N}^3 \mid \forall 1 \le i \le 3 : 1 \le \mu_i \le m\}$
$\widehat{\iota^{\tau}_{p}}$	Index set for the nodal points in the reference element $\hat{\tau}$, 192, 204
$\lVert\cdot\rVert_k$, $\lVert\cdot\rVert_{k,\Omega}$, $\lVert\cdot\rVert_{H^k(\Omega)}$	$H^k(\Omega)$-norm, (2.78), (2.81), (2.85), (2.91), (2.149), (2.162)
$(\cdot,\cdot)_k$, $(\cdot,\cdot)_{k,\Omega}$, $(\cdot,\cdot)_{H^k(\Omega)}$	Associated inner product. Identified with the continuous extension to dual pairings. The complex conjugation is applied to the second argument
$\lvert\cdot\rvert_k$, $\lvert\cdot\rvert_{k,\Omega}$, $\lvert\cdot\rvert_{H^k(\Omega)}$	$H^k(\Omega)$-seminorm, which only contains the highest derivative, (2.79)
$\mu \in \mathbb{N}^d_0$	Multi-index, subject to rules given in (2.67)
ω_d	Surface measure of the unit sphere in $\mathbb{R}^d$
τ	Panel, Definitions 2.2.9 and 4.1.2
$\mathcal{G}$	Surface mesh, Definition 4.1.2
P_S	Poincaré–Steklov operator, (3.127)
$\mathbb{P}^{\Delta}_m$	Space of all polynomials of two variables up to a total maximal degree of $m \in \mathbb{N}_0$ [see (4.23)]

$\mathbb{P}_m^{\square}$	Space of all polynomials of two variables up to a maximal degree of $m \in \mathbb{N}_0$ per component [see (4.67)]
$\mathbb{P}_m$	General term for $\mathbb{P}_m^{\Delta}$ or $\mathbb{P}_m^{\square}$
$\mathbb{Q}_m$	Space of all polynomials of three variables up to a maximal degree of $m \in \mathbb{N}_0$ per component [see (7.8)]
S, D, V, K, K', W	Single layer and double layer potential and associated boundary integral operators, see (3.4), (3.5), Definition 3.1.5, (3.6)
$\mathcal{N}$	Newton potential, (3.8)
$S_{\mathcal{G}}^0, S_{\mathcal{G},\emptyset}^p, S_{\mathcal{G}}^{p,k}, S_{\mathcal{G},\emptyset}^{p,k}$	Boundary element spaces, (4.20), (4.24), Definition 4.1.36
Re, Im	Real part, imaginary part
$\underset{\sim}{Q} := [-1,1]^3$	Reference element for Čebyšev interpolation
$\widehat{Q} = (0,1)^2$	Unit square as reference element
$\widehat{S}$:	Unit triangle (with vertices $(0,0)^{\mathsf{T}}$, $(1,0)^{\mathsf{T}}$, $(1,1)^{\mathsf{T}}$) as reference element
$\widehat{\tau} \in \left\{\widehat{Q}, \widehat{S}\right\}$	General term for the reference element
$\Sigma(\sigma)$	Set of sons of a cluster σ, Definition 7.1.4
$\Sigma(Q)$	Set of eight congruent subcuboids that result when the edges of Q are bisected
$\sigma(T)$	Spectrum of an operator T
$\mathbb{S}_{d-1}$	d-Sphere, surface of the d-dimensional unit sphere
T	Solution operator,73
Z, Z_Ω, Z_+, Z_-	Trace extension operators, Theorem 2.6.11, Notation 2.6.12
$\gamma_0, \gamma_0^+, \gamma_0^-$	Trace operator and one-sided variants, Theorem 2.6.8
$\gamma_1, \gamma_1^+, \gamma_1^-$	Conormal trace operator and one-sided variants, Definition 2.7.6, Remark 2.7.10
$\widetilde{\gamma_1}, \widetilde{\gamma_1}^+, \widetilde{\gamma_1}^-$	Modified conormal trace operator and one-sided variants, (2.107), Definition 2.7.6, Remark 2.7.10
$[u]$	Jump in a function across the boundary, $[u] := \gamma_0^+ u - \gamma_0^- u$
$[\gamma_1 u]$	Conormal jump in a function across the boundary, $[\gamma_1 u] := \gamma_1^+ u - \gamma_1^- u$
σ_Ω	Sign function, $\sigma_\Omega = -1$ for the exterior and $\sigma_\Omega = 1$ for the interior
S_P :	Steklov–Poincaré operator, (3.130)
supp (u)	Support of a function u, (2.74)
$W^{\ell,\infty}(\Omega)$	Sobolev space of functions having all derivatives up to the order ℓ in $L^\infty(\Omega)$,55
$\chi_\tau : \widehat{\tau} \to \tau$	Transformation of the reference element to the panel τ, Definition 4.1.2
$\chi_\tau^{\text{affine}}$	Affine part of the transformation: $\chi_\tau^{\text{affine}}(\hat{\mathbf{x}}) = \mathbf{A}_\tau + \mathbf{m}_\tau \hat{\mathbf{x}}$, Assumption 4.1.6

Index